Nanostructured TiO_2

Nanostructured TiO_2

Synthesis, Photocatalytic Properties, and Applications

Edited by

Jia Hong Pan
Guangxi University

Wan In Lee
Inha University

Detlef W. Bahnemann
St. Petersburg University

WILEY-VCH

Edited By:

Prof. Jia Hong Pan
School of Resources, Environment and Materials
Guangxi University
100 East Daxue Road
Xixiangtang District, Nanning, 530004
Guangxi Zhuang Autonomous Region
People's Republic of China

Prof. Wan In Lee
Department of Chemistry and Chemical Engineering
Inha University
100 Inha-ro, Michuhol-gu
Incheon, 22212
Republic of Korea

Prof. Detlef W. Bahnemann
Photoactive Nanocomposite Materials Laboratory
V.A. Fock Institute of Physics
St. Petersburg University
Ulyanovskaya street 1, Peterhof,
Saint Petersburg, 198504
Russia

Cover Design: Wiley
Cover Image: © Science RF/stock.adobe.com

Typesetting Lumina Datamatics
Printing: CPI Group (UK) Ltd,
Croydon CR0 4YY

Library of Congress Card No.: applied for

British Library Cataloguing-in-Publication Data
A catalogue record for this book is available from the British Library.

Bibliographic information published by the Deutsche Nationalbibliothek
The Deutsche Nationalbibliothek lists this publication in the Deutsche Nationalbibliografie; detailed bibliographic data are available on the Internet at <http://dnb.d-nb.de>.

Print ISBN: 9783527354177
ePDF ISBN: 9783527848331
ePub ISBN: 9783527848324
oBook ISBN: 9783527848348

C9783527354177_011025

Contents

Introduction *ix*
Acknowledgments *xv*

Part I Synthetic Methodologies *1*

1 **Selective Synthesis of TiO_2 Nanoparticles** *3*
Sungjin Park, Jia Hong Pan, Wan In Lee

2 **Synthesis of One-dimensional TiO_2 Nanomaterials** *27*
Jeongho Kim, Wan In Lee, Jia Hong Pan

3 **Synthesis of Two-dimensional TiO_2 Nanostructures** *51*
Sichun Yang, Haijiao Lu, Lianzhou Wang

4 **3D TiO_2 via Self-assembly** *67*
Jie Tian, Mingdeng Wei, Detlef W. Bahnemann, Wan In Lee, Jia Hong Pan

5 **Immobilization of TiO_2 on Various Substrates** *87*
Nishat Khan, Seema Garg, Isha Arora, Gurumurthy Hegde, Suresh Sagadeven, Lalita Ledwani, Amrish Chandra

6 **Surface Modification of TiO_2** *115*
Xintong Zhang, Yingying Li, Rui Wang, Changhua Wang

Part II Photocatalytic Properties *139*

7 **Photocatalytic Fixed Bed Reactions for Efficient Photon Utilization of TiO_2** *141*
Jiazang Chen

8 Carrier Transport and Recombination in Sensitized Nanostructured TiO_2 *159*
James Robert Jennings

9 TiO_2-based Heterojunction Photocatalysts: Type, Characterization, and Activity *185*
Yi Lu, Yi-Xuan Liu, Xiao-Yu Yang

10 Plasmonic Metal–TiO_2 Photocatalysts *207*
Na Zhou, Yan He, Jiahuan Zheng

11 Ecotoxicological and Antibacterial Assessment of Nanostructured TiO_2 *229*
Jiani Liu, Fahim Ullah Khan, Youji Wang

Part III Applications *269*

12 Applications of TiO_2 for Photocatalytic Water Treatment *271*
Muhammed A. Ballim, Xinzhu Pang, Nosipho Moloto, Nathan Skillen, Peter K.J. Robertson, Kathryn Ralphs

13 TiO_2 Photocatalytic Degradation of Antibiotics *301*
Shiping Xu, Wanting Sun, Tianhao Song, Bo Wang

14 Ecotoxicological and Antibacterial Properties of TiO_2 and Processes Based on TiO_2 *343*
Eryk Fernandes, João Lincho, Rui C. Martins, João Gomes

15 The Application of TiO_2 for the Photocatalytic Conversion of the Nitrogen Oxides – NO and NO_2 *361*
Andreas Kafizas

16 Recycling of Heavy Metal Ions from Wastewater by TiO_2 Photocatalysis *389*
Liqiang Yao, Chongchong She, Kun Chen, Jia Hong Pan

17 Solar Fuel Generation *415*
Shuai Wang, Hairong Yue, Patrik Schmuki, Xuemei Zhou

18 Visible-light-induced Selective Oxidation of Organic Molecules over Nanostructured TiO_2 *431*
Yuexin Wang, Fengwei Huang, Xianjun Lang

19 Photoelectrochemical Sensing Applications of Nanostructured TiO_2 *449*
Miao Li, Xiaoqing Qiu

20 Reactor Engineering in Competitive Photocatalytic Performance: Advancements and Challenges *471*
Jéssica Ariane de Oliveira, Daniel Raimundo Sevene, Osmando F. Lopes, Caue Ribeiro

21 TiO_2-based Protective Coatings for Buildings and Monuments *497*
Yijian Cao, Cong Wang, Wanqian Feng, Mara Camaiti

Index *515*

Introduction

Overview of *Nanostructured TiO_2: Synthesis, Properties, and Photocatalytic Applications*

Jia Hong Pan,[1,] Wan In Lee,[2] Detlef W. Bahenmann,[3,4]*

[1]*State Key Laboratory of Featured Metal Materials and Life-Cycle Safety for Composite Structures, School of Resources, Environment and Materials, Guangxi University, Nanning, Guangxi, China*
[2]*Department of Chemistry and Chemical Engineering, Inha University, Incheon, Korea*
[3]*Institut für Technische Chemie, Gottfried Wilhelm Leibniz Universität Hannover, Hannover, Germany*
[4]*Laboratory "Photoactive Nanocomposite Materials", Saint Petersburg State University, Peterhof, Saint Petersburg, Russia*

* To whom all correspondence should be addressed.
E-mail: pan@ncepu.edu.cn (J. H. Pan)

1 TiO_2 Photocatalysis

The Sun emits heat and radiant light to Earth, serving as the most abundant renewable energy resource. Natural photosynthesis is a well-known photochemical process that directly or indirectly powers living organisms in the biosphere by converting solar energy into storable chemical energy. Inspired by this process, heterogeneous photocatalysis mediated by semiconductors has gained significant attention for its potential in efficient solar-to-chemical energy conversion, attracting ongoing interest and research.

TiO_2 is a highly stable inorganic material that is comparatively abundant in the Earth's crust. Known as a versatile and eco-friendly material due to its inert and non-toxic properties, TiO_2 is widely used in our daily lives or industries as an additive in pigments, paints, toothpastes, sun creams, abrasives, pharmaceutical products, and other applications. Since Honda and Fujishima reported photocatalytic water splitting using TiO_2 electrodes in 1972 [1], TiO_2 has been intensively and extensively investigated as a photocatalytic material in various oxidation and reduction reactions. Although various inorganic, organic, and organic–inorganic hybrid semiconductors have been reported, TiO_2 remains the most extensively studied photocatalyst to date because of its exceptional physicochemical properties, including low cost, high quantum efficiency, chemical and photonic stability, favorable band positions, non-toxicity, biocompatibility, and versatility in (photo)catalysis, optoelectronics, sensors, photovoltaics, and biomedical applications.

Figure 1 illustrates the charge carrier generation, trapping, recombination, and interfacial charge transfer during the TiO_2 photocatalysis [2–4]. The process starts with the

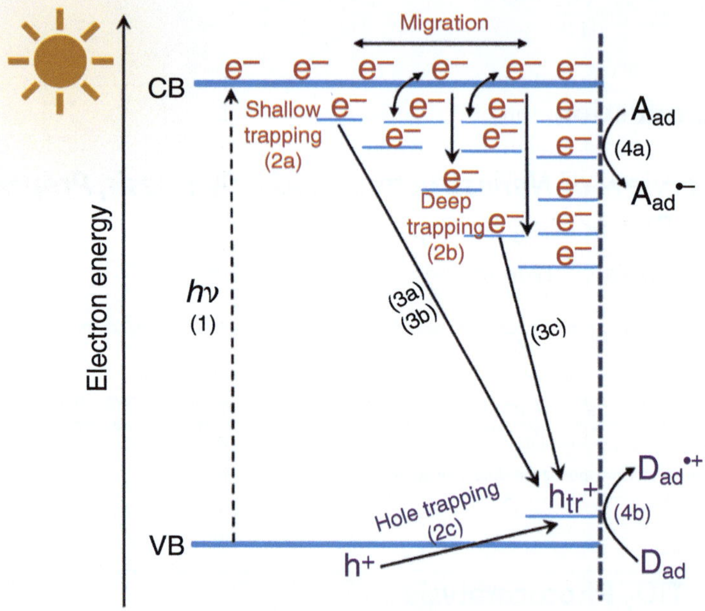

Figure 1 Charge carrier transfer during TiO_2 photocatalysis. Adapted from Ref. 2 with permission from Elsevier.

absorption of photons that possess energy greater than its bandgap. TiO_2 is thus photoexcited, inducing interband transitions and generating electrons (e^-) at the conduction band (CB) and holes (h^+) at the valence band (VB). A large portion of the photogenerated e^-/h^+ pairs is susceptible to recombination within the bulk or TiO_2 surface, resulting in the energy release as light and heat. On the other hand, free charge carriers without recombination can migrate and be trapped at the TiO_2 surface independently. Upon interfacial charge transfer, the trapped e^- and h^+ enable the reduction and oxidation processes, respectively. The underlying reaction can be written as follows [3].

1. Charge carrier generation:
   $$TiO_2 + h\nu \rightarrow TiO_2(e^- + h^+) \quad (1)$$
2. Charge carrier trapping:
   $$e^- + > Ti^{IV}OH \rightarrow \{> Ti^{III}OH\} \text{ (shallow trapping)} \quad (2a)$$
   $$e^- + > Ti^{IV} \rightarrow > Ti^{III} \text{ (deep trapping)} \quad (2b)$$
   $$h^+ + > Ti^{IV}OH \rightarrow \{Ti^{IV}OH^\bullet\}^+ \quad (2c)$$
3. Charge carrier recombination:
   $$e^- + h^+ \rightarrow \text{heat}/h\nu \quad (3a)$$
   $$e^- + \{Ti^{IV}OH^\bullet\}^+ \rightarrow Ti^{IV}OH \quad (3b)$$
   $$h^+ + \{> Ti^{III}OH\} \rightarrow Ti^{IV}OH \quad (3c)$$
   $$h^+ + Ti^{III} \rightarrow Ti^{IV} \quad (3d)$$
   $$\{> Ti^{III}OH\} + \{Ti^{IV}OH^\bullet\}^+ \rightarrow 2Ti^{IV}OH \quad (3e)$$

4. Interfacial charge transfer with acceptor (A) or donor (D) molecules absorbed on the TiO_2 surface:

$$\{> Ti^{III}OH\} + A \rightarrow \{> Ti^{IV}OH\} + A^{\bullet -} \quad (4a)$$

$$\{Ti^{IV}OH^{\bullet}\}^{+} + D \rightarrow \{> Ti^{IV}OH\} + D^{\bullet +} \quad (4b)$$

Optimizing the textural properties of TiO_2 is critical for enhancing its photochemical performance, prompting sustained research interest in the manipulation of TiO_2 nanostructures. Advancements in nanotechnology, particularly in nanomaterial synthesis, have facilitated diverse strategies for tailoring the properties of nanostructured TiO_2. Moreover, various photocatalytic applications of TiO_2 have been developed through the manipulation of electrons (e^-), holes (h^+), and their associated radicals.

2 Crystal Phase of TiO_2

TiO_2 possesses four main crystal phases, i.e., anatase, rutile, brookite, and TiO_2(B), as summarized in Table 1 [5]. Depending on the polymorph, size, and pH, the potential of holes generated from TiO_2 is +1.0 to +3.5 V vs. NHE, while those of electrons are 0.5 to −1.5 V vs. NHE [3, 4]. Anatase and rutile have a tetragonal structure with band gaps of 3.2 and 3.0 eV, respectively, while brookite has an orthorhombic structure with a band gap of 2.96 eV. Rutile is the thermodynamically most stable phase, which is formed above 600 °C, while anatase is a kinetically favorable phase at lower temperatures, and brookite is a metastable phase that often forms at around 500 °C. Monoclinic TiO_2(B) is generally obtained from the proton exchange of

Table 1 Crystal structures of four TiO_2 polymorphs: anatase, rutile, brookite, and TiO_2 (B). Red balls and grey balls are representatives of O and Ti atoms, respectively, and the bright-blue space-filling polyhedron units present octahedral unit constructs from a Ti^{4+} ion and six O^{2-} ions. Adapted from Ref. 5 with permission from Elsevier.

TiO_2 phase	Anatase	Rutile	Brookite	TiO_2(B)
Space group	*I41/amd*	*P42/mnm*	*Pbca*	*C2/m*
Band gap	3.2 eV	3.0 eV	3.0 eV	3.4 eV
Crystal structure	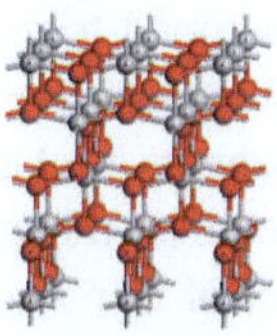	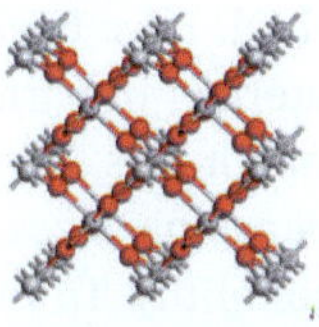	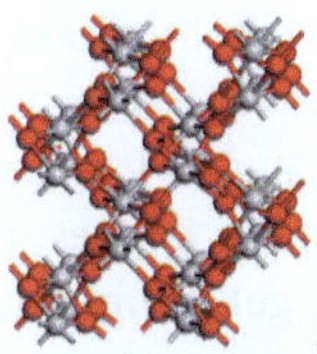	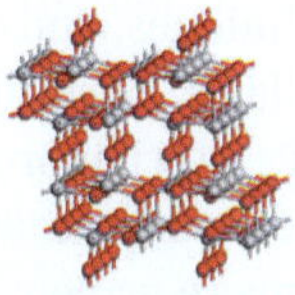
Polyhedral structure	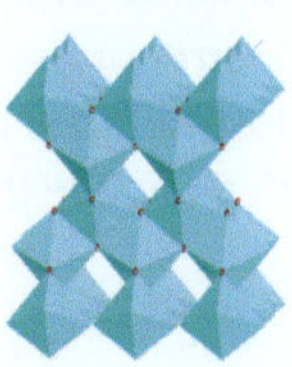	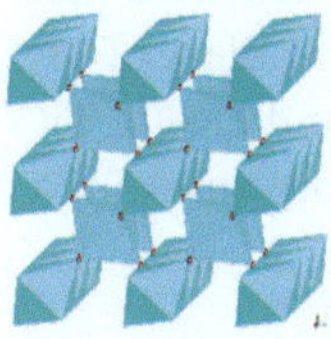	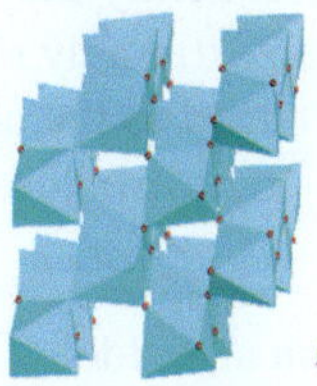	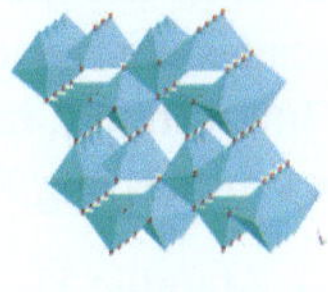

alkaline metal titanates to H_2TiO_3, following a heat treatment at 300–450 °C to induce a phase transformation to TiO_2(B). Unlike the three other key natural polymorphic forms, artificial TiO_2(B) exhibits perovskite-like layered structure.

In semiconductor photocatalysis, anatase is the most widely studied polymorph, showing superior photochemical performances owing to its high electron mobility, electron affinity, and transmittance for visible light. Rutile TiO_2 exhibits relatively low photocatalytic activity. However, under visible light irradiation, rutile with a narrower bandgap might deliver better photocatalytic performances. When designing sunlight-active photocatalysts, the coexistence of anatase and rutile phases in nanostructured TiO_2 can significantly enhance solar energy harvesting. This is the principle behind Aeroxide® P25 TiO_2, which predominantly contains anatase with a smaller proportion of rutile. Brookite is the least studied as a photocatalyst, mainly due to challenges in controlling phase purity. TiO_2(B) is only active in the UV region and exhibits photocatalytic activity comparable to that of anatase.

3 Synthesis of Nanostructured TiO_2

A **nanostructure** refers to a material or object characterized by structural features at the nanoscale, typically with at least one dimension ranging from 1 to 100 nm. This nanoscale confinement imparts a high surface-to-volume ratio to the resulting nanomaterials, significantly enhancing their physicochemical properties and performance compared to bulk materials. Nanostructures can be categorized into four major types based on the degree of spatial confinement:

- Zero-dimensional (0D) nanostructures: All dimensions are on the nanometer scale (e.g., nanoparticles).
- One-dimensional (1D) nanostructures: One dimension is on the nanometer scale, while the other two are larger (e.g., nanorods, nanowires, and nanotubes).
- Two-dimensional (2D) nanostructures: Two dimensions are on the nanometer scale, with the third being larger (e.g., nanosheets and nanoplates).
- Three-dimensional (3D) nanostructures: Porous or hierarchical structures consisting of low-dimensional nanobuilding blocks (NBBs)

A variety of synthesis methods, including the sol-gel process, hydro/solvothermal techniques, and the combustion method, have been developed to enable the controllable fabrication of TiO_2 nanomaterials in various forms of 0D nanoparticles (NPs), 1D nanorods, nanowires, and nanotubes, 2D nanosheets, and 3D superstructures, with variations in phases, dimensions, sizes, porosities, morphologies, and forms (e.g., powder, gel, film, and monolith). To the best of our knowledge, TiO_2 has been the most frequently fabricated transition metal oxide over the past four decades, delivering an extremely rich gallery of TiO_2 nanostructures.

Interestingly, in contrast to the abundance of TiO_2 nanostructures, the precursors for synthesizing TiO_2 nanostructures are quite limited, which include three key types:

- Metallic Ti
- Organic titanium alkoxide
- Inorganic precursors: $TiCl_4$, TiF_4, $Ti(SO_4)_2$, or commercial TiO_2

Clearly, free Ti^{4+} ions seem to be absent in wet-chemical media where sol-gel process is frequently involved. Thus, deep understanding of the physicochemical properties of precursors and their chemical reactions is essential to the wet-chemical synthesis. Moreover, considerable interests have been attracted by the nucleation and growth mechanisms of TiO_2 to reveal the formation of nanostructures. Therefore, Chapters 1–5 of this book systematically address the synthesis strategies of TiO_2 of different dimensions, with a special focus on providing a fundamental insight into the underlying chemical reactions involved, and the accompanied growth models of TiO_2 nuclei, with the ultimate goal of achieving a predictable synthesis of nanostructured TiO_2 with controllable textural properties.

4 Photocatalytic Applications of Nanostructured TiO_2

Continuous research aims to boost the photocatalytic efficiency of nanostructured TiO_2-based photocatalysts, especially in the context of utilizing sunlight and improving their performance for large-scale applications. Photocatalytic applications of nanostructured TiO_2 are vast and diverse, making them a critical material for addressing environmental and energy challenges, as comprehensively addressed in this book.

Environmental purification: TiO_2 nanostructures are highly effective at degrading a wide range of organic pollutants in both atmospheric and aquatic ecosystems. Common atmospheric pollutants include volatile organic compounds (VOCs) and nitrogen oxides (NO_x), while aquatic pollutants include dyes, pesticides, pharmaceuticals, and industrial chemicals. Under sunlight irradiation, photocatalytic oxidation of these contaminants can be conducted by the reactive oxygen species (ROS) generated by excited TiO_2, breaking them down into harmless substances such as CO_2 and water. Additionally, TiO_2 is effective in killing bacteria and viruses in air and water, making it valuable for applications in medical devices, food packaging, and water treatment systems.

Solar fuel generation and chemical synthesis: Nanostructured TiO_2 is widely used in photocatalytic water splitting to produce hydrogen. When exposed to UV light, TiO_2 generates electron–hole pairs that drive the water-splitting reaction, producing hydrogen gas as a clean energy source. TiO_2 nanostructures are also employed for photocatalytic CO_2 reduction, converting it into value-added chemicals such as CO, methane, or methanol. This process not only helps mitigate CO_2 emissions but also produces useful byproducts. Furthermore, TiO_2 is utilized in the photocatalytic synthesis of various organic compounds, offering a sustainable alternative to traditional chemical methods. These photocatalytic processes, including oxidation, reduction, and cross-coupling, operate under mild conditions without the use of harsh reagents or solvents.

Photocatalytic sensors and environmental monitoring: TiO_2-based photocatalytic sensors are used to detect pollutants, such as NO_x and VOCs, in the environment. These sensors leverage the photocatalytic properties of TiO_2 to generate detectable signals in the presence of pollutants, providing real-time monitoring capabilities for environmental quality.

Self-cleaning coating: TiO_2-based photocatalytic coatings show self-cleaning properties when exposed to UV light. Organic dirt, oils, and pollutants on TiO_2 surfaces can be readily degraded or mineralized to maintain cleanliness without manual intervention, which is particularly useful for application in windows, solar panels, and building facades.

References

1 Fujishima, A. and Honda, K. (1972). Electrochemical photolysis of water at a semiconductor electrode. *Nature* 238: 37–38.

2 Qian, R., Zong, H., Schneider, J. et al. (2019). Charge carrier trapping, recombination and transfer during TiO_2 photocatalysis: an overview. *Catalysis Today* 335: 78–90.

3 Hoffmann, M.R., Martin, S.T., Choi, W., and Bahnemann, D.W. (1995). Environmental applications of semiconductor photocatalysis. *Chemical Reviews* 95: 69–96.

4 Schneider, J., Matsuoka, M., Takeuchi, M. et al. (2014). Understanding TiO_2 photocatalysis: mechanisms and materials. *Chemical Reviews* 114: 9919–9986.

5 Ding, Y., Yang, I.S., Li, Z.Q. et al. (2020). Nanoporous TiO_2 spheres with tailored textural properties: controllable synthesis, formation mechanism, and photochemical applications. *Progress in Materials Science* 109: 100620.

Acknowledgments

We would like to express our deepest gratitude to the exceptional team at Wiley for their expertise, unwavering dedication, and invaluable support throughout the entire publication process. In particular, we extend our sincere thanks to Dr. Alice Qian, Elisha Benjamin, Sevanthi Vivekanandhan, and Dhanapriya Velumurthy for their meticulous attention to detail, tireless efforts, and commitment to ensuring the highest quality of presentation for this book. Their professionalism and dedication were pivotal in bringing this project to life.

We are also profoundly grateful to our colleagues and students, whose collaboration, insights, and encouragement have been indispensable over the years. This book would not have been possible without their collective contributions and support. We extend our heartfelt thanks to Yanjie Li and Shuangchao Tang for their invaluable assistance with reference management.

Finally, we would like to express our appreciation to everyone who contributed, directly or indirectly, to this work. Your dedication and support have made this journey truly rewarding.

Jia Hong Pan
Guangxi University, China

Wan In Lee
Inha University, Korea

Detlef W. Bahnemann
St. Petersburg University, Russia
Leibniz Universität Hannover, Germany
Shaanxi University of Science and Technology, China

Part I

Synthetic Methodologies

1

Selective Synthesis of TiO_2 Nanoparticles

Sungjin Park[1], Jia Hong Pan[2], Wan In Lee[1,]*

[1] *Department of Chemistry and Chemical Engineering, Inha University, Incheon, Korea*
[2] *State Key Laboratory of Featured Metal Materials and Life-Cycle Safety for Composite Structures, School of Resources, Environment and Materials, Guangxi University, Nanning, Guangxi, China*

*Corresponding author: wanin@inha.ac.kr

1.1 Introduction

Zero-dimensional titanium dioxide (TiO_2) nanoparticles (NPs) are the most frequently studied nanostructures for various photocatalytic applications. As the particle size decreases, both the specific surface area and surface-to-volume ratio increase significantly [1–4]. More interestingly, quantum confinement effects can occur in TiO_2 NPs. These properties are highly advantageous for photocatalytic applications of TiO_2 NPs, as high-energy photons can drive photocatalytic reactions, and photocatalytic processes can occur efficiently on the TiO_2 surface with large area. Consequently, the photocatalytic performance of TiO_2-based materials is strongly influenced by the size and shape of TiO_2 NPs. In this context, we delve into the synthetic methods for controlling the size and shape of TiO_2 NPs, discussing the principles and mechanisms behind these techniques.

1.2 Synthesis of TiO_2 NPs by Sol–Gel Reaction Under Acidic Condition

Traditionally, nano-sized TiO_2 particles have been formed through an acid-catalyzed sol–gel reaction of titanium alkoxides. The formation process mainly involves an acid-catalyzed hydrolysis step (Eq. 1.1), followed by polycondensation (Eq. 1.2), as described below.

$$\mathrm{Ti(OR)_4 + H_2O \rightarrow (RO)_3Ti{-}OH + R{-}OH} \tag{1.1}$$

$$\mathrm{2[(RO)_3Ti{-}OH] \rightarrow (RO)_3Ti{-}O{-}Ti(OR)_3 + H_2O} \tag{1.2}$$

Nanostructured TiO_2: Synthesis, Photocatalytic Properties and Applications, First Edition.
Edited by Jia Hong Pan, Wan In Lee and Detlef W. Bahnemann

The acidic environment with rich H^+ ions dramatically suppresses the dissociation of water molecules, thereby stabilizing the titanium alkoxide and retarding the hydration reaction.

The molar ratio of H_2O to Ti (denoted as the R-factor) plays a crucial role in this acid-catalyzed sol–gel reaction, as it governs the growth kinetics of the Ti species. When the water content is low, hydrolysis rates are slow, and the excess titanium alkoxides are present in the reaction mixture, the formation of Ti—O—Ti chains is promoted, leading to the development of closely packed, 3D polymeric networks. Thus, at low R-factors (typically $R < 10$), relatively monodispersed aggregates of TiO_2 nanocrystallites are produced. In this case, the growth kinetics of the aggregates are strongly dependent on the colloidal stability in the solution.

In contrast, at high R-factors, the abundance of water accelerates the hydrolysis of titanium alkoxide, favoring the formation of $Ti(OH)_4$. This rapid hydrolysis leads to a large quantity of Ti—OH species, which hinders the formation of 3D polymeric network. As a result, loosely packed first-order sol–gel particles are formed. The rapid hydrolysis causes nucleation and growth to occur almost instantaneously, leading to the precipitation of large aggregates. These initial particles are amorphous, unstable, and result in the formation of a white suspension due to the presence of sizable aggregates.

As the reaction continues under elevated temperatures, chloride or nitrate ions can break up these aggregates into smaller particles. When the temperature goes above 100 °C or a hydrothermal reaction is used, the nucleation and growth of TiO_2 NPs compete with the growth of larger particles. Smaller particles tend to disappear due to dissolution, mutual aggregation or merging with larger particles by Ostwald ripening. As a result, rather monodispersed and crystallized TiO_2 NPs are formed after a long reaction time, as illustrated in Scheme 1.1.

Sol–gel reaction forming TiO_2 nanomaterials can occur even under nonhydrolytic conditions without the addition of water, although the reaction rate is very slow. Typically, titanium halide (TiX_4) such as titanium chloride is used as main Ti-precursor, while metal alkoxides, alcohols, and organic ethers serve as oxygen donor molecules, as illustrated in Eqs. (1.3) and (1.4).

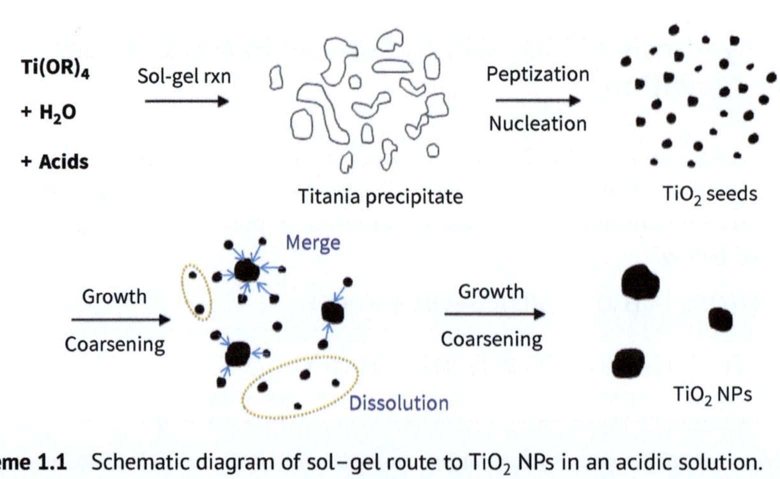

Scheme 1.1 Schematic diagram of sol–gel route to TiO_2 NPs in an acidic solution.

$$TiX_4 + Ti(OR)_4 \rightarrow TiO_2 + 4RX \tag{1.3}$$

$$TiX_4 + 2ROR \rightarrow 2TiO_2 + 4RX \tag{1.4}$$

Apparently, the alkoxide groups in titanium alkoxides, alcohols or ethers can provide oxygen to assist the formation of Ti—O—Ti bridges.

1.2.1 Peptization Process

Peptization refers to the process in which aggregated colloids are separated and redispersed into individual particles during a sol–gel reaction. A typical example of this process is the formation of colloidal TiO_2 NPs through the hydrolysis and condensation of titanium alkoxides. In the presence of a large excess of water (R-factor = 100–200), titanium alkoxides undergo rapid hydrolysis and polycondensation, producing ultrafine primary particles with diameters ranging from 3 to 10 nm. However, these tiny particles tend to agglomerate to minimize surface energy, resulting in the formation of large precipitates of macroscopic dimensions. Consequently, the sol–gel solution becomes opaque with a white color.

By adding acids such as nitric acid (a common peptizing agent with a typical peptization temperature range of 30–90 °C) or hydrochloric acid, redispersion can occur, and the solution becomes translucent after prolonged stirring at an elevated temperature. As shown in Figure 1.1, at same pH condition and with nitric acid as a peptizing agent, TiO_2 NPs showed a gradual decrease in size with the elapse of time. Such deagglomeration and redispersion of TiO_2 are driven by the enhanced electrostatic repulsion among the various primary particles in each agglomerate, as well as the chemical breaking of the bonds that hold the primary particles together.

Thus, two opposing mechanisms simultaneously occur during the peptization process: the fragmentation of agglomerates induced by the peptizing agent, and the reagglomeration of fragments due to insufficient suspension stabilization [5]. When the sol–gel reaction is initiated, precipitation occurs within seconds after mixing the reactants. However, complete peptization is a slow process that may take a few weeks (Figure 1.1). The resulting colloidal TiO_2 NPs remain relatively amorphous, with no significant crystal growth occurring, owing to the relatively low reaction temperature.

According to Matsoukas's report, with the passage of aging time, the resultant colloidal particle size gradually decreases [5]. It is deduced that the colloidal particle size decreases following the first-order kinetics, as described in Eq. (1.5).

$$\frac{D}{D_\infty} = 1 + a\,e^{-\bar{K}_d t}, \tag{1.5}$$

where D is colloidal particle diameter at reaction time (t), D_∞ ultimate colloidal particle diameter, $\overline{K}_d$ the average rate constants for deagglomeration, and a constant.

In addition, the final size of colloidal TiO_2 is strongly dependent on the concentration of Ti-precursor. As shown in Figure 1.1, when the concentration of titanium tetraisopropoxide (TTIP) was varied from 0.0296 to 0.0727 g cm^{-3}, it was

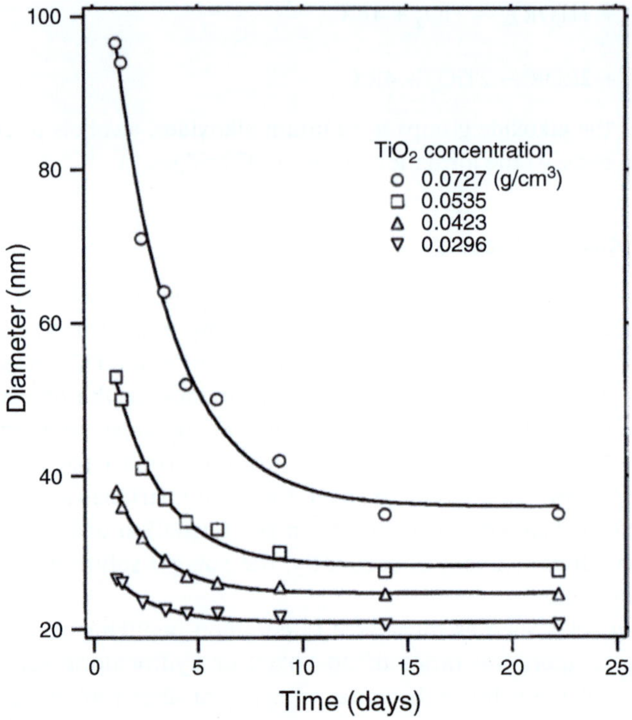

Figure 1.1 Colloidal TiO_2 particle diameter as a function of reaction time for the sol–gel solution with various TTIP concentrations under a constant pH of 1.10. The samples were prepared by diluting a fixed amount of Ti sol in different nitric acid (used as a peptizing agent) solution [5]/with permission of Elsevier.

found that the final particle size of colloidal TiO_2 increases significantly, when the concentration of Ti-precursor increases.

In the peptization process, the sol–gel reaction temperature is also a crucial parameter. Vorkapic and Matsouka prepared TiO_2 NPs derived from sol–gel solutions with various titanium alkoxides (ethoxide, *n*-propoxide, isopropoxide, and butoxide), and monitored the resulting colloidal diameter (*D*) as a function of peptization temperature [6]. As shown in Figure 1.2, the effect of the peptization temperature is quite dramatic. All the peptized TiO_2 NPs exhibited the smallest size at 50 °C, regardless of the titanium alkoxides used. Among the titanium alkoxides, titanium *n*-propoxide provided the smallest colloidal diameter, though the difference was not significant at the optimal temperature of 50 °C.

1.2.2 Coarsening Process

Coarsening, also known as Ostwald ripening, refers to the process in which larger particles grow at the expense of smaller ones. In the wet chemical synthesis, coarsening and aggregation can compete during the nucleation and growth steps, thereby altering the particle size distribution. Coarsening generally occurs at higher

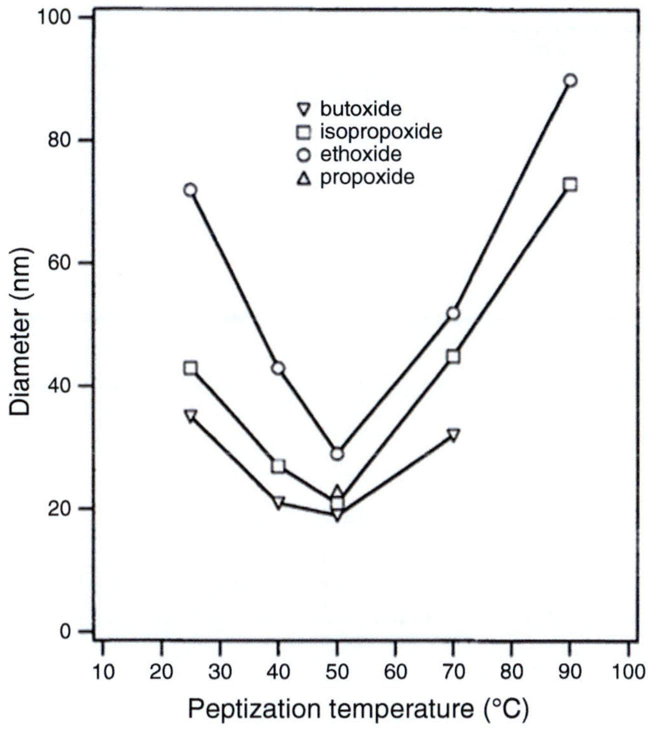

Figure 1.2 Colloidal particle diameter as a function of peptization temperature for sol–gel solutions with various titanium alkoxides [6]/with permission of John Wiley & Sons.

temperatures than those used for peptization, usually above 100 °C or under hydrothermal conditions, while peptization typically is performed at 30–90 °C.

Oskam et al. investigated the growth kinetics of TiO_2 NPs derived from TTIP in the acidic medium [7]. Figure 1.3a shows a transmission electron microscopy (TEM) image of as-prepared anatase TiO_2 particles by peptization of sol–gel solution at 85 °C for 14 h. The average particle radius was 1.6 nm (diameter = 3.2 nm) and the fringe patterns shown in the TEM image suggest that the particles are crystalline. The solution was then further heated to higher temperatures. Figure 1.3b, c compare the particle size of TiO_2 NPs grown at 160 °C and 200 °C for 16 h, respectively, revealing that the particle size increases significantly with the growth temperature.

Figure 1.4a shows the average particle radius (r) versus coarsening time (t) at 160 °C and 200 °C. As the coarsening time progresses, particle radius gradually increases from 1.6 to 2.0 nm and 3.7 nm after 140 h at 160 °C and 125 h at 200 °C, respectively. Figure 1.4b reveals that cubic of particle radius (r^3), which is equivalent to the particle volume, is proportional to coarsening time. Therefore, a longer reaction time in preparing TiO_2 NP leads to larger particle size.

The Lifshitz–Slyozov–Wagner model for coarsening offers a quantitative framework for understanding the time dependence of particle size and size distribution

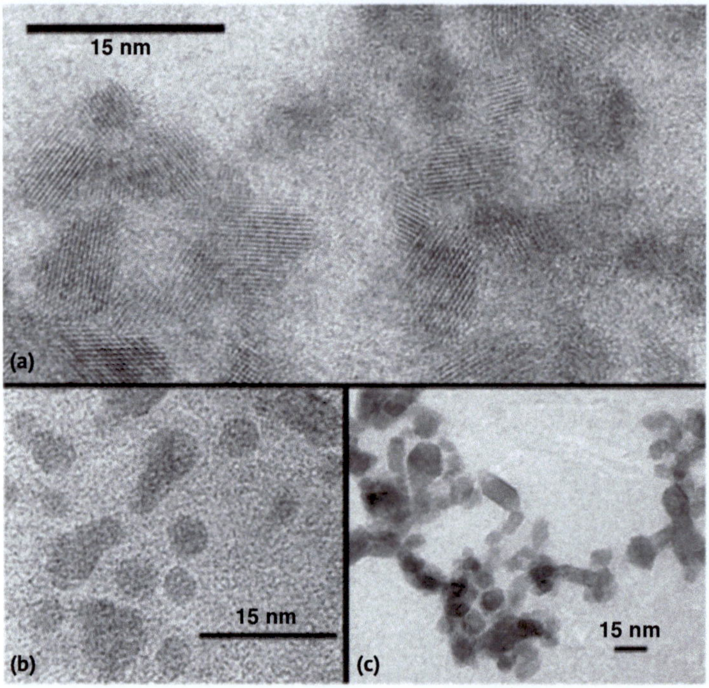

Figure 1.3 (a) High-magnification TEM image of the as-prepared anatase TiO_2 NPs after peptizing reaction at 85 °C for 14 h; (b, c) TEM images of the anatase TiO_2 NPs after 16 h growth at 160 °C and 200 °C, respectively. [7]/American Chemical Society.

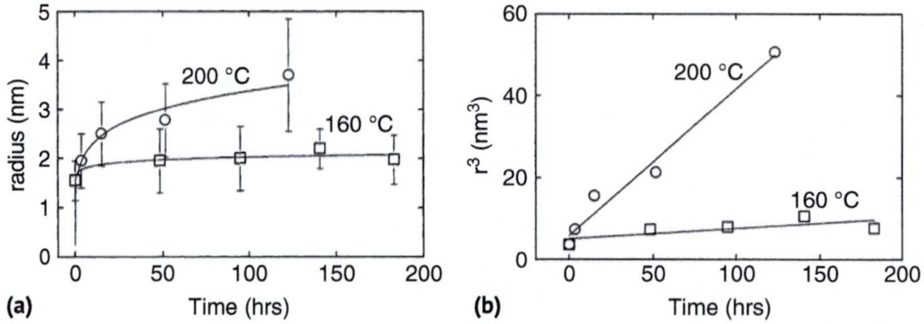

Figure 1.4 Average radius (r) (a) and r^3 (b) of the peptized TiO_2 NPs prepared at 160 °C and 200 °C as a function of growth time [7]/with permission of American Chemical Society.

[7, 8]. The coarsening process is primarily driven by the dependence of the solubility of a solid phase on particle size, as described by the Gibbs–Thomson relation. Assuming spherical NPs, the solubility (c_r) of a particle with radius (r) is expressed by:

$$c_r = c_\infty \exp\left(\frac{2\gamma V_m}{RT}\frac{1}{r}\right), \tag{1.6}$$

where c_∞ is the solubility at a flat surface, γ surface energy of the solid, V_m molar volume, R gas constant, and T absolute temperature. For the case where

$(2\gamma V_m/rRT) < 1$, such that the exponential term in Eq. (1.6) can be linearized [7]. From the assumption that the growth rate is determined by diffusion of the solute from the smaller particles to the larger particles, the following rate law is obtained (Eq. 1.7).

$$\bar{r}^3 - \bar{r}_o^3 = \frac{8\gamma D V_m^2 c_\infty}{9RT} t \tag{1.7}$$

where $\bar{r}$ is the average particle radius at reaction time t and $\bar{r}_o$ the average particle radius at time zero. Equation (1.7) suggests that the average volume of prepared NPs is proportional to coarsening time. Thus, the trend observed in Figure 1.4b agrees with theoretical calculation.

1.2.3 Rutile Phase Formation

Among the various crystallographic phases of TiO_2, the rutile phase is thermodynamically the most stable and forms at higher temperature than 650 °C, whereas the anatase phase is a kinetically more favorable at lower temperatures and on a smaller particle size. In the sol–gel reaction, for example, when TiO_2 NPs are smaller than ~15 nm, the formation of more stable anatase phase is preferred [1]. After a long hydrothermal reaction, larger TiO_2 NPs can be included among the small NPs. Thus, there is a high possibility that these large NPs are of the rutile phase, because a favorable condition for rutile formation has been achieved. For instance, Chae et al. prepared TiO_2 NPs with different particle sizes of 7, 15 and 25 nm by changing the solvent system during hydrothermal reaction under acidic condition. Rutile phase was included only in the 25-nm-sized TiO_2 NPs, whereas 7- and 15-nm-sized TiO_2 NPs showed the pure anatase phase [9].

The formation mechanism of the rutile phase during sol–gel reaction is related to the phase transition from anatase to rutile. Initially, anatase nuclei and ultrafine TiO_2 NPs are formed early in the sol–gel reaction. Over time, these NPs merge to create larger anatase NPs through Ostwald ripening (coarsening process). During this merging step, dislocations or defects are generated at the interface of epitaxially attached anatase NPs. The anatase-to-rutile transformation can be triggered at these interfacial defect sites, forming large rutile NPs with the anatase seeds buried within, as illustrated in Scheme 1.2.

Penn and Banfield investigated the anatase-to-rutile phase transition under hydrothermal conditions [10, 11]. They found that the nucleation of the rutile phase occurs at the twin boundaries of anatase nanocrystals, which are formed by

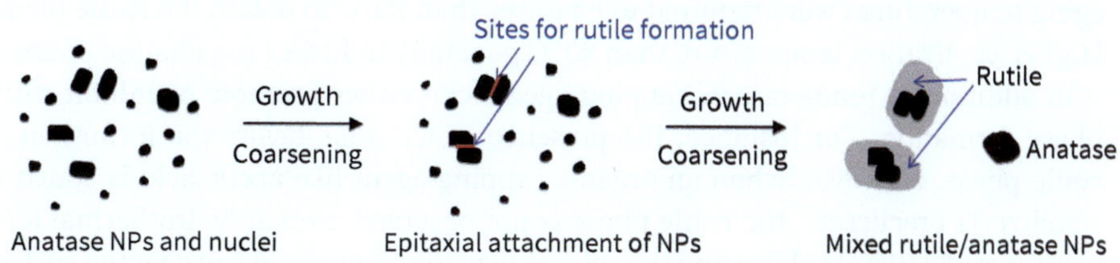

Scheme 1.2 Schematic diagram describing the formation of rutile TiO_2 phase by phase transition from anatase TiO_2 NPs.

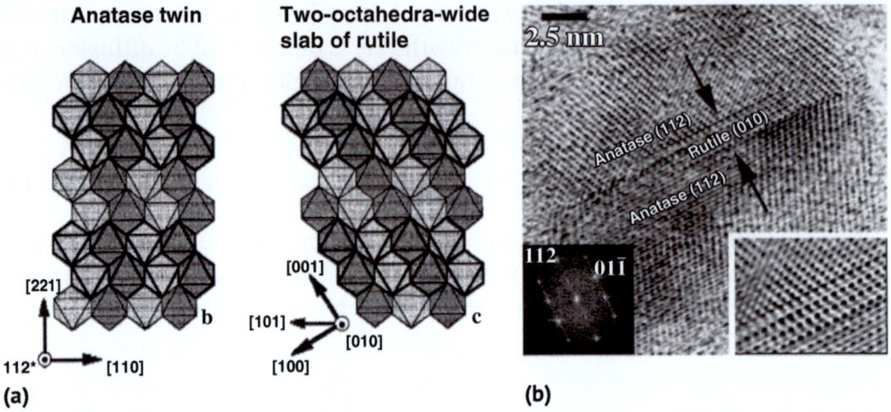

Figure 1.5 (a) Space-filling octahedral structures of the anatase {112} twin (left) and (010) rutile (right). Herein, the bold-outlined octahedra highlights the shared strips between the anatase twin and rutile structures. (b) TEM image exhibiting a rutile nucleus oriented in the (010) plane on the {112} anatase twin surface [10].

the epitaxial attachment of anatase nanocrystals (NCs). Figure 1.5a illustrates their observed space-filling octahedral structures of anatase {112} twins and (010) rutile structures. The nucleation of rutile takes place on the {112} anatase twin surfaces. Figure 1.5b presents a TEM image of rutile nuclei forming at the {112} anatase twin surface, showing that the (010) rutile plane is established between the {112} anatase planes. These interfaces reduce the activation energy barrier for rutile nucleation, thereby promoting the formation of the rutile phase.

Barbé et al. also investigated the rutile phase formation mechanism via sol–gel reaction and hydrothermal process [12]. The peptized TiO_2 NPs were subjected to hydrothermal conditions at 200–250 °C for 12 h. Rutile phase, along with anatase, appeared in TiO_2 NPs reacted at 240–250 °C. The anatase NPs were ~15 nm, while rutile particles were ~200 nm. Rutile formed suddenly at 240 °C, with particles significantly larger than anatase crystallites. This suggests that the anatase-to-rutile transformation involved local rearrangements in anatase aggregates before the rutile formation. Notably, small anatase NPs were buried inside larger rutile crystals, aligning with the proposed rutile formation mechanism (Scheme 1.2).

A highly acidic medium (e.g., $pH < 1$) is favorable for the formation of rutile phase during sol–gel process. For example, Park et al. prepared nanocrystalline rutile TiO_2 NPs by the direct hydrolysis of $TiCl_4$ in water [13]. The obtained 0.5 M TiO_2 sol solution was aged for five days as a peptization process. Interestingly, in this process, aging temperatures were required to be lower than 80 °C to obtain the rutile phase. Higher peptization temperature than 80 °C generally induces pure anatase phase.

In addition to temperature, capping agents can either promote or inhibit rutile phase formation. For instance, the presence of Cl^- ions favors the formation of rutile phase. However, when an organic capping agent like acetic acid is added to stabilize Ti-precursors, the rutile phase is not observed even at hydrothermal temperatures of 250 °C [12]. During the sol–gel reaction, acetic acid can coat the surface of initially formed anatase NPs and prevent their aggregation or epitaxial attachment. Consequently, no suitable sites for rutile growth are formed, blocking the rutile phase from developing. This observation further supports the proposed mechanism for rutile phase formation.

1.2.4 Size Control Depending on Solvent System

The solvent system or Ti-precursor concentration in sol–gel reaction or hydrothermal reaction can influence the size of TiO_2 NP products. By adding alcohol instead of 100% water, the sol–gel reaction can be slowed down, leading to the suppression of the crystal growth rate. Accordingly, the size of the resultant TiO_2 NPs decreases. Also, the Ti-precursor concentration will be an important factor determining the size of TiO_2 NPs. Increase of Ti-precursor concentration can induce fast hydrolysis reaction, thus leading to tremendous increase in the number of nuclei. As a result, only small TiO_2 NPs are formed due to the difficulty in crystal growth for too many nuclei.

Chae et al. synthesized TiO_2 NPs by tuning the TTIP concentration and solvent composition of water and ethanol via solvothermal process [9]. TTIP (1.5–15 mmol) was added dropwise to 150 mL of mixed ethanol and water solution. The pH of the solution was controlled to 0.7 by the addition of nitric acid. After a mild stirring a few hours, the resultant solution was moved to a glass-lined autoclave, where it was subjected to a reaction at 240 °C for 4 h. TiO_2 NPs were then obtained in the form of a colloidal suspension. These NPs were primarily composed of individual particles with minimal aggregation, and their sizes were adjusted in the range of 7–25 nm by altering the reaction parameters. TiO_2 NPs with a size of 25 nm were produced in a water-rich environment $([\text{ethanol}]:[\text{water}] = 1:8)$ and under a low TTIP condition (0.02 M). In contrast, 15 and 7 nm TiO_2 NPs were synthesized using solvent systems of 0.04 M TTIP with ethanol/water = 1:2 and 0.10 M TTIP with ethanol/water = 4:1, respectively, as illustrated in Figure 1.6.

Surface capping agents are often employed to enhance the size uniformity and dispersibility of TiO_2 NPs in solution. For example, Scolan and Sanchez et al. synthesized monodisperse, nonaggregated TiO_2 NPs (1–5 nm size) at 60 °C using acetylacetone and p-toluenesulfonic acid as capping agents [14]. Due to the

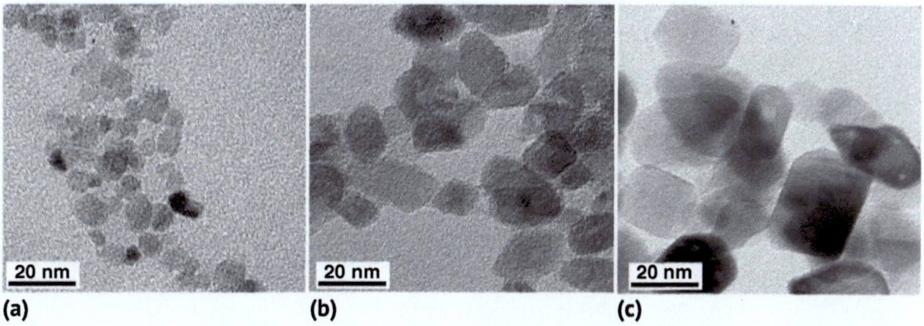

Figure 1.6 TEM images of various TiO_2 NPs prepared under various TTIP concentrations and ethanol/water compositions. (a) 7-nm-sized TiO_2 NPs from 0.10 M TTIP in 4:1 ethanol/water; (b) 15 nm NPs from 0.04 M TTIP in 1:2 ethanol/water; (c) 25 nm NPs from 0.02 M TTIP in 1:8 ethanol/water [9]/American Chemical Society.

successful binding of acetylacetonato ligands to the TiO_2 surface, the obtained TiO_2 NPs remained stable when dispersing in water–alcohol solution at concentrations above 1 M.

1.2.5 Nonhydrolytic Sol–Gel Process

Nonhydrolytic sol–gel process involves reactions between Ti-precursors and oxygen donor molecules, such as metal alkoxides, alcohols, or organic ethers. These reactions lead to the condensation of Ti—Cl and Ti—OR, forming Ti—O—Ti bridges. The Colvin group developed a well-known synthetic method for the nonhydrolytic sol–gel synthesis of uniform, ultrafine TiO_2 NPs under a dry inert gas atmosphere by injecting TTIP into a hot solution (around 300 °C) of TiX_4(X = F, Cl, Br, or I) mixed with trioctylphosphine oxide (TOPO) in heptadecane [15]. TOPO acts as a capping agent, preventing the merging of individual NPs. The average particle size of the resulting TiO_2 NPs was strongly influenced by the halide species in TiX_4. A higher nucleophilicity of the halide leads to smaller anatase NPs. The average particle sizes of the TiO_2 NPs were 9.2 nm for TiF_4, 7.3 nm for $TiCl_4$, 6.1 nm for $TiBr_4$, and 3.8 nm for TiI_4, as shown in Figure 1.7. This suggests that TiX_4 functions both as a crystallization agent and as a reactant in the reactions outlined in Eqs. (1.3) and (1.4). The X^- ion attacks the electrophilic oxygen-bound carbon of the titanium alkoxide, leading to the elimination of alkyl halide and the formation of Ti—O bonds.

Niederberger et al. synthesized crystallized TiO_2 NPs by refluxing $TiCl_4$ and benzyl alcohol for several days, followed by calcination at 450 °C [16]. The nonhydrolytic reaction between $TiCl_4$ and benzyl alcohol, even at temperatures as low as 40 °C, results in anatase phase TiO_2 NPs with high crystallinity and uniformity. By adjusting the temperature and $TiCl_4$ concentration, the particle size can be controlled within the range of 4–8 nm, with higher temperatures and increased $TiCl_4$ concentrations leading to larger particles.

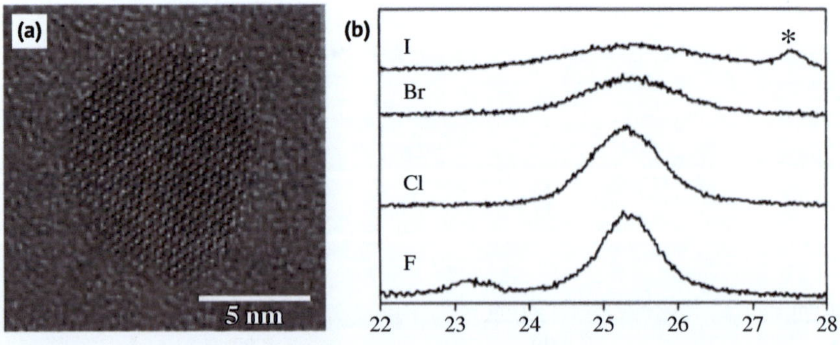

Figure 1.7 (a) TEM image of TiO_2 NPs prepared from $TiCl_4$ and TTIP in a mixed solution of TOPO and heptadecane at 300 °C. (b) X-ray diffraction patterns of the anatase (101) displaying various line broadening depending on the halides. From the Scherrer equation, the particle sizes with TiF_4 were determined to 9.2 nm, 7.3 nm with $TiCl_4$, 6.1 nm with $TiBr_4$, and 3.8 nm with TiI_4 [15]/with permission of American Chemical Society.

1.3 Sol–Gel Reaction in Basic Condition

Highly dispersible anatase TiO_2 NPs of diverse sizes and shapes have been formed by sol–gel reaction under basic condition. Various organic amines or alkylammonium hydroxides have been applied to achieve basic environment. In basic condition, hydrolysis reaction of $Ti(OR)_4$ is expedited due to high OH^- ion concentration in the solution. At the same time, the added amines can play a key role as capping agent or shape controller, because they have good binding affinity toward TiO_2 surface. Therefore, aggregation of TiO_2 NPs can be avoided during coarsening process in sol–gel reaction. As a result, highly crystallized TiO_2 NPs without aggregation can be synthesized. Furthermore, the added organic amines or alkylammonium hydroxides have face-selective binding affinity to the TiO_2 NC. Therefore, TiO_2 NCs can be grown in a certain direction, thus leading to shape control of TiO_2 NCs. It is noteworthy that pure anatase phase without inclusion of rutile phase is formed in general under basic conditions. This is due to a high binding affinity of amine-based capping agents to the surface of TiO_2, which are formed at early stages of sol–gel reaction. Therefore, the merging of individual TiO_2 NCs during the coarsening process is fundamentally blocked, resultantly leading to formation of no defect sites. Therefore, rutile phase cannot be formed due to free from defect sites.

1.3.1 Effect of Primary Amines

Triethanolamine (TEOA) is a common chelating agent that can stabilize TTIP and retard hydrolysis reaction during sol–gel reaction, as shown in Eq. (1.8).

$$2Ti(OR)_4 + 3N(C_2H_4OH)_3 \rightarrow N(C_2H_4O)_3Ti{-}OH_4C_2N(C_2H_4OH)C_2H_4O \\ {-}Ti(OH_4C_2)_3N + 8ROH \tag{1.8}$$

Sugimoto et al. investigated the influence of TEOA on the size and shape of TiO_2 NPs [17]. TTIP was stabilized by TEOA with a molar ratio of 1:2. The mixture was diluted with water and aged at 100 °C for one day and then at 140 °C for three days. In pH 5–10, cuboidal NPs were formed, whereas elongated TiO_2 NPs were formed at pH 11.5, as shown in TEM images of Figure 1.8. The structural evolution can be attributed to the preferential binding affinity to {001} faces of the anatase crystal planes parallel to the *c*-axis of the tetragonal structure under highly alkaline conditions. The analogous facet-selective adsorption and similar shape control to form ellipsoidal particles with a high aspect ratio was also achieved at alkaline conditions

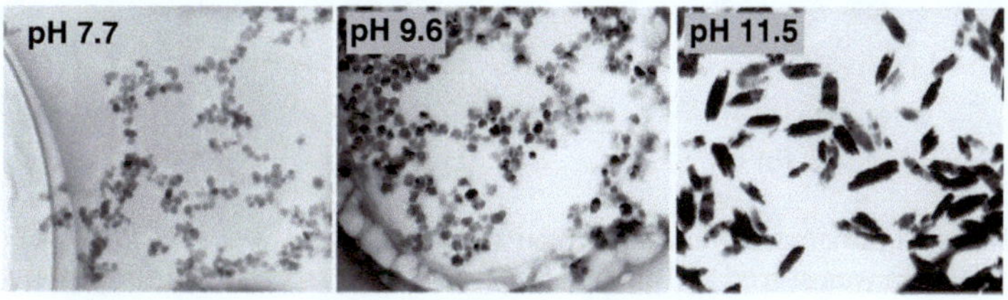

Figure 1.8 Variation of TiO_2 NP shapes depending on the pH in the reactant solution. In all reactions, [TTIP]: [TEOA]= 1:2 [17]/with permission of Elsevier.

and by other primary amines, such as diethylenetriamine, ethylenediamine, trimethylenediamine, and triethylenetetramine [18, 19].

1.3.2 Effect of Secondary and Tertiary Amines

Secondary amines like diethylamine (DEA) and tertiary amines such as trimethylamine (TMA) and triethylamine (TEA) can direct the growth of TiO_2 NPs into ellipsoidal shapes with a low aspect ratio, rather than high aspect ratio ellipsoids. When DEA was introduced at pH values above 11, the TiO_2 structure, was transformed into larger ellipsoidal NPs with a lower aspect ratio (Figure 1.9a, b). In addition, the introduction of TMA and TEA further increased the size of the NPs (Figure 1.9c, d). These results suggest that secondary and tertiary amines promote the crystal growth of TiO_2 in a manner distinct from the effect of TEOA.

Shiu et al. systematically explored the role of TEOA and DEA in a hydrothermal reaction using TTIP as a precursor [20]. The produced TiO_2 single crystals (denoted to HD1, HD2, HD3, HD4, and HD5) exhibited octahedron-like morphologies with the sizes ranging 30–400 nm. TEOA served as a chelating agent and stabilized TTIP during hydrolysis reaction, whereas DEA accelerated the crystal growth. By adjusting hydrothermal reaction time and temperature, growth mechanism and roles of TEOA and DEA were investigated.

Scanning electron microscopy (SEM) images in Figure 1.10 depict the TiO_2 NPs prepared from the two-step sol-hydrothermal process under varying hydrothermal temperatures and reaction times. At a hydrothermal temperature of 230 °C, the particle sizes on average were about 30 nm at 2 h (HD1), 50 nm at 8 h, and 70 nm at 16 h. Increasing the hydrothermal temperature to 250 °C for 12 h expanded the size to over 100 nm. In addition, the TiO_2 nanocrystals converted from a cuboid to the octahedron-like shape, as the temperature increased from 230 to 270 °C. The SEM image in Figure 1.10g shows that larger truncated octahedral nanocrystals (HD3) were synthesized at 270 °C in 16 h. Significantly larger NPs (HD4, shown in Figure 1.10h) were produced by using HD3 as a seed nanocrystal. Subsequently, using HD4 as a seed resulted in even larger truncated octahedral crystals (HD5, shown in Figure 1.10i).

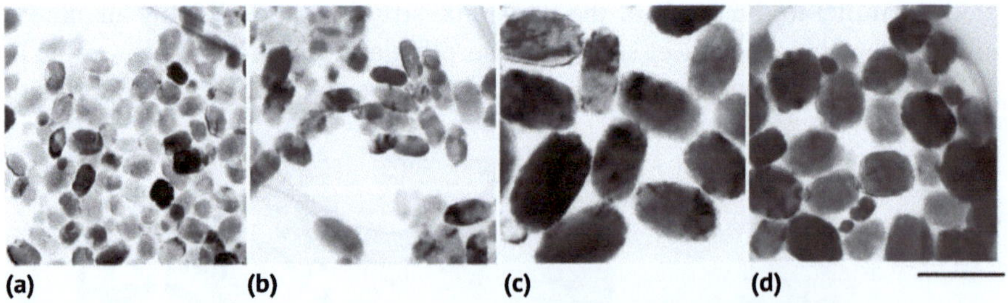

Figure 1.9 TEM images of various TiO_2 NPs in the anatase phase. TTIP/TEOA = 1:2, sol–gel solution was stirred at 100 °C for 24 h, and then reacted at 140 °C in a hydrothermal bomb. For the shape control, various amines were introduced while the solution pH was kept at pH 11.0–11.4. (a) 0.05 M DEA, (b) 0.20 M DEA, (c) 0.50 M TMA, and (d) 0.20 M TEA. The scale bar indicates 200 nm [18]/with permission of Elsevier.

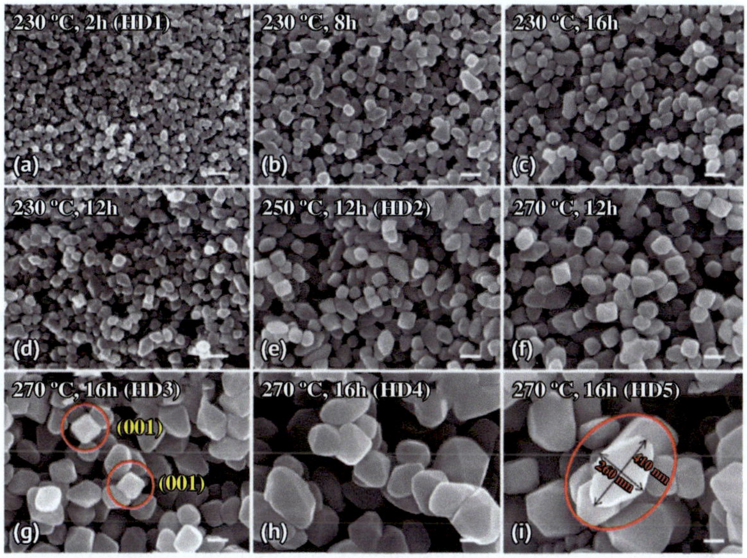

Figure 1.10 SEM images of TiO_2 NCs prepared at various synthetic conditions: (a) 230 °C for 2 h (HD1), (b) 230 °C for 8 h, (c) 230 °C for 16 h, (d) 230 °C for 12 h, (e) 250 °C for 12 h (HD2), (f) 270 °C for 12 h, (g) 270 °C for 16 h (HD3). Large TiO_2 NCs prepared from the additional hydrothermal reaction using HD3 as a seed at 270 °C for 16 h (HD4) (h) and HD4 as a seed at 270 °C for 16 h (HD5) (i). All scale bars represent 100 nm [20]/American Chemical Society.

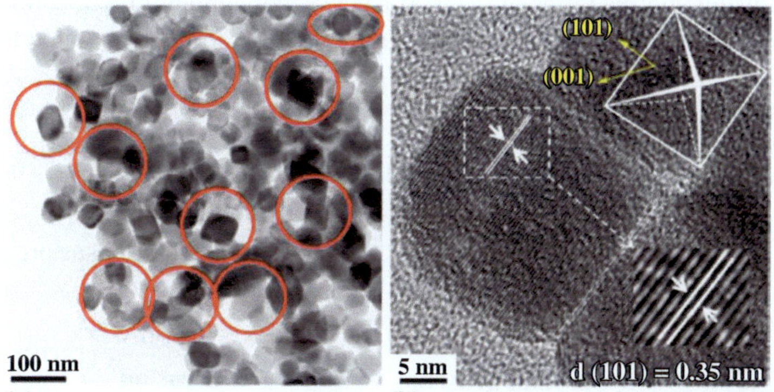

Figure 1.11 (a) TEM images of HD1 NCs displaying a cuboidal shape. The presence of rhombic shaped NCs shown in the red circles suggests formation of octahedron-like NCs. (b) High resolution TEM image of a HD1 NC, indicating a 3D octahedral structure with (101) and (001) plane orientations. The fringe patterns with a distance of 0.35 nm shown in the inset are indicative of the anatase (101) facet [20]/American Chemical Society.

DEA was also used to assist the formation of ellipsoidal TiO_2 NPs [18, 21]. Herein, at initial hydrothermal reaction, small octahedral or cuboid-like NCs (HD1) were formed, as described in TEM images of Figure 1.11. They were then grown to much larger, truncated octahedral nanocrystals (HD3–HD5). Notably, crystal growth is greatly enhanced by the introduction of DEA during hydrothermal reaction. Thus, DEA is considered to be the shape controller as well as the catalyst promoting crystal growth.

As the hydrothermal temperature was raised to 270 °C, the increases in particle size and decrease in surface area of TiO_2 NPs weaken the capping effect of DEA. At this stage, DEA likely facilitated the growth along the [101] facet, resulting in larger truncated octahedral crystals (HD3). During the initial 10 h of the hydrothermal reaction, the growth rate of the [001] facet was initially twice that of the [101] facet. However, after this period, the growth rate of the [001] facet increased to 3–3.5 times faster than that of the [101] facet. This acceleration in growth along the [001] direction may have been influenced by TEOA, which was gradually released from hydrothermal process and inhibited growth along the [001] facet. Consequently, the [001] facet expanded, leading to the formation of truncated octahedral crystals.

As the nanocrystals continued to evolve in a fresh sol solution, DEA once again suppresses the growth along the [101] direction. This allowed further growth along the [001] direction, producing much larger truncated octahedral crystals with a smaller exposed [001] facet area (HD5). The transition from HD1 to HD5 can be explained by a biphasic growth model, which shows the initial expansion of the [001] facet from HD1 to HD3, followed by its contraction from HD3 to HD5, as depicted in Scheme 1.3.

Size and shape of TiO_2 NPs can be tailored through variations in the solvent compositions used in hydrothermal reactions. Sung et al. found that in synthesizing TiO_2 NPs with TTIP as Ti-precursor and TEOA and DEA as amine additives, the solvent system of an ethanol–water mixture instead of pure water could slow down both the sol–gel reaction and the crystal growth during the hydrothermal process [22]. Figure 1.12 shows TEM images of TiO_2 NPs synthesized at 230 °C by varying ethanol/water volume ratios (X_{EtOH}). At a 90% water content ($X_{EtOH} = 0.1$), the particles are elongated NP with diameters around 60 nm and lengths ~100 nm, which

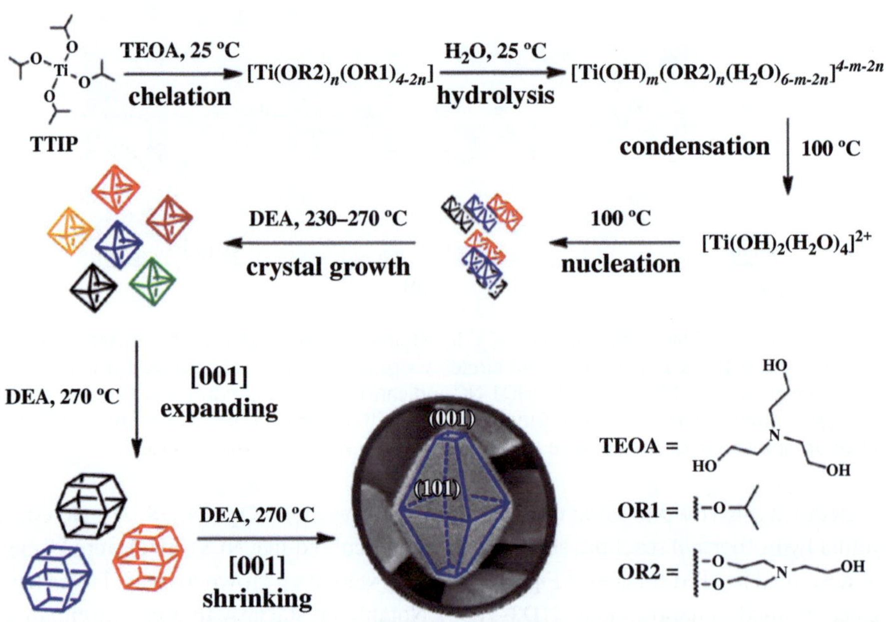

Scheme 1.3 Schematic diagram describing the formation mechanism for the anatase TiO_2 NCs of HD1–HD5 with octahedron-like shape [20]/American Chemical Society.

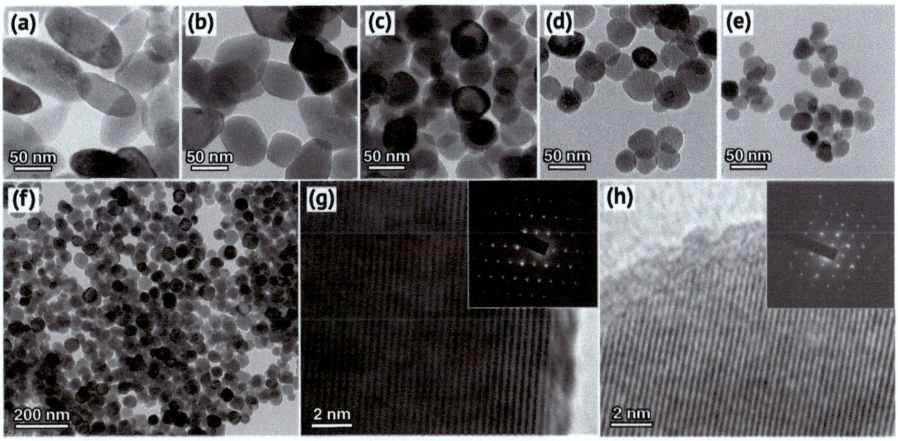

Figure 1.12 TEM images of TiO_2 NPs prepared under X_{EtOH} at 0.1 (a), 0.3 (b), 0.5 (c), 0.7 (d), and 0.85 (e), showing their NP sizes are 100, 65, 50, 40, and 30 nm, respectively. A low magnification TEM image for NP50 (f) and high-resolution images for NP65 (g) and NP50 (h). The insets of (g) and (h) show the selected area electron diffraction (SAED) patterns, obtained from a single NP65 and NP50, respectively. [22]/The Royal Society of Chemistry.

agrees with the previous findings [20]. As the ethanol content increases, there is a noticeable reduction in size and a shift from elongated structure to spherical shape. At $X_{EtOH} = 0.3$, the TiO_2 NPs remain slightly elongated with diameters of 50 nm and lengths of 65 nm (termed NP65), as shown in Figure 1.12b. At higher ethanol content of $X_{EtOH} = 0.5$, 0.7, and 0.85, the NPs take on a uniform spherical shape with diameters decreasing to 50, 40, and 30 nm, respectively, as seen in Figure 1.12c–e. Notably, the TiO_2 NPs with a size of 50 nm (NP50) are highly monodisperse spherical structures and are well separated from each other, as shown in the low-magnification TEM image in Figure 1.12f. Overall, the sizes and shapes of TiO_2 NPs are strongly dependent on X_{EtOH}, and the compositions to achieve spherical morphologies is $X_{EtOH} = 0.5$–0.85. When X_{EtOH} is smaller than 0.5, the NPs become gradually elongated, whereas their diameters do not change significantly.

1.3.3 Effect of Tetramethylammonium Hydroxide (Me_4NOH, TMAH)

Chemseddine and Moritz examined the effect of TMAH on the sol–gel reaction of titanium alkoxides. TMAH catalyzes the sol–gel reaction and introduces an organic cation (Me_4N^+) to TiO_2 NP. That is, Me_4N^+ ions are bound to the surface of the polyanionic anatase clusters, formed in the early sol–gel reaction [23].

Typical synthetic procedure is as follows: 0.333 mL TTIP dissolved in 10 mL anhydrous 2-propanol (the molarity of TTIP: 1.14 μM) was added dropwise to 150 mL aqueous Me_4NOH solution (the molarity of Me_4NOH: 0.26 μM), chilled in an ice bath and vigorously stirred. It was then refluxed at 90–100 °C for 6 h to form TiO_2 NPs. The shape of TiO_2 NP was triangular and monodispersed with a length of ~20 nm, as shown in Figure 1.13a. The as-prepared suspension was transferred to an autoclave and hydrothermally reacted at 175–200 °C for 5 h. Figure 1.13b clearly shows a change in structure from triangle to rectangle (width: 12 nm, length: 28 nm). When

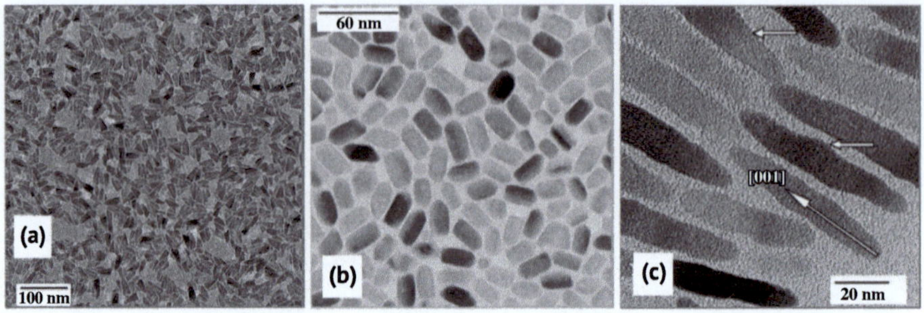

Figure 1.13 TEM images of various TiO_2 NCs. (a) Me_4NOH molarity of 0.26 mM, reaction at 90–100 °C for 6 h. (b) Hydrothermal reaction at 175–200 °C for 5 h using the sample in (a). (c) Me_4NOH molarity of 0.82 mM, reaction at 90–100 °C for 6 h followed by hydrothermal reaction at 200 °C for 5 h [23]/John Wiley & Sons.

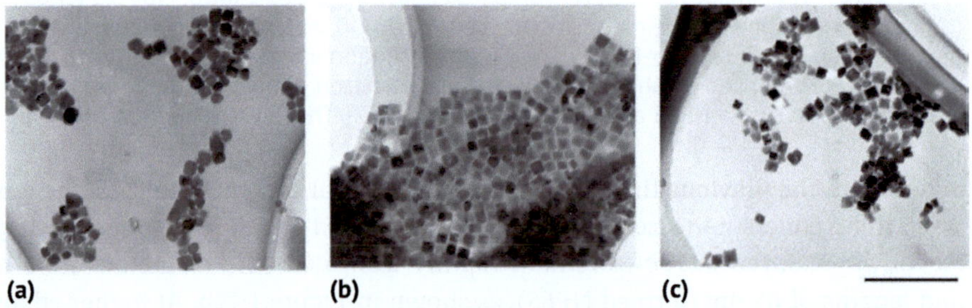

Figure 1.14 Shape control of the anatase TiO_2 NPs using sodium oleate/stearate-assisted 0.10 M TTIP–TEOA (1 : 2 mol ratio) reaction solution at initial pH 10.5. (a) No additive, (b) sodium oleate, and (c) sodium stearate. The scale bar indicates 200 nm [18]/with permission of Elsevier.

the Me_4NOH concentration was increased to 0.82 μM, the obtain TiO_2 NCs were greatly elongated (width: 15–20 nm, length: 76–177 nm), and they were grown to [001] direction, as shown in Figure 1.13c. This confirms that Me_4N^+ cation can stabilize the surface TiO_2 NC and leads to elongation of particles. Therefore, it is deduced that Me_4N^+ cation has strong binding affinity to the [101] facet of anatase nanocrystal. More detailed mechanisms are described in Chapter 2.

1.3.4 Effect of Oleate and Stearate

Sodium oleate and stearate have a considerable effect on the shape of TiO_2 NPs, as found by Sugimoto et al. [24]. These additives, with their high binding affinity for the {001} faces (*c*-faces) and {100} faces (perpendicular to the *c*-faces) of anatase TiO_2, alter the shape of TiO_2 from round-cornered cubes to sharp-edged cubes [24]. This effect is attributed to the reduction in specific surface energies due to the preferential adsorption of oleate or stearate ions on these crystal planes, because the surface free energy of the {001} faces is known to be 1.4 times larger than that for the {101} faces [24].

Figure 1.14 illustrates that the addition of sodium oleate or stearate leads to the formation of sharp-edged cubic TiO_2 NPs. In this experiment, an aqueous solution containing 0.10 M TTIP, 0.20 M TEOA, and 0.02 M sodium oleate (or sodium stearate) at pH 10.5 was aged at 100 °C for 24 h and then hydrothermally reacted at 140 °C for three days. While the size of the TiO_2 NPs remained unchanged, sharp-edged cubic NPs were obtained. In addition, the resulting TiO_2 suspensions became highly transparent, highlighting the role of oleate and stearate as effective colloidal stabilizers. In these sharp-edged cubic particles, oleate and stearate are likely adsorbed preferentially onto the {001} and {100} faces.

At pH 10.5, sodium oleate dissociates into negatively charged carboxyl groups, while TiO_2 surfaces also carry a negative charge. As a result, there is no significant electrostatic driving force for adsorption, suggesting that the adsorption of oleate is primarily governed by van der Waals interactions between the aliphatic groups of oleate and the TiO_2 surface. The face-specific adsorption of oleate may also be influenced by repulsive interactions between the carboxyl groups of oleate and the negatively charged TiO_2 faces with differing charge densities.

1.3.5 Effect of Tetrabutylammonium Hydroxide

Kim et al. used tetrabutylammonium hydroxide (TBAH) to prepare highly dispersible 13-nm-sized TiO_2 NPs via solvothermal reaction [25]. As shown in Figure 1.15, the distribution of colloidal particles, ranging from 12.3 to 13.4 nm, indicated that the particles remained well-separated in the aqueous solution without agglomeration. To prepare the NPs, 2 mmol of TTIP was dissolved in 20 mL of ethanol, and 0.184 mmol of TBAH was slowly added, yielding a molar

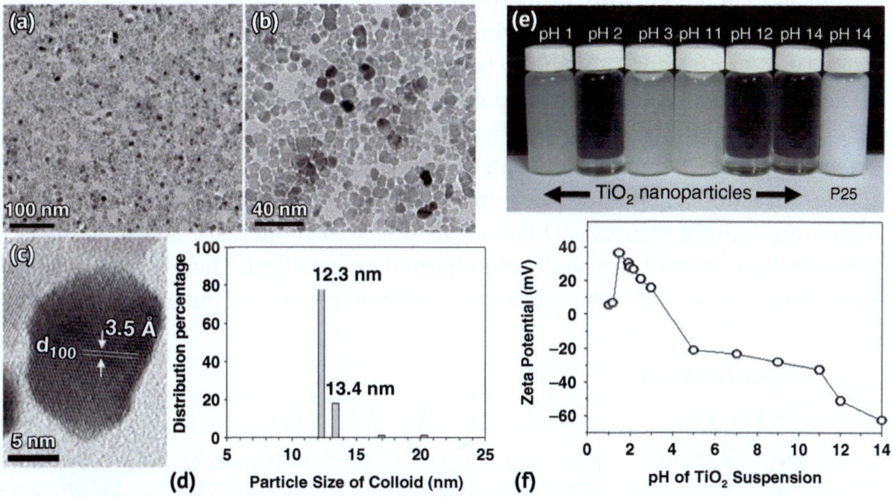

Figure 1.15 TEM images of the as-prepared TiO_2 NPs (a, b) and a high-magnification image for a single NP (c). Size distribution of TiO_2 colloid in water suspension (d). 10 mg sample was suspended in 100 mL water. Change of photographic images (e) and zeta potentials (f) of the TiO_2 water suspensions depending on the pH. To obtain suspension, 0.10 g TiO_2 NPs were dispersed in 150 mL water, and pH of the suspensions were controlled using 1.0 M HCl or 1.0 M TMAH. As a control, Aeroxide® P25 suspension was prepared at the same condition [25]/American Chemical Society.

ratio of 1.0: 0.092: 2.0: 170 (TTIP: TBAH: H_2O: ethanol). After a solvothermal reaction at 240 °C for 6 h, TiO_2 NPs were obtained as a white precipitate.

Initially, their TiO_2 NPs did not disperse well in water or ethanol at pH 7. However, upon adjusting the pH to 1.4–2.3 with 1.0 M HCl, the suspension became highly transparent and remained stable for months. Figure 1.15e shows photographs of TiO_2 suspensions at various pH levels. Similarly, at pH values greater than 12, controlled by adding TMAH, the suspension also became transparent. The zeta potential measurements (Figure 1.15f) indicated that at pH 2, the zeta potential reached +28 mV, a value that ensures the suspension's stability due to electrostatic repulsion between the NPs.

1.4 Shape Control with Capping Agent Under Neutral pH

Numerous metal oxide NPs have been synthesized by reverse micelle method by the aid of capping agents or organic surfactants. For the TiO_2, well-known capping agents are carboxylic acids, organic amines, and organic phosphates due to their high binding affinity toward TiO_2 surface.

In a nonpolar solvent, a capping agent such as carboxylic acid forms a reverse micelle array, as shown in Scheme 1.4. That is, hydrophobic organic chain of carboxylic acid goes outside to mix with nonpolar solvent, while hydrophilic carboxylic group is positioned inside, inducing a spherical array with forming a nanocavity in the center. The added TTIP, regarded to be hydrophilic, is then transported to the nanocavity, as the cavity is in a hydrophilic environment. Subsequently, TTIP in the nanocavity is decomposed and forms the crystallized structure by heating at high temperature or by solvothermal reactions. The produced TiO_2 NPs are fully separated without aggregation due to the capping of carboxylic groups, and the organic group leads TiO_2 NPs to be dissolved in nonpolar solvents.

Oleic acid (OA) is a typical capping agent used for synthesis of uniform TiO_2 NPs. Its carboxylic groups in OA have a good binding affinity to the TiO_2 surface. Typically, OA has comparatively high binding affinity to the anatase {001} faces [24].

One of the important issues for nonhydrolytic sol–gel processes is how to obtain highly crystallized structure. Baek et al. prepared monodispersed 4.5-nm-sized NPs with high crystallinity via a solvothermal process [26]. The high pressure and temperature in solution phase can provide favorable environment for the growth

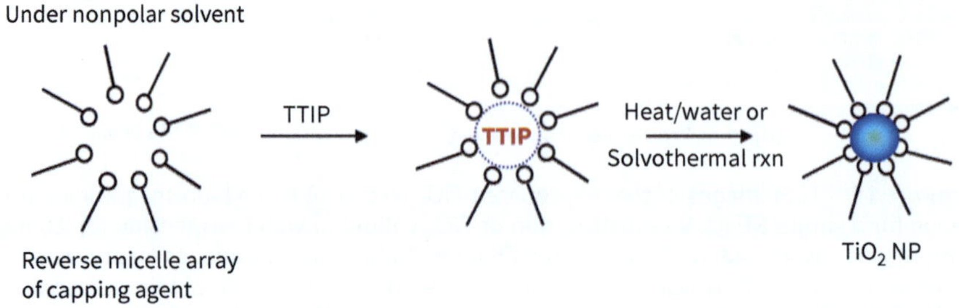

Scheme 1.4 Schematic diagram describing the formation of TiO_2 NPs by reverse micelle method.

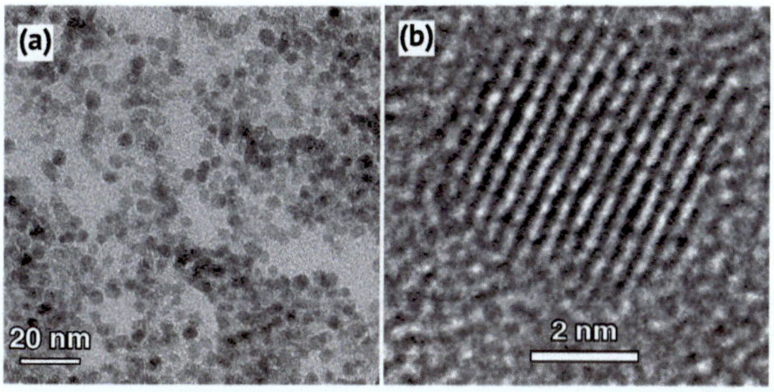

Figure 1.16 Low- and high-magnification TEM image (a and b, respectively) of 4.5–nm-sized monodispersed TiO_2 NPs synthesized at 260 °C in anhydrous condition using oleic acid as capping agent and octyl ether as solvent [26]/John Wiley & Sons.

of TiO_2 nanocrystals. Typically, 0.45 g of TTIP was introduced to a solution containing OA capping agent and 10 mL of octyl ether used as a solvent (molar ratio of TTIP to OA = 1:1). The mixture was stirred for a few hours, and the clear solution was transferred to a glass-lined autoclave. The nonhydrolytic solvothermal reaction was performed at 260 °C for 6 h. During this reaction the TTIP was oxidized by accepting oxygen from octyl ether and isopropoxide to form TiO_2 nuclei. Then the TiO_2 nuclei are crystallized under the solvothermal condition, but their growth is limited due to the surface capping of OA. As a result, well-crystallized 4.5 – nm-sized monodispersed TiO_2 NPs were obtained as a colloidal suspension, as shown in Figure 1.16.

Dinh et al. reported synthesis of TiO_2 NCs with tunable shapes by solvothermal process, employing OA and oleic amine (OM) as capping agent and titanium butoxide (TB) as Ti-precursor [27]. OA and OM preferentially adsorb onto the {001} and {101} faces of anatase TiO_2, respectively, influencing the growth direction of the nanocrystals. OM promotes growth along the [001] direction, although its binding affinity to the {101} faces is weaker compared to that of TMAH [27–29]. The ratio of OA/OM plays a crucial role in the shape of TiO_2 NCs. Specifically, a rhombic structure is formed at a TB: OA: OM ratio of 1: 4: 6, a truncated rhombic shape at 1: 5: 5, and a spherical shape at 1: 6: 4, as shown in Figure 1.17.

The shape of TiO_2 nanocrystals is largely governed by the relative growth rates along the [001] and [101] directions. When the OA/OM ratio is low, hydrolytic reactions dominate in the sol–gel system, leading to rapid NP formation. The high surface energy of the {001} faces, combined with a lower concentration of OA that preferentially binds to these faces, promotes continuous growth along the [001] direction, depleting the {001} faces. Meanwhile, the {101} faces, which have lower surface energy, experience inhibited growth along the [101] direction due to the adhesion of OM to these faces. This results in the formation of rhombic-shaped TiO_2 nanocrystals, as depicted in Scheme 1.5.

As the OA/OM ratio increases, the reaction rate slows down. A higher concentration of OA restricts growth along the [001] direction, as OA preferentially binds to the {001} faces. This binding preserves the surface area of the {001} faces, causing TiO_2 nanocrystals to grow on both the {001} and {101} faces, resulting in truncated rhombic NPs, as shown in Figure 1.17b. When an excess of OA is

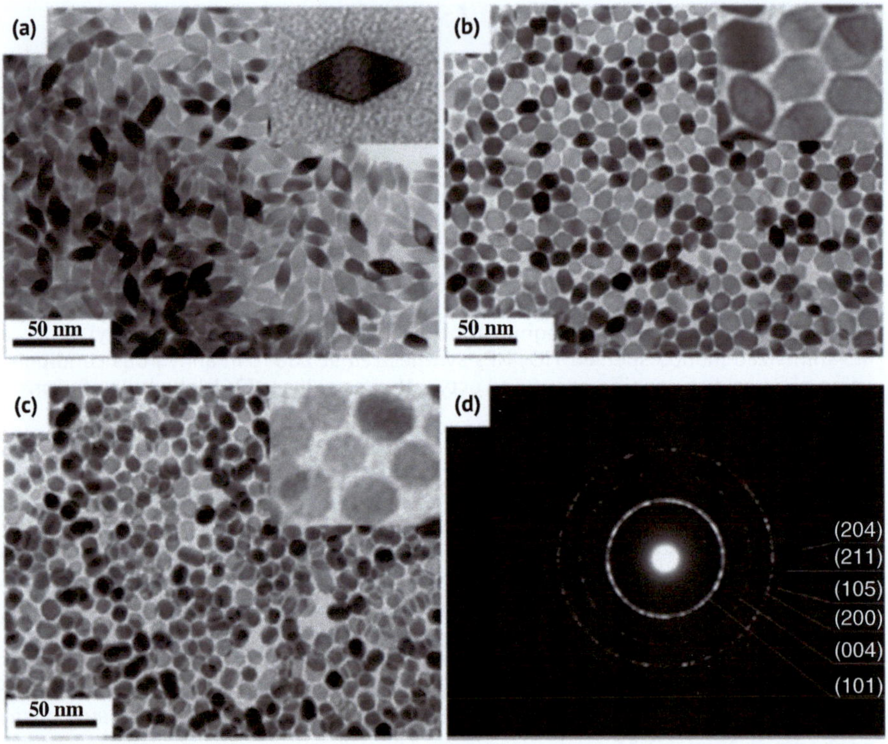

Figure 1.17 TEM images of TiO_2 NPs prepared at various compositions; (a) a rhombic structure at TB: OA: OM = 1: 4: 6; (b) a truncated rhombic structure at TB: OA: OM = 1: 5: 5; (c) a spherical structure at TB: OA: OM = 1: 6: 4. High-magnification image is included in each Inset. (d) Selected area diffraction patterns for the NPs in (b) [27]/American Chemical Society.

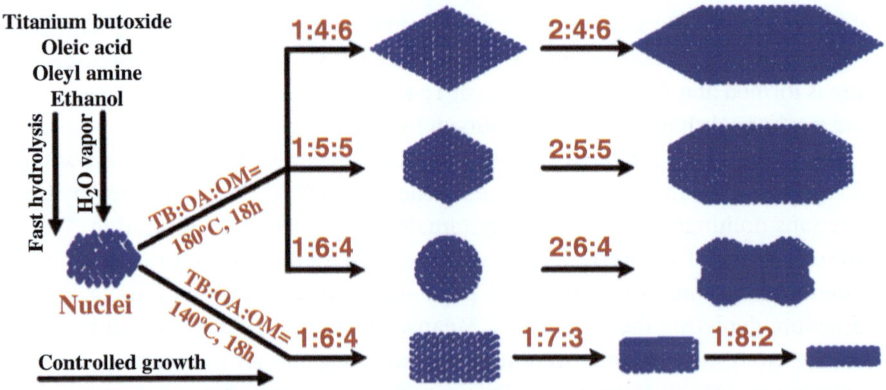

Scheme 1.5 Diagram describing the formation mechanism and shape control of TiO_2 nanocrystals [27]/American Chemical Society.

present (OA: OM ratio = 6: 4), the growth process shifts from hydrolytic to nonhydrolytic condensation. OA adsorbs onto nearly the entire surface of TiO_2, leading to crystal growth primarily on high surface energy faces, such as the {001} faces, and ultimately resulting in the formation of spherical-shaped TiO_2 NCs.

1.5 Synthesis of Extremely Small TiO_2 NPs Using Dendrimer

Extremely small TiO_2 NPs (or quantum dots) can be formed using the dendrimer. The size of NP can be controlled precisely by adjusting the molarity ratio of Ti-precursor and dendrimer. Satoh et al. prepared TiO_2 quantum dots (QDs) using $Ti(acac)Cl_3$ (acac: acetylacetonate) as Ti-precursor and fourth-generation phenylazomethine dendrimer (DPA G4, see Figure 1.18a) [30]. The preparation principle is as follows: when $Ti(acac)cl_3$ is added to DPA G4 and stirred moderately, the Ti-precursors go inside the dendrimer and chemically bind to the tertiary amine groups, as described in Figure 1.18b. Totally, there are 30 tertiary amine groups inside one DPA G4. Therefore, up to 30 $Ti(acac)Cl_3$ molecules can be bound to one DPA G4, as described in Figure 1.18c. The $Ti(acac)Cl_3$ molecules bound inside DPA G4 can form a very small TiO_2 QD through a hydrolysis reaction or a thermal decomposition reaction. DPA G4 exists in the form of a huge cage and is completely separated from the surrounding dendrimers. Therefore, it is predicted that all the $Ti(acac)Cl_3$ molecules existing in one DPA G4 will merge together to form a single TiO_2 QD. As a result, the size of the TiO_2 QD is determined by the equivalence ratio of the added $Ti(acac)Cl_3$ and DPA G4. According to the theoretical value, the diameter of the formed TiO_2 is 1.00 nm when 6 equivalents of $Ti(acac)Cl_3$ are added, 1.23 nm when 14 equivalents are added, and 1.49 nm when 30 equivalents are added. In the actual experiment, the diameters were observed to be 1.47 nm and 1.70 nm when 14 equivalents and

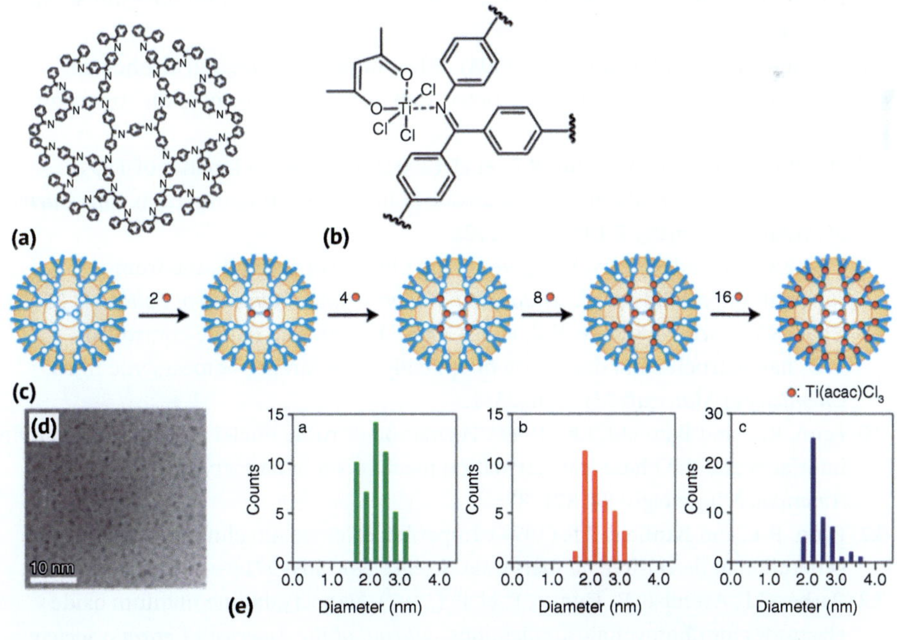

Figure 1.18 (a) Chemical structure of DPA G4 dendrimer; (b) a suggested chemical bonding of $Ti(acac)Cl_3$ onto the secondary amine group in DPA G4; (c) schematic diagram describing gradual complexation of $Ti(acac)Cl_3$ precursors into the dendrimer; (d) TEM image of TiO_2 QDs in DPA G4. Samples were stained by RuO_2; (e) size distribution of TiO_2-DPA G4 complexes depending on the added $Ti(acac)Cl_3$. Adapted from [29].

30 equivalents are added, respectively. The diameter of the TiO_2 NPs increased slightly compared to the theoretical value, which is thought to be the result of some aggregation (Figure 1.18d,e).

The formed TiO_2 exhibits a quantum size effect, and in the case of anatase NPs, the quantum size effect appears at a size of approximately 2 nm or less [29, 30]. That is, the anatase particles with a size of 1.70 nm exhibited a band gap of 3.35 eV, and the particles with a size of 1.47 eV exhibited a band gap of 3.47 eV. According to a previous study, it was reported that no optical blue shift was observed at all when the TiO_2 particle size was 2.1 nm or larger [31], which is correspondent to the results by Satoh et al. [30].

References

1 Chen, X. and Mao, S.S. (2007). Titanium dioxide nanomaterials: synthesis, properties, modifications, and applications. *Chemical Reviews* 107: 2891–2959.

2 Roy, J. (2022). The synthesis and applications of TiO_2 nanoparticles derived from phytochemical sources. *Journal of Industrial and Engineering Chemistry* 106: 1–19.

3 Wang, J., Wang, Z., Wang, W. et al. (2022). Synthesis, modification and application of titanium dioxide nanoparticles: a review. *Nanoscale* 14: 6709–6743.

4 Fujishima, A. and Honda, K. (1972). Electrochemical photolysis of water at a semiconductor electrode. *Nature* 238: 37–38.

5 Vorkapic, D. and Matsoukas, T. (1999). Reversible agglomeration: a kinetic model for the peptization of titania nanocolloids. *Journal of Colloid and Interface Science* 214: 283–291.

6 Vorkapic, D. and Matsoukas, T. (1998). Effect of temperature and alcohols in the preparation of titania nanoparticles from alkoxides. *Journal of the American Ceramic Society* 81: 2815–2820.

7 Oskam, G., Nellore, A., Penn, R.L. et al. (2003). The growth kinetics of TiO_2 nanoparticles from titanium(IV) alkoxide at high water/titanium ratio. *The Journal of Physical Chemistry B* 107: 1734–1738.

8 Lifshitz, I.M. and Slyozov, V.V. (1961). The kinetics of precipitation from supersaturated solid solutions. *Journal of Physics and Chemistry of Solids* 19: 35–50.

9 Chae, S.Y., Park, M.K., Lee, S.K. et al. (2003). Preparation of size-controlled TiO_2 nanoparticles and derivation of optically transparent photocatalytic films. *Chemistry of Materials* 15: 3326–3331.

10 Penn, R.L. and Banfield, J.F. (1999). Formation of rutile nuclei at anatase {112} twin interfaces and the phase transformation mechanism in nanocrystalline titania. *American Mineralogist* 84: 871–876.

11 Penn, R.L. and Banfield, J.F. (1998). Imperfect oriented attachment: dislocation generation in defect-free nanocrystals. *Science* 281: 969–971.

12 Barbé, C.J., Arendse, F., Comte, P. et al. (1997). Nanocrystalline titanium oxide electrodes for photovoltaic applications. *Journal of the American Ceramic Society* 80: 3157–3171.

13 Park, N.G., van de Lagemaat, J., and Frank, A.J. (2000). Comparison of dye-sensitized rutile- and anatase-based TiO_2 solar cells. *The Journal of Physical Chemistry B* 104: 8989–8994.

14 Scolan, E. and Sanchez, C. (1998). Synthesis and characterization of surface-protected nanocrystalline titania particles. *Chemistry of Materials* 10: 3217–3223.

15 Trentler, T.J., Denler, T.E., Bertone, J.F. et al. (1999). Synthesis of TiO_2 nanocrystals by nonhydrolytic solution-based reactions. *Journal of the American Chemical Society* 121: 1613–1614.

16 Niederberger, M., Bartl, M.H., and Stucky, G.D. (2002). Benzyl alcohol and titanium tetrachlorides – a versatile reaction system for the nonaqueous and low-temperature preparation of crystalline and luminescent titania nanoparticles. *Chemistry of Materials* 14: 4364–4370.

17 Sugimoto, T., Zhou, X., and Muramatsu, A. (2003). Synthesis of uniform anatase TiO_2 nanoparticles by gel–sol method 3. Formation process and size control. *Journal of Colloid and Interface Science* 259: 43–52.

18 Sugimoto, T., Zhou, X., and Muramatsu, A. (2003). Synthesis of uniform anatase TiO_2 nanoparticles by gel–sol method 4. Shape control. *Journal of Colloid and Interface Science* 259: 53–61.

19 Sugimoto, T. and Zhou, X. (2002). Synthesis of uniform anatase TiO_2 nanoparticles by the gel–sol method 2. Adsorption of OH^- ions to $Ti(OH)_4$ gel and TiO_2 particles. *Journal of Colloid and Interface Science* 252: 347–353.

20 Shiu, J.W., Lan, C.M., Chang, Y.C. et al. (2012). Size-controlled anatase titania single crystals with octahedron-like morphology for dye-sensitized solar cells. *ACS Nano* 6: 10862–10873.

21 Yun, H.J., Lee, H., Joo, J.B. et al. (2009). Influence of aspect ratio of TiO_2 nanorods on the photocatalytic decomposition of formic acid. *The Journal of Physical Chemistry C* 113: 3050–3055.

22 Sung, S.D., Ojha, D.P., You, J.S. et al. (2015). 50 nm sized spherical TiO_2 nanocrystals for highly efficient mesoscopic perovskite solar cells. *Nanoscale* 7: 8898–8906.

23 Chemseddine, A. and Moritz, T. (1999). Nanostructuring titania: control over nanocrystal structure, size, shape, and organization. *European Journal of Inorganic Chemistry* 1999: 235–245.

24 Jun, Y.W., Casula, M.F., Sim, J.H. et al. (2003). Surfactant-assisted elimination of a high energy facet as a means of controlling the shapes of TiO_2 nanocrystals. *Journal of the American Chemical Society* 125: 15981–15985.

25 Kim, Y.J., Chai, S.Y., and Lee, W.I. (2007). Control of TiO_2 structures from robust hollow microspheres to highly dispersible nanoparticles in a tetrabutylammonium hydroxide solution. *Langmuir* 23: 9567–9571.

26 Baek, S.Y., Chai, S.Y., Hur, K.S. et al. (2005). Synthesis of highly soluble TiO_2 nanoparticle with narrow size distribution. *Bulletin of the Korean Chemical Society* 26: 1333–1334.

27 Dinh, C.T., Nguyen, T.D., Kleitz, F. et al. (2009). Shape-controlled synthesis of highly crystalline titania nanocrystals. *ACS Nano* 3: 3737–3743.

28 Joo, J., Kwon, S.G., Yu, T. et al. (2005). Large-scale synthesis of TiO_2 nanorods via nonhydrolytic sol-gel ester elimination reaction and their application to photocatalytic inactivation of *E. coli*. *The Journal of Physical Chemistry B* 109: 15297–15302.

29 Cozzoli, P.D., Kornowski, A., and Weller, H. (2003). Low-temperature synthesis of soluble and processable organic-capped anatase TiO_2 nanorods. *Journal of the American Chemical Society* 125: 14539–14548.

30 Satoh, N., Nakashima, T., Kamikura, K. et al. (2008). Quantum size effect in TiO_2 nanoparticles prepared by finely controlled metal assembly on dendrimer templates. *Nature Nanotechnology* 3: 106–111.

31 Serpone, N., Lawless, D., and Khairutdinov, R. (1995). Size effects on the photophysical properties of colloidal anatase TiO_2 particles: size quantization versus direct transitions in this indirect semiconductor. *The Journal of Physical Chemistry* 99: 16646–16654.

2

Synthesis of One-dimensional TiO_2 Nanomaterials

Jeongho Kim[1], Wan In Lee[1,], Jia Hong Pan[2,*]*

[1]*Department of Chemistry and Chemical Engineering, Inha University, Incheon, Korea*
[2]*State Key Laboratory of Featured Metal Materials and Life-Cycle Safety for Composite Structures, School of Resources, Environment and Materials, Guangxi University, Nanning, Guangxi, China*

*Corresponding author: wanin@inha.ac.kr; jhpan@gxu.edu.cn

2.1 Introduction

1D TiO_2 nanomaterials have distinct advantages compared with 0D or multidimensional structures in regard to electron and hole transports, as the 1D TiO_2 possesses a highly effective structure for the transport of charges along the axial direction. For instance, anatase TiO_2 nanobelts (NBs) exhibit electric conductivities of $\sim 10^{-7} \Omega^{-1} cm^{-1}$, two orders of magnitude higher than NPs $(\sim 10^{-9} \Omega^{-1} cm^{-1})$ [1, 2]. Therefore, various 1D TiO_2 nanostructures have been synthesized and applied for photocatalysts, solar cells, Li-ion batteries, and gas sensors [3, 4].

The performance of these applications is strongly influenced by the shape, size, morphology, and crystallographic orientation of the 1D TiO_2 structures. Thus, precise control over these parameters is essential for optimizing device performance. This chapter delves into the synthetic strategies and shape control of 1D TiO_2 nanomaterials, including nanorods (NRs), nanotubes (NTs), NBs, nanofibers (NFs), and nanowires (NWs). The fundamental principles of their formation and growth mechanisms are systematically explored, and various studies have demonstrated that the use of 1D TiO_2 nanostructures can improve charge collection efficiency of the solar cells and expedite charge separation that is required for efficient photocatalysts [5, 6].

2.2 TiO_2 NRs

The aspect ratio – defined as the ratio of length to diameter – is a key parameter for evaluating NR morphology. Tailoring the diameter and aspect ratio of TiO_2 NRs is crucial for optimizing their performance in various applications. Wet-chemical

Nanostructured TiO_2: Synthesis, Photocatalytic Properties and Applications, First Edition.
Edited by Jia Hong Pan, Wan In Lee and Detlef W. Bahnemann

synthesis methods are widely used for preparing TiO_2 NRs, and their growth mechanisms are discussed below.

2.2.1 Fundamental Principle of NR Formation

Combined use of strongly binding capping agents (SBCA) and weakly binding capping agents (WBCA) is a common strategy for synthesizing NR structures. SBCA preferentially binds to specific facets of nuclei or small nanocrystals (NCs), while WBCA binds to other facets and dissociate more easily. During solution-phase reactions at elevated temperatures, spherical NCs initially formed evolve into elliptical structures, with merging occurring at WBCA-bound facets. This results in NR growth along the WBCA-bound direction, while SBCA-covered facets remain static, as described in Scheme 2.1. The length of the NRs is proportional to the SBCA concentration, providing a general guideline for synthesizing NRs of metal oxides, chalcogenides, and other materials.

As discussed in Chapter 1, various capping agents have been used in the synthesis of TiO_2 nanostructures, including alkyl amines, alkylammonium hydroxides, carboxylic acids, phosphates, amino acids, and others. Phosphates such as trioctylphosphine oxide (TOPO, $[CH_3(CH_2)_7]_3\ P = O$) are a type of SBCA, but they do not exhibit facet selectivity in general. Alkylammonium hydroxides, like TMAH and TBAH, as well as alkyl amines such as oleic amine, show a strong binding affinity for the {101} facet of anatase TiO_2 NCs, with alkylammonium hydroxides binding more strongly than alkyl amines. Carboxylic acids, such as oleic acid (OA), tend to bind preferentially to the (001) facet of anatase TiO_2, and this binding affinity increases as the alkyl chain length of the carboxylic acids grows.

2.2.2 Oriented Aggregation Mechanism

Oriented aggregation is a key crystal growth pathway in wet-chemical syntheses [7]. Crystal growth occurs through coarsening (Ostwald ripening) or aggregation, with the latter involving either random or oriented aggregation, as illustrated in Figure 2.1a. For anatase TiO_2 NCs, the highly reactive [001] facet connects with other facets, leading to elongated structures such as NRs. The high-resolution transmission electron microscopy (TEM) image in Figure 2.1b reveals that anatase TiO_2 NRs form through the oriented attachment of three quasi-spherical primary

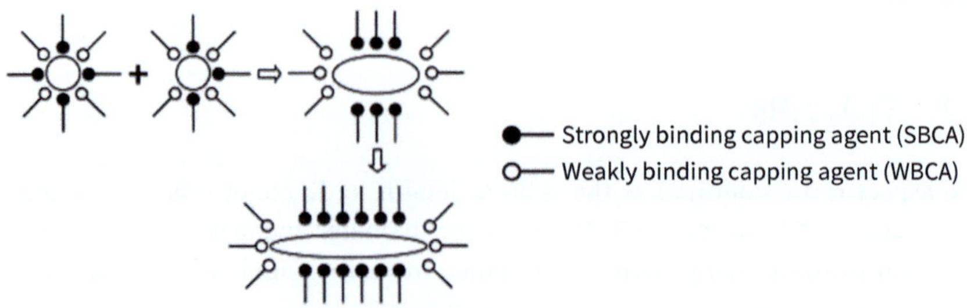

Scheme 2.1 A principle for the formation of NR structures with SBCA and WBCA.

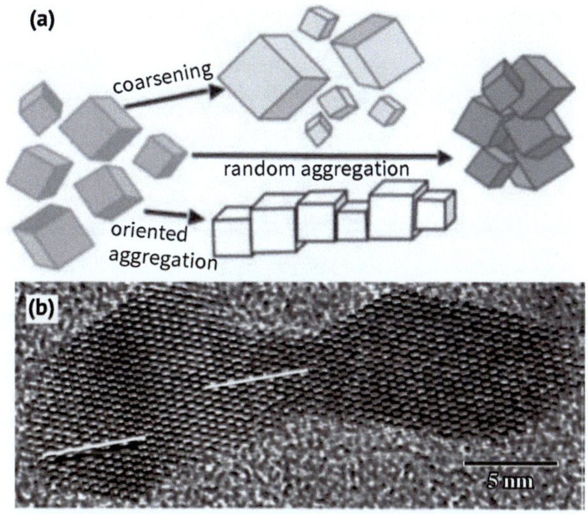

Figure 2.1 (a) Schematic diagram for the formation of TiO_2 NR structure by oriented aggregation mechanism. (b) A high-magnification TEM image of a single TiO_2 NR evolved from the oriented aggregation of three primary TiO_2 NPs. Adapted from [7].

NPs, with [001] facets aligning to create rod-like, single-crystalline nanostructures, despite the nonuniform shape of each individual component.

2.2.3 Selective Growth to the Anatase [100] Direction

2.2.3.1 Fundamental Growth Mechanism

Penn and Banfield et al. demonstrated that the rapid crystal growth of TiO_2 NCs along the [001] direction is driven by the continuously created reactive sites [8]. Figure 2.2a shows a schematic of an anatase NC viewed from the [010] direction. The unshaded octahedra represent potential sites for the preferential attachment of Ti atoms. The (001) face uniquely has the highest number of oxygen atoms arranged in positions that are ideal for forming an octahedral coordination environment, which is key for TiO_2 growth.

Figure 2.2b illustrates the result of new Ti octahedra (depicted with the lightest shading) attaching to these sites. At the sites labeled "A" on the (001) face, Ti attachment generates additional sites that are bound by three oxygen atoms (O^- or OH groups). As a result, the preferential attachment sites on the (001) face continue to generate new sites suitable for further Ti attachment, ensuring a continuous growth process. This ongoing creation of undersaturated sites on the (001) face likely contributes significantly to the rapid extension along the [001] direction.

2.2.3.2 Ester Elimination Reaction

Tiny TiO_2 NRs selectively grown along the [100] direction were synthesized under nonhydrolytic conditions via an ester elimination mechanism. Joo et al. successfully produced a large quantity of TiO_2 NRs with a diameter of 3.4 nm and a length of 38 nm [9]. In a typical synthesis, titanium tetraisopropoxide (TTIP) was introduced into OA, which

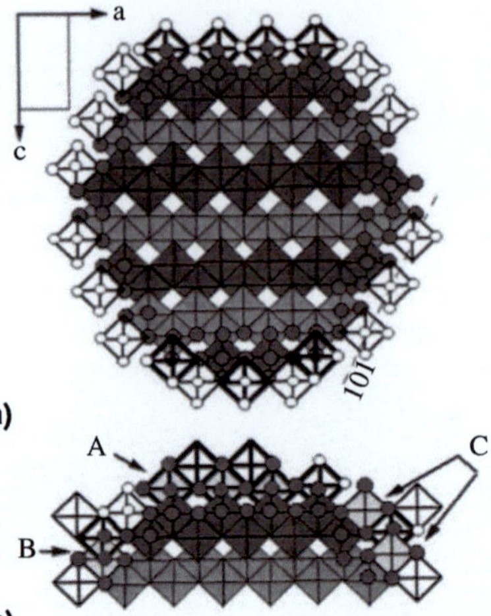

Figure 2.2 Schematic diagrams of an anatase NC projected on the [010] direction (a) and the structure after an attachment of new Ti octahedra (b) [8]/with permission of Elsevier.

served as both the solvent and capping agent. Upon solvothermal reaction at 270 °C for 2 h, the resulting TiO_2 NRs were well capped with oleate anions, enabling their facile redispersion in nonpolar organic solvents such as hexane. As shown in Figure 2.3a, TiO_2 NRs show an aspect ratio exceeding 10. Figure 2.3b further reveals that the single-crystalline anatase nature of TiO_2 NRs and their preferential growth along the [001] direction.

The growth mechanism of TiO_2 NRs can be described by ester elimination reaction. In the nonhydrolytic reaction, direct hydrolysis of titanium alkoxide is not available. Thermal decomposition of titanium alkoxide enables to form TiO_2 and ether, as described in Eq. (2.1), but it will be a very slow reaction. When carboxylic acid such as OA is present in the solution, hydrolysis reaction of titanium alkoxide can be progressed by forming ester, as shown in Eq. (2.2). Then the titanium hydroxide can be used to the growth of TiO_2 by a condensation reaction in Eq. (2.3). In the early stage of nonhydrolytic reaction, the TiO_2 NCs with a small size and isotropic shape are formed. As the reaction proceeds, OA will be adsorbed preferentially on the (001) surface of TiO_2 NCs due to the high binding affinity. Therefore, titanium alkoxide in the solution can be hydrolyzed by reacting with OA on the (001) surface, leading to selective growth of TiO_2 NCs to the [001] direction.

$$\equiv \text{Ti—OR} + \text{RO—Ti} \equiv \ \rightarrow \ \equiv \text{Ti—O—Ti} \equiv + \text{ROR} \tag{2.1}$$

$$\equiv \text{Ti—OR} + \text{HOOC(CH}_2)_7\text{CH=CH(CH}_2)_7\text{CH}_3 \ \rightarrow \ \equiv \text{Ti—OH} + \text{ROOC(CH}_2)_7\text{CH=CH(CH}_2)_7\text{CH}_3 \tag{2.2}$$

$$\equiv \text{Ti—OH} + \text{H—O—Ti} \equiv \ \rightarrow \ \equiv \text{Ti—O—Ti} \equiv + \text{H}_2\text{O} \tag{2.3}$$

In addition, the effect of hexadecylamine (HDA) was examined in the sol–gel reaction between TTIP and OA at an elevated temperature. The amine groups in HDA exhibit strong binding affinities with both the rutile (110) surface and the anatase

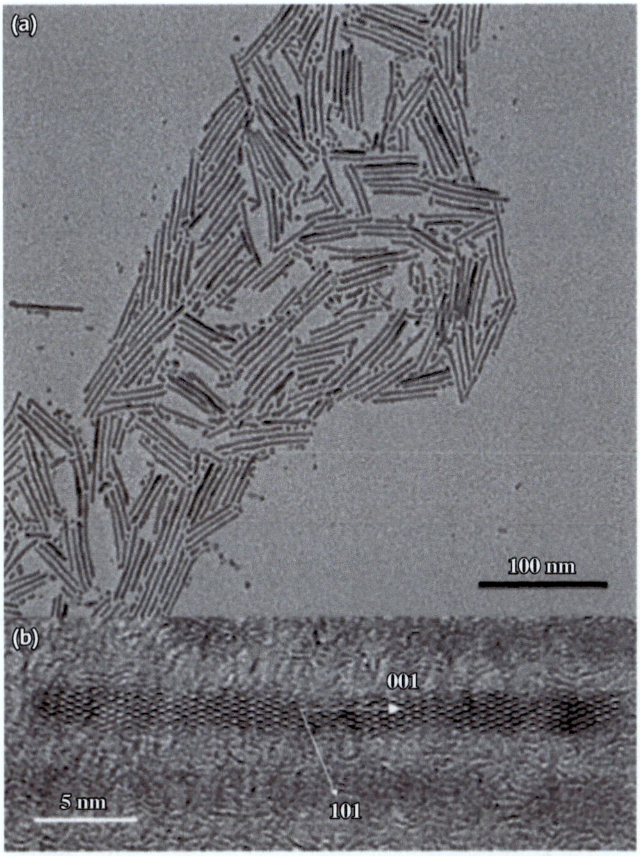

Figure 2.3 TEM image of as-prepared TiO_2 NRs (a) and their high-resolution image (b) [9]/American Chemical Society.

(101) face, because the rutile (110) surface is structurally similar to the anatase (101) face. It was observed that as the concentration of HDA increased, the average diameter of the prepared NRs decreased. As shown in Figure 2.4, the average diameters of the NRs were measured to be 3.4, 2.7, 2.2, and 2.0 nm for HDA concentrations of 0, 1, 5, and 10 mmol, respectively. These findings clearly demonstrate that HDA selectively caps the (101) surface of anatase TiO_2, thereby inhibiting crystal growth along the [101] direction.

2.2.3.3 Growth Mechanism in the Presence of TMAH

Chemsedinne and Moritz [10] investigated the tetramethylammonium hydroxide (Me_4NOH, TMAH)-catalyzed sol–gel reaction of $Ti(OR)_4$. In their method, 1.14 mmol of TTIP in 10 mL of 2-propanol was mixed with 0.83 mmol of TMAH in 150 mL of water. After refluxing at 90–100 °C for 6 h, 1D anatase TiO_2 NRs are formed and displayed a corrugated structure on the (101) surface, as shown in Figure 2.5a.

The Me_4N^+ cation acts as a strongly binding ligand, adhering to and stabilizing the (001) surface of anatase TiO_2, which also helps to control the crystal growth

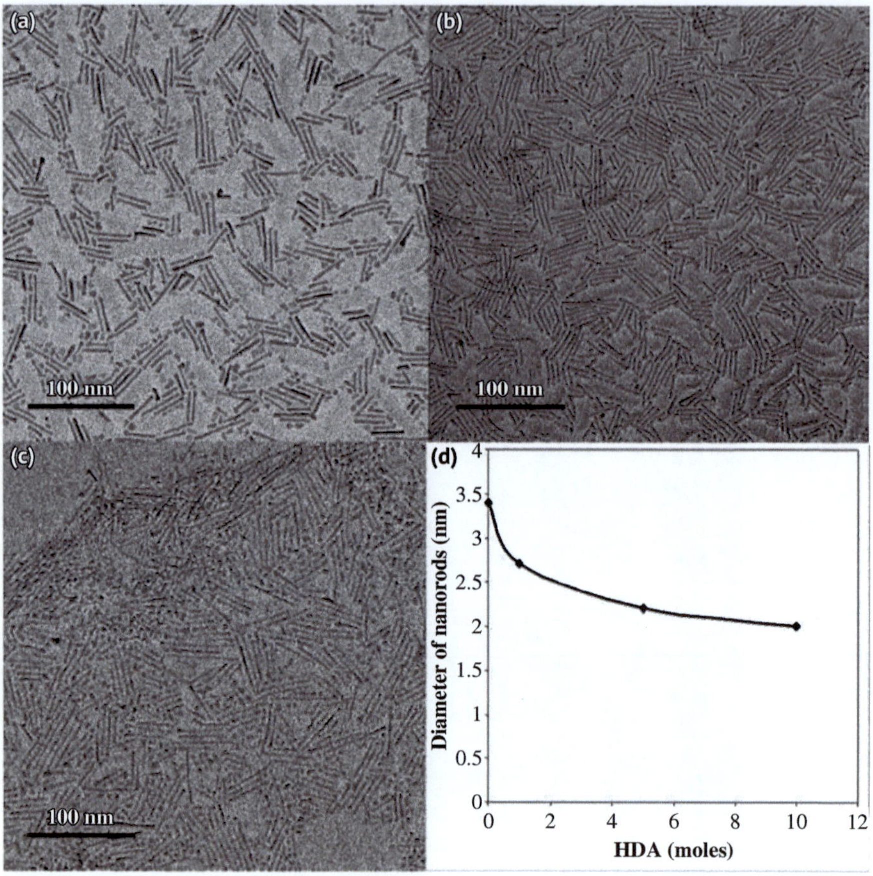

Figure 2.4 TEM images for HDA-dependent various TiO_2 NRs with different diameters and lengths of (a) 2.7 and 28 28 nm, (b) 2.2 and 32 nm, and (c) 2.0 and 39 nm. (d) A plot illustrating the relationship between the diameter of TiO_2 NRs with the dosage of HDA [9]/American Chemical Society.

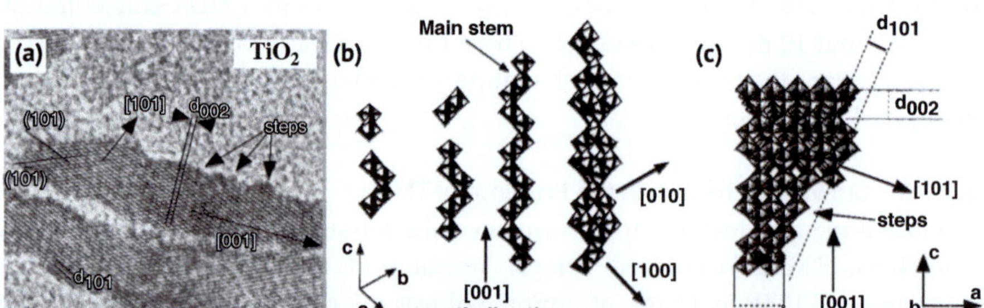

Figure 2.5 (a) High-resolution TEM images of TiO_2 NCs viewed from the [010] plane and grown to the [001] direction. The wrinkled edges in the NCs, indicated by arrows, are formed by the repetitive formation of [101] and [$10\bar{1}$] planes. (b) Suggested mechanism for the growth of TiO_2 NCs. (c) A packing diagram describing the growth of anatase TiO_2 NC, viewed from the [010] direction [10]/with permission of John Wiley & Sons.

of condensed TiO_2 species at the early stages. The linear growth results in skewed chains, corresponding to growth along the [001] direction of the anatase phase, as described in Figure 2.5b. Through an oxolation process, the nuclei can grow in two directions, and the particle shape is determined by the difference in growth rates along the [101] and [001] directions, as shown in Figure 2.5c.

At low concentrations of Me_4N^+, growth is favored along the [101] direction, whereas at high concentrations, growth is enhanced along the [001] direction. By varying the ratio of Ti-precursor to TMAH, the shape of TiO_2 NRs can be fine-tuned, leading to the formation of the corrugated structure on the (101) surface. Specifically, increasing the TMAH concentration promotes the formation of high-aspect-ratio anatase NRs.

2.2.4 Preparation of Anatase TiO_2 NRs with Various Capping Agents

2.2.4.1 Effect of Lauric Acid and TOPO

According to Cheon and Alivisatos groups prepared TiO_2 NRs using lauric acid as a phase selective capping agent binding strongly to the anatase [001] facet, whereas TOPO as a nonselective capping agent [11]. Two capping agents were dissolved in dioctyl ether and heated at 300°C under a vigorous stirring. $TiCl_4$ and TTIP were sequentially introduced. TiO_2 NRs can be formed via the alkyl halide elimination reaction between $TiCl_4$ and TTIP. The reaction was kept for 10–20 mins and then quenched with cold toluene. The concentration of lauric acid was adjusted to manipulate the shape and aspect ratio of 1D TiO_2 nanostructures.

In the absence of lauric acid or with a low dosage of lauric acid, truncated octahedral bipyramid seeds initially formed. As the reaction progressed, bullet-like shapes (Figure 2.6a) and diamond-like shapes (Figure 2.6b) emerged, accompanied by a gradual reduction in the percentage of {001} facets. The diamond shape signifies the complete elimination of {001} facets, leaving only the {101} facets, which are known to be the most thermodynamically stable surfaces of anatase TiO_2.

As the concentration of lauric acid increases, the growth rate along the [001] direction is reduced due to selective binding of lauric acid to the {001} faces. Under this condition, monomers continuously grow further onto both {001} faces and {101} faces of the truncated octahedral bipyramid seeds, in such a way to expand the {101} faces during crystal growth while preserving a constant {001} surface area. When the (101) surface area reaches a critical value under the kinetically driven regime, the formation of a step instead of further homogeneous expansion of (101) surfaces leads to the formation of NR-like structures. The NRs exhibit a zigzag pattern of {101} faces that can be regarded as repeating truncated octahedral bipyramid units.

As the concentration of lauric acid increases, the growth along the [001] direction is suppressed by lauric acid that selectively lands on the {001} facets. The monomers continue to deposit onto both the {001} and {101} facets of the truncated octahedral bipyramid seeds. As a result, {001} facets are preserved while the expansion of the {101} facets during crystal growth. Once the surface area of the {101} facets reaches a critical value under kinetically driven conditions, the formation of steps occurs instead of further homogeneous expansion of the {101} surfaces, resulting in the development of NRs with different length. The shape of TiO_2 NRs shown in Figure 2.6c-e looks

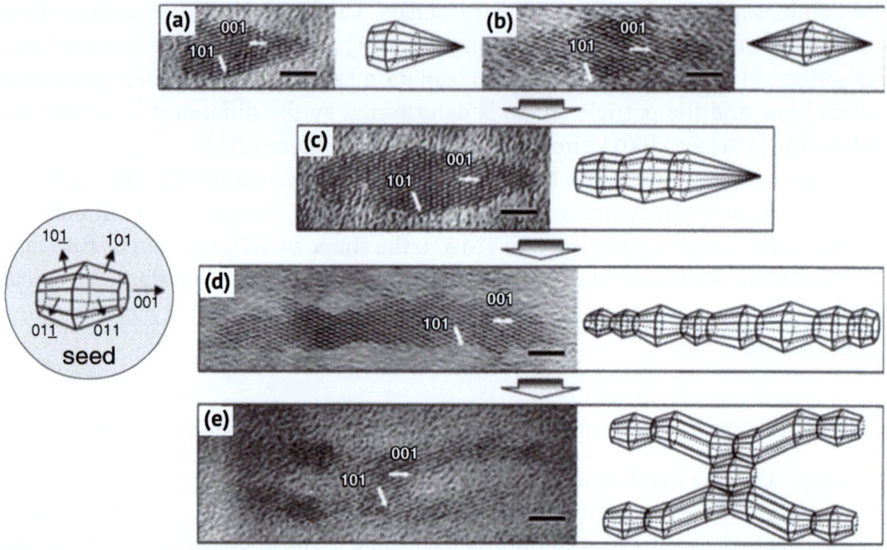

Figure 2.6 High-resolution TEM images and their simulated morphology of bullet (a), diamond (b), short NR (c), long NR (d), and branch nanostructure (e). The axis of elongated TiO_2 NRs is parallel to the anatase c-axis direction. A hexagon-shaped structure truncated by two {001} and four {101} faces is observed in a TiO_2 seed. Consequently, the growth along [101] directions from the initial hexagon structure leads to a branched structure. All scale bars denote 3 nm [11]/with permission of American Chemical Society.

like that of NRs prepared by inter-particular aggregation processes such as oriented attachment, but in reality, the shape control has been achieved by the role of LA, a strong binding capping agent to the anatase [001] facet.

A large excess of lauric acid suppresses the growth along [001] directions, whereas the growth out of {101} faces is favored, generating branched nanostructures (see Figure 2.6e).

2.2.4.2 Effect of Amino Acids

Kanie and Sugimoto prepared TiO_2 NRs with various aspect ratio by using amino acids as shape controller [12]. Synthetic process consists of a first aging of 0.25 M $Ti(TEOA)_2$ (TEOA: triethanolamine) complex with an amino acid for the formation of a $Ti(OH)_4$ gel at 100 °C for 24 h and a second aging for nucleation and growth of anatase TiO_2 particles in the gel at 140 °C for 72 h. In this system, ammonia and primary amines as shape controllers were found to yield ellipsoidal particles by their specific adsorption to the crystal planes parallel to the *c*-axis of a TiO_2 NPs.

By adding amino acids, the anatase TiO_2 particles are elongated, and the aspect ratio is greatly dependent of the kinds of amino acids, as seen in Figure 2.7. Among them, aspartic acid and glutamic acid assist the formation of TiO_2 NRs with higher aspect ratios. The amino groups in amino acids appear to play a key role in their adsorption onto TiO_2 NCs. The adsorption increases with higher pH levels, and lysine adsorbs more strongly onto amorphous TiO_2 compared to glutamic acid [13]. However, the high aspect ratio 1D anatase TiO_2 indicates that these acidic amino acids have a specific affinity for adsorption on crystal planes parallel to the *c*-axis of the tetragonal crystal lattice. This adsorption capability could be stronger than that of neutral or basic amino acids.

2.2.4.3 Effect of DEA and TMAH

Park et al. successfully controlled the length of TiO_2 NRs with TMAH [14]. As shown in Figure 2.8, a significant change in length was only observed when the TMAH concentration reached 39 mM. At 42 mM, the NRs began forming elliptical structures (Figure 2.8b), and as the concentration increased further, their length gradually extended. By setting the TMAH concentration to 42, 45, 49, 53, 59, and

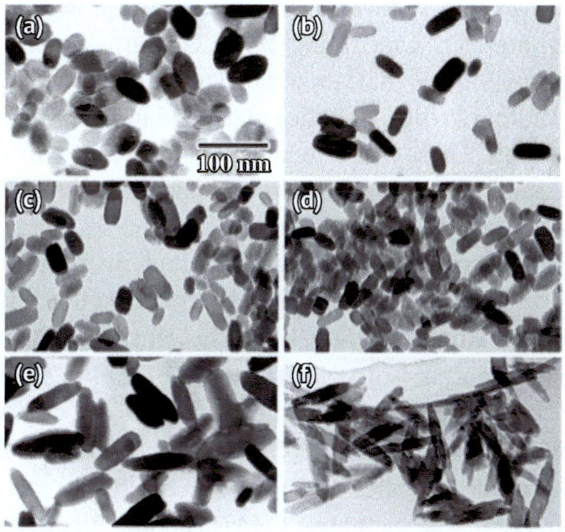

Figure 2.7 TEM images of TiO_2 NRs prepared by sol–gel route without adding any amino acid (a) and with adding 0.020 M of glycine (b), serine (c), lysine (d), aspartic acid (e), and glutamic acid (f) [12]/The Royal Society of Chemistry.

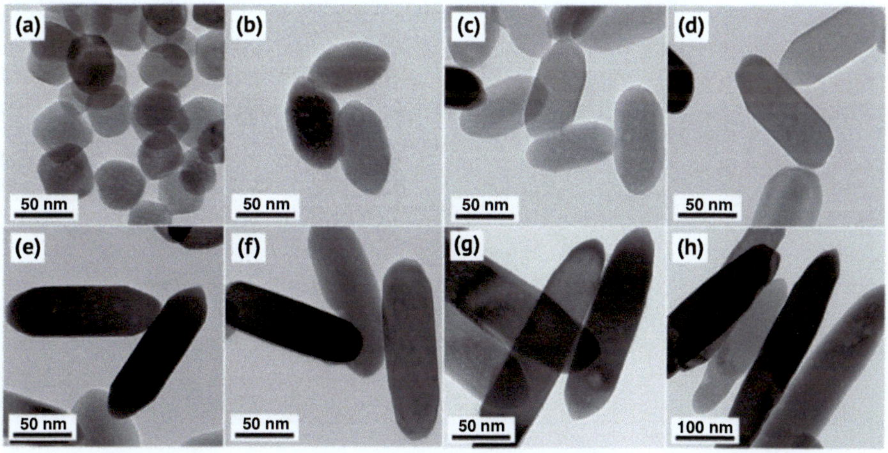

Figure 2.8 TEM images of (a) TiO_2 NPs with sizes of 46 nm, monodispersed TiO_2 NRs with various lengths of (b) 70, 90 (c), 110 (d), 130 (e), 150 (f), and 200 nm (g), and uneven TiO_2 NR (h) [14]/American Chemical Society.

65 mM, the average NR length was precisely controlled at 70, 90, 110, 130, 150, and 200 nm, respectively. Before reaching 53 mM, the average width remained at approximately 46 nm. However, when the concentration increased to 59 and 65 mM, the width slightly increased to 47 and 48 nm, respectively. In contrast, at 73 mM, both the width (> 70 nm) and length (Figure 2.8h) increased significantly. These results indicate that by tuning the TMAH concentration, the length of TiO_2 NRs can be precisely controlled within the range of 70–200 nm, with minimal variation in width.

The crystal characteristics of spindle-like TiO_2 NRs with a length of 110 nm (NR110) were further analyzed by high-resolution TEM, as shown in Figure 2.9. The uniform fringe pattern with a spacing of 0.238 nm, corresponding to the d_{004} lattice plane of anatase TiO_2, confirming that NR110 grows along the [001] direction (Figure 2.9a). Near the rod tips that run parallel to the side facets, the fringe spacing is 0.352 nm, corresponding to the d_{101} lattice plane and indicating that the side facets of NR110 are (101) planes. These facets form an angle of 21.7° with the axial direction of NR110, further confirming that NR110 grows along the [001] direction [15].

The formation of TiO_2 NRs is strongly influenced by the selective binding of TMAH on (101) plane of TiO_2. During the initial hydrothermal reaction, TTIP undergoes hydrolysis and polycondensation to form polyanionic titania nuclei. In the basic environment, the surface of these nuclei becomes Ti—O$^-$, attracting $(CH_3)_4N^+$ ions from TMAH, promoting isotropic growth and forming spherical TiO_2 NPs. As the solvothermal reaction progresses, multiple facets form on the TiO_2, but growth along the [101] direction is inhibited by $(CH_3)_4N^+$ passivation of the (101) plane. The Ti precursor preferentially supplies the highly reactive (001) facet, promoting growth along the [001] direction. The increased TMAH capping further suppresses the (101) plane, and excessive TMAH can bind to other planes, allowing growth in all directions and leading to the expansion of TiO_2 NRs in both length and diameter.

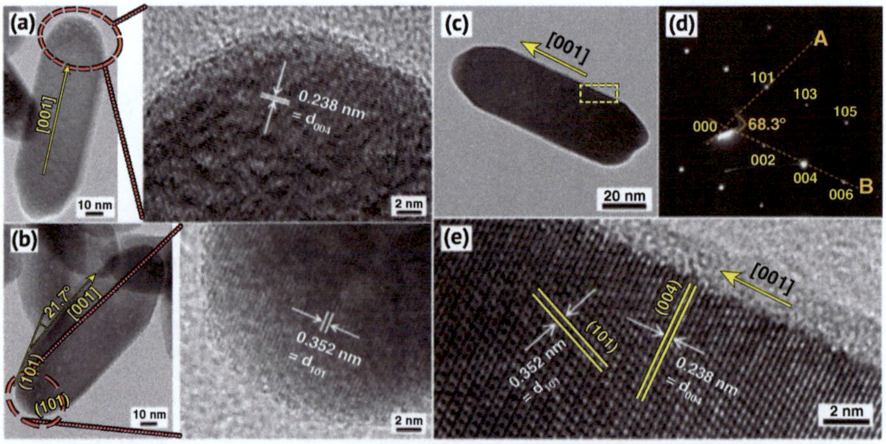

Figure 2.9 (a) TEM (left) and high-resolution TEM (right) micrographs of NR110 with a length of 110 nm and a width of 46 nm showing uniform interspacing of (004) and (b) (101) planes; (c) TEM image, (d) corresponding SAED patterns, and (e) high-magnification image of NR110), acquired from the selected region in Figure 2.9c [14]/American Chemical Society.

2.2.5 Formation of Rutile NRs

Rutile TiO_2 NRs have been prepared under highly acidic conditions with rich Cl^- ions. Huang and Gao reported a simple wet chemical route for preparation of well-dispersed uniform rutile TiO_2 NRs [16]. In a typical synthesis, $TiCl_4$ was dissolved in water to obtain 3 M solution. 20 mL of this solution was added rapidly and stirred vigorously to a 35 mL concentrated HNO_3 solution which was refluxed at 100–250 °C for 6–24 h. Figure 2.10 shows the TEM images of rutile TiO_2 NRs synthesized at different hydrothermal temperature and time. The diameters of the single-crystalline NRs are 4–6 nm and lengths range from 50 to 150 nm. The size of NRs does not appear to be dependent on temperature or reaction time.

Tremel group prepared rutile TiO_2 NRs under a mild condition. They introduced 3-hydroxytyramine hydrochloride into $TiCl_4$ aqueous solution pre-neutralized by Na_2CO_3 [17]. TEM image in Figure 2.11a shows that the resulting 3-hydroxytyramine-capped rutile NRs after hydrothermal process at 150 °C for 13 h exhibit average length and diameter of NRs are 80 nm and 15 nm, respectively. The lattice fringes display a *d*-spacing of 0.32 nm, corresponding to the (110) planes of rutile TiO_2 (Figure 2.11b). The anisotropic growth of the rutile NCs is caused by selective binding of 3-hydroxytyramine and Cl^- ions on the (110) surface of rutile TiO_2 NRs.

2.2.6 Formation of Brookite NR

Bahnemann's group synthesized brookite TiO_2 NRs through the thermal hydrolysis of titanium bis(ammonium lactate) dihydroxide (TALH), as described in Figure 2.12 [18]. Excessive urea serves as an in-situ source of OH^- ions, which are critical for forming brookite-phase TiO_2 NRs. Pure brookite NRs, with diameters of 25 nm and lengths of 150 nm, are selectively formed at urea concentrations above 6.0 M, while anatase NCs coexist at lower urea concentrations. They proposed a

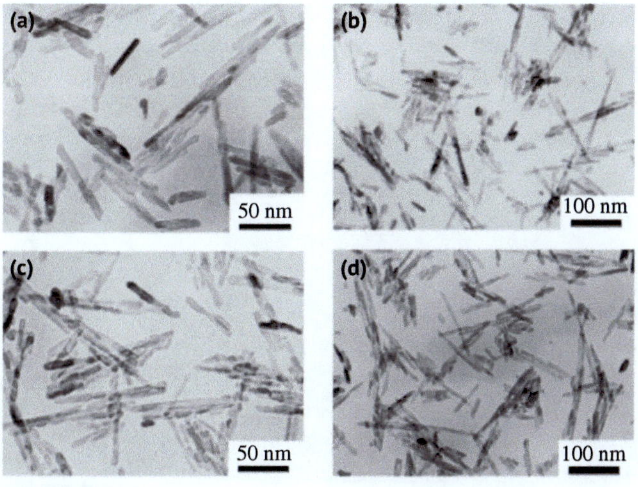

Figure 2.10 TEM images of rutile TiO_2 NRs synthesized at various hydrothermal temperature and time: (a) 110 °C for 12 h, (b) 140 °C for 12 h, (c) 240 °C for 12 h, and (d) 140 °C for 6 h [16]/The Chemical Society of Japan/Oxford University Press.

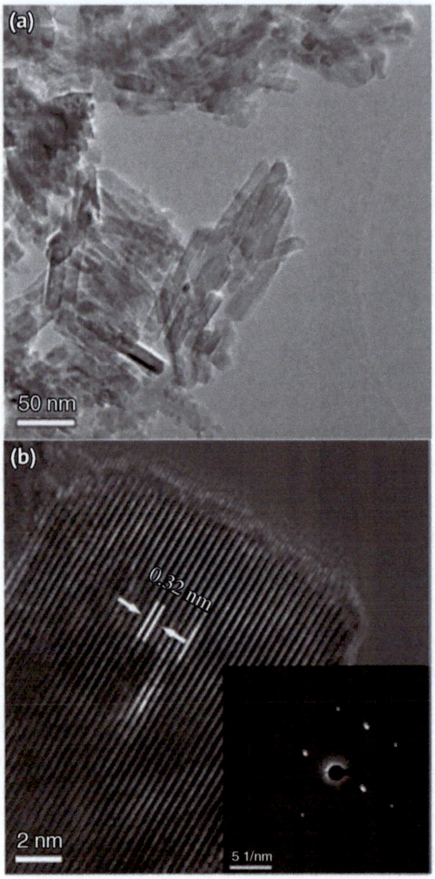

Figure 2.11 Low (a) and high (b) magnification TEM images of rutile TiO_2 NRs capped with 3-hydroxytyramine. The inset in panel b displays the SAED pattern of an individual rutile phase [17]/American Chemical Society.

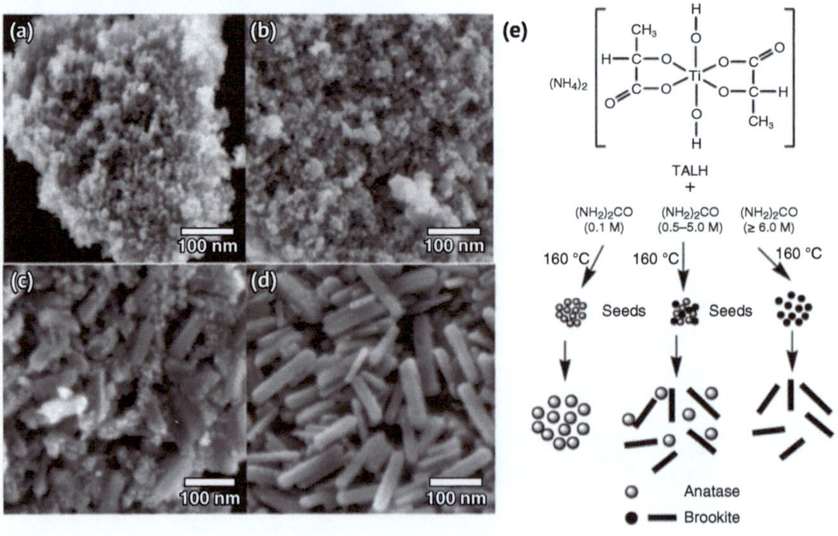

Figure 2.12 SEM images of TiO_2 NRs and NPs obtained by thermal hydrolysis of TALH solution with different urea concentration: 0.1 M (a), 1.0 M (b), 2.0 M (c), and 6.0 M (d), and the formation pathways of anatase NCs, brookite NRs, and their mixtures (e) [18]/American Chemical Society.

seed-assisted growth model, where urea hydrolysis at 160°C generates OH^- ions rapidly, triggering the thermal hydrolysis of TALH and crystallization into brookite NCs. Subsequent growth to NRs does not alter the brookite phase, since the initial seeds remain intact, and crystal growth requires lower activation energy than phase transformation.

2.3 TiO_2 NT Structure

2.3.1 Formation of TiO_2 NTs in NaOH

2.3.3.1 Kasuka Method

Kasuka et al. reported that anatase TiO_2 NPs can be converted into TiO_2 NTs in a 10 M NaOH solution at elevated temperature under hydrothermal or refluxing conditions at 110°C (see Figure 2.13) [19–21]. The corrosive and hot medium breaks the Ti—O—Ti bonds, while Ti—O—Na and Ti—OH bonds generate. Na^+ ions located between the TiO_6 octahedral layers are progressively replaced by water molecules. Because the volume of water molecules is larger than that of Na^+ ions, the interlayer spacing increases, weakening the electrostatic interactions between adjacent TiO_6 octahedral layers. Thus, 2D sodium titanate phase gradually forms in situ on TiO_2 surface, and the layered sodium titanate exfoliates to form nanosheets. To minimize surface energy, the nanosheets curl into NTs.

Even in form of NTs, the presence of sodium titanate significantly lowers the poor photocatalytic activity of NTs. Topo conversion of sodium titanate to crystalline TiO_2 can be achieved by sequent proton exchange and calcination. Proton exchange is conducted by repeated ion exchange in diluted HCl and HNO_3 solution, allowing Na^+ ions in the Ti—O—Na bonds exchanged with H^+ ions [20, 22]. Na^+ ions in the Ti—O—Na bonds are exchanged with H^+ ions, causing exfoliation of the layered structure. The reaction with hydrochloric acid results in the formation of new Ti—O—Ti bonds as Ti—OH bonds react with the acid and water, leading to the transformation into the anatase phase. Meanwhile, laminar TiO_2 is converted into anatase, which then undergo curling to NTs.

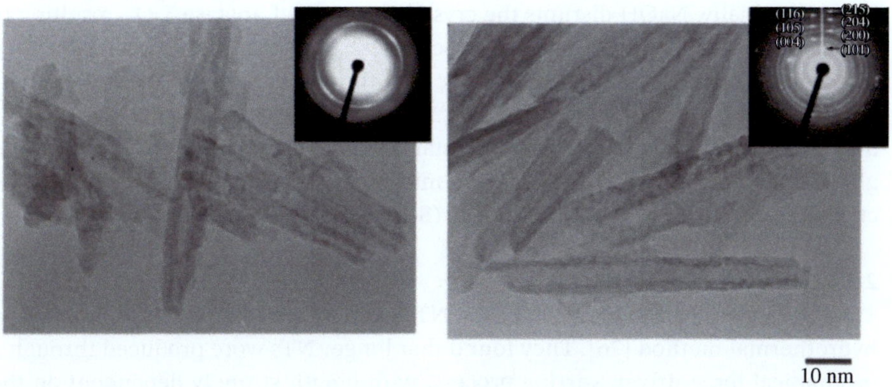

Figure 2.13 TEM images of TiO_2 NTs synthesized through a hydrothermal reaction with TiO_2 particles in 10 M NaOH aqueous solution [20]/John Wiley & Sons.

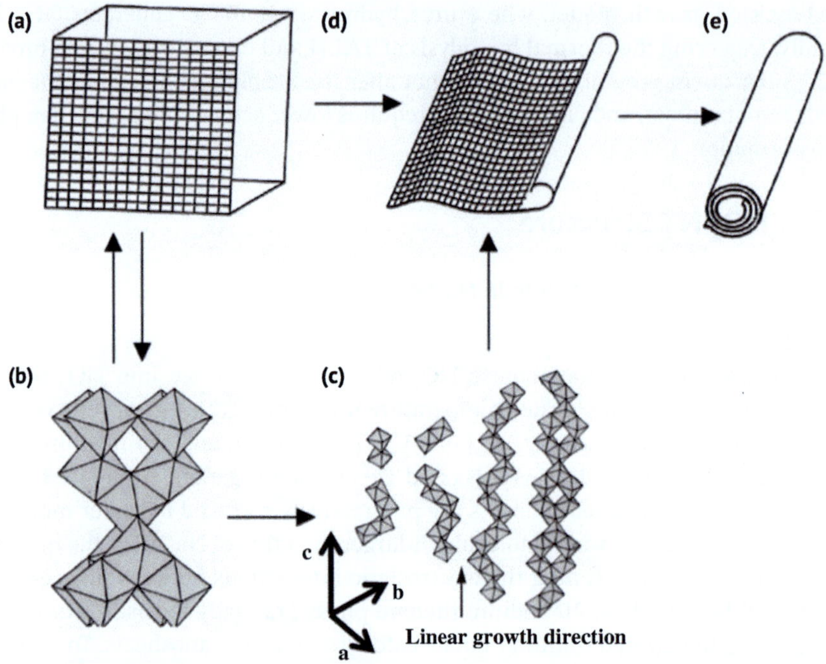

Scheme 2.2 A diagram describing formation mechanism of TiO_2 NT in hydrothermal reaction [25]/with permission of Springer Nature.

The mechanism behind the formation of TiO_2 NTs remains a subject of debate. Du et al. proposed an alternative mechanism for TiO_2 NT formation [22], suggesting that NTs can form during hydrothermal treatment without the need for a hydrochloric acid washing step. Specifically, the TiO_2 NT phase emerges directly during the hydrothermal process. Since then, most studies have supported the idea that NT morphology develops during the alkaline hydrothermal reaction [23, 24]. Additionally, the synthesized NTs predominantly exhibit the anatase phase, rather than the sodium titanate phase [25].

Wang et al. presented a different formation mechanism, as outlined in Scheme 2.2 [25]. Initially, NaOH disrupts the crystal structure of anatase TiO_2, resulting in the formation of free TiO_6 octahedra (Scheme 2.2a,b). These free octahedra then reassemble into edge-sharing structures by forming hydroxyl bridges between Ti ions, creating a zigzag pattern (Scheme 2.2c). Lateral growth occurs along the [001] direction, forming two-dimensional crystal sheets (Scheme 2.2d). To saturate dangling bonds, reduce surface energy, and minimize the surface-to-volume ratio, the crystal sheets curl into anatase TiO_2 NTs (Scheme 2.2e).

2.3.3.2 Elongated TiO_2 NTs

Tang et al. synthesized elongated TiO_2 NTs over 10 μm in length using a stirring hydrothermal method [26]. They found that longer NTs were produced through a mechanical force-driven stirring process, with length strongly dependent on the stirring speed. The continuous stirring force aligned the titanate precursor and

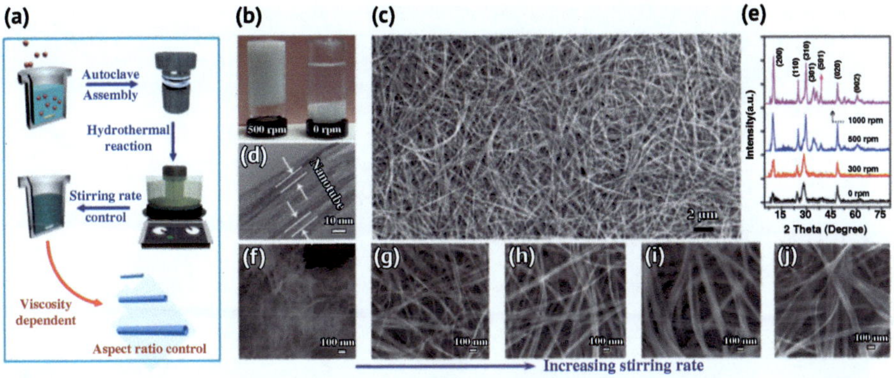

Figure 2.14 Preparation and analysis of titanate NTs showing various aspect ratios. (a) Scheme of the fabrication process. (b) Photographic images of sol solutions produced with stirring at 500 rpm (left) and without stirring (right) after a slow sedimentation. (c) Field emission SEM images of titanate NTs prepared at 500 rpm. (d) TEM image corresponding to c, with the arrow indicating the nanotubular structure. (e) XRD patterns of NTs synthesized under various stirring conditions. (f–j) Field-emission SEM images of NTs derived from hydrothermal reaction at 130 °C for 24 h with stirring speeds of 0, 300, 400, 500, and 1000 rpm, respectively [27]/with permission of John Wiley & Sons.

the formed NTs in the same direction without sedimentation. This process accelerated NC attachment at the NT ends, leading to elongated structures as diffusion and chemical reaction rates increased with the stirring rate. The average length of TiO_2 NTs increases from 0.4 to 30.7 μm (Figure 2.14) as the stirring rate rises from 0 to 500 rpm [27, 28].

2.3.2 TiO_2 Microtubes (MT)

Jeon et al. synthesized a novel TiO_2 microtubes (MTs) with micro-sized pores through a template-free solvothermal reaction of TTIP and tetrabutyl ammonium hydroxide (TBAH) [29]. The shape of TiO_2 nanostructures is highly influenced by the chemical composition of TTIP and TBAH. As shown in Figure 2.15, the prepared TiO_2 MTs in the pure anatase phase, derived from TTIP/TBAH of 0.38 M/0.35 M, exhibit pore diameters of ~1.6 μm and lengths of 15–30 μm. Moreover, the wall of MTs, with a thickness of ~200 nm, is highly crystallized with grain size of ~28 nm, and exhibit a highly porous structure (average pore size on the wall: ~20 nm). As shown in Figure 2. 15f, the structures of MT were not significantly changed by the heat treatment at 600 °C, suggesting that they are physically robust and thermally stable.

The structural evolution was tracked by monitoring the solvothermal intermediates. Initially, TiO_2-based organic–inorganic hybrid NFs form in solution, which then rapidly decompose into nanoparticulate nuclei. These nuclei, with sizes of ~10 nm, spontaneously aggregate to form microcylinders. Subsequently, a small hole forms at the center of the circular plane of the cylindrical structure due to the abrupt release of 1-butene gas originating from TBAH. As the reaction progresses,

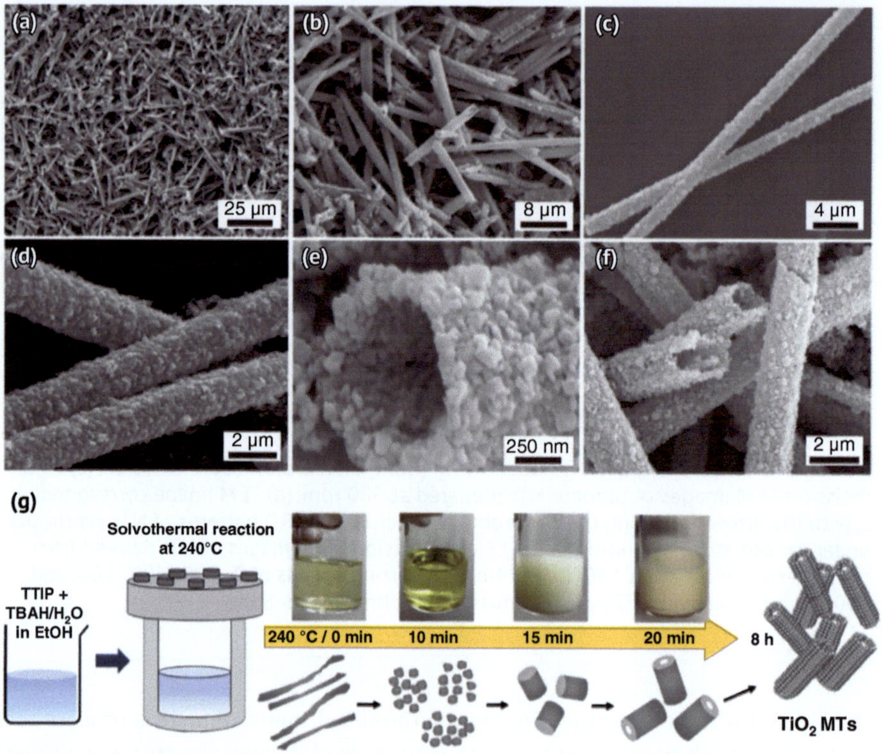

Figure 2.15 SEM images with different magnifications for the as-prepared TiO_2 microtubes upon a solvothermal process at 240 °C for 8 h (a–e) and the MT structure obtained by additional thermal treatment at 600 °C for 1 h (f). (g) A diagram describing structural evolution of TiO_2 microtube during the solvothermal reaction [29]/with permission of Elsevier.

the pore gradually expands, and the wall structure becomes highly crystallized through Ostwald ripening, as illustrated in the schematic diagram in Figure 2.15g.

2.4 TiO_2 NB

Highly crystallized TiO_2 NBs are synthesized by hydrothermal reaction. There are two well-known methods, namely, alkali-hydrothermal process and acid-hydrothermal process [2].

2.4.1 Alkali-Hydrothermal Process

Alkali-hydrothermal process is quite similar to Kasuka method but using a higher hydrothermal temperature. In a typical synthesis, Ti-precursor such as TiO_2 powders is dispersed into a 10 M NaOH aqueous solution. The obtained suspension is hydrothermally heated at 180–200°C for more than 12 h. Sodium titanate NBs are firstly formed, and their composition is tuned and mainly determined by the Ti precursor.

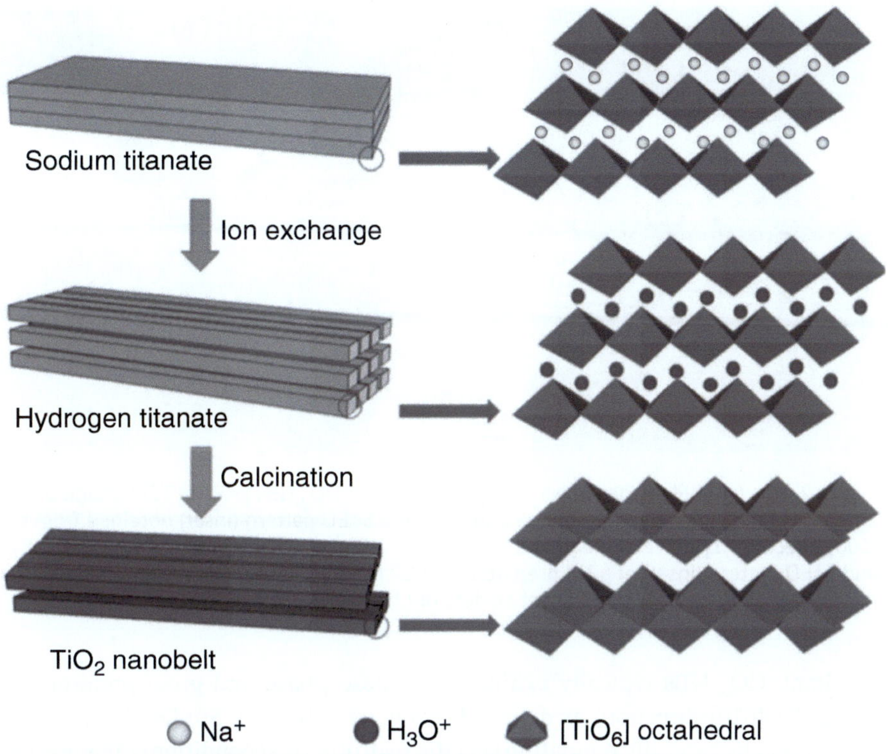

Scheme 2.3 A diagram describing the formation mechanism of titania NBs. The TiO_6 octahedra is composed of one Ti and six oxygen atoms, while Ti atom is located in the body center of the octahedra and six O atoms are in the corners [2]/with permission of John Wiley & Sons.

For example, sodium titanate NBs produced from the TiO_2 powders are $Na_2Ti_3O_7$ or $Na_2Ti_4O_9$. The $Ti(OBu)_4$ and $TiCl_4$ precursors lead to form NB structures of $Na_2Ti_2O_4(OH)_2$, while TiF_4 induces $Na_2Ti_5O_{11}$. Regardless of chemical formular, however, they have a similar layered structure. The formation process of sodium titanate NBs is based on dissolution and recrystallization mechanism [30]. As described in Scheme 2.3, under highly alkaline condition and critical condition, TiO_2 can be dissolved by breaking the Ti—O—Ti bonds, leading to the formation of sodium titanate NBs (Eq. 2.4). The sodium titanate NB is converted to hydrogen titanate NB via proton exchange (Eq. 2.5). Following a calcination, the hydrogen titanate NB can be dehydrated to form crystalline TiO_2 NBs with tunable phase (Eq. 2.6).

$$3TiO_2 + 2NaOH \rightarrow Na_2Ti_3O_7 + H_2O \tag{2.4}$$

$$Na_2Ti_3O_7 + 2HCl \rightarrow H_2Ti_3O_7 + 2NaCl \tag{2.5}$$

$$H_2Ti_3O_7 \rightarrow 3TiO_2 + H_2O \tag{2.6}$$

During these three step procedures, the NB morphologies of $Na_2Ti_3O_7$, $H_2Ti_3O_7$ and TiO_2 are not changed significantly, as shown in Figure 2.16a [31, 32]. The

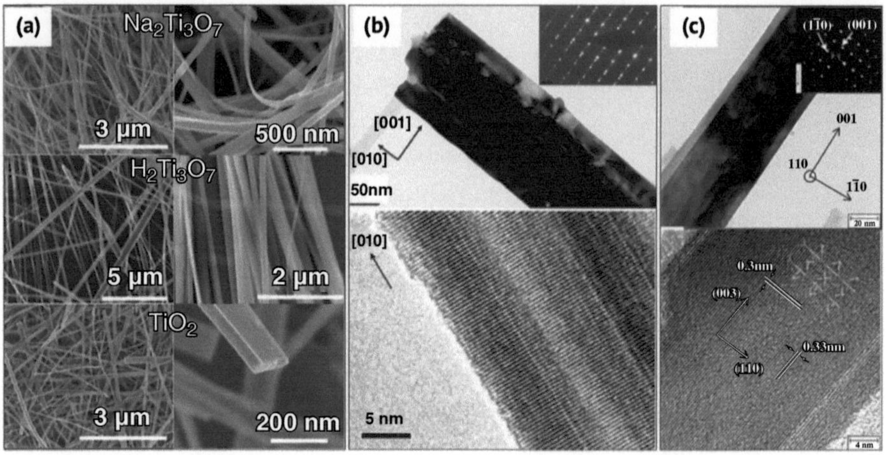

Figure 2.16 (a) SEM images of $Na_2Ti_3O_7$, $H_2Ti_3O_7$ and TiO_2 NBs [32]. (b) TEM image of an anatase NB (top), its magnified image (bottom), and SAED pattern (inset) obtained from the [100] direction [1]. (c) TEM image of a single rutile NB (top), its magnified image (bottom), and SAED pattern (inset) of a NB prepared with 12 M HCl and titanium (IV) butoxide via a hydrothermal process [33]/The Royal Society of Chemistry.

resultant TiO_2 NBs typically exhibit an anatase phase and grow preferentially along the [010] direction, as depicted in Figure 2.16b [1]. Studies have demonstrated that by carefully regulating specific hydrothermal conditions, anatase TiO_2 NBs with dominant (001) facets can also be fabricated through this method [32]. The formation of the belt-like morphology is critically dependent on the process of sodium titanate formation. During the initial stages of the hydrothermal reaction, hexacoordinated $[Ti(OH)_6]^{2-}$ monomers are first generated, which subsequently aggregate to form crystal nuclei. These nuclei further develop into nanosheet structures consisting of layered octahedral $[TiO_6]$ units, with Na^+ ions intercalated between the layers to maintain charge balance. Upon the addition of HCl, the Na^+ ions within the sodium titanate NBs are replaced by larger H_3O^+ ions, leading to an expansion of the interlayer spacing and a reduction in the interlayer attractive forces. Consequently, the laminar sodium titanate undergoes gradual exfoliation, forming hydrogen titanate nanosheets. These nanosheets further fragment into NB structures, thereby alleviating internal stress and minimizing the overall system energy.

2.4.2 Acid-Hydrothermal Process

Acid-hydrothermal process enables to prepare rutile TiO_2 NBs. This route requires precise control for the hydrolysis rate of Ti-precursors in acidic media, leading to the anisotropic growth of rutile TiO_2 NBs along the [001] growth direction (Figure 2.16c) [33]. Zhou et al. monitored the evolution of TiO_2 nanostructures during the acid-hydrothermal process, and rutile TiO_2 NBs can be obtained by hydrothermally treating titanium butoxide in 10 M HCl solution at 180 °C for 2 h [34].

2.5 TiO_2 NFs by Electrospinning

The electrospinning process is a widely employed technique for fabricating NFs with high surface area, porosity, and flexibility. By applying a high-voltage electric field (typically between 10 and 30 kV), a viscous Ti-based sol–gel solution is drawn into fine fibers, enabling precise control over fiber formation from a syringe. In 2003, Li et al. first applied electrospinning to prepare TiO_2 NFs, achieving a diameter of ~53 nm [35]. The electrospinning synthesis involves four key steps: (1) preparation of a Ti sol solution; (2) mixing the sol with a polymer to form a suitable solution for electrospinning; (3) electrospinning the solution to create NFs; and (4) calcining the as-spun NFs to remove organic components and obtain crystalline TiO_2 NFs.

In a typical experiment by Yang et al., TTIP, stabilized in acetic acid, was mixed with ethanol and poly(vinyl pyrrolidone) (PVP) to form a viscous, homogeneous solution [28, 36, 37]. As illustrated in Figure 2.17a, electrospinning the viscous solution onto the grounded substrate under a high voltage of 15 kV forms the organic–inorganic hybrid NFs. Upon organic removal by calcination in air above 500°C, porous and crystalline TiO_2 NFs can be obtained, retaining their fibrous nanostructure, although their diameters decreased and a porous microstructure formed as a result of polymer removal (see Figure 2.17b-e).

2.6 NR or NW Arrays

2.6.1 Rutile NR Arrays

In 2009, Liu and Aydil successfully synthesized rutile TiO_2 NR arrays vertically grown on FTO substrates [38]. Typical procedure is as follows: titanium butoxide was stabilized by concentrated HCl solution in a Teflon-lined autoclave, with pre-cleaned FTO substrates positioned against the Teflon liner's wall. The vessel was closed and kept at 150°C for 20 h. Under these highly acidic conditions, rutile TiO_2 NRs formed on the side of the FTO substrates facing downward in the autoclave.

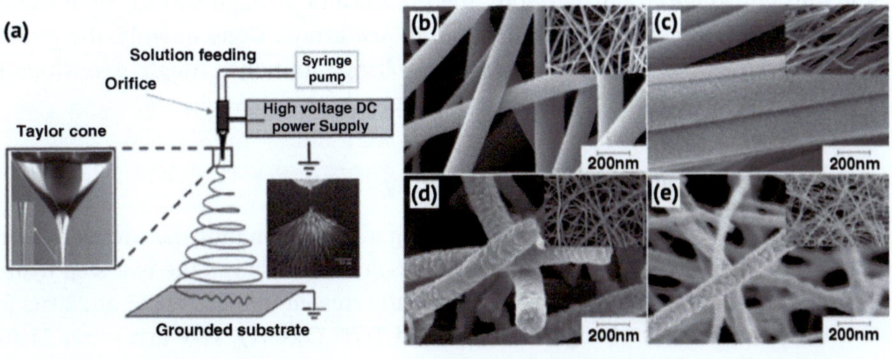

Figure 2.17 (a) A diagram describing the electrospinning process. SEM images of titania NFs prepared by postannealing at 500 °C (b), 600 °C (c), 700 °C (d), and 800 °C (e). Adapted from [36].

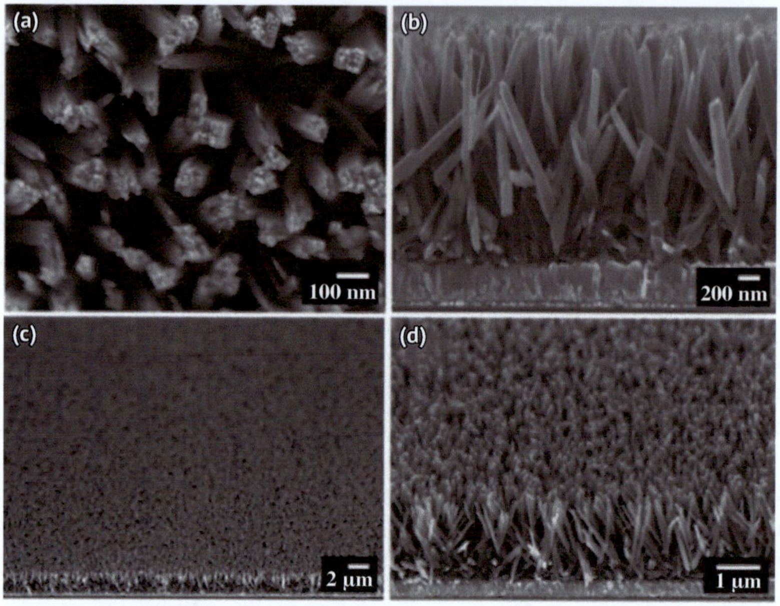

Figure 2.18 SEM images of vertically aligned rutile TiO_2 NRs on an FTO substrate: top view (a) and side views at different magnifications (b–d), prepared by the hydrothermal reaction of $Ti(OBu)_4$ with HCl solution at 150 °C for 20 h [38]/American Chemical Society.

The resulting NRs had an average length of ~1.9 μm and a diameter of ~90 nm. The TiO_2 NR arrays were single-crystalline, with a rutile phase and grew along the [001] direction.

As shown in Figure 2.18, rutile TiO_2 NR arrays were exclusively formed on FTO substrates, keeping the side surfaces smooth. The top surface of TiO_2 NRs featured numerous step edges, exhibiting a tetragonal morphology with square top facets, in accordance with the characteristic growth of rutile phase. The downward-facing orientation of the FTO substrates effectively inhibited free sedimentation, thus promoting preferential growth by facilitating the adsorption of Ti species from the nutrient solution. Furthermore, the presence of a rough surface significantly enhances the formation of well-defined, robust arrays. Consequently, the growth side of a smooth glass substrate must be polished to induce roughness, which in turn fosters improved TiO_2 deposition and growth.

2.6.2 Hierarchical Anatase TiO_2 NW Arrays

Hierarchical anatase TiO_2 NW arrays having a complex nanostructure were fabricated via a layer-by-layer-like method [39]. Scheme 2.4 shows the two-step hydrothermal procedure. Ti foil was hydrothermally treated in 1 M NaOH at 220°C for 24 h. Upon proton exchange, the obtained $H_2Ti_2O_4(OH)_2$ NW arrays on Ti foil was calcined, achieving phase transition to anatase TiO_2. Then, by performing the second-step hydrothermal reaction, short TiO_2 NR branches can be grown on the long TiO_2 NW trunks, forming hierarchical TiO_2 array.

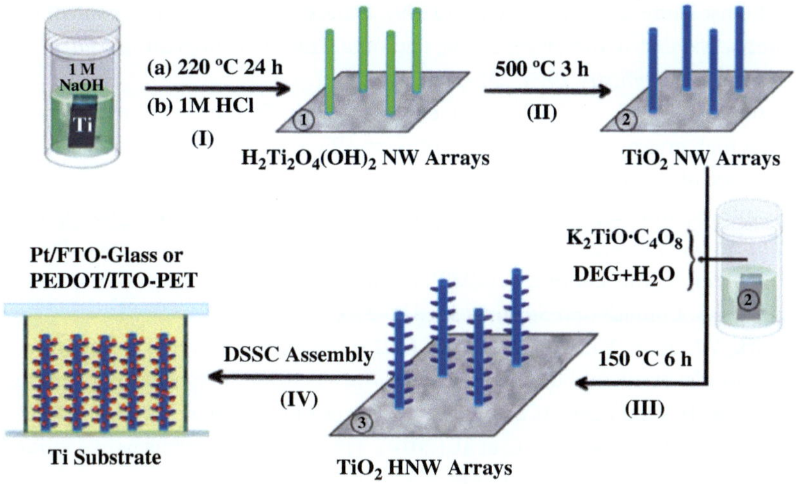

Scheme 2.4 Schematic diagram of hierarchical anatase TiO_2 NW arrays via a two-step hydrothermal process [39]/with permission of The Royal Society of Chemistry.

References

1 Wu, N.Q., Wang, J., Tafen, D. et al. (2010). Shape-enhanced photocatalytic activity of single-crystalline anatase TiO_2 (101) nanobelts. *Journal of the American Chemical Society* 132: 6679–6685.

2 Zhao, Z., Tian, J., Sang, Y. et al. (2015). Structure, synthesis, and applications of TiO_2 nanobelts. *Advanced Materials* 27: 2557–2582.

3 Cargnello, M., Gordon, T.R., and Murray, C.B. (2014). Solution-phase synthesis of titanium dioxide nanoparticles and nanocrystals. *Chemical Reviews* 114: 9319–9345.

4 Tian, J., Zhao, Z., Kumar, A. et al. (2014). Recent progress in design, synthesis, and applications of one-dimensional TiO_2 nanostructured surface heterostructures: a review. *Chemical Society Reviews* 43: 6920–6937.

5 Wang, X., Li, Z., Shi, J. et al. (2014). One-dimensional titanium dioxide nanomaterials: nanowires, nanorods, and nanobelts. *Chemical Reviews* 114: 9346–9384.

6 Lee, K., Mazare, A., and Schmuki, P. (2014). One-dimensional titanium dioxide nanomaterials: nanotubes, *Chemical Reviews* 114: 9385–9454.

7 Penn, R.L., Oskam, G., Strathmann, T.J. et al. (2001). Epitaxial assembly in aged colloids. *The Journal of Physical Chemistry B* 105: 2177–2182.

8 Penn, R.L. and Banfield, J.F. (1999). Morphology development and crystal growth in nanocrystalline aggregates under hydrothermal conditions: insights from titania. *Geochimica et Cosmochimica Acta* 63: 1549–1557.

9 Joo, J., Kwon, S.G., Yu, T. et al. (2005). Large-scale synthesis of TiO_2 nanorods via nonhydrolytic sol-gel ester elimination reaction and their application to photocatalytic inactivation of *E. coli*. *The Journal of Physical Chemistry B* 109: 15297–15302.

10 Chemseddine, A. and Moritz, T. (1999). Nanostructuring titania: control over nanocrystal structure, size, shape, and organization. *European Journal of Inorganic Chemistry* 1999: 235–245.
11 Jun, Y.W., Casula, M.F., Sim, J.H. et al. (2003). Surfactant-assisted elimination of a high energy facet as a means of controlling the shapes of TiO_2 nanocrystals. *Journal of the American Chemical Society* 125: 15981–15985.
12 Kanie, K. and Sugimoto, T. (2004). Shape control of anatase TiO_2 nanoparticles by amino acids in a gel–sol system. *Chemical Communications* 10: 1584–2585.
13 Tentorio, A. and Canova, L. (1989). Adsorption of α-amino acids on spherical TiO_2 particles. *Colloids and Surfaces* 39: 311–319.
14 Park, Y.J., Jeon, Y.I., Yang, I.S. et al. (2023). Selective control of novel TiO_2 nanorods: excellent building blocks for the electron transport layer of mesoscopic perovskite solar cells. *ACS Applied Materials and Interfaces* 15: 9447–9456.
15 Yang, S., Yang, B.X., Wu, L. et al. (2014). Titania single crystals with a curved surface. *Nature Communications* 5: 5355.
16 Huang, Q. and Gao, L. (2003). A simple route for the synthesis of rutile TiO_2 nanorods. *Chemistry Letters* 32: 638–639.
17 Tahir, M.N., Theato, P., Oberle, P. et al. (2006). Facile synthesis and characterization of functionalized, monocrystalline rutile TiO_2 nanorods. *Langmuir* 22: 5209–5212.
18 Kandiel, T.A., Feldhoff, A., Robben, L. et al. (2010). Tailored titanium dioxide nanomaterials: anatase nanoparticles and brookite nanorods as highly active photocatalysts. *Chemistry of Materials* 22: 2050–2060.
19 Kasuga, T., Hiramatsu, M., Hoson, A. et al. (1998). Formation of titanium oxide nanotube. *Langmuir* 14: 3160–3163.
20 Kasuga, T., Hiramatsu, M., Hoson, A. et al. (1999). Titania nanotubes prepared by chemical processing. *Advanced Materials* 11: 1307–1311.
21 Bavykin, D.V., Friedrich, J.M., and Walsh, F.C. (2006). Protonated titanates and TiO_2 nanostructured materials: synthesis, properties, and applications. *Advanced Materials* 18: 2807–2824.
22 Du, G.H., Chen, Q., Che, R.C. et al. (2001). Preparation and structure analysis of titanium oxide nanotubes. *Applied Physics Letters* 79: 3702–3704.
23 Chen, Q., Zhou, W.Z., Du, G.H. et al. (2002). Trititanate nanotubes made via a single alkali treatment. *Advanced Materials* 14: 1208–1211.
24 Pan, X., Zhao, Y., Liu, S. et al. (2012). Comparing graphene-TiO_2 nanowire and graphene-TiO_2 nanoparticle composite photocatalysts. *ACS Applied Materials and Interfaces* 4: 3944–3950.
25 Wang, W., Varghese, O.K., Paulose, M. et al. (2004). A study on the growth and structure of titania nanotubes. *Journal of Materials Research* 19: 417–422.
26 Tang, Y., Zhang, Y., Deng, J. et al. (2014). Mechanical force-driven growth of elongated bending TiO_2-based nanotubular materials for ultrafast rechargeable lithium ion batteries. *Advanced Materials* 26: 6111–6118.
27 Tang, Y.X., Zhang, Y.Y., Deng, J.Y. et al. (2014). Unravelling the correlation between the aspect ratio of nanotubular structures and their electrochemical performance to achieve high-rate and long-life lithium-ion batteries. *Angewandte Chemie* 53: 13488–13492.

28 Ge, M., Cao, C., Huang, J. et al. (2016). A review of one-dimensional TiO_2 nanostructured materials for environmental and energy applications. *Journal of Materials Chemistry A* 4: 6772–6801.

29 Jeon, Y.I., Park, Y.J., Pan J.H. et al. (2022). Template-free synthesis of novel TiO_2 microtube (MT) and N-doped MT and their photocatalytic properties decomposing organic compounds. *Applied Catalysis A: General* 645: 118841.

30 Wu, D., Liu, J., Zhao, X. et al. (2006). Sequence of events for the formation of titanate nanotubes, nanofibers, nanowires, and nanobelts. *Chemistry of Materials* 18: 547–553.

31 Wang, Y.M., Du, G.J., Liu, H. et al. (2008). Nanostructured sheets of Ti–O nanobelts for gas sensing and antibacterial applications. *Advanced Functional Materials* 18: 1131–1137.

32 Pan, F., Wu, K., Li, H.X. et al. (2014). Synthesis of {100} facet dominant anatase TiO_2 nanobelts and the origin of facet-dependent photoreactivity. *Chemistry-A European Journal* 20: 15095–15101.

33 Cha, S.I., Hwang, K.H., Kim, Y.H. et al. (2013). Crystal splitting and enhanced photocatalytic behavior of TiO_2 rutile nano-belts induced by dislocations. *Nanoscale* 5: 753–758.

34 Zhou, W., Liu, X., Cui, J. et al. (2011). Control synthesis of rutile TiO_2 microspheres, nanoflowers, nanotrees and nanobelts via acid-hydrothermal method and their optical properties. *CrystEngComm* 13: 4557–4563.

35 Li, D. and Xia, Y. (2003). Fabrication of titania nanofibers by electrospinning. *Nano Letters* 3: 555–560.

36 Yang, Y., Wang, H., Zhou, Q. et al. (2013). Improved lithium storage properties of electrospun TiO_2 with tunable morphology: from porous anatase to necklace rutile. *Nanoscale* 5: 10267–10274.

37 Zong, H., Xia, X., Liang, Y. et al. (2018). Designing function-oriented artificial nanomaterials and membranes via electrospinning and electrospraying techniques. *Materials Science and Engineering: C* 92: 1075–1091.

38 Liu, B. and Aydil, E.S. (2009). Growth of oriented single-crystalline rutile TiO_2 nanorods on transparent conducting substrates for dye-sensitized solar cells. *Journal of the American Chemical Society* 131: 3985–3990.

39 Liao, J.Y., Lei, B.X., Chen, H.Y. et al. (2012). Oriented hierarchical single crystalline anatase TiO_2 nanowire arrays on Ti-foil substrate for efficient flexible dye-sensitized solar cells. *Energy and Environmental Science* 5: 5750–5757.

3

Synthesis of Two-dimensional TiO_2 Nanostructures

Sichun Yang, Haijiao Lu, Lianzhou Wang**

Nanomaterials Centre, School of Chemical Engineering, and Australian Institute for Bioengineering and Nanotechnology, The University of Queensland, St Lucia, QLD, Australia

*Corresponding authors: l.wang@uq.edu.au; haijiao.lu@uq.edu.au

3.1 Introduction

Low-dimensional materials have attracted significant attention due to their superior thermodynamic stability and excellent physiochemical properties. Among these low-dimensional materials, two-dimensional (2D) materials, characterized by their layered atomic structure, have emerged as promising candidates for various applications among nanomaterials. Graphene, consisting of single-layer carbon atoms, was the first discovered 2D material, and has been widely used in various fields owing to its excellent properties of atomic-scale thickness [1], good flexibility, and high transparency [2]. Inspired by the merits of graphene, researchers have shown particular interest in developing 2D TiO_2.

There are three common crystal structures of TiO_2 including anatase [3, 4], rutile [4, 5] and brookite [6, 7], while all these three crystal structures share octahedral geometry with titanium being bonded to six oxide anions [8]. More importantly, all these crystal structures can be exfoliated to layered structure compounds to obtain 2D TiO_2. The 2D TiO_2 has their stronger planar bonding and weaker interactions between surrounding layers, which hinder the stacking of layers and persist in the single-layer structure. The primary benefit of 2D TiO_2 lies in its ability to minimize the effects of different surface states by providing a single-layered nanosheet (NS) structure without dangling covalent bonds or interplanar interactions. Moreover, the stronger planar bonding of 2D TiO_2 results in outstanding mechanical robustness and high strain-bearing capacity, and their thin layer structure leads to good transparency in the visible range. In addition, the optical properties of 2D TiO_2 can be easily adjusted by tuning the thickness of the layer. The bandgaps of 2D hexagonal-TiO_2 with a single layer, double layers, and triple layers are 2.4, 2.0, and 1.85 eV, respectively [9], which are smaller compared to anatase (3.2 eV), rutile (3 eV), and brookite (3.3 eV). The light absorption in visible range can be

Nanostructured TiO_2: Synthesis, Photocatalytic Properties and Applications, First Edition.
Edited by Jia Hong Pan, Wan In Lee and Detlef W. Bahnemann

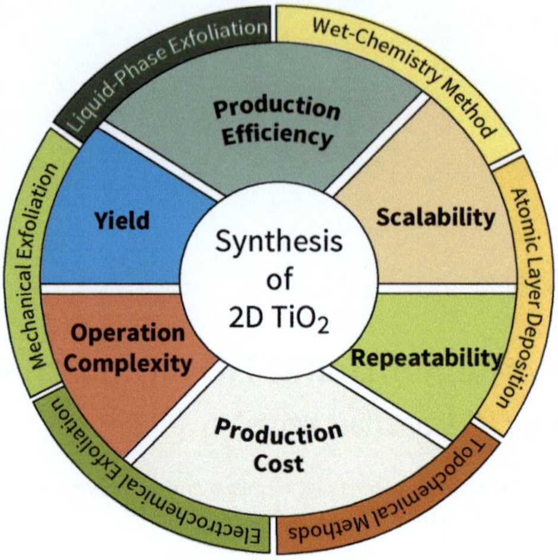

Figure 3.1 Outline for 2D TiO_2 synthesis overview.

improved owing to the narrowed bandgaps in 2D TiO_2, making 2D TiO_2 potential materials for photocatalysts, photoelectrodes, optoelectronic devices, and solar cells [10, 11]. Furthermore, 2D TiO_2 also exhibits good mobility of both electrons and holes, a characteristic closely related to the thickness of the layer. By increasing the thickness to 10 times the original thickness (from 0.5 to 5 nm) of 2D hexagonal-TiO_2, the field mobility of holes increases more than 10 times from 10 to 950 $cm^2\,V^{-1}\,s^{-1}$, both of which are much higher than that of TiO_2 [12]. Attracted by the distinguished mobility of charge carriers and adjustable multiferroic properties, many researchers are exploring the applications of 2D TiO_2 in field-effect transistors, electrothermal, and photovoltaic devices.

This chapter summarizes the state-of-the-art methods for synthesis of 2D TiO_2. Figure 3.1 provides an overview of the main content in this chapter. The most popular synthesis approaches for 2D TiO_2 are discussed, along with the criteria for selecting these methods. Finally, this work delineates the merits and limitations of these approaches, while also offering insights into future research directions for the synthesis of 2D TiO_2.

3.2 Synthesis of 2D TiO_2 Nanostructure

In the past few decades, intensive research and continuous development in the preparation of 2D materials have led to the establishment of mature systematic approaches. Generally, there are two common types of synthesis for 2D materials which are top-down methods and bottom-up methods. These methods are universal to the synthesis of 2D materials including graphene, graphitic carbon nitride, transition metal oxides, MXenes (transition-metal carbides, transition-metal nitrides, and transition-metal carbonitrides), transition-metal dichalcogenides (TMDs), layered double hydroxides, and nonmetal oxides.

3.2.1 Top-down Methods

The top-down method for 2D material synthesis typically involves the exfoliation or delamination of layered bulk materials by applying various approaches to weaken the interlayer electrostatic interactions thereby dismantling the layered structure to obtain single-layered NSs. The starting materials for the top-down approach are usually bulk or powder materials. Various methods are applied to achieve the exfoliation or delamination, including the mechanical exfoliation method, liquid-phase exfoliation, and electrochemical exfoliation.

3.2.1.1 Mechanical Exfoliation

Resulting from the low density of the charge carrier, the interlayer van Der Waals forces in some materials are weak, therefore, allowing the separation of layers in these materials into single-sheet or multilayer 2D materials by applying external mechanical force. A milestone in mechanical exfoliation was the synthesis of monocrystalline graphene films by Novoselov et al. in 2004, for which they were awarded the Nobel Prize in Physics in 2010 [1]. Since then, mechanical exfoliation has been extensively studied and refined for the preparation of 2D materials for decades. For example, 2D monoclinic $CrCl_3$ with single layer and multilayers has been successfully produced by mechanical exfoliating single-crystal $CrCl_3$ in 2020. The 2D monoclinic $CrCl_3$ exhibits better stability than the single-crystal $CrCl_3$, which enables applications in spintronic devices [13].

While 2D NSs produced through mechanical exfoliation inherently contain some defects, they also possess a well-defined crystal structure that enhances their electrical and optical properties. However, defects are sometimes useful for specific applications such as photocatalytic and electrocatalytic reactions as active sites. Despite its ability to generate 2D NSs with varying thicknesses, mechanical exfoliation faces the challenges in controlling the thickness of the resulting sample, and suffers from the issue of low reproducibility. As a result, mechanical exfoliation is limited to lab-scale production of 2D materials due to the low productivity and poor heterogeneity. Moreover, mechanical exfoliation is not applicable to TiO_2 due to the strong electrostatic interaction between layers. Effective exfoliation of TiO_2 can be achieved by introducing species that can replace the interlayer protons, reducing the interactions between host layers and guest layers.

3.2.1.2 Liquid-phase Exfoliation

The liquid exfoliation method has been extensively studied for the synthesis of 2D TiO_2 with single layers or multilayers. Unlike mechanical exfoliation, liquid-phase exfoliation facilitates industrial-scale production of 2D materials due to its high efficiency and production yields. The method involves the introduction of molecules and ions into the solvent, which are then into the interlayers of 3D materials through mechanical shaking or ultrasonication. As these molecules and ions insert between layers, the interlayer interactions gradually diminish. Eventually, this leads to the successful stripping of 2D TiO_2.

For instance, Sasaki's team has applied ion exchange and intercalation reactions on cesium titanates, followed by mechanical shaking to strip layered cesium titanates into single-layered TiO_2 NSs since 1995 [14]. Through years of research, they managed to synthesize single-layered TiO_2 NSs with a film thickness of approximately 1 nm. Furthermore, they successfully adjusted the thickness of the 2D TiO_2 NSs from 1 nm to a few micrometers. Since then, liquid-phase exfoliation has become a widely applied method for synthesizing 2D TiO_2 NSs and found wide applications For example, 2D TiO_2 NSs synthesized via liquid exfoliation method can be applied in sulfur cathodes for lithium–sulfur (Li–S) batteries to prevent the dissolution of lithium polysulfides [15]. The liquid exfoliation was also achieved by intercalation of tetrabutylammonium (TBA^+) cations into the protonic lepidocrocite titanates. The remaining TBA^+ ions in the colloidal suspension of 2D TiO_2 NSs were removed by washing with deionized water. Sulfur species grew on the 2D TiO_2 NSs to form the TiO_2 NSs–sulfur nanocomposite by mixing a colloidal suspension of TiO_2 NSs with $Na_2S\ 9H_2O$ and Na_2SO_3 under vigorous stirring. Ti species on the surface of 2D TiO_2 NSs with high specific surface area can serve as Lewis acids, enabling sufficient interactions with lithium polysulfides and thereby improving the cycling performance of Li–S batteries.

Similarly, 2D anatase NSs can be obtained by exfoliating self-synthesized potassium lithium titanates ($K_{0.8}Ti_{1.73}Li_{0.27}O_4$) [16]. Tetrabutylammonium hydroxide (TBAOH) solution was used to strip acidified protonic titanates along with vigorous mechanical shaking for at least one week. This synthetic process can lead to a high yield of single-layered titanates ($Ti_{0.87}O_2^{0.52-}$) NSs with minimal multilayered protonic titanates.

The liquid-phase exfoliation of 2D TiO_2 usually achieved by mixing parent titanates with excessive surfactant, allowing the intercalation of surfactant to separate the layers. However, the reaction time for both liquid-phase exfoliation and mechanical exfoliation is generally one to two week(s) to ensure complete separation of the layers [17, 18]. Therefore, applying an external driving force more powerful than mechanical shaking is a promising method to reduce the improve the productivity and homogeneity of 2D TiO_2 with desired thickness, specific facets and large specific surface area.

The sono-assisted liquid-phase exfoliation has proved effective for delaminating various materials, including TMDs [19], graphene [19, 20], and TiO_2 [21, 22]. The high-intensity ultrasonic wave with the frequency between 20 kHz and 15 MHz induces formation, vibration, and collapses of bubbles in the solution, which facilitate the intercalation of surfactant and thereby the exfoliation of 2D materials. Utilizing the kinetic energy produced by ultrasonic irradiation, the formation, expansion, and collapsing of bubbles greatly accelerate the contact between surfactant species and interlayer regains of host materials at a specific crystallographic position. This process increases the distance between layers and thus enhances the exfoliation efficiency as well as exposure of specific facets.

For instance, 2D TiO_2 NSs can be synthesized using sono-assisted liquid exfoliation method in 2 h, which is significantly more efficient than liquid-phase exfoliation without external driving force [22]. As shown in Figure 3.2, anatase powder was dispersed in sodium hydroxide (NaOH) solution. This solution was then mixed with a surfactant solution of 1.5 M TABOH under ultrasonic irradiation with a high frequency of 20 kHz for 1 h. After the 2D TiO_2 NSs were fully peeled from the host anatase powder, they were washed with 0.1 M HNO_3 and deionized water until the

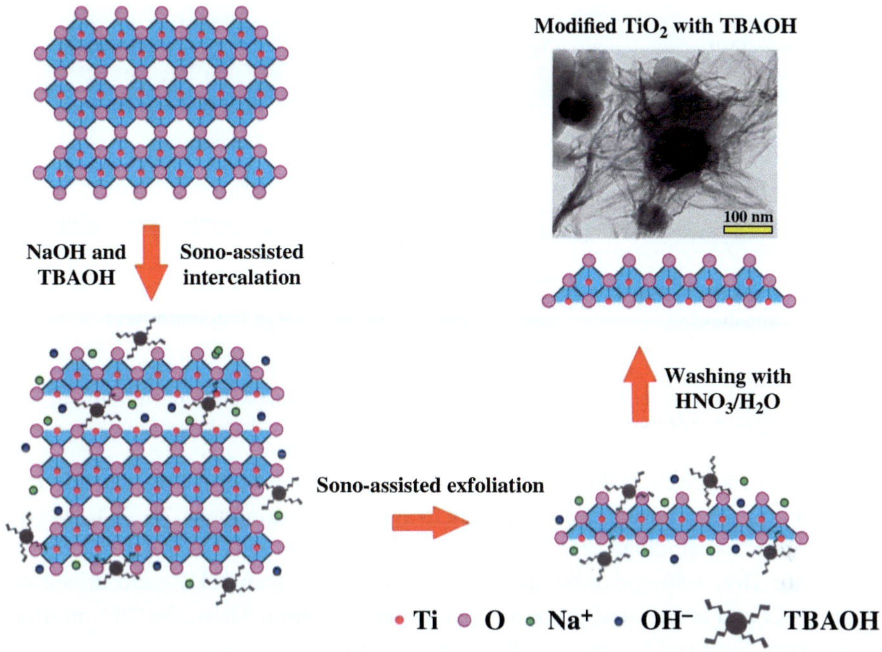

Figure 3.2 The synthesis of TiO_2 NSs from crystalline TiO_2 [22]. Reproduced from Ref. [22]/ with permission of Elsevier.

residual TBA^+ cations and OH^- anions were eliminated. Raman spectra confirmed the exposure of (001) facets of the synthesized TiO_2 NSs, indicated by the decrease in symmetric stretching of Ti—O bonding and the increase in asymmetric and symmetric bending of Ti—O bonding. The exposed (001) facets provide more surface-active sites for the photocatalytic reactions. Additionally, the produced TiO_2 NSs have higher specific surface area and less charge carrier recombination than the pristine anatase TiO_2 powder. However, it was found that defects with detrimental effects on photocatalytic properties were also produced during the ultrasonic process.

TBAOH is a common surfactant used in liquid exfoliation of 2D TiO_2 NSs to maintain the layered structure and control the thickness. The amount of added surfactant can significantly impact the formation of 2D TiO_2 NSs. Figure 3.3 illustrates the synthesis process of 2D TiO_2 NSs, which begins with heating TiO_2 nanoparticles with NaOH to obtain three-dimensional (3D) layered titanates. TBAOH was added to 3D-layered titanates where TBA^+ can attach to the dangling Ti—O bonds with a negative charge enabling the separation of layers and formation of monolayer titanate NSs under ultrasonication. Finally, the surfactants were washed off with strong acids (e.g., HCl and HNO_3) to remove the cations (Na^+ and TBA^+) to obtain TiO_2 NSs. On the other hand, the exclusion of surfactants leads to the production of TiO_2 nanotubes due to the interactions between positively charged Ti and dangling Ti—O bonds with a negative charge, causing the layers to roll up. The as-prepared 2D TiO_2 NSs exhibit better reversible capacity and higher current density than 3D TiO_2 nanotubes and NPs, which implies that NSs are more suitable for electrochemical applications.

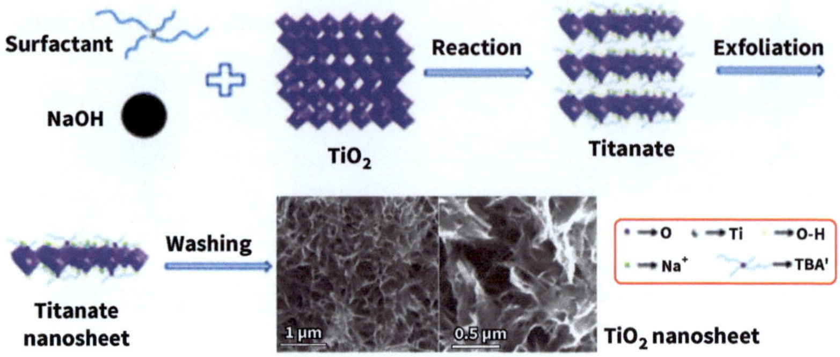

Figure 3.3 Schematics of 2D TiO_2 NSs using liquid exfoliation. Reproduced from Ref. [23]/ The Royal Society of Chemistry.

Taking advantages of high pressure and high temperature, the hydrothermal-assisted liquid exfoliation can be applied to synthesize monolayer 2D TiO_2 NSs with mesoporous morphology [24]. The tetrabutyl titanate (TBOT) was used as the precursor for TiO_2, which can be dissolved in tetrahydrofuran (THF) and mixed with acids (HCl and acetic acid) to obtain TiO_2 monomicelles. Then, the TiO_2 monomicelles were dissolved in a mixed ethanol–glycerol solvent for solvothermal treatment at 100 °C for 10 h. The white precipitation obtained from solvothermal reaction is mesoporous single-layer TiO_2 NSs. The as-produced 2D anatase NSs, with high mesoporosity and crystallinity providing voids for ion adsorption, are favorable candidates for anode materials.

Liquid exfoliation is proved to be an effective and straightforward method to produce 2D TiO_2 NSs with desired thickness and exposure of certain facets. However, it is difficult to precisely control the number of layers of TiO_2 NSs, and the resulted 2D TiO_2 NSs are usually a mixture of a small portion of monolayer TiO_2 NSs and a large portion of multilayer TiO_2 NSs. Furthermore, the surfactant and impurities are difficult to eliminate and some defects induced during ultrasonication may be unfavorable for specific application scenarios. These disadvantages may impede the use of liquid exfoliation method for producing 2D TiO_2 NSs in scenarios that require high-purity TiO_2 NSs with low defects.

3.2.1.3 Electrochemical Exfoliation

Electrochemical exfoliation combines the advantages of both mechanical exfoliation and liquid exfoliation, since it uses electricity as driving force and allows the exfoliation to complete under sonication. The electrochemical exfoliation boosts a higher yield of single-layered 2D TiO_2 than mechanical exfoliation and liquid exfoliation.

Generally, metal sulfides are used as precursors for production of 2D-layered metal oxides through electrochemical exfoliation, assisted by hydrothermal treatment [25, 26]. As shown in Figure 3.4, 2D TiO_2 NSs were synthesized using titanium disulfide (TiS_2) as starting materials using electrochemical exfoliation followed by hydrothermal treatment and oxidation. Initially, the bulk TiS_2 was exfoliated using lithium foil in pouch cells as the counter electrode with the electrode discharged at a

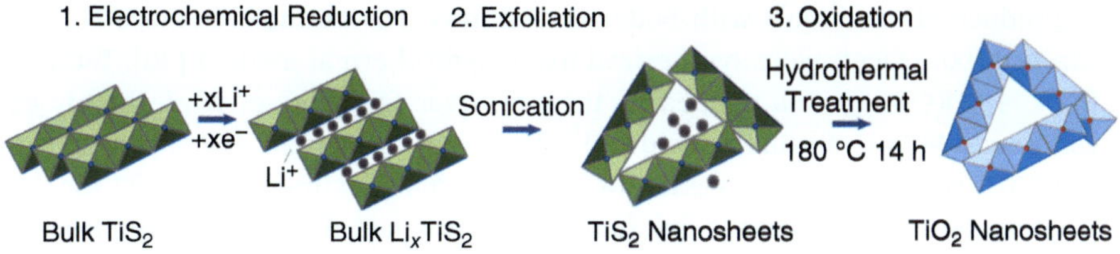

Figure 3.4 Schematics of preparation of TiO_2 NSs by electrochemical exfoliation. Reproduced from [27]/with permission of Springer Nature.

0.05 °C rate to 0.9 V. Then, the electrode was held potentiostatically at 0.9 V for 2 h. Lithiated TiS_2 (Li_xTiS_2) was obtained and then sonicated in an ice-water bath for 1 h to remove the interlayer lithium species and dissociate the TiS_2 NSs. 2D TiO_2 NSs were obtained by oxidizing TiS_2 NSs using hydrothermal method. The TiS_2 NSs solution was purged with oxygen to ensure the dissolved oxygen in deionized water sufficient for oxidation. The TiO_2 NSs prepared using electrochemical phase exfoliation method demonstrated improved photocatalytic performance.

Electrochemical exfoliation is a productive and scalable approach to produce high yields of TiO_2 NSs, especially monolayer TiO_2 NSs, making it a potential industrial-scale production method for single layer TiO_2 NSs. However, it should be noted that this study also reveals that the residual Li carbon black existing in the as-prepared TiO_2 NSs was detrimental for photocatalytic efficiency.

3.2.2 Bottom-up Methods

The bottom-up methods, in contrast to the top-down methods, construct 2D NSs from starting precursors rather than dissociation of bulk materials. Therefore, bottom-up methods offer better flexibility with a wider range of applications. The existing bottom-up methods for TiO_2 NSs synthesis mainly include wet chemistry methods, topochemical methods, and atomic layer deposition (ALD).

3.2.2.1 Wet Chemical Methods

The wet chemistry methods synthesize 2D materials in liquid phase using solutions, suspensions, and emulsions through various chemical reactions. Therefore, 2D NSs obtained from wet chemistry methods are usually colloidal solution. The wet chemistry methods possess excellent scalability, high cost-effectiveness [28], good controllability, and versatility [29], which make them suitable for both lab-scale and large-scale production of 2D materials. In general, wet chemical methods for 2D materials production primarily include template synthesis, self-assembly, and hydro/solvothermal methods.

3.2.2.1.1 Hydro/Solvothermal Methods

Hydrothermal and solvothermal methods use water and nonaqueous solvent as reaction medium, respectively. Both methods include reaction in autoclave vessels where the operating temperature exceeds the boiling point of water/solvent under the high pressure, causing water/solvent to vaporized and facilitating the reaction to produce 2D materials with better crystallinity. For example, anatase NSs with exposed (001) facets were synthesized using hydrothermal method [30]. Tetrabutyl titanate ($Ti(OBu)_4$) was applied as the precursor for anatase, while highly concentrated (48%) hydrofluoric acid (HF) was used for etching. HF was dropped to $Ti(OBu)_4$ slowly in Teflon vessel for hydrothermal treatment at 180 °C for 24 h. Anatase NSs were obtained by further washing off fluorine species from the obtained precipitation. In this process, the thickness of 2D anatase NSs can be controlled by adjusting the amount of HF added. The 2D TiO_2 NSs with oriented (001) facets which are proved to be efficient in photoelectrochemical water-splitting were also synthesized using hydrothermal method and etching with HF [31]. The 2D TiO_2 photoelectrode can also be prepared using a one-pot hydrothermal treatment in solution of diethylene glycol and water which followed by ion exchanges [32]. Although this hydrothermal method can synthesize 2D TiO_2 NSs with controllable thickness, it requires highly corrosive and toxic HF, posing challenges for practical production. Moreover, the stubborn fluorine residuals can cause defects which can be harmful to the physiochemical properties of the produced 2D TiO_2.

3.2.2.1.2 Template Method

In template method, 2D-layered materials, especially graphene oxide (GO), are used as templates to confine the growth of precursor oligomers in 2D space and prevent the expansion in the third dimension. Figure 3.5 illustrates the general process of 2D material synthesis using template method. The liquid-phase precursor oligomers are self-assembled on the template with the assistance of surfactants. Then, they undergo condensation, polymerization, and crystallization in sequence, forming 2D metal oxide NSs with atomic-level thickness. Finally, the targeted 2D materials are obtained by removing the template through high temperature calcination in the air.

GO template is widely used in 2D TiO_2 synthesis due to its 2D structure and ease of removal. As shown in Figure 3.6, 2D porous TiO_2 was synthesized using GO as template. The precursor TBOT was added into GO suspension through vigorous stirring

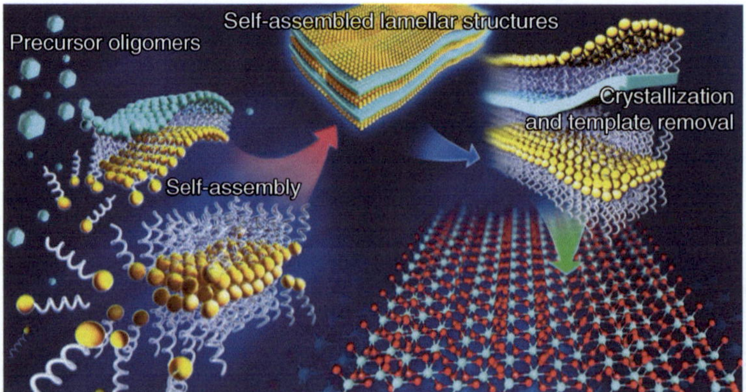

Figure 3.5 The schematics of template synthesis of 2D materials. Reproduced from Ref. [33]/with permission of Springer Nature.

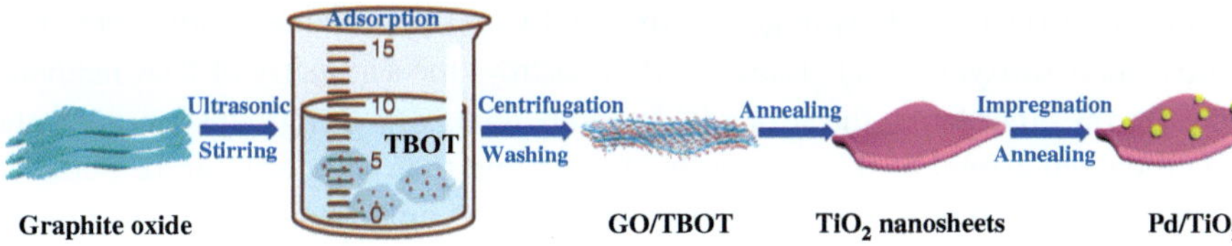

Figure 3.6 Schematic illustration for TiO_2 NSs synthesis using GO template. Reproduced from Ref. [35]/with permission of Elsevier.

and then attached to GO surface due to electrostatic effects. The excessive TBOT was removed by washing with ethanol. At last, GO was removed using calcination in the air at high temperatures (above 500 °C). Interestingly, the template method can not only produce crystal-phase 2D TiO_2 NSs but is also capable of synthesizing 2D amorphous TiO_2 NSs [34]. A similar synthetic procedure except for the usage of titanium tetrachloride as precursor and hydrochloric acid led to the formation of amorphous TiO_2 NSs. They exhibit superior surface-enhanced Raman scattering (SERS) effect due to its improved surface potential than 2D crystalline TiO_2 NSs. While the template method can efficiently produce 2D TiO_2 NSs, only a restricted range of templates have been shown capable of effectively guiding the synthesis of these NSs.

3.2.2.1.3 Template-free Self-assembly Method

In fact, without using a template, amorphous 2D TiO_2 NSs can also be synthesized using hydrogen bonding between the liquid-phase precursor as driving force to guide the self-assembly. Figure 3.7 depicts the synthetic process of a template-free amorphous 2D TiO_2 NS aerogel. The precursor for TiO_2 synthesis is peroxotitanium (PTA) obtained

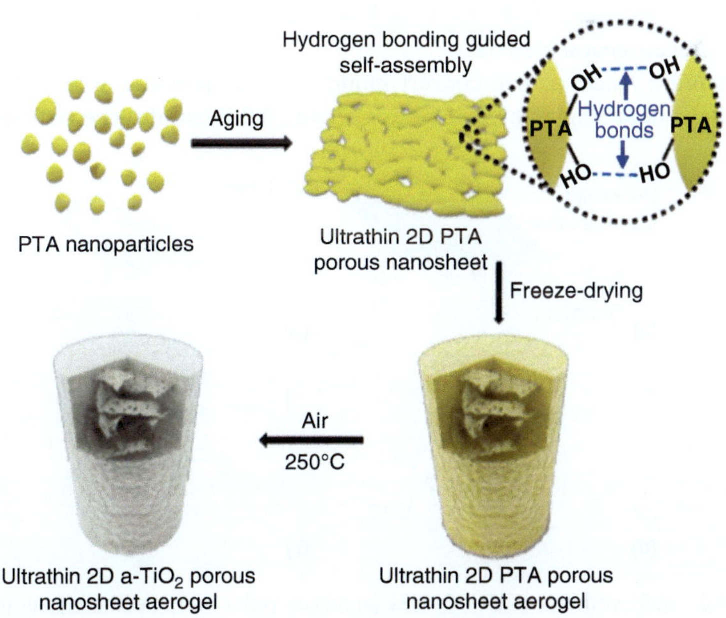

Figure 3.7 Schematic illustration of self-assembled amorphous 2D TiO_2 NS aerogel. Reproduced from [36]/with permission of Springer Nature.

by oxidizing $Ti(OH)_4$ with hydrogen peroxide (H_2O_2). The presence of sufficient O—O groups and hydroxyl (—OH) groups on PTA destroys the symmetry of PTA nanoparticles. On the other hand, the strong hydrogen bonding of —OH groups promotes the anisotropic self-assembly of PTA porous 2D NSs, and then TiO_2 NSs can be obtained by calcining PTA NSs in the air. The as-prepared 2D amorphous TiO_2 NSs possess cation defects which turned out to substantially improve the conductivity. Although the template-free self-assembly method is a useful strategy to prepare 2D TiO_2 NSs, the NSs tend to roll up instead of maintaining as a flat film.

3.2.2.2 Atomic Layer Deposition

The ALD technology, a subclass of chemical vapor deposition for thin film synthesis, is characterized by its self-limiting feature, which ensures one atomic layer is deposited at a time. Therefore, changing the number of deposition cycles can tune the thickness of the deposited layer. Utilizing ALD technique, the structural properties of the materials can be precisely controlled at the atomic level. Thereby, this method is especially useful for 2D materials production which requires uniformity to achieve desired mechanical, physiochemical, optical and electronic features. 2D TiO_2 NSs can be synthesized using ALD as shown in Figure 3.8. The layered polyvinyl alcohol (PVA) used as sacrificial polymer layer was produced by spin coating on silicon wafer. Then dimethylamido titanium mixed with deionized water was used as ALD reactant. By adjusting the ALD cycles, 2D TiO_2 NSs with different thickness can be produced. Then, PVA sacrificial layer can be removed by dissolving in hot water (75 °C). ALD is an efficient way to synthesize high-quality 2D TiO_2 NSs with precise control of thickness. However, the operating cost for ALD is significantly higher than other approaches, posing challenges for cost-effective large-scale production.

3.2.2.3 Topochemical Method

Topochemical reactions can convert monomers into polymer crystals by synthesizing individual single polymer chains. Figure 3.9 illustrates the synthesis of 2D

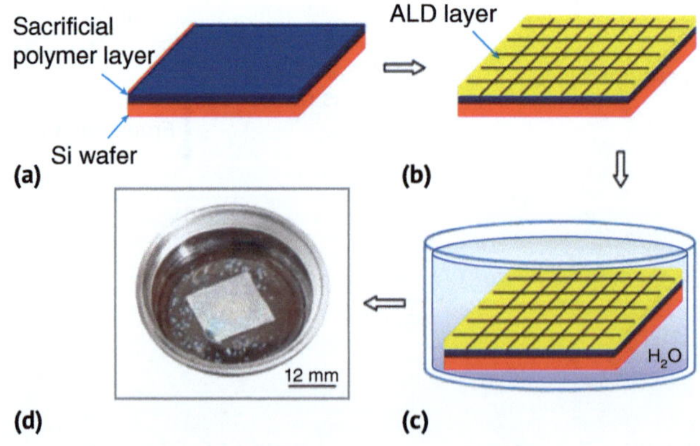

Figure 3.8 ALD synthesis of 2D TiO_2 NSs. (a) Silicon wafer coated with polymer layer by spin coating, (b) ALD-coated TiO_2 layered on sacrificial layer and then scratched into small pieces by razor edge, (c) dissolution of sacrificial layer, (d) formed TiO_2 nanosheets. Reproduced from Ref. [37]/with permission of Elsevier.

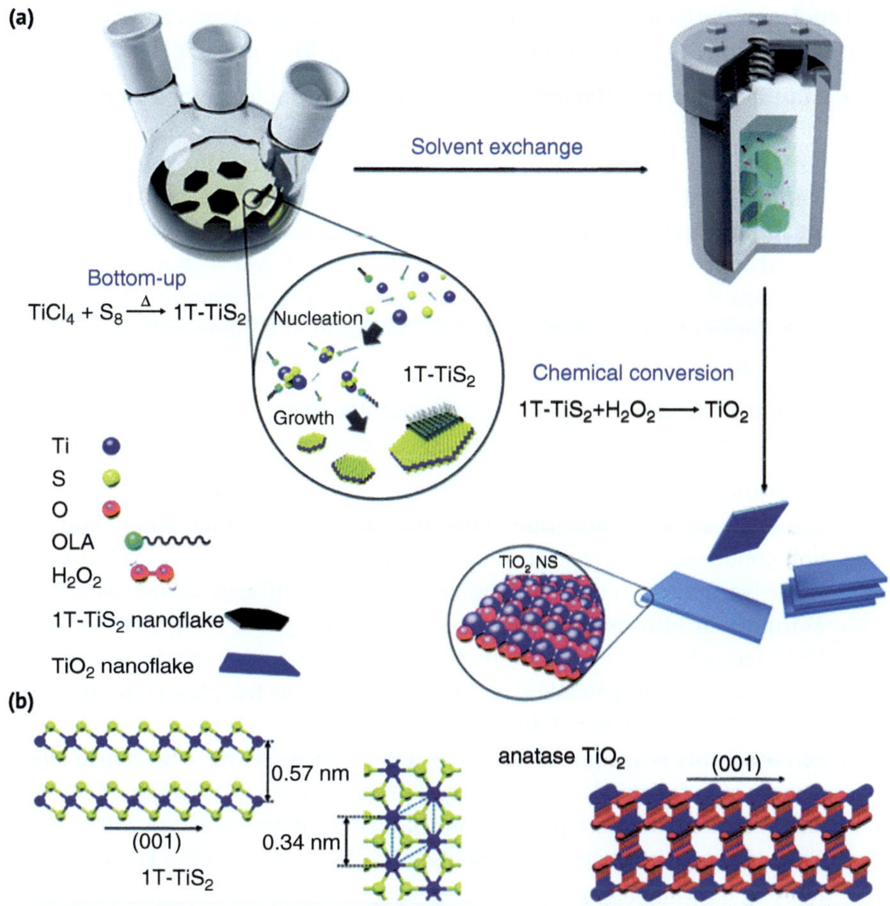

Figure 3.9 Scheme of topochemical synthesis of 2D TiO_2 NSs. (a) Bottom-up method for precursor synthesis followed by desulfurization process through hydrothermal reactions, (b) omission of Cl atoms [38]/The Royal Society of Chemistry/CC BY 3.0.

TiO_2 NSs using topochemical reactions. First, 1T-TiS_2 NSs were produced through a topochemical reaction between titanium tetrachloride ($TiCl_4$) and oleylamine-sulfur in Argon atmosphere at 300 °C. Then, H_2O_2 was then added to oxidize 1T-TiS_2 NSs dispersed in ultrapure water to 2D anatase NSs through hydrothermal treatment. In this way, topochemical method successfully builds up 2D anatase NSs without usage of template or toxic surfactant. However, it can be delicate and time consuming to optimize the experimental parameters.

3.3 Conclusion

This chapter provides a comprehensive summary of various synthesis methods for 2D TiO_2 nanostructures including both top-down and bottom-up methods. However, each method has its own advantages and disadvantages, necessitating a method selection tailored to specific requirements.

Top-down methods mainly consist of liquid-phase exfoliation and electrochemical exfoliation both including delamination of bulk TiO_2 into 2D TiO_2 NSs:

- **Liquid-phase exfoliation** is an effective and simple method to produce 2D TiO_2 NSs with desired thickness and exposure of specific facets. However, the uniformity and number of layers are difficult to control. Moreover, the contaminations from residual surfactant and impurities can cause detrimental defects for specific applications.
- **Electrochemical exfoliation** gives better control over thickness of the layers and requires less production time. However, it shares common disadvantages with liquid-phase exfoliation, such as inclusion of contaminations and impurities and introduction of defects.

Bottom-up methods majorly include wet chemistry, ALD, and topochemical approach, which all start from various liquid precursors to build up the 2D TiO_2 NSs.

- **Hydro/solvothermal synthesis** produces high-quality, crystallized 2D TiO_2 in a relatively short time due to the high-pressure and high-temperature reaction conditions. However, it is inevitable that HF is used which is toxic and corrosive, and the remaining fluorine species are difficult to eliminate.
- **Template methods** are proven to be effective in synthesizing uniform 2D TiO_2 NSs. Removing templates can increase the complexity of synthesis and risk of introducing detrimental defects. Moreover, only few templates have been demonstrated to be capable of directing the synthesis of 2D NSs.
- **Self-assembly** is a template-free method to produce 2D TiO_2 NSs. Due to intrinsic features of self-assembly, the as-obtained structure is always rolled- up 2D TiO_2 NSs.
- **ALD** enables the most precise control of the thickness of as-synthesized 2D TiO_2. However, the operating cost of ALD is high and the yield is relatively low, limiting its scalability for industrial-scale production.
- **Topochemical methods** successfully build up 2D TiO_2 NSs without usage of templates or toxic surfactant. However, it can be delicate and time consuming to optimize the experimental parameters.

Future advancements in 2D TiO_2 synthetic methods should focus on enhancing the quality, uniformity, efficiency, and scalability of the produced 2D TiO_2. The following suggestions can be considered to optimize the current methods:

- **Developing more templates and precursors:** investigating more feasible templates which are easy to remove and cause less defects; explore more useful precursors for more uniform and defect-free 2D TiO_2 NSs.
- **Improving the sustainability:** substituting toxic and corrosive chemicals that used in synthesis with nontoxic and environmental-friendly chemicals; optimizing the synthetic procedures to prepare 2D TiO_2 at room temperature and atmospheric pressure to reduce the energy consumption.
- **Developing standard protocols for various synthesis methods:** some synthesis methods for 2D TiO_2 synthesis are mature and well demonstrated over past decades. However, there is a lack of standardized and detailed protocols for the procedures. Researchers in the same field should all contribute to developing the protocols to improve consistency and reproducibility.

References

1 Novoselov, K.S., Geim, A.K., Morozov, S.V., et al. (2004). Electric field effect in atomically thin carbon films. *Science* **306** (5696): 666–669.

2 Bertolazzi, S., Bondavalli, P., Roche, S., et al. (2019). Nonvolatile memories based on graphene and related 2D materials. *Advanced Materials* **31** (10): 1806663.

3 Xiao, M., Zhang, L., Luo, B., et al. (2020). Molten-salt-mediated synthesis of an atomic nickel Co-catalyst on TiO_2 for improved photocatalytic H_2 evolution. *Angewandte Chemie* **132** (18): 7297–7301.

4 Hu, Y., Pan, Y., Wang, Z., et al. (2020). Lattice distortion induced internal electric field in TiO_2 photoelectrode for efficient charge separation and transfer. *Nature Communications* **11** (1): 2129.

5 Liu, G., Wang, L., Yang, H.G., et al. (2010). Titania-based photocatalysts—crystal growth, doping and heterostructuring. *Journal of Materials Chemistry* **20** (5): 831–843.

6 Liu, G., Yang, H.G., Sun, C., et al. (2009). Titania polymorphs derived from crystalline titanium diboride. *CrystEngComm* **11** (12): 2677–2682.

7 Wu, T., Niu, P., Yang, Y., et al. (2019). Homogeneous doping of substitutional nitrogen/carbon in TiO_2 plates for visible light photocatalytic water oxidation. *Advanced Functional Materials* **29** (25): 1901943.

8 Wang, L. and Sasaki, T. (2014). Titanium oxide nanosheets: graphene analogues with versatile functionalities. *Chemical Reviews* **114** (19): 9455–9486.

9 Grimes, C.A. and Mor, G.K. (2009). *TiO_2 Nanotube Arrays: Synthesis, Properties, and Applications*. Springer Science & Business Media.

10 Zhang, W., He, H., Li, H., et al. (2021). Visible-light responsive TiO_2-based materials for efficient solar energy utilization. *Advanced Energy Materials* **11** (15): 2003303.

11 Wang, L. (2008). Titanium oxide based photocatalysts: from research to applications. *Recent Patents on Materials Science* **1** (2): 165–175.

12 Zhang, B.Y., Xu, K., Yao, Q., et al. (2021). Hexagonal metal oxide monolayers derived from the metal–gas interface. *Nature Materials* **20** (8): 1073–1078.

13 Kazim, S., Alì, M., Palleschi, S., et al. (2020). Mechanical exfoliation and layer number identification of single crystal monoclinic $CrCl_3$. *Nanotechnology* **31** (39): 395706.

14 Sasaki, T., Watanabe, M., Michiue, Y., et al. (1995). Preparation and acid-base properties of a protonated titanate with the lepidocrocite-like layer structure. *Chemistry of Materials* **7** (5): 1001–1007.

15 Patil, S.B., Kim, H.J., Lim, H.-K., et al. (2018). Exfoliated 2D lepidocrocite titanium oxide nanosheets for high sulfur content cathodes with highly stable Li–S battery performance. *ACS Energy Letters* **3** (2): 412–419.

16 Zhang, Y., Wu, W., Zhang, K., et al. (2016). Raman study of 2D anatase TiO_2 nanosheets. *Physical Chemistry Chemical Physics* **18** (47): 32178–32184.

17 Maluangnont, T., Matsuba, K., Geng, F., et al. (2013). Osmotic swelling of layered compounds as a route to producing high-quality two-dimensional materials. A comparative study of tetramethylammonium versus tetrabutylammonium cation in a lepidocrocite-type titanate. *Chemistry of Materials* **25** (15): 3137–3146.

18 Choy, J.-H., Lee, H.-C., Jung, H., et al. (2002). Exfoliation and restacking route to anatase-layered titanate nanohybrid with enhanced photocatalytic activity. *Chemistry of Materials* **14** (6): 2486–2491.

19 Shen, J., He, Y., Wu, J., et al. (2015). Liquid phase exfoliation of two-dimensional materials by directly probing and matching surface tension components. *Nano Letters* **15** (8): 5449–5454.

20 Tapia, J.I. and Quintana, M. (2016). Carbon nanostructures produced by liquid phase exfoliation of graphite in the presence of small organic molecules. *Open Material Sciences* **3** (1): 76–82.

21 Zhu, Y., Li, H., Koltypin, Y., et al. (2001). Sonochemical synthesis of titania whiskers and nanotubes. *Chemical Communications* 24: 2616–2617.

22 Thasirisap, E., Vittayakorn, N., and Seeharaj, P. (2017). Surface modification of TiO_2 particles with the sono-assisted exfoliation method. *Ultrasonics Sonochemistry* **39**: 733–740.

23 Leng, M., Chen, Y., and Xue, J. (2014). Synthesis of TiO_2 nanosheets via an exfoliation route assisted by a surfactant. *Nanoscale* **6** (15): 8531–8534.

24 Lan, K., Liu, Y., Zhang, W., et al. (2018). Uniform ordered two-dimensional mesoporous TiO_2 nanosheets from hydrothermal-induced solvent-confined monomicelle assembly. *Journal of the American Chemical Society* **140** (11): 4135–4143.

25 Zeng, Z., Sun, T., Zhu, J., et al. (2012). An effective method for the fabrication of few-layer-thick inorganic nanosheets. *Angewandte Chemie International Edition* **36** (51): 9052–9056.

26 Zeng, Z., Tan, C., Huang, X., et al. (2014). Growth of noble metal nanoparticles on single-layer TiS_2 and TaS_2 nanosheets for hydrogen evolution reaction. *Energy & Environmental Science* **7** (2): 797–803.

27 Jing, H., Cheng, Q., Weller, J.M., et al. (2018). Synthesis of TiO_2 nanosheet photocatalysts from exfoliation of TiS_2 and hydrothermal treatment. *Journal of Materials Research* **33** (21): 3540–3548.

28 Lin, Y.-C., Dumcenco, D.O., Huang, Y.-S., et al. (2014). Atomic mechanism of the semiconducting-to-metallic phase transition in single-layered MoS_2. *Nature Nanotechnology* **9** (5): 391–396.

29 Pottathara, Y.B. et al. (2019). Synthesis and processing of emerging two-dimensional nanomaterials. In: *Nanomaterials Synthesis* (ed. Y. Beeran Pottathara, et al.), 1–25. Elsevier.

30 Yang, X.H., Li, Z., Liu, G., et al. (2011). Ultra-thin anatase TiO_2 nanosheets dominated with {001} facets: thickness-controlled synthesis, growth mechanism and water-splitting properties. *CrystEngComm* **13** (5): 1378–1383.

31 Wang, X., Liu, G., Wang, L., et al. (2011). TiO_2 films with oriented anatase {001} facets and their photoelectrochemical behavior as CdS nanoparticle sensitized photoanodes. *Journal of Materials Chemistry* **21** (3): 869–873.

32 Butburee, T., Bai, Y., Wang, H., et al. (2018). 2D porous TiO_2 single-crystalline nanostructure demonstrating high photo-electrochemical water splitting performance. *Advanced Materials* **30** (21): 1705666.

33 Sun, Z., Liao, T., Dou, Y., et al. (2014). Generalized self-assembly of scalable two-dimensional transition metal oxide nanosheets. *Nature Communications* **5** (1): 3813.

34 Wang, X., Shi, W., Wang, S., et al. (2019). Two-dimensional amorphous TiO_2 nanosheets enabling high-efficiency photoinduced charge transfer for excellent SERS activity. *Journal of the American Chemical Society* **141** (14): 5856–5862.

35 Wang, D., Yang, J., Bao, L., et al. (2021). Pd nanocrystal sensitization two-dimension porous TiO_2 for instantaneous and high efficient H_2 detection. *Journal of Colloid and Interface Science* **597**: 29–38.

36 Qi, Y., Zeng, X., Xiao, L., et al. (2022). An invisible hand: hydrogen bonding guided synthesis of ultrathin two-dimensional amorphous TiO_2 nanosheets. *Science China Materials* **65** (11): 3017–3024.

37 Edy, R., Zhao, Y., Huang, G.S., et al. (2016). TiO_2 nanosheets synthesized by atomic layer deposition for photocatalysis. *Progress in Natural Science: Materials International* **26** (5): 493–497.

38 Zarattini, M., Dun, C., Isherwood, L.H., et al. (2022). Synthesis of 2D anatase TiO_2 with highly reactive facets by fluorine-free topochemical conversion of 1T-TiS_2 nanosheets. *Journal of Materials Chemistry A* **10** (26): 13884–13894.

4

3D TiO_2 via Self-assembly

Jie Tian[1], Mingdeng Wei[2], Detlef W. Bahnemann[3], Wan In Lee[4,], Jia Hong Pan[1,*]*

[1] *State Key Laboratory of Featured Metal Materials and Life-Cycle Safety for Composite Structures, School of Resources, Environment and Materials, Guangxi University, Nanning, Guangxi, China*
[2] *Fujian Key Laboratory of Electrochemical Energy Storage Materials, Fuzhou University, Fuzhou, Fujian, China*
[3] *International Joint Center of Nature-Inspired Carbon Neutrality Solutions, Saint Petersburg State University, Peterhof, Saint Petersburg 198504, Russia*
[4] *Department of Chemistry and Chemical Engineering, Inha University, Incheon, South Korea*

*Corresponding authors: wanin@inha.ac.kr; jhpan@gxu.edu.cn

4.1 Introduction

The photocatalytic performance of TiO_2 hinges critically upon several key factors including its inherent crystal phase and crystallinity. However, for sustained and practical applications, careful attention must also be paid to the microstructure (e.g., active nanobuilding blocks (NBBs), surface area, and porosity) and morphology [1]. Nanosized TiO_2 particles, characterized by the quantum size effect, have been widely studied due to their expansive surface area and superior dispersion capabilities. Nonetheless, a notable drawback arises from their propensity to aggregate when suspended, leading to a rapid decline in active sites and photocatalytic efficiency. Furthermore, the challenges associated with the separation, recovery, and reusability of nanosized particles present additional hurdles in the context of practical applications. Moreover, the management of nanoparticles becomes increasingly intricate, given escalating concerns surrounding their biocompatibility and toxicity. To effectively tackle these technical challenges, two main strategies have been developed: the creation of 3D TiO_2 structures with porous configurations and the assembly of low-dimensional TiO_2 into well-defined hierarchical superstructures [2]. Thus, 3D TiO_2 shows merits in structural diversity: (i) the pore–wall structure and well-assembled NBB delivering a large number of active centers for photocatalytic reactions, (ii) enhanced light harvesting properties due to the efficient light reflection and scattering in the pore channels or voids among NBBs, and (iii) easy reclaiming and recovery after each cycle of photocatalytic process [3–5].

Nanostructured TiO_2: Synthesis, Photocatalytic Properties and Applications, First Edition.
Edited by Jia Hong Pan, Wan In Lee and Detlef W. Bahnemann

According to the classification by the International Union of Pure and Applied Chemistry (IUPAC), microporous, mesoporous, and macroporous materials show different pore diameters of <2, 2–50, and > 50 nm. The template method, employing both hard templates such as porous SiO_2 and carbon, as well as soft templates like surfactants, has emerged as a general approach for fabricating porous TiO_2-based photocatalysts characterized by tunable pore size, high surface areas, precisely defined NBBs, and controllable morphologies. From application point view, mesoporous TiO_2 is of particular interest due to its comprehensive advantages. It shows superior mechanical stability compared to macroporous structure, as well as enhanced crystallinity, better accessibility in pore–wall structure and more suitable grain size compared to typically low-crystallinity microporous TiO_2 [6, 7].

Template method has been widely explored for mesoporous TiO_2, inspired by the discovery of mesoporous SiO_2 sieves, that is, MCM family in the early 1990s. Depending on the type of template chosen, strategies of hard template route and soft template have been explored. Additional, self-template method has been developed for high-yield synthesis. Thermal treatment is generally used to induce phase transition and topological transformation of the well-defined Ti-containing compounds to porous TiO_2. In both template and self-template synthesis paradigms, the sol–gel process serves as a cornerstone for the deposition of hard templates or facilitates the cooperative assembly between the Ti-precursor and soft templates. This method provides a robust foundation for the predictable fabrication of porous photocatalysts with tailored pore–wall structures and photocatalytic properties, thereby enhancing the photocatalytic performances in various applications.

In addition to strategies aimed at generating porous structures, self-organization via wet-chemical routes have significantly advanced for the synthesis of robust 3D TiO_2 consisting of well-attached NBBs over the past two decades. Inspired by natural biomineralization, such bottom-up synthesis is conducted in solution, where crystalline or amorphous TiO_2 nucleates and grows, initially producing low-dimensional (0–2D) TiO_2 nanostructures. These diminutive TiO_2 entities, characterized by heightened surface energy, exhibit spontaneous attraction, stacking, and agglomeration, thereby self-organizing into well-defined assemblies comprising low-dimensional 0–2D NBBs. Consequently, they demonstrate structural diversity across various scales, ranging from the nano to the bulk level. In contrast to the random aggregation of low-dimensional TiO_2 into dense bulk structures, 3D TiO_2 architectures retain active sites within the highly exposed 0–2D TiO_2 NBBs, resulting in a significantly enhanced structural stability. The repulsion between adjacent NBBs promotes the formation of hierarchical structures with abundant accessible pores and voids, facilitating efficient mass transportation and exchange. Notably, the spherical morphology of these structures, typically in the (sub)micron range, enhances light harvesting by effectively reflecting and scattering light through the pores and voids, thereby ensuring efficient and long-lasting photocatalysis. Moreover, the larger size of 3D TiO_2 (sub)microspheres, compared to the tiny 0–2D nanostructures, offers notable advantages in terms of recovery and reuse for practical applications.

This chapter begins by summarizing the template synthesis of porous TiO_2 with crystalline frameworks. We then highlight the self-template strategy, especially recent advancements in the sol–gel preparation of hydrous TiO_2 colloidal spheres

(HTCS), marking a significant milestone in the development of self-organization theory, and summarize their self-template crystallization with selective crystal phase and interior structure. Finally, we discuss the one-pot hydro/solvothermal synthesis of hierarchical spheres under the operation of self-organization mechanism, with a focus on controlling low-dimensional NBBs and elucidating the self-organization pathway and underlying mechanism.

4.2 Template Method

The sacrificial template strategy for the preparation of 3D TiO_2 generally employs the easy-fabricated and cost-effective solids or supermolecules as robust molds to direct the organization of TiO_2 NBBs in form of molecules, oligomers, and nanoparticles through wetting, sorption, deposition, complexation, grafting, and ion exchange, followed with the confined sol–gel reaction and calcination process. Control of the interface between the Ti-precursors and hard or soft templates is essential to the successful creation of porous structure. The desired architecture of porous TiO_2 is attained by selectively removing the template. Notably, the textural characteristics of the resulting 3D TiO_2 are primarily dictated by the properties and structure of the template. Two principal template types have emerged based on their properties: the hard template and soft template.

4.2.1 Hard Template Method

Porous silica, carbon, polymer, alumina, xerogels, resins, wood, and cellulose fibers have been employed as hard templates, yielding various morphologies such as foam, tube, fiber, and bead. The underlying sol–gel process and host–guest chemistry are essential to induce the Ti-precursors wetting the surface and permeating into the pore channels. After the polycondensation of Ti source, a slow heating process can be conducted for TiO_2 crystallization with high interconnection, which is stable for template removal by corrosive agent and high-temperature calcination [3].

Porous silica has been widely used as a hard template. However, microporous silica is unsuitable due to the challenge of accommodating the larger size of oligomeric Ti-precursors within its small pores. Macroporous TiO_2 inverse opals can be synthesized utilizing the self-assembled colloidal SiO_2 photonic crystals on different substrates and as hard template. The pore size and geometry of macroporous TiO_2 can be adjusted by tuning the diameter and packing properties of monodisperse SiO_2 spheres. Therefore, highly ordered mesoporous SiO_2 possessing interconnected pores with size in range of 2–50 nm is an ideal hard template for the loading of Ti-precursors and subsequently the confined growth of TiO_2 nanocrystallites. A diverse array of mesoporous silica architectures such as MCM, MSU, SBA, KIT, and FDU offer tunable template for synthesizing mesoporous TiO_2 with predictable morphology, mesophase (2D hexagonal (*p6mm*) and 3D cubic (*Im3m*, or bi-continuous *Ia3d*)), and porous structure parameter (pore size and surface area) [8].

Challenges remain in preparing mesoporous TiO_2 using hard templates. These include (i) the rapid hydrolysis and polycondensation of Ti-precursors, which can be mitigated by adding stabilizing agents like acetylacetone and triethanolamine to the sol solution to prevent bulky precipitate formation; (ii) the need for repeated impregnation of TiO_2 precursors to achieve high loading within the template's pore channels and to ensure strong interconnections in the TiO_2 scaffold after template removal; and (iii) difficulties in template removal. Complete removal of conventional SiO_2 templates typically requires environmentally and user-unfriendly fluorine-containing compounds. An alternative is to use NaOH solution to dissolve SiO_2 at elevated temperatures, though this can leave residual content. For carbon templates, complete removal requires calcination at ultra-high temperatures (e.g., >600 °C), which might lead to the collapse of the mesoporous rutile-rich TiO_2 scaffold [7, 8].

4.2.2 Soft Template Method

The soft template method harnesses surfactant micelles as templates. Various surfactants, including CTAB, PEG, Brij 56, and block copolymers (e.g., $EO_{20}PO_{70}EO_{20}$, $EO_{102}PO_{70}EO_{102}$, PEO–PHB (KLE), and PEO–PIB) have been employed to directly synthesize porous TiO_2 [6]. Unlike hard templates, which typically remain inert and undergo minimal change before removal, surfactants readily dissolve in the sol solution and actively engage in the sol–gel process. They facilitate the formation of a hybrid mesostructure between TiO_2 oligomers and surfactant through mechanisms such as hydrogen bonding, electrostatic attraction, hydrophilic–hydrophilic interaction, and electronic attraction [9]. Upon calcination, organic surfactants are burned out, leaving behind porous TiO_2. Given that uncontrolled crystal growth of TiO_2 during calcination could potentially eliminate the pores, establishing a robust inorganic–organic hybrid is crucial for preserving the porous structure. In this context, the key in successful synthesis of mesoporous TiO_2 using surfactant as a soft template lies in the control the interaction between block copolymer with Ti oligomers. That is, it is essential to manipulate the hydrolysis of TiO_2 precursors to self-assemble with surfactant to form rigid mesostructure before the hydrolysis and polycondensation of TiO_2 alone [7, 10, 11].

Among various surfactants, block copolymer is of particular interest. During phase separation, block copolymer with immiscible blocks is able to generate various highly ordered and periodical mesophases with hexagonal, cubic, gyroidal, and lamellar equilibrium morphologies, depending on the nature of block copolymer and the experimental condition. Moreover, block copolymer template method has coupled with the so-called evaporation-induced self-assembly (EISA) process, offering an effective and reproducible technique for the facile synthesis of highly organized mesoporous TiO_2 in forms of powders and films, as shown in Figure 4.1 [9, 11]. The synthesis derives from a homogeneous sol solution containing a Ti-precursor and block copolymer. The sol can be casted in a petri dish (to yield powders) or coated onto substrates (for film formation). Upon the solvent evaporation, the concentration of the surfactant exceeds the critical micelle concentration, which energetically triggers the phase separation of the block copolymer template and guides the supramolecular self-assembly, resulting in the formation of a highly organized mesostructure [12–15]. Moreover, an aging post-treatment by exposing to the mesostructured hybrid to a humid condition is imperative. Such aging process allows to complete hydrolysis of Ti species and to promote inorganic–organic crosslinking, The fixed robust mesostructure, comprised of well hybridized Ti

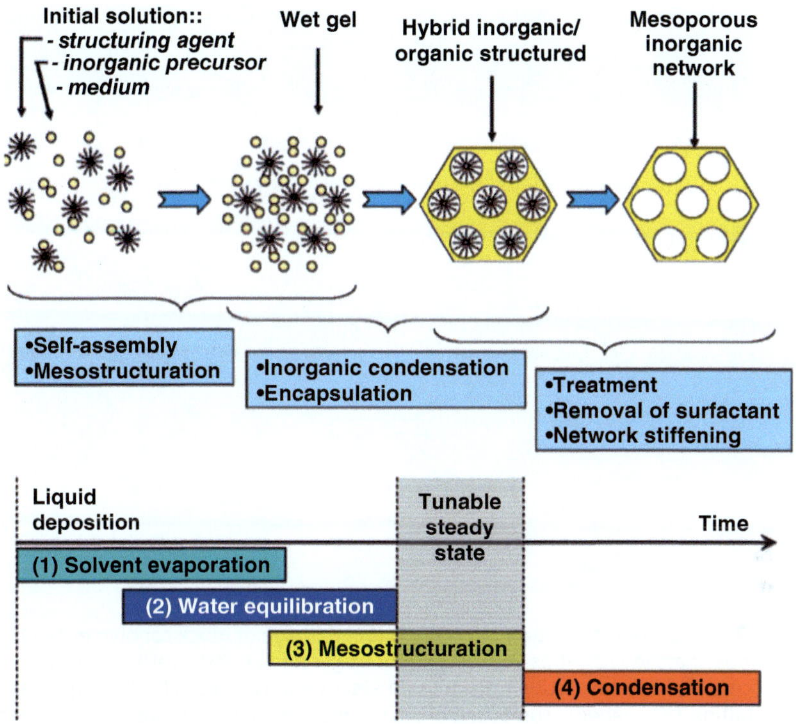

Figure 4.1 Schematic illustration of the EISA process using surfactant as soft template and the underlying multistep structural evolution. Reproduced from [9], Copyright {2008} American Chemical Society.

precursors and block copolymer, show enhanced thermal stability. [3, 11]. Although the block copolymer can be largely removed and phase transition to anatase occurs at 350 °C, a calcination temperature over 450 °C is need to fully crystallize the amorphous hybrid mesostructure into mesoporous inorganic framework consisting of interconnected anatase TiO_2 nanocrystallites. Moreover, the complete elimination of surfactants and residual organics inevitably results in a significant volume contraction (40–60% by volume). Thus, it is imperative to maintain the thickness of the casted films at a level thin enough, ideally less than 1.0 mm, and to utilize a low ramping rate (e.g., 1°C min^{-1}) during the calcination process to prevent the distortion or collapse of the mesoporous structure frequently caused by internal stress within the TiO_2 scaffold [11, 16]. Figure 4.2 shows the representative electronic microscopy images for mesoporous TiO_2 films with anatase frameworks.

The EISA process is predominantly utilized in the film preparation, although its effectiveness in particle processing remains constrained. To address this limitation, several granulation technologies, such as spray pyrolysis, electrospraying, and electrospinning, have been integrated with the EISA process. This integration enables the continuous and large-scale production of mesoporous TiO_2 with diverse shapes, including fibers, tubes, and spheres. Moreover, this combined approach facilitates the generation of well-defined morphologies alongside the mesoporous microstructure, thereby augmenting its potential for practical applications [5].

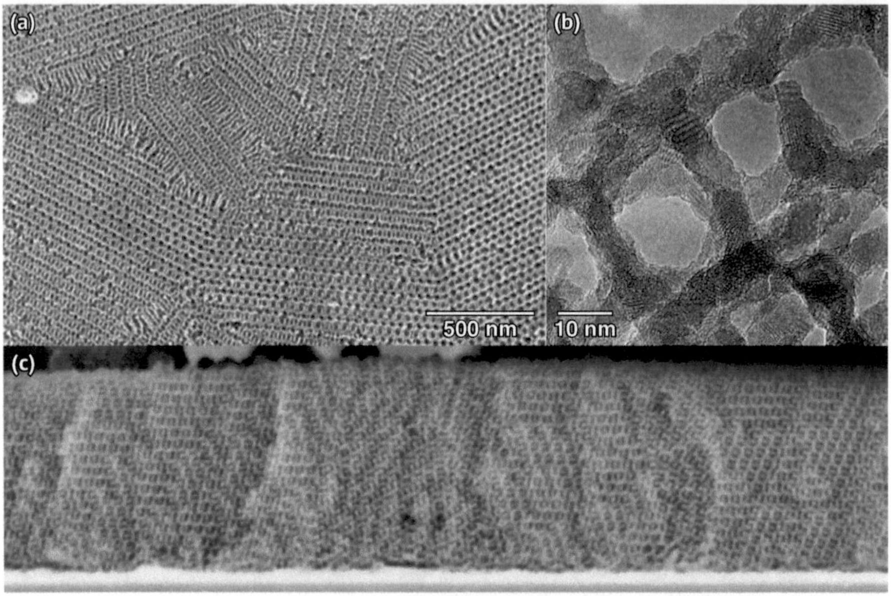

Figure 4.2 (a) Scanning electron microscopy (SEM) image of block copolymer-templated mesoporous anatase TiO_2 films with bicontinuous gyroid network synthesized via EISA process. (b) High-resolution transmission electron microscopy (TEM) image of the polycrystalline TiO_2 network composed of ~ 10 nm sized anatase grains. (c) Cross-sectional SEM image of periodically mesoporous TiO_2 films (image width, 3 mm) [16], Copyright {2009} American Chemical Society.

4.3 Self-Template Route

Although the sacrificial template method enables controlled synthesis of porous TiO_2, its yield is also low, and the process is costly since the removed template is not reclaimable. To improve TiO_2 productivity, self-template approach has been developed to enable straightforward construction of porous TiO_2. Notably, "self-template" sometimes refers to an intermediate phase or structural transition of TiO_2, particularly in structural transformations from solid to hollow TiO_2 spheres during solvo- or hydrothermal processes (see Section 4.4). Here, we define it as an independent product with well-defined composition and structure, where TiO_2 can succeed the architecture of the self-template through additional steps, such as heat treatment. Thus, in this section, we examine two key self-templates – HTCS and titanium metal–organic frameworks (MOFs) – to discuss strategies for designing self-templates and converting them into porous TiO_2 with targeted textural properties.

4.3.1 Hydrous TiO_2 Colloidal Spheres

4.3.1.1 Self-Organization Theory from HTCS

Amorphous HTCS were reported in 1982 by Barringer and Bowen. Their synthesis approach is based on sol–gel chemistry through the direct hydrolysis of titanium alkoxide [$Ti(OR)_4$, where R represents an alkyl group] in ethanol, containing a small amount of water at ambient conditions. N_2 sorption analysis in HTCS revealed these HTCS feature a large surface area exceeding 300 m^2/g, and pore size distribution analysis showed HTCS are highly porous, possessing abundant micropores and mesopores (< 3 nm) derived from the voids between densely packed NBBs, that is, nanosized amorphous TiO_2 hydrates [4, 5, 17–19].

Before this pioneering work, colloidal SiO_2 spheres had been widely synthesized using the Stöber method since the 1960s [20], via the hydrolysis of tetraethyl orthosilicate (TEOS) in an ethanol–water mixture containing ammonia as a hydrolytic catalyst. The widely referenced LaMer model [21–23], which describes a process of instantaneous nucleation followed by diffusion-controlled growth, was frequently applied to explain the multistage formation of colloidal SiO_2 spheres, as well as other materials [5, 24, 25]. Typically, in the first successful synthesis of monodisperse TiO_2 spheres by Matijević's group in 1977 – achieved through the forced hydrolysis of $TiCl_4$ in an ethanol–water solvent – the LaMer model was still used to explain the formation of their rutile-phase TiO_2 spheres [26]. This reliance on the LaMer model to explain HTCS formation likely arose from the high similarities in the synthesis methods for SiO_2 and TiO_2 colloidal spheres and in the development in the concentrations of solute Ti-precursor in sol solution as a function of time [25].

However, the difference in porosity between nonporous SiO_2 colloidal spheres and the highly porous HTCS suggests a fundamental divergence in formation mechanisms. In the 1990s, Zukoski's group proposed the self-organization theory, which was based on studies of HTCS [27–30]. According to this theory, the distinctive porous microstructure and spherical morphology of HTCS result from the self-organization of nanosized TiO_2 nuclei, that is, each sphere forms independently through the spontaneous aggregation of hydrous TiO_2 nuclei via an exact 0D → 3D self-assembly pathway.

The primary difference in the formation mechanisms between SiO_2 and TiO_2 colloidal spheres stems from the distinct sol–gel kinetics of silicon and titanium alkoxides. TEOS, used for SiO_2 synthesis, undergoes relatively slower hydrolysis, whereas moisture-sensitive titanium alkoxides, such as $Ti(OR)_4$, exhibit much faster hydrolysis kinetics. In sol–gel media, $Ti(OR)_4$ rapidly hydrolyzes and condense, which greatly hinders the diffusion-controlled growth according to the LaMer model. Instead, the as-formed small TiO_2 hydrates are prone to self-assembly to (sub)microspheres or nanospheres to minimize the high surface energy, as shown in Figure 4.3a. Actually in 1997, Kim's group revisited the findings of Matijević's work from 20 years earlier and found that these 3D spheres are composed of self-organized 0D rutile nanograins [26, 31].

4.3.1.2 Structure-Directing Agents for HTCS Synthesis

In the early 21st century, considerable progress has been achieved in developing structure-directing agents (SDAs) for HTCS preparation, marked by notable enhancements in reproducibility, yield, and quality attributes such as particle size uniformity, monodispersity, and sphericity. Different SDAs, such as salts [32], NH_3 [33], polymers [32], amines (methylamine (MA) [34, 35], dodecylamine (DDA) [10, 36], and hexadecylamine (HDA) [37, 38]), ethylene glycol (EG) [39, 40], and organic acid [41].

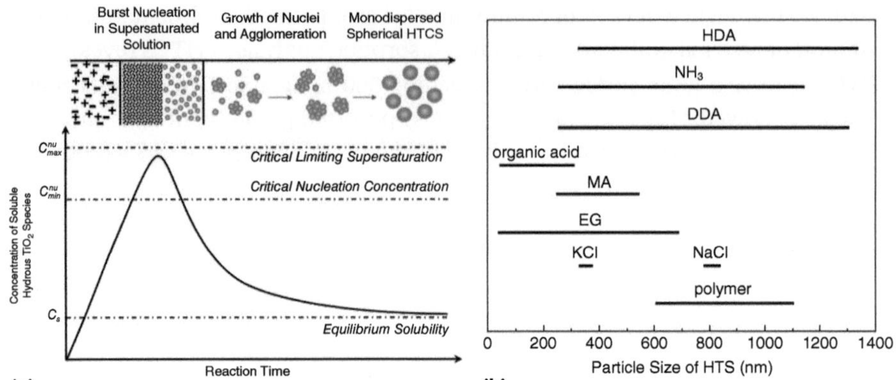

(a) (b)

Figure 4.3 (a) The formation pathway of monodisperse HTCS via self-assembly process and the corresponding variation in the concentration of soluble TiO_2 hydrates in the sol–gel medium; (b) particle size ranges of uniform HTCS prepared using different SDAs at ambient conditions. Adapted from [19] with permission of Elsevier.

Various SDAs, including salts (e.g., NaCl, KCl), organic acids, amines (e.g., methylamine, dodecylamine, hexadecylamine), and ammonia, have been integrated into the $Ti(OR)_4$—EtOH—H_2O system to facilely modulate the agglomeration dynamics of TiO_2 hydrates through orchestrating the underlying driving forces [5, 19, 41, 42].

This intricate control mechanism enables the production of HTCS through spherical aggregation among hydrous TiO_2 nuclei or nanoparticles. SDAs play dual roles, precipitation prompter and morphological controller, in formation of HTCS. SDAs can accelerate the hydrolysis and condensation of $Ti(OR)_4$ to TiO_2 nuclei, and their subsequent self-organization to spheres, leading to expedited precipitation and a shortened induction timc. Moreover, SDAs direct the self-organization of nanosized TiO_2 hydrates to form monodisperse spheres through the introduction of additional driving forces. For instance, the inclusion of salts elevates the surface potential of TiO_2 hydrates, augmenting their mutual electrostatic forces; ammonia enhances an interaction through hydrogen bonding; and amines facilitate the self-assembly process by hybridizing with TiO_2 hydrates, forming organic–inorganic hybrid mesostructures in HTCS [4, 33, 41].

The meticulous selection and utilization of SDAs are essential for achieving precise control over the synthesis process and ensuring the quality attributes of HTCS. Among SDAs, the particularly notable are amines for their efficacy in synthesizing (sub)microspheres of highly uniform HTCS with adjustable spherical diameters. Amines exhibit versatility and robustness, facilitating the facile and reproducible self-assembly synthesis of uniform HTCS across a wide range of amine concentrations. In contrast, NH_3 and salts require precise control of dosage and the utilization of high-purity Ti-precursors and solvents due to their relatively lower self-organizing capability.

SDAs also play a crucial role in controlling the particle size of HTCS. Previously, the molar ratio of water to Ti-precursors has been widely used to regulate particle size, with higher water content leading to faster hydrolysis and smaller particle sizes. Moreover, each type of SDA appears to have a limitation in the particle size range it can generate, with an associated optimal size range, as summarized in Figure 4.3b.

4.3.1.3 Crystallization of HTCS Self-Template by External Field Enhancements

The development of HTCS not only provides new insights into the bottom-up self-organization mechanism for hierarchical spheres of transitional metal oxides via wet-chemistry routes but also enables the large-scale synthesis of well-crystallized mesoporous TiO_2 spheres with the aid of external field enhancements by direct calcination, hydro/solvothermal process followed with calcination, and UV irradiation.

Direct calcination of HTCS results in the results in the undesirable fusion of NBBs, leading to the loss of the porous structure and yielding a small surface area, typically around $\approx 10\ m^2/g$. To address this issue, a hydro/solvothermal post-treatment prior to calcination is essential. This post-treatment is crucial for stabilizing porous spheres with partially or fully crystallized TiO_2 NBBs. The intermediate pre-crystallized spheres exhibit enhanced thermal stability during subsequent calcination processes, enabling the preservation of the porous structure with a significantly larger surface area, typically $\approx 100\ m^2/g$ [38, 41, 43, 44]. The resulting mesoporous TiO_2 submicrospheres consist of closely packed anatase nanograins with sizes ranging from 10 to 15 nm, as shown in Figure 4.4a–d. Their mesopores are thus formed from the voids between the nanograins. To reduce the typical hydro/solvothermal reaction time from 8–24 hours to less than 1 hour, microwave-assisted synthesis has been employed [45, 46].

UV irradiation can induce the crystallization of amorphous HTCS at room temperature. This gentle crystallization process preserves a large surface area, exceeding 300 m^2/g. Notably, adjusting the wavelength (λ) of UV light and the type of dispersant allows for precise control over the crystal phase, facilitating the selective formation of anatase, rutile, or their biphasic mixtures. However, the achievement of high crystallinity is impeded by the presence of crystallized TiO_2 on the surface, which can obstruct UV penetration and hinder further internal crystallization [47].

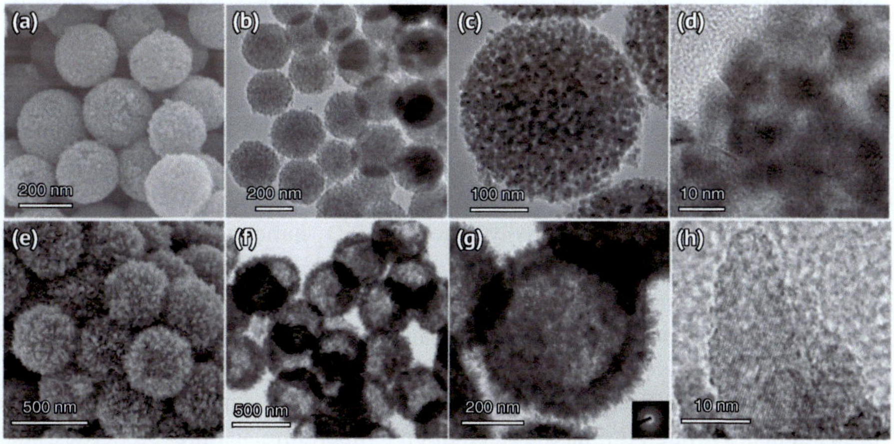

Figure 4.4 SEM and TEM images of mesoporous anatase TiO_2 spheres consisting of anatase nanocrystallites with solid (a–d) and hollow (e–h) interiors nanograins synthesized using HTCS as self-template. Adapted from Ref. [19] with permission of Elsevier.

Apart from the direct phase transition of HTCS to crystallized solid spheres, design of appropriate HTCS for the selective etching is crucial to obtain hollow spheres, for which an ideal self-template should possess significant differences in physicochemical properties along the radial direction from the center to surface. That is, TiO_2 hydrates inside should be more dissoluble than those in the surface region, and the etching should be targeted to the interior when corrosive agents is the penetrated into the porous HTCS [19, 48]. Basically, the primary TiO_2 hydrates located in the core are rapidly agglomerated with other "outer" TiO_2 rather than grow bigger. Thus, they are relatively small and more densely packed, thereby having a higher mean curvature and surface energy as compared to those in the vicinity of surface. As a result, they present relatively high reactivity and mobility toward dissolution in the presence of etchants [48–51]. Thus, mesoporous TiO_2 hollow spheres can be synthesized using amorphous HTCS the self-template with the corrosive fluoride and ammonia as etchants under the critical condition by hydro/solvothermal process. As shown in Figure 4.4e–h, the presence of fluoride induces selective interior hollowing and promotes the preferential growth of single-crystal anatase TiO_2 nanorods with dominant (101) facets [5, 19, 33].

4.3.2 Titanium MOFs

MOF community has been rapidly developed since the early 1990s, attracting research fields from coordination chemistry, host–guest chemistry, nano engineering, gas storage and separation, (photo)catalysis, membrane technology, optoelectronics, sensing, energy storage and conversion, biomedical science, etc. MOFs have a metal center of di- or trivalent ions of 3p metals, 3d transition metals, or lanthanides that have weak interaction with organic ligands. Direct calcination of MOFs induces the oxidization to TiO_2 and subsequent crystallization without destroying the origin morphology, since the limited amount of Ti sources in the highly porous MOFs, using MOF self-template, enables to genera mesoporous structure consisting of close-packed nanosized TiO_2 grains by effectively preventing the rapid crystal growth from the crystal aggregation and fusion.

Ti-based zeolitic imidazolate framework-8 (Ti-ZIF-8) has been widely reported as self-template [52]. It can be synthesized by solvothermal process of a mixed solution of *N,N*-dimethylmethanamide and methanol containing 1,4-benzenedicarboxylic acid-stabilized titanium isopropoxide at 150 °C [42, 53, 54]. The cake-like mesoporous TiO_2 can be obtained by calcining Ti-ZIF-8 over 380 °C, with the octahedron morphology well reserved. The polycrystalline TiO_2 nanograins build up a mesoporous structure with a surface area of $\approx 147\ m^2\ g^{-1}$ and an average pore size of ≈ 10 nm, as illustrated in Figure 4.5 [42].

Up to now, various Ti-based MOFs, including NH_2-MIL-125 [55, 56], MIL-167 [56], and NTU-9 [57], have been used as self-template for the facile synthesis of mesoporous TiO_2 with different morphologies, and their photocatalytic properties have been studied. Through tuning the calcination temperature, textural properties, such as crystal phase, surface area, oxygen vacancy, and grain size can be optimized to produce photocatalytically more active TiO_2. As demonstrated by Kampouri et al. [56], their optimum NH_2-MIL-125-derived TiO_2 with anatase and rutile phases by

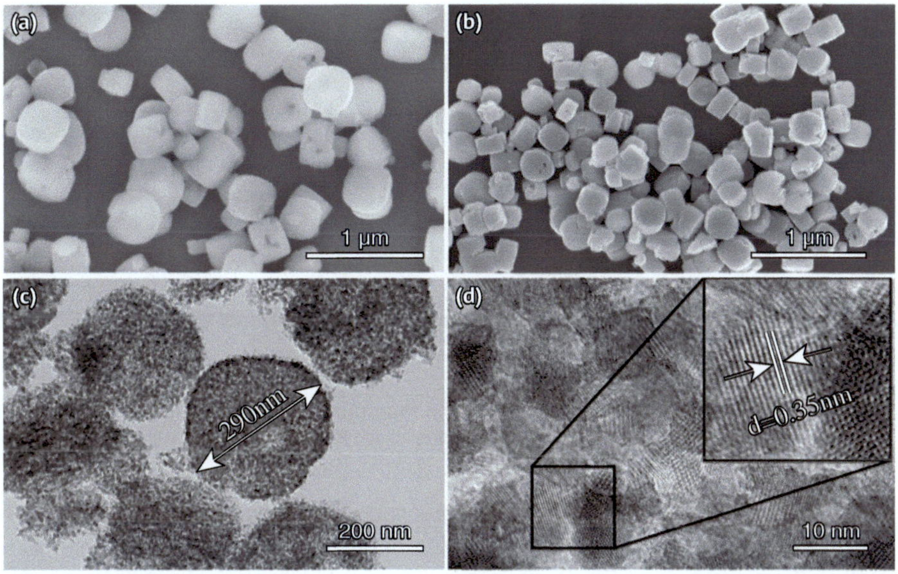

Figure 4.5 SEM images of (a) cakelike MIL-125 self-templates with a diameter of 400 nm, and (b) their calcined product, porous TiO_2 with particle size 250–350 nm. (c) Low- and (d) high-resolution TEM images of porous TiO_2 consisting of anatase nanocrystallites with lattice fringe is 0.35 nm [42], Copyright {2020} American Chemical Society.

calcination at 600°C notably outperforms the commercial Aeroxide P25 and the conventionally synthesized TiH_4O_4- and MIL-167-derived TiO_2 in photocatalytic hydrogen generation.

4.4 3D Hierarchical Porous TiO_2 Spheres by Solvo/Hydrothermal Process

Following the discovery of HTCS formation, the self-origination process has been widely operated in wet-chemical synthesis, and hydro/solvothermal process provides a facile and straightforward synthesis of 3D TiO_2 crystalline spheres by subjecting the reaction system with the dispersed Ti-precursors to critical conditions, resulting in nucleation, growth, self-assembly, and crystallization into porous spheres. Various driving forces, such as electrostatic repulsive force, van der Waals attractive force, hydrogen bonding, repulsion, steric force, and hydrophobic–hydrophobic or organic–inorganic interactions, orchestrate the self-assembly process. Leveraging these varied driving forces is crucial for achieving optimal textural properties of NBBs and 3D sphere, including uniformity, size, shape, porosity, and surface area. Moreover, hollow interiors might be generated due to the crystal fusion, dissolution–recrystallization, or Ostwald ripening under the critical condition.

To date, alcohol and organic acids are frequently employed as organic solvents to leverage nonequilibrium states and critical solvent conditions. In such settings, nonhydrolytic reactions between the Ti-precursor and organic solvent can

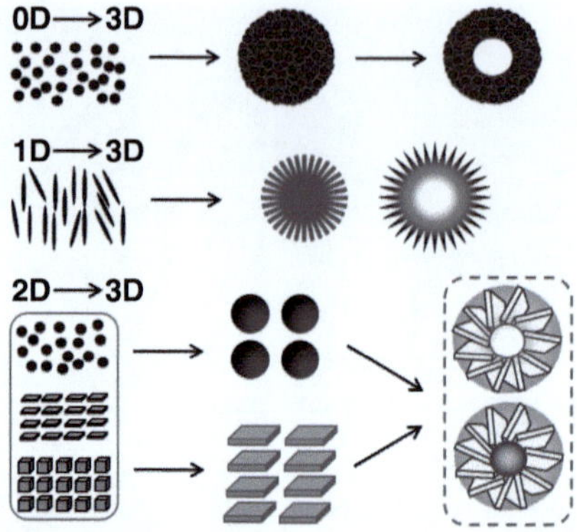

Figure 4.6 Self-organization pathways to the formation of 3D TiO_2 spheres by assembling low-dimensional (0D, 1D, and 2D) NBBs. Adapted from Ref. [3] with permission from Elsevier.

directly initiate TiO_2 crystallization. Moreover, water, either generated during the solvothermal reaction or introduced beforehand, acts as an oxygen source, facilitating the formation of Ti—O—Ti bonds. The gradual hydrolysis of TiO_2, enabled by the controlled formation of NBBs, promotes subsequent agglomeration and self-organization into spherical structures.

The self-organized hierarchical superstructures generally show spherical in shape and consisting of well-arranged low-dimensional NBBs. Figure 4.6 schematically illustrates the formation of 3D hierarchical TiO_2 spheres via different self-organization pathways (0/1/2D → 3D) with diverse low-dimensional NBBs, that is, 0D nanoparticles, 1D nanowires, nanorods, or nanoribbons, and 2D nanosheets or nanoplates [3, 5]. The dimensions of these NBBs are primarily determined by the crystal phase of TiO_2 and the synthetic conditions employed. Most of these ordered self-assembly configurations optimize NBB exposure mirroring analogous natural formations observed in flora and fauna such as pomegranates, sea urchins, flowers, and desert roses.

4.4.1 0D→3D Pathway

0D is the most popular nanostructure of TiO_2 in wet-chemical synthesis, and the self-assembly through 0D → 3D is also the most common pathway. Like the formation of 3D spherical HTCS from 0D hydrous TiO_2 nuclei, 0D NBBs readily form through the controlled hydrolysis of titanium species under specific conditions, following with their self-organization to 3D spheres. Mechanically, 0D → 3D pathway involves the forced hydrolysis and burst nucleation of Ti sources, followed with spontaneous agglomeration of the as-formed 0D nanocrystallites to 3D spherical geometry to minimize their high surface energy. Notably, the kinetics of spontaneous agglomeration of 0D TiO_2 entities generally is energetically, and the 3D reconstruction occurs immediately after the nucleation of Ti-precursors and before the crystallization of TiO_2 NBBs, particularly under critical conditions.

An intriguing phenomenon widely observed in 0D → 3D self-assembly during the hydro/solvothermal process is that a hollow reconstruction frequently occurs after the formation of solid spheres. Akin to the structure difference in the hydrous TiO_2 NBBs in HTCS, the as-hydrothermal TiO_2 solid spheres exhibit low crystallinity and radial divergence from the center to the surface. Each solid sphere comprised of aggregated TiO_2 NNBs that remain loosely packed and contains rich inorganic–liquid interfaces. With the further proceeding of hydro/solvothermal process and the crystal growth of TiO_2 NBBs, mechanisms like "dissolution and recrystallization" or Ostwald ripening may facilitate symmetric hollowing at the interface between the inorganic TiO_2 and residual liquid from the solvent and soluble intermediates. Therefore, such hollow reconstruction is often linked to the presence of corrosive agents in the reaction medium, such as fluoride ions (F^-) and amines, which play a crucial role in facilitating the repeated dissolution–recrystallization cycles of "weak" TiO_2, ultimately contributing to the formation of hollow structure [5, 50, 58].

4.4.2 1D → 3D Pathway

In the 1D → 3D self-organization, capping ligands (for anatase) and ions (e.g., Cl^- for 1D rutile) have been utilized to adjust the surface energy through adsorption or interaction, so that the anisotropic crystal growth of 1D nanostructure can initiate due to the significant difference in surface free energy among the crystal faces. The resultant hierarchical sphere is formed through the oriented aggregation of 1D anatase and rutile nanocrystallites along the radial direction, which results from the surface energy mineralization and the repulsion between the 1D TiO_2 NBBs.

Anatase nanocrystallites as NBBs and in a shape of 1D nanorods are mainly derived from bipyramidal crystal structure with dominant (101) facets [33]. Selecting proper chelating agents is essential to guide the formation of 1D nanostructures, and the available include diethylene glycol, glycerol, isopropanol, hydrogen peroxide, F^- ions, acetic acid (HAc), ammonia, and amine [5, 59, 60].

Significant progress has been made in the synthesis of nanoporous TiO_2 microspheres composed of 1D anatase nanorods, derived from a binary system of titanium butoxide as Ti source and HAc as solvent. In this process, HAc can complex with the distributed Ti source through ligand exchange and substitution, resulting in a metastable six-coordinated structure, $Ti_6O_6(Ac)_6(OBu)_6$, with the gradual release of butyl alcohol. The Ti-complex exhibits slower hydrolysis kinetics, guiding a preferential growth. Meanwhile, the generating butyl alcohol may react with the HAc solvent, yielding H_2O through a slow esterification reaction, which facilitates the hydrolysis and condensation of Ti species. During the condensation process, 1D Ti-complex nanorods, temporarily stabilized by HAc, are formed, and orientally attach, forming hierarchical Ti-complex spheres. Upon calcination, 1D anatase-phase TiO_2 is obtained with the remained urchin-like superstructure [61–63].

Wider success has been made in the self-assembly of rutile-phase TiO_2 spheres through 1D → 3D assembly. The 1D nanostructure results from a preferred growth along the [110] direction of rutile. Interestingly, Cl^- ions show effective for the promotion of rutile phase formation at mild temperatures. They can be adsorbed selectively on the lowest-surface-energy (110) plane of rutile TiO_2, suppressing but accelerating the growth of rutile along the [110] and [100] directions, respectively [6]. Therefore, it is not surprised that $TiCl_4$ and $TiCl_3$ have been widely used as Ti-precursor for the formation of 1D rutile NBBs and their assembly to spherical structure [64, 65]. Moreover, chlorides, such as NaCl and HCl, can be used to assist non-$TiCl_x$ precursors for the anisotropic growth of 1D rutile. Apart from Cl^- ions, the reaction media and the additives (e.g., surfactants or SDAs) were also found to have influences on the growth of 1D nanobuilding units and the final structure of rutile TiO_2 spheres.

To facilely control the hydrolysis rate and self-organization, Fang et al. [66] employed dual chelating agents of ethylene glycol and urea. The hydrolysis rate of TTIP in the different solution is in the order of TTIP > TTIP + EG > TTIP + EG + urea, leading to the directional growth of crystals in the 1D feature of nanorods, nanoribbons, and nanowires, respectively [66].

4.4.3 2D→3D Pathway

In anatase TiO_2, the ultrathin 2D layered structure is related to the dominant high-energy facet (001) facet. Excessive F^- ions can efficiently reduce the surface free energy of (001) facets, inducing a preferential growth of single-crystalline anatase TiO_2 nanosheets with well-defined (001) facets. Facilitated by the mutual repulsions among 2D NBBs under the one-pot hydro/solvothermal process of a mixture of titanium butoxide, HF aqueous solution and ethanol, the flower-like spheres can be assembled from the interconnected nanoflakes, through which each nanoflakes presents maximized exposure to the surface [67].

A solvothermal system of TTIP–diethylenetriamine–IPA–H_2O has been developed to synthesize solid or hollow TiO_2 microspheres with selective 1D and 2D anatase TiO_2. Herein diethylenetriamine acts as a fluorine-free capping agent to direct the preferential growth along the [101] direction with the maintenance of (001) facets without water [68]. However, water in this quaternary system is beneficial to the generation of (101) facets in single-crystal anatase TiO_2 NBBs, and the interior hollowing of microspheres. Therefore, a sequent evolution in spherical morphology occurs, from solid to sphere-in-sphere and finally to hollow spheres during the solvothermal crystallization process [60]. Also, a TTIP–1,4-butanediamine (BDA, capping agent and N dopant source)–IPA ternary system has been developed for synthesizing hierarchical core-shell N-doped TiO_2 spheres consisting of nanosheets with exposed (116) high-energy facets [69].

4.5 Conclusion

This chapter reviews the progress in the synthesis of 3D TiO_2, focusing on porous and spherical TiO_2 with crystalline frameworks, and highlights the evolution of synthesis strategies. We have summarized various template-based approaches for creating porous TiO_2 spheres, including hard templates, soft templates, and self-templates. A key emphasis is placed on the advances in the sol–gel preparation of uniform HTCS with controllable particle size and porous structure, which represent a significant milestone in self-organization theory for inorganic materials. HTCS also serve as a promising self-template for the subsequent large-scale synthesis of mesoporous TiO_2 spheres with tunable crystal phases and customizable microstructures. Finally, we examine the one-pot hydro/solvothermal synthesis of hierarchical TiO_2 spheres, focusing on the role of self-organization mechanisms in guiding the formation of low-dimensional NBBs. Understanding these self-organization pathways and their underlying mechanisms offers substantial potential for the rational design of spherical TiO_2 assemblies. Further research is needed to refine these methods and explore their applications in semiconductor photocatalysis.

References

1 Reghunath, S., Pinheiro, D., and Kr, S.D. (2021). A review of hierarchical nanostructures of TiO_2: advances and applications. *Applied Surface Science Advances* 3: 100063.

2 Gatou, M.A., Syrrakou, A., Lagopati, N., and Pavlatou, E.A. (2024). Photocatalytic TiO_2-based nanostructures as a promising material for diverse environmental applications: a review. *Reactions* 5: 135–194.

3 Pan, J.H., Dou, H., Xiong, Z. et al. (2010). Porous photocatalysts for advanced water purifications. *Journal of Materials Chemistry* 20: 4512–4528.

4 Pan, J.H., Wang, Q., and Bahnemann, D.W. (2014). Hydrous TiO_2 spheres: an excellent platform for the rational design of mesoporous anatase spheres for photoelectrochemical applications. *Catalysis Today* 230: 197–204.

5 Ding, Y., Yang, I.S., Li, Z. et al. (2020). Nanoporous TiO_2 spheres with tailored textural properties: controllable synthesis, formation mechanism, and photochemical applications. *Progress in Materials Science* 109: 100620.

6 Lan, K., Wei, Q., and Zhao, D. (2022). Versatile synthesis of mesoporous crystalline TiO_2 materials by monomicelle assembly. *Angewandte Chemie International Edition* 61: e202200777.

7 Wang, A., Ma, Y., and Zhao, D. (2024). Pore engineering of porous materials: effects and applications. *ACS Nano* 18: 22829–22854.

8 Yu, H., Liu, M., Lu, Q. et al. (2024). Mesoporous metal oxides. *The Journal of Physical Chemistry C* 128: 19945–19961.

9 Sanchez, C., Boissière, C., Grosso, D. et al. (2008). Design, synthesis, and properties of inorganic and hybrid thin films having periodically organized nanoporosity. *Chemistry of Materials* 20: 682–737.

10 Pan, J.H., Cai, Z., Yu, Y., and Zhao, X.S. (2011). Controllable synthesis of mesoporous F–TiO_2 spheres for effective photocatalysis. *Journal of Materials Chemistry* 21: 11430–11438.

11 Pan, J.H., Zhao, X.S., and Lee, W.I. (2011). Block copolymer-templated synthesis of highly organized mesoporous TiO_2-based films and their photoelectrochemical applications. *Chemical Engineering Journal* 170: 363–380.

12 Dendooven, J., Goris, B., Devloo-Casier, K. et al. (2012). Tuning the pore size of ink-bottle mesopores by atomic layer deposition. *Chemistry of Materials* 24: 1992–1994.

13 Pan, J.H., Chai, S.Y., Lee, C. et al. (2007). Controlled formation of highly crystallized cubic and hexagonal mesoporous SnO_2 thin films. *The Journal of Physical Chemistry C* 111: 5582–5587.

14 Pan, J.H. and Lee, W.I. (2005). Selective control of cubic and hexagonal mesophases for titania and silica thin films with spin-coating. *New Journal of Chemistry* 29: 841–846.

15 Pan, J.H. and Lee, W.I. (2006). Preparation of highly ordered cubic mesoporous WO_3/TiO_2 films and their photocatalytic properties. *Chemistry of Materials* 18: 847–853.

16 Crossland, E.J.W., Kamperman, M., Nedelcu, M. et al. (2009). A bicontinuous double gyroid hybrid solar cell. *Nano Letters* 9: 2807–2812.

17 Barringer, E.A. and Bowen, H.K. (1982). Formation, packing, and sintering of monodisperse TiO_2 powders. *Journal of the American Ceramic Society* 65: C-199–C-201.

18 Barringer, E.A. and Bowen, H.K. (1985). High-purity, monodisperse TiO_2 powders by hydrolysis of titanium tetratethoxide. 2. Aqueous interfacial electrochemistry and dispersion stability. *Langmuir* 1: 420–428.

19 Ma, D., Schneider, J., Lee, W.I., and Pan, J.H. (2021). Controllable synthesis and self-template phase transition of hydrous TiO_2 colloidal spheres for photo/electrochemical applications. *Advances in Colloid and Interface Science* 295: 102493.

20 Stöber, W., Fink, A., and Bohn, E. (1968). Controlled growth of monodisperse silica spheres in the micron size range. *Journal of Colloid and Interface Science* 26: 62–69.

21 Whitehead, C.B., Özkar, S., and Finke, R.G. (2019). LaMer's 1950 model for particle formation of instantaneous nucleation and diffusion-controlled growth: a historical look at the model's origins, assumptions, equations, and underlying sulfur sol formation kinetics data. *Chemistry of Materials* 31: 7116–7132.

22 Whitehead, C.B., Özkar, S., and Finke, R.G. (2021). LaMer's 1950 model for particle formation of instantaneous nucleation and diffusion-controlled growth: a historical look at the model's origins, assumptions, equations, and underlying sulfur sol formation kinetics data. *Materials Advances* 2: 186–235.

23 Lamer, V.K. and Dinegar, R.H. (1950). Theory, production and mechanism of formation of monodispersed hydrosols. *Journal of the American Chemical Society* 72: 4847–4854.

24 Thanh, N.T.K., Maclean, N., and Mahiddine, S. (2014). Mechanisms of nucleation and growth of nanoparticles in solution. *Chemical Reviews* 114: 7610–7630.

25 Ghimire, P.P. and Jaroniec, M. (2021). Renaissance of Stöber method for synthesis of colloidal particles: new developments and opportunities. *Journal of Colloid and Interface Science* 584: 838–865.

26 Matijević, E., Budnik, M., and Meites, L. (1977). Preparation and mechanism of formation of titanium dioxide hydrosols of narrow size distribution. *Journal of Colloid and Interface Science* 61: 302–311.

27 Bogush, G.H. and Zukoski Iv, C.F. (1991). Studies of the kinetics of the precipitation of uniform silica particles through the hydrolysis and condensation of silicon alkoxides. *Journal of Colloid and Interface Science* 142: 1–18.

28 Look, J.L., Bogush, G.H., and Zukoski, C.F. (1990). Colloidal interactions during the precipitation of uniform submicrometre particles. *Faraday Discussions of the Chemical Society* 90: 345–357.

29 Look, J.L. and Zukoski, C.F. (1992). Shear induced aggregation during the precipitation of titanium alkoxides. *Journal of Colloid and Interface Science* 153: 461–482.

30 Look, J.L. and Zukoski, C.F. (1995). Colloidal stability and titania precipitate morphology: influence of short-range repulsions. *Journal of the American Ceramic Society* 78: 21–32.

31 Park, H.K., Kim, D.K., and Kim, C.H. (1997). Effect of solvent on titania particle formation and morphology in thermal hydrolysis of $TiCl_4$. *Journal of the American Ceramic Society* 80: 743–749.

32 Eiden-Assmann, S., Widoniak, J., and Maret, G. (2004). Synthesis and characterization of porous and nonporous monodisperse colloidal TiO_2 particles. *Chemistry of Materials* 16: 6–11.

33 Pan, J.H., Wang, X.Z., Huang, Q. et al. (2014). Large-scale synthesis of urchin-like mesoporous TiO_2 hollow spheres by targeted etching and their photoelectrochemical properties. *Advanced Functional Materials* 24: 95–104.

34 Mine, E., Hirose, M., Nagao, D. et al. (2005). Synthesis of submicrometer-sized titania spherical particles with a sol–gel method and their application to colloidal photonic crystals. *Journal of Colloid and Interface Science* 291: 162–168.

35 Yu, I.G., Kim, Y.J., Kim, H.J. et al. (2011). Size-dependent light-scattering effects of nanoporous TiO_2 spheres in dye-sensitized solar cells. *Journal of Materials Chemistry* 21: 532–538.

36 Pan, J.H., Shen, C., Ivanova, I. et al. (2015). Self-template synthesis of porous perovskite titanate solid and hollow submicrospheres for photocatalytic oxygen evolution and mesoscopic solar cells. *ACS Applied Materials & Interfaces* 7: 14859–14869.

37 Chen, Y., Huang, F., Chen, D. et al. (2011). Effect of mesoporous TiO_2 bead diameter in working electrodes on the efficiency of dye-sensitized solar cells. *ChemSusChem* 4: 1498–1503.

38 Chen, D., Cao, L., Huang, F. et al. (2010). Synthesis of monodisperse mesoporous titania beads with controllable diameter, high surface areas, and variable pore diameters (14–23 nm). *Journal of the American Chemical Society* 132: 4438–4444.

39 Jiang, X., Herricks, T., and Xia, Y. (2003). Monodispersed spherical colloids of titania: synthesis, characterization, and crystallization. *Advanced Materials* 15: 1205–1209.

40 Pal, M., García Serrano, J., Santiago, P., and Pal, U. (2006). Size-controlled synthesis of spherical TiO_2 nanoparticles: morphology, crystallization, and phase transition. *The Journal of Physical Chemistry C* 111: 96–102.

41 Zhao, T., Qian, R., Tang, Y. et al. (2020). Controllable synthesis and crystallization of nanoporous TiO_2 deep-submicrospheres and nanospheres via an organic acid-mediated sol–gel process. *Langmuir* 36: 7447–7455.

42 Zhao, X., Zhao, J., He, J. et al. (2020). Porous anatase TiO_2 nanocrystal derived from the metal–organic framework as electron transport material for carbon-based perovskite solar cells. *ACS Applied Energy Materials* 3: 6180–6187.

43 Yu, S., You, J.S., Yang, I.S. et al. (2016). Tailoring of nanoporous TiO_2 spheres with 100–200 nm sizes for efficient dye-sensitized solar cells. *Journal of Power Sources* 325: 7–14.

44 Zhao, T., Qian, R., Zhou, G. et al. (2021). Mesoporous WO_3/TiO_2 spheres with tailored surface properties for concurrent solar photocatalysis and membrane filtration. *Chemosphere* 263: 128344.

45 Ma, D., Lai, L., de León, C.P. et al. (2022). Microwave-assisted self-template synthesis of mesoporous anatase TiO_2 spheres for non-aqueous Al-ion batteries: textural property optimization and enhanced reversible Al^{3+} storage. *Sustainable Materials and Technologies* 32: e00419.

46 Ma, D., Li, J., Yang, J. et al. (2021). Solid-state self-template synthesis of Ta-doped $Li_2ZnTi_3O_8$ spheres for efficient and durable lithium storage. *iScience* 24: 102991.

47 Ma, D., Li, K., and Pan, J.H. (2020). Ultraviolet-induced interfacial crystallization of uniform nanoporous biphasic TiO_2 spheres for durable lithium-ion battery. *ACS Applied Energy Materials* 3: 4186–4192.

48 Tsai, M.C., Lee, J.Y., Chen, P.C. et al. (2014). Effects of size and shell thickness of TiO_2 hierarchical hollow spheres on photocatalytic behavior: an experimental and theoretical study. *Applied Catalysis B: Environmental* 147: 499–507.

49 Pan, J.H., Bai, Y., and Wang, Q. (2015). Reconstruction of colloidal spheres by targeted etching: a generalized self-template route to porous amphoteric metal oxide hollow spheres. *Langmuir* 31: 4566–4572.

50 Pan, J.H., Zhang, X., Du, A.J. et al. (2008). Self-etching reconstruction of hierarchically mesoporous F-TiO_2 hollow microspherical photocatalyst for concurrent membrane water purifications. *Journal of the American Chemical Society* 130: 11256–11257.

51 Cao, L., Chen, D., and Caruso, R.A. (2013). Surface-metastable phase-initiated seeding and Ostwald ripening: a facile fluorine-free process towards spherical fluffy core/shell, yolk/shell, and hollow anatase nanostructures. *Angewandte Chemie - International Edition* 52: 10986–10991.

52 Yang, Y., Li, M., Chen, L. et al. (2024). MIL-125 and its derivatives based photoelectrodes for photoelectrochemical applications. *Progress in Solid State Chemistry* 75: 100476.

53 Dou, J., Li, Y., Xie, F. et al. (2016). Metal–organic framework derived hierarchical porous anatase TiO_2 as a photoanode for dye-sensitized solar cell. *Crystal Growth & Design* 16: 121–125.

54 Lv, T., Xiao, B., Zhou, S. et al. (2021). Rich oxygen vacancies, mesoporous TiO_2 derived from MIL-125 for highly efficient photocatalytic hydrogen evolution. *Chemical Communications* 57: 9704–9707.

55 Crespí Sánchez, N., Turnes Palomino, G., and Palomino Cabello, C. (2021). TiO_2 derived from NTU-9 metal-organic framework as highly efficient photocatalyst. *Materials Science and Engineering: B* 273: 115424.

56 Kampouri, S., Ireland, C.P., Valizadeh, B. et al. (2018). Mixed-phase MOF-derived titanium dioxide for photocatalytic hydrogen evolution: the impact of the templated morphology. *ACS Applied Energy Materials* 1: 6541–6548.

57 Zhuang, H., Wang, L., Shao, S. et al. (2024). Photocatalysis-promoted tumor ferroptosis enabled by MOF-derived black TiO_2. *Chemical Engineering Journal* 496: 154204.

58 Guo, R., Bao, Y., Kang, Q. et al. (2022). Solvent-controlled synthesis and photocatalytic activity of hollow TiO_2 microspheres prepared by the solvothermal method. *Colloids and Surfaces A: Physicochemical and Engineering Aspects* 633: 127931.

59 Wu, D., Zhu, F., Li, J. et al. (2012). Monodisperse TiO_2 hierarchical hollow spheres assembled by nanospindles for dye-sensitized solar cells. *Journal of Materials Chemistry* 22: 11665–11671.

60 Pan, J.H., Han, G., Zhou, R., and Zhao, X.S. (2011). Hierarchical N-doped TiO_2 hollow microspheres consisting of nanothorns with exposed anatase {101} facets. *Chemical Communications* 47: 6942–6944.

61 Ye, J., Liu, W., Cai, J. et al. (2011). Nanoporous anatase TiO_2 mesocrystals: additive-free synthesis, remarkable crystalline-phase stability, and improved lithium insertion behavior. *Journal of the American Chemical Society* 133: 933–940.

62 Liao, J.Y., Lei, B.X., Kuang, D.B., and Su, C.Y. (2011). Tri-functional hierarchical TiO_2 spheres consisting of anatase nanorods and nanoparticles for high efficiency dye-sensitized solar cells. *Energy & Environmental Science* 4: 4079.

63 Liao, J.Y., He, J.W., Xu, H. et al. (2012). Effect of TiO_2 morphology on photovoltaic performance of dye-sensitized solar cells: nanoparticles, nanofibers, hierarchical spheres and ellipsoid spheres. *Journal of Materials Chemistry* 22: 7910.

64 Bai, X., Xie, B., Pan, N. et al. (2008). Novel three-dimensional dandelion-like TiO_2 structure with high photocatalytic activity. *Journal of Solid State Chemistry* 181: 450–456.

65 Park, K.S., Min, K.M., Jin, Y.H. et al. (2012). Enhancement of cyclability of urchin-like rutile TiO_2 submicron spheres by nanopainting with carbon. *Journal of Materials Chemistry* 22: 15981.

66 Fang, W.Q., Zhou, J.Z., Liu, J. et al. (2011). Hierarchical structures of single-crystalline anatase TiO_2 nanosheets dominated by {001} facets. *Chemistry - A European Journal*: 1423–1427.

67 Liu, S., Yu, J., and Jaroniec, M. (2010). Tunable photocatalytic selectivity of hollow TiO_2 microspheres composed of anatase polyhedra with exposed {001} facets. *Journal of the American Chemical Society* 132: 11914–11916.

68 Chen, J.S., Tan, Y.L., Li, C.M. et al. (2010). Constructing hierarchical spheres from large ultrathin anatase TiO_2 nanosheets with nearly 100% exposed (001) facets for fast reversible lithium storage. *Journal of the American Chemical Society* 132: 6124–6130.

69 Chen, Q., Ren, B., Zhao, Y. et al. (2014). Template-free synthesis of core–shell TiO_2 microspheres covered with high-energy {116}-facet-exposed N-doped nanosheets and enhanced photocatalytic activity under visible light. *Chemistry - A European Journal* 20: 17039–17046.

5

Immobilization of TiO_2 on Various Substrates

Nishat Khan[1], Seema Garg[1,], Isha Arora[1], Gurumurthy Hegde[2], Suresh Sagadeven[3], Lalita Ledwani[4], Amrish Chandra[5,*]*

[1] *Department of Chemistry, Amity Institute of Applied Sciences, Amity University, Noida, Uttar Pradesh, India*
[2] *Department of Chemistry and Centre for Advanced Research and Development (CARD), CHRIST (Deemed to be University), Bengaluru, Karnataka, India*
[3] *Nanotechnology & Catalysis Research Centre, University of Malaya, Kuala Lumpur, Malaysia*
[4] *Manipal University Jaipur, Dehmi Kalan, Jaipur-Ajmer Expressway, Jaipur, Rajasthan, India*
[5] *Lexgin IP LLP, T21/1602, Paras Tierea, Sector 137, Noida, Uttar Pradesh, India*

*Corresponding authors: sgarg2@amity.edu; amrish.chandra@sharda.ac.in

5.1 Introduction

Photocatalysis has made significant progress as a water treatment method, becoming affordable for use in developing countries due to the availability of semiconductors with high degradation efficiency and low production costs. One of the major challenges in the industrial application of photocatalytic water treatment is the immobilization of semiconductors, which is essential to reduce separation and filtration costs.

Ultraviolet (UV)-active titanium dioxide (TiO_2) has been the most widely used photocatalyst in semi-industrial water purification, both in immobilized form and in suspension. While TiO_2 offers several advantages, addressing immobilization challenges is critical for its effective use. Moreover, when immobilizing visible-light-active photocatalysts, it is essential to minimize defects during the process to maintain their efficiency. Various photocatalysts have been developed for photocatalytic applications, but their practical use is often limited by challenges such as ineffective separation and recovery from aqueous media during wastewater treatment. Immobilizing photocatalysts on a suitable substrate is a promising solution to overcome these limitations. Immobilized photocatalysts offer two key advantages: (i) they enable rapid separation and easy recovery for reuse and (ii) they prevent aggregation of the photocatalyst, thereby enhancing photocatalytic efficiency. Photocatalysts have been immobilized on a range of substrates, including activated carbon fiber, zeolite, stainless steel, silica, and glass. Among these, alumina (Al_2O_3)-based ceramic fiber stands out due to its unique properties, making it an ideal

Nanostructured TiO_2: Synthesis, Photocatalytic Properties and Applications, First Edition.
Edited by Jia Hong Pan, Wan In Lee and Detlef W. Bahnemann

support for photocatalyst immobilization. These fibers offer excellent shape and size adaptability, flexibility, high thermal conductivity, and superior tensile strength. In addition to their mechanical robustness, alumina-based ceramic fibers exhibit resistance to acid and alkali attacks at elevated temperatures and provide a large surface area for photocatalyst attachment. Their high thermal conductivity (32 W/m · K) facilitates the absorption and transfer of incident light to the immobilized photocatalyst, enhancing charge carrier activation and photogeneration. Moreover, the elasticity and wear resistance of ceramic fibers allow for custom shaping, making them highly versatile for practical applications.

TiO_2 is a well-established photocatalyst widely utilized in wastewater and air treatment, as well as solar energy conversion, due to its cost effectiveness, non-toxic nature, high stability, and excellent catalytic efficiency under light irradiation. However, the removal and recycling of powdered TiO_2 suspended in treated water significantly increase operational costs and can cause secondary contamination. To address this issue, extensive efforts have been made to immobilize TiO_2 nanoparticles onto various substrates, such as glass, stainless steel plates, and polymers, using methods like sol–gel synthesis, thermal treatment, chemical vapor deposition (CVD), and electrophoretic deposition. The adhesion between TiO_2 nanoparticles and substrates typically relies on physical adsorption, electrostatic interactions, hydrogen bonding, or chemical bonding.

Literature highlights the successful immobilization of TiO_2 on Al_2O_3-based ceramic paper using TiO_2 as a binder [1]. Similarly, researchers [2, 3] have demonstrated the immobilization of TiO_2 on carbon fiber. While ceramic fibers and other substrates have been investigated as effective supports for photocatalysts, much of the existing research focuses on TiO_2 immobilization, with ceramic fiber being highlighted as a promising substrate for various photocatalysts [1]. In most cases, solutions containing the photocatalyst are used, and the semiconductor particles must be removed from the treated solution. However, in one study, the catalyst immobilization step was eliminated by using poly(vinyl alcohol) (PVA), which positively influenced the polymer's crosslinking behavior [4].

Heterogeneous photocatalysis, classified as an advanced oxidation process, meets the requirements for effective pollutant degradation [5]. For TiO_2-coated surfaces to be viable for practical applications, they must meet several critical criteria: strong adhesion between the catalyst and the substrate to ensure stability, retention of photocatalytic activity despite the layered coating, an expanded surface area to enhance contaminant adsorption and degradation efficiency, and the ability of the substrate to withstand photocatalytic reactions and prolonged radiation exposure. Key characteristics of immobilized catalyst films include thickness, porosity, quality, and specific surface area. Although immobilized catalysts are generally more cost effective than suspended catalysts, they suffer from a reduced surface area and limitations in mass transfer [6]. Catalysts play a critical role in photocatalytic processes for the removal of organic pollutants from water. Among them, TiO_2 has gained significant attention and undergone extensive research due to its diverse and attractive properties. TiO_2 is characterized by its chemical and biological inertness, ease of fabrication, environmental and human safety, and stability in photocatalytic processes.

TiO_2 photocatalysts find applications in a variety of fields, including self-cleaning surfaces, water purification, air purification, anticorrosion coatings, and photocatalytic lithography. The primary objective of this study is to investigate the use of TiO_2 for the photodegradation of organic pollutants. Literature extensively documents large-scale systems where TiO_2 in powdered form is dispersed in aqueous solutions and subjected to UV radiation [6]. While this reactor design offers a substantial surface area, it poses challenges during the extraction of the catalyst from the final product.

Immobilizing the catalyst can eliminate the need for a costly and labor-intensive separation process, thereby reducing operational expenses. Extensive research has been conducted on various immobilization methods using a wide range of substrates, including glass, silica gel, metals, ceramics, polymers, thin films, fibers, zeolites, alumina clays, activated carbon, cellulose, and reactor walls. Among these methods, the sol–gel technique has garnered significant interest as a deposition method.

While the application of TiO_2 in macroscale systems is well documented, its use in microreactors is less explored. However, TiO_2 in microreactors offers several advantages, such as a higher surface-to-volume ratio and reduced diffusion rates. Therefore, this study focuses on investigating the performance and potential of TiO_2 in microreactor systems [7].

The primary objective of this study was to evaluate various methods of TiO_2 immobilization, with a particular focus on identifying techniques suitable for microreactor fabrication. Microreactors were constructed using different binding approaches to analyze the photocatalytic activity of the immobilized films. The degradation of methylene blue (MB), a model organic pollutant, was used as a benchmark for photocatalytic performance.

Traditionally, reactor designs employ either a TiO_2 slurry or an immobilized film. Slurry reactors, such as those utilizing commercially available Aeroxide P25, offer a substantial surface area due to their high surface-to-volume ratio. However, they present significant challenges, including contamination of the treated solution, clogging of filter layers, and difficulty in separating fine particles, leading to increased operational costs. Immobilizing TiO_2 on solid supports in the form of particles or thin films eliminates the need for filtration and reduces economic expenses [8].

5.2 Immobilization Techniques

Various immobilization techniques have been developed for depositing colloidal TiO_2 onto silicon surfaces.

5.2.1 Liquid-phase Deposition

Pipetting and Spin Coating (Turn Coating): The effects of sintering temperatures, steric stabilization, suspension pH, spin-coating parameters, and P25 concentration on the surface morphology, film quality, and structural stability were evaluated using tape testing and scanning electron microscopy (SEM). The immobilized films were subjected to sintering at temperatures ranging from 400 to 600 °C to enhance adhesion to the substrate.

Substrates for Immobilization: Common substrates used for TiO_2 immobilization include glass, silica gel, activated carbon, ceramics, polymers, fibers, alumina clays, and metals. These substrates must possess specific properties such as: strong adhesion to TiO_2 particles, resistance to high temperatures, large surface area to enhance pollutant retention, glass and silica are attractive due to their transparency, while activated carbon, with its high porosity, significantly enhances catalytic activity [9].

Sol–Gel Method: The sol–gel method is a widely used technique due to its affordability and simplicity. It involves transforming a colloidal solution of TiO_2 into a viscous gel and subsequently into a rigid solid. Oxygen bridges form between the substrate and TiO_2 particles during this process, ensuring strong adhesion. The homogeneous sol solution can be used as precursor solution for the subsequent coating process, that is, dip coating, spin coating, or spread coating. To make thicker films, the sol–gel process can utilize commercially available TiO_2 nanopowders such as Aeroxide P25, instead of precursors like $TiCl_4$ [10].

5.2.2 Gas-phase Deposition

Alternatively, facile and scalable gas-phase deposition methods have been developed. CVD involves exposing the substrate to a volatile precursor in an inert atmosphere, where the precursor decomposes to form a TiO_2 layer. This method allows for uniform coatings on various surfaces. Sputtering, heat treatment, and electrophoretic deposition are also employed for creating stable films [11]. The stability of colloidal TiO_2 particles is influenced by the balance of attractive (van der Waals) and repulsive (electrostatic) forces. The surface charge of TiO_2 particles depends on the pH, with neutral charge occurring at the point of zero charge. At lower pH levels, incomplete protonation leads to aggregation due to van der Waals forces. Steric stabilization can prevent aggregation by introducing polymers such as PVA. The lyophobic segments of the polymer adhere to TiO_2 particles, while the lyophilic segments extend into the dispersion medium, creating a steric barrier [12]. During sintering, PVA decomposes at around 190–200 °C, creating porosity in the film. Higher PVA content and larger particle sizes increase pore size. Smaller TiO_2 particles offer a higher surface area and better adhesion, resulting in mechanically stable films [13]. Superior surface coverage and stabilized films exhibit more uniform surface coverage compared to un-stabilized films.

Smaller TiO_2 particles improve adhesion, increasing film durability. Improved Photocatalytic Efficiency: Increased contact area between smaller particles and the substrate enhances pollutant degradation. Reduced Surface Cracking: Steric stabilization prevents grain enlargement during sintering, minimizing surface defects [14].

This study demonstrates that immobilizing TiO_2 using techniques such as spin coating, pipetting, and the sol–gel method provides effective solutions for photocatalytic applications in microreactors. By optimizing parameters such as sintering temperature, suspension pH, and steric stabilization, durable and efficient TiO_2

films can be fabricated, offering significant advantages over traditional slurry-based systems.

5.3 Techniques for Immobilizing TiO_2 on to Various Substrates

Photocatalytic reactions are widely recognized as an efficient method for degrading environmental contaminants. TiO_2 is a particularly effective photocatalyst due to its non-toxic nature and environmental safety. Despite its potential, the widespread industrial application of TiO_2 in wastewater treatment facilities remains limited. One of the primary barriers is the high overall cost associated with the technique.

While biodegradation processes continue to dominate wastewater treatment, they have significant drawbacks, including their inability to effectively degrade certain xenobiotics. Moreover, these processes can introduce hazardous genetic material into the environment, posing potential risks to human health. In contrast, the use of immobilized photocatalysts offers a promising alternative by reducing operational costs and simplifying the photocatalytic process.

Immobilizing TiO_2 on various substrates can lead to durable coatings that enhance photocatalytic performance. Glass, in particular, is an ideal substrate due to its favorable optical properties. While it is generally accepted that immobilized catalyst systems are less efficient than slurry-based catalytic suspensions, several studies demonstrate that TiO_2 immobilized on suitable substrates can outperform bare TiO_2 in the photocatalytic degradation of pharmaceuticals and dyes.

Recent research has focused on enhancing the photocatalytic activity of TiO_2 by modifying it with noble metals or non-metallic atoms. These modified catalysts exhibit significant activity under visible light irradiation, broadening the scope of TiO_2 applications beyond UV light.

The objective of this study is to develop a highly efficient catalytic system by immobilizing a commercially available TiO_2 photocatalyst on a commercially accessible substrate [1, 15–17]. The goal was to achieve a system with both high photocatalytic activity and long-term stability, while also prioritizing simplicity in the preparation process.

Coating TiO_2 onto different substrates is a widely employed strategy to enhance its durability, reusability, and catalytic efficiency. The choice of substrate depends on the specific application and desired properties. TiO_2-coated substrates have found applications in diverse fields, including photocatalysis, sensors, and heterogeneous catalysis.

5.3.1 Immobilization of TiO_2 on Glass or Quartz Surfaces Using the Sol–Gel Method

The sol–gel method is a versatile technique for immobilizing TiO_2 on glass or quartz surfaces. This process involves hydrolysis and condensation reactions of titanium alkoxides, resulting in the formation of a gel that adheres firmly to the substrate [18] (Figure 5.1).

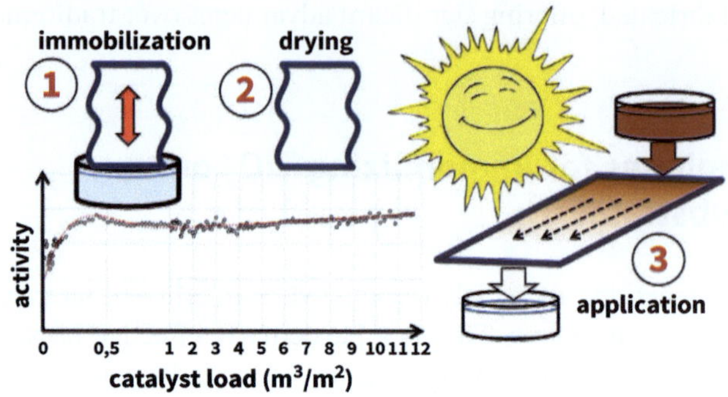

Figure 5.1 Illustration of immobilization of TiO_2 on glass or quartz surfaces. Adapted from Ref. [19].

5.3.1.1 Procedure

1. **Preparation of the Titanium Alkoxide Precursor:** It begins with a titanium alkoxide precursor, commonly titanium tetra-isopropoxide (TTIP).
2. **Hydrolysis and Gel Formation:** Induce hydrolysis by adding water to the precursor while carefully regulating the pH to facilitate the formation of a stable titanium oxide gel.
3. **Application of the Gel:** Evenly apply the gel onto the glass or quartz substrate to ensure consistent coverage.
4. **Heat Treatment:** Heat the coated substrate to remove excess solvent and organic residues, promoting the formation of a robust TiO_2 layer.

5.3.1.2 Thin Film Production via Sol–Gel

The sol–gel method is particularly effective for producing thin coatings of TiO_2. This process begins with:

1. **Sol Preparation:** Hydrolysis and condensation of titanium alkoxides (e.g., TTIP) in the presence of alcohol and catalytic agents to form a sol.
2. **Film Deposition:** The deposition is typically performed using the dip-coating technique, followed by thermal treatment to form a nanocrystalline TiO_2 thin film [18].

5.3.1.3 Advantages of the Sol–Gel Method

1. **Precise Control:** It allows meticulous regulation of film thickness.
2. **Excellent Adherence:** It provides strong adhesion between the TiO_2 film and the substrate.
3. **Scalability:** it is suitable for mass production of TiO_2 coatings.

5.3.2 Dip-coating Process for TiO_2 Immobilization

The dip-coating technique involves submerging a substrate into a TiO_2 sol, followed by a controlled withdrawal, resulting in a uniform coating. The sol consists of TiO_2 nanoparticles dispersed in a liquid medium as shown in Figure 5.2 [20]. Procedure: Disperse TiO_2 nanoparticles in a liquid to form the sol. Immerse the glass or quartz

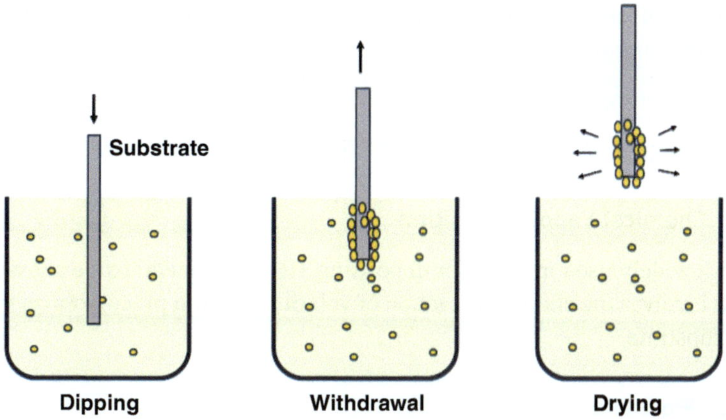

Figure 5.2 Immobilization of TiO_2 by dip coating method. Adapted from Ref. [20].

substrate completely in the sol, ensuring full coverage. Retrieve the substrate at a regulated pace, allowing excess solvent to evaporate, leaving a thin and uniform TiO_2 coating on the surface. Apply heat to the coated substrate to enhance adhesion and stabilize the TiO_2 film.

5.3.2.1 Advantages of the Dip-coating Method

1. **Efficiency and Cost-effectiveness:** This process is both efficient and economical for large-scale applications.
2. **Applicability for Large Surface Areas:** It is particularly well suited for applying coatings over wide areas, making it ideal for industrial applications.
3. **Thickness Control:** The thickness of the TiO_2 coating can be adjusted by regulating the withdrawal speed, providing flexibility in film deposition.

5.4 Electrospun/Organically Modified Montmorillonite Nanofibrous Membrane

In this study, we propose using an **electrospun/organically modified montmorillonite (O-MMT) nanofibrous membrane** as a substrate for TiO_2 immobilization. The membrane is first coated with TiO_2 sol–gel before immobilizing laccase (Lac). The resulting composite nanofibrous material integrates several functionalities:

1. **Adsorption Capabilities:** The O-MMT and electrospun nanofibrous membrane provide superior adsorption properties.
2. **Enzymatic Degradability:** Lac enhances the enzymatic degradation of contaminants.
3. **Photocatalytic Degradability:** TiO_2 contributes to photocatalytic degradation, allowing the composite material to target a wide range of pollutants.

This composite system combines the adsorption capacity of O-MMT, the electrospun nanofibrous membrane's structural properties, and the degradation capabilities of both Lac and TiO_2, offering an innovative approach for environmental remediation.

5.5 Metal Oxide Substrates (Alumina, Silica, etc.)

5.5.1 Chemical Vapor Deposition

CVD is a widely used method for depositing TiO_2 onto metal oxide substrates. This process involves the chemical reaction of volatile titanium precursors on the surface of the substrate.

5.5.1.1 Procedure

Introduce volatile titanium precursors and oxygen-containing reactants into a controlled environment. Facilitate chemical reactions on the surface of the metal oxide substrate, leading to the deposition of TiO_2. Regulate temperature and other parameters to achieve the desired film thickness and quality (see Figure 5.4).

5.5.2 Atomic Layer Deposition

Atomic layer deposition (ALD) provides precise control over the thickness of TiO_2 films by exposing the substrate to alternating precursor gases in a cyclical manner. This method enables the deposition of TiO_2 layer by layer, ensuring high uniformity and conformity to the substrate's surface.

5.5.2.1 Procedure

Alternate the exposure of the metal oxide substrate to titanium precursors and a reactive oxygen source. Self-limiting surface reactions allow for controlled deposition of TiO_2 in a layer-by-layer manner. Regulate temperature and other parameters to obtain the desired film thickness and quality. ALD is known for its high precision, making it ideal for applications requiring thin, conformal coatings on complex substrates.

5.6 Microplasma Printer for TiO_2 Deposition

The microplasma printer, shown in two-dimensional schematic views (Figure 5.1) [19], is another method used for depositing TiO_2 in CVD and ALD modes. In CVD mode, the printer supplies TTIP and N_2/O_2 plasma concurrently onto substrates such as glass and crystalline silicon (c-Si). In ALD mode, the printer alternates between exposing the substrate to the TTIP precursor and N_2/O_2 plasma.

The microplasma system is equipped with needle electrodes, and the plasma gas input and exhaust lines are positioned at the anterior and posterior ends of the device, respectively. The exhaust lines are connected to a rotary pump, and a needle valve regulates the pumping speed, preventing disruption to the gas flow. The diagrammatic illustration for the difference of ALD and CVD is shown in Figure 5.3 [21].

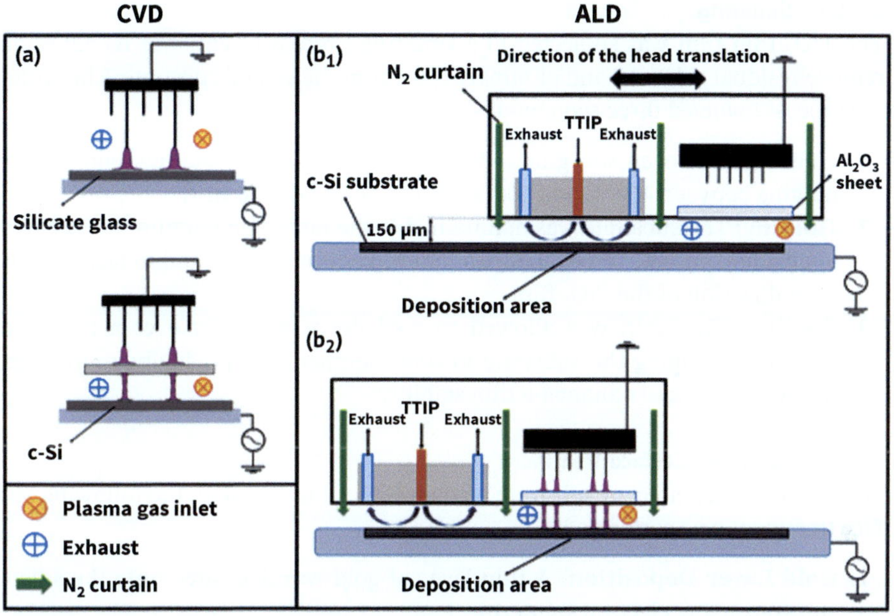

Figure 5.3 2-D Microplasma printer overview [21]/John Wiley & Sons/CC BY 3.0.

Additional purging steps are performed both before and during plasma ignition when operating in ALD mode. A supporting animation of the ALD process using the microplasma device is available.

5.7 Polymeric Substrates

TiO_2 nanoparticles may be applied to polymer substrates using in situ polymerization. This approach is especially beneficial for applications that need adaptability. Suspension preparation: 2-propanol and water were mixed with TiO_2 powder (85% Anatase and 15% Rutile) and PVA. After magnetic stirring for three hours, acetic acid was added to the solution to adjust its pH to a range of 2.8–6.6. An pH meter was used to measure the pH. After the pH was adjusted, the suspension was allowed to magnetically stir overnight at 50 °C. After that, it was exposed to sonication for 10–60 min [22].

5.7.1 Substrate Coating

After rinsing with water and ethanol and washing with nitric acid for 1 h (Merck), the silicon wafer de4-inch wafer, p-type (100) was manually cut into square chips. Two distinct methods were employed to apply a coating on these chips: pipetting or spin coating. The solution was uniformly applied to completely cover the whole substrate in both pipetting and spin coating processes. The substrate was allowed to dry after pipetting, until a significant portion of the solvent had evaporated. The chip was subjected to spin coating for 50 s at a speed ranging from 500 to 2000 rpm, resulting in the deposition of a thin layer on the substrate.

5.7.1.1 Sintering

The TiO_2 thin film was subjected to a sintering process to enhance its adhesion, remove residual solvents, and eliminate PVA from the coated substrate. The sintering process involved three sequential steps:

1. **Heating:** The oven was gradually heated to the target sintering temperature, ranging between 400 °C and 600 °C, over a period of **230 min**.
2. **Holding:** The substrate was maintained at the target temperature for **120 min** to ensure complete removal of residual solvents and PVA, and to facilitate the crystallization of the TiO_2 film.
3. **Cooling:** The oven was allowed to cool down naturally over a period of **230 min**, bringing the substrate to room temperature gradually, minimizing thermal stress and enhancing film stability.

5.7.1.2 Substrate Surface Treatment

To investigate the effect of substrate type on TiO_2 film adhesion, the following surface treatment was performed:

1. **Gold Layer Deposition:** A thin layer of gold was deposited onto the silicon substrate using electron sputtering.
2. **TiO_2 Coating:** The gold-coated silicon substrate was then coated with the TiO_2 suspension, followed by sintering under the conditions to evaluate the adhesion and structural integrity of the TiO_2 film [18].

5.8 Carbon-Based Substrates (Activated Carbon, Carbon Nanotubes, etc.)

5.8.1 Chemical Vapor Deposition

CVD can be utilized to coat carbon-based materials such as activated carbon or carbon nanotubes (CNTs) with a uniform TiO_2 layer (see Figure 5.4) [23]. This method involves the chemical reaction of volatile titanium precursors with oxygen-containing reactants on the surface of the carbon substrate, resulting in the deposition of TiO_2 (Figure 5.2) [1].

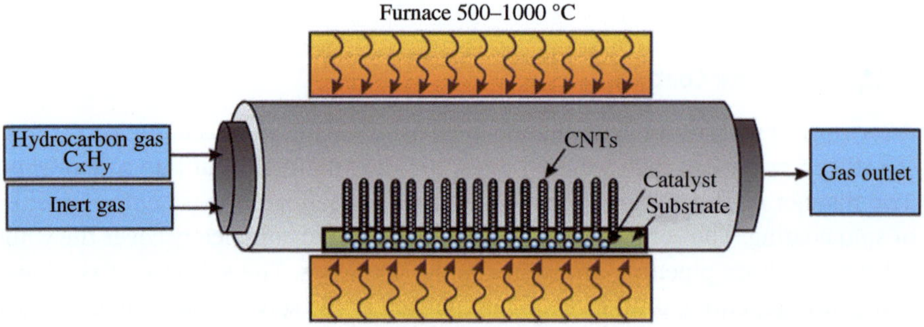

Figure 5.4 CVD method for coating carbon nanotubes to catalyst substrate [23]/Springer Nature/CC BY 4.0.

5.8.2 Hydrothermal Synthesis

TiO_2 can also be deposited on carbon-based substrates via hydrothermal synthesis, which involves immersing the substrate in a hydrothermal solution and exposing it to high temperature and pressure.

Procedure: Place the carbon-based substrate into a hydrothermal reactor containing a TiO_2 precursor solution. Maintain elevated temperature and pressure for a specified duration to promote crystallization. The TiO_2 layer forms on the substrate via hydrothermal crystallization.

5.9 Ceramic Substrates

5.9.1 Spray Pyrolysis

This technique involves spraying a precursor solution onto a ceramic substrate and subjecting it to pyrolysis, leading to the formation of a stable TiO_2 coating.

Procedure: Dissolve or disperse TiO_2 nanoparticles in a compatible solvent to create a precursor solution. Spray the solution evenly over the ceramic substrate. Allow the solvent to evaporate, leaving behind a thin TiO_2 film. Subject the substrate to pyrolysis at moderate temperatures (100–500 °C), promoting film adhesion and stability.

Advantages: Produces films with large surface area and excellent adhesion. Compatible with low-cost substrates and chemicals. Suitable for applications in the glass industry, solar cells, and other areas. Enables precise control over film characteristics by adjusting parameters such as substrate temperature, airflow pressure, and precursor solution molarity [24].

5.10 Textile Substrates

5.10.1 Pad-Dry-cure Method

This method is used to immobilize TiO_2 nanoparticles on textile substrates.

Procedure: Prepare a TiO_2 suspension or sol. Submerge the textile substrate in the TiO_2 solution, ensuring uniform coverage. Remove excess solution by applying pressure. Allow the coated textile to dry completely. Cure the dried textile in an oven or under UV radiation to permanently attach the TiO_2 nanoparticles.

5.10.2 Layer-by-layer Deposition

This method involves depositing alternating layers of TiO_2 and polymers with opposing charges onto textile substrates.

Procedure: Sequentially deposit layers of TiO_2 and polymers with opposite charges onto the substrate. Repeat the process to achieve a multilayered coating with precise control over thickness and composition.

5.10.3 Considerations for TiO_2 Immobilization

When selecting an immobilization method for TiO_2, it is essential to consider: The properties of the substrate, such as flexibility, thermal stability, and chemical compatibility. The intended application, including environmental conditions and required performance. TiO_2 particle size, shape, and surface area, as these factors significantly influence photocatalytic efficiency and overall performance [25].

5.10.4 Spin-Coating Procedure

Spin Coating is a widely used technique for depositing thin, uniform films of TiO_2 on a variety of substrates.

Procedure: Place a small amount of TiO_2 sol onto the center of the substrate using a precise application method. Rotate the substrate at high speeds, causing the sol to spread evenly over the surface due to centrifugal force. Allow the coating to dry briefly to remove excess solvent. Apply heat to the coated substrate to eliminate any remaining solvent and promote the formation of a thin, uniform TiO_2 layer.

Advantages: Provides precise control over the film thickness by adjusting spin speed and sol viscosity. Enables uniform coating on complex and small surfaces. Ensures high consistency and reproducibility in the deposited film.

5.10.5 Plasma-assisted Deposition

It is a technique that integrates plasma treatment with deposition methods like sol–gel or dip-coating to enhance film properties.

Procedure: Introduce a plasma environment during the deposition process to activate the substrate surface. Deposit the TiO_2 layer using sol–gel, dip-coating, or spin-coating techniques while maintaining the plasma environment, as a diagrammatic illustration of spin coating (dip coating) is shown in Figure 5.5. Optimize process parameters such as gas composition, pressure, and plasma power to achieve desired surface modifications. Apply heat treatment if necessary to enhance the film's structural and chemical properties.

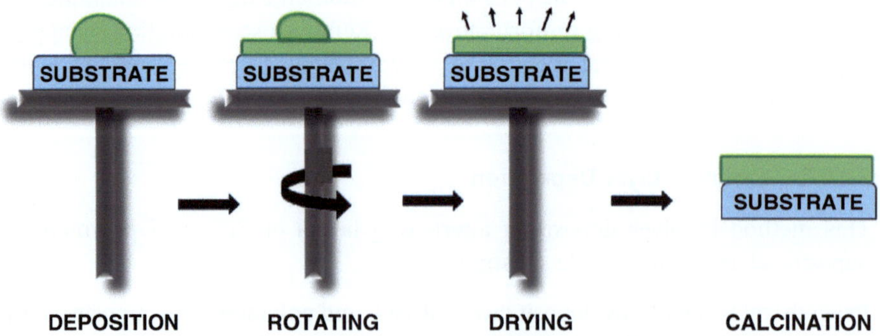

Figure 5.5 Immobilization by spin-coating. Adapted from Ref. [27].

Advantages: Improves film adhesion by modifying the substrate surface. Enhances film density, stability, and mechanical properties. Alters surface characteristics, such as wettability and surface energy, to improve performance.

5.11 Critical Parameters for TiO_2 Immobilization

When immobilizing TiO_2 on glass, quartz, or other substrates, careful tuning of the following parameters is essential to achieve optimal film quality:

1. **Precursor Concentration:** Controls the viscosity of the sol and influences the film thickness and uniformity.
2. **pH of the Sol:** Affects the hydrolysis and condensation reactions during sol-gel synthesis, influencing film morphology.
3. **Drying Conditions:** Determines the rate of solvent evaporation, which can affect the uniformity and adhesion of the film.
4. **Heat Treatment:** Facilitates the removal of residual solvents and organic components, enhancing the crystallinity, adhesion, and stability of the TiO_2 layer.

Additionally, **substrate surface preparation** is critical to ensure proper adhesion and film uniformity. This typically involves: Cleaning the substrate with water, ethanol, or nitric acid to remove contaminants. Drying the substrate under controlled conditions to avoid surface impurities.

By carefully controlling these parameters, it is possible to produce high-performance TiO_2 coatings with tailored properties for specific applications [24–26].

5.12 TiO_2-immobilized Biodegradable Polymers for Enhanced Photocatalytic Activity

Recent advancements in photocatalysis have explored the integration of TiO_2 with biodegradable polymers to create environmentally friendly and efficient photocatalysts. These biodegradable polymers offer several advantages over traditional non-biodegradable substrates, primarily due to their sustainability and potential for safe degradation in the environment. Both synthetic and natural biodegradable polymers have been investigated, but research has shown that combining the two can produce more stable and durable photocatalytic structures. Despite this, only a few studies have successfully integrated natural and synthetic biodegradable polymers for photocatalytic applications [28].

5.12.1 Importance of Immobilization Techniques

The immobilization method plays a crucial role in determining the photocatalytic efficiency of TiO_2. The method must be carefully selected based on: The type of substrate (biodegradable polymer). The specific pollutant targeted for degradation. The potential effect on the photocatalytic activity of TiO_2.

5.12.2 Common Immobilization Techniques for TiO_2 Include

1. **Sol–Gel Process:** Hydrolysis of titanium alkoxides at room temperature to form nano-sized TiO_2 particles.
   Advantages: Simple, cost-effective, and scalable.
   Limitations: Requires high-temperature calcination (~300 °C) for effective crystallization, which limits its use with low-thermal-stability polymers.
   Future Scope: Research is needed to develop thermally stable polymers compatible with sol–gel processing.
2. **3D Printing:** Allows for the creation of custom structures with precise geometry for TiO_2 immobilization.
3. **Hydrothermal Treatment:** Enhances TiO_2 crystallization and adhesion to polymeric surfaces.
4. **Plasma Treatment:** Modifies the surface properties of the polymer, improving TiO_2 adhesion and stability.
5. **Electrospinning and Electrospraying:** Produce nanofibrous polymer–TiO_2 composites with high surface areas.

5.12.3 Photocatalytic Performance Evaluation

The performance of TiO_2-immobilized biodegradable polymers is influenced by various factors:

1. **Light Source:** The selection of a light source (UV, visible light, or sunlight) is critical as it determines the generation of hydroxyl radicals (OH•) responsible for pollutant degradation.
2. **Pollutant Concentration:** Higher pollutant concentrations reduce the availability of hydroxyl radicals, limiting photocatalytic efficiency.
3. **pH of the Solution:** The degradation efficiency varies based on the pH of the environment:
   Acidic Conditions: Suitable for certain pollutants.
   Neutral pH: Often preferred for general wastewater treatment.
   Alkaline Conditions: Beneficial for specific organic compounds.
4. **Temperature:**
   Moderate Temperature: Enhances photocatalytic reactions by increasing molecular collisions and TiO_2 surface interactions.
   High Temperature: Can degrade the polymer substrate, reducing photocatalytic efficiency due to changes in crystallinity and surface area.

5.12.4 Impact of Immobilization on Photocatalytic Efficiency

The immobilization of TiO_2 within polymer matrices often results in a reduction in photocatalytic activity compared to TiO_2 suspended in aqueous solutions. This is primarily due to: a decrease in the available surface area for light interaction and reduced mobility of TiO_2 particles, limiting their interaction with pollutants.

However, the integration of TiO_2 with biodegradable polymers offers benefits such as:

1. **Reusability:** Immobilized TiO_2 can be recovered and reused more easily than suspended TiO_2.
2. **Environmental Safety:** Biodegradable polymers reduce the risk of environmental contamination by TiO_2 nanoparticles.

Examples of TiO_2–Polymer Composites for Photodegradation:

Our literature survey highlights several innovative examples of TiO_2-immobilized biodegradable polymer composites that have shown promise for the degradation of various pollutants:

1. **TiO_2–Chitosan Composites:** Chitosan, a natural polysaccharide, has been used as a support for TiO_2 due to its biocompatibility and adsorption properties. Photodegradation of dyes and pharmaceutical compounds has been reported with high efficiency.
2. **TiO_2–Poly(Lactic Acid) (PLA) Composites:** PLA is a synthetic biodegradable polymer with good thermal stability. TiO_2–PLA composites have demonstrated effective degradation of organic pollutants under UV and visible light.
3. **TiO_2–Cellulose Composites:** Cellulose-based materials offer high surface area and excellent mechanical properties. TiO_2–cellulose composites have been used for the photodegradation of pesticides and heavy metal contaminants.

5.12.5 Future Research Directions

Development of New Biodegradable Polymers: Exploring polymers with higher thermal stability to expand the range of immobilization techniques, such as sol–gel processing.

Optimization of Immobilization Techniques: Investigating methods that preserve or enhance the photocatalytic activity of TiO_2 while improving its stability and reusability.

Hybrid Composites: Combining natural and synthetic biodegradable polymers to create robust, multifunctional photocatalytic materials.

Advanced Characterization: Utilizing advanced analytical techniques to study the structural, thermal, and photocatalytic properties of TiO_2–polymer composites.

The development of TiO_2-immobilized biodegradable polymers represents a promising avenue for sustainable photocatalytic applications, with potential for significant environmental benefits through efficient pollutant degradation [28, 29].

5.13 Polymers (Natural and Biodegradable)

5.13.1 Chitosan, Often Known as CS

Synthesis and Characterization Methods of Chitosan-Based TiO_2 Photocatalysts: **Chitosan (CS)**, a linear polysaccharide derived from the deacetylation of chitin found in the exoskeletons of crustaceans and arthropods, is widely recognized for its nontoxic, biodegradable, and biocompatible properties. Due to its abundant availability, low cost, and customizable features, CS has been extensively explored for environmental remediation applications, especially in combination with TiO_2 as a composite photocatalyst. CS possesses amino ($-NH_2$) and hydroxyl (—OH) functional groups, which act as coordination sites for binding with metals, metal oxides, and other nanomaterials. These groups not only facilitate the immobilization of TiO_2 nanoparticles but also enhance pollutant removal through selective adsorption and complex formation.

5.13.2 Chitosan–TiO_2 (CS/TiO_2) Composite Synthesis Techniques

1. **Cross-linking with Alkaline Agents**
   Process: CS films are treated with TiO_2 nanoparticles, followed by cross-linking using alkaline agents like NaOH to enhance TiO_2 adhesion.
   Advantages: Enhances chemical stability and improves photocatalytic performance by creating a strong polymeric network.
2. **Incorporation of Organic/Inorganic Nanofillers**
   Nanofillers: Carbon dots (CDs), graphene oxide (GO), CNTs, and other metal oxides.
   Function: Act as co-catalysts and support materials, enhancing TiO_2dispersion, charge separation, and light absorption.
   Example: CS integrated with montmorillonite (Mt) clay forms hydrogen bonds between CS's hydroxyl groups and Mt's Si—O—Si units, creating an effective adsorbent layer for TiO_2.
3. **Microwave Solvothermal Method**
   Process: Combines TiO_2 with ZrO_2 using carboxymethylated CS as the support.
   Advantages: Produces composites with enhanced hydrophilicity, functional group density, and specific surface area, essential for interactions with dyes and pollutants.
4. **3D Printing of CS Scaffolds**
   Method: CS is 3D-printed into scaffolds, followed by TiO_2 nanoparticle embedding. The scaffolds are treated with ammonia solution to neutralize and stabilize the network.
   Outcome: Provides precise geometric control, mechanical strength, and flexibility for multiple photocatalytic cycles.
5. **Electrospinning for Nanostructured Fibers**
   Parameters: Applied voltage: 20 kV, tip-to-collector distance: 10 cm, flow rate: 0.25 mL h^{-1}
   Outcome: The inclusion of nanomaterials such as SWCNTs (single-walled CNTs) and Fe_3O_4 increases fiber diameter and enhances thermal stability, as demonstrated by thermogravimetric analysis.

5.13.3 Photocatalytic Applications of CS–TiO_2 Composites

1. **Pollutant Degradation Efficiency:** Tetracycline hydrochloride, 2,4-dichlorophenol, Reactive Blue 4, and Reactive Red 15 pollutants were degraded. Most pollutants are degraded within 2–3 h. For compounds like tetracycline hydrochloride, degradation may take up to 240 min.

2. **Agitation Speed Effect:** Increased agitation reduces the boundary layer thickness, improving mass transfer and enhancing photocatalytic activity.
3. **Photocatalytic Efficiency Under Visible Light: By Incorporation of CDs: It** enhances visible light absorption, leading to efficient electron–hole pair separation and increased surface area. It shows higher photocatalytic efficiency in the degradation of organic dyes.
4. **Layer-by-layer Fabrication of CS–TiO_2 Scaffolds:** Three-layer scaffolds showed the highest photocatalytic activity due to better UV penetration and effective surface interaction with pollutants. Fifteen-layer scaffolds exhibited lower activity due to deep nanoparticle incorporation, limiting their exposure to UV light.

5.13.4 Photocatalytic Mechanism of CS–TiO_2 Composites

The photocatalytic activity of CS–TiO_2 composites is driven by the generation of reactive oxygen species (ROS) such as hydroxyl radicals (OH•) and superoxide radicals (O_2^-•) under light irradiation. The mechanism involves:

1. **Photon Absorption:** TiO_2 absorbs UV or visible light, promoting electrons from the valence band to the conduction band, creating electron–hole pairs.
2. **Charge Separation and Transfer:** CS acts as a supporting matrix, facilitating the transfer of electrons to the pollutant surface. Nanofillers like CDs and CNTs enhance charge separation, reducing recombination.
3. **Reactive Species Formation:** Photogenerated holes react with water to produce hydroxyl radicals (OH•), which are highly reactive and responsible for breaking down organic pollutants.
4. **Pollutant Degradation:** ROS oxidize organic pollutants, leading to their mineralization into harmless byproducts like CO_2 and H_2O.

5.13.5 Advantages of CS–TiO_2 Composites

1. **Biodegradability:** Eco-friendly disposal and reduced environmental impact.
2. **High Adsorption Capacity:** Amino and hydroxyl groups in CS enhance pollutant adsorption.
3. **Cost Effectiveness:** Abundant availability of chitosan and simple fabrication methods.
4. **Reusability:** CS–TiO_2 scaffolds exhibit excellent mechanical strength and flexibility, allowing for multiple photocatalytic cycles.

5.13.6 Future Research Directions

1. **Optimization of Nanofiller Content:** Investigate the optimal concentration of nanofillers like CNTs, CDs, and metal oxides for maximum photocatalytic efficiency.
2. **Exploration of Hybrid Materials:** Combine CS with other biodegradable polymers to enhance mechanical and thermal properties.

3. **Advanced Characterization Techniques:** Employ surface morphology and structural analysis methods (e.g., SEM, TEM, XRD) to understand the interaction between CS and TiO_2.
4. **Pilot-scale Applications:** Test CS–TiO_2 composites in real-world wastewater treatment scenarios to evaluate scalability and long-term performance.

In conclusion, CS–TiO_2 composites offer a promising solution for environmental cleanup due to their efficient photocatalytic activity, sustainability, and cost-effectiveness. Further research into optimizing their synthesis and expanding their applications could lead to significant advancements in photocatalysis and environmental remediation.

5.13.7 Synthesis and Characterization Methods of Cellulose-based TiO_2 Photocatalysts

Cellulose, a naturally abundant, nontoxic, biodegradable, and biocompatible polysaccharide, has garnered significant attention as a sustainable material for various applications, including environmental remediation. Its structure is composed of linear chains of β–(1 → 4)-linked D-glucose units, with numerous hydroxyl (—OH) groups capable of forming strong hydrogen bonds. These properties make cellulose and its derivatives highly suitable for use as reinforcing agents in photocatalytic systems, particularly when combined with TiO_2.

Cellulose-based materials offer several advantages for environmental cleanup, such as high mechanical strength, chemical stability, and enhanced pollutant-binding capabilities. Various techniques have been employed to synthesize cellulose-supported TiO_2 photocatalysts, optimizing their performance for pollutant degradation, dye removal, and heavy metal adsorption.

5.13.7.1 Synthesis Techniques for Cellulose–TiO_2 Composites

Sol–Gel Process: The sol–gel method involves the hydrolysis and condensation of titanium alkoxides, followed by the incorporation of cellulose fibers to form a hybrid composite.

Application: Used to fabricate 3D porous cellulose–TiO_2 nanocomposites with hierarchical architectures, enhancing photocatalytic activity through increased surface area and efficient pollutant interaction.

Phase-Inversion Technique:

Regenerated cellulose derived from recycled newspaper is combined with N-doped TiO_2 nanorods to form nanocomposite films. This is achieved through a phase-inversion approach, where a cellulose solution is precipitated into a solid matrix in the presence of TiO_2.

5.13.7.2 Characterization: FTIR Spectroscopy

FTIR confirms strong hydrogen bonding interactions between the hydroxyl groups of cellulose and the TiO_2 nanorods.

1. **Application:**
   These nanocomposites exhibit excellent photocatalytic properties for dye degradation and pollutant removal from water.
2. **Tape Casting and Suction Filtration:**
   Yu et al. [30] demonstrated the fabrication of triple-layered Au-TiO_2 nanoparticles embedded into porous cellulose membranes using tape casting and suction filtration.
3. **Layered Structure:**
   1. **Bottom Layer (Cellulose):** Provides structural support, filtration, and thermal insulation.
   2. **Intermediate Layer (Au Nanoparticles):** Enhances photo-adsorption and facilitates electron–hole separation.
   3. **Top Layer (TiO_2 Nanoparticles):** Serves as the active photocatalytic surface.
   4. **Outcome:** The multilayered design significantly improves the photocatalytic degradation of organic pollutants.
   5. **Layer-by-layer Assembly**:
      Multilayered films of TiO_2 nanoparticles and phosphor-tungstic acid (HPW) are fabricated using cellulose phosphate (CP) as a polyelectrolyte substrate. No additional pH adjustment is required. Rapid and uniform deposition of TiO_2 nanoparticles onto the cellulose substrate.

5.13.7.3 Characterization: SEM

Reveals the porous structure and uniform nanoparticle distribution. **Application:** Demonstrated enhanced photocatalytic degradation of stearic acid (SA) and crystal violet (CV) under UV light.

5.13.7.4 Forward Osmosis (FO) Applications

1. **Process:** Cellulose triacetate membranes are employed as a support matrix for TiO_2 deposition, enhancing the hydrophilicity and performance of forward osmosis systems.
2. **Application:** Primarily used for desalination of produced water and simultaneous removal of hazardous substances, showcasing the versatility of cellulose-based photocatalysts in water treatment technologies.

5.13.7.5 Characterization of Cellulose–TiO_2 Composites

1. **FTIR Spectroscopy:** Confirms the presence of functional groups and interactions between cellulose and TiO_2, particularly through hydrogen bonding.
2. **SEM Analysis:** Visualizes the surface morphology, porosity, and uniform distribution of TiO_2 nanoparticles within the cellulose matrix.
3. **X-Ray Diffraction (XRD):** Identifies the crystalline phases of TiO_2 (anatase or rutile) and evaluates the structural properties of the composite material.
4. **Thermogravimetric Analysis:** Assesses the thermal stability of the composite, indicating improvements due to the incorporation of TiO_2 nanoparticles.

5.13.7.6 Photocatalytic Efficiency of Cellulose-based TiO_2 Composites

1. **Pollutant Degradation**
   The photocatalytic efficiency of cellulose-supported TiO_2 composites has been evaluated across various pollutants, demonstrating excellent degradation rates. The pollutants studied include organic dyes such as MB, rhodamine B (RhB), methyl orange (MO), crystal violet (CV), reactive brilliant red K-2BR, and cationic red X-GRL, with degradation efficiencies consistently exceeding 90%. These studies were conducted under UV, sunlight, and visible light, with irradiation times ranging from 10 to 180 min.

   Additional organic contaminants such as phenol, benzene, toluene, ethylbenzene, and xylene were also examined under UV and visible light, yielding degradation efficiencies between 70% and 90% after 180–360 min of exposure.
2. **Reusability and Stability:** The reusability of cellulose-based photocatalytic membranes was tested over three to six cycles. The composites demonstrated excellent photocatalytic stability and maintained high efficiency across multiple cycles, indicating their durability and potential for long-term applications.
3. **Structural Influence on Efficiency:** The superior photocatalytic activity of these cellulose-based materials is attributed to several structural factors:
   a. Hierarchical Porous Network: The 3D porous architecture of the cellulose matrix provides a large surface area, facilitating pollutant adsorption and interaction with photocatalytic sites.
   b. Uniform Distribution of Catalysts: A well-dispersed distribution of TiO_2 nanoparticles throughout the cellulose matrix enhances light absorption and photocatalytic activity.
   c. Synergistic Interactions: The combination of cellulose and TiO_2, along with other active components like Au or phosphor-tungstic acid, results in synergistic effects that improve electron-hole separation and overall photocatalytic performance.

5.13.7.7 Challenges in Catalyst Dispersion

In cases where dimethyl sulfoxide (DMSO) was used as a solvent during the preparation of cellulose–TiO_2 composites, uneven distribution of TiO_2 nanoparticles was observed, leading to reduced photocatalytic efficiency. This highlights the need for optimized dispersion techniques to enhance performance.

5.13.7.8 Optimization of Sol–Gel Parameters

The sol–gel process used to synthesize TiO_2 plays a critical role in determining the photocatalytic properties of the final composite. Several factors influence the structure, size, and crystallinity of TiO_2 particles:

a. Temperature and Time: Higher temperatures expedite gelation and enhance crystallinity, leading to larger particle sizes and denser structures. However, they can reduce porosity and surface area. Lower temperatures slow down hydrolysis and condensation reactions, resulting in smaller, more amorphous particles with higher porosity and surface area.

b. Humidity: High humidity can hinder hydrolysis, disrupt gel formation, and lead to nonuniform particle sizes. It is advisable to conduct the sol–gel process in controlled environments with low humidity levels to ensure uniformity.
c. Agitation Rate: An optimal stirring speed (~500 rpm) ensures even dispersion of precursors, leading to uniform particle size, enhanced surface area, and improved mechanical properties. Higher stirring rates reduce particle size and increase crystallinity.
d. Chemical Addition Sequence: The order of chemical addition influences hydrolysis kinetics, particle stability, and agglomeration. Adding acid before water produces smaller particles, while the reverse results in larger particles. Chelating agents introduced before titanium alkoxide precursors lead to more uniform and smaller particles.
e. Injection Rate: A slower injection rate during the sol–gel process results in smaller particles with higher crystallinity, porosity, surface area, and mechanical strength. Gradual addition allows for controlled nucleation and growth.
f. Aging Process: Aging enhances cross-linking between TiO_2 particles, leading to improved stability, mechanical strength, and photocatalytic activity. Prolonged aging, however, can reduce porosity and surface area, potentially decreasing reactivity. Optimized aging times (e.g., 7 days at 25 °C or 1 h at 160–180 °C) balance stability and porosity.

5.13.7.9 Impact of Sol–Gel Parameters on Photocatalytic Performance

Optimizing the sol–gel synthesis parameters significantly impacts the photocatalytic efficiency of cellulose-TiO_2 composites:

1. Elevated Temperatures: Promote larger particles and higher crystallinity but reduce surface area.
2. Low Temperatures: Yield smaller particles with higher porosity and surface area, enhancing photocatalytic performance.
3. Controlled Humidity: Ensures uniform particle size and prevents hydrolysis disruptions.
4. Optimal Agitation: Leads to uniform particle distribution, higher surface area, and improved mechanical properties.
5. Precise Chemical Addition and Injection Rates: Ensure controlled nucleation, particle size, and crystallinity.
6. Aging: Improves stability, strength, and photocatalytic activity, though prolonged aging may reduce porosity.

5.13.7.10 Future Recommendations

a. Employ advanced dispersion techniques to ensure uniform distribution of TiO_2 nanoparticles.
b. Optimize sol–gel parameters to achieve the desired balance between particle size, surface area, and crystallinity.
c. Conduct further studies on the scalability and long-term stability of cellulose-based photocatalytic membranes for industrial applications.

5.13.7.11 TiO_2 Gel Formation

This section describes how to create TiO_2 gels using the sol–gel method, focusing on the selection of materials, reaction conditions, and techniques. The process involves several key steps, including the preparation of the sol, hydrolysis, aging, and gelation.

1. **Preparation of the Sol (Precursor Solution):**
   Solvent: Ethanol is chosen as the solvent because it has high polarity and low viscosity. These properties help the TTIP to dissolve easily and prevent particle clumping.
   Precursor: TTIP is added slowly to the ethanol while stirring at a speed of 500 rpm. This step takes 10 min to ensure proper mixing.
   Additives: Diethanolamine (DEA) is added to stabilize the solution, with a molar ratio of DEA:TTIP = 1:2, if necessary.
2. **Hydrolysis Process:**
   The next step is hydrolysis, where 250 mL of water mixed with an acid (with a pH of 1.5) is added gradually over a period of 240 min using a dropping funnel. The solution is stirred continuously during this time.
   After hydrolysis, the mixture is stirred overnight until a gel with a viscous and clear or slightly translucent consistency forms.
3. **Aging and Gelation:**
   If the gel does not form after overnight stirring, it can be aged at room temperature or heated between 30 °C and 60 °C to speed up the evaporation process and gel formation.
   Once the gel has formed, it can be stored at 4 °C for up to one month.

5.13.7.12 Rationale for Material Selection

1. **TTIP:** TTIP is used as a precursor because of its high reactivity and ability to easily undergo hydrolysis, making it ideal for the gel formation.
2. **Ethanol:** Ethanol is selected as the solvent due to its ability to dissolve TTIP effectively, and it helps prevent particle aggregation because of its high polarity and low viscosity.
3. **HNO_3 (Nitric Acid):** Nitric acid is used as a mild acid for hydrolysis. It is an inorganic acid that does not react with other chemicals in the process, avoiding unwanted side reactions.

5.13.7.13 Coating Process for Photocatalytic Microreactors

The process of applying TiO_2 coatings to flat silicon surfaces is essential for creating photocatalytic microreactors. Two main coating methods were compared: pipetting and spin coating.

1. **Coating Methods:**
   1. **Pipetting:** In this method, TiO_2 sol is applied drop by drop to the substrate. However, the resulting surface is less uniform, and the mechanical stability is lower compared to spin coating.
   2. **Spin Coating:** This technique provides a more uniform distribution of the TiO_2 solution across the substrate. It also allows for faster solvent

evaporation, leading to a more consistent coating with better mechanical properties.

2. **Sintering:**
   1. The films are sintered at different temperatures (400 °C, 500 °C, and 600 °C) to improve their stability. Higher sintering temperatures help the films adhere better to the substrate, but they also lead to the formation of larger TiO_2 particles and agglomerations.
   2. When the sintering temperature is raised, the TiO_2 particles tend to grow, and their crystallization changes from anatase to rutile, which may lower the photocatalytic activity due to the loss of surface area.
3. **Substrate Bonding:**
   1. The process of bonding polydimethylsiloxane (PDMS) to the coated silicon substrate was tested using plasma oxidation and optical adhesives. Plasma oxidation and adhesive bonding were not successful, but partial curing of the PDMS before bonding worked better.
   2. Even though this method showed some success, leakage was still observed in some cases [29].

5.13.7.14 Film Characterization and Mechanical Stability

SEM Analysis: SEM was used to examine the surfaces of the coated films. Spin-coated films were found to be more uniform and consistent in appearance compared to pipetted films.

Mechanical Stability: Spin-coated films were more mechanically stable than pipetted films, likely due to the better dispersion of the TiO_2 particles and the higher quantity of PVA used in the spin-coated recipes. The use of PVA helped stabilize the colloidal solution and prevented the formation of cracks in the coating.

5.13.7.15 Challenges and Recommendations

1. **Mechanical Stability:** While sintering temperatures above 500 °C improved the mechanical properties of the films, they also led to increased particle size and reduced surface area. This can decrease the photocatalytic efficiency because the larger particles have less surface area for reactions.
2. **Adhesion Issues:** There were problems in bonding PDMS to TiO_2-coated substrates, likely due to the TiO_2 layer preventing the formation of Si—O—Si linkages. Trying different bonding methods, such as partial curing of PDMS before application, seemed to be the most effective solution.
3. **Future Research Needs:**
   1. There is a need for further study on film thickness, porosity, crystal structure, and surface area, as these factors significantly impact the performance and durability of the films.
   2. More experiments on the degradation of MB in microreactors are necessary to optimize the coating formulations and the sintering conditions.
   3. Developing theoretical models to validate the experimental results will help refine the process.

5.13.7.16 Applications and Future Directions

1. **Photocatalysis and Environmental Remediation:** TiO_2 films prepared by the sol–gel method show potential for photocatalytic applications, such as breaking down pollutants like MB in water.
2. **Emerging Trends:**
   1. New precursors and additives are being explored to tailor the properties of TiO_2 for specific uses.
   2. Novel reaction conditions, such as microwave heating or continuous flow reactors, could improve the efficiency and reproducibility of the sol–gel process.
   3. Integrating TiO_2 with other materials like graphene or metal nanoparticles could lead to advanced composites with enhanced properties.

This research demonstrates the importance of controlling the sol–gel synthesis process and the coating methods to optimize TiO_2 films for photocatalytic applications. It also highlights the need for further studies to fully understand and refine the characteristics of TiO_2 coatings and their practical applications in environmental and industrial processes [31].

5.14 Future Perspective and Discussions

The issue of low binding strength, caused by incomplete curing, suggests a need to explore alternative bonding methods. While there is considerable research on binding PDMS to silicon, the primary focus should now be on overcoming the challenges posed by the TiO_2 coating during the bonding process. The use of theoretical modeling can help assess whether light can effectively reach the catalyst surface in situations where the microreactor's walls are completely covered, or when it is illuminated from behind. This analysis is crucial for understanding how well light penetrates through both the TiO_2 coating and the silicon substrate. Additionally, a new approach could involve flushing the TiO_2 suspension into the microreactor after the bonding process to potentially coat the entire microreactor uniformly trial settings, the ability of photocatalysts to be reused several times is essential. To ensure that TiO_2 nanoparticles do not wash away during extended use, strong adhesion to the substrate is critical. Chemical bonding is the most effective method for securing TiO_2 nanoparticles, and thus, a key area of future research will be how to achieve this bonding while maintaining the photocatalytic activity of TiO_2 over multiple cycles [32].

5.15 Conclusion

The decrease in the total surface area of TiO_2 catalysts often leads to a reduced rate of chemical reactions. As a result, composite photocatalysts typically show lower photocatalytic activity compared to pure TiO_2 when suspended in aqueous solutions. However, the swelling power of a crosslinked matrix, whether achieved through chemical or physical processes, is higher. This enhanced swelling allows the embedded TiO_2 particles to have better contact with organic pollutants in the solution, thereby improving photocatalysis. This method provides a direct approach for creating an immobilized framework with excellent photocatalytic efficiency over several cycles. This is typically achieved through a combination of solution-casting and heat treatment or thermal methods.

The study also examined the effect of immobilizing TiO_2 through chemical bonding and evaluated the recyclability of TiO_2 composite films over 25 cycles. It was found that the arrangement conditions significantly influence the photocatalytic activity of the material, with immobilization altering the behavior of the TiO_2 photocatalyst. Various substrates, including glass, metal oxides, carbon-based materials, textiles, cellulose, and polymeric substrates, have been widely used as matrices for immobilizing TiO_2 photocatalysts due to their excellent processability and cost-effectiveness. These materials are highly suitable for enhancing the stability and reusability of TiO_2 in photocatalytic applications.

References

1 Yadav, M., Garg, S., Chandra, A., and Hernadi, K. (2019). Immobilization of green BiOX (X = Cl, Br and I) photocatalysts on ceramic fibers for enhanced photocatalytic degradation of recalcitrant organic pollutants and efficient regeneration process. *Ceramics International* 45 (14): 17715–17722. https://doi.org/10.1016/j.ceramint.2019.05.340.

2 Chen, P., Gu, L., Xue, X., Li, M., and Cao, X. (2010). Engineering the growth of TiO_2 nanotube arrays on flexible carbon fibre sheets. *Chemical Communications* 46 (32): 5906–5908. https://doi.org/10.1039/C0CC00869A.

3 Guo, W., Zhang, F., Lin, C., and Wang, Z.L. (2012). Direct growth of TiO_2 nanosheet arrays on carbon fibers for highly efficient photocatalytic degradation of methyl orange. *Advanced Materials* 24 (35): 4761–4764. https://doi.org/10.1002/adma.201201075.

4 Hegedűs, P., Szabó-Bárdos, E., Horváth, O., Szabó, P., and Horváth, K. (2017). Investigation of a TiO_2 photocatalyst immobilized with poly(vinyl alcohol). *Catalysis Today* 284: 179–186. https://doi.org/10.1016/j.cattod.2016.11.050.

5 Sohn, Y., Huang, W., and Taghipour, F. (2017). Recent progress and perspectives in the photocatalytic CO_2 reduction of Ti-oxide-based nanomaterials. *Applied Surface Science* 396: 1696–1711. https:/doi.org/10.1016/j.apsusc.2016.11.240.

6 Shehzad, N., Tahir, M., Johari, K., Murugesan, T., and Hussain, M. (2018). A critical review on TiO_2 based photocatalytic CO2 reduction system: strategies to improve efficiency. *Journal of CO_2 Utilization* 26: 98–122. https://doi.org/10.1016/j.jcou.2018.04.026.

7 Guo, R.T., Xia, C., Bi, Z.X., Zhang, Z.R., and Pan, W.G. (2023). Recent progress of photothermal effect on photocatalytic reduction of CO_2. *Fuel Processing Technology* 241: 107617. https://doi.org/10.1016/j.fuproc.2022.107617.

8 Markovskaya, D.V., Lyulyukin, M.N., Zhurenok, A.V., and Kozlova, E.A. (2021) New composite photocatalysts based on the solid solutions of cadmium sulfide,

zinc sulfide, titania, and platinum for the photocatalytic reduction of carbon dioxide with water vapor under visible light. *Kinetics and Catalysis* 62 (4): 488–495. https://doi.org/10.1134/S002315842104008X.

9 Ainali, N.M., Kalaronis, D., Evgenidou, E., Bikiaris, D.N., and Lambropoulou, D.A. (2021). Insights into biodegradable polymer-supported titanium dioxide photocatalysts for environmental remediation. *Macromolecule* 1: 201–233. https://doi.org/10.3390/macromol1030015.

10 Feng, X., Pan, F., Zhao, H. et al. (2018). Atomic layer deposition enabled MgO surface coating on porous TiO_2 for improved CO_2 photoreduction. *Applied Catalysis B: Environmental* 238: 274–283. https://doi.org/10.1016/j.apcatb.2018.07.027.

11 Matthews, R.W. (1987). Photooxidation of organic impurities in water using thin films of titanium dioxide. *Journal of Physical Chemistry* 91 (12): 3328–3333. https://doi.org/10.1021/j100296a044.

12 Fujishima, A., Rao, T.N., and Tryk, D.A. (2000). Titanium dioxide photocatalysis. *Journal of Photochemistry and Photobiology C: Photochemestry Reviews* 1 (1): 1–21. https://doi.org/10.1016/S1389-5567(00)00002-2.

13 Mohamad Idris, N.H., Rajakumar, J., Cheong, K.Y. et al. (2021). Titanium dioxide/polyvinyl alcohol/cork nanocomposite: a floating photocatalyst for the degradation of methylene blue under irradiation of a visible light source. *ACS Omega* 6 (22): 14493–14503. https://doi.org/10.1021/acsomega.1c01458.

14 Sapalidis, A.A. (2020). Porous polyvinyl alcohol membranes: preparation methods and applications. *Symmetry* 12: 960. https://doi.org/10.3390/sym12060960.

15 Yadav, M., Garg, S., Chandra, A., Gläser, R., and Hernadi, K. (2020). Green BiOI impregnated 2-dimensional cylindrical carbon block: a promising solution for environmental remediation and easy recovery of the photocatalyst. *Separation and Purification Technology* 240: 116628. https://doi.org/10.1016/j.seppur.2020.116628.

16 Adams, M., Skillen, N., McCullagh, C., and Robertson, P.K.J. (2013). Development of a doped titania immobilised thin film multi tubular photoreactor. *Applied Catalysis B: Environmental* 130: 99–105. https://doi.org/10.1016/j.apcatb.2012.10.008.

17 Veréb, G., Ambrus, Z., Pap, Z., Mogyorósi, K., Dombi, A., and Hernádi, K. (2014). Immobilization of crystallized photocatalysts on ceramic paper by titanium(IV) ethoxide and photocatalytic decomposition of phenol. *Reaction Kinetics, Mechanisms and Catalysis* 113 (1): 293–303. https://doi.org/10.1007/s11144-014-0734-y.

18 Yang, H., Yang, B., Chen, W., and Yang, J. (2022). Preparation and photocatalytic activities of TiO_2-based composite catalysts. *Catalysts* 12: 1263. https://doi.org/10.3390/catal12101263.

19 Adamek, E., Baran, W., Ziemiańska-Błaszczyk, J., and Sobczak, A. (2019). Immobilisation of TiO_2-P25 on a glass fibre mat: preparation, photocatalytic activity and stability. *Solar Energy* 188: 1232–1242. https://doi.org/10.1016/j.solener.2019.07.034.

20 Navidpour, A.H., Xu, B., Ahmed, M.B., and Zhou, J.L. (2024). Immobilization of TiO_2 and ZnO by facile surface engineering methods to improve semiconductor performance in photocatalytic wastewater treatment: a review. *Materials Science in Semiconductor Processing* 179: 108518. https://doi.org/10.1016/j.mssp.2024.108518.

21 Aghaee, M., Verheyen, J., Stevens, A.A.E., Wilhelmus, K.M.M., and Creatore, M. (2019). TiO_2 thin film patterns prepared by chemical vapor deposition and atomic layer deposition using an atmospheric pressure microplasma printer. *Plasma Processes and Polymers* 16: 1900127. https://doi.org/10.1002/ppap.201900127.

22 Kwon, J.M., Kim, Y.H., Song, B.K., Yeom, S.H., Kim, B.S., and Im, J.B. (2006). Novel immobilization of titanium dioxide (TiO_2) on the fluidizing carrier and its application to the degradation of azodye. *Journal of Hazardous Materials* 134 (1–3): 230–236. https://doi.org/10.1016/j.jhazmat.2005.11.024.

23 Zaytseva, O. and Neumann, G. (2016). Carbon nanomaterials: production, impact on plant development, agricultural and environmental applications. *Chemical and Biological Technologies in Agriculture* 3: 17. https://doi.org/10.1186/s40538-016-0070-8.

24 Anderson, C. and Bard, A.J. (1995). Improved photocatalyst of TiO_2/SiO_2 prepared by a sol-gel synthesis. *Journal of Physical Chemistry* 99 (24): 9882–9885. https://doi.org/10.1021/j100024a033.

25 Linsebigler, A.L., Lu, G., and Yates, J.T. (1995). Photocatalysis on TiO_2 surfaces: principles, mechanisms, and selected results. *Chemical Reviews* 95 (3): 735–758. https://doi.org/10.1021/cr00035a013.

26 Mouele, E.S.M., Ngqoloda, S., Pescetelli, S. et al. (2020). Spin coating immobilisation of C-N-TiO_2 co-doped nano catalyst on glass and application for photocatalysis or as electron transporting layer for perovskite solar cells. *Coatings* 10 (11): 1029. https://doi.org/10.3390/coatings10111029.

27 Sadek, O., Touhtouh, S., Dahbi, A., and Hajjaji, A. (2023). Photocatalytic degradation of methylene blue on multilayer TiO_2 coatings elaborated by the sol-gel spin-coating method. *Water, Air, & Soil Pollution* 234: 698. https://doi.org/10.1007/s11270-023-06558-4.

28 Tinoco Navarro, L.K. and Jaroslav, C. (2023). Enhancing photocatalytic properties of TiO_2 photocatalyst and heterojunctions: a comprehensive review of the impact of biphasic systems in aerogels and xerogels synthesis, methods, and mechanisms for environmental applications. *Gels* 9: 976. https://doi.org/10.3390/gels9120976.

29 Bessekhouad, Y., Robert, D., and Weber, J.V. (2003). Preparation of TiO_2 nanoparticles by sol-gel route. *International Journal of Photoenergy* 5 (3): 153–157. http://dx.doi.org/10.1155/S1110662X03000278.

30 Yu et al. (2020). A simple strategy to design 3-layered Au-TiO2 dual nanoparticles immobilized cellulose membranes with enhanced photocatalytic activity. *Carbohydrate Polymers* 1:231:115694. 10.1016/j.carbpol.2019.115694

31 Jerczynski, K., Lipinska, M., Raj, W. et al. (2022). Effect of hybrid TiO_2 nanoparticles with controlled morphology on rheological properties of poly(styrene-co-acrylonitrile) nanocomposites. *Materials Today Chemistry* 26: 101189. http://dx.doi.org/10.1016/j.mtchem.2022.101189.

32 Tian, S., Feng, Y., Zheng, Z., and He, Z. (2023). TiO_2-based photocatalytic coatings on glass substrates for environmental applications. *Coatings 13*: 1472. https://doi.org/10.3390/coatings13081472.

6

Surface Modification of TiO_2

Xintong Zhang, Yingying Li, Rui Wang, Changhua Wang*

Key Laboratory of UV-Emitting Materials and Technology, Northeast Normal University, Changchun, China

*Corresponding author: xtzhang@nenu.edu.cn

6.1 Introduction

Surface modification is an often-adopted strategy to make nanostructured TiO_2 photocatalysts more efficient [1]. This is self-evident because TiO_2 surface is the only place for the photocatalytic chemical conversions initiated by photogenerated charge carriers, no matter forward or back reaction. Surface modification thereby holds the promise to extend the lifetime of photogenerated charge and modulate the reaction pathways to the desired direction. Surface modification can even do more, for instance, make the only ultraviolet-sensitive TiO_2 respond to the visible light, or modulate surface electronic states to make the surface-trapped electrons and holes more reactive.

Surface modification of TiO_2 photocatalysts can be traced back more than four decades to early observation of the significant reactivity increase resulted from partial platinization of TiO_2 powders for photodecarboxylation reactions [2]. In the seminal work, partial platinized TiO_2 powders were treated as short-circuited photoelectrochemical cells in which reduction sites on Pt were separated from oxidation sites on TiO_2 to promote photocatalytic processes. Ever since that work, surface modification has been recognized as one of the most effective means to enhance photocatalysis over TiO_2. Various strategies or concepts, including surface doping, cross-interface charge transfer, multiple-electron reduction/oxidation, proton-coupled electron transfer, have been adopted into surface modification of TiO_2, and various physical, chemical methods or both have been employed to modify TiO_2 surface [3]. Knowledge of photoelectrochemistry, catalysis, surface physical chemistry, and semiconductor physics is combined together to reach an understanding on the mechanism of surface modification. All these efforts make the surface modification of TiO_2 be active until now.

In this chapter, we attempt to make a summary on the surface modification methods for nanostructured TiO_2 thus far, of which a major proportion are extracted from the published works of our own. We will start with a brief summary on the various surface modification methods, and then introduce some recent advancements on the

Nanostructured TiO_2: Synthesis, Photocatalytic Properties and Applications, First Edition.
Edited by Jia Hong Pan, Wan In Lee and Detlef W. Bahnemann

modulation of surface electronic states as well as single-atom cocatalyst modification. Lastly, we will give a brief outlook for the future direction of TiO_2 surface modification.

6.2 General Strategies of Surface Modification

In this section, we will give a brief summary on the general strategies employed to modify the surface of nanostructured TiO_2 photocatalysts and their efficacy to enhance the photocatalytic performance. The works summarized herein were conducted on TiO_2; however, it is no doubt that the strategies also work well for other semiconductor photocatalysts.

6.2.1 Modification with Noble Metal Cocatalysts

Modification of TiO_2 with noble metal cocatalysts, such as Pt, Pd, Ru, Au, and Ag, has been considered as the most effective means to enhance the photocatalytic performance of TiO_2. It is generally believed that at the interface of TiO_2 and noble metal (often in the form of nanoparticles) an ohmic junction is formed, thus leading to the enhanced charge separation; the excellent catalytic properties of noble metal particles facilitate the interfacial charge transfer and modulate the reaction pathway, working like cocatalysts. Various chemical or physical methods, including photodeposition, hydrothermal synthesis, sputtering, atomic layer deposition, chemical vapor deposition, electrodeposition, and colloidal synthesis, have been developed for the modification of TiO_2 with noble metal cocatalysts [4–10]. Some of the works are listed in Table 6.1. Selection of the deposition method and the control over loading amount, composition, and size of the noble metal nanoparticles, are crucial for optimizing the photocatalytic performance of the modified TiO_2.

Pt was reported to be the most effective cocatalyst to date for water splitting and photocatalytic reforming of biomass to H_2 under anaerobic conditions [11], whereas their contribution to photocatalytic oxidation reactions in aerobic environments is less pronounced. Pd, a noble metal cocatalyst comparable to Pt in enhancing photocatalytic reactions but at a lower cost, exhibits similar efficacy in both aerobic and anaerobic conditions [12]. To optimize reaction activity, many studies have adopted the strategy of using Pt–Pd alloys, exploring the synergistic effects of both noble metals [13]. Ru, as a cocatalyst, is notable for its economic advantage being the cheapest among the platinum group elements, making it appealing for industrial photocatalytic applications. It uniquely supports both hydrogen evolution reaction in its metallic form and oxygen evolution reaction in oxide form (RuO_2), synergistically enhancing the overall efficiency of water splitting reactions [14]. Au and Ag are distinguished by their localized surface plasmon resonance (SPR) that enhances light absorption and catalytic efficiency across visible to near-infrared spectra [15]. Au cocatalysts can enhance photocatalytic degradation of organic pollutants and CO oxidation on TiO_2 under visible light irradiation [16]. Ag nanoparticles exhibit a unique antibacterial property due to the SPR-induced surface oxidation, which is important for photocatalytic purification and disinfection [17]. Additionally, the photothermal properties of noble metals, particularly Au and Ag, have been

Table 6.1 Representative works of TiO_2 modification with noble metal nanoparticles

Method	Noble metal	Preparation detail	References
Photodeposition	Pt	TiO_2 was dispersed in a solution containing $H_2PtCl_6 \cdot 6H_2O$. Then, the suspension was irradiated with a Hg lamp.	[4]
Impregnation	Pt	Pt precursor was added in TiO_2 solution. The solution was heated to 343 K until completed dried.	[5]
Sputtering	Pt	Pt was deposited using a tilted-target sputtering method under a fixed target angle, distance, pressure, power and time.	[6]
Atomic layer deposition (ALD)	Pt	TiO_2 substrate was loaded into the reactor. $MeCpPtMe_3$ was used as Pt source. Adjust ALD parameters for deposition.	[7]
Chemical vapor deposition (CVD)	Pt	Pt (IV) acetylacetonate was used as Pt source and put in chamber. Adjust CVD parameters for deposition.	[8]
Electrodeposition	Pt	Platinum was deposited from a H_2PtCl_6 solution on TiO_2 in a conventional three-electrode cell.	[9]
Colloidal deposition	Au	TiO_2 was added to an Au colloidal suspension.	[10]

harnessed in photothermal catalysis due to their SPR, where the localized heat generated accelerates reaction kinetics [18].

It should be noted that high cost and scarcity of noble metal limit their practical use in photocatalytic applications. Therefore, research efforts are devoted to explore alternative, more cost-effective cocatalysts or to optimize the use of noble metals through precise control of their loading amount. In this regard, single-atom noble metal is a promising way for the surface modification of TiO_2 [19].

6.2.2 Modification with Sensitizers

Surface modification with dye molecules, metal complexes, semiconductor quantum dots or in the form of surface complex have been extensively investigated in the past decades for the purpose of sensitizing TiO_2 into the visible light region. The sensitizing mechanism is believed to be photo-induced charge injection, that is, the visible light-excited sensitizers inject electrons to the conduction band of TiO_2, afterwards the electrons and the oxidized sensitizers initiate reactions at or close to the TiO_2 surface [20].

Sensitizers are commonly adsorbed on TiO_2 surface from solution by electrostatic interaction, metal–ligand bonds or covalent bonds. The loading number of sensitizers is determined by the specific surface area of TiO_2 due to the monolayered adsorption. TiO_2 can be colored with 7,7,8,8-tetracyanoquiondimenthane (TCNQ), although both pristine materials do not show absorption in the visible region. Density functional theory calculations and some experimental test show that interfacial charge-transfer transitions appeared on the TiO_2–TCNQ surface complex [21].

Alternatively, toluene 2,4-diisocyanate (TDI) was chemically adsorbed on TiO_2 by a surface chemical reaction [22], the resultant TiO_2–TDI exhibited a good photostability and a high photocatalytic performance for the degradation of phenol, 2,4-dichlorophenol, fluorescein and methyl orange.

Kisch et al. conducted pioneering research on modifying TiO_2 with platinum (IV) chloride complexes by the grinding method for photocatalytic degradation of 4-chlorophenol(4-CP) [23]. Under the excitation of UV or visible light, the Pt—Cl bond cleavage generated a Pt^{III} species and a weakly bound chlorine atom. The electron was then transferred from Pt^{III} into the conduction band (CB) of TiO_2 regenerates Pt^{IV}, while the chlorine atom abstracted an electron from 4-CP to regenerate the sensitizer, followed the electron transfer on CB of TiO_2 to 4-CP. This charge transfer ligand-to-metal (CTLM) process led to an increased photocatalytic activity for 4-chlorophenol degradation and mineralization. Similar phenomena were observed in a variety of metal complex, such as Rh(III), Au(III); however, their charge transfer paths are different, which affects their efficacy. For Rh(III) chloride complexes, there was electron transfer from rhodium to TiO_2 but without the cleavage of Rh—Cl bond, which made it ineffective in cyanuric acid decomposition. By choosing appropriate metal complex, a desired modification effect can be obtained to meet specific application fields.

6.2.3 Modification with MO_x Cluster

Metal oxide clusters (MO_x), especially transition metal (M = Cu, Ni, Fe, Co, Cr) oxide clusters, were considered as the low-cost substitute to noble metal nanoparticles for TiO_2 modification[1]. MO_x clusters can extract photogenerated electrons or holes from TiO_2 and catalyze the multielectron reaction such as water oxidation and oxygen reduction, dependent on the chemical composition. MO_x clusters can also function as sensitizers by the mechanism of cross-interface charge transfer. The grafting of MO_x clusters TiO_2 surface was often carried out by chemical methods, like the impregnation method [24], the sol–gel method [25], and the chemisorption-calcination method [26]. Reaction conditions, such as precursor concentrations, reaction time and temperature, can significantly influence the size, crystallinity, and loading amount of the MO_x clusters, thereby impacting the photocatalytic performance.

Irie et al. first modified TiO_2 with CuO_x, CrO_x, and FeO_x, by the impregnation method [24]. They considered that electrons were photo-excited from the valence band (VB) of TiO_2 to the surface-grafted MO_x clusters, or from the d orbitals of MO_x clusters to the conduction band (CB) of TiO_2. The net result is the photo-induced charge separation by the cross-interface charge transfer and the photocatalytic chemical conversions thereafter. In the case of CuO_x/TiO_2, electrons in the VB of TiO_2 are transferred to CuO_x, leaving holes in the VB of TiO_2 and the reduced CuO_x clusters. The reduced CuO_x clusters will catalyze multielectron oxygen reduction reaction, together with the oxidation function of VB holes, reaching a quantum efficiency of 8.8% for photocatalytic decomposition of 2-propanol under visible light excitation. The same mechanism also works well for the FeO_x-modified TiO_2, accounting for its photocatalytic activity under visible light [26, 27].

We conducted a comparative study on the effect of the surface-grafted CrO_x, CoO_x, and CuO_x on the photocatalytic mineralization of acetaldehyde under visible light ($\lambda > 400$ nm), using Pt cocatalyst as a reference, as shown in Figure 6.1 [28].

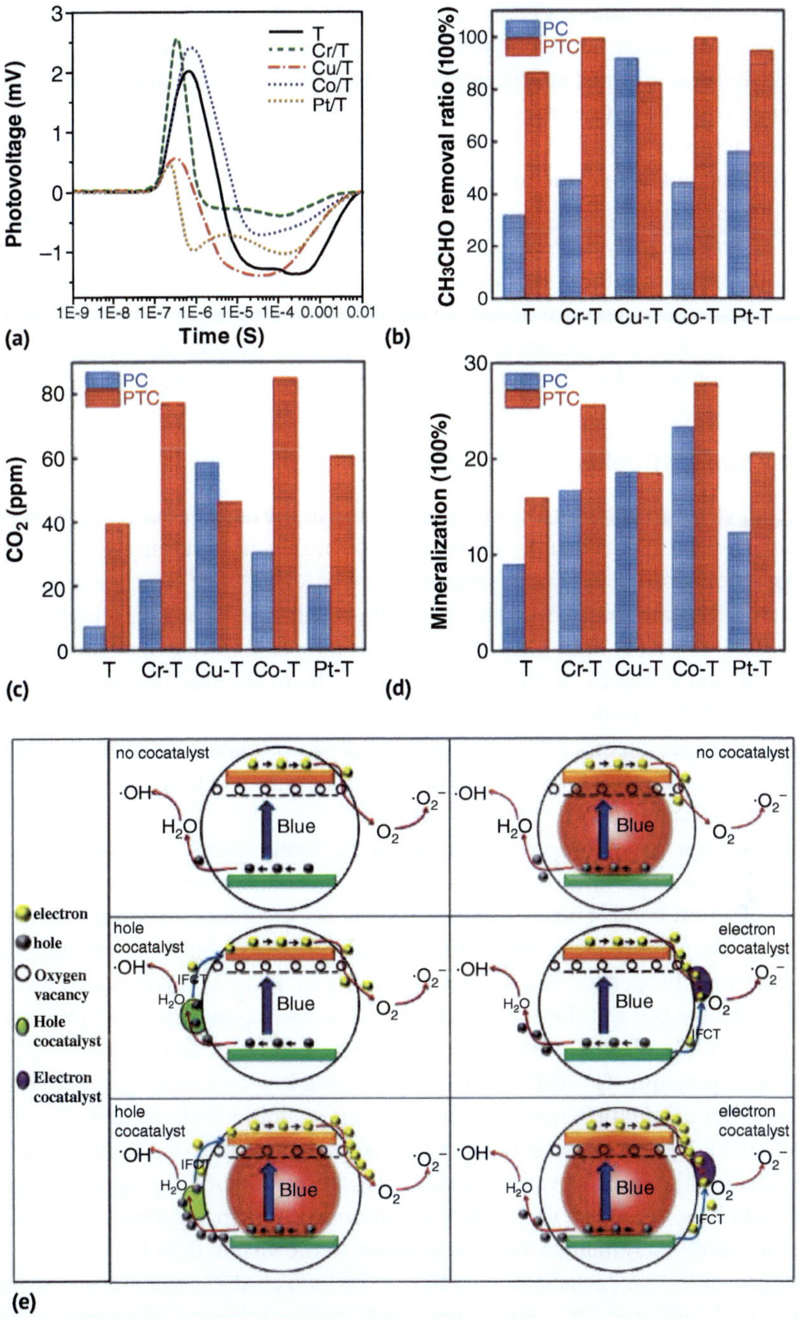

Figure 6.1 (a) TPV spectra of different samples. Catalytic activity in terms of (b) acetaldehyde removal, (c) CO_2 generation, and (d) mineralization yield over different samples. Pristine TiO_2, CrO_x/TiO_2, CuO_x/TiO_2, CoO_x/TiO_2 and Pt/TiO_2 are named T, Cr/T, Cu/T, Co/T, and Pt/T. PC stands for photocatalysis and PTC stands for photothermocatalysis. (e) The mechanism explanation for the difference of catalytic activity under blue light between pristine TiO_2 and cocatalyst/TiO_2. Reproduced from Ref. [28]/with permission of Elsevier.

By transient photovoltage (TPV) measurements, CoO_x and CrO_x were identified as hole-type cocatalysts, and CuO_x and Pt were electron-type cocatalysts. Using gaseous acetaldehyde as a model substrate, we observed the obvious mineralization of acetaldehyde with the MO_x-grafted anatase TiO_2 under the blue light irradiation; at the elevated temperature of 333 K, the mineralization rate was remarkably increased. It should be noted that the anatase TiO_2 before modification is only UV sensitive. Moreover, it is interesting to find that at room temperature CuO_x and Pt, that is, the electron-type cocatalyst, performed better than CoO_x and CrO_x, the hole-type cocatalyst, for promoting the mineralization of acetaldehyde under blue light; however, at the elevated temperature of 333 K, the hole-type cocatalyst, here CoO_x and CrO_x showed superior performance to the electron-type cocatalyst, suggesting the involvement of additional photothermal synergistic pathway by the hole-type MO_x cocatalyst.

6.2.4 Surface Doping Modification

In recent years, many efforts have been devoted to the defect effect on TiO_2 photocatalysis. The defect was usually introduced by foreign atom doping and self-doping. Foreign atom doping by metal cations (Fe^{3+}, Cu^{2+}, Zn^{2+}, Co^{2+}) or non-metal anions (N^{3-} and F^-) can create new energy levels within the bandgap, allowing for the absorption of visible light [1a]. The foreign atom doping can be achieved by both chemical and physical methods, such as the hydrothermal method, the sol–gel method, and the sputtering method. Selection of dopant ions and their concentrations are crucial factors affect the photocatalytic performance of TiO_2. Excessive doping can lead to the formation of new recombination centers, reducing the photocatalytic efficiency [29].

The concept of self-doping is proposed in recent years, that is, the intrinsic defects include Ti^{3+} and oxygen vacancies, whose defect levels are located below the conduction band. These defects, acting as color centers, would not narrow the bandgap, but provide a channel in the mid-gap for visible light absorption and alter the electronic properties and surface chemistry of TiO_2. Many methods have been used to realize the self-doping of TiO_2, such as hydrogen treatment [30], vacuum treatment [31], plasma treatment [32], chemical reduction reaction [33], etc. Fang et al. prepared Ti^{3+} self-doped TiO_2 by using low-cost $NaBH_4$ as the reductant via one-step calcination method and one-step solvothermal method. The activity under visible light is improved without weakening the catalytic activity under ultraviolet light [34]. We, however, reported that a simple ethanol impregnation followed with mild heat treatment (150–400°C) introduced oxygen vacancies onto TiO_2 surface and made TiO_2 nanoparticles responsive to the visible light [35].

Oxygen-excessive surface, in contrast to the oxygen-deficient surface, is another type of self-doping of TiO_2. Such type of self-doping relies on an oxygen-rich preparation conditions, to form oxygen interstitial or Ti vacancy defects. In such case, defect levels were introduced into the band gap, at the position above valence band. Due to the large radius of oxygen anion, it is obviously more difficult to form oxygen-excessive surface. However, there did have several reports on the oxygen-excessive TiO_2, prepared from peroxo titanate precursors. For example, we reported that TiO_2 microspheres prepared by the ultrasonic pyrolysis method, using peroxo titanate as the precursor, exhibited a red-shifted absorption edge and photocatalytic activity under visible light ($\lambda > 400$ nm) [36].

It is worth noting that the MO_x cocatalyst and the self-doping can have a remarkable synergistic effect on TiO_2 photocatalysis. Taking FeO_x–TiO_{2-x} as an example, we prepared the material by impregnating TiO_2 into the ethanol solution of ferric ethoxide, followed with mild heating [37]. The FeO_x–TiO_{2-x} sample heat-treated at 423 K exhibited the optimal photocatalytic performance for β-naphthol degradation with rate constant 6.0, 2.7, and 3.9 times higher than that of TiO_2, TiO_{2-x}, and FeO_x–TiO_2, respectively. In another study, CuO_x cocatalysts were grafted on oxygen-excessive TiO_{2+x} microsphere to catalyze the multiple-electron reduction process. The CuO_x–TiO_{2+x} sample achieved a photocatalytic activity 8.6 times higher than TiO_{2+x}, and 11 times higher than CuO_x–TiO_2 [36]. The mechanism of the photocatalytic process is elucidated in Figure 6.2.

The synergy of the MO_x cocatalysts and the surface oxygen vacancies can be more significant in the case of photothermocatalysis. We have reported a strategy to co-modify TiO_2 with oxygen vacancies and CoO_x nanoclusters for enhanced photothermocatalytic reduction of CO_2 with water [38]. As shown in Figure 6.3, the TiO_{2-x}/CoO_x material exhibits prominently enhanced activity for the yield of CH_4 and CO under ultraviolet irradiation at elevated temperature of 393 K, which is 111.3 and 13.2 times greater yield of CH_4 and CO, respectively, than the conventional photocatalytic process at 298 K, and 175.1 and 2.9 times greater yield of CH_4 and CO, respectively, than the pristine TiO_2 under the same photothermocatalytic conditions, as shown in Figure 6.3. Control experiments over singly modified TiO_2 and doubly modified TiO_2 by different preparation history, together with high-resolution transmission electron microscope (HRTEM), electron spin resonance (ESR), and transient photovoltage measurements reveal the synergistic effect of oxygen vacancies and surface-grafted CoO_x on the photothermocatalytic reduction of CO_2 to CH_4, that is, oxygen vacancies at TiO_2 surface facilitate the adsorption and reduction of CO_2 and the dispersion of CoO_x nanoclusters, whereas

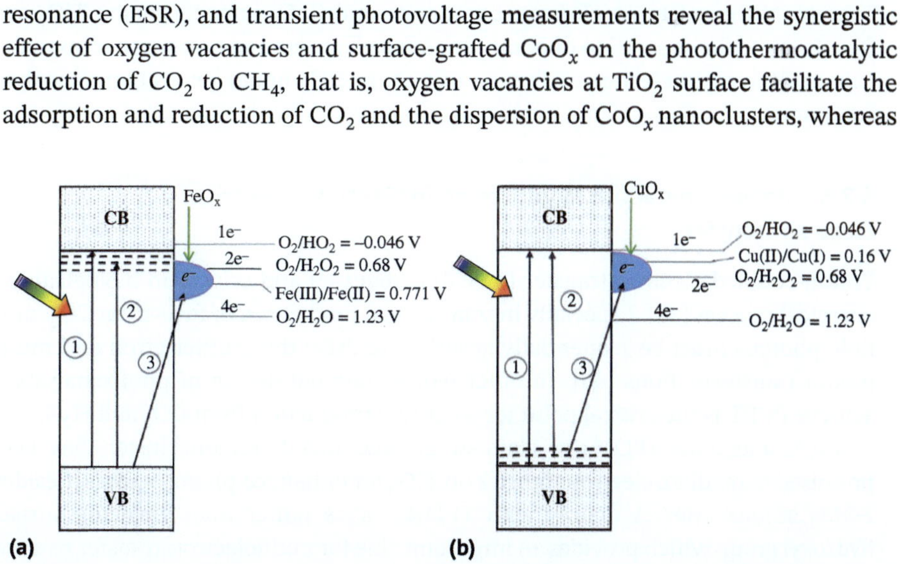

Figure 6.2 (a) FeO_x modified TiO_2 with oxygen vacancies, in which ② is the electron excited from VB to oxygen vacancies and ③ is the interfacial charge transfer with FeO_x, and (b) CuO_x modified oxygen-rich TiO_2, in which ② is the electron excited from oxygen excess defects to CB and ③ is the interfacial charge transfer with CuO_x. Adapted from [36].

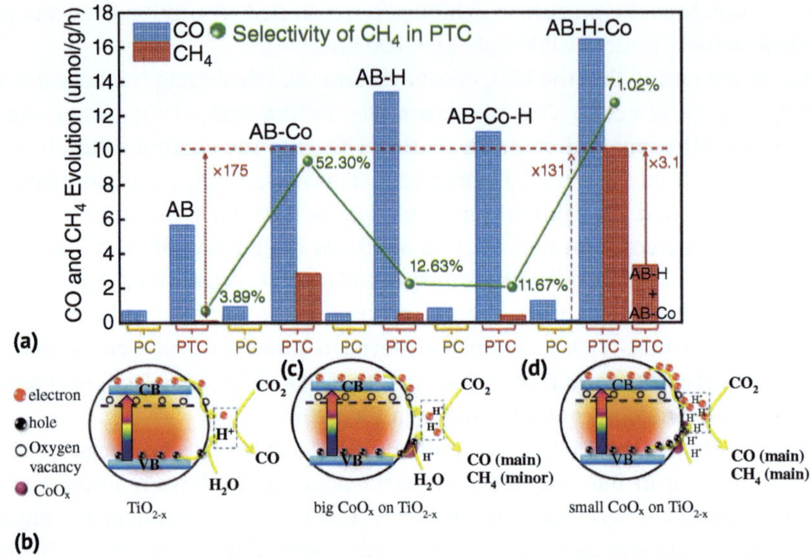

Figure 6.3 (a) Photocatalytic and photothermal catalytic CO_2 conversion for a series of as-prepared samples. In this work, TiO_2 was the nanotube composed of anatase and TiO_2(B), which named AB. CoO_x grafted on AB named AB-Co. H represents the sample was annealing calcined at 450°C for 2 h in N_2/H_2(9:1) mixture. PC stands for photocatalysis and PTC stands for photothermocatalysis. Scheme of photothermocatalytic reaction over three typical samples: (b): TiO_{2-x}; (c): big CoO_x clusters modified TiO_{2-x}; and (d): small CoO_x clusters modified TiO_{2-x}. Reproduced from Ref. [38]/with permission of Elsevier.

surface-grafted CoO_x clusters facilitate the hole trapping and the oxidation of H_2O. Thereby, the coexistence of oxygen vacancies and CoO_x nanoclusters at TiO_2 surface promote the separation of photogenerated electrons and holes, and remarkably enhance the eight-electron reduction of CO_2 to CH_4 under photothermocatalytic conditions.

6.2.5 Polyoxometallate Modification for Proton-coupled Electron Transfer

Proton-coupled electron transfer (PCET) is required and critical in biological and chemical process [39]. Especially in solar-driven production of fuels from CO_2, multiple photons must be sequentially absorbed to drive the multielectron and multiproton transformations [40]. In other words, rational design of photocatalysts to achieve PCET is the critical point for solar fuel production from CO_2 and H_2O.

Polyoxometallate (POM), a class of anionic metal–oxygen cluster, has been promised to mediate electron transfer on TiO_2 for enhanced photocatalysis. Besides, POMs anions interact mainly with Ti-OH_2^+ sites rather than replacing surface hydroxyl group which provides an important clue for multielectron-transfer reaction via PCET. The methods to integrate POMs on TiO_2 surface include the impregnation method, the sol–gel method, as well as physical mixing. Recently, we developed an ultrasonic spray pyrolysis method to integrate TiO_2 with Keggin-type POMs and employed the composite photocatalysts to achieve photocatalytic CO_2 reduction [41].

Three kinds of POMs, including phosphotungstic acid (HPW), tungstosilicic acid (HSiW), and phosphomolybdic acid (HPMo), were mixed respectively with TiO_2 powders (Degussa P25) in water to produce stable dispersion. The POM–TiO_2 porous microspheres were then prepared by ultrasonic spraying the respective dispersion into a heated furnace. The method is very simple, and very suitable for the preparation of nanocomposite materials from functional blocks. Our successful examples include TiO_2–WO_3, TiO_2–rGO, TiO_2–WO_3–Pt, and MoS_2–rGO [42].

The photocatalytic activity of the three POM–TiO_2 was comparatively investigated by CO_2 reduction. It was found that among the three kinds of POM–TiO_2 samples, HPW–TiO_2 photocatalysts performed the best in terms of CO and CH_4 production, as shown in Figure 6.4. A smaller amount of CH_4 could be formed with the HSiW–TiO_2 photocatalyst, while no CH_4 could be detected in the case of HPMo–TiO_2. Comprehensive analysis including transient photovoltage, surface photovoltage, and photochromic phenomena revealed that energy level match between POMs and the surface mid-gap levels of TiO_2 played a decisive role in the CO_2 conversion and the product selectivity. The better energy level match of HPW

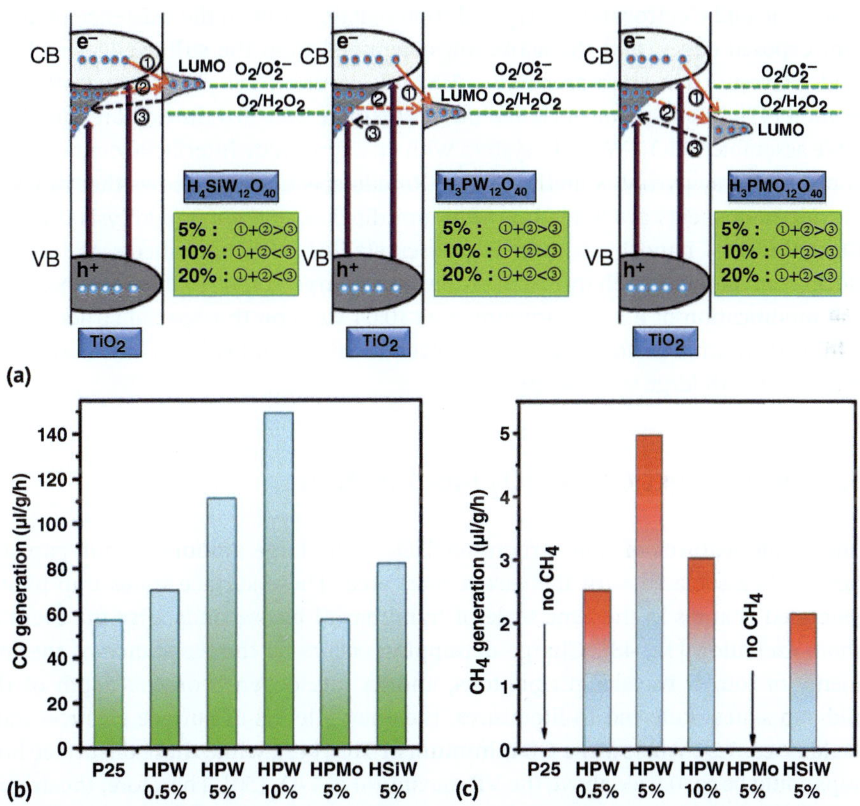

Figure 6.4 (a) Illustration of charge transfer and separation at different POM–TiO_2 interface. The signal ">" and "<" means a higher and lower probability of combined process of 1 and 2 than that of process 3, respectively. Photocatalytic production yield of (b) CO and (c) CH_4 over different samples after reaction for 5.5 h. Reproduced from Ref. [41]/with permission of Elsevier.

to mid-gap level of TiO_2, as well as the appropriate loading amount, facilitated the coupling of multiple-electron transfer to multiple-proton transfer, and thus favored the formation of CH_4.

6.2.6 Two-dimensional Materials Modification

Two-dimensional (2D) materials, such as transition metal dichalcogenides (TMDs), graphene, and layered double hydroxides (LDHs), exhibit unique physical and chemical properties attributed to their novel structural features. Modifying 2D materials on TiO_2 surface can improve the photocatalytic activity by providing a great number of active sites and promoting charge separation and transfer. Deposition 2D materials on TiO_2 surface can achieved by physical or chemical vapor deposition [43], photoreduction deposition [44], ultrasonic treatment [45], and hydrothermal reaction [46].

2D materials have an excellent charge transport properties which can enhance charge separation and prolong the lifetime of photogenerated carriers, boosting the photocatalytic activity. MoS_2 has been reported as promising catalyst for both photocatalytic and electrocatalytic H_2 evolution reactions due to the existence of abundant exposed edges, and the active sites stemmed from the sulfur edges of MoS_2 crystal layers. Yuan et al. prepared a 2D MoS_2 layer on TiO_2 by the hydrothermal method [47]. The maximum H_2 evolution rate increased 36.4 times than pure TiO_2.

We assembled TiO_2–WO_3–Pt system with an appropriate interfacial contact using ultrasonic spray pyrolysis method [42c]. Raman spectroscopy shows that tungsten trioxide nanosheets are loaded on titanium dioxide, and spray pyrolysis does not affect the WO_3 nanosheet layer. Highly crystalline WO_3 nanosheets as building blocks can maximize charge separation and charge transfer. It is worth noting that modification of a lower amount of Pt (0.04 wt%) on this special structure can exhibit an optimal performance than conventional amount (1 wt%), due to the WO_3 nanosheet with large surface area.

6.3 New Direction of TiO_2 Modification

One of the features of nanostructured TiO_2 is the large amount of mid-gap surface states associated with the vast surface area. These surface states trap photogenerated charges in the time scale of hundreds of picoseconds after the event of photo-excitation [1a, 48]. Charge detrapping occurs by the assistance of thermal energy or longer wavelength photons, and its rate depends on the depth of the mid-gap states. Referred to literatures, the energy levels of surface electron traps could be ca. 1.2 eV below the CB minimum of TiO_2 [49], while those of surface hole traps could be ca. 0.9 eV above the VB maximum of TiO_2 [50]. Therefore, the driving force for chemical conversions could be significantly lowered after trapping the photogenerated electrons and holes at surface, especially when the surface is heavily doped with foreign elements or self-doped. This issue encourages us to explore new means to modulate the surface electronic states of TiO_2 and to activate the trapped charge carriers for more efficient photocatalysis.

6.3.1 Solution Plasma Modification

Solution plasma, that is, plasma generated in a solution by applying a high voltage pulse to metal electrodes at normal temperature and pressure [51], is a novel and unique technology for the surface modification of catalysts. This process produces multiple physical fields, including electric fields, thermal fields (10^4–10^5 K), light fields, liquid annealing, etc., and various active species with densities up to 10^{17}–10^{20} cm^{-3} (e^-_{aq}, ·H, ·OH, ·O, O_2^-, etc.) [52]. These physicochemical effects can synergistically impact on the surface of catalysts for modification. More importantly, the high density and strong internal pressure of the solution environment lead to weaker kinetic energy of the plasma [53], which facilitates the gentle surface modification of catalysts, without changing the bulk properties.

We utilized alcohol plasma technology to form a thin amorphous layer on the surface of TiO_2 (here Degussa P25) [54]. The subsequent characterizations confirmed that the surface amorphous layer passivated surface O_Vs, then reduced the deep-energy-level electron traps and raised the energy level of electron traps, eventually boosting the reduction capacity of photogenerated charges. HRTEM images distinctly revealed that a surface amorphous layer with a thickness of ~0.7 nm was introduced on the TiO_2 surface (Figure 6.5a). Moreover, the Ti-$L_{2,3}$ edges of TiO_2 collected in the surface region shifted to the lower energy (Figure 6.5b). This indicated that pristine TiO_2 contained several O_Vs, and these were mainly located on the surface. Differently, the Ti-$L_{2,3}$ edges of alcohol-plasma-treated TiO_2 barely shifted with a scan from the bulk to the surface, manifesting its O_Vs were evenly distributed (Figure 6.5c). These results

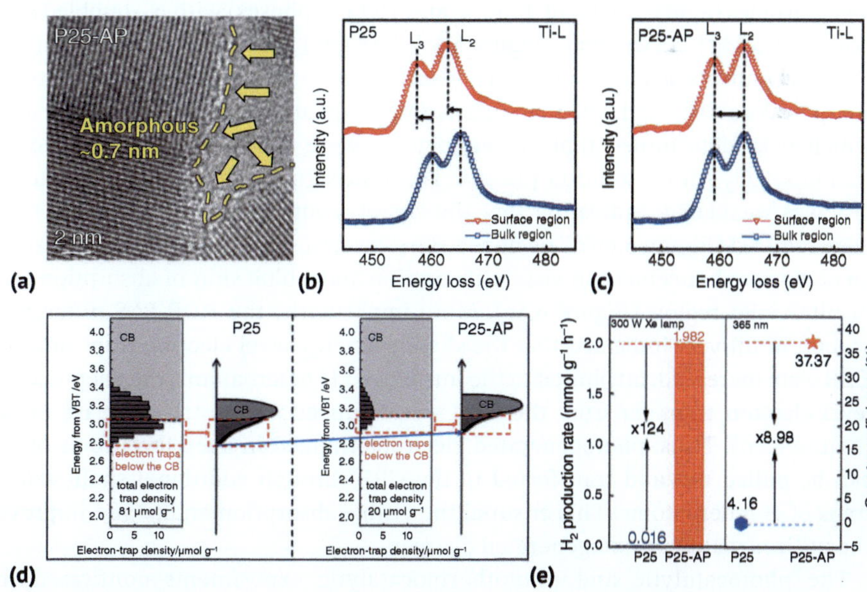

Figure 6.5 (a) HRTEM images of alcohol-plasma-treated P25. EELS spectra of Ti-$L_{2,3}$ for (b) P25 and (c) alcohol-plasma-treated P25. (d) RDB-PAS spectroscopy of P25 and alcohol-plasma-treated P25. (e) Photocatalytic hydrogen evolution rate and AQE of P25 and alcohol-plasma-treated P25. Reproduced from [54]/with permission of American Chemical Society.

showed that the surface O_vs of alcohol-plasma-treated TiO_2 were decreased, ascribed to the passivation effect of the surface amorphous layer.

The decrease in surface O_vs affected the distribution of electron traps in the band gap. The alcohol-plasma-treated TiO_2 showed a decreased density of electron traps from 81 μmol g^{-1} in pristine TiO_2 to 20 μmol g^{-1}, as measured by the reverse double-beam photoacoustic spectroscopy (RDB-PAS, Figure 6.5d). The energy level of electron traps was simultaneously uplifted after alcohol plasma treatment (Figure 6.5d). This verified that the electron-donating ability of alcohol-plasma-treated TiO_2 was raised. Thus, the surface amorphization of TiO_2 benefited the charge separation and the reduction ability of photoinduced electrons, which would boost the catalytic reaction. Consequently, the hydrogen evolution rate of alcohol-plasma-treated TiO_2 was increased by ca. 124 times compared with pristine TiO_2. And the apparent quantum efficiency (AQE) of it loaded with 0.3 wt.% Pt reached 37.37% at 365 nm (Figure 6.5e). Therefore, the deep-energy-level traps in the forbidden band can be eliminated through the surface amorphization for TiO_2, and then enhanced the electron-donating ability of TiO_2, thus facilitating the photocatalytic reduction reaction.

In another effort, we intended to make the O_vs-rich TiO_2 more efficient by introducing H shallow levels with solution plasma. The O_vs, as mentioned above, is deep trap for electrons [55]; it can make TiO_2 colored, but with low photocatalytic performance. H heteroatoms, however, are shallow donors in TiO_2, lying above the O_vs level in the band gap [56]. As shown in Figure 6.6, the incorporation of H shallow levels to O_vs-rich TiO_2 might help the extraction of electrons deeply trapped at O_vs through H bridge, and make the visible sensitive TiO_{2-x} photocatalyst more efficient (Figures 6.6g,h). We treated the as-synthesized rich-O_vs TiO_2-AB nanoparticles (a mixed TiO_2 of anatase and TiO_2(B) phases) with N_2-bubbled solution plasma [32]. As shown in Figure 6.6a, the signal peak at 6.15 ppm assigned to hydroxyl radicals adsorbed on the TiO_2 surface was increased, and two additional sharp peaks at 0.87 and 0.82 ppm referred to as signals of Ti—H or O—H appeared, which both determined that H heteroatoms were successfully doped into the rich-O_vs TiO_2-AB by solution plasma. The ESR signal at $g = 2.003$ attributed to Ovs was weakened, indicating that the O_vs of solution-plasma-treated TiO_2-AB were reduced (Figure 6.6b). The doped H heteroatoms and the reduced O_vs caused an obvious enhancement in visible absorption and a blue shift of absorption edge in ultraviolet region (Figure 6.6c). More importantly, the RDB-PAS differential spectra intuitively indicated that the shallow-energy-level electron traps near the CBB were increased, attributed to the introduced H heteroatoms; the deep-energy-level electron traps far from the CBB were reduced due to the decrease in O_vs (Figure 6.6d). Thus, photogenerated electrons trapped in deep-level traps of O_vs can be pulled out and transferred to the CBB through additional shallow-level traps of H heteroatoms. This ensured the visible absorption ability and improved the utilized yield of photogenerated charges.

The photocatalytic and photothermocatalytic experiments confirmed the enhancement of catalytic performance vis abovementioned electron state regulation. Irradiated with solar light, the photothermal catalysis of CO and CH_4 yield over solution-plasma-treated TiO_2-AB were respectively increased by 299.9 and 108.5 times compared to that over TiO_2-AB in photocatalysis (Figure 6.6e).

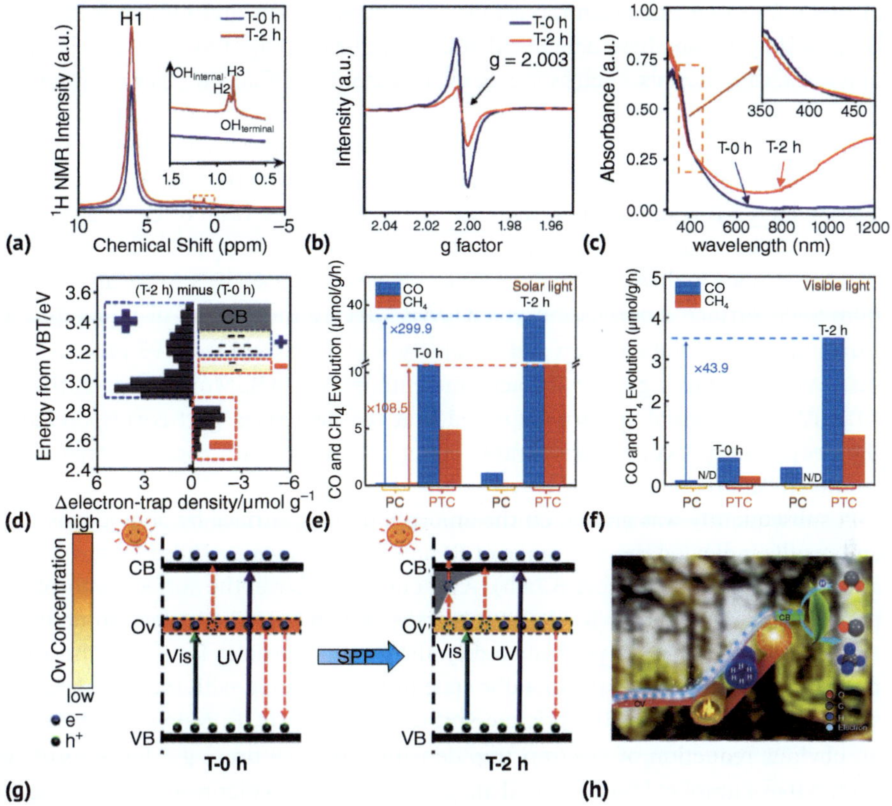

Figure 6.6 (a) ^{1}H NMR spectra; (b) ESR spectra; (c) ultraviolet–visible absorption spectra of TiO_2-AB and solution-plasma-treated TiO_2-AB. (d) The difference value of electron trap density between TiO_2-AB and solution-plasma-treated TiO_2-AB. Photo- and photothermal CO_2 conversion rate over TiO_2-AB and solution-plasma-treated TiO_2-AB samples under (e) solar light and (f) visible light irradiation. (g) Schematic illustration of the electron–hole separation mechanism for TiO_2-AB and solution-plasma-treated TiO_2-AB during photocatalysis driven by UV and visible light irradiation. (h) Schematic illustration of electron transfer from O_v to CB under solar light irradiation with thermal assist. [32]/John Wiley & Sons/CC BY 4.0.

Moreover, solution-plasma-treated TiO_2-AB displayed the 3.51 μmol g^{-1} h^{-1} of CO yield and 1.16 μmol g^{-1} h^{-1} of CH_4 yield under visible light (Figure 6.6f). Such surface electron state engineering was also applicable to other oxides, such as ZnO, WO_3, and Ta_2O_5. It also worked well for the commercial N-doped TiO_2 (Sumitomo Chemicals), doubled its photocatalytic activity under the visible light [57].

6.3.2 Ultrasonic Modification

Ultrasonics can generate intense thermal field and high-pressure force field in liquid [58]. The periodic compression and expansion of ultrasonic wave generate countless cavitation bubble around the probe, and then these grow and break. This process instantly induces a local microzone filled with shock waves and/or micro-jets derived from the collapse of bubbles in solution [59]. The temperature and pressure of this microzone can reach thousands of degrees Celsius and hundreds of atmospheres [60]. These physicochemical effects can be employed to synthesize a range of nanosized materials, such as metal, oxide, carbide, sulfide, etc., as well as modify the materials surface.

We selected TiO_2-AB as a model catalyst and introduced an amorphous surface by ultrasonic process [61]. The XPS spectra indicated that the surface amorphization of TiO_2-AB resulted in an increase in the peak area ratio of O—H and Ti—O from 0.33 for TiO_2-AB to 0.40 for surface-amorphized TiO_2-AB. This suggested that the surface hydroxyl content of TiO_2-AB was increased by surface amorphization. Such surface modification of TiO_2-AB affected the density of mid-gap electron states. The total electron states density was decreased from 93 μmol g^{-1} for TiO_2-AB to 59 μmol g^{-1} for surface amorphous TiO_2-AB. More importantly, the RDB-PAS differential spectra displayed that the decreased part corresponded to the deep-energy-level electron traps, being consistent with the aforementioned results in Section 6.3.1.

Pt subsequently was grafted on the amorphous TiO_2 surface by photodeposition, and results indicated the size range of Pt nanoparticles was reduced on the surface amorphous TiO_2-AB (Figure 6.7a,b). As mentioned above, the surface amorphization reduced deep-energy-level electron traps and enhanced the electron-donating ability of TiO_2 photocatalyst. Coherently, the Pt nanoparticles loaded on the amorphous surface showed a fully metallic state of Pt^0 compared with the mixed valence state of Pt^0 and Pt^{2+} on the pristine surface (Figure 6.7c,d). Moreover, there was an obvious reduction of electron trap densities from 59 μmol g^{-1} for amorphous TiO_2-AB to 4 μmol g^{-1} for Pt-loaded amorphous TiO_2-AB (Figure 6.7e,f). During the RDB-PAS test, holes were annihilated by methanol, while electrons were captured by electron traps in the forbidden band, which in turn recorded the distribution and density of electron traps by detecting the photoacoustic signals of trapped electrons. Therefore, the reduction of electron trap densities indicated that the loaded Pt nanoparticles extracted and stored photogenerated electrons. This enhanced the charge separation efficiency.

The photothermal CO_2 conversion rate of Pt-loaded amorphous TiO_2-AB showed a 71.08-fold increase than that in photocatalysis, and a 155-fold enhancement compared with TiO_2 in photocatalysis (Figure 6.7g,h). Furthermore, the selectivity of CH_4 was significantly improved from 33.92% to 87.46% (Figure 6.7h). Interestingly, the Pt-loaded catalysts produced additional H_2 in photothermal catalytic process. By investigating the by-product H_2, the smaller sized Pt nanoparticles produced less H_2 (Figure 6.7h). This suggested that smaller Pt nanoparticles, in synergy with the amorphous surface of TiO_2, promoted the splitting of H_2 molecule to ·H, and thus accelerated the hydrogenation of CO to CH_4. We deduced that the smaller sized Pt nanoparticles would increase the contact interface with the surface amorphous TiO_2-AB, on which H_2 splitting occurred. Therefore, the ultrasonic processing induced surface amorphization of TiO_2, and then reduced the size and valence state of grafted Pt nanoparticles. Subsequently, the yield and selectivity of CH_4 in photothermal catalytic CO_2 conversion were boosted by the combination of smaller sized Pt nanoparticles and amorphous surface.

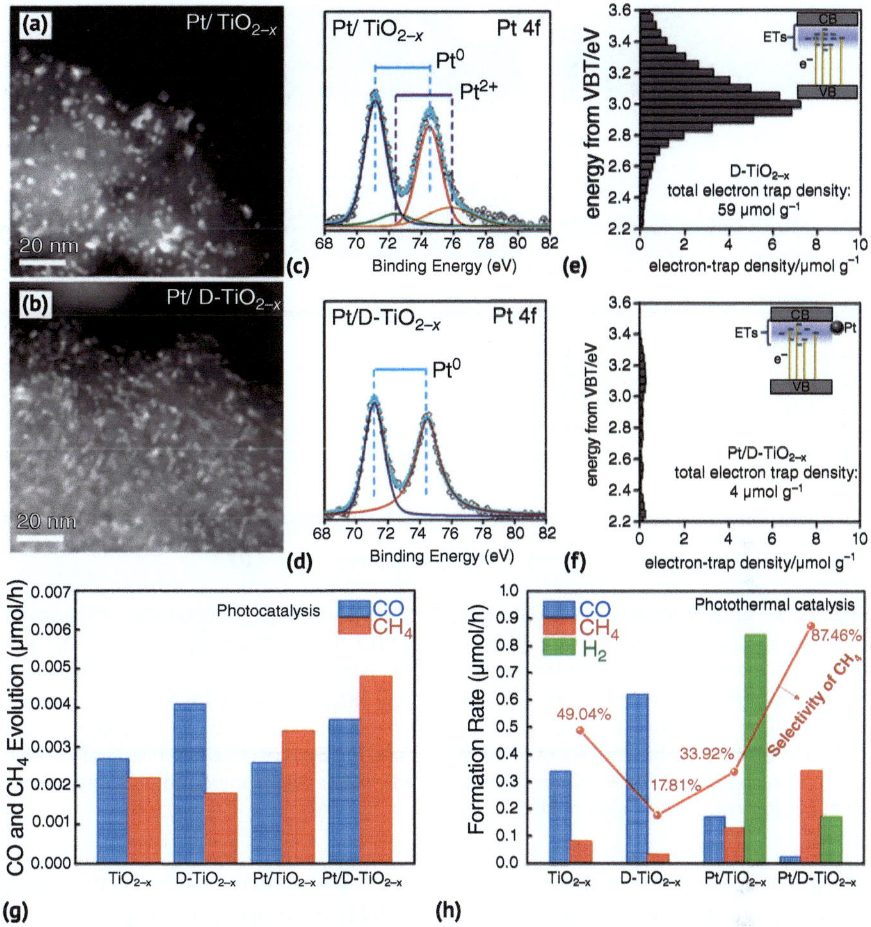

Figure 6.7 (a) and (b) HAADF-STEM images of Pt-loaded TiO_2-AB and Pt-loaded surface amorphous TiO_2-AB. XPS spectra of the Pt 4f doublet states of (c) Pt-loaded TiO_2-AB and (d) Pt-loaded surface amorphous TiO_2-AB. RDB-PAS results of (e) surface amorphous TiO_2-AB and (f) Pt-loaded surface amorphous TiO_2-AB. (g) Photocatalytic and (h) photothermal catalytic CO_2 conversion for TiO_2-AB and surface amorphous TiO_2-AB with and without Pt loading. Reproduced from [61]/with permission of The Royal Society of Chemistry.

6.3.3 Single-atom Cocatalyst

Compared with clusters, single-atom cocatalysts have a higher utilization rate of atoms, stronger metal-support interaction, and unique electronic structure for excellent catalytic performance. In recent years, single-atom metals overcome the problem of precious metal dosage, and have attracted sustained interest from researchers [62]. Unfortunately, the more abundant and cheaper nonprecious single-atom metals, still have received less attention.

We recently utilized ethanol solution impregnation method to successfully graft atomically dispersed MnO_x on TiO_2 microspheres (Figure 6.8a), thereby boosting the deep oxidation of NO_x under mild photocatalytic conditions [63]. The HAADF-STEM images and statistics plot of Mn species demonstrated that

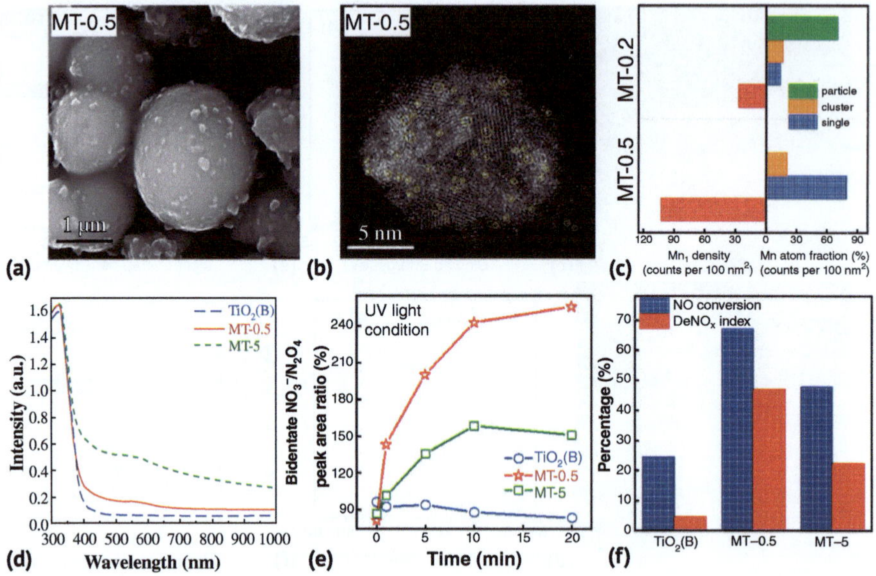

Figure 6.8 (a) SEM images and (b) HAADF-STEM images of MT-0.5. (c) Statistical diagram of Mn atom number of different distribution forms of MT-0.5 and MT-2 samples in 100 nm^2 area. (d) UV–vis spectra of Mn species loaded TiO_2(B) microsphere samples. (e) The peak area ratio of bidentate NO_3^- (located at 1294 cm^{-1}) and N_2O_4(2 NO_2) (located at 1735 cm^{-1}) upon the in situ FTIR of NO adsorption under photocatalytic reaction condition at the time point of 0, 1, 5, 10, and 20 min for the samples of MT-0, MT-0.5 and MT-5, respectively. (f) histogram of NO conversion and $DeNO_x$ index for Mn species loaded TiO_2(B) microsphere samples. Reproduced from Ref. [63]/with permission of Elsevier.

the grafted Mn species were ultra-dispersed and nearly in the form of single atoms (Figure 6.8b,c). Due to the d–d electron transition of grafted Mn^{3+}, Mn-loaded TiO_2 samples presented a characteristic absorption peak at 457–650 nm (Figure 6.8d), which improved visible absorption. Atomically dispersed MnO_x species would act as the part of photogenerated hole trap, and thus promote the charge separation of photogenerated carriers in TiO_2, better than MnO_x nanoparticles-loaded TiO_2.

In-situ FTIR analyses illustrated that the monoatomically dispersed MnO_x-loaded TiO_2(B) sample exhibited excellent low concentration NO_x abatement performance, which provided a rational mechanism for the NO deep oxidation. N_2O_4 was the dimer of toxic byproduct NO_2 produced by the chemical adsorption and polymerization of NO since the dark adsorption process ($2NO + O_2 \rightarrow 2NO_2 \leftrightarrow N_2O_4$). The bidentate NO_3^- was identified as the stable oxidation state product for NO in the photocatalytic reaction process. These two vital intermediates in NO_x abatement were detected on in-situ FTIR and quantified as shown in Figure 6.8e. As a result, the ultra-dispersed MnO_x-loaded TiO_2(B) showed the highest peak area ratio of bidentate NO_3^- and N_2O_4 compared with MnO_x nanoparticles-loaded TiO_2 and TiO_2, indicating monoatomically dispersed Mn species facilitated the formation of stable bidentate NO_3^-. Correspondingly, the NO conversion and $DeNO_x$ index over monoatomically dispersed MnO_x-loaded TiO_2(B) increased remarkably to 67% and 47%, respectively (Figure 6.8f). Thus, the well-designed modification of single atom cocatalyst on TiO_2 catalyst holds great potential to boost photocatalytic efficiency and optimize catalytic paths for target valuable products.

6.4 Conclusion

In this chapter, various surface modification strategies to enhance the photocatalytic performance of TiO_2 are discussed, including the incorporation of noble metal cocatalysts, the introduction of sensitizers, modification with MO_x clusters, surface doping, and the modification with 2D materials. These strategies not only significantly enhance the light absorption capabilities of TiO_2 but also promote charge separation and transfer, thereby greatly improving its potential applications in solar fuel production and environmental purification.

Specifically, the addition of noble metal cocatalysts effectively enhances the efficiency of charge separation; the use of sensitizers broadens the light response range of TiO_2; MO_x cluster modification further enhances catalytic efficiency by providing additional active sites, while surface doping and the modification with 2D materials offer new pathways for photocatalysis through the adjustment of TiO_2 electronic structure and surface characteristics. The development of these modification techniques not only reflects the innovative progress in the field of photocatalysis within material science but also provides a significant theoretical and experimental foundation for the design and development of new, efficient photocatalysts.

While current research on TiO_2 photocatalysis mainly focuses on enhancing its response to visible light, the efficiency of these surface-modified TiO_2 under visible light remains limited. UV light, as part of the solar spectrum, possesses high energy characteristics that make TiO_2 photocatalytic processes effective, especially demonstrating unique advantages in certain environmental purification and disinfection applications. We have demonstrated here that methods like solution plasma processing and ultrasonic processing as well as MnO_x single-atom modification are effective to modulate surface electronic states of TiO_2 for more efficient UV photocatalysis. Moving forward, the enhancement of UV photocatalytic activity is still indispensable for TiO_2 photocatalysis to achieve more successful applications over an extended period.

In conclusion, precise surface modification and functionalization of TiO_2 not only significantly improve its photocatalytic performance but also allow for the customization of catalyst properties according to specific application needs. These advancements not only propel the development of TiO_2 photocatalysis technology but also provide robust technical support for addressing energy and environmental issues. Future research should continue to explore new modification strategies, while not overlooking the potential in the field of UV catalysis either, with the aim of discovering more efficient, stable, and practically applicable photocatalytic systems.

Acknowledgments

Part of works involved in this chapter was supported by the Natural Science Foundation of China (52273236, U22A2078, 91833303, 51102001) and Jilin Province Science and Technology Development Project (20220201073GX).

References

1 a) Fujishima, A., Zhang, X., and Tryk, D.A. TiO_2 photocatalysis and related surface phenomena (2008). *Surface Science Reports* 63: 515; b) Wang, C., Zhang, X., and Liu, Y. Promotion of multi-electron transfer for enhanced photocatalysis: A review focused on oxygen reduction reaction (2015). *Applied Surface Science* 15: 28–45.

2 Kraeutler, B. and Bard, A.J. Heterogeneous photocatalytic decomposition of saturated carboxylic acids on titanium dioxide powder. Decarboxylative route to alkanes (1978). *Journal of American Chemical Society* 100: 5985.

3 a) Hoffmann, M.R., Martin, S.T., Choi, W. et al. Environmental applications of semiconductor photocatalysis (1995). *Chemical Reviews* 95: 69–96; b) Parka, H., Park, Y., Kimb, W. et al. Surface modification of TiO_2 photocatalyst for environmental applications (2013). *Journal of Photochemistry and Photobiology C: Photochemistry Reviews* 15: 1–20; c) Schneider, J., Matsuoka, M., Takeuchi, M. et al. Understanding TiO_2 Photocatalysis: Mechanisms and Materials (2014). *Chemical Reviews* 114: 9919–9986.

4 Bamwenda, G.R., Tsubota, S., Nakamura, T. et al. Photoassisted hydrogen production from a water-ethanol solution: a comparison of activities of Au-TiO_2 and Pt-TiO_2 (1995). *Journal of Photochemistry and Photobiology A: Chemistry* 89, 177.

5 DeRita, L., Dai, S., Lopez-Zepeda, K. et al. Catalyst Architecture for Stable Single Atom Dispersion Enables Site-Specific Spectroscopic and Reactivity Measurements of CO Adsorbed to Pt Atoms, Oxidized Pt Clusters, and Metallic Pt Clusters on TiO_2 (2017). *Journal of American Chemical Society* 139: 14150–14165.

6 Wang, W.-N., An, W.-J., Ramalingam, B. et al. Size and Structure Matter: Enhanced CO_2 Photoreduction Efficiency by Size-Resolved Ultrafine Pt Nanoparticles on TiO_2 Single Crystals (2012). *Journal of American Chemical Society* 134: 11276.

7 Zhou, Y., King, D.M., Liang, X. et al. Optimal preparation of Pt/TiO_2 photocatalysts using atomic layer deposition (2010). *Applied Catalysis B:Environmental* 101: 54–60.

8 Shah, S.I., Li, W., Huang, C.-P. et al. Study of Nd^{3+}, Pd^{2+}, Pt^{4+}, and Fe^{3+} dopant effect on photoreactivity of TiO_2 nanoparticles (2002). *The Proceedings of the National Academy of Sciences* 99: 6482–6486.

9 Ismail, A.A., Bahnemann, D.W., Rathousky, J. et al. Multilayered ordered mesoporous platinum/titania composite films: does the photocatalytic activity benefit from the film thickness? (2011). *Journal of Materials Chem*istry 21: 7802–7810.

10 Rogers, S.M., Catlow, C.R.A., Chan-Thaw, C.E. et al. Tailoring gold nanoparticle characteristics and the impact on aqueous-phase oxidation of glycerol (2015). *ACS Catalysis* 5: 4377–4384.

11 Christoforidis, K.C. and Fornasiero, P. Photocatalytic hydrogen production: a rift into the future energy supply (2017). *ChemCatChem* 9: 1523–1544.

12 Chiarello, G.L., Bernareggi, M., and Selli, E. Redox Dynamics of Pt and Cu Nanoparticles on TiO_2 during the Photocatalytic Oxidation of Methanol under

Aerobic and Anaerobic Conditions Studied by In Situ Modulated Excitation X-ray Absorption Spectroscopy (2022). *ACS Catalysis* 12: 12879–12889.

13 a) Caudillo-Flores, U., Muñoz-Batista, M.J., Fernández-García, M. et al. Bimetallic Pt-Pd co-catalyst Nb-doped TiO_2 materials for H_2 photo-production under UV and Visible light illumination (2018). *Applied Catalysis B:Environmental.* 238: 533–545, b) Huang, S.-Y., Ganesan, P., and Popov, B.N. Electrocatalytic activity and stability of titania-supported platinum–palladium electrocatalysts for polymer electrolyte membrane fuel cell (2012). *ACS Catalysis* 2: 825–831.

14 a) Rozman, N., Nadrah, P., Cornut, R. et al. TiO_2 photocatalyst with single and dual noble metal co-catalysts for efficient water splitting and organic compound removal (2021). *International Journal of Hydrogen Energy* 46: 32871–32881; b) Kondratenko, E.V., Mul, G., Baltrusaitis, J. et al. Status and perspectives of CO_2 conversion into fuels and chemicals by catalytic, photocatalytic and electrocatalytic processes (2013). *Energy Environmental Science* 6: 3112–3135.

15 a) Tanaka, A., Sakaguchi, S., Hashimoto, K. et al. Preparation of Au/TiO_2 with Metal Cocatalysts Exhibiting Strong Surface Plasmon Resonance Effective for Photoinduced Hydrogen Formation under Irradiation of Visible Light (2013). *ACS Catalysis* 3: 79–85; b) Ge, M., Cao, C., Li, S. et al. In situ plasmonic Ag nanoparticle anchored TiO_2 nanotube arrays as visible-light-driven photocatalysts for enhanced water splitting (2016). *Nanoscale* 8: 5226–5234.

16 a) Lin, S.D., Bollinger, M., and Vannice, M.A. Low temperature CO oxidation over Au/TiO_2 and Au/SiO_2 catalysts (1993). *Catalysis Letters* 17: 245–262; b) Tada, K., Maeda, Y., Koga, H. et al. TiO_2 Crystal Structure Dependence of Low-temperature CO Oxidation Catalyzed by Au/TiO_2 (2018). *Chemistry Letters* 47, 200–203; c) Fu, A., Chen, X., Tong, L. et al. Remarkable Visible-Light Photocatalytic Activity Enhancement over Au/p-type TiO_2 Promoted by Efficient Interfacial Charge Transfer (2019). *ACS Applied Materials Interfaces* 11: 24154–24163.

17 a) Sharma, V.K., Yngard, R.A., and Lin, Y.O. Silver nanoparticles: green synthesis and their antimicrobial activities (2009). *Advances in Colloid and Interface Science* 145: 83–96; b) Akhavan Lasting antibacterial activities of Ag–TiO_2/Ag/a-TiO_2 nanocomposite thin film photocatalysts under solar light irradiation (2009). *Journal of Colloid and Interface Science* 336: 117–124.

18 a) Furube, A. and Hashimoto, S. Insight into plasmonic hot-electron transfer and plasmon molecular drive: new dimensions in energy conversion and nanofabrication (2017). *NPG Asia Materials* 9: e454; b) Shi, X., Ueno, K., Oshikiri, T. et al. Enhanced water splitting under modal strong coupling conditions (2018). *Nature Nanotechnology* 13: 953–958; c) Gurbatov, S.O., Modin, E., Puzikov, V. et al. Black Au-Decorated TiO_2 Produced via Laser Ablation in Liquid (2021). *ACS Applied Materials Interfaces* 13: 6522–6531.

19 a) Zhang, Z., Zhu, Y., Asakura, H. et al. Thermally stable single atom Pt/m-Al_2O_3 for selective hydrogenation and CO oxidation (2017). *Nature Communications* 8: 16100; b) Kaushik, S., Wu, D., Zhang, Z. et al. Universal Synthesis of Single-Atom Catalysts by Direct Thermal Decomposition of Molten Salts for Boosting Acidic Water Splitting (2024). *Advanced Materials* 2401163; c) Hwang, I., Mazare, A., Will, J. et al. Inhibition of H_2 and O_2 Recombination: The Key to a Most Efficient

Single-Atom Co-Catalyst for Photocatalytic H_2 Evolution from Plain Water (2022). *Advanced Functional Materials* 32: 2207849.

20 Zhang, G., Kim, G., and Choi, W. Visible light driven photocatalysis mediated via ligand-to-metal charge transfer (LMCT): an alternative approach to solar activation of titania (2014). *Energy Environmental Science* 7: 954–966.

21 a) Jono, R., Fujisawa, J., Segawa, H. et al. Theoretical Study of the Surface Complex between TiO_2 and TCNQ Showing Interfacial Charge-Transfer Transitions (2011). *The Journal of Physical Chemistry Letters* 2: 1167–1170; b) Jono, R., Awai, F., Kondo, T. et al. Basic Principle of Photoinduced Charge Separation for TiO_2–TCNQ Surface Complex Revealed by a Multibody Model (2020). *The Journal of Physical Chemistry C* 124: 13535–13540.

22 Chen, F., Zou, W., Qu, W. et al. Photocatalytic performance of a visible light TiO_2 photocatalyst prepared by a surface chemical modification process (2009). *Catalysis Communications* 10: 1510–1513.

23 a) Macyk, W., Burgeth, G., and Kisch, H. Photoelectrochemical properties of platinum(iv) chloride surface modified TiO_2 (2003). *Photochemical Photobiological Sciences* 2: 322–328; b) Macyk, W., and Kisch, H. Photosensitization of Crystalline and Amorphous Titanium Dioxide by Platinum(IV) Chloride Surface Complexes (2001). *Chemistry A European Journal* 7: 1862–1867.

24 a) Irie, H., Miura, S., Kamiya, K. et al. Efficient visible light-sensitive photocatalysts: grafting Cu (II) ions onto TiO_2 and WO_3 photocatalysts (2008). *Chemical Physics Letters* 457: 202–205; b) Qiu, X., Miyauchi, M., Sunada, K. et al. Hybrid Cu_xO/TiO_2 Nanocomposites As Risk-Reduction Materials in Indoor Environments (2012). *ACS Nano* 6: 1609–1618; c) Miyauchi, M., Irie, H., Liu, M. et al. Visible-Light-Sensitive Photocatalysts: Nanocluster-Grafted Titanium Dioxide for Indoor Environmental Remediation (2016). *The Journal of Physical Chemistry Letters* 7: 75–84.

25 Li, D., Fernández, Y., Ola, O. et al. On the impact of Cu dispersion on CO_2 photoreduction over Cu/TiO_2 (2012). *Catalysis Communications* 25: 78–82.

26 Tada, H., Jin, Q.L., Nishijima, H. et al. Titanium (IV) dioxide surface-modified with iron oxide as a visible light photocatalyst (2011). *Angewandte Chemie International Edition* 50: 3501–3505.

27 Murakami, N., Chiyoya, T., Tsubota, T. et al. Switching redox site of photocatalytic reaction on titanium (IV) oxide particles modified with transition-metal ion controlled by irradiation wavelength (2008). *Applied Catalysis A: General* 348: 148–152.

28 Peng, X., Wang, C., Li, Y. et al. Revisiting cocatalyst/TiO_2 photocatalyst in blue light photothermalcatalysis (2019). *Catalysis Today* 335: 286–293.

29 Kong, L., Wang, C., Zheng, H. et al. Defect-Induced Yellow Color in Nb-Doped TiO_2 and Its Impact on Visible-Light Photocatalysis (2015). *The Journal of Physical Chemistry C* 119: 16623–16632.

30 Wang, G., Wang, H., Ling, Y. et al. Hydrogen-Treated TiO_2 Nanowire Arrays for Photoelectrochemical Water Splitting (2011). *Nano Letters* 11(7): 3026–3033.

31 Xing, M., Zhang, J., Chen, F. et al. An economic method to prepare vacuum activated photocatalysts with high photo-activities and photosensitivities (2011). *Chemical Communications* 47: 4947–4949.

32 Yu, F., Wang, C., Li, Y. et al. Enhanced Solar Photothermal Catalysis over Solution Plasma Activated TiO_2 (2020). *Advanced Science* 7: 2000204.

33 Ou, G., Xu, Y., Wen, B. et al. Tuning defects in oxides at room temperature by lithium reduction (2018). *Nature Communications* 9: 1302.

34 a) Xing, M., Fang, W., Nasir, M. et al. Self-doped Ti^{3+}-enhanced TiO_2 nanoparticles with a high-performance photocatalysis (2013). *Journal of Catalysis*. 297: 236–243; b) Fang, W., Xing, M., Zhang, J. et al. A new approach to prepare Ti^{3+} self-doped TiO_2 via $NaBH_4$ reduction and hydrochloric acid treatment (2014). *Applied Catalysis B:Environmental* 160–161: 240–246.

35 Kong, L., Jiang, Z., Wang, C. et al. Simple Ethanol Impregnation Treatment Can Enhance Photocatalytic Activity of TiO_2 Nanoparticles under Visible-Light Irradiation (2015). *ACS Applied Materials Interfaces* 7: 7752–7758.

36 Kong, L., Zhang, X., Wang, C. et al. Synergic effects of Cu_xO electron transfer co-catalyst and valence band edge control over TiO_2 for efficient visible-light photocatalysis (2017). *Chinese Journal of Catalysis* 38: 2120–2131.

37 Kong, L., Wang, C., Wan, F. et al. Synergistic effect of surface self-doping and Fe species-grafting for enhanced photocatalytic activity of TiO_2 under visible-light (2017). *Applied Surface Science* 396: 26–35.

38 Li, Y., Wang, C., Song, M. et al. TiO_{2-x}/CoO_x photocatalyst sparkles in photothermocatalytic reduction of CO_2 with H_2O steam (2019). *Applied Catalysis B:Environmental* 243: 760–770.

39 Parada, G.A., Goldsmith, Z.K., Kolmar, S. et al. Concerted proton-electron transfer reactions in the Marcus inverted region (2019). *Science* 364: 471–475.

40 Dempsey, J.L. How a highly driven reaction hits the brakes (2019). *Science* 364: 436–437.

41 Zheng, H., Wang, C., Zhang, X. et al. Control over energy level match in Keggin polyoxometallate-TiO_2 microspheres for multielectron photocatalytic reactions (2018). *Applied Catalysis B:Environmental* 234: 79–89.

42 a) Yang, J., Zhang, X., Liu, H. et al. Heterostructured TiO_2/WO_3 porous microspheres: preparation, characterization and photocatalytic properties (2013). *Catalysis Today* 201: 195–202; b) Yang, J., Zhang, X., Li, B. et al. Photocatalytic activities of heterostructured TiO_2-graphene porous microspheres prepared by ultrasonic spray pyrolysis (2014). *Journal of Alloys and Compounds* 584: 180–184; c) Zheng, H., Wang, C., Zhang, X. et al. Ultrasonic spray pyrolysis assembly of a TiO_2–WO_3–Pt multi-heterojunction microsphere photocatalyst using highly crystalline WO_3 nanosheets: less is better (2016). *New Journal of Chem*istry 40: 3225–3232.

43 a) Guo, L., Yang, Z., Marcus, K. et al. MoS_2/TiO_2 heterostructures as nonmetal plasmonic photocatalysts for highly efficient hydrogen evolution (2018). *Energy Environmental Science* 11: 106–114; b) Wang, J. and Zhang, W.-D. Modification of TiO_2 nanorod arrays by graphite-like C_3N_4 with high visible light photoelectrochemical activity (2012). *Electrochimica Acta* 71: 10–16.

44 Ho, W., Yu, J.C., Lin, J. et al. Preparation and Photocatalytic Behavior of MoS_2 and WS_2 Nanocluster Sensitized TiO_2 (2004). *Langmuir* 20: 14, 5865–5869.

45 a) Williams, G., Seger, B., and Kamat, P.V. TiO_2-Graphene Nanocomposites. UV-Assisted Photocatalytic Reduction of Graphene Oxide (2008). *ACS Nano*

2: 1487–1491; b) Su, T., Hood, Z.D., Haguib, M. et al. Monolayer Ti_3C_2Tx as an Effective Co-catalyst for Enhanced Photocatalytic Hydrogen Production over TiO_2 (2019). *ACS Applied Energy Materials* 2: 4640–4651.

46 a) Liang, Y., Wang, H., Casalongue, H.S. et al. TiO_2 Nanocrystals Grown on Graphene as Advanced Photocatalytic Hybrid Materials (2010). *Nano Research* 3(10): 701–705; b) Zhou, W., Yin, Z., Du, Y. et al. Synthesis of Few-Layer MoS_2 Nanosheet-Coated TiO_2 Nanobelt Heterostructures for Enhanced Photocatalytic Activities (2013). *Small* 9(1): 140–147; c) Qiu, B., Xing, M., and Zhang, J. Mesoporous TiO_2 Nanocrystals Grown in Situ on Graphene Aerogels for High Photocatalysis and Lithium-Ion Batteries (2014). *Journal of American Chemical Society* 136: 5852–5855.

47 Yuan, Y.-J., Ye, Z.-J., Lu, H.-W. et al. Constructing Anatase TiO_2 Nanosheets with Exposed (001) Facets/Layered MoS_2 Two-Dimensional Nanojunctions for Enhanced Solar Hydrogen Generation (2016). *ACS Catalysis* 6: 532–541.

48 Tamaki, Y., Hara, K., Katoh, R. et al. Femtosecond Visible-to-IR Spectroscopy of TiO_2 Nanocrystalline Films: Elucidation of the Electron Mobility before Deep Trapping (2009). *The Journal of Physical Chemistry C* 113: 11741–11746.

49 Di Valetin, C., Pacchioni, G., and Selloni, A. Reduced and n-Type Doped TiO_2: Nature of Ti^{3+} Species (2009). *The Journal of Physical Chemistry C* 113: 20543–20552.

50 Zhu, M., Mi, Y., Zhu, G. et al. Determination of Midgap State Energy Levels of an Anatase TiO_2 Nanocrystal Film by Nanosecond Transient Infrared Absorption – Excitation Energy Scanning Spectra (2013). *The Journal of Physical Chemistry C* 117: 18863–18869.

51 Saito, N., Bratescu, M.A., and Hashimi, K. Solution plasma: A new reaction field for nanomaterials synthesis (2018). *Japanese Journal of Applied Physics* 57: 0102A4.

52 a) Wang, R., Wang, C., Xing, Y. et al. Solution Plasma for Surface Design of Advanced Photocatalysts (2023). *Catalysts* 13: 1124; b) Namihira, T., Sakai, S., Yamaguchi, T. et al. Electron temperature and electron density of underwater pulsed discharge plasma produced by solid-state pulsed-power generator (2007). *IEEE Transactions on Plasma Science* 35: 614–618.

53 Dack, M.R.J. The importance of solvent internal pressure and cohesion to solution phenomena (1975). *Chemical Society Reviews* 4: 211–229.

54 Wang, R., Che, G., Wang, C. et al. Alcohol plasma processed surface amorphization for photocatalysis (2022). *ACS Catalysis* 12: 12206–12216.

55 Cronemeyer, D.C. Infrared Absorption of Reduced Rutile Ti Single Crystals (1959). *Physical Review* 113: 1222–1226.

56 a) Valentin, C.D., Pacchioni, G., and Selloni, A. Reduced and n-Type Doped TiO_2: Nature of Ti^{3+} Species (2009). *The Journal of Physical Chemistry C* 113: 20543–20552; b) Guo, Y., Chen, S., Yu, Y. et al. Hydrogen-Location-Sensitive Modulation of the Redox Reactivity for Oxygen-Deficient TiO_2 (2019). *Journal of American Chemical Society* 141: 8407–8411.

57 Yu, F., Wang, C., Wang, R. et al. Solution plasma engineering the surface of nitrogen doped TiO_2 for photothermal catalysis (2023). *Applied Surface Science* 624: 157119.

58 Suslick, K.S. Sonochemistry (1990). *Science* 247: 1439–1445.
59 Bhangu, S.K. and Ashokkumar, M. Theory of sonochemistry (2016). *Topics in Current Chemistry* 374: 56.
60 Liu, L. and Corma, A. Metal catalysts for heterogeneous catalysis: from single atoms to nanoclusters and nanoparticles (2018). *Chemical Reviews* 118: 4981–5079.
61 Yu, F., Wang, C., Ma, H. et al. Revisiting Pt/TiO_2 photocatalysts for thermally assisted photocatalytic reduction of CO_2 (2020). *Nanoscale* 12: 7000–7010.
62 a) Hejazi, S., Mohajernia, S., Osuagwu, B. et al. On the Controlled Loading of Single Platinum Atoms as a Co-Catalyst on TiO_2 Anatase for Optimized Photocatalytic H_2 Generation (2020). *Advanced Materials* 32: 1908505; b) Zhou, X., Hwang, I., Tomanec, O. et al. Advanced photocatalysts: pinning single atom co-catalysts on titania nanotubes (2021). *Advcanced Functional Materials* 31: 2102843; c) Hwang, I., Mazare, A., Will, J. et al. Inhibition of H_2 and O_2 Recombination: The Key to a Most Efficient Single-Atom Co-Catalyst for Photocatalytic H_2 Evolution from Plain Water (2022). *Advcanced Functional Materials* 32: 2207849; d) Wang, Y., Denisov, N., Qin, S. et al. Stable and Highly Active Single Atom Configurations for Photocatalytic H_2 Generation (2024). *Advanced Materials* 2400626. e) Qin, S., Denisov, N., Kim, H. et al. Photocatalytic H_2 Generation: Controlled and Optimized Dispersion of Single Atom Co-Catalysts Based on Pt-TCPP Planar Adsorption on TiO_2 (2024). *Angewandte Chemie International Edition* 63: e202316660; f) Yoo, J.E., Lee, K., Altomare, M. et al. Self-organized arrays of single-metal catalyst particles in TiO_2 cavities: a highly efficient photocatalytic system (2013). *Angewandte Chemie International Edition* 52: 7514–7517.
63 Ma, H., Li, D., Li, Y. et al. Atomically dispersed Mn sites on TiO_2 (B) microspheres enables efficient photocatalytic abatement of NO_x (2024). *Separation and Purification Technology* 337: 126357.

Part II

Photocatalytic Properties

7

Photocatalytic Fixed Bed Reactions for Efficient Photon Utilization of TiO_2

*Jiazang Chen**

School of Engineering, Hangzhou Normal University, Hangzhou, China

*Corresponding author: chenjiazang@hznu.edu.cn

7.1 Introduction

The rapid updates in value-added oxidation half-reaction integrated with hydrogen evolution (HE) encourage us to address certain issues related to the practical application of photocatalysis [1, 2]. Photocatalyst immobilization, which allows a straightforward separation from the process stream and simplifies the installation and operation in practical application [3], encounters certain dissenting voices. This is because that, by shrinking the photocatalyst originally dispersed in slurry suspension into thin coating, the probability for the target collision of the reactant(s) with reactive sites should be severely reduced. This will lower the volume activity of the reaction [4, 5]. Another drawback to photocatalyst immobilization is that the time constants for occurring mass transport of the reactant(s) in the pores of the formed photocatalyst films is far larger than the widely believed femtosecond-millisecond timescale for all the photoinduced electronic processes and steps occurred in photocatalytic reaction (Figure 7.1) [6–14]. Regarding these issues, texturing the internal walls and introducing transparent supports that can furthest expose the active surface of the immobilized photocatalysts are the most common (Figure 7.2) [15–18]. Despite of these efforts, the photocatalytic photon utilization of these configurations is still far lower than that of dispersed counterparts.

Unlike imitating the slurry suspension, an opposite strategy that anchors photocatalyst onto the smooth internal walls of the light incident side of the reactor can surprisingly exhibit superiority over dispersed counterparts in photocatalytic photon utilization [19]. This promising result encourages us to gain the insight into photoinduced electronic processes and explore the applications of the formed photocatalytic fixed bed reactions (PFBRs) [20–26]. In this chapter,

Nanostructured TiO_2: Synthesis, Photocatalytic Properties and Applications, First Edition.
Edited by Jia Hong Pan, Wan In Lee and Detlef W. Bahnemann

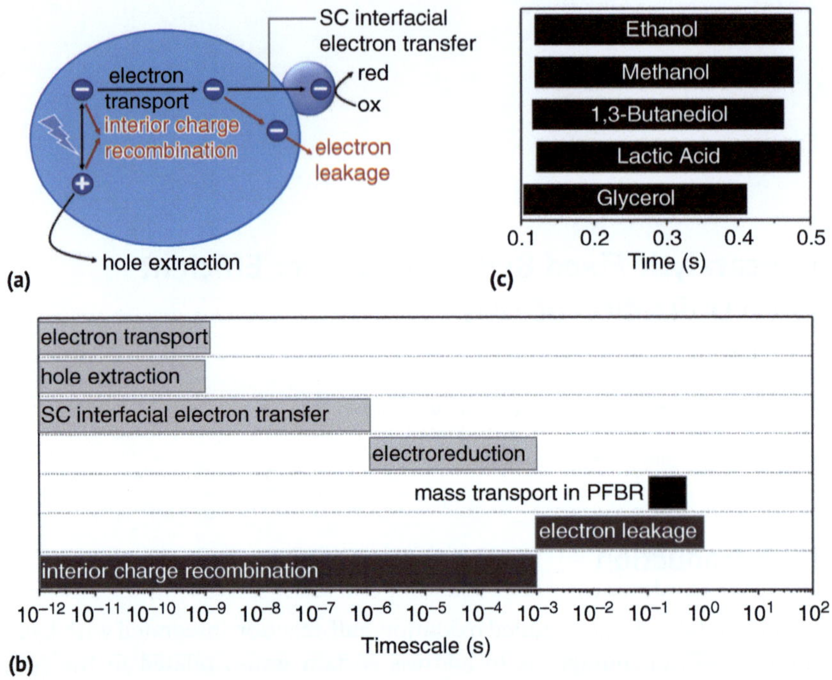

Figure 7.1 Schematic diagram for the photoinduced electronic processes occurring in photocatalytic reaction (a) and literature reported timescales (b). Replotted from Ref. [26]/ Elsevier (b). For comparison, the timescales for occurring mass transport of certain common organic compounds in the pores of the immobilized photocatalyst films (c) are also shown. Reproduced from [19]/with permission of American Chemical Society.

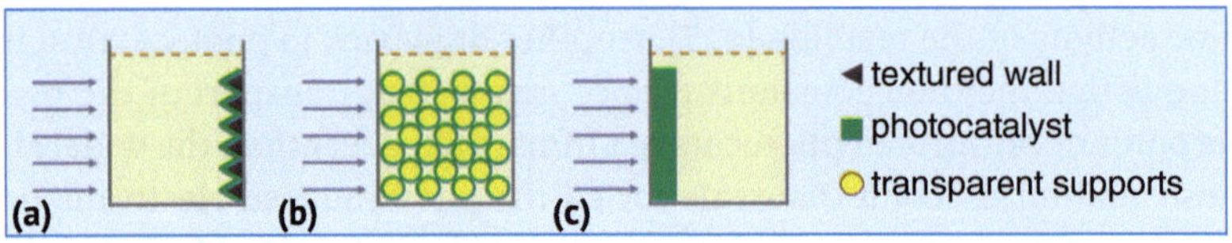

Figure 7.2 Schematic diagrams for immobilization of photocatalysts on textured internal wall of the reactor (a), transparent support (b), and smooth internal wall of light incidence side (c). Reproduced from [19]/with permission of American Chemical Society.

we will cognize certain key electronic processes occurring in actual photocatalytic reaction and discuss some important features that endow the PFBR with efficient photon utilization.

7.2 Cognition of Time Consumption for Photoinduced Electronic Processes

7.2.1 Light-intensity-dependent Electronic Process

In photocatalytic reaction, the photoinduced electronic processes exhibit a power-law dependence on the effective incident photons [27–30]. The index of this relation

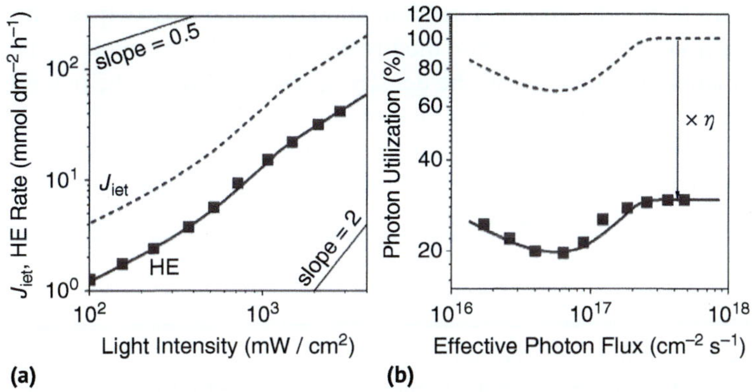

Figure 7.3 Light-intensity-dependent SC interfacial electron transfer and the correlated photocatalytic HE rate (a) and photon utilization (b). Since SC interfacial electron transfer dominates photocatalytic reaction, the simulated curves for this interfacial electronic process can be generally parallel to the plots of photocatalytic behaviors with a constant distance in the double-logarithm coordinates. The electron utilization can be estimated from the constant distance. Reproduced from [20]/with permission of American Chemical Society.

is also sensitively light-intensity-dependent and usually deviates from unity. For example, photoinduced semiconductor-cocatalyst (SC) interfacial electron transfer rate that can be described as current (J_{iet}) can be given approximately by (Figure 7.3) [28, 30]:

$$J_{iet} = k_{iet} \cdot (A \cdot P_{inc})^{\alpha} = \frac{k_{iet}}{(A \cdot P_{inc})^{1-\alpha}} \cdot A \cdot P_{inc} = k_{iet}' \cdot A \cdot P_{inc}, \text{ (with } 0.5 \leq \alpha \leq 2) \quad (7.1)$$

in which k_{iet} is the pseudo-rate-constant after separating the variables of incident photons (P_{inc}) and absorption ratio of photons (A) from the spectrum. However, this pseudo-rate-constant is also dependent on the incident photons [31]. Since SC interfacial electron transfer mainly contains the parallel connection of trap-assisted charge recombination and thermionic emission [30], J_{iet} can be given approximately by:

$$J_{iet} \approx J_{cr} + J_{te} \quad (7.2a)$$

in which J_{cr} and J_{te} are, respectively, the rates for photoinduced trap-assisted charge recombination and thermionic emission and can be given by [30, 31]:

$$J_{cr} = k_{cr} \cdot (A \cdot P_{inc})^{\beta}, \text{ (with } \beta \to 0.5) \quad (7.2b)$$

and

$$J_{te} = k_{te} \cdot (A \cdot P_{inc})^{\gamma}, \text{ (with } \gamma \to 2) \quad (7.2c)$$

Numerically and physically, trap-assisted charge recombination mainly accounts for SC interfacial electron transfer under weak illumination (Figure 7.3), because most of the photogenerated electrons can hardly overcome the very high potential barrier [30, 32, 33]. Under intense irradiation, however, thermionic emission plays major role to this interfacial electronic process [28, 30]. The nearly quadratic growth in J_{te} and J_{iet} (Figure 7.3) indicates that the photogenerated charges mainly determined by the absorption of incident photons not only participate into the interfacial electronic processes but also modify the kinetics. For example, by intensifying the irradiation, the increase in the density of electron lowers the potential barrier for SC interfacial electron transfer by thermionic emission [28, 31].

By further intensifying the irradiation, the height of potential barrier for thermionic emission can be minimized; SC interfacial electron transfer rate therefore can generally increase linearly with the incident photons (Figure 7.3) [20, 28]. Therefore, Eq. (7.1) can be given approximately by:

$$J_{\text{iet}} \approx k_{\text{te-line}} \cdot A \cdot P_{\text{inc}} \approx A \cdot P_{\text{inc}} \tag{7.2d}$$

in which $k_{\text{te-line}}$ can be nearly unity.

The index (α) shown in Eq. (7.1) increases from 0.5 to 2 with the light intensity and, eventually decrease to nearly unity under very intense irradiation that is rarely used in photocatalytic reaction (Figure 7.3) [28, 30]. The nonlinearity makes the characteristic time constants for occurring these electronic processes sensitively depends on the light intensity.

It should be pointed out that photocatalytic reaction rates also exhibit power-law dependence on the incident photons with the index that can be the same to α shown in Eq. (7.1) [28, 30, 31]. The simulated curves for SC interfacial electron transfer can be generally parallel to the plots of photocatalytic behaviors with a constant distance in the double-logarithm coordinates (Figure 7.3). This is because SC interfacial electron transfer with very high potential barrier (activation energy) dominates photocatalytic reaction [28, 30, 32]. The difference between the simulated curve and the experimental plot can be described as electron utilization (η) that is mainly determined by the side/back reaction occurring at the reduction cocatalyst [28]. To further cognize the rate-determining step and electron utilization, next, we will discuss the time consumption for SC interfacial electron transfer in actual photocatalytic reaction and photon utilization correlated with photocatalyst configuration.

7.2.2 Time Constant Estimation for SC Interfacial Electron Transfer

For certain instrumental methods based on ultrafast pulse laser, the photon flux of incident beam is evidently higher than that of the irradiation used in actual photocatalytic reaction [27]. For example, the photon flux of the laser in a pulse duration (in the level of 10 ps) can be as high as 10^{23}–10^{26} cm^{-2} s^{-1}, which is 5–10 orders of magnitude larger than 10^{16}–10^{18} cm^{-2} s^{-1} of the common irradiation used in photocatalytic reaction [27]. Because of light-intensity-dependence, the kinetic information obtained from these instrumental methods cannot reflect the real electronic processes occurring in actual photocatalytic reaction. For the same reason, the time constants for the electronic processes driven by such very dense photon beam should be far smaller than that in actual working condition.

To obtain the kinetic information that can reflect the real electronic processes occurring in actual photocatalytic reaction, monitoring the decay of open-circuit potential (OCP) from a biased value has recently been proposed [28, 34]. The potential or the irradiation biased on the photocatalyst (semiconductor) can be the same or similar to the actual working condition of the photocatalytic reaction. During the decay of OCP, the kinetic information like the time constants for transfer of electrons from semiconductor to the solution occurring in actual working condition can be reflected. This interfacial electronic process contains parallel connection of three channels that are electron leakage, SC-solution (SCS) interfacial electron transfer for target reaction, and depriving electrons from cocatalyst for the side/ back reaction. By manipulating the electron transfer channel(s), we can obtain the time constants from the OCP decay behaviors (Figure 7.4):

$$\tau_{nC} = \frac{k_B T}{q}\left(\frac{\mathrm{d}E}{\mathrm{d}t}\right)^{-1}, \text{ (with } n = 1,\ 2,\ 3) \tag{7.3}$$

in which k_B is the Boltzmann constant, T is the temperature, q is the elementary charge, t is the time along the decay of potential (E), and n represents number of channel(s).

When the cocatalyst is absent, the direct transfer of electrons from semiconductor to the solution described as single-channel electron transfer can reflect the electron leakage. Thus, the time constants (τ_{leak}) for electron leakage can be:

$$\tau_{leak} = \tau_{1C} \tag{7.4a}$$

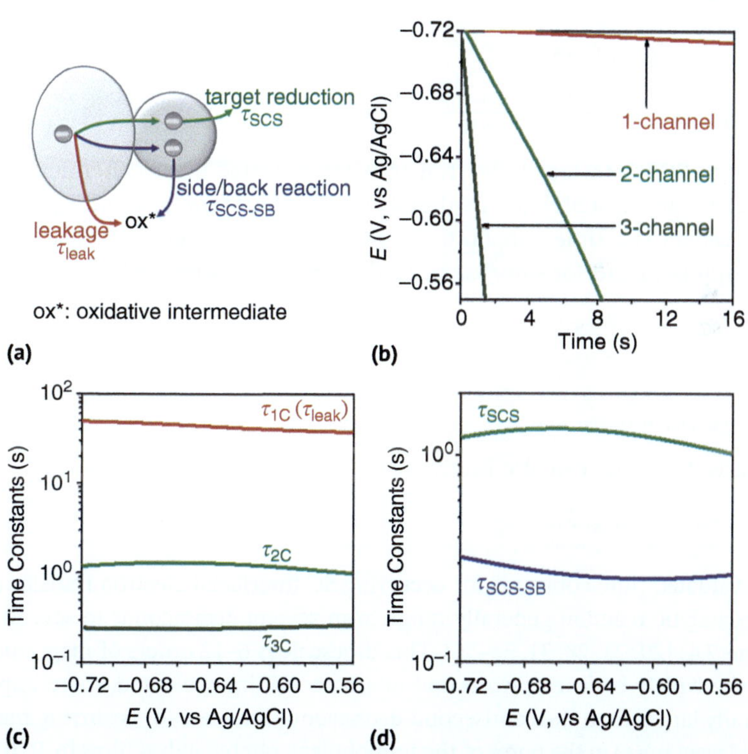

Figure 7.4 Schematic diagram for electron transfer (a). By manipulating the electron transfer channel(s), OCP decay with one, two, and three channels can be obtained (b). By taking the derivation of OCP plots with time, the time constants for electron transfer with one (τ_{1C}), two (τ_{2C}), and three (τ_{3C}) channels can be obtained (c). The time constants of electron transfer through each channel (a) can be obtained by deconvoluting the parallel connection (d). Reproduced from [28]/with permission of American Chemical Society.

By loading cocatalyst onto the semiconductor, time constants (τ_{2C}) for occurring the dual-channel electron transfer can be obtained by monitoring the OCP decay from a biased potential in dark. The time constants (τ_{SCS}) of SCS interfacial electron transfer for target reduction can be given by:

$$\frac{1}{\tau_{2C}} = \frac{1}{\tau_{1C}} + \frac{1}{\tau_{SCS}} \tag{7.4b}$$

Under irradiation, side/back reaction also occurs. The side/back reaction that mainly accounts for the lowering of electron utilization is intrinsically the deprivation of electrons from reduction cocatalyst by intermediates. These intermediates formed by reacting substance with photogenerated holes are highly reactive. By considering the side/back reaction that competes with target reduction, the relationship for time constants of each interfacial electronic process can be:

$$\frac{1}{\tau_{3C}} = \frac{1}{\tau_{2C}} + \frac{1}{\tau_{SCS\text{-}SB}} = \frac{1}{\tau_{1C}} + \frac{1}{\tau_{SCS}} + \frac{1}{\tau_{SCS\text{-}SB}} \tag{7.4c}$$

in which $\tau_{SCS\text{-}SB}$ is the time constants for side/back reaction.

Since τ_{1C} is evidently larger than τ_{SCS} and $\tau_{SCS\text{-}SB}$, according to Eq. (7.4b), we have:

$$\tau_{SCS} \approx \tau_{2C} \tag{7.4d}$$

Similarly, Eq. (7.4c) can be:

$$\frac{1}{\tau_{3C}} = \frac{1}{\tau_{2C}} + \frac{1}{\tau_{SCS\text{-}SB}} \approx \frac{1}{\tau_{SCS}} + \frac{1}{\tau_{SCS\text{-}SB}} \tag{7.4e}$$

Although SCS interfacial electron transfer contains electrochemical steps, the time constants of which are negligibly small. Therefore, τ_{SCS} and $\tau_{SCS\text{-}SB}$ can generally reflect the time constants for SC interfacial electron transfer for target reduction (τ_{SC}) and for side/back reaction ($\tau_{SC\text{-}SB}$), respectively:

$$\tau_{SC} \approx \tau_{SCS} \approx \tau_{2C} \tag{7.5a}$$

and

$$\tau_{SCS\text{-}SB} \approx \tau_{SC\text{-}SB} \tag{7.5b}$$

Similarly, Eq. (7.4c) can also be given by:

$$\frac{1}{\tau_{3C}} \approx \frac{1}{\tau_{SC}} + \frac{1}{\tau_{SC\text{-}SB}} \tag{7.5c}$$

The evaluated time constants for occurring SC interfacial electron transfer in actual photocatalytic reaction generally range from several deciseconds to several seconds (Figure 7.4) [20, 23, 28, 31, 34–36]. This timescale is 6–12 orders of magnitude larger than the widely believed picosecond-microsecond timescale and, more importantly, generally larger than the centisecond-decisecond timescale for occurring mass transport of reactant(s) in the pores of the immobilized photocatalyst films [6, 9, 10, 19, 37]. This new cognition indicates that the long-distance mass transport does not severely slows photocatalytic reaction. PFBR formed by immobilizing photocatalyst would not severely lower the photon utilization. With this information in hand, next, we will discuss the configuration for immobilizing photocatalyst that crucially influences the photon utilization.

7.3 Immobilized Pt/TiO_2 Photocatalyst

Since the timescale for occurring mass transport of reactant(s) in the formed pores is generally smaller than that for SC interfacial electron transfer (Figures 7.1 and 7.4), immobilization of photocatalyst would not severely slow the photocatalytic reaction. Regarding this, we here immobilize the platinum loaded TiO_2 (Pt/TiO_2) photocatalyst onto the smooth internal wall of the light incident side of the reactors to evaluate the photocatalytic performance. The photocatalytic HE rate of water splitting for immobilized Pt/TiO_2 photocatalyst is only ~58% of that for the dispersed counterpart (Figure 7.5) [20]. By introducing methanol into the reaction, the HE rate ratios of the immobilized Pt/TiO_2 photocatalyst to dispersed counterpart are generally higher than the case of water splitting and can even increase to ~86% under intense irradiation (Figure 7.5) [19]. One obvious difference between these reactions is that the introduced substance (methanol) can easily be oxidized by the photogenerated holes.

In photocatalytic reaction, the deprivation of electrons from reduction cocatalyst by the oxidative intermediates also plays a crucial role to the effective separation of charge carriers and photon utilization [28]. This is because the intermediates formed by reacting the substance with photogenerated holes are highly reactive prior to evolving into final product(s). The electron deprivation that can be called side/back reaction occurs on a relatively less negative potential [28], as compared with the competitive target reduction. In the restrained space (e.g., pores) formed by photocatalyst immobilization, the probability for the collision of intermediates with the cocatalyst should be increased [20]. This will severely lower the electron utilization. Comparing with the intermediates generated from methanol, the reactivity of oxygen-containing species like hydroxyl radicals (OH) generated in water splitting is high [20]. The deprivation of electrons in water splitting should be more severe than the case containing methanol.

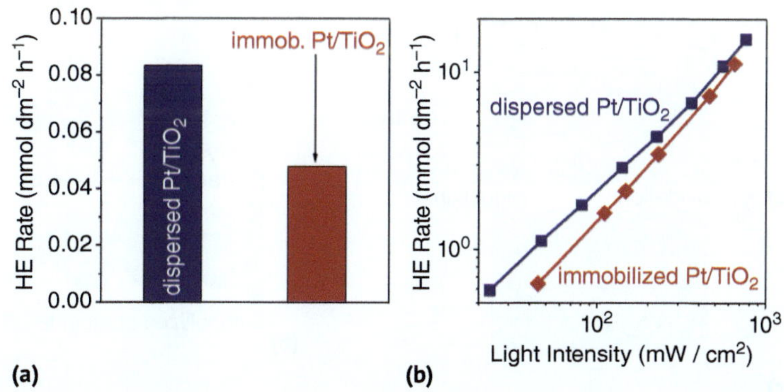

Figure 7.5 Photocatalytic HE behaviors over dispersed and immobilized Pt/TiO_2 in the absence (a) and presence of methanol (b). Reproduced from [20]/with permission of American Chemical Society (a). Reproduced from [19]/with permission of American Chemical Society (b).

7.4 Configuration Features of Immobilized TiO_2|Pt Photocatalyst for Efficient PFBR

7.4.1 Spatial Separation of Oxidation and Reduction Half-reactions

Efficient photocatalysis requires to promote the effective charge separation that includes hole extraction and SC interfacial electron transfer and the followed target electrocatalytic reduction. In addition to the very high potential barriers encountered by SC interfacial electron transfer [32], electron deprivation from reduction cocatalyst is also crucial to the effective separation of charge carriers [28], as mentioned above (e.g., Figure 7.3). Generally, the increase in the size of semiconductor that can increase the average distance between the generation sites of the oxidative intermediates and the reduction cocatalyst can alleviate the electron deprivation [28]. This is because the increase in the mass transport distance of intermediates will prolong the time for this step. During the mass transport, some of the intermediates will evolve into final products with relative weak reactivity.

Porous film made of semiconductor nanostructures with efficient electron transport can be treated as a macroscale material. Hypothesize that the photocatalyst can be formed by anchoring the semiconductor nanostructures like TiO_2 nanoparticles onto the smooth internal walls of the reactor following with loading cocatalysts (e.g., platinum) onto the outer layer of the formed porous films, the effective size that influences the side/back reaction can be increased. In this immobilized photocatalyst (TiO_2|Pt) configuration, the effective size can be mainly the thickness of the porous semiconductor films (Figure 7.6).

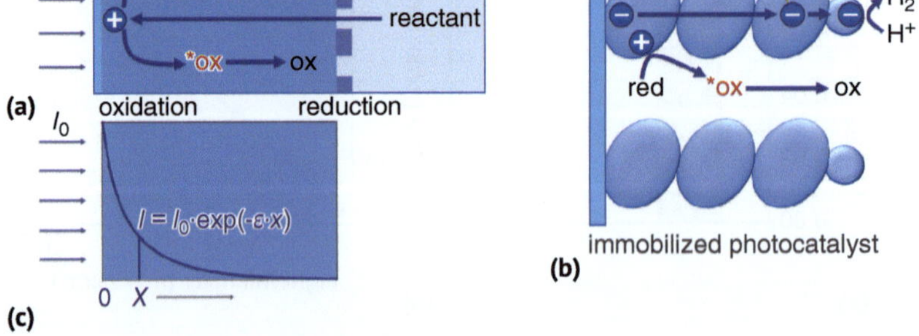

Figure 7.6 Schematic diagrams for configuration of PFBR (a), photoinduced electronic processes and steps occurring in PFBR (b), and the attenuation of incident light along the position of the film (c). Reproduced from [19]/with permission of American Chemical Society (a). Reproduced from [20]/with permission of American Chemical Society (b).

According to the attenuation law, the light intensity at a given position (x) of the semiconductor film can be (Figure 7.6):

$$I = I_0 \cdot \exp(-\varepsilon \cdot x) \tag{7.6}$$

in which I_0 is the intensity of incident light, and ε is the light extinction coefficient of the semiconductor film. For a porous TiO_2 film, ε can be generally $>10^4$ cm^{-1}. This indicates that the incident light undergoes a rapid attenuation along the incident direction (Figure 7.6). Therefore, most of the incident photons can be absorbed by the very thin superficial part of the film near the incident walls. The oxidation half-reaction therefore mainly occurs on the very restrained space that absorbs most of the incident photons. On the other hand, the reduction half-reaction occurs on the outer layer of the semiconductor films that is far always from the generation sites for occurring hole extraction. These two half-reactions can be spatially separated mainly by the semiconductor films. To occur side/back, the photogenerated intermediates should travel a very long distance to reach the reduction cocatalyst (Figure 7.6). During the transport, considerable percentage of intermediates can evolve into the final product(s). By virtue of the spatial separation, the undesired electron deprivation can be effectively alleviated. PFBR formed by immobilizing photocatalyst (e.g., TiO_2|Pt) take obvious superiority over the slurry suspension in reaction rates and photon utilization (Figure 7.7) [19, 20].

The alleviation of side/back reaction not only save the electrons from deprivation, but also increases the density of electrons in the semiconductor (Figure 7.7). This is because suppressing the reaction that occurred at a relatively positive potential can upshift the electronic Fermi level of the semiconductor [20]. The increase in the density of electrons will flatten the potential barrier for thermionic emission and further promote SC interfacial electron transfer. Therefore, the immobilized photocatalyst with configuration like TiO_2|Pt can deliver nearly constant photon utilization along the variation of light intensity (Figure 7.7).

7.4.2 Estimation of Utilization of Electrons at Reduction Cocatalyst

As mentioned above, the immobilized photocatalyst like TiO_2|Pt with spatial separation of oxidation and oxidation half-reactions can alleviate the side/back reaction. To make a step further, we here try to quantitatively analyze the utilization of photogenerated electrons (η) that transferred to the reduction cocatalyst from the semiconductor. Since the potential barrier for SC interfacial electron transfer by thermionic emission under very intense irradiation can be minimized (Eq. 7.2d), photocatalytic photon utilization that can be described by gross quantum yield (GQY) therefore can be the utilization of photogenerated electrons at reduction cocatalyst:

$$\text{GQY} \approx \eta \cdot \frac{J_{\text{iet}}}{A \cdot P_{\text{inc}}} \tag{7.7}$$

Since the target reduction half-reaction and side/back reaction that compete with each other and occur in parallel, the time constants for occurring these two reactions can be used to evaluate the electron utilization:

$$\eta \approx \frac{1}{1 + \tau_{\text{SCS}}/\tau_{\text{SCS-SB}}} \tag{7.8}$$

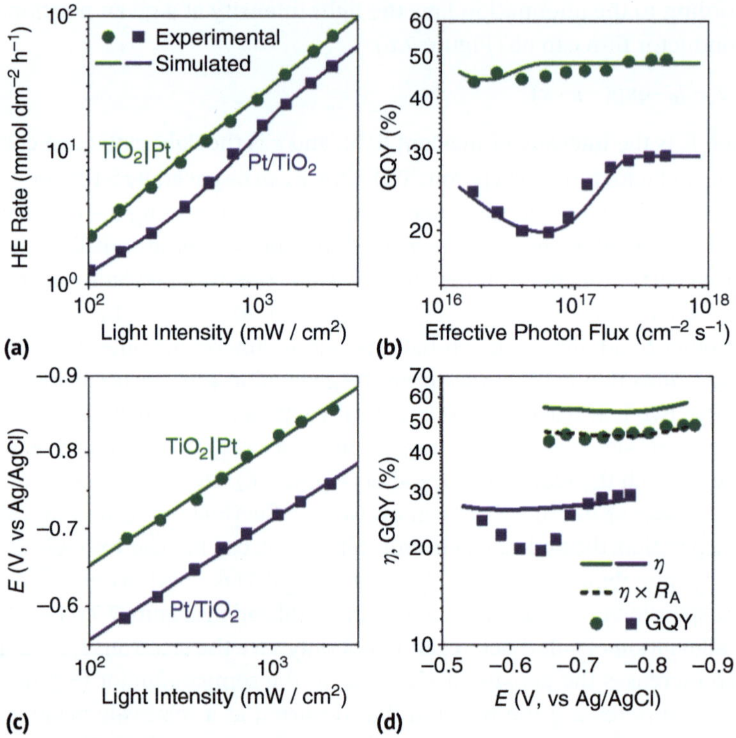

Figure 7.7 Simulated and experimental photocatalytic HE rate (a) and photon utilization (b) over the dispersed (Pt/TiO_2) and in PFBR ($TiO_2|Pt$). By suppressing the side/back reaction, the photoinduced equilibrium potential for $TiO_2|Pt$ can be more negative than that for Pt/TiO_2 (c). Although the TiO_2 film in $TiO_2|Pt$ photocatalyst cannot completely absorb all the incident photons (R_A: ~84%), the very high electron utilization resulted from effective spatial separation of oxidation and reduction half-reactions endows PFBR with high and nearly constant photon utilization (d). Reproduced from [20]/with permission of American Chemical Society.

By acquiring the time constants (Figure 7.4), the evaluated electron utilization (η) for PFBR can be evidently higher than that of slurry suspension (Figure 7.7).

7.4.3 Light Absorption

In photocatalytic reaction, the reactant(s) and/or the pollutant(s) dissolved in liquid media will usually absorb incident photons from a wide range of wavelength. Although the light extinction coefficient is generally not very high (e.g., $<1\ cm^{-1}$), the very low concentration of photocatalyst (semiconductor) dispersed in the liquid media make the target absorption very adverse. This competitive absorption of light comes to be a spiny issue to the efficient utilization of incident light [23].

For simplicity, we first discuss the case that the liquid does not absorb incident photons. The light extinction coefficient of the photocatalyst (semiconductor) dispersed in slurry suspension exhibits a power-law dependence on the concentration with an index less than but close to unity [31]:

$$\varepsilon = \varepsilon_0 \cdot C_{cat}{}^{\vartheta}, \text{ (with } \vartheta < 1) \tag{7.9a}$$

or can be given approximately by:

$$\varepsilon \approx \varepsilon_0 \cdot C_{cat} \tag{7.9b}$$

in which C_{cat} is the volume concentration of photocatalyst (semiconductor) dispersed in the slurry suspension. This means that light extinction coefficient of the dispersed photocatalyst (semiconductor) increases almost linearly with the concentration. The absorption (A_{ph}) of incident photons at a given position (x) can be (Figure 7.8):

$$A_{ph}(x) = A \cdot P_{inc} \cdot \varepsilon \cdot \exp(-\varepsilon \cdot x) \approx A \cdot P_{inc} \cdot \varepsilon \cdot \exp(-\varepsilon_0 \cdot C_{cat} \cdot x) \tag{7.9c}$$

By integrating the light absorption from the incident side to the given position, we have:

$$A_{ph}(x)|_0^x = A \cdot P_{inc} \cdot [1 - \exp(-\varepsilon \cdot x)] \approx A \cdot P_{inc} \cdot [1 - \exp(-\varepsilon_0 \cdot C_{cat} \cdot x)] \tag{7.9d}$$

Clearly, the increase in the concentration of photocatalyst (semiconductor) can reduce the required depth for absorbing a given percentage of incident photons (Figure 7.8).

If the liquid (e.g., wastewater) can absorb the incident photons with an extinction coefficient of ε_w, the absorption (A_w) of incident photons integrating from the incident side to a given position (x) can be given approximately by (Figure 7.9):

$$A_w(x)|_0^x \approx A \cdot P_{inc} \cdot \{1 - \exp[-(1 - C_{cat}) \cdot \varepsilon_w \cdot x]\} \tag{7.10}$$

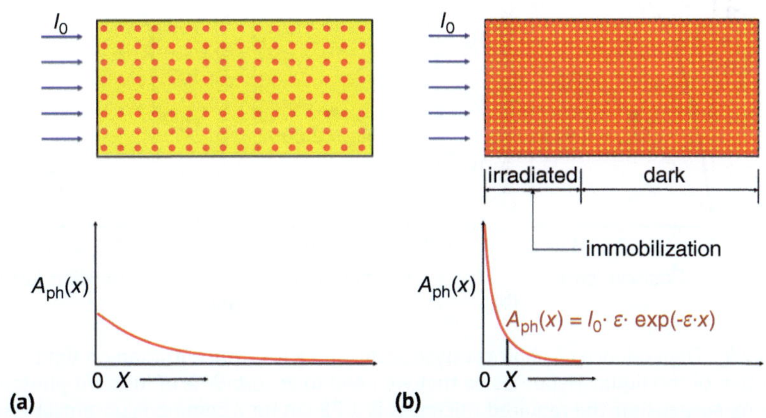

Figure 7.8 When the concentration of photocatalyst (semiconductor) is low, the tail of light absorption can be long (a). By increasing the concentration, most of the incident photons can be absorbed in a narrow region near the incident side (b). Hypothesize that the solid concentration is extremely high, by fixing the photocatalyst (semiconductor) in the irradiated region and removing the rest (dark), the light absorption can be treated as the situation of an immobilized photocatalyst (b).

The extinction coefficient of the slurry suspension (ε_{ss}) can be given approximately by:

$$\varepsilon_{ss} \approx \varepsilon_0 \cdot C_{cat} + (1 - C_{cat}) \cdot \varepsilon_w \quad (7.11a)$$

The absorption of incident photons (A_{ss}) integrating from the incident side to a given position of the slurry suspension can be given by (Figure 7.9):

$$A_{ss}(x)|_0^x \approx A \cdot P_{inc} \cdot [1 - \exp(-\varepsilon_{ss} \cdot x)] \quad (7.11b)$$

Hypothesize an extreme case that the volume concentration can be very high, for example, ~0.4. The rest (i.e., ~0.6) that filled with liquid media in photocatalytic reaction can be imagined as porosity of an immobilized photocatalyst (Figure 7.8). In such very high C_{cat}, the light extinction coefficient can be very high and the required depth to absorb certain percentage of incident photons can be effectively reduced. On the other hand, the increase in C_{cat} simultaneously means a lowering of "concentration" of liquid. By reducing the "concentration" and thickness of the liquid, the competitive absorption by liquid media can be nearly eliminated (Figure 7.9).

The light absorption in PFBR formed by anchoring semiconductor porous film onto the smooth substrate can be very similar to the situation with extremely high concentration of photocatalyst (semiconductor). By immobilizing the photocatalyst (semiconductor) near the light incident side of the reactor and removing the solid that is almost dark from the dispersion, it can be treated as immobilized photocatalyst (semiconductor) porous films (Figure 7.8). This means that PFBR can effectively address the issue of competitive light absorption by the reactant(s) and/or pollutant(s).

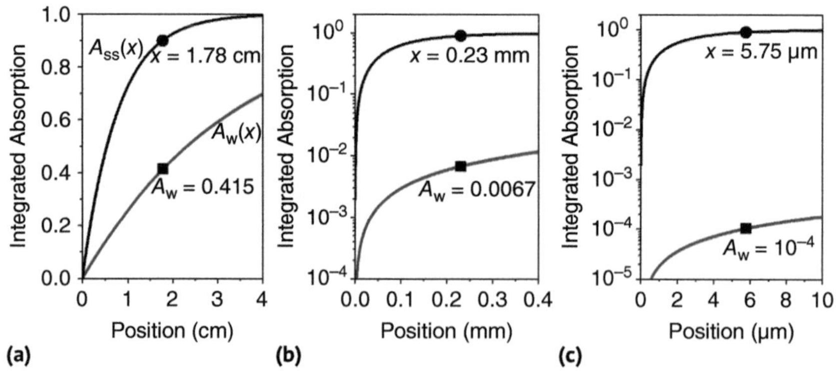

Figure 7.9 Dependence of photocatalyst (semiconductor) concentration on light absorption of the liquid. Hypothesize that we need to absorb 90% of incident photons by the slurry suspension, the required thickness is 1.78 cm for a common concentration of photocatalyst (e.g., by dispersing ~40 mg photocatalyst into 100 mL liquid, $C_{cat} = 0.0001$, a). In this situation, 41.5% of the incident photons will be absorbed by the liquid (a). By increasing the solid concentration ($C_{cat} = 0.01$), the required thickness reduced to 0.23 mm and the photon loss reduced to 0.67% (b). In an extreme case with $C_{cat} = 0.4$, the required thickness of the slurry suspension is only 5.75 μm and the photons absorbed by liquid is only 0.01% (c). For this prediction, the extinction coefficients adopted for the solid (ε_0) and liquid (ε_w) are respectively 10 000 and 0.3 cm^{-1}.

7.4.4 Long-distance Electron Transport

According to the features of PFBR, the electrons generated in the photon incident side of the semiconductor film should transport to another side for occurring SC interfacial electron transfer and the followed electrocatalytic reduction. During this procedure, electrons suffer from intergrain scattering, multiple trapping/detrapping events, and leakage to the solution [33, 38, 39]. Intergrain electron scattering accounts for the major lowering of electron mobility in the nanostructured films [38, 39]. Although trapping/detrapping also occurs in slurry suspension even with ideal monodispersed photocatalyst [33], the evident increase in transport distance in PFBR undoubtedly increases the probability for such undesired events. These will increase the time consumption for electron transport and, more seriously, increase the probability to occur electron leakage [33]. To realize effective spatial separation of oxidation and reduction half-reactions, efficient transport of electrons through the semiconductor film is required.

To our knowledge, the oxide semiconductors like TiO_2 and ZnO with efficient electron transport can be used to fabricate PFBR with efficient photon utilization [19, 20, 33]. This is because the fact that, for example, the resistance for electron transport can be evidently reduced by biasing the films [19, 33]. Besides, the relatively low density of trap states in these oxides can reduce the trapping/detrapping events during the electron transport and suppress the electron leakage [32, 33, 39]. These make the oxides like TiO_2 to be the most widely adopted semiconductor to transport electrons in photoelectrochemical cells (e.g., sensitized films).

We here take TiO_2 films made of P25 powders (Degussa) as an example to demonstrate the features of electronic conductivity. As estimated from impedance spectra, the electron transport resistance of TiO_2 film at the potentials of the actual photocatalytic reactions can be generally 1–10 Ω cm^2 (Figure 7.10), which is 3–4 orders of magnitude lower than that for electron leakage [19]. The evaluated diffusion coefficients of electrons in TiO_2 film in such condition are generally in the level of 10^{-4}– $10^{-3}cm^2s^{-1}$ (Figure 7.10), which is 10–1000 times higher than that of mass transport. The time constants for survival of electrons in TiO_2 film are generally in the level of 10^1 s. By these virtues, the effective transport length of electrons in TiO_2 film reaches 10^2–10^3 microns, which are generally >10 times larger than the thickness (Figure 7.10). The time constants for transport of electrons through the film can be generally < 10 ms (Figure 7.10). This can save the electrons from loss during the transport and endow the PFBR with effective spatial separation of oxidation and reduction half-reactions. By these virtues, the photocatalytic HE rate of PFBR can be maximally 1.74 times higher than that of slurry suspension [19]. By adopting phase-pure anatase TiO_2 synthesized by hydrothermal treatment, the photocatalytic reaction rate and photon utilization of the immobilized photocatalyst can be maximally ~2.3 times higher that of dispersed counterpart in HE carried out in methanol aqueous solutions (Figure 7.7) [20]. In a scalable pilot reaction integrated HE with nonoxidative dehydrogenation of ethanol, HE rate reaches as high as 0.262 mol/h with a quantum yield of 65.2% and the acetaldehyde selectivity can be 96.4% in the PFBR equipped with a 320-W low-pressure mercury lamp [19].

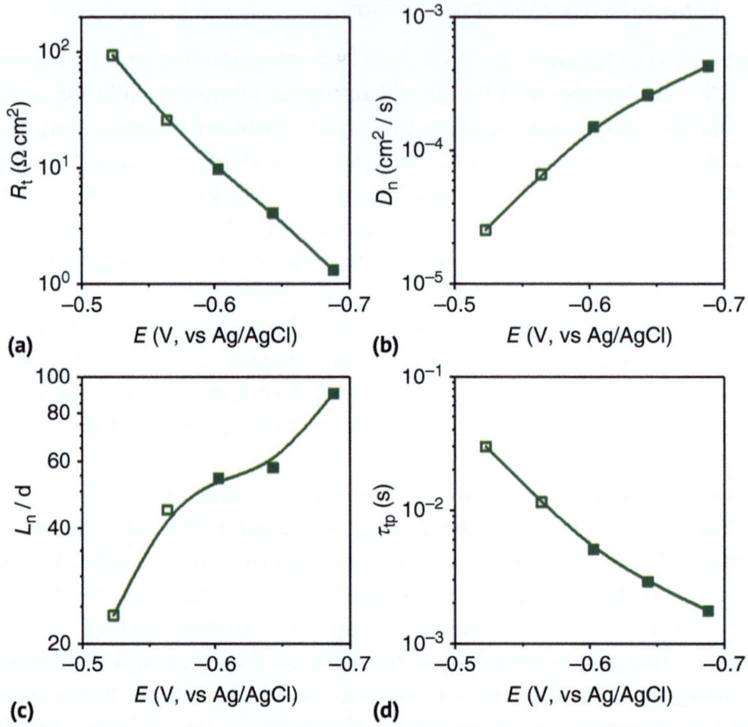

Figure 7.10 The electronic features reflecting the actual working condition (filled squares) of TiO_2 film in PFBR. The electron transport resistance (a), electron diffusion coefficient (b), the ratios of electron diffusion length to the film thickness (c), and the time constants for transport through the film (d) are evaluated from impedance spectra. The characteristic parameters (open squares) at the potential less negative than the actual photocatalytic reaction are also shown. Reproduced from [19]/with permission of American Chemical Society.

PFBE formed by immobilizing semiconductors like CdS and C_3N_4 that cannot form highly conductive films only deliver inferior photocatalytic photon utilization, as compared with dispersed counterparts [19]. Because of poor electron transport, for these semiconductors, more efforts should be devoted to realize acceptable conductivity.

7.5 Applications of PFBR

PFBR formed by anchoring semiconductor onto the smooth internal wall of the light incident side of the reactor with TiO_2|Pt configuration is superior to slurry suspension with dispersed photocatalyst, as demonstrated above in HE integrated with nonoxidative dehydrogenation of alcohols dissolved in water (e.g.: Figure 7.7) [19, 20]. These exciting results encourage us to explore more valuable reactions. By applying PFBR based on TiO_2|Pt photocatalyst in nonoxidative dehydrogenation of anhydrous methanol, a quantum yield of 89.9% for HE with acetaldehyde selectivity of 95.5% can be realize by introducing protons into the system [40]. This is because the potential barriers for SC interfacial electron transfer that mainly determined by the contact barrier to the cocatalyst flattens with increase in proton concentration, while the position of conduction band at the semiconductor surface does not evidently change. The very high quantum yield also indicates that the side/back can also be effectively eliminated in the PFBR.

Application of PFBR to oxidize CO in H_2 stream for lossless purification of hydrogen can be a paragon [21, 35]. In photocatalytic hydrogen purification, in addition to the desired CO oxidation [41, 42],

$$\cdot OH + CO \rightarrow CO_2 + \cdot H \tag{7.12a}$$

one of the most concerned issues is the reaction of H_2 with photogenerated $\cdot OH$ [42],

$$\cdot OH + H_2 \rightarrow H_2O + \cdot H \tag{7.12b}$$

This reaction (Eq. 7.12b) will undesirably consume H_2 since the hydrogen radicals ($\cdot H$) are usually not able to form molecular hydrogen in liquid media with disperse photocatalyst, due most like to the low probability for the target collision [21].

In PFBR, the plentiful molecular hydrogen in the gaseous ambient could bridge the transport of hydrogen radicals ($\cdot H$) with low activation energy [43],

$$\cdot H + H{-}H \rightarrow H{-}H + \cdot H \tag{7.13a}$$

The resulted hydrogen exchange can recombine the hydrogen radicals into the molecules,

$$\cdot H \rightarrow \frac{1}{2}H_2 \tag{7.13b}$$

By considering the associated H_2 generation resulted from proton and the photogenerated electron (e^-) that are the products accompanied by the generation of $\cdot OH$,

$$H^+ + e^- \rightarrow \frac{1}{2}H_2 \tag{7.13c}$$

the total reaction occurring in PFBR can be [21]:

$$CO + H_2O \xrightarrow{h\nu} CO_2 + H_2 \tag{7.14}$$

The photocatalytic lossless hydrogen purification is similar to water-gas-shift reaction in form but the reverse reaction can be minimized. In a polit PFBR equipped with Mg-doped-TiO_2|Cu photocatalyst, the CO in the effluent stream can be <1 ppm by feeding H_2 stream with a flow rate of 1 L min^{-1} [21, 35].

7.6 Conclusion

By cognizing the timescales for occurring electronic processes in actual photocatalytic reaction, efficient photocatalytic fixed bed reactors can be formed. By anchoring TiO_2 onto the smooth internal wall of the light incident side of the reactors following with depositing cocatalyst onto the outer layer of the formed porous semiconductor film, the immobilized photocatalyst with TiO_2|cocatalyst configuration takes obvious advantages over dispersed counterpart, not only in simplification of installation and operation but also in photon utilization. Like the situations of HE and lossless hydrogen purification, PFBRs will shine brilliance in more applications.

Acknowledgments

The author acknowledges Hangzhou Normal University for financial support.

References

1 Xiong, L. and Tang, J. (2021). Strategies and challenges on selectivity of photocatalytic oxidation of organic substances. *Advanced Energy Materials* 11: 2003216.

2 Chen, Y., Xie, H., Ma, M. et al. (2024). Simultaneous H_2 fuel evolution and value-added organic transformation from a one-dimensional noble-metal-free photocatalyst with spatially separated catalytic sites. *Applied Catalysis B: Environmental* 342: 123433.

3 Miranda-García, N., Suárez, S., Sánchez, B. et al. (2011). Photocatalytic degradation of emerging contaminants in municipal wastewater treatment plant effluents using immobilized TiO_2 in a solar pilot plant. *Applied Catalysis B: Environmental* 103: 294–301.

4 Rincón, G.J. and La Motta, E.J. (2019). A fluidized-bed reactor for the photocatalytic mineralization of phenol on TiO_2-coated silica gel. *Heliyon* 5: e01966.

5 Wood, D., Shaw, S., Cawte, T. et al. (2020). An overview of photocatalyst immobilization methods for air pollution remediation. *Chemical Engineering Journal* 391: 123490.

6 Schneider, J. and Curti, M. (2023). Spectroscopic and kinetic characterization of photogenerated charge carriers in photocatalysts. *Photochemical & Photobiological Sciences* 22: 195–217.

7 Chen, D., Li, F., and Ray, A.K. (2001). External and internal mass transfer effect on photocatalytic degradation. *Catalysis Today* 66: 475–485.

8 Friedmann, D., Mendive, C., and Bahnemann, D. (2010). TiO_2 for water treatment: Parameters affecting the kinetics and mechanisms of photocatalysis. *Applied Catalysis B: Environmental* 99: 398–406.

9 Schneider, J. Matsuoka, M. Takeuchi, M. et al. (2014). Understanding TiO_2 photocatalysis: Mechanisms and materials. *Chemical Reviews* 114: 9919–9986.

10 Li, R. and Li, C. (2017). Photocatalytic water splitting on semiconductor-based photocatalysts.In: *Advances in Catalysis, Vol. 60* (ed. C. Song), 1–57. Academic Press.

11 Hisatomi, T., Takanabe, K., and Domen, K. (2015). Photocatalytic water-splitting reaction from catalytic and kinetic perspectives. *Catalysis Letters* 145: 95–108.

12 Bahnemann, D.W., Hilgendorff, M., and Memming, R. (1997). Charge carrier dynamics at TiO_2 particles: Reactivity of free and trapped holes. *The Journal of Physical Chemistry B* 101: 4265–4275.

13 Furube, A., Asahi, T., Masuhara, H. et al. (2001). Direct observation of a picosecond charge separation process in photoexcited platinum-loaded TiO_2 particles by femtosecond diffuse reflectance spectroscopy. *Chemical Physics Letters* 336: 424–430.

14 Leung, D.Y.C., Fu, X., Wang, C. et al. (2010). Hydrogen production over titania-based photocatalysts. *ChemSusChem* 3: 681–694.

15 Vescovi, T., Coleman, H.M., and Amal, R. (2010). The effect of pH on UV-based advanced oxidation technologies-1,4-dioxane degradation. *Journal of Hazardous Materials* 182: 75–79.

16 Marques, A.C., Vale, M., Vicente, D. et al. (2021). Porous silica microspheres with immobilized titania nanoparticles for in-flow solar-driven purification of wastewater. *Global Challenges* 5: 2000116.

17 Alexiadis, A. and Mazzarino, I. (2005). Design guidelines for fixed-bed photocatalytic reactors. *Chemical Engineering and Processing: Process Intensification* 44: 453–459.

18 Bideau, M., Claudel, B., Dubien, C. et al. (1995). On the "immobilization" of titanium dioxide in the photocatalytic oxidation of spent waters. *Journal of Photochemistry and Photobiology A: Chemistry 91*: 137–144.

19 Wei, X., Liu, H., Gao, S. et al. (2022). Photocatalyst: To be dispersed or to be immobilized? The crucial role of electron transport in photocatalytic fixed bed reaction. *The Journal of Physical Chemistry Letters* 13: 9642–9648.

20 Wei, X. and Chen, J. (2023). Revealing the efficient and constant photon utilization of photocatalytic fixed bed reaction. *The Journal of Physical Chemistry Letters* 14: 5386–5389.

21 Liu, H., Wei, X., Fang, Y. et al. (2023). Self-promoted H_2 formation: The feasibility of photoinduced CO removal for lossless hydrogen purification. *The Journal of Physical Chemistry Letters* 14: 2087–2091.

22 Gao, S., Ge, T., Li, B. et al. (2023). Electrostatic induction promotes photocatalytic contaminant mineralization in saline wastewater. *Catalysis Science & Technology* 13: 6631–6634.

23 Xiang, H., Wang, Z., and Chen, J. (2022). Revealing the role of elementary doping in photocatalytic phenol mineralization. *Chinese Journal of Structural Chemistry* 41: 2209069–2209073.

24 Fang, Y., Wei, X., Liu, H. et al. (2023). Homogeneous interfacial electron transfer promotes photoinduced hole extraction for phenol mineralization. *Catalysis Science & Technology* 13: 3495–3498.

25 Luo, T., Wang, Z., Wei, X. et al. (2022). Surface enriching promotes decomposition of benzene from air. *Catalysis Science & Technology* 12: 2340–2345.

26 Yang, H. and Chen, J. (2024). Photocatalytic fixed bed reactions for contaminant mineralization. *Journal of Environmental Chemical Engineering* 12: 112765.

27 Ge, T. and Chen, J. (2024). Time constant estimation and alleviation of interior charge recombination for photocatalytic reaction guided by correlation with photoelectrochemical behaviors. *The Journal of Physical Chemistry Letters* 15: 1241–1245.

28 Wei, X., Gao, S., Liu, H. et al. (2023). Three-channel electron transfer for estimating time constants correlated with photocatalytic photon utilization. *The Journal of Physical Chemistry Letters* 14: 3721–3726.

29 Ge, T. and Chen, J. (2023). Evaluating a second time scale for hole extraction in an actual photocatalytic reaction: The method. *The Journal of Physical Chemistry Letters* 14: 7477–7482.

30 Wang, Z., Qiao, W., Yuan, M. et al. (2020). Light-intensity-dependent semiconductor-cocatalyst interfacial electron transfer: A dilemma of sunlight-driven photocatalysis. *The Journal of Physical Chemistry Letters* 11: 2369–2373.

31 Wei, X., Xiang, H., Xu, Y. et al. (2023). Photocatalytic modeling for maximizing utilization of real-time changing sunlight and rationalizing evaluation. *AIChE Journal* 69: e17985.

32 Wang, Z., Xue, N., and Chen, J. (2019). Semiconductor-cocatalyst interfacial electron transfer dominates photocatalytic reaction. *The Journal of Physical Chemistry C* 123: 24404–24408.

33 Chen, J., Tao, H.B., and Liu, B. (2017). Unraveling the intrinsic structures that influence the transport of charges in TiO_2 electrodes. *Advanced Energy Materials* 7: 1700886.

34 Xiang, H., Wang, Z., and Chen, J. (2021). Revealing and facilitating the rate-determining step for efficient sunlight-driven photocatalysis. *The Journal of Physical Chemistry Letters* 12: 7665–7670.

35 Liu, H., Huang, X., and Chen, J. (2023). Surface electronic state modulation promotes photoinduced aggregation and oxidation of trace CO for lossless purification of H_2 stream. *Chinese Journal of Catalysis* 51: 49–54.

36 Xu Y., Wang, Z., Xiang, H., Yang, D., Wang, J., and Chen, J. (2022). Revealing the role of electronic doping for developing cocatalyst-free semiconducting photocatalysts. *The Journal of Physical Chemistry Letters* 13: 2039–2045.

37 Hoffmann, M.R., Martin, S.T., Choi, W. et al. (1995). Environmental applications of semiconductor photocatalysis. *Chemical Reviews* 95: 69–96.

38 Chen, J., Yang, H.B. Tao, H.B. et al. (2016). Surface rutilization of anatase TiO_2 nanorods for creation of synergistically bridging and fencing electron highways. *Advanced Functional Materials* 26: 456–465.

39 Chen, J., Yang, H.B. Miao, J. et al. (2014). Thermodynamically driven one-dimensional evolution of anatase TiO_2 nanorods: One-step hydrothermal synthesis for emerging intrinsic superiority of dimensionality. *Journal of the American Chemical Society* 136: 15310–15318.

40 Jia, K., Wei, X., Xu, Y., Wang, Z, and Chen, J. (2023). Asymmetric potential barrier lowering promotes photocatalytic nonoxidative dehydrogenation of anhydrous methanol. *Applied Catalysis A: General* 650: 119009.

41 Hoare, D.E. (1962). Reactions of gaseous hydroxyl radicals. *Nature* 194: 283–283.

42 Ung, A.Y.-M. and Back, R.A. (1964). The photolysis of water vapor and reactions of hydroxyl radicals. *Canadian Journal of Chemistry* 42: 753–763.

43 Christensen, H. and Sehested, K. (1983). Reaction of hydroxyl radicals with hydrogen at elevated temperatures. Determination of the activation energy. *The Journal of Physical Chemistry* 87: 118–120.

8

Carrier Transport and Recombination in Sensitized Nanostructured TiO_2

James Robert Jennings[1,2,*]

[1] *Faculty of Science, Universiti Brunei Darussalam, Bandar Seri Begawan, Brunei Darussalam*
[2] *Optoelectronic Device Research Group, Universiti Brunei Darussalam, Bandar Seri Begawan, Brunei Darussalam*

*Corresponding author: james.jennings@ubd.edu.bn

8.1 Introduction

Nanostructured titanium dioxide is an undisputed workhorse in the fields of photochemistry and photovoltaics, primarily due to its favorable bandgap, excellent chemical stability, nontoxicity, and low cost. After the first demonstration of reasonably efficient solar cells employing dye-sensitized nanostructured TiO_2 by O'Regan and Gratzel [1], research interest in dye-sensitized and the related quantum-dot-sensitized solar cells (DSSCs and QDSSCs, respectively) exploded, resulting in the publication of over 33 000 scholarly works mentioning "sensitized solar cells" (SSCs) according to the Elsevier Scopus database at the time of writing, ~21 000 of which have been published in the last 10 years, with most also containing the keyword "TiO_2" or similar (the remainder being SSCs employing other metal oxides, especially ZnO and SnO_2).

Since the significant breakthroughs in power conversion efficiency (PCE) for alkylammonium lead iodide perovskite solar cells (PSCs) since 2012 [2, 3], many variants of which also employ nanostructured TiO_2 electron transport layers [4], much of the DSSC research community began to focus their attention on this promising new class of photovoltaic materials and interest in conventional DSSCs has waned somewhat (the peak in annual publications about "dye-sensitized solar cells" occurred in 2014). It is also noteworthy that the record certified PCE of organic solar cells (OSCs) surpassed that of DSSCs by a significant margin in 2017–2018 through advances in ternary blend and tandem devices [5–7]. However, despite advances in competing "emerging" photovoltaic technologies, significant new research into more traditional DSSCs continues, albeit at a more moderate pace compared with the previous decade.

Nanostructured TiO_2: Synthesis, Photocatalytic Properties and Applications, First Edition.
Edited by Jia Hong Pan, Wan In Lee and Detlef W. Bahnemann

In particular, persistent research efforts have yielded high PCEs for DSSCs under both natural sunlight and indoor conditions with artificial lighting. These advances were largely driven by the development of novel panchromatic dye systems and innovative redox shuttles [8–11], as well as through enhanced understanding of the interaction between dyes and redox mediators [12–15]. Under standard 1 Sun AM 1.5G illumination, PCEs of up to 13% (certified by the Fraunhofer Institute for Solar Energy Research) and 15.2% (noncertified) have been achieved using nanostructured TiO_2 cosensitized with organic dyes [16, 17].

Another major advantage of DSSCs is their insensitivity to the angle of incident light, which allows them to efficiently utilize diffuse light. This characteristic makes them particularly suitable for applications where sunlight is not direct, such as in indoor settings or during overcast weather. Studies have shown that DSSCs can perform well under low intensity artificial lighting, such as fluorescent lighting, in many cases outperforming traditional silicon-based solar cells, with PCEs of up to 36% being reported under these conditions [11, 18–20].

Although perhaps unable to compete in terms of raw PCE, DSSCs offer other advantages over rival solar technologies, such as relatively low toxicity materials (especially compared with Pb-based PSCs), ease of fabrication (no vacuum deposition steps are required, unlike most PSCs and OSCs), and great flexibility in material choices and design, which make them attractive for specific applications [21, 22]. DSSCs also benefit from tunable aesthetics thanks to a wide array of dyes to choose from, allowing for customizable colors and semi-transparency, which are advantageous in building-integrated photovoltaics [23, 24], portable electronic devices [25], and other niche consumer applications such as power-generating clothing [26, 27]. Therefore, while the rise of PSCs and OSCs has shifted some focus away from DSSCs and other sensitized nanostructured TiO_2 systems in recent years, the latter still possess unique advantages that warrant continued research and development.

The purpose of this chapter is to review selected research on SSCs, with an emphasis on advances in fundamental understanding of charge transport and recombination processes. This chapter does not aim to provide an exhaustive overview of research in the field. Instead, it will provide the author's own perspective and highlight particularly significant work conducted over the past fifteen years to facilitate a broader discussion on the current state of understanding of sensitized nanostructured TiO_2 in terms of transport and recombination mechanisms.

8.2 Overview of SSCs

In this section, a brief overview of SSCs based on nanostructured TiO_2 is given. Figure 8.1 shows the typical device configuration of DSSCs and QDSSCs, which share many common features, differing mainly in the type of sensitizer (dye, quantum dot, or conformal semiconductor thin film), electrolyte (typically I^-/I_3^-, $Co^{2+/3+}$ or $Cu^{+/2+}$ electrolytes for DSSCs or S^{2-}/S_n^{2-} for QDSSCs).

The nanostructured TiO_2 electrodes used in SSCs like DSSCs and QDSSCs typically comprise sintered TiO_2 nanocrystals or more ordered structures such as nanotube arrays [28, 29]. These nanostructured electrodes are either prepared by

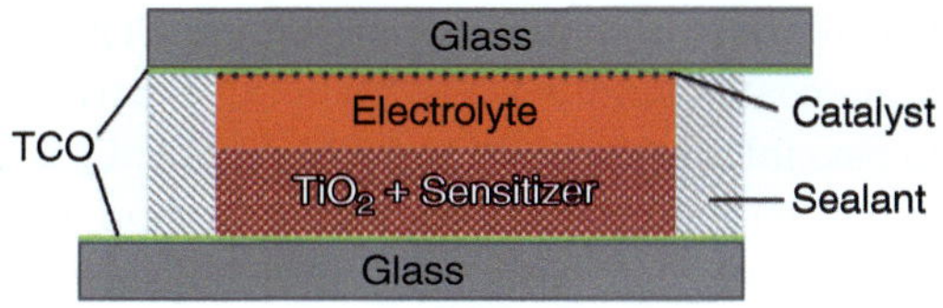

Figure 8.1 Schematic showing the typical structure of a sensitized nanostructured TiO_2 solar cell.

deposition of nanoparticles (often synthesized via sol–gel methods) [30] onto an electrically conductive substrate, or by direct anodization of a titanium metal substrate [31]. For photoelectrochemical applications, substrates coated with transparent conducting oxides (TCOs) are often preferred, with fluorine-doped tin oxide (FTO) being the most common, allowing for illumination through the substrate, which in the case of DSSCs tends to yield superior performance compared with illumination of the opposite side of the electrode through the electrolyte. Adhesion of the nanostructured TiO_2 to the TCO and interparticle connections can be enhanced through sintering. In the case of nanotube electrodes, the initially amorphous films are typically crystallized by annealing before visible-light sensitization by adsorption of a dye or semiconductor quantum dots.

A notable feature of nanostructured TiO_2 electrodes is the presence of two characteristic length scales, namely (i) the thickness of the electrode (a crucial parameter for conventional compact semiconductor electrodes) and (ii) the minimum nanostructure feature size (e.g., particle diameter or nanotube wall thickness), which, by definition, is on the nanometer scale (typically 10–100 nm). For nanostructured TiO_2 electrodes, feature sizes are small relative to the Debye length. As a result, no substantial band bending can occur between the interior and the surface of the particles when they are contacted by an electrolyte [32, 33]. This has significant implications for charge transfer across the semiconductor–electrolyte interface, as electric fields do not play a dominant role. The separation, interfacial transfer, and transport of charge carriers is instead governed by kinetics and diffusion, rather than by carrier drift. For instance, the electron quasi-Fermi level (QFL) throughout the entire electrode thickness can usually be controlled by applying an externally controlled (using a potentiostat) potential at the electrode contact, a behavior distinct from bulk semiconductor crystals where external bias primarily affects band bending near the interface [33].

Another key characteristic of nanostructured TiO_2 electrodes is their very high microscopic surface area, corresponding to roughness factors of up to 1000 for nanostructured TiO_2 electrodes in high-efficiency SSCs. This is in stark contrast with compact, planar semiconductor electrodes, where the microscopic area nearly equals to the geometric projected area. The large microscopic surface area in nanostructured TiO_2 electrodes is crucial to SSC performance because it permits a high sensitizer loading while maintaining only a thin layer (ideally monolayer coverage). This enables good light harvesting by electrodes only ~10 μm thick, while still offering a large semiconductor/sensitizer/electrolyte interfacial area for efficient charge separation following light absorption.

A typical DSSC comprises a porous, nanostructured TiO_2 layer supported by a transparent substrate coated with a TCO. The TCO electrically contacts the TiO_2 and allows light to pass into the cell. The sensitizer in DSSCs is a molecular dye, initially ruthenium bipyridyl complexes in original incarnations but now more usually an organic dye [1, 9, 34], which is adsorbed to the TiO_2 surface. This dye-sensitized electrode performs the function of light-harvesting in the solar cell and participates in charge transport. In QDSSCs, the dye is replaced with a semiconductor quantum dot, with a wide range of semiconductors having been tested including InP, PbS, CdS, CdSe, $CdSe_xTe_{1-x}$, $CuInS_2$, alloyed Zn–Cu–In–Se, and alkyl ammonium lead iodide perovskite, the latter of which currently holds the efficiency record for QDSSCs, with ZnCuInSe alloy QDs holding the record for Pb-free, nonperovskite QDSSCs [35, 36]. The sensitized TiO_2, which forms the photoanode, is sealed to a cathode coated with a catalyst that is suitable for the employed redox mediator (typically platinum for cobalt or iodine based mediators or a metal sulfide for polysulfide electrolytes) using a polymer sealant. A concentrated electrolyte containing a redox couple, usually a trisbipyridyl cobalt (II/III) complex or the I_3^-/I^- mediator for DSSCs [37], or a polysulfide electrolyte for QDSSCs, fills the TiO_2 pores and the gap between the electrodes. Electrolyte additives like *tert*-butylpyridine, lithium salts, and guanidinium thiocyanate are also essential for optimum performance, and their precise mechanisms of action remain an active area of research.

Upon illumination, the dye or QD sensitizer absorbs light, becomes excited, and then injects electrons into the TiO_2 conduction band. The timescale of injection is on the order of femtoseconds to picoseconds, depending on the type of sensitizer [38]. The oxidized sensitizer is then re-reduced ("regenerated") by the reduced half of the electrolyte redox mediator, which is subsequently oxidized. A conductive pathway between the electrodes enables electron transfer to the cathode, where reduction of the oxidized mediator completes the circuit, sustaining a photocurrent. Electron transport through the TiO_2 layer occurs via diffusion, with the photogenerated electrons eventually reaching the contact to be extracted [39]. The basic operating principles of all SSCs share some common steps, namely charge carrier generation and injection, charge carrier transport, and charge carrier recombination. Figure 8.2 illustrates these steps schematically for a typical sensitized TiO_2 system.

After injection into the TiO_2, electrons navigate through the nanostructured TiO_2 film to reach the conductive substrate, which is typically made from FTO. Electron transport in nanostructured TiO_2 is fundamentally influenced by its morphological and structural characteristics, including particle size, surface area, and crystallographic phase. The nanostructured nature of TiO_2 leads to a high surface area-to-volume ratio, offering ample sites for dye adsorption, but this also introduces a high concentration of surface states that act as electron traps. Trap states may also occur in the bulk of the nanostructured TiO_2 due to structural disorder (e.g., Urbach tails) or the presence of oxygen vacancies or impurities [41]. These trap states have a profound influence on carrier dynamics in sensitized TiO_2 by providing sites for temporary localization of carriers and either assisting or impeding transport depending on their nature and distribution, as well as acting as recombination centers. Indeed, the interplay between electron transport and recombination is a critical aspect of the

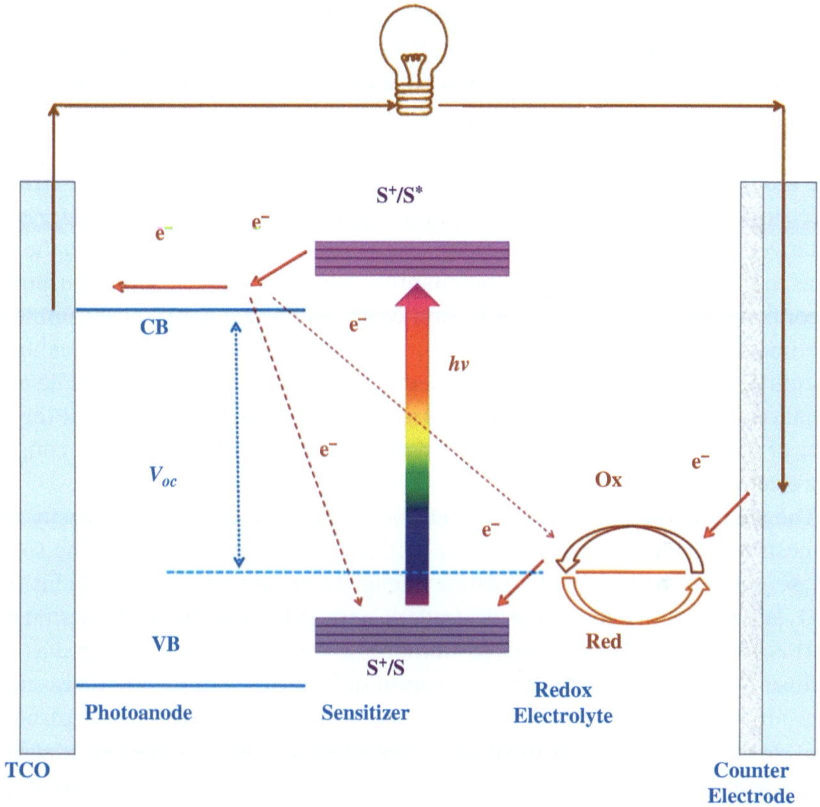

Figure 8.2 Illustration of the operating principles of a sensitized nanostructured TiO_2 solar cell. Reproduced from Ref. [40]/with permission of Elsevier.

functionality of nanostructured TiO_2. Recombination, the process by which electrons recombine with the oxidized sensitizer and/or redox mediator, competes with the desired charge collection at the FTO electrode, thus reducing the device PCE. Understanding and controlling these recombination pathways is a key challenge in improving the performance of SCCs based on nanostructured TiO_2.

8.3 Fundamentals of Carrier Transport and Recombination in Sensitized TiO_2

8.3.1 The Diffusion Model

The diffusion model is one of the most popular frameworks for describing the steady-state behavior of sensitized nanostructured TiO_2 solar cells. First applied to DSSCs by Södergren et al., this model has since been adopted and adapted by many researchers [42–45]. Although it is now clear that the original model is vastly oversimplified in many aspects (especially in its assumption of negligible recombination between TiO_2 electrons and oxidized sensitizer species), it is of historical importance remains instructive for gaining a more quantitatively grounded intuition about the main factors governing the performance of SSCs. For these reasons, a brief overview of the basic diffusion model will be presented in this section, while extensions and limitations of the model will be presented in subsequent sections.

The porous TiO_2 layer, sensitizer, and electrolyte are considered a single phase in the model, assuming the nanoscale mixture is homogeneous in terms of structure, electrical properties, and light absorption. Light absorption by the sensitizer is characterized using an effective absorption coefficient, electron injection from the sensitizer to the TiO_2 using an injection quantum efficiency, and electron transport and recombination by an electron diffusion coefficient and a constant recombination lifetime or rate constant, respectively. It is implicitly assumed that recombination occurs between TiO_2 electrons and acceptor species in the electrolyte, while recombination with the oxidized sensitizer is assumed negligible on account of the rapid rate of sensitizer regeneration by the donor (reduced half of the redox couple) in the electrolyte.

The small particle size, high dielectric constant, and low doping density of the nanostructured TiO_2, along with the high density of mobile ions in the solution phase, make it reasonable to assume negligible *macroscopic* electric fields in the TiO_2 layer perpendicular to the electrode substrate. The treatment of electron transport is greatly simplified by this assumption, allowing it to be considered as solely diffusion-driven, without the need to account for drift as needed for conventional p–n junction solar cells. Additionally, electron transfer from the nanostructured TiO_2 to the oxidized redox mediator is presumed to follow a rate law that is first order with respect to free electron concentration and zero order in oxidized mediator concentration due to its relatively high background levels. This assumption, along with the constant I_3^- concentration throughout the cell under operating conditions, justifies using a constant, first-order lifetime or rate constant for electron recombination with the oxidized redox mediator.

Based on these assumptions, the following electron continuity equation can be formulated to describe the steady-state behavior of electrons in the sensitized nanostructured TiO_2 electrode:

$$\frac{dn_c(x)}{dt} = 0 = I_0 \alpha e^{-\alpha x} + D_0 \frac{\partial^2 n_c(x)}{\partial x^2} - \frac{\left(n_c(x) - n_{eqm}\right)}{\tau_0} \tag{8.1}$$

Here, $n_c(x)$ represents the conduction band (or "free") electron concentration at a position x in the TiO_2 (with the origin, $x = 0$, being the FTO substrate), I_0 is the flux of incident photons, α is the absorption coefficient, D_0 and τ_0 are the diffusion coefficient and lifetime of free electrons, and n_{eqm} is the equilibrium electron concentration in the dark. The term "free" is used to describe electrons in the conduction band as opposed to those in bandgap trap states, which will be discussed later.

To solve Eq. (8.1), we require two boundary conditions, the first of which is $dn_c(x)/dx_{x=d} = 0$ with d the thickness of the electrode. This so-called reflecting boundary condition arises from the requirement that there can be no net electron flux at the electrolyte side of the TiO_2 layer. Various boundary conditions can be applied at the TCO/TiO_2 interface. At short-circuit, $n_c(0)$ can be set to n_{eqm}, assuming no free energy change is needed for electron transfer between TiO_2 and TCO. Under open-circuit conditions, the reflecting boundary condition can also be applied at $x = 0$ to prevent electron extraction at the TCO. To account for a finite electron extraction rate from TiO_2 to the TCO, an extraction rate constant, k_{ext}, can be introduced, yielding $D_0 dn_c(x)/dx_{x=0} = k_{ext}(n_c(x)_{x=0} - n_{eqm})$ as the boundary condition. Here, k_{ext}, with units of velocity, can be interpreted as the speed at which electrons cross the TiO_2/TCO interface, with an upper limit of approximately 10^7 cm s^{-1}, corresponding to the thermal velocity of free electrons at room temperature.

To construct the current density–voltage (j–V) characteristics of an SSC using $n_c(x)$ from the solution to Eq. (8.1) (which can be found elsewhere) [42–46], k_{ext} is continuously adjusted between 0 (open-circuit) and a sufficiently high value (e.g., 10^7 cm s^{-1}, short-circuit) to achieve diffusion-limited extraction. The photocurrent density, j, is then calculated from:

$$j = qD_0 \frac{dn_c(x)}{dx}\bigg|_{x=0} \tag{8.2}$$

where q is the elementary charge. The cell voltage is determined by:

$$n_c(x)_{x=0} = n_{eqm} e^{\frac{qV}{k_B T}}, \tag{8.3}$$

where the cell voltage, V, represents the electrode potential of the TCO under the TiO_2 ($x = 0$) relative to the counter electrode, which itself is equilibrated with the electrolyte redox potential. Under high currents, an additional potential drop (iR drop) occurs due to the non-zero series resistance of the TCO.

An alternative approach to calculating the net current density involves adding the photocurrent from solving Eq. (8.1) with fixed k_{ext} to the dark current, which is obtained by setting $I_0 = 0$ and applying Eq. (8.3) as a boundary condition at the TCO—TiO_2 contact. Though both approaches are mathematically equivalent, the latter fixes k_{ext} as a constant related to the TCO/semiconductor interface properties, isolating it from effects of an externally applied bias.

The electron diffusion length, L_n, naturally emerges from solving Eq. (8.1) and is given by:

$$L_0 = \sqrt{D_0 \tau_0} \tag{8.4}$$

This length represents the distance an electron travels, on average, before recombining. For efficient SSCs, L_0 should significantly exceed the thickness of the semiconductor layer, d, to ensure collection of all injected electrons. While this simple model assumes the electron diffusion coefficient and lifetime do not depend on electron concentration, thus predicting constant electron diffusion length, experimental findings point to a diffusion length that varies with electron concentration and/or light intensity [47–49]. This indicates that more detailed models incorporating nonlinear recombination are needed for a complete description of the diffusion length in sensitized nanostructured TiO_2. However, the linear model described here provides a useful foundation for discussing the general behavior of SSCs.

The diffusion model makes a number of important predictions. When $L_0 \gg d$, the concentration profile of electrons in the semiconductor should be relatively flat

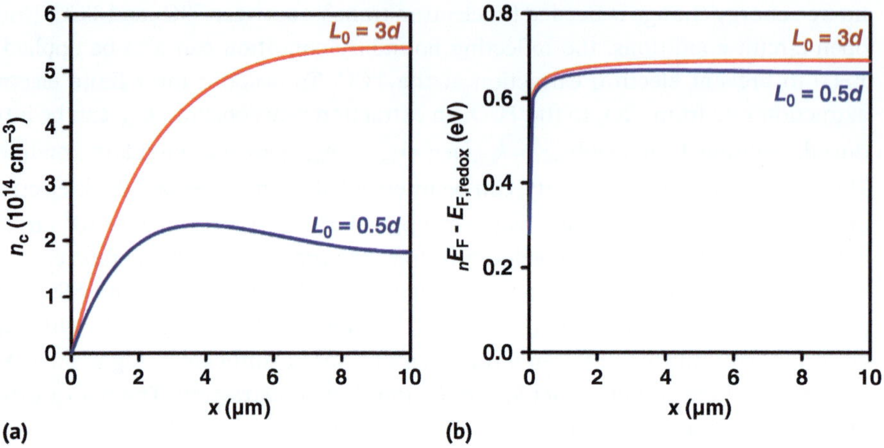

Figure 8.3 Simulated electron concentration (a) and QFL (b) profiles predicted by the diffusion model (Eq. 8.1). Parameters used: $I_0 = 10^{18}$ cm^2 s^{-1}, $\alpha = 4000$ cm^{-1}, $D_0 = 0.4$ cm^2 s^{-1}, $E_c - E_{redox} = 1$ eV, and $N_c = 10^{22}$ cm^{-3}.

at open circuit, and this also applies for $L_0 < d$ if light is weakly absorbed (low α). At short circuit, the electron collection efficiency should be nearly 100%, provided $L_0 \gg d$. When L_0 is comparable to or less than d, charge collection depends heavily on α and the direction of illumination. Illustrative electron concentration and QFL profiles under illumination at short circuit, for $L_0 = 0.5d$ and $L_0 = 3d$, are shown in Figure 8.3.

It is evident from Figure 8.3 that if $L_0 < d$, not all of the sensitized TiO_2 contributes to photocurrent production, as electrons photogenerated beyond the diffusion length recombine before reaching the contact at $x = 0$. The predicted QFL profile at short-circuit has been verified in experiments by Lobato et al., who incorporated a sensor electrode made of titanium metal on the electrolyte facing side of the TiO_2 film ($x = d$) and found agreement with the diffusion model predictions [50–52].

8.3.2 The Multiple Trapping Model

Nanostructured TiO_2 is known to have a significant concentration of electron traps within the forbidden bandgap [53]. The energetic distribution of the trap states is approximately exponential, with the concentration of traps increasing toward the edge of the conduction band. Electron transport and recombination in the TiO_2 are perturbed by trapping/detrapping involving these states, which has a considerable influence on the non-steady-state or "dynamic" characteristics of SSCs.

The multiple trapping model extends the diffusion model by explaining how the transport of electrons, as observed in photocurrent relaxation and other dynamic experiments, seems to accelerate as the light incident intensity increases [46]. This model is derived from theories formulated by Tiedje and Rose describing similar behaviors in amorphous silicon [54], and it has been widely used to model DSSC and QDSSC behavior by various researchers [33, 36]. The multiple trapping model posits that electrons move randomly in the conduction band where they are considered "free." However, this random walk is intermittently interrupted when the electrons are trapped in states lower in energy than the conduction band edge, rendering them immobile or "trapped." The trapped electrons remain localized until they are thermally "detrapped" (released) and return to the conduction band, allowing them to continue their random walk. The process of trapping/detrapping occurs repeatedly during electron transit through the TiO_2 film, hence the term "multiple trapping." Consequently, the average time for an electron to traverse the TiO_2 film is much longer than if there were no trapping. The electron diffusion coefficient determined by photocurrent transients or frequency-domain methods such as impedance spectroscopy has become known as the "chemical" diffusion coefficient due to its connection with the chemical potential of electrons, and it frequently exhibits an exponential dependence on cell voltage. In the multiple trapping model, this dependency is interpreted as a direct consequence of the exponentially distributed traps in the TiO_2 bandgap.

It is crucial to recognize that under ideal conditions, trapping and detrapping primarily influence the transient behavior of electrons. Under the steady-state conditions relevant to solar cell operation, the local rates of trapping/detrapping in the TiO_2 layer must be equal if trapped electrons do not transfer to the electrolyte, thus the trapped electron concentration must remain constant in time. Therefore, under steady-state conditions, trap states in the bulk of the TiO_2 (i.e., states not communicating with the electrolyte) cannot in principle affect steady-state parameters like the diffusion length of electrons or the charge collection efficiency; it is sufficient to only consider free electrons to formulate these parameters. This situation of multiple trapping is depicted in Figure 8.4. It is also important to note that if the bulk trap state concentration is high enough, so the average distance between traps is small enough, electron transport may occur by hopping between traps [55, 56].

The time-dependent electron continuity equation for a sensitized nanostructured TiO_2 solar cell, incorporating multiple trapping processes, can be expressed as:

$$\frac{\partial n_c(x,t)}{\partial t} = I_0 \alpha e^{-\alpha x} + D_0 \frac{\partial^2 n_c(x,t)}{\partial x^2} - \frac{\left(n_c(x,t) - n_{eqm}\right)}{\tau_0} + k_{detrap} f N_{t,0} - k_{trap} n_c(x,t)(1-f) \tag{8.5}$$

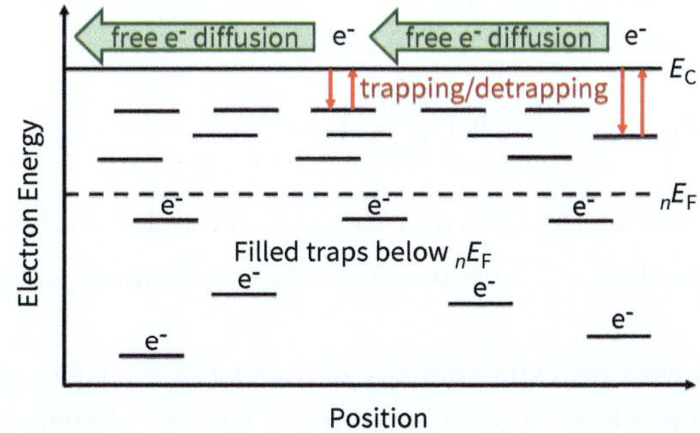

Figure 8.4 Schematic demonstrating the basic principles of the multiple trapping model in nanostructured TiO_2. For simplicity, only transport through the bulk TiO_2 is shown and electron exchange with the electrolyte via surface states is omitted.

where k_{trap} and k_{detrap} are rate constants for trapping/detrapping, respectively, and f is the fractional trap state occupancy. The rate constant for trapping can be written as:

$$k_{trap} = \sigma_n v_{th} N_{t,0} \tag{8.6}$$

where σ_n is the trap capture cross-section, v_{th} is the free electron thermal velocity, and $N_{t,0}$ is the trap state concentration. The rate constant for detrapping is proportional to the probability with which electrons are thermally detrapped. In the case of a single trap state energy, k_{detrap} can be written in terms of k_{trap} according to Shockley-Read-Hall statistics [57]:

$$k_{detrap} = k_{trap} \frac{N_c}{N_{t,0}} \exp\left(\frac{E_t - E_c}{k_B T}\right) \tag{8.7}$$

where N_c is the effective density of conduction band states, E_c is the conduction band edge energy, and E_t is the trap state energy. While expressions can be derived for multiple trapping with other trap state distributions (such as an exponential distribution), the reader is referred elsewhere for a more detailed coverage [58–60].

Equation (8.5) implies that dynamic characteristics of SSCs will be affected by trapping/detrapping of electrons, which makes it more complicated to determine the true diffusion coefficient and lifetime of free electrons from transient and frequency-domain techniques. However, at steady state, trapping/detrapping rates must be equal at any point x in the TiO_2, causing the last two terms in Eq. (8.5) to cancel out, simplifying it to the electron continuity equation without traps (Eq. 8.1). This suggests that the steady-state properties of the SSCs can be explained without explicit consideration of trapping. While this is mostly true for predicting the free electron concentration and photocurrent, it is important to note that trapped electrons must still be present in the TiO_2. It is possible to calculate the trapped electron concentration by recognizing that under equilibrium conditions, the trapped and free electron concentrations can be linked by a shared QFL, under the assumption that exchange of electrons between trap states and the oxidized mediator (or oxidized sensitizer if regeneration is inefficient) is negligibly slow compared with exchange between trap states and the conduction band. The local QFL relative to the electrolyte redox level can be derived from the free electron concentration $n_c(x)$ (obtained from the solution to Eq. 8.1) using:

$${}_nE_F(x) - E_{F,redox} = \frac{k_B T}{q} \ln\left(\frac{n_c(x)}{n_{eqm}}\right) \tag{8.8}$$

where ${}_nE_F(x)$ is the electron QFL and $E_{F,redox}$ is the redox energy of the electrolyte. Note that n_{eqm}, the dark equilibrium electron concentration, occurs when ${}_nE_F(x) = E_{F,redox}$.

Assuming a known trap state density of states (DOS), $g_t(E)$, the excess (relative to the dark) trapped electron concentration, n_t, can be calculated from the electron QFL using:

$$n_t = \int_{E_{F,redox}}^{E_c} f(E) g_t(E)\, dE \tag{8.9}$$

The trap state DOS is often approximated as an exponential function for nanostructured TiO_2:

$$g_t(E_t) = \frac{N_{t,0}}{k_B T_c} \exp\left(\frac{E_t - E_c}{k_B T_c}\right) \tag{8.10}$$

where T_c is a characteristic temperature that describes the trap distribution width. This empirical expression has been used to explain various electrochemical measurements of nanostructured TiO_2. When T_c greatly exceeds T, T can be set to zero in the Fermi–Dirac occupancy function (the so-called zero-Kelvin approximation), reducing $f(E)$ to a step function with a step at $E = {}_nE_F$. In this approximation, with a DOS described by Eq. (8.10), the concentration of trapped electrons is approximately:

$$n_t \approx N_{t,0}\exp\left(\frac{{}_nE_F - E_c}{k_B T_c}\right) - N_{t,0}\exp\left(\frac{E_{F,redox} - E_c}{k_B T_c}\right) \tag{8.11}$$

As a result, the total concentration of electrons in the TiO_2 is significantly larger than would be expected if only free electrons were considered. Moreover, the dependence of the total electron concentration on QFL is primarily governed by Eq. (8.11) and not Eq. (8.8).

8.3.3 The Quasi-static Approximation

If the equilibrium of an SSC is disturbed, such as by a short light pulse, the timescale over which free carriers (and, consequently, the photocurrent, and photovoltage) relax is not only influenced by D_0 and τ_0, but also by the terms in Eq. (8.5) describing trapping/detrapping. This makes interpretating the time constants found in transient photocurrent and photovoltage measurements more difficult.

Bisquert and Vikhrenko showed that a simple quasi-static approximation can greatly assist in interpretating the time constants found in such dynamic experiments when studying SSCs [61]. They showed that when trapping/detrapping is very rapid compared with other relaxation processes involving free carriers (i.e., conduction band transport and recombination), free and trapped carriers remain in quasi-equilibrium with each other, even as a perturbation pushes the system away from equilibrium. The approximation that invokes this concept is called the *quasi-static approximation*. Under quasi-static conditions, the following relationship connecting the rates of change of trapped and free carrier concentrations can be used:

$$\frac{\partial n_c}{\partial t} = \frac{\partial n_c}{\partial n_t}\frac{\partial n_t}{\partial t} \tag{8.12}$$

Under this approximation, Fickian diffusion can still be used to describe local electron transport, but with a modified diffusion coefficient, termed the chemical diffusion coefficient, D_n:

$$D_n = D_0 \frac{1}{\left(1 + \frac{\partial n_t}{\partial n_c}\right)} \tag{8.13}$$

where $\partial n_t/\partial n_c$ represents the change in trapped electron concentration as free electron concentration varies at equilibrium. If multiple trapping is to significantly impact the measured diffusion coefficient, $\partial n_t/\partial n_c$ must be much greater than 1, which permits simplification of Eq. (8.13) to:

$$D_n \approx D_0 \frac{\partial n_c}{\partial n_t} \tag{8.14}$$

For an exponential trap state distribution, $\partial n_t/\partial n_c$ can be derived from Eqs. (8.8) and (8.11), leading to the following expression:

$$D_n = D_0 \frac{N_c^{\beta}}{\beta N_{t,0}} n_c^{1-T/T_c} \tag{8.15}$$

If n_c linearly depends on I_0 (as predicted by both the simple diffusion model and the multiple trapping model for recombination first order in n_c), Eq. (8.15) predicts that D_n will vary I_0 raised to a power $1 - T/T_c$, which is commonly observed in experiments. The steps used to derive Eq. (8.15) are valid only when the QFL remains constant throughout the TiO_2 layer, and this has implications for time constants derived from dynamic techniques when the QFL is not uniform across the TiO_2 layer, such as photocurrent transients measured under short-circuit conditions. These nuances will be discussed in more detail later in this chapter.

Under the assumption of first-order recombination, the electron lifetime, τ_n, can similarly be expressed as:

$$\tau_n = \tau_0 \frac{\beta N_{t,0}}{N_c^{\beta}} n_c^{T/T_c-1} \tag{8.16}$$

It is worth emphasizing that the electron lifetime and chemical diffusion coefficient from the multiple trapping model are theoretical constructs that allow interpretation of time constants extracted from small-amplitude dynamic measurements within a simple linear framework. They are only indirectly related to the steady-state behavior of SSCs under practical device operating conditions. Moreover, the actual values of the free electron lifetime and diffusion coefficient cannot be extracted from τ_n and D_n without knowing several other, often difficult to measure, parameters (i.e., N_c, $N_{t,0}$, and T_c). For these reasons, caution should be exercised when using τ_n and D_n values obtained in dynamic experiments to explain steady-state SSC performance.

When recombination occurs via states other than the conduction band, τ_0 becomes dependent on electron concentration, rendering Eq. (8.18) invalid. Detailed discussions of experimental approaches for measuring the electron lifetime in DSSCs and QDSSCs can be found elsewhere [62, 63].

8.3.4 Electron–Redox Mediator Recombination

The most elementary model for recombination in SSCs supposes that recombination arises from electrons in the TiO_2 conduction band being transferred to acceptor species in the electrolyte, namely the oxidized half of the redox couple such as I_3^-, a Co^{3+}, or Cu^{2+} complex, or S_n^{2-} ions, depending on the redox couple employed.

Recombination of electrons with oxidized sensitizers after electron injection is assumed to be negligible on account of near quantitative regeneration by the redox mediator. While this assumption is likely true for efficient cells at short circuit or under low light intensities, it may not be valid under more realistic operating conditions (e.g., at 1 Sun illumination and/or the maximum power point), as discussed below in Section 8.3.5. In this model, since electron transfer initiates from only one energy state (the bottom edge of the conduction band) toward a distribution of acceptor states in solution that is independent of electron concentration, the rate constant for electron removal does not depend on electron concentration. Additionally, if the acceptor concentration is much higher than the free electron concentration (typically the case for the concentrated electrolyte solutions used), the recombination rate equation can be expressed in pseudo-first-order form as:

$$\frac{dn_c}{dt} = \frac{(n_c - n_{eqm})}{\tau_0} \tag{8.17}$$

This equation is the same as the recombination term in the previously described diffusion model. At open circuit with steady, spatially uniform photogeneration (experimentally achievable by using continuous, very weakly absorbed light), a flat profile of electron concentration is expected through the TiO_2 layer. Under such conditions, the electron continuity equation can be written in a simplified form as:

$$\frac{dn_c}{dt} = \frac{I_0(1 - e^{-\alpha d})}{d} - \frac{(n_c - n_{eqm})}{\tau_0} = 0 \tag{8.18}$$

The open-circuit photovoltage, V_{oc}, of an SSC is the difference in QFL between the sensitized TiO_2 and the redox electrolyte. Therefore, the free electron concentration depends on open-circuit photovoltage according to:

$$n_c = n_{eqm}\exp\left(\frac{qV_{oc}}{k_B T}\right) \tag{8.19}$$

It follows from Eqs. (8.18) and (8.19) that V_{oc} should vary linearly with $\log(I_0)$, so that semi-logarithmic plots exhibit slopes of ~59 mV/decade at room temperature (when illumination is bright enough to achieve $n_c \gg n_{eqm}$, which is most practical cases). While nearly ideal slopes of 59 mV/decade are occasionally observed for DSSCs [64], significantly greater slopes are far more common. The reasons for these so-called "non-ideal" slopes have been heavily debated. For example, some researchers suggest that recombination through the FTO substrate leads to deviations from ideal behavior and linearity [65, 66]. Additionally, some researchers propose that the mechanism of recombination is more complicated than implied by Eq. (8.17), suggesting that recombination may also occur via bandgap trap states near the surface in addition to via the conduction band [67, 68]. Other hypotheses include unpinning of the semiconductor energy levels due to a non-negligible change in potential drop across the electric double layer at the semiconductor–electrolyte interface upon charging the TiO_2, meaning that Eq. (8.19) may not hold, even if Eq. (8.17) is valid [14, 69].

A careful analysis of V_{oc} versus I_0 data while varying electrolyte Li^+ concentration found that the shape and average slope of V_{oc}–I_0 plots is influenced by electrolyte

Li^+ concentration [70]. It was already well known that increasing Li^+ concentration results in a decrease in V_{oc} due to specific adsorption of Li^+ ions positively charging the TiO_2 surface and causing a downward shift in the semiconductor energy levels. However, it had not previously been appreciated that the slopes of the semi-logarithmic V_{oc} versus I_0 plots decreased as the concentration of Li^+ increased, with ideal slopes of ~59 mV/decade (consistent with a reaction order of unity with respect to conduction band electrons) observed for 1 M Li^+. These results align well with a recombination model first elaborated by Salvador et al., who suggested that electron transfer to the acceptors in solution occurs both from the conduction band and from surface states within the bandgap [68]. A key feature of this model is that electron transfer follows Marcus–Gerischer electron transfer theory [71], meaning that electrons are transferred from the semiconductor to acceptor states with a Gaussian distribution in energy. A good fit of this model to the data was obtained using a surface state distribution that was independently measured by impedance spectroscopy. Figure 8.5 shows plots of V_{oc} versus I_0 for DSSCs with electrolytes containing various Li^+ concentrations, along with fits to the aforementioned model [68, 70].

These results indicate that recombination with electrolyte acceptors via the conduction band occurs in the Marcus inverted region, in which increasing the charge transfer driving force decreases the recombination rate constant. Another element altering the V_{oc} versus I_0 plots is the positive shift in the conduction band edge with increasing Li^+, which causes the concentration of conduction band electrons at a particular V_{oc} to increase. As Li^+ concentration increases and the band edge is further lowered toward the electrolyte redox level, the fraction of recombination via the conduction band increases until ultimately, at 1 M Li^+, transfer via the conduction band totally dominates the recombination process.

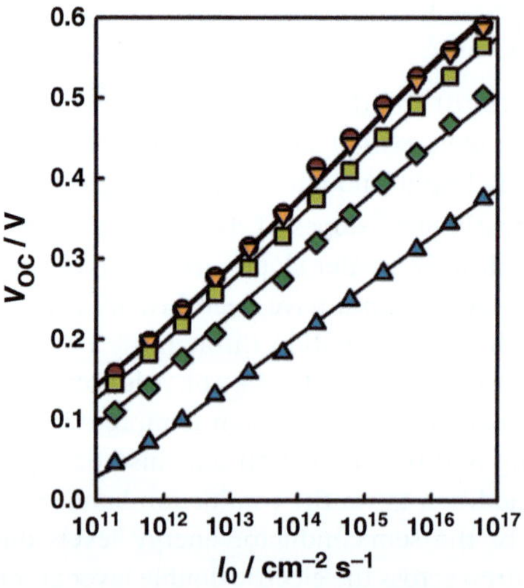

Figure 8.5 Plots of V_{oc} versus I_0 for DSSCs containing varying electrolyte Li^+ concentrations: 0 M (red), 1 mM (orange), 10 mM (yellow), 100 mM (green), and 1 M (cyan). Solid lines represent fits using the model of Salvador et al. described in the main text. Reproduced from [70]/with permission of American Chemical Society.

The hypothesis that recombination from the TiO_2 conduction band (or "shallow" states near the conduction band) is in the Marcus inverted region, and that displacing the conduction band alters the relative importance of recombination via traps states has been further explored by in theoretical work by Gonzalez-Vazquez et al. [72]. These authors used random walk numerical simulations in conjunction with Marcus–Gerischer theory to model recombination in DSSCs. It was found that the aforementioned model is a plausible explanation for experimental observations that L_0 depends strongly on V_{oc} when the energy of the conduction TiO_2 band edge is far above the electrolyte redox level but is independent of V_{oc} when the conduction band edge is lowered [49]. This change in recombination behavior is illustrated in Figure 8.6. More recently, several researchers have investigated the recombination mechanism in DSSCs employing cobalt-based electrolytes, finding evidence for both Marcus normal and inverted region behavior depending on the type of dye employed [73–75].

An alternative, more empirical model for the observed non-ideal variation of V_{oc} with I_0 has recently been proposed and tested by Barr and Meyer [76]. These authors consider the exponentially distributed trap states in the TiO_2 mentioned previously, together with the experimentally determined (and well-known) [62] dependence of electron lifetime on the total TiO_2 electron concentration to explain the variation of V_{oc} with I_0. Their analysis provides compelling evidence that the recombination reaction is first order with respect to *total* TiO_2 electron concentration, which is consistent with Ti^{3+} (often cited as the chemical origin of trap states in nanostructured TiO_2) being the reactive species involved in the recombination reaction. These authors also argue that molecular I_2 is most likely to be the electron acceptor in I_3^-/I^- based DSSCs, making the overall recombination reaction

$$Ti^{3+} + I_2 \rightarrow Ti^{4+} + I_2^- \tag{8.20}$$

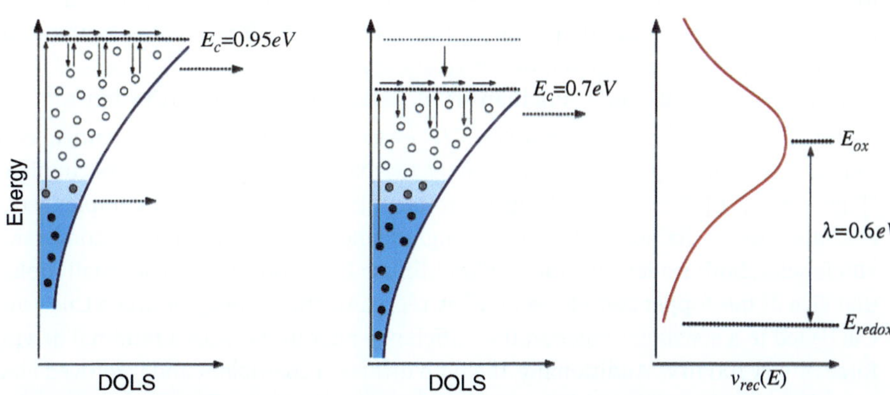

Figure 8.6 Diagram illustrating how recombination behavior changes with shifts in the QFL (represented by varying levels of blue shading) as the conduction band moves to lower energy. For a band edge of 0.95 eV above the redox level (left), recombination occurs via shallow and deep states, and the electron diffusion length increases with increasing QFL. For a band edge of 0.70 eV (right), recombination primarily occurs from shallow traps, leading to a constant diffusion length, independent of the QFL. Reproduced from [72]/with permission of American Chemical Society.

On first sight, these findings may appear to be in conflict with the previously mentioned recombination models based on Marcus–Gerischer theory, in which a transition from recombination via the conduction band and/or shallow traps to recombination via deeper, exponentially distributed trap states as the conduction band energy decreases relative to the electrolyte redox level is given as the main reason for the increasingly non-ideal slopes of V_{oc}–I_0 characteristics as Li^+ concentration is lowered. However, an essential component of the model of Barr and Mayer is an experimentally determined intensity dependent (and thus V_{oc} dependent) electron lifetime, which is also predicted by models based on Marcus–Gerischer theory with an exponential distribution of traps. Such models, like that proposed by Salvador et al. in ref. [68] and applied in ref. [70], involve an electron transfer rate constant that depends on the energy of the TiO_2 state involved in the transfer, and the *apparent* subunity reaction order is just a manifestation of the continuously varying rate constant as the QFL moves through the TiO_2 DOS (and electrolyte acceptor energy levels), leading to a voltage-dependent (and thus intensity dependent) rate constant. The reaction order with respect to electron concentration *for transfer from a single TiO_2 energy state* is of course unity in the Marcus–Gerischer model, which is fully consistent with the results of Barr and Meyer. In other words, the subunity reaction order (which is not actually a constant) in the models based on Marcus–Gerischer theory is merely an emergent property of these models rather than a fundamental assumption.

In another recent study by Sampaio et al., the importance of interfacial electric fields on the recombination process was investigated [77]. These authors took advantage of the recently identified Stark shift in the absorption spectrum of the sensitizing dye to gauge the magnitude of the interfacial electric field and study its effect on the recombination rate. The interfacial electric field was experimentally controlled by varying the size of the electrolyte cation, with a field strength of 1.1 MV cm^{-1} for Na^+, increasing though Li^+ and Mg^{2+} before reaching 2.2 MV cm^{-1} for Ca^{2+}. The recombination rate constant was found to be directly proportional to this interfacial field strength, providing the first direct evidence that recombination with I_3^- is likely to be hindered by electrostatic repulsion.

In another recent study, Pradhan et al. compared copper- ($[Cu(dmp)_2]^{1+/2+}$), cobalt- ($[Co(bpy)_3]^{2+/3+}$), and iodine (I_3^-/I^-)-based electrolytes under identical conditions using the organic LEG4 sensitizer to analyze charge transfer mechanisms (Figure 8.7) [78]. The authors focused on the potential of the copper complex redox mediator due to its ability to achieve higher photovoltage and power conversion efficiencies, both under full sun and low/diffuse light conditions. The small molecular size of the copper complexes and low reorganization energy between Cu(I) and Cu(II) led to a sensitizer regeneration efficiency near unity, with a minimal driving force of just 100 mV. Additionally, the high diffusion coefficient and improved electron diffusion length provided superior charge collection efficiency compared to cobalt-based systems. Copper electrolytes also addressed mass transport limitations more effectively than their cobalt counterparts.

A very detailed study of recombination between TiO_2 electrons and oxidized copper-based redox mediators was recently conducted by Saygili et al. [79]. Their investigation focused on four copper complexes, namely $[Cu(dmby)_2]^{2+/1+}$,

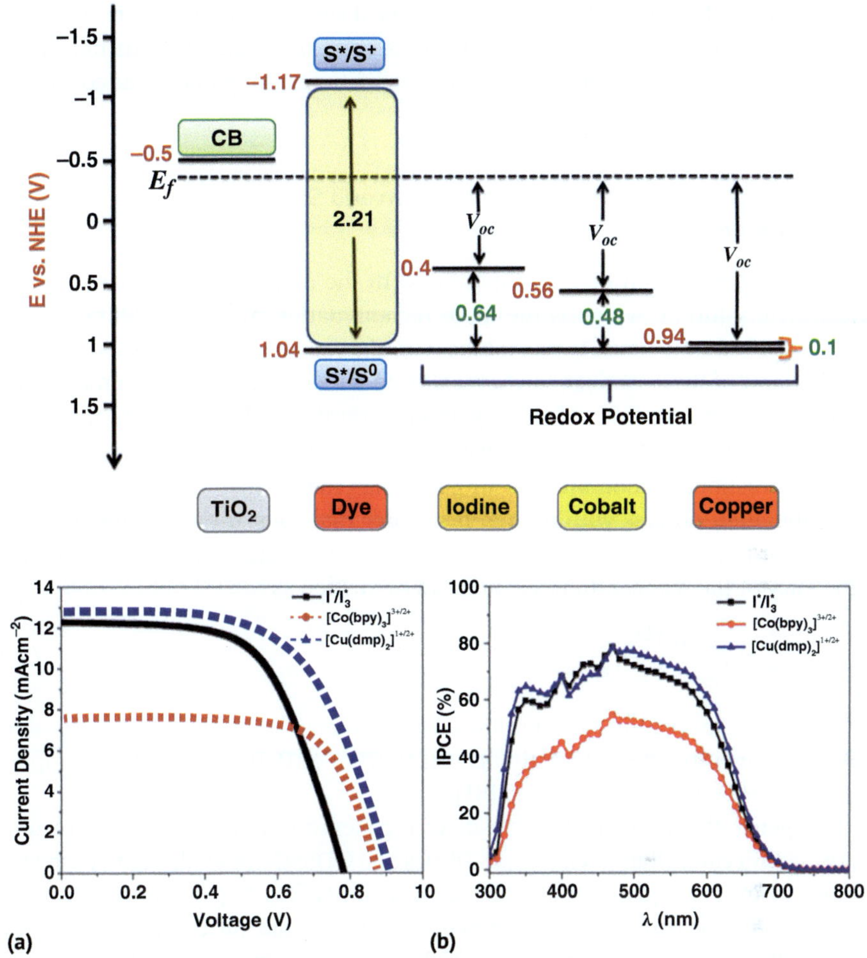

Figure 8.7 Energy level diagram of LEG4-sensitized nanostructured TiO_2 with three different redox electrolytes: I^-/I_3^-, $[Co(bpy)_3]^{2+/3+}$ and $[Cu(dmp)_2]^{1+/2+}$ (top panel) together with the corresponding current–voltage characteristics and incident photon-to-current efficiency (IPCE) spectra (bottom panel). Reproduced from [78]/with permission of The Royal Society of Chemistry.

$[Cu(tmby)_2]^{2+/1+}$, $[Cu(eto)_2]^{2+/1+}$, and $[Cu(dmp)_2]^{2+/1+}$, paired with different organic dye sensitizers (D5, D35, and D45) that varied in their ability to block recombination. Experimental results, supplemented with density functional theory computations, highlighted the low reorganization energies ($\lambda = 0.31$–0.34 eV) of the Cu(I) complexes, which allowed for efficient dye regeneration at minimal driving force (approximately 0.1 eV) in the Marcus normal region. However, for the Cu(II) species, if 4-tert-butylpyridine (an additive used to increase the V_{oc} of almost all high-efficiency DSSCs) was present in the electrolyte, recombination kinetics were altered due to the formation of penta-coordinated complexes with higher reorganization energies ($\lambda = 1.23$–1.40 eV), leading to rapid charge recombination in the

Marcus normal regime rather than the expected slow recombination in the Marcus inverted regime. Despite these shifts, the copper redox species demonstrated higher recombination resistance and longer electron lifetimes compared to the cobalt redox mediator used as a ref. [79].

8.3.5 Electron–Sensitizer Recombination and Sensitizer Regeneration

Besides recombination of TiO_2 electrons with the oxidized redox mediator in the electrolyte solution, which is the main recombination route at short circuit, it is also possible for electrons to recombine with the sensitizer. After injection from the excited state of the sensitizer, the now oxidized sensitizer undergoes a process called regeneration, where it is reduced by an electron donor, D, present in the electrolyte with pseudo-first-order rate constant k_{reg}. For effective sensitizer regeneration, the regeneration reaction must occur faster than the competing electron–sensitizer recombination process with pseudo-first-order rate constant k_{rec}, where the semiconductor transfers electrons back to the oxidized sensitizer. The overall quantum efficiency of the regeneration reaction can be written as:

$$\eta_{reg} = \frac{k_{reg}[D]}{k_{rec} + k_{reg}[D]} \tag{8.21}$$

where [D] is the donor concentration. Iodide remains the most widely used and effective donor for DSSCs [80], though Co(III/II), Fe(III/II), and Cu(II/I) complexes have also been extensively studied [81].

It is generally accepted that sensitizer regeneration is near quantitative under short-circuit conditions in champion DSSCs and QDSSCs, as evidenced by the observation of internal quantum efficiencies near unity [82, 83]. Transient absorption spectroscopy measurements performed ex-situ (i.e., not on complete operational solar cells) have also indicated very high sensitizer regeneration efficiencies [84, 85]. However, it has been shown that under open-circuit conditions or at the maximum power point, the efficiency of dye regeneration is substantially less than unity [86–89]. In one such study, the efficiency of dye regeneration was deliberately altered by varying the iodide concentration in the electrolyte while holding all other experimental variables constant. The cation concentration and ionic strength of the electrolyte were maintained constant while varying I^- by substituting it with electrochemically inert ClO_4^- ions, which are similar in size to I^- ions. A detailed analysis of steady-state (*j*–*V* and IPCE) and dynamic (impedance spectroscopy) data revealed a drop in photocurrent that could only be explained by incomplete dye regeneration [86].

A potential criticism of the aforementioned study is that no "direct" observation of dye regeneration by transient absorption spectroscopy was made. This issue was later addressed by Feng et al. where the dye regeneration efficiency in a complete DSSC was measured under open-circuit conditions over a wide range of light intensities and thus V_{oc} values [89]. For comparison, measurements were also performed at several points along the *j*–*V* characteristic under 1 Sun equivalent illumination, namely at short circuit, the maximum power point, and open circuit. Consistent with the previous study involving varying I^- concentration, clear evidence of decreased dye regeneration and increased electron–dye recombination was found as the TiO_2 electron concentration increased with increasing light intensity. This study also yielded information about the reaction order of electron–dye recombination in a DSSC with a normal electrolyte as well as for an "inert" electrolyte without added I^-. Similar conclusions were are drawn in a separate study by Robson et al., who used transient absorption spectroscopy to investigate regeneration of D–π–A sensitizers with iodide- and cobalt-based electrolytes [88].

Further insights into recombination dynamics in nanostructured TiO_2 sensitized with ruthenium-based sensitizers were gained by Brigham et al. using kinetic isolation conditions (Ostwald isolation) to determine the reaction order for electron–dye recombination [90]. By applying an external bias to maintain known populations of oxidized sensitizer, S^+, and TiO_2 electrons, $TiO_2(e^-)$, they were able to isolate and study the interfacial charge recombination process. Using pulsed laser excitation, they tracked the $TiO_2(e^-)|S^+ \rightarrow TiO_2|S$ reaction spectroscopically, revealing that the recombination follows a first-order rate law with respect to both the oxidized sensitizer and TiO_2 electrons under reverse bias conditions. However, under forward bias conditions, which are more applicable to operating conditions for a solar cell, the recombination behavior became more complex. The researchers proposed a mechanism involving a pre-equilibrium diffusional encounter between $TiO_2(e^-)$ and S^+, suggesting that diffusion limits recombination when the electron concentration in TiO_2 is low, while electron transfer dominates when the electron concentration is larger [90].

More recently, regeneration in DSSCs using copper-based electrolytes has gained significant attention due the promising power conversion efficiencies achievable using these newer redox mediators, particularly under full sunlight and low/diffuse light conditions [11, 78, 79]. Copper-based redox mediators offer the advantage of achieving higher photovoltage without sacrificing photocurrent, making them a highly attractive alternative to traditional electrolytes. One of the key factors behind this success is the small reorganization energy between Cu(I) and Cu(II), combined with the small molecular size of copper complexes [79]. This enables near 100% regeneration efficiency with minimal driving force, as low as 100 mV.

8.4 Conclusion

This chapter has provided an overview of carrier transport and recombination in sensitized nanostructured TiO_2 solar cells, with a particular emphasis on DSSCs. The diffusion model, although simplified, provides a foundational understanding of electron transport and recombination in these systems. Extensions of this model, such as the multiple trapping model, offer deeper insights into how trapping and detrapping events influence electron transport and recombination dynamics. These models are crucial for predicting the behavior of electrons in nanostructured TiO_2 and for identifying strategies to mitigate losses due to recombination.

Recent developments, particularly the realization that sensitizer regeneration is not 100% efficient at the maximum power point, as well as the use of copper-complex-based electrolytes, have demonstrated the potential for further enhancing the performance of SSCs. Copper-based electrolytes in particular have shown the ability to deliver higher photovoltage without sacrificing photocurrent, primarily due to their small reorganization energy and efficient charge transfer properties. This innovation, along with ongoing research into the role of surface states and interfacial charge dynamics, suggests that SSCs will continue to evolve and find niche applications.

Despite competition from other emerging photovoltaic technologies such as perovskite solar cells, SSCs maintain unique advantages, including their tunable aesthetic properties, flexibility in material choices, and the potential for high efficiency under nonideal lighting conditions. These features make them attractive for specific applications, such as building-integrated photovoltaics and portable electronic devices. Moving forward, further improvements in the understanding of charge transport mechanisms and recombination pathways, combined with the development of novel materials, are likely to lead to even more efficient and commercially viable SSCs.

References

1 O'Regan, B. and Grätzel, M. (1991). A low-cost, high-efficiency solar cell based on dye-sensitized colloidal TiO_2 films. *Nature* 353: 737–740. https://doi.org/10.1038/353737a0.

2 Lee, M.M., Teuscher, J., Miyasaka, T. et al. (2012). Snaith, efficient hybrid solar cells based on meso-superstructured organometal halide perovskites. *Science* (80–.) 338: 643–647. https://doi.org/10.1126/science.1228604.

3 Liu, M., Johnston, M.B., and Snaith, H.J. (2013). Efficient planar heterojunction perovskite solar cells by vapour deposition. *Nature* 501: 395–398. https://doi.org/10.1038/nature12509.

4 Fatima, Q., Haidry, A.A., Zhang, H. et al. (2024). A critical review on advancement and challenges in using TiO_2 as electron transport layer for perovskite solar cell. *Materials Today Sustainability* 27: 100857. https://doi.org/10.1016/j.mtsust.2024.100857.

5 Cui, Y., Yao, H., Gao, B. et al. (2017). Fine-tuned photoactive and interconnection layers for achieving over 13% efficiency in a fullerene-free tandem organic solar cell. *Journal of the American Chemical Society* 139: 7302–7309. https://doi.org/10.1021/jacs.7b01493.

6 Xiao, Z., Jia, X., and Ding, L. (2017). Ternary organic solar cells offer 14% power conversion efficiency. *Science Bulletin* 62: 1562–1564. https://doi.org/10.1016/j.scib.2017.11.003.

7 Meng, L., Zhang, Y.X., Wan, C. et al. (2018). Organic and solution-processed tandem solar cells with 17.3% efficiency. *Science* (80–.) 361: 1094–1098. https://doi.org/10.1126/science.aat2612.

8 Kakiage, K., Aoyama, Y., Yano, T. et al. (2015). Highly-efficient dye-sensitized solar cells with collaborative sensitization by silyl-anchor and carboxy-anchor dyes. *Chemical Communications* 51: 15894–15897. https://doi.org/10.1039/c5cc06759f.

9 Ji, J.M., Zhou, H., Eom, Y.K. et al. (2020). 14.2% efficiency dye-sensitized solar cells by co-sensitizing novel thieno[3,2-b]indole-based organic dyes with a promising porphyrin sensitizer. *Advanced Energy Materials* 10: 1–12. https://doi.org/10.1002/aenm.202000124.

10 Feldt, S.M., Gibson, E.A., Gabrielsson, E.L. et al. (2010). Hagfeldt, design of organic dyes and cobalt polypyridine redox mediators for high-efficiency dye-sensitized solar cells. *Journal of the American Chemical Society* 132: 16714–16724. https://doi.org/10.1021/ja1088869.

11 Zhang, D., Stojanovic, M., Ren, Y. et al. (2021). A molecular photosensitizer achieves a V oc of 1.24 V enabling highly efficient and stable dye-sensitized solar cells with copper(II/I)-based electrolyte. *Nature Communications* 12: 2–11. https://doi.org/10.1038/s41467-021-21945-3.

12 Liu, Y., Jennings, J.R., Huang, Y. et al. (2011). Cobalt redox mediators for ruthenium-based dye-sensitized solar cells: a combined impedance spectroscopy and near-IR transmittance study. *The Journal of Physical Chemistry C* 115: 18847–18855. https://doi.org/10.1021/jp204519s.

13 Mosconi, E., Yum, J.H., Kessler, F. et al. (2012). Cobalt electrolyte/dye interactions in dye-sensitized solar cells: a combined computational and experimental study. *Journal of the American Chemical Society* 134: 19438–19453. https://doi.org/10.1021/ja3079016.

14 Liu, Y., Jennings, J.R., Zakeeruddin, S.M. et al. (2013). Heterogeneous electron transfer from dye-sensitized nanocrystalline TiO_2 to [Co(bpy)3]3+: insights gained from impedance spectroscopy. *Journal of the American Chemical Society* 135: 3939–3952. https://doi.org/10.1021/ja311743m.

15 Pashaei, B., Shahroosvand, H., Graetzel, M. et al. (2016). Influence of ancillary ligands in dye-sensitized solar cells. *Chemical Reviews* 116: 9485–9564. https://doi.org/10.1021/acs.chemrev.5b00621.

16 Ren, Y., Zhang, D., Suo, J. et al. (2023). Hydroxamic acid pre-adsorption raises the efficiency of cosensitized solar cells. *Nature* 613: 60–65. https://doi.org/10.1038/s41586-022-05460-z.

17 Green, M.A., Dunlop, E.D., Yoshita, M. et al. (2024). Solar cell efficiency tables (version 64). *Progress in Photovoltaics: Research and Applications* 32: 425–441. https://doi.org/10.1002/pip.3831.

18 Venkatesan, S., Cho, Y.-S., Liu, I.-P. et al. (2021). Novel architecture of indoor bifacial dye-sensitized solar cells with efficiencies surpassing 25% and efficiency ratios exceeding 95%. *Advanced Optical Materials* 9: 2100936. https://doi.org/10.1002/adom.202100936.

19 Michaels, H. Rinderle, M. Freitag, R. et al. (2020). Dye-sensitized solar cells under ambient light powering machine learning: towards autonomous smart sensors for the internet of things. *Chemical Science* 11: 2895–2906. https://doi.org/10.1039/c9sc06145b.

20 Venkatesan, S. Hsu, T.-H. Teng, H. et al. (2023). Dye-sensitized solar cells with efficiency over 36% under ambient light achieved by cosensitized tandem structure. *Solar RRL* 7: 2300220. https://doi.org/10.1002/solr.202300220.

21 Muñoz-García, A.B., Benesperi, I., Boschloo, G. et al. (2021). Dye-sensitized solar cells strike back. *Chemical Society Reviews* 50: 12450–12550. https://doi.org/10.1039/d0cs01336f.

22 Freitag, M. and Boschloo, G. (2017). The revival of dye-sensitized solar cells. *Current Opinion in Electrochemistry* 2: 111–119. https://doi.org/10.1016/j.coelec.2017.03.011.

23 Heiniger, L.-P., O'Brien, P.G., Soheilnia, N.Y. et al. (2013). See-through dye-sensitized solar cells: photonic reflectors for tandem and building integrated photovoltaics. *Advanced Materials* 25: 5734–5741. https://doi.org/10.1002/adma.201302113.

24 Tai, Q. and Yan, F. (2017). Emerging semitransparent solar cells: materials and device design. *Advanced Materials* 29. https://doi.org/10.1002/adma.201700192.

25 Li, G., Sheng, L., Li, T. et al. (2019). Engineering flexible dye-sensitized solar cells for portable electronics. *Solar Energy* 177: 80–98. https://doi.org/10.1016/j.solener.2018.11.017.

26 Hashemi, S.A., Ramakrishna, S., and Aberle, A.G. (2020). Recent progress in flexible-wearable solar cells for self-powered electronic devices. *Energy & Environmental Science* 13: 685–743. https://doi.org/10.1039/c9ee03046h.

27 Ke, H., Gao, M., Li, S. et al. (2023). Advances and future prospects of wearable textile- and fiber-based solar cells. *Solar RRL* 7. https://doi.org/10.1002/solr.202300109.

28 Ohsaki, Y., Masaki, N., Kitamura, T. et al. (2005). Dye-sensitized TiO_2 nanotube solar cells: fabrication and electronic characterization. *Physical Chemistry Chemical Physics* 7: 4157–4163. https://doi.org/10.1039/b511016e.

29 Jennings, J.R., Ghicov, A., Peter, L.M. et al. (2008). Dye-sensitized solar cells based on oriented TiO_2 nanotube arrays: transport, trapping, and transfer of electrons. *Journal of the American Chemical Society* 130: 13364–13372. https://doi.org/10.1021/ja804852z.

30 MacWan, D.P., Dave, P.N., and Chaturvedi, S. (2011). A review on nano-TiO_2 sol-gel type syntheses and its applications. *Journal of Materials Science* 46: 3669–3686. https://doi.org/10.1007/s10853-011-5378-y.

31 Roy, P., Berger, S., and Schmuki, P. (2011). TiO_2 nanotubes: synthesis and applications. *Angewandte Chemie International Edition* 50: 2904–2939. https://doi.org/10.1002/anie.201001374.

32 Bisquert, J., Garcia-Belmonte, G., and Fabregat-Santiago, F. (1999). Modelling the electric potential distribution in the dark in nanoporous semiconductor electrodes. *Journal of Solid State Electrochemistry* 3: 337–347. https://doi.org/10.1007/s100080050164.

33 Peter, L. (2009). "Sticky electrons" transport and interfacial transfer of electrons in the dye-sensitized solar cell. *Accounts of Chemical Research* 42: 1839–1847. https://doi.org/10.1021/ar900143m.

34 Yella, A., Lee, H.W., Tsao, H.N. et al. (2011). Porphyrin-sensitized solar cells with cobalt (II/III)-based redox electrolyte exceed 12 percent efficiency. *Science* (80–.) 334: 629–634. https://doi.org/10.1126/science.1209688.

35 Pan, Z., Rao, H., Mora-Seró, I. et al. (2018). Quantum dot-sensitized solar cells. *Chemical Society Reviews* 47: 7659–7702. https://doi.org/10.1039/c8cs00431e.

36 Aqoma, H., Lee, S.-H., Imran, I.F. et al. (2024). Alkyl ammonium iodide-based ligand exchange strategy for high-efficiency organic-cation perovskite quantum dot solar cells. *Nature Energy* 9: 324–332. https://doi.org/10.1038/s41560-024-01450-9.

37 Wu, J., Lan, Z., Lin, J. et al. (2015). Electrolytes in dye-sensitized solar cells. *Chemical Reviews* 115: 2136–2173. https://doi.org/10.1021/cr400675m.

38 Ardo, S. and Meyer, G.J. (2009). Photodriven heterogeneous charge transfer with transition-metal compounds anchored to TiO_2 semiconductor surfaces. *Chemical Society Reviews* 38: 115–164. https://doi.org/10.1039/b804321n.

39 Bisquert, J. (2008). Physical electrochemistry of nanostructured devices. *Physical Chemistry Chemical Physics* 10: 49–72. https://doi.org/10.1039/b709316k.

40 Ye, M., Wen, X., Wang, M. et al. (2015). Recent advances in dye-sensitized solar cells: from photoanodes, sensitizers and electrolytes to counter electrodes. *Matererial Today* 18: 155–162. https://doi.org/10.1016/j.mattod.2014.09.001.

41 Choudhury, B. and Choudhury, A. (2014). Oxygen defect dependent variation of band gap, Urbach energy and luminescence property of anatase, anatase-rutile mixed phase and of rutile phases of TiO_2 nanoparticles. *Physica E: Low-Dimensional Systems and Nanostructures* 56: 364–371. https://doi.org/10.1016/j.physe.2013.10.014.

42 Sodergren, S., Hagfeldt, A., Olsson, J. et al. (1994). Theoretical models for the action spectrum and the current-voltage characteristics of microporous semiconductor films in photoelectrochemical cells. *The Journal of Physical Chemistry* 98: 5552–5556. https://doi.org/10.1021/j100072a023.

43 Peter, L.M. (2007). Characterization and modeling of dye-sensitized solar cells. *The Journal of Physical Chemistry C* 111: 6601–6612. https://doi.org/10.1021/jp069058b.

44 Rudra, S., Sarker, S., and Kim, D.M. (2019). Review on simulation of current–voltage characteristics of dye-sensitized solar cells. *Journal of Industrial and Engineering Chemistry* 80: 516–526. https://doi.org/10.1016/j.jiec.2019.08.030.

45 Rahmatian, M. and Sayyaadi, H. (2024). Steady and transient modeling of dye-sensitive solar cells: the impact of electrode thickness and dye specifications. *Energy Conversion and Management X*. 24: 100709. https://doi.org/10.1016/j.ecmx.2024.100709.

46 Dloczik, L., Ileperuma, O., Lauermann, I. et al. (1997). Dynamic response of dye-sensitized nanocrystalline solar cells: characterization by intensity-modulated photocurrent spectroscopy. *The Journal of Physical Chemistry B* 101: 10281–10289. https://doi.org/10.1021/jp972466i.

47 Barnes, P.R.F., Liu, L., Li, X. et al. (2009). O'Regan, re-evaluation of recombination losses in dye-sensitized cells: the failure of dynamic relaxation methods to correctly predict diffusion length in nanoporous photoelectrodes. *Nano Letters* 9: 3532–3538. https://doi.org/10.1021/nl901753k.

48 Bisquert, J. and Mora-Seró, I. (2010). Simulation of steady-state characteristics of dye-sensitized solar cells and the interpretation of the diffusion length. *The Journal of Physical Chemistry Letters* 1: 450–456. https://doi.org/10.1021/jz900297b.

49 Jennings, J.R., Li, F., and Wang, Q. (2010). Reliable determination of electron diffusion length and charge separation efficiency in dye-sensitized solar cells. *The Journal of Physical Chemistry C* 114: 14664–14674. https://doi.org/10.1021/jp105486k.
50 Lobato, K., Peter, L.M., and Würfel, U. (2006). Direct measurement of the internal electron quasi-Fermi level in dye sensitized solar cells using a titanium secondary electrode. *The Journal of Physical Chemistry B* 110: 16201–16204. https://doi.org/10.1021/jp063919z.
51 Lobato, K. and Peter, L.M. (2006). Direct measurement of the temperature coefficient of the electron quasi-Fermi level in dye-sensitized nanocrystalline solar cells using a titanium sensor electrode. *The Journal of Physical Chemistry B* 110: 21920–21923. https://doi.org/10.1021/jp064538o.
52 Jennings, J.R. and Peter, L.M. (2007). A reappraisal of the electron diffusion length in solid-state dye-sensitized solar cells. *The Journal of Physical Chemistry C* 111: 16100–16104. https://doi.org/10.1021/jp076457d.
53 Berger, T., Anta, J.A., and Morales-Flórez, V. (2012). Electrons in the band gap: spectroscopic characterization of anatase TiO_2 nanocrystal electrodes under Fermi level control. *The Journal of Physical Chemistry C* 116: 11444–11455. https://doi.org/10.1021/jp212436b.
54 Tiedje, T. and Rose, A. (1981). A physical interpretation of dispersive transport in disordered semiconductors. *Solid State Communication* 37: 49–52. https://doi.org/10.1016/0038-1098(81)90886-3.
55 Bisquert, J. (2007). Hopping transport of electrons in dye-sensitized solar cells. *The Journal of Physical Chemistry C* 111: 17163–17168. https://doi.org/10.1021/jp077419x.
56 Wang, Q., Zhang, Z., Zakeeruddin, S.M. et al. (2008). Enhancement of the performance of dye-sensitized solar cell by formation of shallow transport levels under visible light illumination. *The Journal of Physical Chemistry C* 112: 7084–7092. https://doi.org/10.1021/jp800426y.
57 Shockley, W. and Read, W.T. (1952). Statistics of the recombinations of holes and electrons. *Physical Reviews* 87: 835–842. https://doi.org/10.1103/PhysRev.87.835.
58 Bisquert, J. (2008). Beyond the quasistatic approximation: impedance and capacitance of an exponential distribution of traps. *Physical Review B—Condensed Matter and Materials Physics* 77: 235203. https://doi.org/10.1103/PhysRevB.77.235203.
59 Bisquert, J. (2015). *Nanostructured Energy Devices: Equilibrium Concepts and Kinetics*. Boca Raton: CRC Press. https://books.google.com/books?id=5YvaBAAAQBAJ&lr=&source=gbs_navlinks_s.
60 Bisquert, J. (2018). Nanostructured energy devices: foundations of carrier transport. *MRS Bulletin* 43: 713. https://doi.org/10.1557/mrs.2018.228.
61 Bisquert, J. and Vikhrenko, V.S. (2004). Interpretation of the time constants measured by kinetic techniques in nanostructured semiconductor electrodes and dye-sensitized solar cells. *The Journal of Physical Chemistry B* 108: 2313–2322. https://doi.org/10.1021/jp035395y.
62 Bisquert, J., Fabregat-Santiago, F., Mora-Seró, I. et al. (2009). Electron lifetime in dye-sensitized solar cells: theory and interpretation of measurements. *The Journal of Physical Chemistry C* 113: 17278–17290. https://doi.org/10.1021/jp9037649.
63 Mora-Seró, I., Giménez, S., Fabregat-Santiago, F. et al. (2009). Recombination in quantum dot sensitized solar cells. *Accounts of Chemical Research* 42: 1848–1857. https://doi.org/10.1021/ar900134d.

64 Peter, L.M. (2007). Dye-sensitized nanocrystalline solar cells. *Physical Chemistry Chemical Physics* 9: 2630–2642. https://doi.org/10.1039/b617073k.

65 Cameron, P.J., Peter, L.M., and Hore, S. (2005). How important is the back reaction of electrons via the substrate in dye-sensitized nanocrystalline solar cells? *The Journal of Physical Chemistry B* 109: 930–936. https://doi.org/10.1021/jp0405759.

66 Cameron, P.J. and Peter, L.M. (2005). How does back-reaction at the conducting glass substrate influence the dynamic photovoltage response of nanocrystalline dye-sensitized solar cells? *The Journal of Physical Chemistry B* 109: 7392–7398. https://doi.org/10.1021/jp0407270.

67 Bisquert, J., Zaban, A., Greenshtein, M. et al. (2004). Determination of rate constants for charge transfer and the distribution of semiconductor and electrolyte electronic energy levels in dye-sensitized solar cells by open-circuit photovoltage decay method. *Journal of the American Chemical Society* 126: 13550–13559. https://doi.org/10.1021/ja047311k.

68 Salvador, P. Hidalgo, M.G. Zaban, A. et al. (2005). Illumination intensity dependence of the photovoltage in nanostructured TiO_2 dye-sensitized solar cells. *The Journal of Physical Chemistry B* 109: 15915–15926. https://doi.org/10.1021/jp051515l.

69 Mandal, D. and Hamann, T.W. (2016). Charge distribution in nanostructured TiO_2 photoanode determined by quantitative analysis of the band edge unpinning. *ACS Applied Materials & Interfaces* 8: 419–424. https://doi.org/10.1021/acsami.5b09200.

70 Jennings, J.R. and Wang, Q. (2010). Influence of lithium ion concentration on electron injection, transport, and recombination in dye-sensitized solar cells. *The Journal of Physical Chemistry C* 114: 1715–1724. https://doi.org/10.1021/jp9104129.

71 Memming, R. (2015). Electron transfer theories. In: *Semiconductor Electrochemistry* (ed. R. Memming), 127–168. Weinheim: WILEY-VCH. https://doi.org/10.1002/9783527688685.ch6.

72 Gonzalez-Vazquez, J.P., Oskam, G., and Anta, J.A. (2012). Origin of nonlinear recombination in dye-sensitized solar cells: interplay between charge transport and charge transfer. *The Journal of Physical Chemistry C* 116: 22687–22697. https://doi.org/10.1021/jp306517g.

73 Feldt, S.M., Lohse, P.W., Kessler, F. et al. (2013). Regeneration and recombination kinetics in cobalt polypyridine based dye-sensitized solar cells, explained using Marcus theory. *Physical Chemistry Chemical Physics* 15: 7087–7097. https://doi.org/10.1039/c3cp50997d.

74 Gao, W., Liang, M., Tan, Y. et al. (2015). New triarylamine sensitizers for high efficiency dye-sensitized solar cells: recombination kinetics of cobalt(III) complexes at titania/dye interface. *Journal of Power Sources* 283: 260–269. https://doi.org/10.1016/j.jpowsour.2015.02.121.

75 Soman, S., Pradhan, S.C., Yoosuf, M. et al. (2018). Probing recombination mechanism and realization of Marcus normal region behavior in DSSCs employing cobalt electrolytes and triphenylamine dyes. *The Journal of Physical Chemistry C* 122: 14113–14127. https://doi.org/10.1021/acs.jpcc.8b01325.

76 Barr, T.J. and Meyer, G.J. (2017). Evidence for first-order charge recombination in dye-sensitized solar cells. *ACS Energy Letters* 2: 2335–2340. https://doi.org/10.1021/acsenergylett.7b00569.

77 Sampaio, R.N., O'Donnell, R.M., Barr, T.J. et al. (2014). Electric fields control TiO_2(e-) + I3- → charge recombination in dye-sensitized solar cells. *The Journal of Physical Chemistry Letters* 5: 3265–3268. https://doi.org/10.1021/jz5016444.
78 Pradhan, S.C., Hagfeldt, A., and Soman, S. (2018). Resurgence of DSCs with copper electrolyte: a detailed investigation of interfacial charge dynamics with cobalt and iodine based electrolytes. *Journal of Materials Chemistry A* 6: 22204–22214. https://doi.org/10.1039/C8TA06948D.
79 Saygili, Y., Stojanovic, M., Michaels, H. et al. (2018). Effect of coordination sphere geometry of copper redox mediators on regeneration and recombination behavior in dye-sensitized solar cell applications. *ACS Applied Energy Materials* 1: 4950–4962. https://doi.org/10.1021/acsaem.8b00957.
80 Boschloo, G. and Hagfeldt, A. (2009). Characteristics of the iodide/triiodide redox mediator in dye-sensitized solar cells. *Accounts of Chemical Research* 42: 1819–1826. https://doi.org/10.1021/ar900138m.
81 Sun, Z., Liang, M., and Chen, J. (2015). Kinetics of iodine-free redox shuttles in dye-sensitized solar cells: interfacial recombination and dye regeneration. *Accounts of Chemical Research* 48: 1541–1550. https://doi.org/10.1021/ar500337g.
82 Liu, Y., Jennings, J.R., Parameswaran, M. et al. (2011). An organic redox mediator for dye-sensitized solar cells with near unity quantum efficiency. *Energy & Environmental Science* 4: 564–571. https://doi.org/10.1039/c0ee00519c.
83 Wang, P., Yang, L., Wu, H. et al. (2018). Stable and efficient organic dye-sensitized solar cell based on ionic liquid electrolyte. *Joule* 2: 2145–2153. https://doi.org/10.1016/j.joule.2018.07.023.
84 Clifford, J.N., Palomares, E., Nazeeruddin, K.M. et al. (2007). Dye dependent regeneration dynamics in dye sensitized nanocrystalline solar cells: evidence for the formation of a ruthenium bipyridyl cation/iodide intermediate. *The Journal of Physical Chemistry C* 111: 6561–6567. https://doi.org/10.1021/jp067458t.
85 Daeneke, T., Mozer, A.J., Uemura, Y. et al. (2012). Dye regeneration kinetics in dye-sensitized solar cells. *Journal of the American Chemical Society* 134: 16925–16928. https://doi.org/10.1021/ja3054578.
86 Jennings, J.R., Liu, Y., and Wang, Q. (2011). Efficiency limitations in dye-sensitized solar cells caused by inefficient sensitizer regeneration. *The Journal of Physical Chemistry C* 115: 15109–15120. https://doi.org/10.1021/jp2053053.
87 Li, F., Jennings, J.R., and Wang, Q. (2013). Determination of sensitizer regeneration efficiency in dye-sensitized solar cells. *ACS Nano* 7: 8233–8242. https://doi.org/10.1021/nn403714s.
88 Robson, K.C.D., Hu, K., Meyer, G.J. et al. (2013). Atomic level resolution of dye regeneration in the dye-sensitized solar cell. *Journal of the American Chemical Society* 135: 1961–1971. https://doi.org/10.1021/ja311640f.
89 Anderson, A.Y., Barnes, P.R.F., Durrant, J.R. et al. (2011). Quantifying regeneration in dye-sensitized solar cells. *The Journal of Physical Chemistry C* 115: 2439–2447. https://doi.org/10.1021/jp1101048.
90 Brigham, E.C. and Meyer, G.J. (2014). Ostwald isolation to determine the reaction order for TiO_2(e-)|S+↔ TiO_2|S charge recombination at sensitized TiO_2 interfaces. *The Journal of Physical Chemistry C* 118: 7886–7893. https://doi.org/10.1021/jp501814r.

9

TiO_2-based Heterojunction Photocatalysts: Type, Characterization, and Activity

Yi Lu[1,2,], Yi-Xuan Liu[1], Xiao-Yu Yang[1,2]*

[1]*State Key Laboratory of Advanced Technology for Materials Synthesis and Processing, Wuhan University of Technology, Wuhan, China*
[2]*National Energy Key Laboratory for New Hydrogen-Ammonia Energy Technologies, Foshan Xianhu Laboratory, Foshan, China*

*Corresponding author: yilu@whut.edu.cn

9.1 Introduction

TiO_2 has garnered substantial research attention owing to its affordability, nontoxicity, and chemical inertness [1]. However, the limited photo response to ultraviolet (UV) light and the rapid recombination of charge carriers hinder the efficient harnessing of solar energy and the effective utilization of generated photocarriers. To address these limitations, strategies such as the incorporation of co-catalysts [2], doping with impurities [3], and the fabrication of heterojunctions [4] have been employed to enhance light absorption and carrier utilization. Notably, meticulously engineered heterojunctions can markedly improve UV to visible light absorption, facilitate the separation of electron–hole pairs, and even bolster the redox capabilities of TiO_2, thereby optimizing its performance in solar energy conversion and photocatalytic applications [5].

The performance of heterojunctions is significantly influenced by the interfaces they form when two semiconductors come into contact. Heterojunctions, derived from semiconductor physics, are categorized based on their composition of two distinct single-crystal semiconductors. These classifications include antipode heterojunctions, such as p–n junctions, and homotypic heterojunctions, which encompass n–n and p–p junctions, primarily used in photocatalytic applications. However, as photocatalytic technology has progressed, the traditional concept was sufficient for identifying the composition of heterojunctions, it failed to provide a clear differentiation between their photocatalytic mechanisms. In response, a new generation of heterojunctions has been developed, including p–n heterojunctions, straddling alignment (type-I), staggered alignment (type-II), broken alignment (type-III), Z-scheme, and S-scheme configurations [6]. The interplay and categorization of

Nanostructured TiO_2: Synthesis, Photocatalytic Properties and Applications, First Edition.
Edited by Jia Hong Pan, Wan In Lee and Detlef W. Bahnemann

these heterogeneous junctions remain confusing, particularly in delineating the direct Z-scheme from type-II and S-scheme junctions. Furthermore, the photocatalytic mechanisms of each heterojunction type have not been exhaustively explored, for instance, p–n junctions exhibit a multiplicity of mechanisms beyond a singular pathway. It is important to recognize that heterojunctions can be conceptualized from multiple vantage points, suggesting that there are likely intersections rather than mere parallelisms among the different types. In this chapter, we propose a classification system for heterojunctions, with a particular focus on type-II and Z-Scheme/S-Scheme configurations. We also provide a comprehensive summary of the characterization techniques and the functional roles of heterojunctions within photocatalytic systems. Our goal is to offer a deeper and more nuanced perspective on these pivotal components, enhancing the understanding of their significance in the field of photocatalysis.

9.2 Classification of Heterojunction

9.2.1 P–N Heterojunction

When the p-type semiconductor is contacted with an n-type semiconductor, the electrons transfer from n-type to p-type until their Fermi level aligns [7]. This results in electron accumulation in the p-type material and positive charge buildup in the n-type region. Such charge redistribution induces energy band bending and creates an internal electric field oriented from the n-type to the p-type semiconductor. Upon light exposure, photogenerated electrons are transferred from the conduction band (CB) of the p-type material to the n-type counterpart, while photogenerated holes move from the valence band (VB) of the n-type to the p-type semiconductor, thereby enhancing the separation efficiency of photoinduced charge carriers.

9.2.2 Type-I, Type-II, and Type-III Heterojunctions

As illustrated in Figure 9.1, the energy band alignment of a standard type- I heterojunction demonstrates that semiconductor I (SC I) completely encompasses the conduction and VB positions of semiconductor II (SC II). This configuration facilitates the migration of charge carriers, with electrons moving from the more negative CB of SC I and holes transferring from the more positive VB to the corresponding bands within the narrower bandgap of SC II. However, the photogenerated charges accumulate on the same semiconductor which cannot achieve effective spatial charge separation.

The typical type-II heterojunction presents staggered alignment, upon illumination, the electrons transfer from the more negative CB of SC I to the corresponding bands of SC II, while holes migrate from more positive VB of SC II to SC I. Compared with type-I heterojunction, electron and hole in type-II heterojunction accumulate in CB and VB of two semiconductors respectively, achieving more efficient electron–hole pairs spatial separation, which induces the most widely investigation among conventional heterojunctions. However, the redox potential ability of the

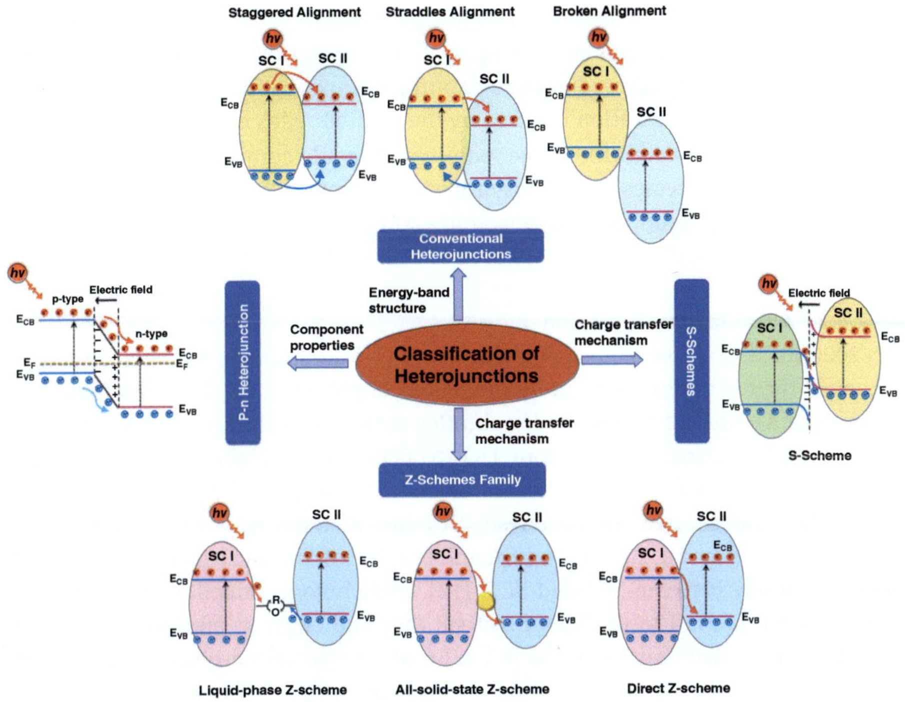

Figure 9.1 Classification of heterojunctions.

heterojunction materials decreases because of charge carriers move from a high energy level to a low energy level.

As for the typical type-III heterojunction, the bandgap of two materials is broken alignment, so there is no charge transfer in type-III heterojunction, which is less favorable for photocatalytic applications.

9.2.3 Z-Scheme Family

Due to the diminished redox capabilities of type-II heterojunctions, coupled with the electrostatic repulsion that hinders the transfer of electrons from SC I to SC II and the movement of holes from SC II to SC I, the feasibility of type-II heterojunctions has been questioned from both thermodynamic and kinetic perspectives [8]. Consequently, the concept of the Z-scheme heterojunction was introduced to preserve the redox potential of semiconductors. The Z-scheme heterojunction family encompasses a variety of configurations, including liquid-phase Z-schemes, all-solid-state junctions, and direct Z-scheme heterojunctions [9].

9.2.3.1 Liquid-phase Z-scheme

In 1979, Bard proposed a semiconductor model of photosynthesis from the biological photosynthetic system [10]. This model was composed of two semiconductors and a suitable redox couple, the higher oxidation potential of SC I caused water oxidation and the higher reduction potential of SC II caused hydrogen generation while the redox couple consumed the photogenerated electrons from SC I and holes generated from SC II, thus promoting spatially separation of redox reaction sites. However, the use of redox couples can confine its application to the solution phase and decrease light utilization induced by the colored redox couple solution such as Fe^{3+}/Fe^{2+} couple. Furthermore, the redox couple could compete with the reduction and oxidation processes in photosynthesis, leading to backward reactions. The moderate redox potentials of these couples could decrease the photogenerated carrier concentration, thus diminishing the overall reaction efficiency [9a].

9.2.3.2 All-solid-state Z-scheme

In 2006, an all-solid-state Z-scheme CdS-Au-TiO_2 nanojunction has been discovered to avoid the restriction induced by the redox couples [11]. As shown in Figure 9.2, they used Au instead of ionic redox shuttles as the electron mediator, and the photogenerated electron transfer from TiO_2 to CdS via Au to promote charges separation and restrict the photocorrosion of CdS. The typical all-solid-state Z-scheme typically composes two photocatalysts (SCI and SCII, respectively) and an electron mediator (such as Au, Ag, Bi, and Cu and graphene), the electron mediator should contact closely between SCI and SCII [12]. The photogenerated electrons transfer from the CB of SCII to the VB of SCI via the solid electron mediator under light irradiation, recombining with the holes. Similar to the liquid-phase Z-scheme, the high oxidation and reduction abilities are kept due to the left holes on SC II as well as electrons on SC I [12, 13]. Significantly, the selection of a conductor mediator should have an appropriate work function adapted to the Fermi levels of SC I and SC II to perform favorable energy band bending at the interface between the conductor mediator and semiconductors [14].

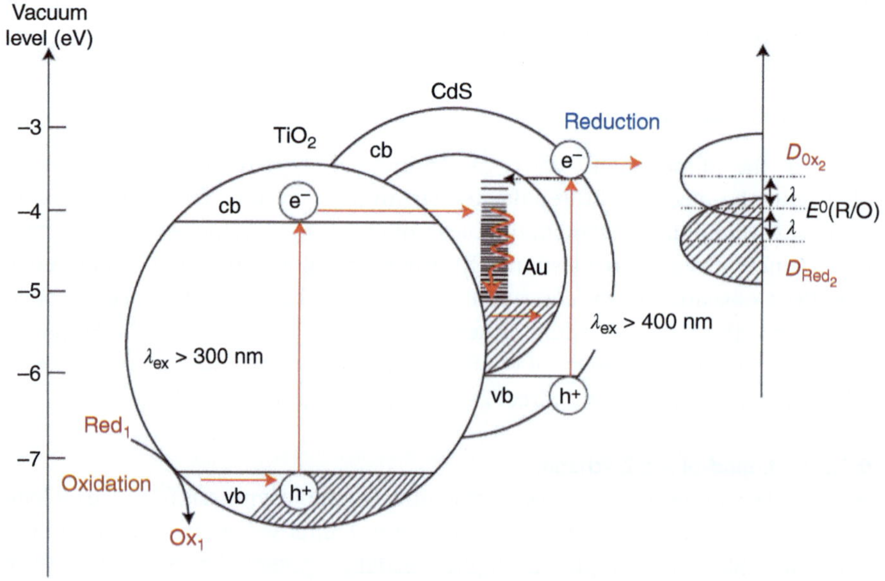

Figure 9.2 Charge transfer in the photoexcited CdS-Au-TiO_2 system.

Additionally, the intimate contact interface formed between the conductor mediator and the semiconductors exhibits low electrical resistance, facilitating the formation of an Ohmic contact instead of a Schottky barrier, especially when compared to the photo-deposition method. Within this composite architecture, the CB of Photosystem II (PS II) and VB of Photosystem I (PS I) are believed to form a low-resistance Ohmic junction via the solid-state mediator, thereby minimizing impedance to charge carrier transport and facilitating efficient electron–hole migration.

9.2.3.3 Direct Z-scheme

The widespread implementation of all-solid-state Z-scheme heterojunctions faces significant challenges due to the substantial expenses associated with noble metal utilization and the intricate fabrication processes required for precise control of conductor particles functioning as mediators rather than cocatalysts. Consequently, researchers have developed an alternative approach by constructing mediator-free direct Z-scheme systems through the strategic combination of two semiconductors with appropriately aligned band structures [13b, 15]. In this configuration, electrostatic forces facilitate the migration of photogenerated electrons from the CB of the SC II to the VB of the SC I, which is abundant in photogenerated holes. This charge transfer mechanism preserves carriers with superior redox capabilities, maintaining electrons with elevated reduction potential and holes with enhanced oxidation potential for subsequent photocatalytic reactions.

Furthermore, interfacial band bending emerges upon contact between the two semiconductor materials, accompanied by the establishment of an internal electric field oriented from SC II toward SC I. This built-in potential facilitates the recombination of photoinduced electrons from CB of SC II with holes in VB of SC I. Crucially, the development of highly efficient direct Z-scheme systems necessitates precise optimization of several parameters: the relative quantity of each semiconductor component, their structural configuration, and the quality of interfacial contact between the two materials [4a, 15a].

The outstanding Z-scheme heterojunction photocatalyst should have the following points: (1) high specific surface area to maximize catalytic activity; (2) strong chemical stability to ensure longevity and resistance to degradation; (3) low interface resistance to facilitate efficient charge transfer; (4) large contact area of redox components to enhance reaction kinetics; (5) wider light response range to capture a wider spectrum of the solar irradiation, etc. Consequently, careful evaluation of the constituent materials' physicochemical characteristics prior to integration is crucial for optimal system design. Maximizing photocatalytic efficiency requires establishing intimate interfacial contact between the oxidative and reductive components within the Z-scheme heterojunction architecture. Current fabrication strategies for these photocatalytic systems primarily encompass four methodologies: (1) deposition-precipitation techniques, (2) ion-exchange processes, (3) hydrothermal and solvothermal approaches, and (4) electrostatic self-assembly protocols.

The direct Z-scheme architecture typically exhibits higher interfacial resistance compared to its second-generation all-solid-state counterpart, primarily attributed to the unmediated interaction between semiconductor components. This challenge can be mitigated through interfacial engineering strategies that enhance contact quality and electrical connectivity. Computational modeling reveals that controlled introduction of defects at the semiconductor interface may promote ohmic behavior, as these structural imperfections function as effective charge carrier mediators [4a]. Experimental evidence further demonstrates that such defect sites operate as recombination hubs, creating multiple intermediate states within the bandgap that facilitate electron–hole pair annihilation [16].

Nevertheless, the Z-scheme heterostructure confronts several unresolved issues: (1) Due to its inherent charge carrier transport mechanism, the efficiency of electron and hole utilization is capped at 50%. (2) The performance of the Z-scheme heterojunction is highly dependent on the tightness of contact between its oxidized and reduced components, necessitating as close an interface as possible between these elements in the Z-scheme photocatalyst. (3) Enhancing the performance and selectivity of the Z-scheme photocatalyst for photocatalytic CO_2 conversion is essential, given the complexity of the multistep reaction process and the inherent stability of CO_2 molecules; (4) The current comprehension of CO_2 photocatalytic conversion through Z-scheme heterojunctions remains underdeveloped, highlighting the need to better utilize advanced characterization methods to elucidate the electronic transport mechanisms within these systems. (5) Additionally, the strategies for designing Z-scheme photocatalysts require further refinement to identify optimal pairs of oxidation and reduction components, ensuring enhanced photocatalytic efficiency.

9.2.4 S-scheme

Traditional heterojunction systems, such as type-II heterojunctions and the Z-scheme family, have encountered fundamental limitations in their applicability. For instance, traditional Z-scheme photocatalysts exhibit significant limitations, primarily stemming from their heterogeneous combinations and inherent structural imperfections. To address these defects, novel concepts have emerged that provide a more nuanced understanding of photocatalytic mechanisms. One such concept is the step-scheme (S-scheme), which was introduced by Fu et al. [17]. This innovative addition to the heterojunction family has since become a significant focus within the research community. The S-scheme presents a promising alternative to the conventional Z-scheme, effectively addressing the shortcomings associated with earlier iterations of the Z-scheme framework, which has addressed the inherent challenges associated with various photocatalytic mechanisms, including those of type-II heterojunctions, traditional Z-schemes, and all-solid-state Z-scheme heterojunctions [8a]. This innovative system offers a promising avenue for overcoming the limitations inherent in these configurations.

Figure 9.1 provides a conceptual overview of the S-scheme heterojunction, which is instrumental in the synthesis of photocatalysts. This configuration improves the separation and migration of charge carriers, leading to superior photocatalytic performance. In detail, the S-scheme heterojunction is formed by combining two n-type semiconductor photocatalysts: SC I for oxidation and SC II for reduction (Figure 9.1). In this structure, electrons are placed in the CB of SC II and the VB of

SC I. The electric field generated at the interface of SC I and SC II acts as a critical driver for efficient charge separation.

In traditional configurations, the oxidation SC I typically exhibits a high work function and a low Fermi level, while the reduction SC II possesses a low work function and a high Fermi level. Upon contact between SC I and SC II, the Fermi level of SC I shifts upward, whereas that of SC II shifts downward, eventually aligning at an equilibrium level. This alignment causes the CB and VB of SC I to bend downward, while those of SC II bend upward. As a result, an electric field forms at the interface, promoting the efficient transfer of electrons from SC II to SC I. When exposed to light, the S-scheme heterojunction enables the recombination of excess electrons in the CB of SC I and surplus holes in the VB of SC II at their interface, where they are effectively neutralized. Simultaneously, the primary electrons in the CB of SC II and the main holes in the VB of SC I actively participate in redox reactions, driven by the internal electric field. This mechanism highlights a key advantage of the S-scheme over the type-II heterojunction: while the latter often experiences reduced redox efficiency due to the accumulation of electrons in the CB of SC I and holes in the VB of SC II, the S-scheme heterojunction enhances redox activity by leveraging the built-in electric field to optimize charge separation and reaction dynamics.

As shown in Figure 9.1, the charge transfer mechanism in the S-scheme system closely resembles that of the direct Z-scheme heterojunction, incorporating key factors such as Fermi level alignment (work function), the presence of a built-in electric field, and band bending. Consequently, the novelty of the S-scheme heterojunction becomes ambiguous when considering the charge transfer mechanism. Moreover, many researchers have overlooked the fundamental relationship linking the S-scheme to the direct Z-scheme. Since 2020, the pieces of literature indicate a preference for studying the direct Z-scheme over the S-scheme. As a result, the latter remains in its early developmental stages, with both its theoretical framework and practical applications facing numerous obstacles that necessitate further exploration and validation over time.

It is worth mentioning that Xu et al. have made a significant contribution by standardizing the selection of semiconductors in S-scheme heterojunctions [8a]. Previously, in a more restricted context, oxidation photocatalysts (OPs) and reduction photocatalysts (RPs) were predominantly considered as two distinct classes of n-type semiconductors. Early classifications divided semiconductor photocatalysts into two categories-RPs and OPs-based on their respective energy-band structures [8b]. It is important to note that RPs typically possess higher band positions than OPs. The formation of S-scheme heterojunctions involves pairing one RP with one OP, where the RP features a higher Fermi level than the OP. Consequently, both OPs and RPs can exist as either p-type or n-type semiconductors. These diverse combinations are illustrated in Figure 9.3, paving the way for innovative approaches to designing and optimizing S-scheme heterojunctions in the future.

The S-scheme heterojunction has exhibited exceptional capabilities that were not achievable with earlier photocatalyst generations, including type-II heterojunctions and Z-scheme variants. To develop an S-scheme photocatalyst optimized for targeted reactions, a deep comprehension of charge-transfer mechanisms and the role

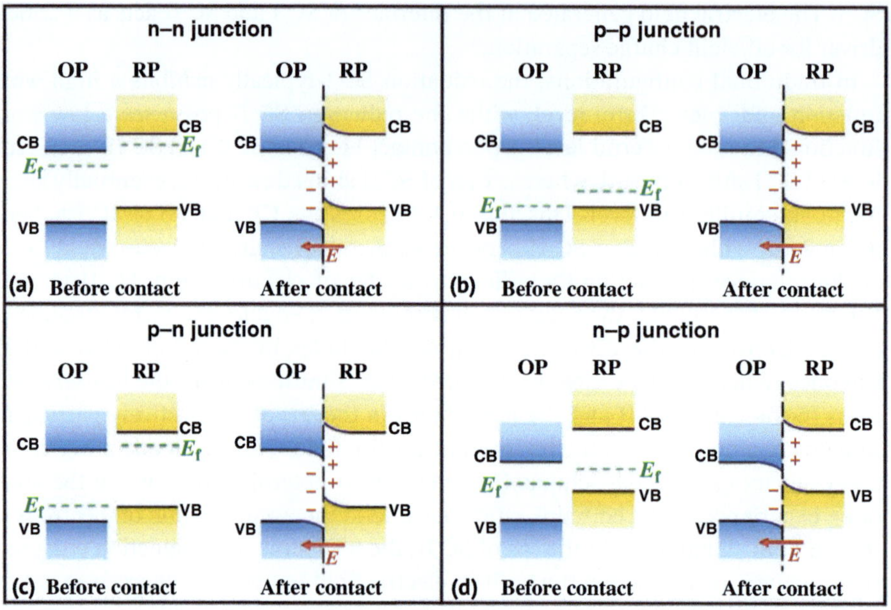

Figure 9.3 The interfacial contact in different types of S-scheme heterojunctions: (a) n–n junction; (b) p–p junction; (c) p–n junction; and (d) n–p junction.

of co-catalysts within the S-scheme framework is crucial. Depositing co-catalysts on the oxidation and reduction sites of the S-scheme heterojunction markedly improves the separation efficiency of electron–hole pairs. Additionally, incorporating co-catalysts can lower the activation energy required for photocatalytic reactions, further enhancing overall performance. Incorporating suitable dopants into the photocatalyst can broaden the Fermi level gap, facilitating alignment, and altering its structure to enhance solar energy absorption capabilities. Expanding the interfacial contact between two phases, particularly through a two-dimensional face-to-face configuration, can increase the spatial charge transfer channel, thereby Maximizing the charge separation.

At present, S-scheme photocatalysts are still in the early stages of development. A primary constraint of these photocatalysts is the powder form. Therefore, there is a pressing need to broaden the scope of innovative S-scheme heterostructures to address environmental and energy-related photocatalytic challenges in a sustainable manner.

9.3 Characterization Technologies of S-scheme Heterojunction

The charge transfer process in S-scheme heterojunction can be characterized by various technologies, such as in situ-irradiated X-ray photoelectron spectroscopy (ISIXPS), radicals-dependent experiments, density functional theory (DFT) calculations, surface photovoltage (SPV), and selective metal (oxide) photo-deposition.

9.3.1 In Situ-irradiated X-ray Photoelectron Spectroscopy

In situ-irradiated XPS analysis can characterize the charge dynamics and surface chemical environment, making it invaluable for confirming the presence of an S-scheme charge transfer mechanism. He et al. used in situ-irradiated XPS analysis to detect the charge transfer mechanism in the TiO_2/Bi_2O_3 S-scheme photocatalyst [18]. As illustrated in Figure 9.4, when Bi_2O_3 and TiO_2 interact, the higher Fermi level of Bi_2O_3 causes electrons to spontaneously migrate from Bi_2O_3 to TiO_2 at their interface. This electron transfer results in the formation of a depletion layer on the Bi_2O_3 side and an accumulation layer on the TiO_2 side, establishing an internal electric field directed from Bi_2O_3 to TiO_2. Consequently, the binding energy of Ti in TiO_2 decreases as it gains electrons, while the binding energy of Bi increases due to electron loss. Under light irradiation, the binding energy of Bi decreases, reflecting an increase in electron density at Bi sites, whereas the binding energy of Ti rises, indicating electron depletion at TiO_2 sites. These observations confirm the transfer of photogenerated electrons from the CB of TiO_2 to the VB of Bi_2O_3, aligning perfectly with the S-scheme mechanism.

9.3.2 Radical-dependent Experiments

During photocatalytic process, photogenerated electrons from CBs and holes from VBs may have the potential to interact with adsorbed oxygen and water, generating the superoxide anion radicals ($\cdot O_2^-$) and hydroxyl radicals ($\cdot OH$), which could provide insight into the charge-transfer mechanism based on the knowledge that

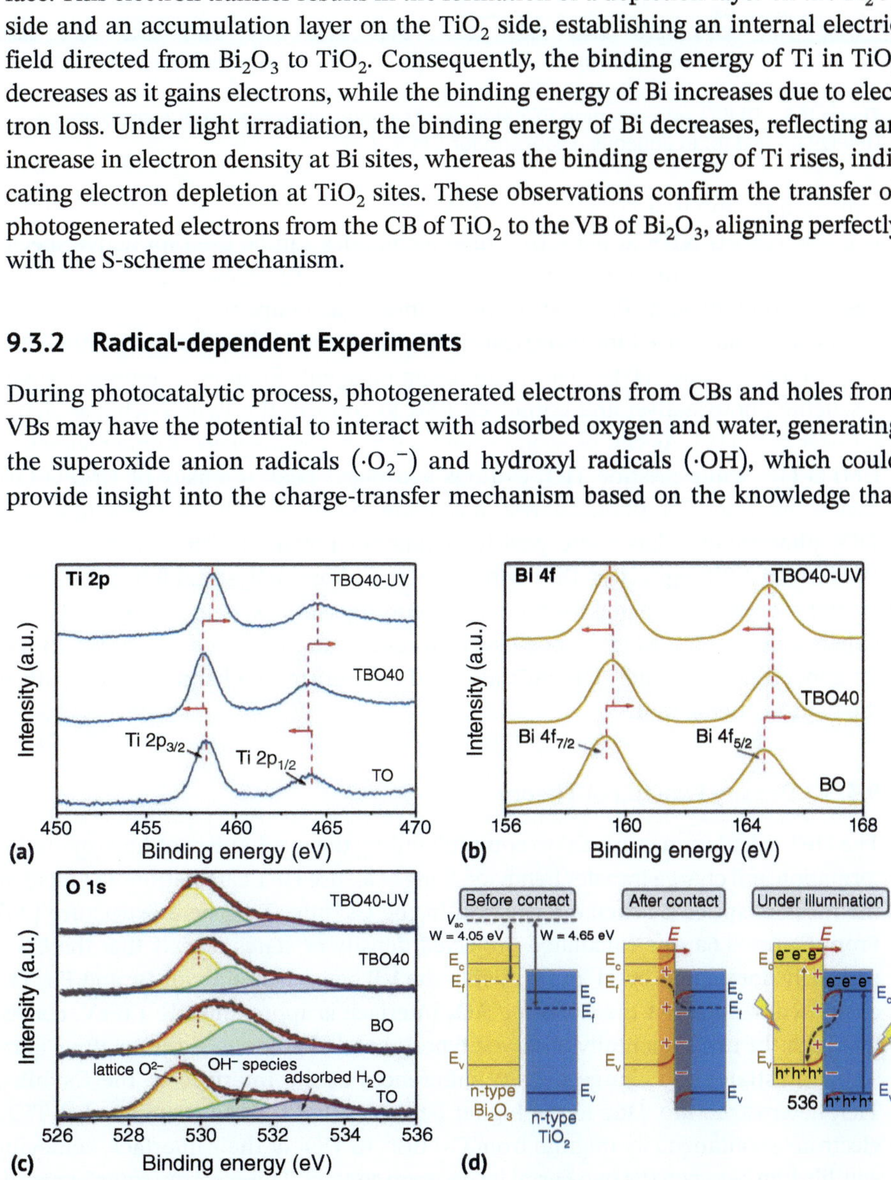

Figure 9.4 XPS spectra of: (a) Ti 2p, (b) Bi 4f, (c) O 1s spectra under light irradiation, (d) charge-transfer processes of samples with or without light irradiation.

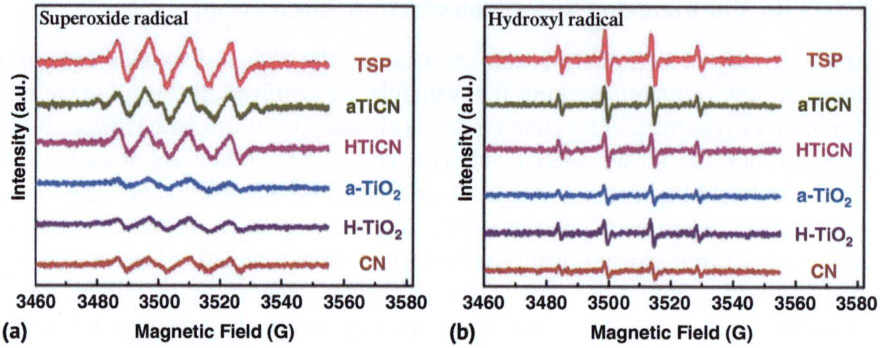

Figure 9.5 DMPO spin-trapping ESR signals of samples (a) in methanol dispersion for DMPO-O_2^- and (b) in aqueous dispersion for DMPO-OH.

reduction potential of $O_2/\cdot O_2^-$ is −0.33 V, and the oxidation potential of $H_2O/\cdot OH$ is 2.37 V (versus NHE at pH = 0). Three techniques can be used for active species trapping experiments: ex situ and in situ electron spin resonance (ESR), photoluminescence probing and UV–vis spectrophotometric assessment.

ESR was widely used for investigating $\cdot O_2^-$ and ·OH radicals, with 5,5-dimethyl-1-pyrroline N-oxide (DMSO) as a spin-trapping agent. Ruan et al. reported a twin S-scheme photocatalyst and employed ESR to confirm the twin S-scheme charge transfer path [19]. As can be seen in Figure 9.5a,b, CN displays stronger signals of DMPO-O_2^- while pristine TiO_2 exhibits a stronger peak intensity of DMPO-OH, suggesting that CN photocatalyst has more negative reduction potential and TiO_2 photocatalyst has more positive oxidation potential. The DMPO-O_2^- and DMPO-OH ESR signals of the twin S-scheme photocatalyst exhibit significantly greater intensity compared to those of the control samples, demonstrating that the twin S-scheme photocatalyst possesses strong oxidation and reduction capabilities for generating $\cdot O_2^-$ and · OH radicals, further confirming charge transfer path of the twin S-scheme in it.

9.3.3 Density Functional Theory

The DFT calculation is crucial to comprehending the mechanism of heterojunction formation and charge transfer behavior. Song et al. use DFT calculations to elucidate the mechanism of Z-scheme formation in the $CsPbBr_3/TiO_2$ heterostructure [20]. From Figure 9.6a, the calculated projected density of states reveals that the CB is primarily dominated by Ti and O, while the VB consists mainly of Br and Pb. The computed bandgap of the $CsPbBr_3/TiO_2$ interface is approximately 1.6 eV, consistent with the experimentally observed type-II band alignment. Additionally, Figure 9.6b,c illustrates the charge density differences at the interface of the $CsPbBr_3/TiO_2$ heterostructure. Due to the higher Fermi level of $CsPbBr_3$ compared to TiO_2, electrons spontaneously migrate from $CsPbBr_3$ to TiO_2 at their interface, achieving equilibrium between the two Fermi levels upon contact. This electron transfer results in the formation of a depletion layer on the $CsPbBr_3$ side and an accumulation layer on the TiO_2 side, generating an internal electric field. In contrast, $CsPbBr_3$ has more

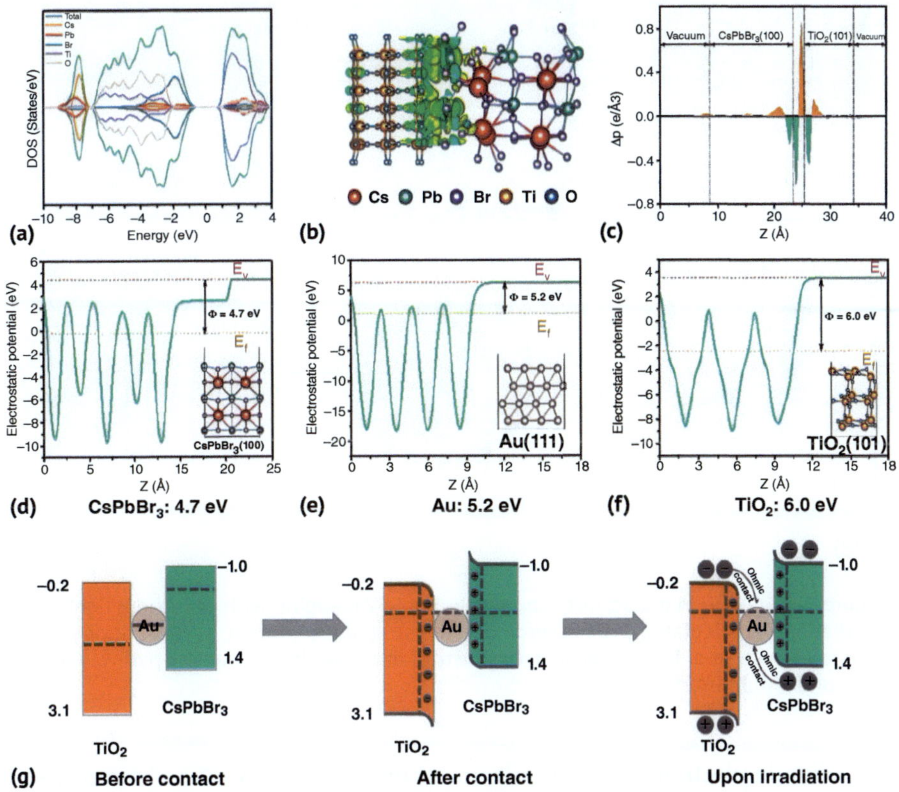

Figure 9.6 (a) Calculated DOS for the interface of $CsPbBr_3/TiO_2$. (b) Charge density difference at the interface of $CsPbBr_3/TiO_2$. The yellow and green regions represent electron accumulation and electron depletion, respectively. (c) Calculated charge density difference at the interface of $CsPbBr_3/TiO_2$. Work functions of (d) $CsPbBr_3$, (e) Au, and (f) TiO_2. (g) Z-scheme formation mechanism after irradiation.

negatively positioned CB edge potential, the photogenerated electrons would spontaneously migrate from $CsPbBr_3$ to TiO_2 under light irradiation. The $CsPbBr_3/TiO_2$ heterostructure converts to the all-solid-state Z-scheme heterostructure after the addition of Au. As shown in Figure 9.6d–f, the different work functions of $CsPbBr_3$, Au, and TiO_2 with Au as a mediator formed an all-solid-state Z-scheme and constructed a charge transfer pathway (Figure 9.6g).

9.3.4 Surface Photovoltage

Surface photovoltage (SPV) is a powerful method to verify the carrier migration mechanism by mapping the surface potential of heterojunction photocatalysts. Photovoltage is usually quantified through capacitance–voltage (C–V), kelvin probe force microscopy (KPFM), and electrostatic voltmeter techniques. For instance, Moon et al. applied KPFM analysis to evaluate the spatial charge separation of their sample under irradiation (Figure 9.7a–c) [21]. As shown in Figure 9.7d, the accumulation of photogenerated electrons leads to a reduction in SPV. Conversely,

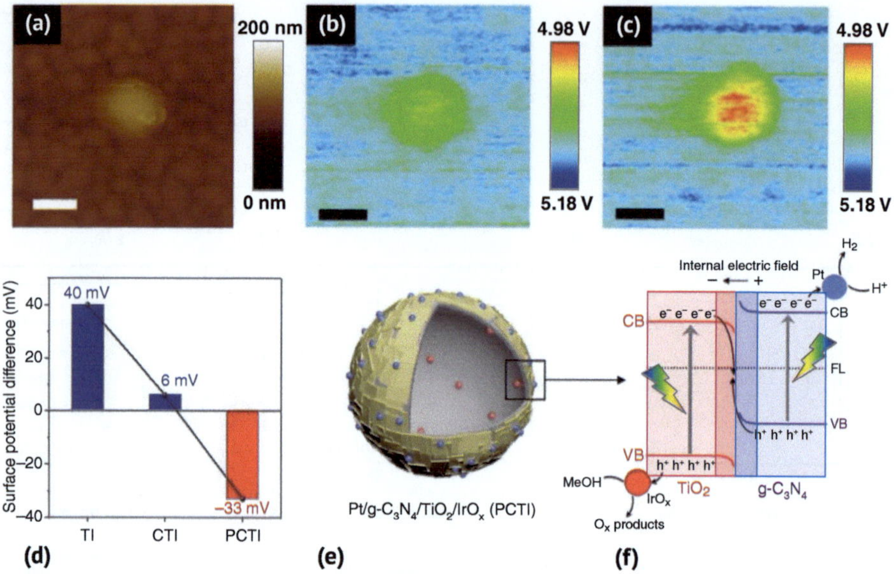

Figure 9.7 (a) KPFM image. (b, c) Corresponding SPV mapping with or without light irradiation. (d) Calculated ΔSPV according to the structure evolution. (e) Illustration of the physical structure of PCTI. (f) Schematic illustration of the charge-transfer pathway within PCTI under light irradiation.

a decline in electron density results in an increase in SPV. The observed rise in SPV for TiO_2/IrO_x following light exposure suggests that electrons are transferred from the outer TiO_2 shell to the inner IrO_x particles. However, when TiO_2/IrO_x is combined with g-C_3N_4, a notable decrease in ΔSPV is observed under light irradiation compared to TiO_2/IrO_x alone. This reduction reflects a decline in the inward electron migration, attributed to the Z-scheme charge transfer mechanism and the accumulation of photogenerated electrons on the outermost g-C_3N_4 layer (Figure 9.7e, f). Notably, PCTI demonstrates a unique negative SPV shift under illumination, signifying that photogenerated electrons migrate to the external g-C_3N_4 and are subsequently captured by the Pt cocatalysts.

9.3.5 Selective Metal (Oxide) Photo-deposition

The photo-deposition method usually be applied in loading cocatalysts. In recent years, this technique has been employed to verify the charge transfer process in heterojunction photocatalysts. During the photoreduction process, if the heterojunction is S-scheme, the electron will transfer from the CB of SC I to VB of SC II, precious metal ions, such as Pt should be photo-deposited almost on SC II (Figure 9.8) [22]. In a type II heterojunction, electrons migrate from the CB of SC II to the CB of SC I, accumulating on SC I and resulting in the deposition of metal onto SC I. Conversely, in an S-scheme heterojunction, holes remain available in the VB of SC I during the photooxidation process, causing metal oxide nanoparticles to

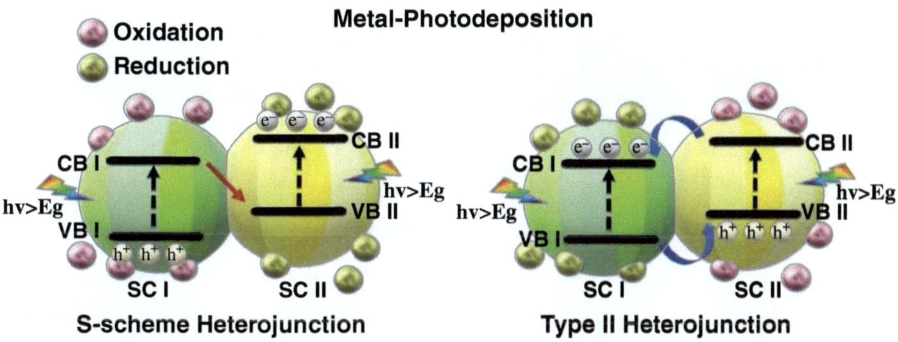

Figure 9.8 Schematic representation of charge transfer in S-scheme and type II heterojunction.

predominantly deposit on SC I. In contrast, for a type II heterojunction, holes accumulate in the VB of SC II, leading to the deposition of metal oxide nanoparticles primarily on SC II.

9.4 Roles of Heterojunctions

Photocatalysts should have high catalytic activity, wide light utilization and long stability simultaneously to achieve large-scale application, which can be difficult to obtain by a single semiconductor. The well-designed heterojunction is a satisfactory method to overcome the drawback of a single semiconductor by improving stability, prolonging the lifetime of photogenerated charges, promoting charges separation, accelerating electron transfer and broadening the light adsorption, thus, realizing high catalytic efficiency.

TiO_2 has been widely used as a protection layer in photo-/photoelectro-catalysis to prevent semiconductors against dissolution such as $BiVO_4$. As shown in Figure 9.9a,b, Yang's group coated amorphous TiO_2 on $BiVO_4$ by atomic layer deposition [23]. It is reported that the presence of the amorphous TiO_2 layer can remarkably suppress the dissolution of $BiVO_4$. Lee et al. also demonstrated the use of a TiO_2 layer could keep $BiVO_4$ stable in strongly basic media (Figure 9.9c) [24]. From the fascinating charge transport mechanism, heterojunctions also can be designed to help hinder photo-corrosion in the photocatalytic process. As shown in Figure 9.9d, Lu et al. fabricated a hierarchical CdS/mesoporous TiO_2/graphene ternary photocatalyst using amorphous Ti(IV) domains that act as efficient hole capture centers which have a more negative potential to trap holes generated from VB of CdS, preventing CdS oxidation [25]. As shown in Figure 9.9e even after 80 h, CdS/mesoporous TiO_2/G remained at 82% of the initial photoactivity while the photoactivity of pure CdS only kept 4% activity. Recently, TiO_2 has usually been fabricated into heterojunctions for resisting seawater corrosion due to its high chemical stability. Yang et al. fabricated an H-TiO_2/Cu_2O heterojunction to elevate the stability for photo-driven seawater splitting [26]. It has been confirmed that the improved corrosion resistance in seawater is attributed to the high concentration of surface hydroxyl groups present on

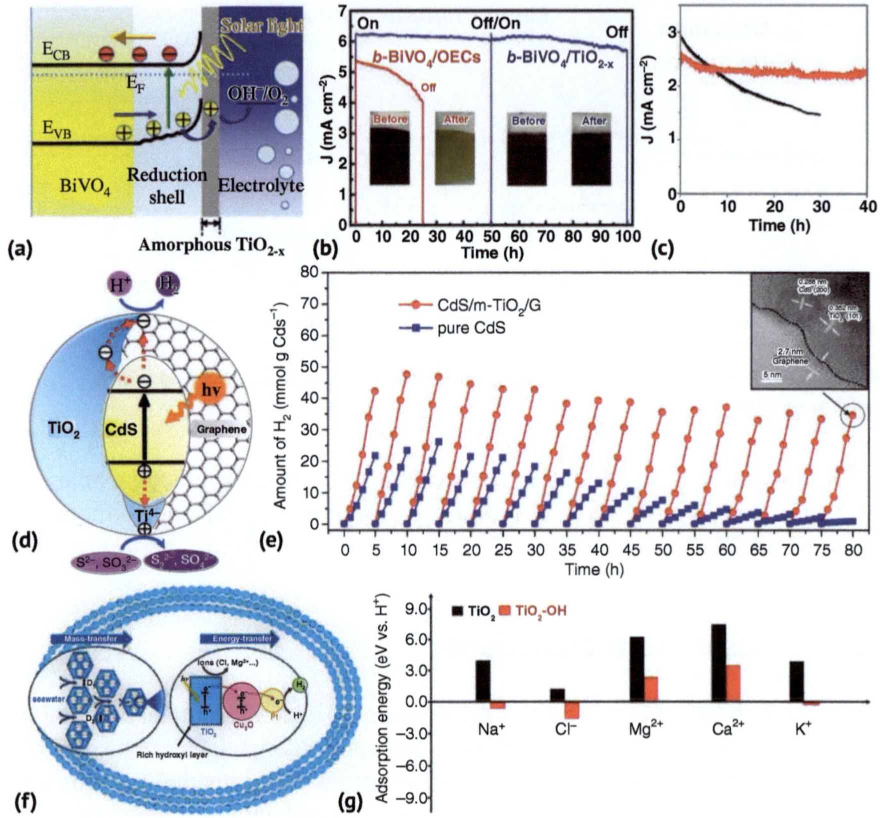

Figure 9.9 (a) Charge-transfer processes of the *b*- $BiVO_4/TiO_{2-x}$ photoanode. (b) Chronoamperometry curves of b- $BiVO_4/TiO_{2-x}$ and b-$BiVO_4$/FeOOH/NiOOH photoanodes. (c) Chronoamperometry curves at 0.6 V versus RHE of $BiVO_4$ (black) and $BiVO_4/TiO_2$ (red). (d) Schematic description of the charge transfer in the CdS/m-TiO_2/G structure. (e) Photostability for the H_2 evolution of CdS/m-TiO_2/G and CdS. The inset is an HRTEM image of CdS/m-TiO_2/G after 80-h photocatalytic H_2 evolution test. (f) Schematic description of oriented mass and energy transfer in H-TiO_2/Cu_2O for photo-driven seawater splitting. (g) Calculated adsorption energies of main ions in seawater vs H^+ toward TiO_2-OH and TiO_2.

TiO_2 and direct mass- and energy-transfer induced by the well-designed heterojunction, as shown in Figure 9.9f,g.

A single semiconductor should have suitable band edges that straddle the reduction and oxidation potentials of water splitting. Moreover, the photocatalysts should possess a strong ability of charge transport and separation to achieve high solar to hydrogen efficiency. However, only a few photocatalysts can realize overall water splitting and most of them show low efficiency at present. In contrast, heterojunction photocatalysts capable of outstanding charge separation and transport apply to efficient overall water splitting [27]. Zuo et al. constructed a Z-scheme heterojunction by integrating TiO_2 on $ZnIn_2S_4$ with a well-matched energy band (Figure 9.10) [13b]. The internal electric field induced by energy-band bending impels the photogenerated electrons migrating from E_{CB} of TiO_2 tor to E_{VB} of

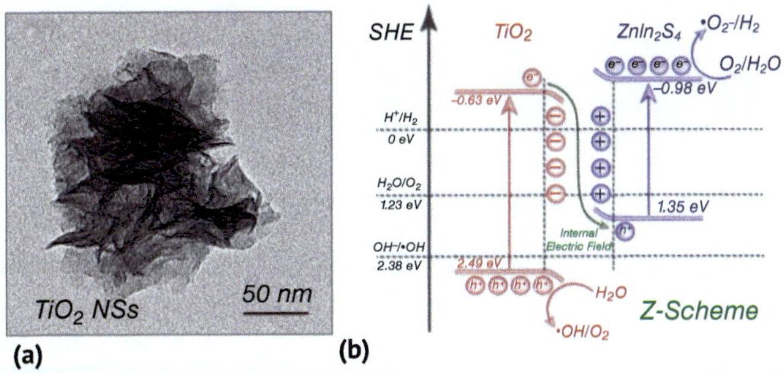

Figure 9.10 (a) TEM images of ultrathin TiO_2 nanosheets. (b) Schematic illustration of direct Z-scheme TiO_2-$ZnIn_2S_4$ heterojunction.

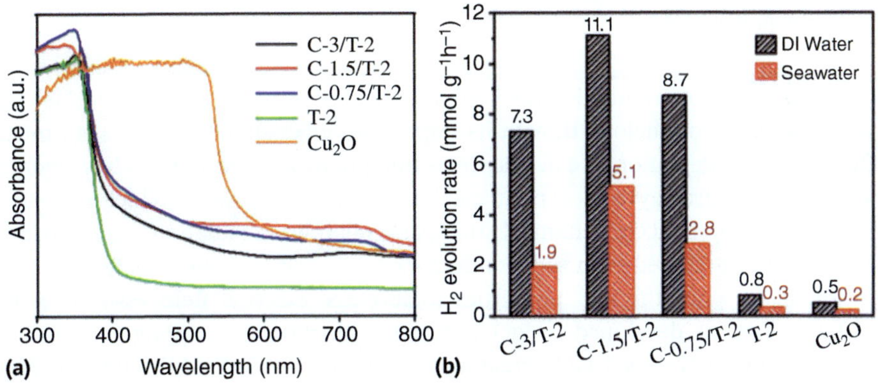

Figure 9.11 (a) UV–vis DRS spectra. (b) Hydrogen production rate constants of Cu_2O, TiO_2 and Cu_2O/TiO_2 catalysts.

$ZnIn_2S_4$ to consume the useless holes in $ZnIn_2S_4$, retaining strong redox potential to conduct water reduction and oxidation. This Z-scheme TiO_2-$ZnIn_2S_4$ heterojunction showed greatly improved overall water splitting ability without any cocatalyst and sacrificial agent.

The initial step in any photocatalytic reaction involves the absorption of photons from incident light. However, only 5% of the sun spectrum is ultraviolet, 43% is visible, and 52% is infrared irradiation. Therefore, visible-even infrared-light-driven water splitting for a single semiconductor material is required to improve the solar to hydrogen efficiency. Since the conventional TiO_2 has a large band gap which limits its applicability under visible irradiation conditions. Lv et al. constructed a direct Z-scheme heterojunction by coupling mesoporous TiO_2 with Cu_2O for promoting seawater splitting [15a]. As shown in Figure 9.11, Cu_2O/TiO_2 heterojunction has a significantly highly red-shifted in comparison with TiO_2 from UV–vis diffuse reflectance spectroscopy. This shift is primarily attributed to the strong interfacial connection between Cu_2O and TiO_2, which enhances its photocatalytic performance, achieving hydrogen production rates of 11 mmol h^{-1} g^{-1} in water and 5.1 mmol h^{-1} g^{-1} in seawater, respectively.

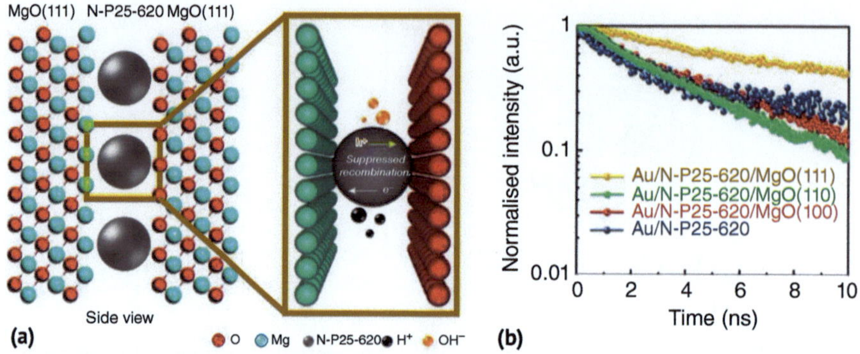

Figure 9.12 (a) Schematic illustration of the local electric field of polar MgO (111) nanocrystal-assisted photocatalytic water splitting. (b) Time-resolved photoluminescence measurements of Au/N-P25-620/MgO (111), Au/N-P25-620/MgO (110), Au/N-P25-620/MgO (100), and Au/N-P25-62.

Constructing heterojunctions can enhance charge separation at the interface between two semiconductors, though its impact on bulk separation remains limited. Recently, the integration of external fields into photocatalytic systems has emerged as a promising strategy to boost photocatalytic efficiency. These fields include electric fields, magnetic fields, mechanical stress, and microwave radiation. Consequently, TiO_2 can be combined with semiconductors exhibiting piezoelectric, ferroelectric, magnetic, or photothermal properties to leverage external field assistance. For instance, Li et al. developed a MgO-TiO_2 heterojunction by coupling polar-faceted MgO (111) with N-doped TiO_2, creating a localized electric field through the MgO nanostructure's exposed (111) polar facets to improve charge carrier separation, as illustrated in Figure 9.12a [28]. As depicted in Figure 9.12b, the photogenerated carrier lifetime increased from 2.56 to 5.76 ns due to the localized electric field generated by MgO, achieving a quantum efficiency of 81.8% at 437 nm.

Apart from adding external fields, engineering vacancies are another effective method to promote charge separation and transport. As described above, introducing vacancies can simplify the formation of intimate contact and act as an interfacial mediator to accelerate electron migration at the interface in Z/S-scheme heterojunction. In addition, vacancies such as O- and Ti-vacancies can act as electron donor and acceptor type centers which propel electron transfer from O-vacancy to Ti-vacancy. Therefore, combining heterojunction architecture with vacancy engineering to form a directed charge transfer channel from the inner lattice to the surface could greatly improve charge utilization. For example, Lu et al. designed a TiO_2/carbon/graphene ternary spatially heterojunction with titanium-defect to achieve inside-out charge transfer (Figure 9.13a,b) [29]. As shown in Figure 9.13c, they proposed a cascade effect of charges in the spatial heterojunctions, wherein the electron migrates from the inner lattice to the Ti-vacancies, then transfers to CQDs, and finally arrives at the surface of rGO sheets. The integrated V_{Ti}@CQDs@rGO spatial heterojunction exhibited a superior H_2 production rate of 31.9 $\mu mol\ h^{-1}$, which is over 3.2-fold in comparison with the random surface heterojunctions (Figure 9.13d).

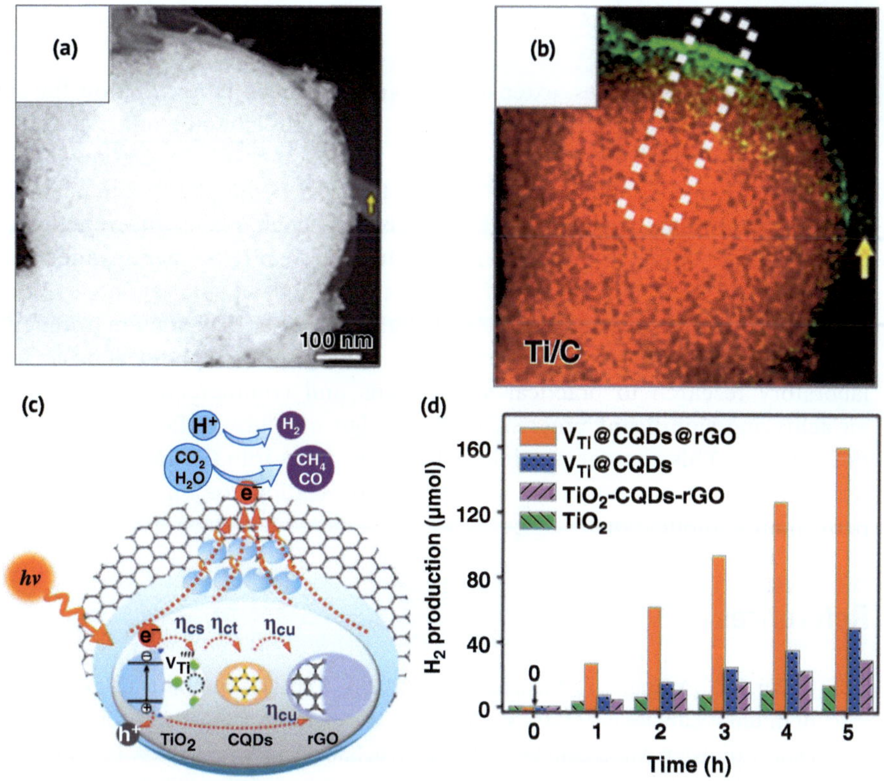

Figure 9.13 (a) HAADF-STEM image. (b) EDS elemental map. (c) Schematic description of the spatial electron-transfer cascade mechanism. (d) Photocatalytic H_2 production over time.

The surface reaction with holes–electrons is an important part of the water splitting, slow water oxidation–reduction kinetics at the catalyst/solution junction can cause additional surface recombination. To enhance the surface reaction kinetics, cocatalysts such as transition metal dichalcogenides, phosphorous metal, metal oxides, metal hydroxides, etc. have been introduced to decrease activation energy and provide more active sites [30]. Besides, the charge recombination can be further restrained driven by the energy band bending between cocatalysts and electrolytes, and lower the reaction barrier by changing the reaction pathway. Ai et al. engineered a cobalt phosphate water oxidation cocatalyst on the TiO_2 nanowire arrays photoanode [31]. This created a strong interfacial connection between cobalt phosphate and the electrolyte, causing upward band bending. As a result, electrons were depleted in the space charge region, while excess holes were captured by the cocatalyst. This process facilitated the transfer of photogenerated holes to the photoanode surface, preventing their accumulation and reducing recombination. Additionally, when combined with TiO_2, the photogenerated holes from the VB of TiO_2 were utilized by cobalt phosphate to drive the cyclic catalytic reactions of $Co^{2+}/^{3+} \rightarrow Co^{4+} \rightarrow Co^{2+}/^{3+}$, alongside water oxidation. This mechanism effectively lowered the overpotential for the oxygen evolution reaction.

9.5 Conclusion

Over the past few decades, extensive research has been conducted on the synthesis of diverse heterojunction photocatalysts. This chapter offers a concise evaluation of various heterojunction types, such as p–n heterojunctions, type-I, type-II, and type-III heterojunctions, as well as Z-scheme and S-scheme configurations. Additionally, we highlight the distinct mechanisms and functions of these heterojunctions in photocatalysis, aiming to inspire further advancements in this critical area. The pursuit of sustainable and clean solar energy solutions is central to addressing the intertwined challenges of energy and environmental sustainability. However, the transition of heterojunction photocatalysts from laboratory research to practical applications and commercialization depends on achieving significant progress in the design of highly efficient heterojunction systems. This chapter provides valuable insights into the rational design of semiconductor heterojunctions and their optimization for high photocatalytic performance applications.

References

1 a) Wu, S.M., Wang, Y.T., Xiao, S.T. et al. (2022). A spatial homojunction of titanium vacancies decorated with oxygen vacancies in TiO_2 and its directed charge transfer. *Nanoscale* 14: 13373; b) Wu, S.M., Liu, X.L., Lian, X.L. et al. (2018). Homojunction of Oxygen and Titanium Vacancies and its Interfacial n–p Effect. *Advanced Materials* 30: 1802173; c) Lu, Y., Liu, Y.X., He, L. et al. (2020). Interfacial co-existence of oxygen and titanium vacancies in nanostructured TiO_2 for enhancement of carrier transport. *Nanoscale* 12: 8364.

2 a) Chen, R., Meng, L., Xu, W. et al. (2024). Cocatalysts-Photoanode Interface in Photoelectrochemical Water Splitting: Understanding and Insights. *Small* 20: 2304807; b) Okazaki, M., Suganami, Y., Hirayama, N. et al. (2020). Site-selective deposition of a cobalt cocatalyst onto a plasmonic Au/TiO_2 photoanode for improved water oxidation. *ACS Applied Energy Materials* 3: 5142.

3 a) Rani, B.J., Praveenkumar, M., Ravichandran, S. et al. (2019). Ultrafine M-doped TiO_2 (M = Fe, Ce, La) nanosphere photoanodes for photoelectrochemical water-splitting applications. *Materials Characterization* 152: 188; b) Yang, L., Wang, C., Li, Y. et al. (2024). Frustrated Lewis Pairs on Zr Single Atoms Supported N-Doped TiO2-x Catalysts for Electrochemical Nitrate Reduction To Ammonia. *Advanced Functional Materials* n/a: 2401094.

4 a) Xing, X., Zhu, H., Zhang, M. et al. (2018). Interfacial oxygen vacancy layer of a Z-scheme BCN-TiO_2 heterostructure accelerating charge carrier transfer for visible light photocatalytic H_2 evolution. *Catalysis Science and Technology* 8: 3629; b) Dong, Y., Chen, S.Y., Lu, Y. et al. (2018). Hierarchical MoS_2@TiO_2 heterojunctions for enhanced photocatalytic performance and electrocatalytic hydrogen evolution. *Chemistry-An Asian Journal* 13: 1609.

5 a) You, Y., Chen, S., Zhao, J. (2024). Rational design of S-scheme heterojunction toward efficient photocatalytic cellulose reforming for H_2 and Formic acid in pure water. *Advanced Materials* 36: 2307962; b) Ahmadi, M., Alavi, S.M., and Larimi, A. (2023). Pt-Cu@Bi_2MoO_6/TiO_2 photocatalyst for CO_2 reduction. *Inorganic Chemistry* 62: 20372.

6 a) Zhou, S., Shen, Q., Yang, F.L. et al. (2023). Engineering cuboctahedral N-doped C-coated p-CuO/n-TiO_2 heterojunctions toward high-performance photocatalytic cross-dehydrogenative coupling. *Nanoscale* 15: 13313; b) Meng, X., Wang, S., Zhang, C. et al. (2022). Boosting hydrogen evolution performance of a CdS-Based photocatalyst: in situ transition from type I to Type II heterojunction during photocatalysis. *ACS Catalysis* 12: 10115; c) Zhang, J., Zhou, D., Dong, S. et al. (2019). Respective construction of Type-II and direct Z-scheme heterostructure by selectively depositing CdS on {001} and {101} facets of TiO_2 nanosheet with CDots modification: A comprehensive comparison. *Journal of Hazardous Materials* 366: 311; d) Low, J., Dai, B., Tong, T. et al. (2019). In situ irradiated X-ray photoelectron spectroscopy investigation on a direct Z-scheme TiO_2/CdS composite film photocatalyst. *Advanced Materials* 31: 1802981; e) Zhang, L., Zhang, J., Yu, H. et al. (2022). Emerging S-Scheme Photocatalyst. *Advanced Materials* 34: 2107668.

7 Liu, Y., Niu, H., Gu, W. et al. (2018). In-situ construction of hierarchical CdS/MoS_2 microboxes for enhanced visible-light photocatalytic H_2 production. *Chemical Engineering Journal* 339: 117.

8 a) Xu, Q., Zhang, L., Cheng, B. et al. (2020). S-Scheme Heterojunction Photocatalyst. *Chem* 6: 1543; b) Zhang, L., Zhang, J., Yu, H. et al. (2022). *Advanced Materials* 34: e2107668.

9 a) Maeda, K. (2013). Z-scheme water splitting using two different semiconductor photocatalysts. *ACS Catalysis* 3: 1486; b) Zhang, G., Wang, Z., and Wu, J. (2021). Construction of a Z-scheme heterojunction for high-efficiency visible-light-driven photocatalytic CO_2 reduction. *Nanoscale* 13: 4359.

10 Bard, A.J. (1979). Photoelectrochemistry and heterogeneous photo-catalysis at semiconductors. *Journal of Photochemistry and Photobiology* 10: 59.

11 Tada, H., Mitsui, T., Kiyonaga, T. et al. (2006). All-solid-state Z-scheme in CdS-Au-TiO_2 three-component nanojunction system. *Nature Materials* 5: 782.

12 Xia, X., Song, M., Wang, H. et al. (2019). Latest progress in constructing solid-state Z scheme photocatalysts for water splitting. *Nanoscale* 11: 11071.

13 a) Liu, H., Geng, W., Jin, C.J. et al. (2019). Silica coating with well-defined micro-nano hierarchy for universal and stable surface superhydrophobicity. *Chemical Physics Letters* 730: 594; b) Zuo, G., Wang, Y., Teo, W.L. et al. (2021). Direct Z-scheme TiO_2-$ZnIn_2S_4$ nanoflowers for cocatalyst-free photocatalytic water splitting. *Applied Catalysis B: Environmental.* 291: 120126.

14 Xia, M., Pan, L., Liu, Y. et al. (2023). Efficient Cu_2O photocathodes for aqueous photoelectrochemical CO_2 reduction to formate and syngas. *Journal of the American Chemical Society* 145: 27939.

15 a) Lv, S., Wang, Y., Zhou, Y. et al. (2021). Oxygen vacancy stimulated direct Z-scheme of mesoporous Cu_2O/TiO_2 for enhanced photocatalytic hydrogen production from water and seawater. *Journal of Alloys and Compounds* 868: 159144; b) Lin, S., Ren, H., Wu, Z. et al. (2021). Direct Z-scheme WO_3- nanowire-bridged TiO_2 nanorod arrays for highly efficient photoelectrochemical overall water splitting. *Journal of Energy Chemistry* 59: 721.

16 Zhao, Y., Linghu, X., Shu, Y. et al. (2022). Classification and catalytic mechanisms of heterojunction photocatalysts and the application of titanium dioxide (TiO_2)-based heterojunctions in environmental remediation. *Journal of Environmental Chemical Engineering* 10: 108077.

17 Fu, J., Xu, Q., Low, J. et al. (2019). Ultrathin 2D/2D WO_3/g-C_3N_4 step-scheme H_2-production photocatalyst. *Applied Catalysis B: Environmental.* 243: 556.

18 He, B., Wang, Z., Xiao, P. et al. (2022). Cooperative coupling of H_2O_2 production and organic synthesis over a floatable polystyrene-sphere-supported TiO_2/Bi_2O_3 S-scheme photocatalyst. *Advanced Materials* 34: 2203225.

19 Ruan, X., Huang, C., Cheng, H. et al. (2023). A twin S-scheme artificial photosynthetic system with self-assembled heterojunctions yields superior photocatalytic hydrogen evolution rate. *Advanced Materials* 35: 2209141.

20 Song, W., Chong, K.C., Qi, G. et al. (2024). Unraveling the Transformation from Type-II to Z-Scheme in Perovskite-Based Heterostructures for Enhanced Photocatalytic CO_2 Reduction. *Journal of the American Chemical Society* 146: 3303.

21 Moon, H.S., Hsiao, K.C., Wu, M.C. et al. (2023). Spatial separation of cocatalysts on Z-scheme organic/inorganic heterostructure hollow spheres for enhanced photocatalytic H_2 evolution and in-depth analysis of the charge-transfer mechanism. *Advanced Materials* 35: 2200172.

22 Salazar-Marín, D., Oza, G., Real, J.A.D. et al. (2024). Distinguishing between type II and S-scheme heterojunction materials: A comprehensive review. *Applied Surface Science* 19: 100536.

23 Tian, Z., Zhang, P., Qin, P. et al. (2019). Novel Black $BiVO_4/TiO_{2-x}$ photoanode with enhanced photon absorption and charge separation for efficient and stable solar water splitting. *Advanced Energy Materials* 9: 1901287.

24 Lee, D., Kvit, A., and Choi, K.S. (2018). Enabling solar water oxidation by $BiVO_4$ photoanodes in basic media. *Chemistry of Materials* 30: 4704.

25 Lu, Y., Cheng, X., Tian, G. et al. (2018). Hierarchical CdS/m-TiO_2/G ternary photocatalyst for highly active visible light-induced hydrogen production from water splitting with high stability. *Nano Energy* 47: 8.

26 Xiao, S.T., Wu, S.M., Wu, L. et al. (2023). Confined heterojunction in hollow-structured TiO_2 and its directed effect in photodriven seawater splitting. *ACS Nano* 17: 18217.

27 a) Jiang, Y., Sun, Z., Chen, Q. et al. (2022). Fabrication of 0D/2D TiO2 Nanodots/g-C_3N_4 S-scheme heterojunction photocatalyst for efficient photocatalytic overall water splitting. *Applied Surface Science* 571: 151287; b) Lin, S., Ren, H., Wu, Z. et al. (2021). Direct Z-scheme WO_{3-x} nanowire-bridged TiO_2 nanorod arrays for highly efficient photoelectrochemical overall water splitting. *Journal of Energy Chemistry* 59: 721.

28 Li, Y., Peng, Y.K., Hu, L. et al. (2019). Photocatalytic water splitting by N-TiO_2 on MgO (111) with exceptional quantum efficiencies at elevated temperatures. *Nature Communications* 10: 4421.

29 Lu, Y., Liu, X.L., He, L. et al. (2020). Spatial Heterojunction in Nanostructured TiO_2 and Its Cascade Effect for Efficient Photocatalysis. *Nano Letters* 20: 3122.

30 Yang, J., Wang, D., Han, H. et al. (2013). Roles of Cocatalysts in Photocatalysis and Photoelectrocatalysis. *Accounts of Chemical Research* 46: 1900.

31 Ai, G., Mo, R., Li, H. et al. (2015). Cobalt phosphate modified TiO_2 nanowire arrays as co-catalysts for solar water splitting. *Nanoscale* 7: 6722.

10

Plasmonic Metal–TiO_2 Photocatalysts

Na Zhou, Yan He, Jiahuan Zheng*

School of Materials Science and Engineering, Shijiazhuang Tiedao University, Shijiazhuang, Hebei, P.R. China

*Corresponding author: chmzhouna@stdu.edu.cn

10.1 Surface Plasmon Resonance

Metal nanoparticles (NPs) are widely utilized due to their unique localized surface plasmon resonance (LSPR). The generation of LSPR by metal NPs is depicted in Figure 10.1. When light interacts with metal NPs smaller than the light wavelength, coherent oscillation of electrons on the conduction band is induced by interaction with an electromagnetic field. The resonant frequency is highly dependent on the shape, size, interparticle distance, and the nature of the surrounding medium, indicating that the LSPR can be adjusted by modifying the factors [1–7].

10.2 The Enhanced Photocatalytic Mechanism Based on LSPR

Plasmonic metal NPs have been utilized to enhance the solar energy conversion activities of semiconductors due to the LSPR [8–13]. To elucidate the enhanced processes, various mechanisms such as direct electron transfer (DET, also known as hot electron injection), plasmon-induced resonant energy transfer (PIRET), or resonant photon scattering, have been proposed based on the heterostructure of metal and semiconductor. The mechanisms are illustrated detailed in the following sections.

10.2.1 DET Mechanism

The DET or hot electron injection mechanism involves a heterostructure in which the semiconductor surface directly interfaces with metal NPs [14]. As shown in Figure 10.2a, upon UV light irradiation, the semiconductors within the composite NPs are excited, leading to the generation of electron–hole pairs. Electrons from

Nanostructured TiO_2: Synthesis, Photocatalytic Properties and Applications, First Edition.
Edited by Jia Hong Pan, Wan In Lee and Detlef W. Bahnemann

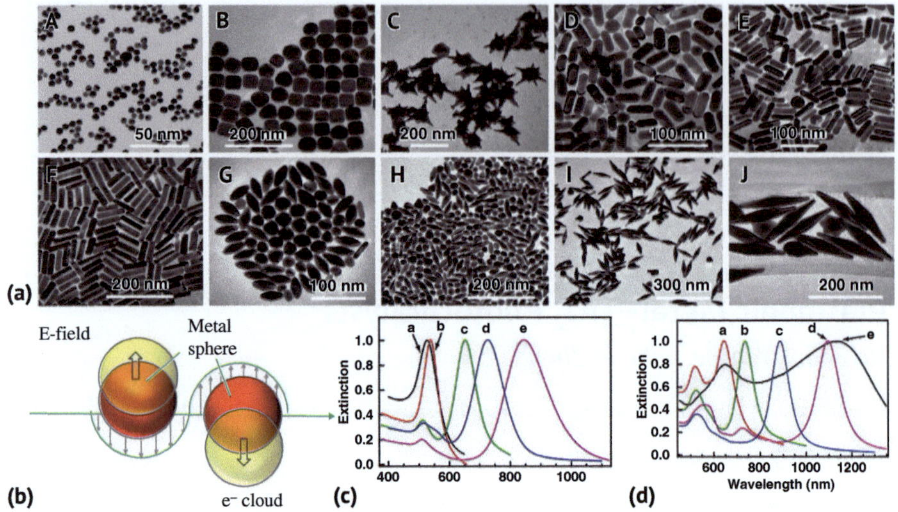

Figure 10.1 (a) The TEM images of metal NPs with different shapes. (b) Scheme for generation of coherent oscillation by metal NPs induced by light. (c, d) The normalized extinction spectra of Au NPs of different shapes and sizes. (c) Spectra a and b correspond to nanospheres and nanocubes, and spectra c–e correspond to nanorods with different aspect ratios. (d) Spectra a–e correspond to nanobipyramids with different aspect ratios. Reproduced from [7]/American Chemical Society.

n-type semiconductor, possessing a higher Fermi level, may transfer to metal NPs with a lower Fermi level, leaving behind holes on the semiconductors. The electron transfer process can effectively reduce recombination of electrons and holes in the semiconductor, thereby enhancing the trapping efficiency of electrons or holes by reactants. To validate this transfer mechanism, Ag@TiO_2 core–shell nanocomposites were subjected to UV light irradiation and exhibited a blue shift in their surface plasmon band due to accumulation of electrons on the metal NP surface. Furthermore, addition of electron acceptors resulted in a shift back of the surface plasmon band (Figure 10.2c–e) [15].

The electron transfer path is different under visible-light illumination. As shown in Figure 10.2b, visible light can directly excite metal NPs rather than semiconductors, leading to the generation of hot electrons on the metal NPs that may cross the Schottky barrier at the interface of metal NPs and semiconductors, and subsequently move to the conduction band of semiconductors [5, 8, 16–18]. According the DET mechanism, adjustment of the LSPR of metal NPs enables metal–semiconductor nanocomposites to utilize solar light in the visible to near-infrared region for enhanced harvesting of solar energy and higher conversion efficiency. For instance, Au nanorods were fabricated on the surface of TiO_2 single crystal to facilitate the water oxidation under the near-IR light irradiation. The photocurrent spectrum (IPCE) under different wavelengths of light irradiation matched with the LSPR bands shape of Au nanorods, indicating mediation of photocatalytic action by excitation of metal NPs (as shown in Figure 10.3a) [19]. Ultrafast spectra in Au@ZnO nanocomposites also confirmed DET (Figure 10.3b). Compared with pure Au metal, long-lived transient species were observed in the transient absorption spectra

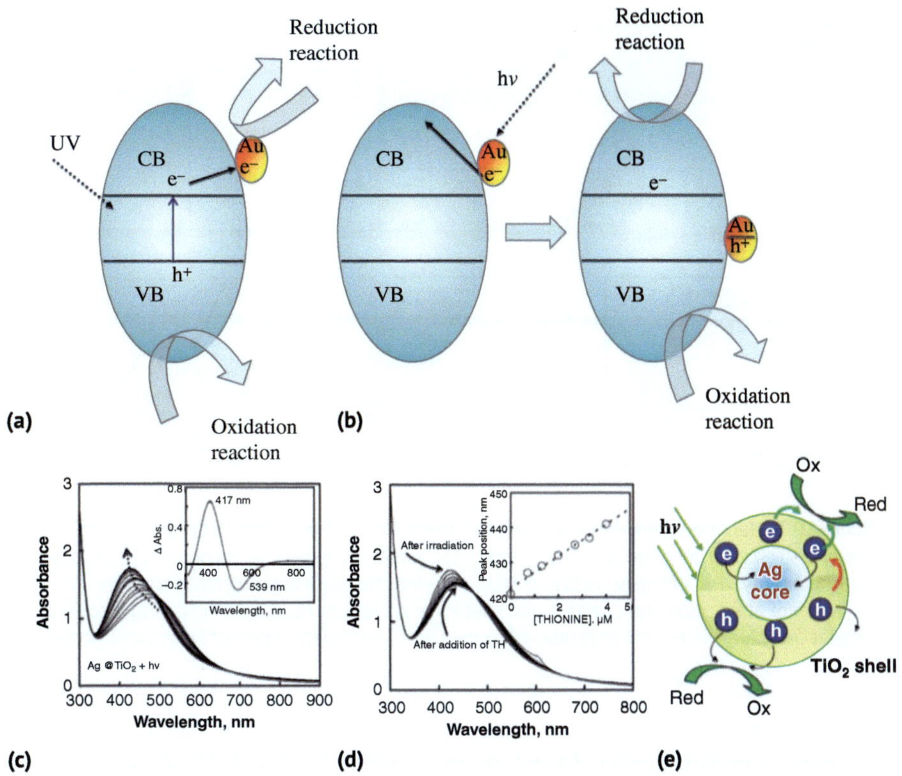

Figure 10.2 (a) Scheme for DET mechanism under UV light irradiation. (b) DET mechanism by visible light irradiation. (c) Absorption spectra of Ag@TiO_2 in ethanol at different UV irradiation time. The direction of the arrow indicates the spectral shift observed during the UV irradiation. (d) The addition of different amount of thionine dye into pre-UV-irradiated Ag@TiO_2 suspension. (e) The mechanism to illustrate the enhanced photocatalytic of TiO_2 by metal core. Reproduced from [15]/with permission of American Chemical Society.

of Au@ZnO nanocomposites, providing direct evidence for hot electron transfer from metal to semiconductor. Further evidence for DET was demonstrated by faster decay dynamics at a single wavelength of Au NR@ZnO [20, 21].

10.2.2 The Plasmon-induced Resonance Energy Transfer Mechanism

The electron transfer between metal and semiconductor is commonly observed in metal–semiconductor nanocomposites. However, several studies have also demonstrated that plasmonic metals can still enhance solar energy conversion efficiency when isolated from semiconductor by insulators, such as SiO_2 and Al_2O_3 [22, 23]. In this scenario, the presence of insulator blocks DET, leading to the proposal and validation of PIRET or near-field enhancement mechanism. As depicted in Figure 10.4a, when the LSPR of metal NPs overlaps with the absorption band of a semiconductor, electromagnetic field-mediated plasmonic energy transfer occurs from the LSPR dipole of the metal to the transition dipole of the semiconductor [24].

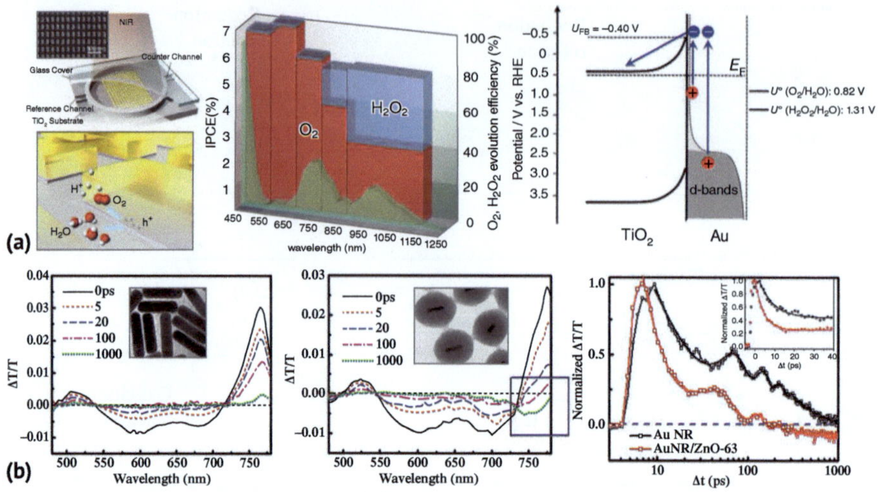

Figure 10.3 (a) The scheme for the three-electrode PEC with Au–NRs–TiO_2 as working electrode. The relation of generated O_2 and H_2O_2 between the photocurrent and IPCE spectrum. And the energy level of the Au NRs and TiO_2, showing the process for electron transfer. (b) Transient absorption and pump-probe spectra of Au NRs or Au@ZnO, indicating electron transfer from Au NRs to ZnO. Reproduced from [19, 20]/with permission of American Chemical Society.

The plasmon-induced resonance energy transfer differs from the radiatively enhanced mechanism of SPR-mediated local electromagnetic field, as it is a nonradiative process induced by the relaxation of the localized surface plasmon dipole. This process can contribute to the generation of electron–hole pairs in the semiconductor without photon emission. It is important to note that the strength of PIRET dependent to the distance between the dipoles of metal NPs and semiconductor. The PIRET process has potential to enhance solar energy conversion efficiency by more electron–hole pairs [25]. Transient absorption spectroscopy and calculations have been used to study the

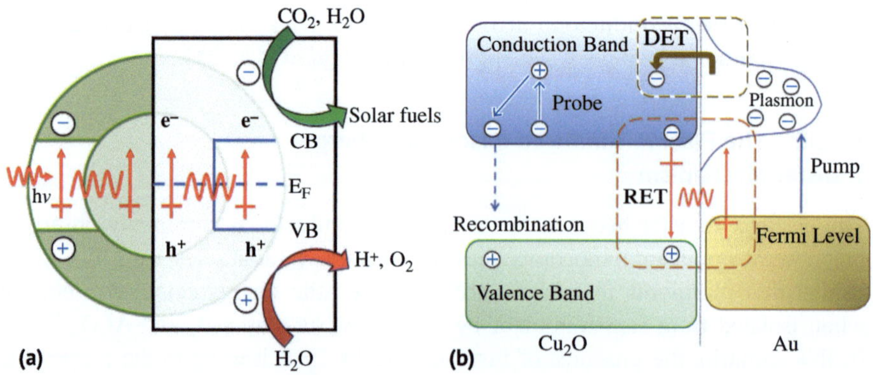

Figure 10.4 (a) The increased the charge separation processes by local electromagnetic field enhancement (LEMF) or resonant energy transfer (RET). (b) The DET and PIRET occurring in Au@Cu_2O nanocomposites. Reproduced from [24]/with permission of American Chemical Society.

PIRET and DET processes in Au@Cu_2O and Au@SiO_2@Cu_2O nanocomposites, indicating that the carriers were created in both of Au@Cu_2O or Au@SiO_2@Cu_2O nanocomposites. And the PIRET mechanism dominated the enhanced carries generation in the Au@SiO_2@Cu_2O (Figure 10.4b).

10.2.3 Resonant Photon Scattering Mechanisms

It is well established that metal NPs demonstrated size-dependent light scattering properties, resulting in distinct characteristic colors. Moreover, larger metal NPs may induce strong scattering [26]. The phenomenon of light scattering facilitates the reflection of light between NPs and increases the average path of photons, thereby enhancing the likelihood of photon absorption by metal–semiconductor nanocomposites. Notably, plasmonic NPs with a diameter exceeding ~50 nm typically exhibit pronounced scatting effects. Silver (Ag) NPs have been employed to enhance the charge carrier generation in organic photovoltaic films [27]. And the X-channel photoinduced absorption spectra confirmed the absorption of organic film was enhanced by Ag NPs through either the light scattering or the near field absorption enhancement effects of metal NPs.

10.3 The Strategies for Synthesis of Plasmonic Metal–TiO_2 Nanocomposites

The photocatalytic performance of metal–TiO_2 (M–TiO_2) nanocomposites is significantly influenced by the synthesis conditions, methods, and morphologies. Various techniques have been developed to incorporate plasmonic metal NPs with TiO_2, including sol–gel, co-precipitation, hydrothermal, and photodeposition methods. Each synthesis method has its distinct advantages and disadvantages in terms of performance and preparation. Therefore, the selection of appropriate methods should be based on the design and structure of M–TiO_2. In the subsequent section, we will elaborate on the commonly used methods.

10.3.1 The Sol–Gel Method for Hydrolyzing TiO_2 on the Surface of Plasmonic Metal NPs

The sol–gel method, a wet chemical process commonly employed for the preparation of metal oxide materials through hydrolysis and condensation of precursors of metal ions, is utilized to prepare M–TiO_2 nanocomposites. In this approach, metal NPs are initially synthesized and subsequently functionalized with surfactants to absorb the titanium precursor. The TiO_2 precursor is then hydrolyzed on the surface of plasmonic metals via a sol–gel method.

Noncentrosymmetric Au@TiO_2 photocatalysts were synthesized using a single-step sol–gel route by dissolving gold NPs and titanium precursors in isopropanol and ammonia solution (Figure 10.5). Gold NPs of diameters of 30, 50, and 70 nm were synthesized using a seed growth method with citrate reduction.

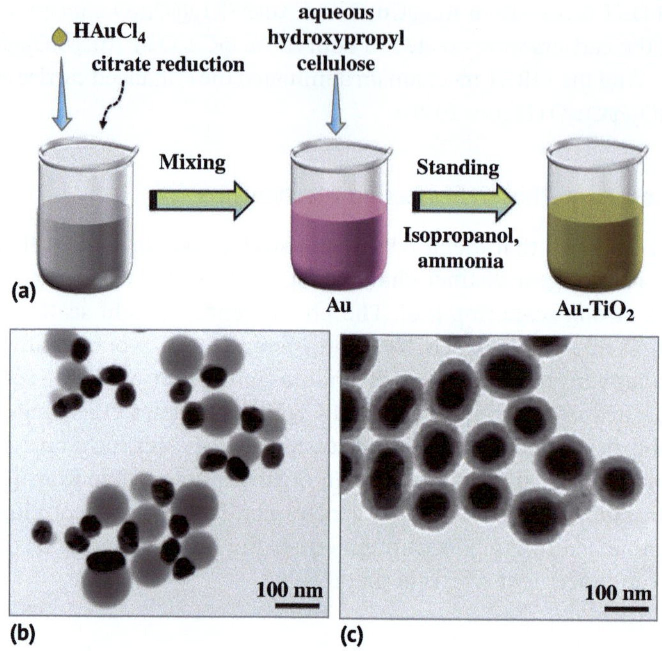

Figure 10.5 (a) Schematic illustration of the preparation of Au–TiO_2; (b, c) TEM images of Janus (b) and core–shell; (c) $Au_{50\ nm}$–TiO_2 nanostructures. Reproduced from [28]/ John Wiley & Sons.

The resulting gold colloids (500 mL) were mixed with a hydroxypropyl cellulose solution (20 mL, 0.5 wt.%) and stirred overnight. Subsequently, a solution containing the gold NPs (100 mL) was combined with isopropanol (400 mL) and ammonia solution (12.5 mL, 25%) to prepare Au–TiO_2 nanocomposites. Isopropanol was used as a homogenizer due to its compatibility with both water and the TiO_2 precursor, addressing the issue of poor miscibility between water and the TiO_2 precursor [28].

Triple core–shell nanocomposites of Ag@SiO_2@TiO_2 were also synthesized using a sol–gel method (Figure 10.6). The pre-prepared Ag@SiO_2 NPs were modified with 3-aminopropyltrimethoxysilane (APTMS) prior to the hydrolysis of titanium tetraisopropoxide (TTIP) in an ethanol solution. Subsequently, a layer of amorphous TiO_2 shell was coated onto the surface of Ag@SiO_2 to achieve the desired triple core–shell structure. The final nanocomposites were obtained after calcination at 450 °C in air [29].

10.3.2 Deposition Metal NPs on TiO_2 by Reduction

Metal ions can be directly reduced on the surface of TiO_2 through chemical or photo reduction. However, this method results in uncontrollable and limited spherical shapes of metal NPs. Despite these limitations, it remains one of the most commonly utilized approaches due to its straightforward procedures.

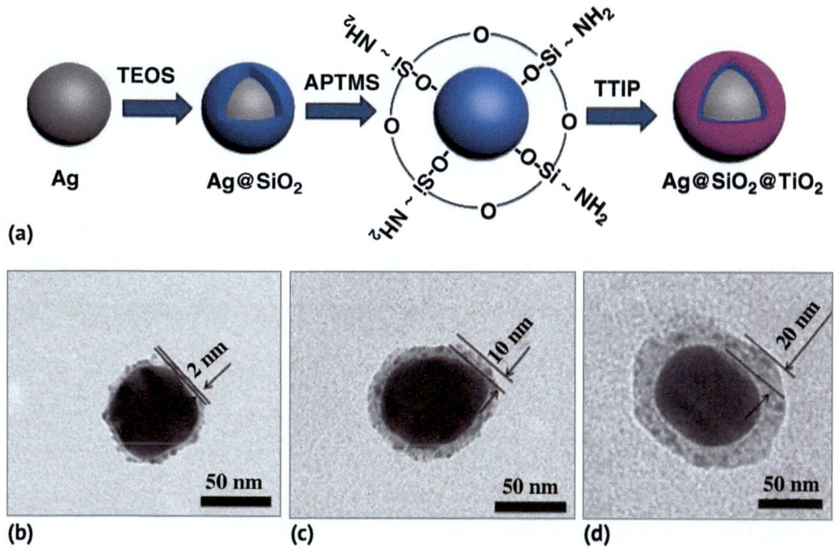

Figure 10.6 (a) Schematic illustration of the preparation of Ag@SiO_2@TiO_2 triple core–shell NPs; (b–d) TEM images of Ag@SiO_2@TiO_2 with TiO_2 shell thicknesses of 2, 10, and 20 nm shell, respectively. Reproduced from [29]/The Royal Society of Chemistry.

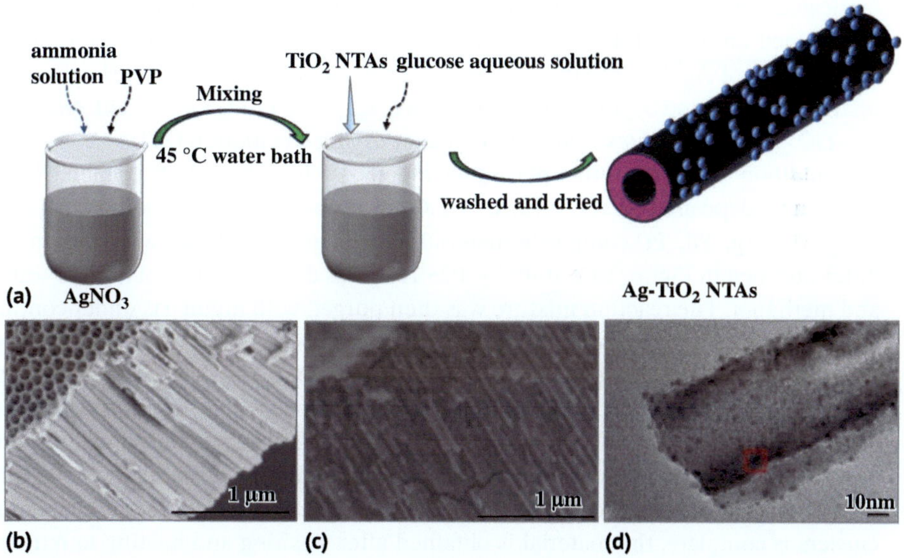

Figure 10.7 (a) Schematic diagram of the procedures for preparation of Ag–TiO_2 NTAs; (b) SEM images of the pristine TiO_2 NTAs; (c) SEM images of Ag–TiO_2 NTAs; (d) TEM of Ag–TiO_2 NTAs. Reproduced from [30]/The Royal Society of Chemistry.

Vertically aligned TiO_2 nanotube arrays (NTAs) were decorated with plasmonic silver NPs through an ultrasonic-assisted in situ reduction approach (Figure 10.7). The growth solution, containing ammonia and polyvinylpyrrolidone, was thoroughly mixed with the assistance of ultrasonic waves. Subsequently, the pre-prepared

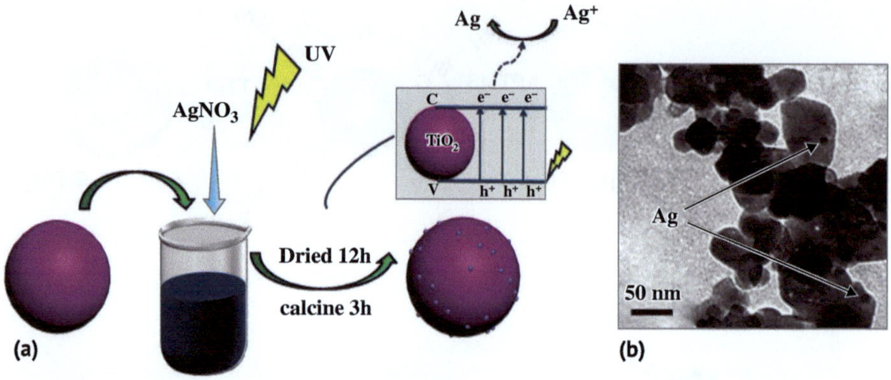

Figure 10.8 (a) Schematic diagram for the synthesis of Ag–TiO_2 using photoreduction method; (b) TEM image of the Ag–TiO_2. Reproduced from [31]/Taylor & Francis.

TiO_2 NTAs were immersed into the growth solution followed by the addition of glucose. The resulting nanocomposites were collected after washing and drying [30].

The photoreduction of metal ions on the surface of semiconductor is commonly employed for the preparation of metal–semiconductor nanocomposites, wherein the reduction occurs as a result of electron excitation in the conduction band of semiconductor induced by light with sufficient energy. Shokri et al. successfully fabricated Ag–TiO_2 NPs using a photo deposition method (Figure 10.8). A predetermined amount of TiO_2 is added to 100 mL of deionized water, followed by introduction of $AgNO_3$ into the TiO_2 suspension. After adjustment of pH value, the mixture is irradiated by ultraviolet light (30 W, $\lambda_{max} = 254$ nm) for 3 h. The resulting Ag–TiO_2 nanocomposites were obtained after drying calcination [31].

In addition to Ag NPs, other plasmonic metals, such as Au, Pd, Pt, or metal alloy, also can be deposited on the surface of TiO_2 by a photoreduction method. M–Au–TiO_2 (M = Ag, Pd, Pt) composite materials were synthesized following the procedures outlined in Figure 10.9. Initially, P25 TiO_2 was dispersed in a mixture of water and methanol. The reaction mixture was then purged with argon gas under continuous stirring for 30 min. Subsequently, $HAuCl_4 \cdot 4H_2O$ was added to the mixture and exposed to a 400 W ultraviolet lamp for 30 min to obtain Au–TiO_2 nanocomposites. Following this step, $H_2PtCl_6 \cdot H_2O$ or another metal precursor is introduced into the same reaction mixture and subjected to ultraviolet light for an additional 30 min. The presence of both Pt and Au in the solution facilitates the formation of bimetallic (or alloy) NPs on the surface of Au–TiO_2. Once the photodeposition of metal nanoclusters is complete, the material is obtained after washing and heating to remove any unreacted substance [32].

10.3.3 Attaching Plasmonic Metal NPs to TiO_2 NPs by Linkers

Approaches involving the reduction or deposition of metal NPs onto the surface of TiO_2 may be limited due to the irregular shape of the metal NPs, resulting in uncontrollable LSPR. Metal NPs can be attached to pre-prepared TiO_2 NPs using molecular linkers. This method allows for the separate synthesis of TiO_2 NPs and metal NPs under their respective optimal conditions, so it becomes possible to combine

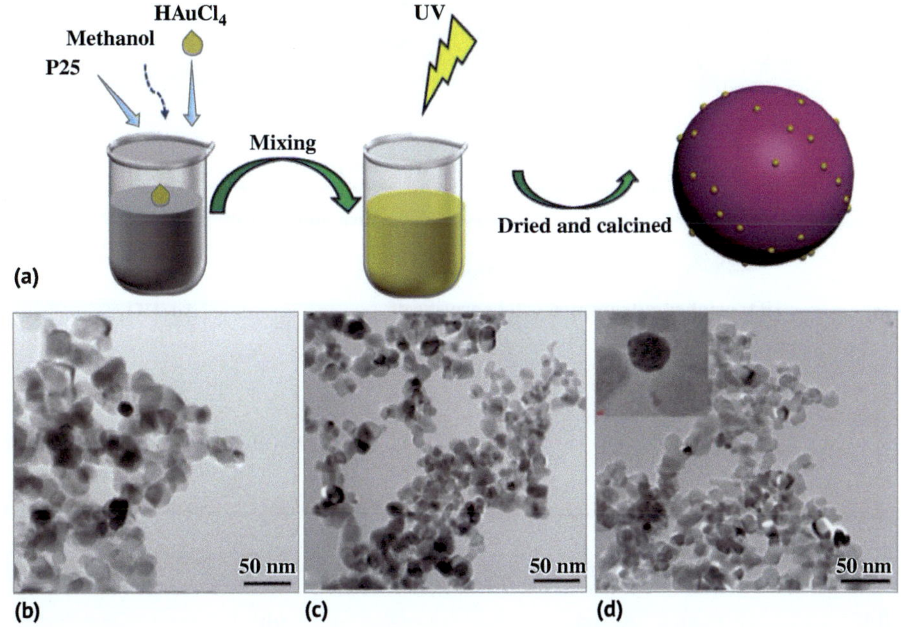

Figure 10.9 (a) Schematic diagram of the procedures for preparation of M–TiO_2; (b–d) TEM images of, (b) Au–TiO_2, (c) Pt–TiO_2, and (d) Pt–Au–TiO_2. Reproduced from [32]/The Royal Society of Chemistry.

TiO_2 nanocomposites of any shape with a variety of metal NPs in order to achieve the most optimized enhancement.

The Au NPs were deposited onto TiO_2 nanocrystals using a ligand-exchange approach to produce Au–TiO_2 nanocomposites. TiO_2 NPs with various shapes, including octahedra, truncated octahedra, and sheets, were initially synthesized through a hydrothermal method using different precursors or adjusting the precursor ratios (Figure 10.10). The diverse shapes of TiO_2 offer the opportunity to customize the exposed facets in order to enhance the photocatalytic performance of the nanocomposites. Subsequently, the pre-prepared Au NPs were attached to the surface of TiO_2 sample using 3-mercaptopropionic acid (MPA). The detailed procedures are as follows: 10 mg of TiO_2 NPs are dispersed in 5 mL of an aqueous solution containing 150 μL of MPA and 150 μL of $NH_3 \cdot H_2O$ (28%) via ultrasonication. Following this initial dispersion, 400 μL of an aqueous dispersion containing Au NPs $(1.64\ mg\ mL^{-1})$ is added under ultrasonication. The mixture is then allowed to settle before undergoing multiple centrifugation and water washing steps to remove any unbound substances. Finally, the sample is annealed at 200 °C to remove the MPA and enhance the contact between Au and TiO_2, thus improving the material's properties [33]. In fact, a variety of metal NPs with diverse LSPR can be attached on TiO_2 in order to fully optimize its photocatalytic efficiency, highlighting the advantages of this approach.

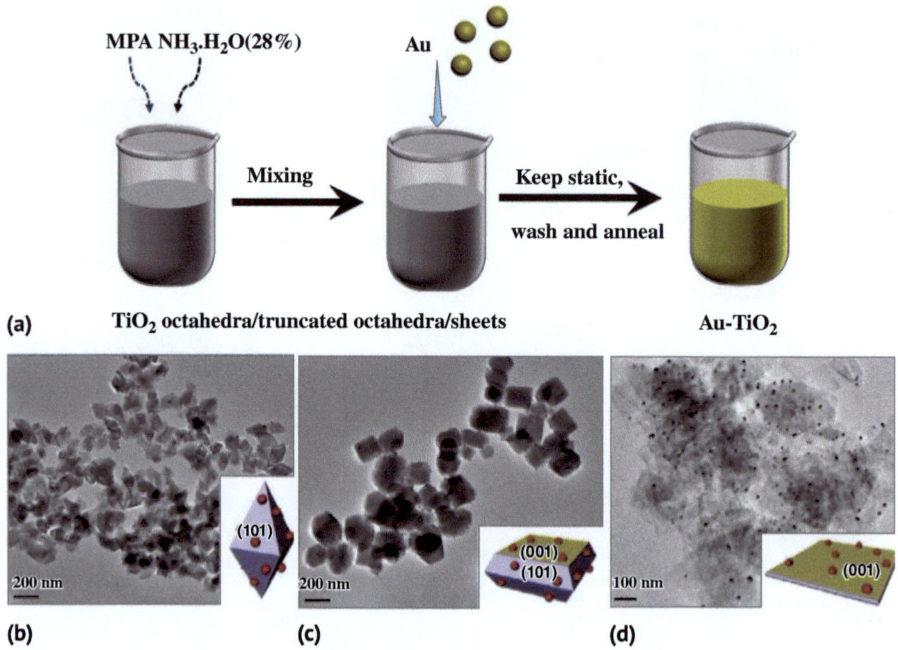

Figure 10.10 (a) Schematic diagram of the approach to synthesize Au–TiO_2; (b–d) TEM image of the Au–TiO_2 with the shapes of octahedra, truncated octahedra, and sheets, respectively. Reproduced from [33]/with permission of Elsevier.

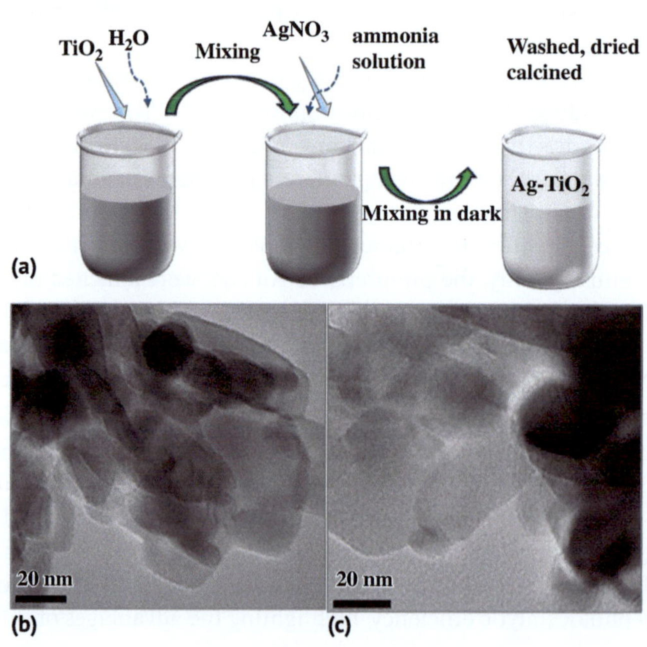

Figure 10.11 (a) Schematic illustrations of the approach to synthesize of single atom Ag–TiO_2. TEM images of (b) TiO_2 and (c) Ag–TiO_2 (0.23 wt.%). Reproduced from [34]/The Royal Society of Chemistry.

10.3.4 Loading Single Metal Atom on the Surface of TiO_2

A novel photocatalyst, consisting of single Ag atoms loaded onto anatase-type TiO_2 NPs (Ag–TiO_2), was synthesized as depicted in Figure 10.11. The TiO_2 NPs were initially dispersed in water with the aid of sonication to ensure uniform dispersion. This dispersion was then slowly added into a solution containing $AgNO_3$ drop by drop. The combined solution was stirred continuously for 2 h in a dark environment to prevent any light-induced reactions. Subsequent to the stirring process, the resultant white solution underwent centrifugation and washing to remove any impurities and excess reagents. The collected white product was air-dried and subsequently calcined, yielding the final Ag–TiO_2 sample [34].

10.4 Applications of Plasmonic Metal–TiO_2 Photocatalysts

Environmental pollution and the energy crisis are two of the most pressing issues that need to be addressed. Photocatalytic techniques offer effective solutions for treating pollution and harnessing solar energy. TiO_2 photocatalysts, as a traditional and widely used semiconductor, have been applied in various areas such as the degradation of organic pollutants, hydrogen evolution, and reduction of CO_2. Consequently, plasmonic metal–TiO_2 nanocomposites have also been extensively utilized to achieve higher levels of photocatalytic efficiency.

10.4.1 Photodegradation of Organic Pollutions

Ag NPs were photodeposited on the surface of TiO_2 nanorices (TNRs) to synthesize TNRs–Ag nanocomposites. The photocatalytic efficiency of TNRs–Ag nanocomposites was assessed through degradation of Acid Blue 92 (AB92) or Crystal Violet (CV). Under visible light irradiation, the degradation of AB92 reached 72.06% in 90 min when using TNR–Ag, compared to 56.34% for TNRs alone (Figure 10.12). Consistent with the characterization results, the presence of silver NPs (Ag NPs) was found to enhance the photocatalytic properties of TNRs. The enhancement was also observed in the photodegradation of CV. And the enhancement is attributed to a DET mechanism as described previously. Excitation by light with appropriate energy levels leads to excitation of the TiO_2 nanorice, causing electrons to move to the conduction band while leaving behind holes in the valence band. Electrons on the conduction band may then transfer to the Ag NPs, reducing recombination of electron–hole pairs. Furthermore, absorption-induced surface plasmonic resonance in Ag NPs may further contribute to improving the photocatalytic efficiency of TNRs [35].

Ag@SiO_2@TiO_2 nanocomposites with a triple core–shell nanostructure were synthesized through the hydrolysis of TiO_2 on the surface of Ag@SiO_2 shell. The thickness of SiO_2 interlayer and TiO_2 shell was tailored by varying the amounts of precursors. And the results for photocatalytic degradation of methylene blue demonstrated that Ag@SiO_2@TiO_2 nanocomposites with a 2 nm SiO_2 interlayer and

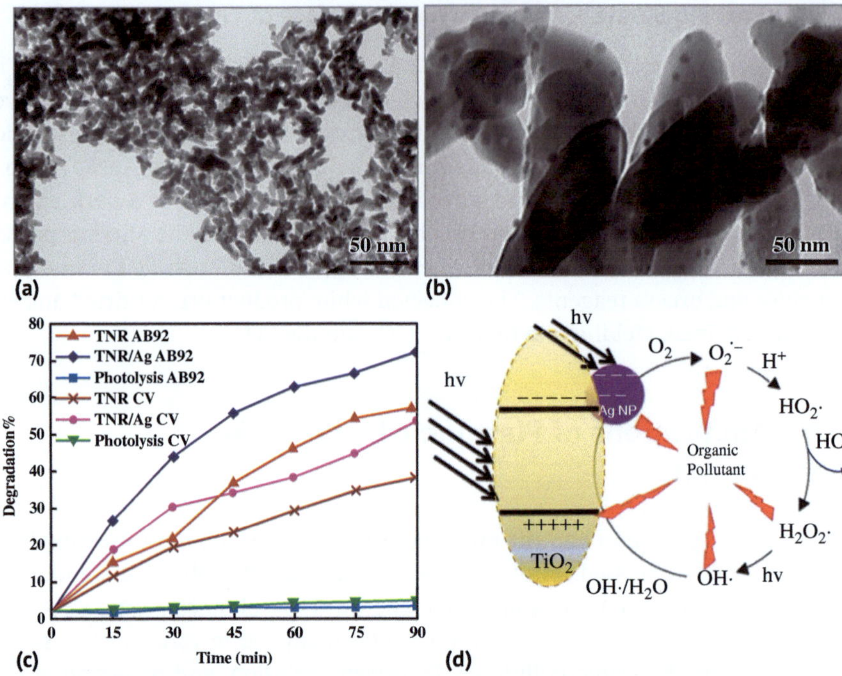

Figure 10.12 TEM image of TNR–Ag before photocatalytic tests (a) and (b). (c) Photocatalytic degradation of AB92 and CV using TNRs or TNRs–Ag samples. (d) Schematic possible degradation mechanism of AB92. Reproduced from Ref. [35]/with permission of Elsevier.

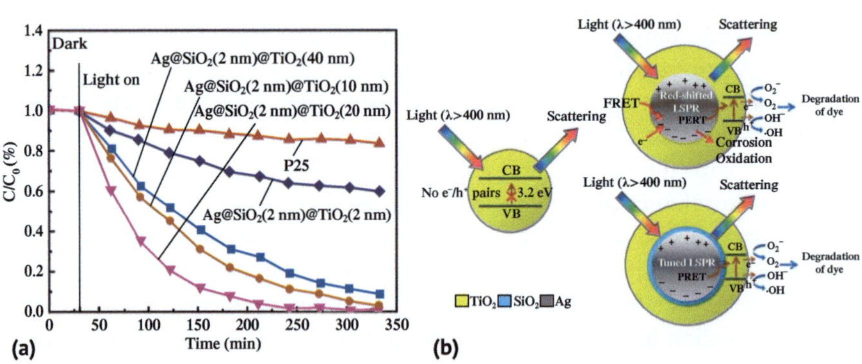

Figure 10.13 (a) Photocatalytic degradation of methylene blue with various photocatalysts; (b) mechanism of the photodegradation process of pure TiO_2, $Ag@TiO_2$, and $Ag@SiO_2@TiO_2$ during the irradiation of visible light. Reproduced from ref. [29]/with permission of The Royal Society of Chemistry.

a 20 nm TiO_2 shell exhibited superior photocatalytic performance, approximately 31 times higher than P25. A proposed mechanism involving plasmon-induced resonance energy transfer from Ag to TiO_2 and scattering by Ag NPs elucidated this enhancement (Figure 10.13) [29]. Upon irradiation, an electric field generated around Ag NPs within a specific range may enhance the absorption of TiO_2, leading to an increased generation of electron–hole pairs. Notably, the overlapping of the LSPR of Ag NPs with the absorption spectrum of TiO_2 suggests coupling between these two components, facilitating resonance-induced energy transfer from Ag NPs to TiO_2. Furthermore, scattering due to the large size of Ag NPs may also contribute to enhanced light absorption by TiO_2, thereby improving its photocatalytic efficiency.

10.4.2 Photocatalytic Hydrogen Evolution

M–Au–TiO_2 (M = Ag, Pd, Pt) composite materials were synthesized using a facile photodeposition method. The results of their photocatalytic hydrogen generation indicated that Pt–Au–TiO_2 demonstrated the highest hydrogen (H_2) production rate at approximately 1.3 ± 0.07 mmol/h/g, with an apparent quantum yield (AQY) of 6.4%. And more, the photocatalytic performance of Pt–Au–TiO_2 maintained well even after undergoing more than 25 cycles, each lasting 5 h, for a total of 125 h. The enhance was attributed to the synergistic interaction between Au, Pt, and TiO_2. Under one sun illumination, photo-excitation occurs due to the SPR of gold as a result of the Au–TiO_2 or Pt–Au–TiO_2 systems. Following photoexcitation, charge separation took place due to the Schottky barrier formed between gold and TiO_2 in the Au–TiO_2 system. Furthermore, in the Pt–Au–TiO_2 system, effective charge separation and storage occur due to the presence of Pt. The Schottky barrier height for Pt–Au was anticipated to be between that of gold (1.3 eV) and platinum (1.9 eV) when these metals were present on the surfaces of TiO_2. However, equilibration of the Fermi level occurring within the composite not only electronically integrated but also aligned energy levels of the Au's surface plasmon resonance state, the conduction band of TiO_2, and the Fermi energy of platinum at the interfaces. This equilibration facilitated charge separation and storage significantly easier, leading

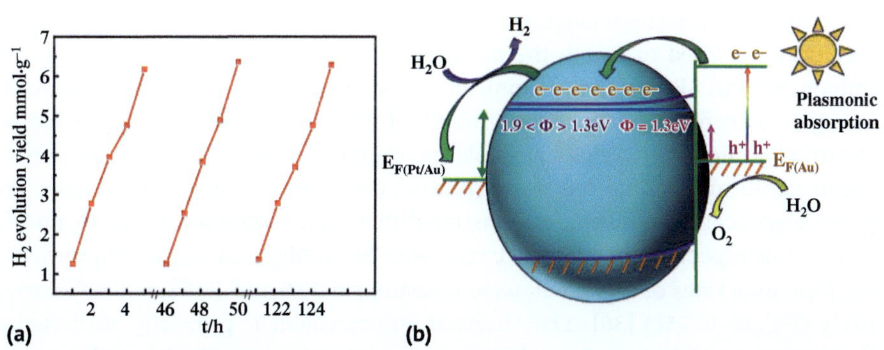

Figure 10.14 (a) Photocatalytic H_2 evolution activity of $Pt_{0.5}$–Au_1–TiO_2 under one sun conditions with aqueous methanol solution and the data for cycle use; (b) schematic energy-band diagram for $Pt_{0.5}$–Au_1–TiO_2 heterojunctions and DET at the Au–Pt alloy surfaces, leading to electrons and holes with longer lifetime for catalytic reactions. Reproduced from [32]/with permission of The Royal Society of Chemistry.

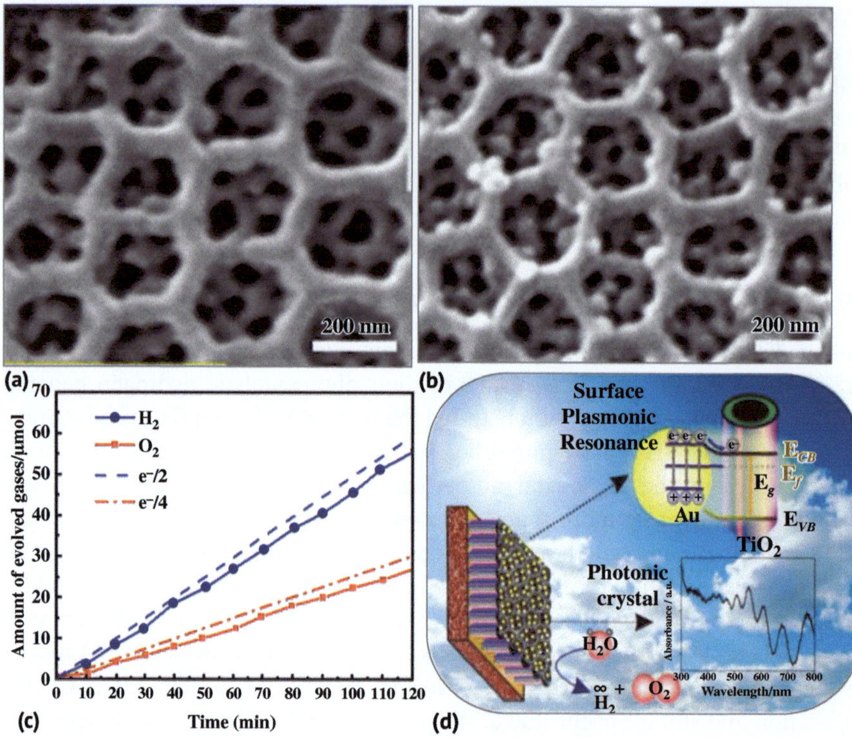

Figure 10.15 SEM images of TiO_2 NTPC (a) and Au–TiO_2 NTPC (b); (c) H_2 and O_2 evolution illumination of simulated solar light for Au–TiO_2 NTPC electrode; (d) the enhanced mechanism based on hot electron transfer from Au to the conduction band of TiO_2. Reproduced from [36]/with permission of American Chemical Society.

to an extended lifetime of charge carriers, which was beneficial for water splitting (Figure 10.14) [32].

The Au nanocrystals were assembled with TiO_2 tube-based photonic crystal (NTPC) substrates in order to create plasma photocatalytic composites that are responsive to visible light (Figure 10.15). The selection of the Au nanocrystals was specifically tailored to match the bandgap of the TiO_2, allowing for significantly influence of Au SPR on the photonic crystal substrate. This composite design was anticipated to greatly enhance the strength of the gold SPR, thereby improving hot electron injection from Au nanocrystals into the conduction band of TiO_2 and resulting in significantly enhanced water splitting performance under visible light. The photo-electrocatalytic results demonstrated that the formation rates of H_2 and O_2 reached the expected stoichiometric ratio with full sunlight irradiation. Specifically, the formation rates of H_2 and O_2 were measured at 27.9 and 13.6 μmol h^{-1}, respectively (Figure 10.15c) [36]. The observed enhancement is primarily attributed to the generation of hot electrons by Au under illumination, which are then transferred to the conduction band of TiO_2 for hydrogen evolution. The Au NPs effectively function as a sensitizer, enhancing the ability of TiO_2 to harness a greater proportion of visible light.

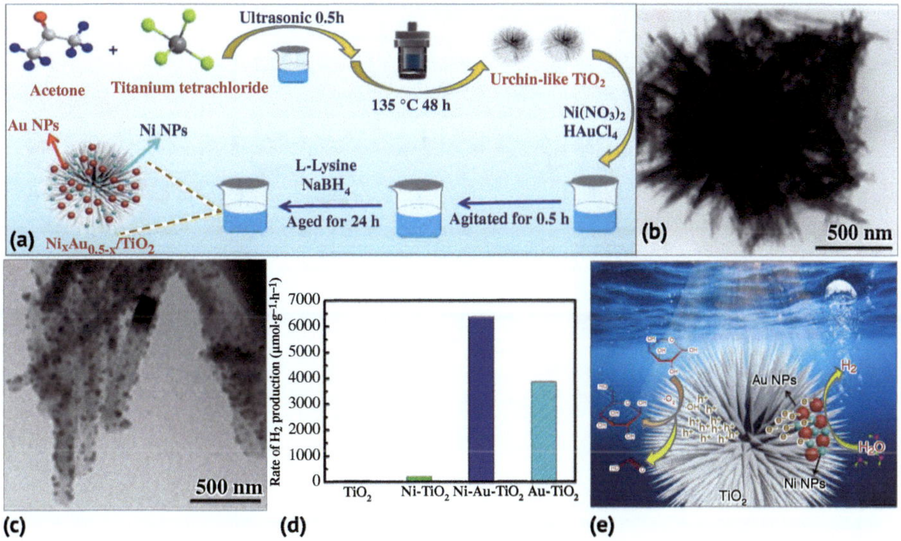

Figure 10.16 (a) The approach for synthesis of Ni_x–$Au_{0.5\text{-}x}$–TiO_2; (b) and (c) TEM images of $Ni_{0.05}$–$Au_{0.45}$–TiO_2; (d) rate of H_2 production with various photocatalysts; (e) the mechanism of H_2 evolution by glucose photoreforming with $Ni_{0.05}$–$Au_{0.45}$–TiO_2 catalyst. Reproduced from [37]/with permission of Springer Nature.

Ni and Au NPs were loaded on the surface of urchin-like TiO_2 through direct reduction of $Ni(NO_3)_2$ and $HAuCl_4$ with $NaBH_4$ (Figure 10.16) [37]. In comparison to pure TiO_2, bimetallic Ni and Au decorated TiO_2 composites exhibited enhanced photocatalytic hydrogeneration efficiency during the photoreforming of glucose. The H_2 production rate of the optimized $Ni_{0.05}$–$Au_{0.45}$–TiO_2 reached 6391.86 $\mu mol \cdot h^{-1} \cdot g^{-1}$, which was 118.57 times higher than that of pure TiO_2. The excellent H_2 evolution activity of $Ni_{0.05}$–$Au_{0.45}$–TiO_2 can be attributed not only to the urchin-like structure of TiO_2 itself but also the synergetic effect resulting from the presence of Ni and Au NPs. This enhancement is primarily attributed to the following two reasons: First, the LSPR effect of Au NPs broadens the light absorption of $Ni_{0.05}$–$Au_{0.45}$–TiO_2 in the visible light range, leading to an increased generation of electron–hole pairs for photoreforming of glucose reaction under visible irradiation. Second, a Schottky contact is formed between TiO_2 and either Ni NPs or Au NPs, where electrons on the conduction band of TiO_2 could transfer to Ni and Au NPs for H^+ reduction, thereby reducing the recombination of electrons and holes and prolonging the lifetime of electrons for efficient H_2 production.

10.4.3 Photocatalytic CO_2 Reduction

The TiO_2 nanowires were decorated with plasmonic Au–Ag alloy NPs using a simple photodeposition methods. The photocatalytic CO_2 reduction of Au–Ag–TiO_2 nanowires exhibit superior selectivity and higher efficiency compared to TiO_2 nanowires. The main product of CO_2 reduction with H_2 by Au–Ag–TiO_2 nanowires is CO, reaching 1813 $\mu mol \cdot g^{-1} \cdot h^{-1}$ at a selectivity of 98%, resulting in a photocatalytic

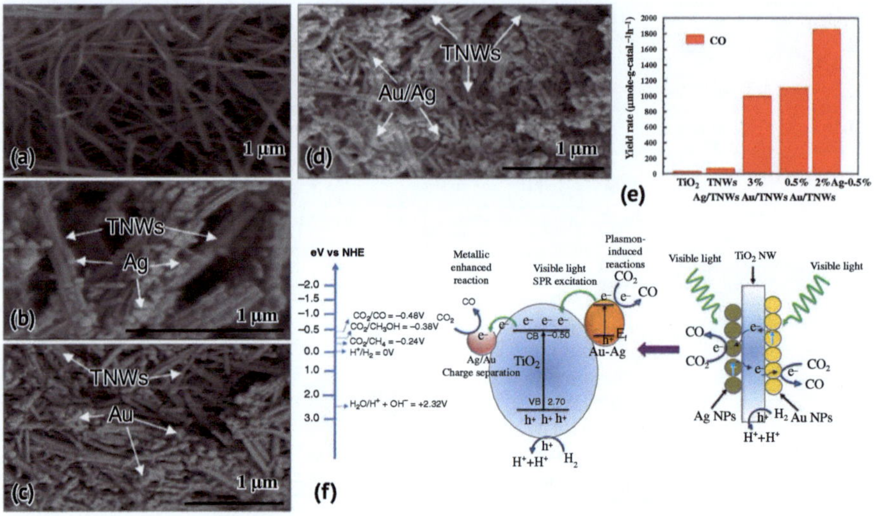

Figure 10.17 SEM images of TiO_2 nanowires (a), Ag–TiO_2 nanowires (b), Au–TiO_2 nanowires (c), and Au–Ag–TiO_2 nanowires (d); (e) yield rate of CO with different photocatalysts during CO_2 reduction with H_2; (f) proposed photocatalytic process and enhanced mechanism for CO_2 reduction with Au–Ag–TiO_2 catalysts. Reproduced from [38]/with permission of Elsevier.

CO_2 reduction rate 78 times higher than pure TiO_2. The enhanced photocatalytic activities achieved by co-depositing Au–Ag NPs can be attributed to the synergistic effected arising from LSPR effect of Au–Ag alloy. Once excitation, hot electrons generated on the surface of plasmonic metals can transfer to the conduction band of TiO_2 nanowires through the interface between metal and TiO_2. Simultaneously, electrons generated by the TiO_2 may transfer to the metal NPs in order to reduce mitigate electron–hole recombination and thereby extend the lifetime of electrons for CO_2 reduction (Figure 10.17) [38].

In additional to enhancing the photocatalytic efficiency of CO_2 reduction, metal NPs were reported to influence the selectivity of the reduced products. Homogeneous ligand-free metal NPs, including Ag, Cu, and Au, were deposited on the surface of TiO_2 through a chemical reduction method. The photocatalytic CO_2 reduction activities with water as an electron donor revealed that deposition of Group XI metals, including Cu, Au, and Ag, exhibited high selectivity toward CO production. Conversely, the deposition of metals in Group X, such as Pt and Pd showed high selectivity toward hydrocarbons like methane and ethane. Notably, Ag–TiO_2 demonstrated superior activity as a photocatalyst for producing CO at a rate of 5.2 $\mu mol \cdot g^{-1} \cdot h^{-1}$, while Pd–$TiO_2$ was active in methane production at a rate of 2.4 $\mu mol \cdot g^{-1} \cdot h^{-1}$ [39].

The photochromic single atom Ag–TiO_2 photocatalyst exhibited enhanced efficiency and high selectivity for the photocatalytic conversion of CO_2 to CH_4 (Figure 10.18). The optimal rate for photocatalytic CO_2 reduction to CH_4 reached 46.0 $\mu mol \cdot g^{-1} \cdot h^{-1}$ with an electron selectivity of 91%. The improved photocatalytic efficiency and selectivity were attributed to the incorporation of a single Ag

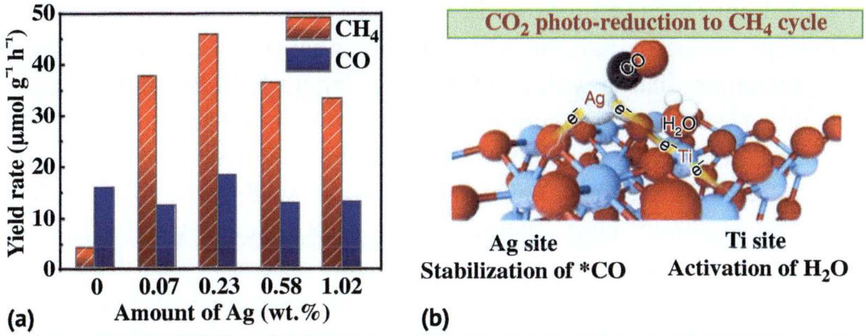

Figure 10.18 (a) Performance of photocatalytic CO_2 reduction on single atom Ag–TiO_2 with varying amounts of Ag loading; (b) the photocatalytic CO_2-to-CH_4 conversion cycle over the Ag–Ti active site on Ag–TiO_2. Reproduced from [34]/with permission of The Royal Society of Chemistry.

atom, which was demonstrated to capture photogenerated electrons form excited TiO_2, thereby prolonging their lifetime for CO_2 reduction. Furthermore, the high selectivity towards CH_4 was ascribed to the presence of Ag and Ti sites, which can stabilize the key intermediates such as *CO and *CHO in the production of CH_4 [34].

Most of the reported plasmonic metal–TiO_2 photocatalysts for CO_2 reduction have been designed based on the mechanism of DET enhancement, while the impact of resonance energy transfer induced by LSPR to the photocatalytic CO_2 reduction was rarely addressed. In order to investigate the LSPR effect of plasmonic metals on Pt–TiO_2, Au NPs with sizes of 4, 8, 18, and 26 nm, as well as their corresponding Au@SiO_2 core–shell NPs, (the corresponding TEM images shown in Figure 10.19 a-c)

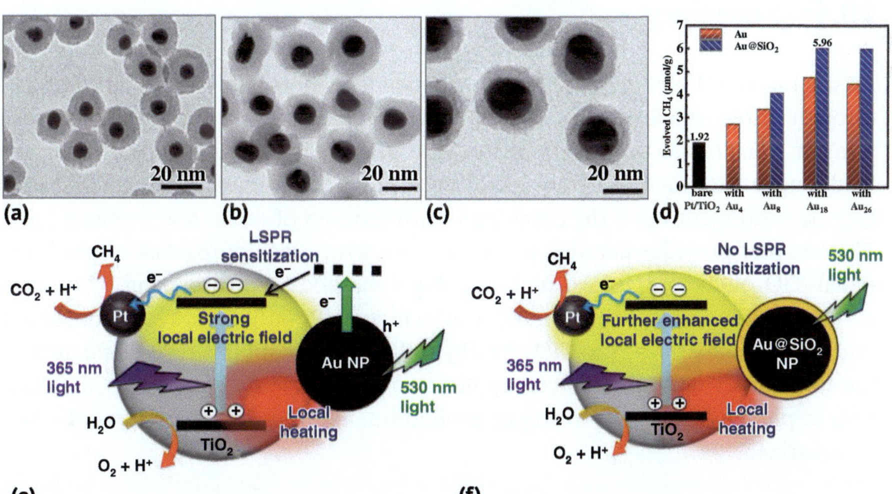

Figure 10.19 (a–c) TEM images of $Au_8@SiO_2$ (a), $Au_{18}@SiO_2$ (b), $Au_{26}@SiO_2$ (c); (d) amounts of CH_4 evolved under coirradiation of 365 and 530 nm wavelength of light catalyzed by Au–Pt–TiO_2 or Au@SiO_2–Pt–TiO_2 catalysts with varying sizes of Au NPs; (e, f) proposed mechanism presenting a photocatalytic process for CO_2 reduction on Au–Pt–TiO_2 (e) or Au@SiO_2–Pt–TiO_2 (f) catalysts under visible light irradiation 365 or 530 nm. Reproduced from [40]/with permission of Elsevier.

were loaded onto the pre-prepared Pt–TiO_2 surface for catalyzing CO_2 conversion. The findings revealed that Au NPs did not enhance the catalytic efficiency of Pt–TiO_2 under irradiation with a 365 nm wavelength light due to their inability to be activated by such light. In contrast, Au NPs were found to significantly enhance the photocatalytic CO_2 reduction of Pt–TiO_2 photocatalysts under the irradiation of 530 nm light (Figure 10.19e), which corresponds well the LSPR of Au NPs, indicating that the enhancement was induced by LSPR effect of Au NPs. The enhancement can be attributed to several factors (Figure 10.19e): First, the local electric field induced by Au NPs can promote the light absorption of TiO_2, resulting in increased generation of electron–hole pairs. Second, the local-heat generated by the oscillation of electrons in excited Au NPs may improve catalytic activity of Pt–TiO_2 due to elevated temperature. Additionally, hot electrons from Au NPs can transfer to the conduction band of TiO_2, leading to an increase in available electrons for catalytic reactions. The use of Au@SiO_2-loaded Pt–TiO_2 yielded similar results, with no enhancement of the photocatalytic activity under 365 nm light irradiation, but a significant increase under 530 nm light irradiation. The enhancement was attributed to the local electric field induced by Au NPs due to the blocking of electron transfer by SiO_2 insulator (Figure 10.19f). The comparison of catalytic rates in Figure 10.19d also indicated that the enhancement from Au@SiO_2 was greater than that from Au NPs, suggesting that PIRET effect played a more important than DET effect in the photocatalytic CO_2 conversion reaction. It is worth noting that both Au NPs and Au@SiO_2 exhibited size-dependent plasmonic effects on the enhancement of CO_2 reduction reaction [40].

10.5 Summary

Plasmonic metal NPs are utilized to enhance the photocatalytic activities of TiO_2 through three mechanisms, DET between metal and semiconductor, energy transfer from metal to semiconductor due to the LSPR induced electromagnetic field, and scattering effect from metal NPs. The predominant enhancement effect may be contingent upon the combined configuration of metal and semiconductor. This chapter provides detailed illustration for synthesis approaches of plasmonic metal–TiO_2 nanocomposites, such as sol–gel method for hydrolyzing TiO_2 on the surface of plasmonic metal NPs, deposition metal NPs on TiO_2 by reduction, attaching plasmonic metal NPs to TiO_2 NPs by linkers, and loading single metal atom on the surface of TiO_2. Furthermore, applications of plasmonic metal–TiO_2 photocatalysts in photodegradation, hydrogen evaluation, and CO_2 reduction have also been summarized.

References

1 Chou, C.-H. and Chen, F.-C. (2014). Plasmonic nanostructures for light trapping in organic photovoltaic devices. *Nanoscale* 6 (15): 8444–8458. https://doi.org/10.1039/C4NR02191F.

2 Zhang, X., Kong, X., Lv, Z. et al. (2013). Bifunctional quantum dot-decorated Ag@ SiO_2 nanostructures for simultaneous immunoassays of surface-enhanced Raman scattering (SERS) and surface-enhanced fluorescence (SEF). *Journal of Materials Chemistry B* 1 (16): 2198–2204. https://doi.org/10.1039/c3tb20069h.

3 Liao, C., Tang, L., Gao, X. et al. (2015). Bright white-light emission from Ag/SiO_2/ CdS-ZnS core/shell/shell plasmon couplers. *Nanoscale* 7 (48): 20607–20613. https://doi.org/10.1039/c5nr05526a.

4 Yuan, K., Qin, R., Yu, J. et al. (2020). Effects of localized surface plasmon resonance of Ag nanoparticles on luminescence of carbon dots with blue, green and yellow emission. *Applied Surface Science* 502: 144277. https://doi.org/10.1016/j.apsusc.2019.144277.

5 Rycenga, M., Cobley, C.M., Zeng, J. et al. (2011). Controlling the synthesis and assembly of silver nanostructures for plasmonic applications. *Chemical Reviews* 111 (6): 3669–3712. https://doi.org/10.1021/cr100275d.

6 Li, X., Cao, Y., Li, S. et al. (2020). The preparation of plasmonic Au@SiO_2 NPs and its application in polymer solar cells. *Materials Letters* 268: 127599. https://doi.org/10.1016/j.matlet.2020.127599.

7 Chen, H., Kou, X., Yang, Z. et al. (2008). Shape- and size-dependent refractive index sensitivity of gold nanoparticles. *Langmuir* 24 (10): 5233–5237. https://doi.org/10.1021/la800305j.

8 Zarick, H.F., Hurd, O., Webb, J.A. et al. (2014). Enhanced efficiency in dye-sensitized solar cells with shape-controlled plasmonic nanostructures. *ACS Photonics* 1 (9): 806–811. https://doi.org/10.1021/ph500281v.

9 Zhou, N., Polavarapu, L., Gao, N. et al. (2013). TiO_2 coated Au/Ag nanorods with enhanced photocatalytic activity under visible light irradiation. *Nanoscale* 5 (10): 4236–4241. https://doi.org/10.1039/C3NR00517H.

10 Pan, Y., Deng, S., Polavarapu, L. et al. (2012). Plasmon-enhanced photocatalytic properties of Cu_2O nanowire–Au nanoparticle assemblies. *Langmuir* 28 (33): 12304–12310. https://doi.org/10.1021/la301813v.

11 Lee, J.E., Bera, S., Choi, Y.S., and Lee, W.I. (2017). Size-dependent plasmonic effects of M and M@SiO_2 (M = Au or Ag) deposited on TiO_2 in photocatalytic oxidation reactions. *Applied Catalysis B: Environmental* 214: 15–22. https://doi.org/10.1016/j.apcatb.2017.05.025.

12 Zhou, N., López-Puente, V., Wang, Q. et al. (2015). Plasmon-enhanced light harvesting: applications in enhanced photocatalysis, photodynamic therapy and photovoltaics. *RSC Advances* 5 (37): 29076–29097. https://doi.org/10.1039/c5ra01819f.

13 Wang, B., Zhu, X., Li, S. et al. (2018). Ag@SiO_2 core-shell nanoparticles embedded in a TiO_2 mesoporous layer substantially improve the performance of perovskite solar cells. *Nanomaterials* 8 (9): 701.

14 Clavero, C. (2014). Plasmon-induced hot-electron generation at nanoparticle/metal-oxide interfaces for photovoltaic and photocatalytic devices. *Nature Photonics* 8 (2): 95–103. https://doi.org/10.1038/nphoton.2013.238.

15 Hirakawa, T. and Kamat, P.V. (2005). Charge separation and catalytic activity of Ag@TiO_2 core–shell composite clusters under UV–irradiation. *Journal of the*

American Chemical Society 127 (11): 3928–3934. https://doi.org/10.1021/ja042925a.

16 Liu, Z., Hou, W., Pavaskar, P. et al. (2011). Plasmon resonant enhancement of photocatalytic water splitting under visible illumination. *Nano Letters* 11 (3): 1111–1116. https://doi.org/10.1021/nl104005n.

17 Ingram, D.B. and Linic, S. (2011). Water splitting on composite plasmonic-metal/semiconductor photoelectrodes: evidence for selective plasmon-induced formation of charge carriers near the semiconductor surface. *Journal of the American Chemical Society* 133 (14): 5202–5205. https://doi.org/10.1021/ja200086g.

18 Talebi, H. and Emami, F. (2022). Broadband plasmonic absorption enhancement of perovskite solar cells with embedded Au@SiO_2@graphene core–shell nanoparticles. *Semiconductor Science and Technology* 37 (5): 055002. https://doi.org/10.1088/1361-6641/ac4ef5.

19 Nishijima, Y., Ueno, K., Kotake, Y. et al. (2012). Near-infrared plasmon-assisted water oxidation. *The Journal of Physical Chemistry Letters* 3 (10): 1248–1252. https://doi.org/10.1021/jz3003316.

20 Zhou, N., Zhu, H., Li, S. et al. (2018). Au nanorod/ZnO core–shell nanoparticles as nano-photosensitizers for near-infrared light-induced singlet oxygen generation. *The Journal of Physical Chemistry C* 122 (14): 7824–7830. https://doi.org/10.1021/acs.jpcc.7b12463.

21 Zhou, N., Yan, R., Wang, X. et al. (2021). Tunable thickness of mesoporous ZnO-coated metal nanoparticles for enhanced visible-light driven photoelectrochemical water splitting. *Chemosphere* 273: 129679. https://doi.org/10.1016/j.chemosphere.2021.129679.

22 Cushing, S.K. and Wu, N. (2013). Plasmon-enhanced solar energy harvesting. *The Electrochemical Society Interface* 22 (2): 63. https://doi.org/10.1149/2.F08132if.

23 Kholmicheva, N., Royo Romero, L., Cassidy, J., and Zamkov, M. (2018). Prospects and applications of plasmon-exciton interactions in the near-field regime. *Nanophotonics* 8 (4): 613–628. https://doi.org/10.1515/nanoph-2018-0143.

24 Cushing, S.K., Li, J., Meng, F. et al. (2012). Photocatalytic activity enhanced by plasmonic resonant energy transfer from metal to semiconductor. *Journal of the American Chemical Society* 134 (36): 15033–15041. https://doi.org/10.1021/ja305603t.

25 Li, J., Cushing, S.K., Meng, F. et al. (2015). Plasmon-induced resonance energy transfer for solar energy conversion. *Nature Photonics* 9 (9): 601–607. https://doi.org/10.1038/nphoton.2015.142.

26 Atwater, H.A. and Polman, A. (2010). Plasmonics for improved photovoltaic devices. *Nature Materials* 9 (3): 205–213. https://doi.org/10.1038/nmat2629.

27 Kulkarni, A.P., Noone, K.M., Munechika, K. et al. (2010). Plasmon-enhanced charge carrier generation in organic photovoltaic films using silver nanoprisms. *Nano Letters* 10 (4): 1501–1505. https://doi.org/10.1021/nl100615e.

28 Seh, Z.W., Liu, S., Low, M. et al. (2012). Janus Au-TiO_2 photocatalysts with strong localization of plasmonic near-fields for efficient visible-light hydrogen generation. *Advanced Materials* 24 (17): 2310–2314. https://doi.org/10.1002/adma.201104241.

29 Zhang, X., Zhu, Y., Yang, X. et al. (2013). Enhanced visible light photocatalytic activity of interlayer-isolated triplex Ag@SiO_2@TiO_2 core–shell nanoparticles. *Nanoscale* 5 (8): 3359–3366. https://doi.org/10.1039/c3nr00044c.

30 Ge, M.-Z., Cao, C.-Y., Li, S.-H. et al. (2016). In situ plasmonic Ag nanoparticle anchored TiO_2 nanotube arrays as visible-light-driven photocatalysts for enhanced water splitting. *Nanoscale* 8 (9): 5226–5234. https://doi.org/10.1039/c5nr08341a.

31 Shokri, M., Jodat, A., Modirshahla, N., and Behnajady, M.A. (2013). Photocatalytic degradation of chloramphenicol in an aqueous suspension of silver-doped TiO_2 nanoparticles. *Environmental Technology* 34 (9): 1161–1166. https://doi.org/10.1080/09593330.2012.743589.

32 Melvin, A.A., Illath, K., Das, T. et al. (2015). M–Au/TiO_2(M = Ag, Pd, and Pt) nanophotocatalyst for overall solar water splitting: role of interfaces. *Nanoscale* 7 (32): 13477–13488. https://doi.org/10.1039/c5nr03735b.

33 Wang, A., Wu, S., Dong, J. et al. (2021). Interfacial facet engineering on the Schottky barrier between plasmonic Au and TiO_2 in boosting the photocatalytic CO_2 reduction under ultraviolet and visible light irradiation. *Chemical Engineering Journal* 404: 127145. https://doi.org/10.1016/j.cej.2020.127145.

34 Ban, C., Wang, Y., Feng, Y. et al. (2024). Photochromic single atom Ag/TiO_2 catalysts for selective CO_2 reduction to CH_4. *Energy & Environmental Science* 17 (2): 518–530. https://doi.org/10.1039/d3ee02800c.

35 Zarepour, M.A. and Tasviri, M. (2019). Facile fabrication of Ag decorated TiO_2 nanorices: highly efficient visible-light-responsive photocatalyst in degradation of contaminants. *Journal of Photochemistry and Photobiology A: Chemistry* 371: 166–172. https://doi.org/10.1016/j.jphotochem.2018.11.007.

36 Zhang, Z., Zhang, L., Hedhili, M.N. et al. (2012). Plasmonic gold nanocrystals coupled with photonic crystal seamlessly on TiO_2 nanotube photoelectrodes for efficient visible light photoelectrochemical water splitting. *Nano Letters* 13 (1): 14–20. https://doi.org/10.1021/nl3029202.

37 Eqi, M., Shi, C., Xie, J. et al. (2022). Synergetic effect of Ni-Au bimetal nanoparticles on urchin-like TiO_2 for hydrogen and arabinose co-production by glucose photoreforming. *Advanced Composites and Hybrid Materials* 6 (1): 5. https://doi.org/10.1007/s42114-022-00580-6.

38 Tahir, M., Tahir, B., and Amin, N.A.S. (2017). Synergistic effect in plasmonic Au/Ag alloy NPs co-coated TiO_2 NWs toward visible-light enhanced CO_2 photoreduction to fuels. *Applied Catalysis B: Environmental* 204: 548–560. https://doi.org/10.1016/j.apcatb.2016.11.062.

39 Jin, L., Shaaban, E., Bamonte, S. et al. (2021). Surface basicity of metal@TiO_2 to enhance photocatalytic efficiency for CO_2 reduction. *ACS Applied Materials & Interfaces* 13 (32): 38595–38603. https://doi.org/10.1021/acsami.1c09119.

40 Bera, S., Lee, J.E., Rawal, S.B., and Lee, W.I. (2016). Size-dependent plasmonic effects of Au and Au@SiO_2 nanoparticles in photocatalytic CO_2 conversion reaction of Pt/TiO_2. *Applied Catalysis B: Environmental* 199: 55–63. https://doi.org/10.1016/j.apcatb.2016.06.025.

11

Ecotoxicological and Antibacterial Assessment of Nanostructured TiO_2

Jiani Liu[1,2], *Fahim Ullah Khan*[1,2], *Youji Wang*[1,2,*]

[1] *International Research Center for Marine Biosciences, Ministry of Science and Technology, Shanghai Ocean University, Shanghai, China*
[2] *Key Laboratory of Exploration and Utilization of Aquatic Genetic Resources, Ministry of Education, Shanghai Ocean University, Shanghai, China*

*Corresponding author: Youji Wang (youjiwang2@gmail.com)

11.1 Introduction

Nanostructured TiO_2 is a metal oxide nanomaterial with a diameter of less than 100 nm. TiO_2 is an emerging high-performance material that can be derived from natural environments such as ash, desert dust, and aerosols [1], but produced with large scale for commercialization [2]. The TiO_2 is mainly found bounded in an inert matrix of various chemical compositions in their pristine form. Gradual degradation of this matrix will result in the release of TiO_2 into the surrounding environments, which would affect the environmental safety [3]. Due to its specific physicochemical properties, including antibacterial, ultraviolet (UV) absorption, and photocatalytic properties [4], it has wide-ranged applications in various fields such as cosmetics, sunscreens, ceramics, coatings, food, environment, energy, and medical applications [5]. The increasing application of TiO_2 in consumer products has accelerated environmental, health and safety concerns globally [6–9]. The TiO_2 exposure in the environment may occur during its entire life cycle from production to use and disposal [10]. Likewise, market data suggest a similar global production growth of TiO_2 with a market share approximating US$ 15.76 billion in 2018 and a compound annual growth rate of 8.7% from 2019 to 2025 [11].

In recent years, TiO_2 has garnered significant attention due to its distinctive properties and wide-ranged applications, with one of its most crucial applications being in the realm of photocatalysis [12]. TiO_2 presents in three crystal forms (anatase, rutile, brookite), among which only the rutile and anatase forms exhibit photocatalytic properties [13]. Anatase has a higher number of industrial uses as compared to rutile and brookite. However, it is considered most toxic form of TiO_2 [4].

Nanostructured TiO_2: Synthesis, Photocatalytic Properties and Applications, First Edition.
Edited by Jia Hong Pan, Wan In Lee and Detlef W. Bahnemann

Moreover, of all the phases mentioned above, rutile is stable at high temperature (for bulk TiO_2), while anatase and brookite are common in nano-scale natural and synthetic samples. Anatase is considered to be the most stable phase when the size of the nanoparticles (NPs) is below 50 nm. The transformation from one phase to the other is dependent on the processing temperature [14]. Research indicates that the rutile crystal lattice, owing to a higher number of defects, can generate more oxygen vacancies for electron capture, thus diminishing the recombination of photocatalytically produced holes and electrons compared to the anatase form [15]. The discovery of water photo-splitting on TiO_2 electrodes by Fujishima and Honda [16] marked the inception of extensive research into the photocatalytic properties of TiO_2, generating widespread interest [17]. TiO_2 can absorb light energy, leading to the generation of electron–hole pairs that react with surrounding molecules (Figure 11.1), initiating diverse chemical reactions [18]. In most scenarios, photocatalytic reactions involving TiO_2 primarily occur in air and aqueous solutions, as the formation of radical groups hinges on the interaction of oxygen or water molecules with photo-generated electrons and holes. Under light exposure with energy equal to or exceeding the band-gap energy of n-type semiconductor TiO_2, if electron–hole pairs remain uncaptured by electron acceptors or donors during their migration process, some pairs may recombine, releasing energy in the form of heat or other forms [19]. Carey et al. [20] pioneered the use of TiO_2 photocatalysts for the degradation of highly toxic polychlorinated biphenyls (PCBs), marking the initiation of TiO_2 photocatalyst applications in environmental protection. Upon exposure to UV light, TiO_2 can completely degrade most organic pollutants into harmless inorganic substances [21]. The US Environmental Protection Agency has identified nine major categories of organic pollutants (totaling 114 substances) that are effectively treatable using photocatalytic methods [22]. Numerous studies have underscored that TiO_2 not only degrades hazardous substances like phenol in wastewater and pesticides in soil but also demonstrates photocatalytic effects in sewage treatment, air purification, and self-cleaning

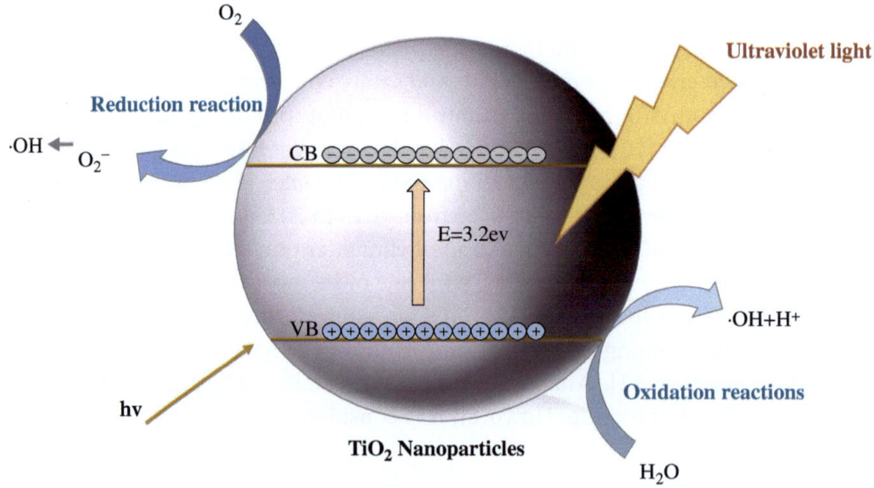

Figure 11.1 Mechanism of TiO_2 photocatalysis.

applications [23]. Moreover, it has utility in photocatalytic coatings for air purification, solar fuel generation, cosmetics, etc., showcasing its broad-ranging applications in environmental protection [24].

The antibacterial properties of TiO_2 have garnered much attention in recent years. These properties are influenced by intrinsic factors such as morphology, size, chemical composition, source, and nanostructure of NPs [25, 26]. TiO_2 exhibits remarkable antibacterial activity distinct from its bulk properties [27, 28]. It was reported that sunlight irradiation, humic acid treatment, and their dynamic interactions all help in the significant cytotoxic effects of TiO_2 on bacterial assemblages [29]. Due to their photosensitive nature, TiO_2 could further inhibit cell viability of bacterial assemblages under light condition in comparison with dark condition [29]. Specifically, the antibacterial effectiveness of TiO_2 largely hinges on its photocatalytic performance, which is significantly impacted by its morphology, structure, and texture characteristics [30]. The production of more $OH^{\cdot}$ radicals in photocatalytic reactions can profoundly impact bacterial cell walls and membranes [31, 32]. Additionally, TiO_2 finds application in specific personal care products like cosmetics and sunscreens due to its antibacterial and UV-absorbing properties [33]. In such products, it plays a role in preventing microbial contamination and growth. Coatings or filters containing TiO_2 can exhibit antibacterial properties by capturing and deactivating bacteria, viruses, and other pathogens [34].

Ecotoxicology is the scientific discipline that investigates the toxicity and potential impacts of chemicals on organisms within an ecosystem. When examining the ecotoxicology of TiO_2, the primary focus lies in its effects on biodiversity, species interactions, and ecological functions in the environment [35]. However, the extensive utilization of TiO_2 has raised concerns regarding its potential impacts on both human health and environment [36]. The toxic mechanism of TiO_2 involves multiple pathways, including physical interactions, oxidative stress, inflammation, genotoxicity, and cell death pathways [37, 38]. At the cellular level, TiO_2 can disrupt cell structure and signal transduction pathways through physical interactions, leading to cellular dysfunction or cell death [39, 40]. Exposure to TiO_2 at the tissue level may trigger inflammatory responses, resulting in oxidative damage and structural alterations in organs, particularly affecting organs such as the lungs, liver, and kidneys [41, 42]. Additionally, TiO_2 may induce organ-specific toxicity, such as pulmonary fibrosis [43, 44], hepatotoxicity [42], nephrotoxicity [45, 46], and neurotoxicity [47]. Chronic exposure to TiO_2 can potentially lead to immunosuppression, endocrine disruption, and carcinogenicity, thereby impacting overall health [48–50]. Moreover, exposure to TiO_2 may raise concerns about developmental and reproductive toxicity [51, 52]. Furthermore, TiO_2 exhibits ecotoxicity [53], potentially causing adverse effects on aquatic organisms, soil microorganisms, and plants [53–55]. Previous studies demonstrated that TiO_2 can increase bioavailability and bioaccumulation of heavy metal in various biological systems (such as mammalian cells, zebrafish, and carb) or can even exert synergistic genotoxicity effects in interaction with heavy metals [56–58]. This indicates indirect genotoxic effects of TiO_2 that may occur in biological systems, which require further research investigation to ensure the safe utilization of TiO_2 [59].

In order to thoroughly evaluate the ecological risks, photocatalytic properties, and antibacterial capabilities of TiO_2, it is crucial to conduct detailed research into its behavior, interactions, and potential toxicity within the environment. This approach will enhance our understanding of TiO_2's potential impacts on ecosystems, thus ensuring its sustainable and responsible utilization.

11.2 Ecotoxicological Effects of TiO_2

11.2.1 Toxicity Mechanisms and Modes of Action of TiO_2

In recent years, significant strides have been made in ecotoxicology research concerning TiO_2, particularly in unraveling its toxicity mechanisms and modes of action (Figure 11.2). The toxicity of TiO_2 on organisms can be condensed into three main aspects: First, under light conditions, TiO_2 generates reactive oxygen species (ROS) due to electron–hole pair induction, which possess potent oxidative capabilities, capable of damaging cell membranes, proteins, and DNA, leading to intracellular oxidative stress responses and eventual cell death [60, 61]. Second, owing

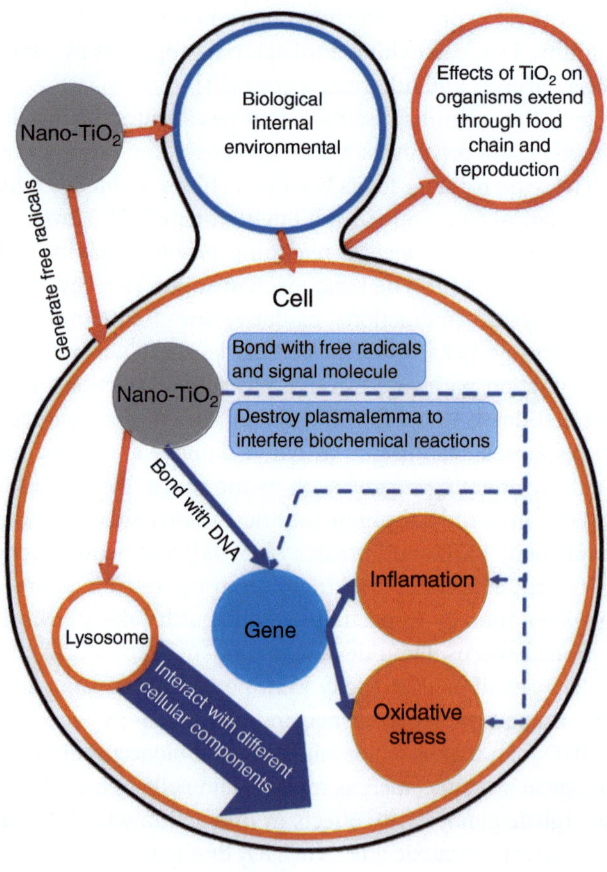

Figure 11.2 The likely adverse outcome pathways underlying the effects of TiO_2. The dotted lines represent suspected pathway. The solid lines represent demonstrated pathways. Luo et al., 2020/John Wiley & Sons.

to its substantial surface area, TiO_2 can adhere to cells through electrostatic forces, causing cell wall damage and lipid peroxidation of cell membranes [62]. Third, TiO_2 can directly compromise cell membrane integrity through mechanical actions, leading to cytoplasm leakage and subsequent attachment of TiO_2 to intracellular organelles and biomolecules, thereby interfering with their normal functions [63, 64]. Additionally, several studies have explored other toxicity mechanisms of TiO_2. For instance, TiO_2 may also impact cell signaling pathways, gene expression, and protein synthesis, further amplifying its toxic effects [65]. Concentration dependent genotoxic effects of TiO_2 have been reported [66]. Moreover, TiO_2 with different morphologies were reported to induce epigenomic toxicity, resulting in decreased global DNA methylation and altered expression levels of methylation-related genes and proteins [67]. In summary, advancements in the ecotoxicology research of TiO_2 highlight its toxicity mechanisms, primarily revolving around high surface activity, oxidative stress, cellular uptake and accumulation, and genotoxicity. The ongoing exploration of TiO_2 toxicity mechanisms and modes of action is deepening, providing crucial references and guidance for better comprehension and evaluation of the impacts of nanosized titanium dioxide on the environment and organisms.

11.2.2 The Toxicity Impact of TiO_2 on Aquatic Organisms

The products containing TiO_2 are frequently introduced to various aquatic bodies (such as river and sea), causing a significant volume of TiO_2 to enter aquatic ecosystems. The photocatalytic nature of TiO_2 would act as a significant redox agent in the environment that might have detrimental effects on biological systems [69]. Release of TiO_2 into water could exert adverse effects on overall ecosystem health, especially when the concentration is large enough to have some detrimental effect on aquatic animals and that it readily accumulates in drinking water [70]. TiO_2 exerts pronounced toxic effects on aquatic organisms, with particular emphasis on its impact on fish within the field of ecotoxicology. First, regarding behavior and physiology, exposure to TiO_2 can induce aberrant behaviors in fish. Studies demonstrate significant alterations in swimming behaviors of zebrafish larvae following TiO_2 exposure, including changes in average speed, maximum speed, and activity level [71]. Second, TiO_2 influences the growth and development of fish. Research underscores the notable toxicity of TiO_2 on zebrafish embryos, resulting in abnormal embryo development and growth inhibition. Exposure to TiO_2 diminishes the survival rate of zebrafish embryos, retards development, and may even induce organ malformations [71]. Moreover, zebrafish larvae exposed to TiO_2 demonstrate premature hatching, while organisms subjected to UV radiation exhibit impaired development [72]. TiO_2 is also implicated in organ damage in fish. Studies indicate that TiO_2 presence may impair fish liver cells, disrupting their physiological functions and causing oxidative stress-induced cellular damage [73]. Carp exposure to TiO_2 leads to histopathological alterations in liver and gill tissues [73]. At the biochemical level, TiO_2 exposure can modulate the activity of specific enzymatic reactions. Yeo and Kang [74] observed that zebrafish exposure to TiO_2 not only induces embryo abnormalities but also enhances the activity of enzymes such as glutathione reductase, glutathione S-transferase, and catalase.

Research has demonstrated that TiO_2 may pose potential toxic effects on shrimp, especially those inhabiting freshwater or marine ecosystems (Figure 11.3). Exposure to elevated levels of TiO_2 can lead to reduced behavioral activity, slowed growth rates, and potential reproductive complications in shrimp [75]. Additionally, TiO_2 hampers shrimp growth, disrupts tissue structure and the expression of immune-related genes, and induces dysbiosis in the gut microbiota [75]. Moreover, bivalves in marine environments are also susceptible to varying degrees of toxicity from TiO_2 exposure (Figure 11.3). High concentrations of TiO_2 can cause multiple adverse effects on bivalves. Physiologically, bivalves may experience respiratory system disturbances following TiO_2 exposure, resulting in altered respiratory rates [76, 77]. Furthermore, feeding behavior can be affected, with TiO_2 exposure leading to reduced feeding rates or altered feeding behavior in bivalves [78]. These behavioral changes may negatively impact bivalve growth, leading to decreased growth or developmental irregularities [79]. Additionally, TiO_2 exposure can influence bivalve behavior, inhibiting the attachment strength of bivalve foot threads and reducing gene expression levels encoding foot proteins, resulting in sluggish or unstable swinging movements. Such behavioral alterations may disrupt normal bivalve activities and foraging behaviors [63]. Furthermore, TiO_2 exposure may affect bivalve reproduction. Studies suggest that exposure to high TiO_2 concentrations can cause reproductive issues in bivalves, including significantly reduced fertilization success rates. This exposure leads to notably slower sperm swimming speeds, limits gamete collisions, significantly reduces gamete fusion, and exacerbates gamete DNA damage [80]. Additionally, TiO_2 exhibits immunotoxicity toward bivalves, as evidenced by in vitro experiments

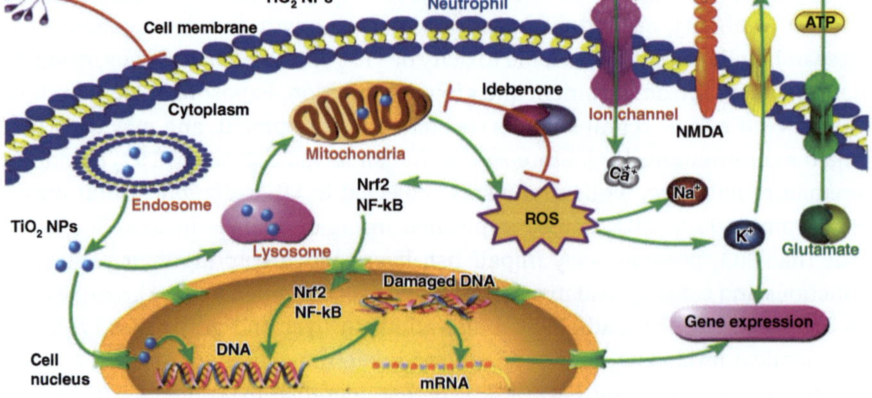

Figure 11.3 Schematic diagram of the toxicity of TiO_2 to invertebrates and vertebrates. Hou et al., 2019/with permission of Elsevier.

with TiO_2 suspensions stimulating immune and inflammatory responses in bivalve hemocytes [81].

11.2.3 The Toxic Impact of TiO_2 on Algae and Plants

The impact of TiO_2 on algae is a significant concern in ecological studies (Figure 11.4). Studies have shown that TiO_2 inhibits the photosynthesis and respiration of freshwater green algae [82], which affects their cellular pigments and subsequently influences their growth and metabolic activity [83]. This inhibition can lead to oxidative stress in cell membranes [84], with substantial implications for aquatic ecosystems [63]. For instance, when TiO_2 concentration exceeds 1 mg/L, the photosynthetic activity of Volvox algae significantly decreases [85]. Furthermore, TiO_2 may impact algal growth rate. Studies propose that the encapsulation or adsorption of TiO_2 can result in inadequate light exposure for algal cells, leading to growth inhibition [86], and exposure to low concentrations of TiO_2 can reduce algal population density [87]. Beyond its effects on photosynthetic activity and growth rate, TiO_2 may also alter the cell structure and metabolism of algae. Some studies indicate that TiO_2 particles cause cell wall damage in algae [88], influencing critical physiological processes like nutrient absorption [89]. Additionally, TiO_2 can induce changes in carbon allocation, leading to a decrease in the proportion of carbohydrates and lipids [90].

$nTiO_2$ displays varying levels of toxicity across different plant species (Figure 11.4). Exposure to elevated concentrations of TiO_2 can result in leaf shedding, decreased chlorophyll levels, and reduced growth rates in plants, thereby impacting their

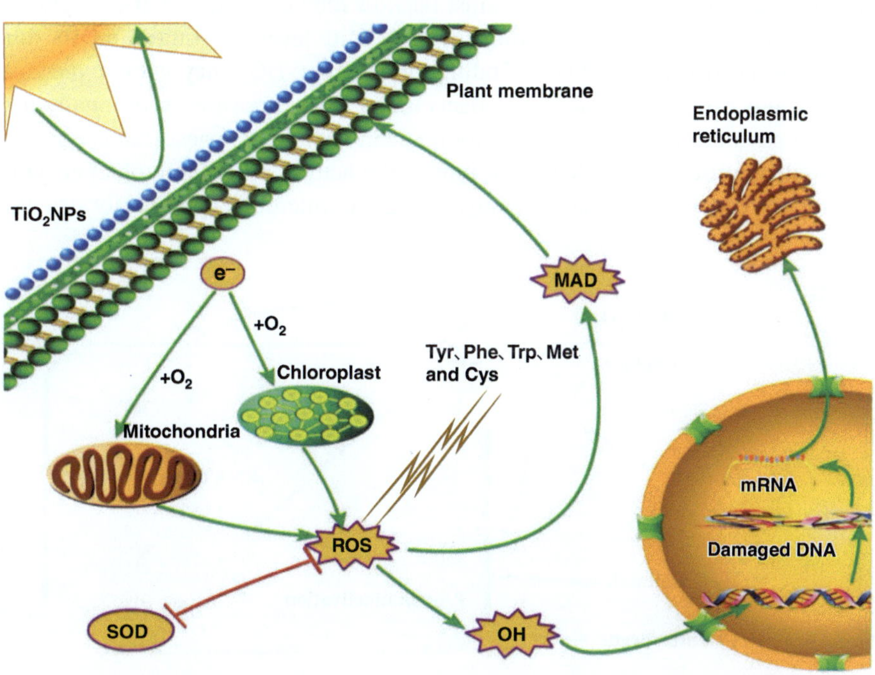

Figure 11.4 Schematic diagram of the toxicity of TiO_2 to algae and plants. [62]/with permission of Elsevier.

photosynthesis and growth processes [62]. Additionally, TiO_2 particles measuring 100 nm induce growth inhibition and DNA damage in onion roots, along with DNA damage in tobacco leaves [91]. The toxic mechanisms of TiO_2 on algae and plants seem to involve ROS and free radicals that directly harm chlorophyll molecules, leading to lipid peroxidation in cell membranes, elevated levels of malondialdehyde, decreased chlorophyll a content, reduced photosynthetic efficiency, and diminished superoxide dismutase activity. This process exacerbates lipid peroxidation and membrane harm [92].

11.2.4 The Toxicity Impact of TiO_2 on Microorganisms

$nTiO_2$ have been observed to exert toxic effects on various microorganisms (Figure 11.5), with *Escherichia coli* showing particular sensitivity characterized by a typically low lethal concentration (LC50) [93]. The LC50, a measurement unit indicating the concentration at which half of the test organisms die within a specified period (typically 48 h), serves as a gauge for substance toxicity. High sensitivity implies that *E. coli* may experience significant effects, potentially resulting in mortality, even at lower concentrations of TiO_2. The bactericidal mechanism of TiO_2 involves interacting with the bacterial surface, leading to membrane damage and impaired metabolic functions [94]. Moreover, the toxicity of TiO_2 can be influenced by environmental factors. For example, exposure to light can trigger the generation of ROS by TiO_2, thereby intensifying its harmful impact on bacteria, whereas its toxicity may diminish under dark conditions [95, 96]. TiO_2 NPs display heightened antibacterial activity against *Bacillus subtilis* and *E. coli* under light conditions compared to dark conditions, with activity levels 2.5 times and 1.8 times higher, respectively [97]. These findings imply that TiO_2 may exhibit increased toxicity and stronger antibacterial effects under light exposure. Additionally, TiO_2 has the potential to impact soil microorganisms. Studies suggest that exposure to elevated concentrations of TiO_2 can inhibit the activity of soil microbial enzymes, leading to reduced microbial diversity, changes in microbial community structure,

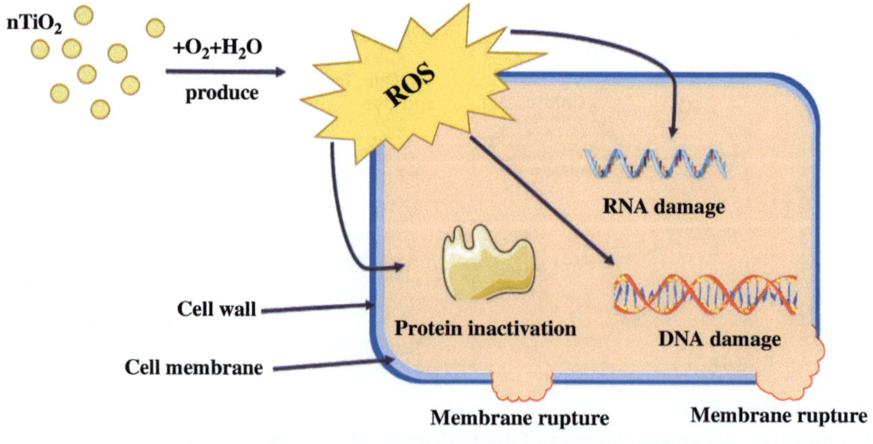

Figure 11.5 Schematic diagram of the toxicity of TiO_2 to microorganisms.

and potential disruptions to beneficial microbial populations in the soil [98]. These effects can further influence nutrient cycling in the soil and the overall stability of ecosystems.

11.2.5 The Toxicity Impact of TiO_2 on Mice and Humans

$nTiO_2$ also demonstrates toxicity in model organisms like mice. Studies have revealed that prolonged exposure to TiO_2 can induce immune system instability in mice, leading to cytokine imbalance or heightened inflammatory responses [99]. Moreover, TiO_2 is capable of triggering neurotoxicity in mice by stimulating ROS production in brain microglial cells, resulting in rapid neuronal damage within complex brain tissue [47]. Additionally, TiO_2 exposure may impact the reproductive system of mice [100]. Some investigations have reported abnormal structures or functional disorders in the reproductive organs of mice exposed to high doses of TiO_2, causing reduced reproductive capacity or breeding difficulties [101]. Furthermore, TiO_2 can traverse the placental barrier in pregnant mice, leading to neurotoxicity in their offspring [102]. Exposure to TiO_2 inhibits rat placental and fetal development on GD 18 (Gestation Day 18), resulting in elevated levels of THs (T3 and T4) in the placenta and impaired placental function [103]. Additionally, TiO_2 exhibits other toxic effects in mice, including increased interleukin (IL) 6 and tumor necrosis factor (TNF) α levels in the liver, along with elevated nuclear factor (NF) κB mRNA levels [102]. Shukla et al. [104] observed significant DNA damage in mouse liver cells due to heavy accumulation of TiO_2, causing DNA damage and cell apoptosis through intrinsic pathways.

$nTiO_2$ presents potential risks to human health. It is commonly used as a food additive, with studies detecting the deposition of TiO_2 in postmortem spleen samples [105]. This suggests that TiO_2 can enter and accumulate in the human body through various exposure routes, including the digestive system. Exposure to Ag/N-TiO_2 NPs can trigger increased production of ROS, activate the ERK1/2 MAP kinase pathway, and release pro-inflammatory mediators such as CXCL1, GM-CSF, and MIF in human lung epithelial cells, thereby exacerbating lung inflammation [106]. In human HepG2 liver cells, TiO_2 could induce a concentration-dependent increase in oxidative DNA damage and micronucleus formation, which was linked with increased ROS generation, and oxidative stress [66]. Both anatase and rutile crystalline forms of TiO_2 induced intracellular ROS generation and DNA strand break in HepG2 cells [66, 107, 108]. However, both forms induced DNA damage to a different extent, that is, anatase form resulted in severe DNA strand break, while rutile form stimulated mRNA expression of *p21, gadd45a*, and *mdm2* genes that are linked with the DNA repair mechanism [107]. Moreover, TiO_2 can traverse the blood–testis barrier, leading to inflammation, cytotoxicity, and alterations in gene expression that contribute to damage in the male reproductive system [109]. Studies focusing on human cells demonstrate that TiO_2 has the capability to penetrate cell membranes, enter cells, and disrupt their structure and function, potentially causing cell damage, death, and other adverse effects. Therefore, further investigation is necessary to fully comprehend the potential health risks associated with TiO_2 exposure [59].

11.2.6 The Antidotal Effects of TiO_2

Additionally, TiO_2 can impact the biotoxicity of various substances. Hartmann et al. [110] conducted a comparative study on the effects of 10, 300 nm, and P25 rutile-type TiO_2 on Cd^{2+} toxicity in *Chlamydomonas reinhardtii*, revealing that all three types of TiO_2 reduced the biotoxic inhibition of Cd^{2+} on *C. reinhardtii*. Deng et al. [111] observed that high concentrations of TiO_2 could alleviate the toxic effects of Tris (2-chloropropyl) phosphate (TCPP) on the respiratory metabolism and gut microbiota of bivalves. Furthermore, TiO_2 has a promotive effect on plant growth TiO_2 has been found to enhance photosynthesis and nitrogen metabolism in spinach, thereby promoting its growth [112]. Shyla and Natarajan [113] observed that TiO_2, under visible light and UV radiation, could increase the activity of superoxide dismutase in germinating soybeans, thereby enhancing their antioxidant capacity and promoting seed germination and seedling growth.

11.2.7 The Accumulation of TiO_2 in Organisms

The prolonged presence of TiO_2 in the environment can result in its bioaccumulation within organisms. This bioaccumulation is facilitated through biomagnification along the food chain, contributing to the overall increase of TiO_2 levels in the environment and amplifying its enduring impact on ecosystems. Research has highlighted that aquatic organisms like algae and plankton ingest TiO_2, entering the food chain [114, 115], where it progressively accumulates and eventually affects the well-being of higher trophic level organisms [116]. The transference of TiO_2 within aquatic ecosystems has spurred concerns. Some studies indicate the absence of biomagnification of TiO_2 within aquatic food chains, evidenced by no observed biomagnification in the transfer from copepods to zebrafish ($BMF = 0.009 - 0.024$) [117], as well as its transformation from sandworms to juvenile flounder ($BMF = 0.30 - 0.33$) [118]. Zhu et al. [117] demonstrated that TiO_2 can indeed move from copepods to zebrafish through dietary intake, highlighting dietary exposure as the primary route for high trophic level aquatic organisms. The enduring ecological ramifications of TiO_2 encompass its accumulation within environmental organisms, potentially disrupting ecosystem stability and posing risks to population viability and ecological integrity [119].

11.3 The Antimicrobial Properties of TiO_2

11.3.1 The Antimicrobial Mechanisms of TiO_2

The potent antibacterial properties of TiO_2 make it highly promising for a wide range of applications in medical, food, water treatment, and other sectors. Initially, TiO_2 demonstrates photocatalytic activity, causing bacterial decomposition under light exposure, leading to its antibacterial effect. The antimicrobial efficacy of TiO_2 has been documented against several pathogens such as *E. coli, Staphylococcus aureus, Pseudomonas aeruginosa*, and *Aspergillus niger* [120]. Bacteria are eradicated upon contact with TiO_2 photocatalysts under light, resulting in the generation of hydroxyl radicals (·OH) and ROS on the TiO_2 surface during illumination. These radicals deactivate microorganisms by oxidizing the unsaturated phospholipid components present on microbial cell membranes [121]. Under the influence of an electric field, electrons and holes separate, migrate to various positions on the particle surface, undergo a series of reactions, and capture dissolved oxygen on the TiO_2 surface, forming O_2 (Figure 11.5). The produced superoxide anion radicals react with numerous organic substances, including those within bacterial cells, leading to the formation of CO_2 and H_2O [122]. Concurrently, holes oxidize OH and H_2O molecules adsorbed on the TiO_2 surface into ·OH. The bactericidal potency of ·OH against *E. coli* is estimated to be approximately 1000–10 000 times higher than that of common disinfectants like chlorine, ozone, and chlorine dioxide [123, 124]. OH can initiate a chain reaction by attacking unsaturated bonds in organic substances or extracting H atoms, ultimately leading to bacterial decomposition [125, 126]. The impact of ROS on bacterial cells is multifaceted, disrupting cell wall integrity, altering macromolecular structures to reduce adhesion, significantly affecting cell integrity, gene expression, as well as impeding phosphate uptake and cell communication [127]. Titanium dioxide generates ROS under UV light, inducing cell death, and has been utilized in photodynamic therapy for treating various diseases, ranging from psoriasis to cancer. Moreover, the micro-nanostructured nature of TiO_2 offers a larger surface area, enhancing direct contact opportunities with bacteria. TiO_2 particles can penetrate bacterial cells due to their small size and high specific surface area, disrupting cell structure and function, ultimately leading to bacterial death [128]. Furthermore, TiO_2 can release a certain amount of titanium ions ($Ti4^+$) in water. When trace amounts of Ti^{4+} reach microbial cell membranes, which are negatively charged, they adhere firmly due to Coulombic attraction. Metal ions penetrate the cell wall and react with thiol groups (–SH), causing proteins to coagulate, disrupting enzyme activity, and impeding the cells' ability to divide and reproduce, ultimately resulting in cell death [129]. $Ti4^+$ ions may also disrupt microbial electron transfer systems, respiratory systems, and substance transport systems [130]. The application of TiO_2 photocatalysis in deep water treatment significantly reduces bacterial counts without causing mutagenic effects postconsumption, meeting the standards for safe drinking water [24, 131]. Incorporating TiO_2 into coatings can yield antibacterial, antifouling, deodorizing, and self-cleaning coatings. These coatings are suitable for application in bacteria-rich and easily replicating environments like hospital wards, operating rooms, and home bathrooms, effectively eliminating harmful bacteria such as *E. coli* and *S. aureus*, thus preventing infections [132].

When utilized as an antibacterial agent, TiO_2 photocatalysts also exhibit significant drawbacks, particularly the necessity for a specific UV light environment. When exposed to light, TiO_2 can produce photoinduced electrons and holes with redox capabilities [94]. As explained in the preceding photocatalytic mechanism, these photoinduced electrons can undergo reduction by oxygen, leading to the generation of potent oxidative species such as ·OH, OOH^-, and H_2O_2. These highly oxidative substances, combined with hydroxyl radicals formed due to the capture of photoinduced holes on the surface, can degrade organic matter within living organisms [125]. Consequently, TiO_2 photocatalysts also exhibit antibacterial, antifungal, and antiviral properties. Research indicates the potential for enhancing the antibacterial attributes of TiO_2 by incorporating antibacterial metals like silver, zinc, or copper into its structure [133, 134]. Such modifications can partially alleviate the dependence of TiO_2's antibacterial efficacy on light conditions.

11.3.2 The Purification and Decontamination Effects of TiO_2 in Different Environments

The antibacterial effectiveness of TiO_2 varies across different environments. In natural conditions, when exposed to UV light, TiO_2 demonstrates a potent photocatalytic effect. This effect relies on active sites present on the surface of TiO_2 to adsorb and stimulate oxygen molecules, initiating a cascade of redox reactions and producing ROS like hydroxyl radicals and superoxide radicals [135]. These ROS possess strong oxidizing properties capable of disrupting bacterial membrane structures and DNA, ultimately causing bacterial death. Furthermore, incorporating TiO_2 into disinfectants and cleaning products can enhance their antibacterial properties, effectively targeting various bacteria, viruses, and fungi, thereby enhancing the efficiency of disinfection and cleaning procedures [124].

nTiO_2 possesses water purification capabilities primarily due to the redox abilities of electrons and holes created upon light excitation. These redox processes utilize cost-effective oxidants like oxygen (O_2) from the atmosphere, enabling photocatalytic oxidation to break down both organic and inorganic pollutants in water [136]. In aquatic environments, particularly under UV light exposure, TiO_2 effectively hinders the proliferation of bacteria and other microorganisms [95]. This efficacy stems mainly from its photocatalytic activity, where generated ROS disrupt and eradicate microorganisms [137]. Moreover, TiO_2 contributes to the purification of organic compounds in water [138, 139], including halogenated derivatives, dye wastewater, pesticide wastewater, surfactants, oily wastewater, and papermaking wastewater. This purification is achieved through adsorption and photocatalytic degradation techniques targeting organic pollutants in water [140]. Grzechulska et al. [141] utilized a blend of TiO_2 and KOH postcalcination as a photocatalyst, successfully degrading oil pollution entirely under UV light within 2 h. Despite the promising potential of powdered TiO_2 photocatalysts in water treatment, there are notable challenges and unresolved issues. Notably, TiO_2 photocatalysts are wide-bandgap semiconductors that necessitate high-energy UV light sources for activation, posing a challenge in maximizing solar utilization efficiency for photocatalysts in TiO_2 catalyst research [142].

nTiO_2 finds application in indoor air purifiers. Owing to the presence of specific airborne substances like ammonia, hydrogen sulfide, methyl mercaptan, and acetaldehyde [143], TiO_2 contributes to deodorization by breaking down these organic compounds [144]. When exposed to UV light, TiO_2 coatings can decompose formaldehyde emitted from newly constructed materials and adhesives, acetaldehyde from tobacco smoke, mercaptans from household dust, as well as oils and organic surface pollutants [145]. Nano titanium dioxide employs photocatalysis to oxidize and disintegrate these adsorbed substances, resulting in decreased air concentrations and alleviation or elimination of environmental discomfort [144, 146]. Moreover, Stauff et al. [147] illustrated that the incorporation of TiO_2 into polylactic acid (PLA) to form PLA/TiO_2 composite films effectively reduces benzene levels in indoor air to 5 ppm.

11.3.3 The Antibacterial Applications of TiO_2

nTiO_2 is widely utilized across diverse fields due to its exceptional antibacterial properties. TiO_2 demonstrates a broad-spectrum antibacterial efficacy, effectively combatting microorganisms such as bacteria, viruses, and fungi [148]. TiO_2 developed in various forms (e.g., sphere, tube, cylinder, fiber, sheet, whisker, wire, and rod) have found extensive applications in advanced biomedical contexts [149]. TiO_2 plays a significant role in the production of wound dressing materials. Chronic wounds, characterized by persistent infections, often harbor pathogens like *S. aureus* capable of forming bacterial biofilms [150]. However, due to the substantial risks associated with antibiotics, such as resistance, recurrent infections, compromised healing, and delayed hypersensitivity reactions, local antibiotic application is discouraged [151]. Consequently, there is a rising trend in developing nano-enhanced polymer matrices (nanofilled polymer composites) for nano-composite skin grafts and wound dressings, avoiding the incorporation of antibiotics into biomaterial structures [152]. TiO_2 has been instrumental in the development of antibacterial skin grafts and wound dressings [153]. Research indicates that incorporating TiO_2 into polymer matrices not only enhances their mechanical properties but also exhibits antibacterial effects against various Gram-positive and Gram-negative bacteria [153]. The cytotoxicity associated with TiO_2 can be mitigated by controlling the slow release of titanium ions from composite biomaterials, thereby providing antibacterial protection and expediting wound healing [154, 155].

In recent years, heightened by outbreaks of infectious diseases, there has been a noticeable increase in public concern regarding health and safety within domestic environments. Everyday life at home poses several potential health risks, prompting the adoption of antibacterial ceramics as a viable solution. These ceramics incorporate or apply antibacterial agents on their surfaces, with TiO_2, standing out as an exceptional antibacterial material. Its application on ceramic surfaces creates an environmentally friendly and self-cleaning effect, effectively suppressing the proliferation of pathogenic bacteria to acceptable levels [156]. Consequently, antibacterial tiles find utility in critical settings like hospital restrooms and operating rooms, where they efficiently curb the spread of pathogens, thereby mitigating the risk of cross-contamination [157].

The antibacterial properties of TiO_2 have garnered significant attention in research. However, several unresolved issues and challenges persist, necessitating further exploration and resolution. Specifically, there is a need for heightened focus on the stability and durability of TiO_2 in prolonged and widespread applications. This is particularly crucial in intricate settings like organisms or outdoor environments, where understanding the potential influence of various factors on TiO_2's antibacterial efficacy and its long-term performance maintenance is paramount. Moreover, a comprehensive evaluation of the potential ecological toxicity and human health impacts associated with TiO_2 is imperative. Despite its proven antibacterial efficacy, ongoing debates and inquiries persist regarding TiO_2

accumulation in the environment, potential ecological risks, and the health hazards posthuman exposure. Therefore, future research endeavors should comprehensively tackle these challenges through rigorous experimental and theoretical studies, and gain a deeper understanding of the complex between TiO_2's antibacterial properties and the prevailing challenges, thus offering more reliable support and guidance for its practical applications.

11.4 The Impact of TiO_2 on Health

nTiO_2 produces ROS when exposed to UV light, leading to oxidative stress [60, 61]. Upon entry into the human body, TiO_2 has the potential to induce inflammation, cytotoxicity, immunotoxicity, and other effects [158]. This nanotoxicity poses a risk of harm to organs such as the liver, kidneys, lungs, and skin [159, 160]. TiO_2 particles with a diameter below 20 nm can trigger a more pronounced inflammatory response in human cells [161]. Additionally, at low concentrations, TiO_2 can act as an immunomodulator by provoking inflammatory responses through specific interactions with immune system cells [162]. The primary pathways of potential TiO_2 exposure include three routes: (i) Dermal exposure – typically through the use of cosmetics and sunscreen products; (ii) Inhalation exposure – primarily occurring in occupational and manufacturing settings; and (iii) Oral exposure – mainly through the consumption of food (Figure 11.6). With regard to the fact that TiO_2 are extensively used in many industries, the risk of exposure increases; hence, their potential effect on the human health should be explored in more detail [14].

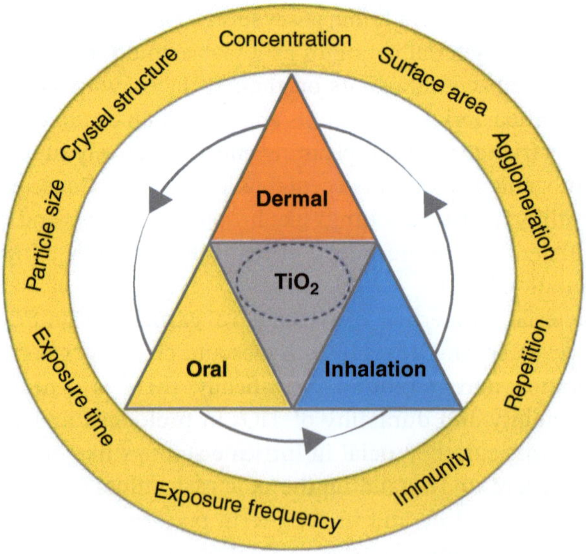

Figure 11.6 Risk management strategies of human exposure to TiO_2 NPs via different exposure routes. [159]/MDPI/CC BY 4.0.

11.4.1 The Skin Contact Risk of TiO_2

$nTiO_2$ finds widespread use in cosmetics and sunscreens, generally regarded as safe for skin application. Its mechanism involves reflecting and scattering UV light, thereby shielding the skin from UV radiation damage. The skin, with its extensive surface area, comprises the epidermis and dermis layers, where the stratum corneum forms the uppermost layer of the epidermis [163]. Under normal conditions, TiO_2 particles larger than 100 nm do not permeate skin tissue. Only a fraction of nano-sized particles may breach the stratum corneum, failing to reach viable epidermal or dermal cells. When emulsions containing TiO_2 are applied to excised human skin, they predominantly reside on the stratum corneum's surface. Similarly, in in vitro experiments using static diffusion cells on porcine skin with larger TiO_2 particles (ranging from 80 to 200 nm or up to 1 μm), no penetration was observed [164]. Schulz et al. [165] explored rutile-type TiO_2 penetration into the skin, finding some particles entering hair follicles but not reaching the dermis. Other studies concur that TiO_2 typically does not penetrate the skin [166]. Notably, applying TiO_2-containing sunscreen on live miniature pigs daily for 4 weeks revealed elevated titanium levels in the epidermis, particularly in the stratum corneum and near hair follicles, with minimal particles in suspected contaminated skin [160]. Moreover, UVB-induced skin damage did not markedly increase TiO_2 sunscreen penetration [167]. Overall, evidences suggest that TiO_2 largely remains confined to the stratum corneum and does not breach deeper skin layers or enter systemic circulation, irrespective of skin health status [168]. However, the specific risks related to skin contact with TiO_2 may vary based on particle size, concentration, and exposure conditions, necessitating further research and evaluation to ensure its safety and reliability.

11.4.2 The Inhalation Risk of TiO_2

Inhalation is considered the primary route through which NPs enter the human body [169]. The inhalation risk associated with TiO_2 primarily stems from its nanoscale properties. The lungs, comprising approximately 2300 km of airways and 300 million alveoli, are shielded by a viscous mucus layer on the airway epithelium, which acts as a relatively robust barrier [170]. Upon inhalation, TiO_2 particles are transported to various lung tissues, capillaries, airways, and alveoli, and subsequently disseminated to organs such as the heart, liver, and nervous system [171]. Due to their exceedingly small size, inhaled TiO_2 can penetrate alveoli and potentially breach the alveolar wall into the bloodstream, thereby reaching other organs [172]. It is estimated that approximately 10% of insoluble particles persist in human lungs due to their very slow clearance rate, leading to lung deposition [173]. TiO_2 particles provoke the respiratory system, potentially causing irritation to lung tissues and eliciting inflammatory responses in the respiratory tract, resulting in symptoms such as coughing, sore throat, or breathing difficulties [159]. Prolonged or high-dose exposure to TiO_2 may induce lung inflammation and fibrosis, wherein normal lung tissue is replaced by scar tissue, thereby impairing lung function [174], consequently leading to lung inflammation, fibrosis, and oxidative damage [158, 175]. Studies on lung toxicity suggest that in addition to particle size and surface area, crystal structure and surface treatment are also critical factors [176]. Research indicates that the lung toxicity of rutile-type TiO_2 surpasses that of anatase-type TiO_2 [177]. Moreover, the analysis of particles in lung tissues of workers exposed to inhalable TiO_2 revealed the presence of particles such as TiO_2, SiO_2, and other minerals, which persisted even years after exposure cessation [159]. Absorption of TiO_2 in liver, spleen, kidney, and lung tissues may lead to renal toxicity and liver damage [178]. Even after extended release, TiO_2 can continue to impact organisms in the environment through biological pathways. Furthermore, TiO_2 may exert an influence on the immune system, as some studies have reported that prolonged or high-concentration exposure may activate immune cells and trigger inflammatory responses, thereby adversely affecting immune function [179].

11.4.3 The Edible Risk of TiO_2

$nTiO_2$ has been classified as a Group 2B carcinogen by the International Agency for Research on Cancer, despite its widespread use as a food additive [180]. It is present in various food products such as pastries, candies, chewing gum, yogurt, and coffee creamers, with concentrations ranging from 0.164 to 2.64 $\mu g\ mg^{-1}$, as detected in chewing gum [4]. The estimated daily ingestion of TiO_2 by adults in the UK is up to 0.36 $mg\ kg^{-1}$ [181], while children who consume a significant amount of sweets may ingest TiO_2 at a rate up to 24 times higher than adults [4]. Additionally, postmortem studies on spleen samples from 84-year-old Caucasians have revealed a Ti content of 0.08 ± 0.1 mg Ti/kg, with more than 24% of particles measuring smaller than 100 nm [105], suggesting the potential for TiO_2 to enter the human body via the gastrointestinal tract and accumulate within it.

The primary mode of TiO_2 entry into the food supply is through oral exposure, where it traverses the gastrointestinal tract, enters the circulatory system, and interacts with various components such as plasma proteins, clotting factors, platelets, red blood cells, and white blood cells [182]. These interactions can potentially affect the distribution, metabolism, and excretion of TiO_2 [183]. TiO_2 exhibits a distribution pattern across different tissues and organs, with studies indicating a hierarchy of distribution as liver > kidney > spleen > lung > brain > heart [184]. However, contrasting findings suggest a different trend, where TiO_2 accumulates predominantly in the spleen, followed by the liver, kidney, and lungs in descending order [185]. Studies employing radiolabeled TiO_2 have demonstrated its presence in various organs such as the liver, lung, kidney, brain, spleen, uterus, and skeleton seven days after a single dose was administered to rats [186]. These findings imply that prolonged oral exposure to TiO_2 may lead to its accumulation in specific organs, thus posing potential risks to individuals exposed to TiO_2 orally [168]. Upon oral ingestion, nano-sized TiO_2 particles may penetrate the gastrointestinal mucosa. Investigations using an ex vivo model of porcine oral mucosa have revealed rapid interaction between TiO_2 and the mucous layer, penetration of the oral epithelium, and impact on the physiological balance of buccal/sublingual cells within the oral cavity [187]. In vivo studies in rats have yielded varying results, indicating either extremely low absorption [186], no significant absorption [188], or negligible absorption of TiO_2 through the gastrointestinal tract [189]. While in vitro/ex vivo oral mucosal models suggest potential penetration of TiO_2, its actual absorption after oral ingestion, whether in rats or humans, appears to be minimal or even negligible. Nonetheless, it is essential to note that the concentrations used in these studies greatly surpass typical real-world exposure levels, and the actual absorption scenario in the human body remains unclear.

The impact of TiO_2 on health risks has received significant attention. It is crucial to delve deeper into the long-term accumulation of TiO_2 in the human body and its potential implications for biological longevity. Risk assessments related to the inhalation and dermal exposure pathways of TiO_2 are also necessary, particularly for individuals exposed to TiO_2 in occupational settings or daily life. Furthermore, emphasis should be placed on assessing the effects of TiO_2 on the human immune and endocrine systems, as well as the potential for immunotoxicity and endocrine disruption. This will help better evaluate the actual health risks of TiO_2 and provide a more scientific basis and guidance for its safe use across various application domains.

11.5 The Medical Potential of TiO_2

11.5.1 The Application of TiO_2 in Medical Devices and Products

The antibacterial properties of TiO_2 coated surfaces have enabled them to have applications in hospitals, considering that traditional disinfection methods are not as effective as photocatalytic approaches [190]. The use of TiO_2 in medical devices and products is extensive due to its antibacterial properties, biocompatibility, and photocatalytic activity. The escalating risk of healthcare-associated infections has prompted significant attention toward antibacterial strategies for medical devices. Coating device surfaces with antibacterial active metals is a crucial approach. TiO_2 can be integrated into the production of surgical knives, dental tools, endoscopes, and other devices [191]. Coating these tools with TiO_2 facilitates their application as antibacterial coatings or surface modifications in medical devices. This coating generates ROS that disrupt the cell walls and membranes of microorganisms, thereby inhibiting bacterial and microbial growth [192, 193], leading to antibacterial effects [194], that help mitigate the risk of healthcare-associated infections. Owing to their nanostructure composed of TiO_2 nanotube (NT) walls, TiO_2 NTs are well suited for biomedical applications, including antibacterial surface coatings [195]. TiO_2 is also suitable for the production of artificial joints and implants, owing to its antibacterial properties, excellent biocompatibility, and chemical stability [128]. Additionally, it exhibits good cell compatibility, which promotes osteoblast diffusion, proliferation, and differentiation [196, 197]. The porous structure of titanium dioxide NTs supports bone regeneration and repair, making TiO_2 suitable for incorporation into medical implant materials such as artificial hip joints and dental implants [198].

The addition of TiO_2 to medical implants can improve their biocompatibility and mechanical properties, while also mitigating inflammation around the implants. For example, the incorporation of TiO_2 into artificial hip joints can enhance their

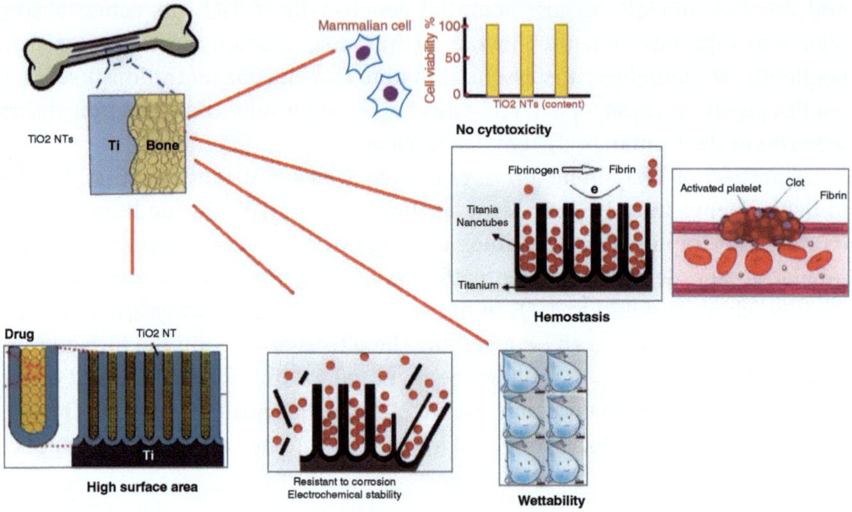

Figure 11.7 The main properties of TiO_2 nanostructure make it suitable for implant and other in vivo applications. [128]/Dove Medical Press Ltd/CC BY.

wear resistance and minimize damage to the surrounding tissues [199] (Figure 11.7). Furthermore, the osteoinductive properties of TiO_2 render it a suitable material for bone repair and regeneration, as it can stimulate the growth and differentiation of bone cells, thereby accelerating the healing of fractures [153].

Ti is extensively utilized in dentistry for dental implants owing to its exceptional chemical and mechanical properties. However, implant failure in titanium-based dental implants can occur due to infection. TiO_2 NTs not only exhibit antibacterial properties but also enhance the osteogenic activity of implants [200, 201]. Liu et al. [56] have demonstrated that in combination with zinc at appropriate concentrations, TiO_2 NTs can improve the effectiveness of implants by capitalizing on their cell compatibility and antibacterial functions. Furthermore, TiO_2 is incorporated into dental filling materials and oral care products to enhance their antimicrobial and cleaning capabilities, as TiO_2 shows inhibitory effects on bacteria responsible for causing dental caries [202]. Choi et al. [203] revealed that titanium-based dental implants coated with a multilayer TiO_2 nano-network can enhance cell adhesion, proliferation, and mobility.

$nTiO_2$ is utilized in the production of medical textiles including masks, surgical gowns, and medical bed sheets, providing antimicrobial and deodorizing functions to mitigate cross-infection and odor issues. Notably, TiO_2 fibers in masks demonstrate effective barriers against the permeation of microorganisms and particles, thereby reducing the risk of disease transmission [204]. In a study by Li et al. [205], a multifunctional textile incorporating TiO_2 and Ag NPs exhibited biocidal activity against common Gram-positive and Gram-negative bacteria, emphasizing its potential antibacterial properties. The prevention of microbial infection is a crucial consideration in the development of antimicrobial wound healing materials, while uncontrolled bleeding at wound sites significantly contributes to infections in accidents and surgical cases. Hence, the development of materials with combined hemostatic and antimicrobial properties is imperative. Research has shown that the combination of metal oxides like TiO_2 with natural polymers holds promise for addressing both infection and hemorrhage [206].

Archana et al. [154] fabricated nanocomposite films comprising TiO_2 loaded with chitosan-pectin, exhibiting high antibacterial activity against a broad range of bacteria and enhanced hemostatic capability attributed to the presence of TiO_2. Furthermore, the introduced nanocomposite demonstrated nontoxicity in both in vitro and in vivo studies.

11.5.2 The Application of TiO_2 in Drug Delivery Systems and Therapy

In recent years, various nanostructured systems have been developed for delivering chemotherapeutic drugs to effectively target tumor cells. However, most delivery systems still lack specificity in targeting tumor cells and have poorly controlled drug release processes, leading to severe side effects [128] (Figure 11.8). The utilization of nanoscale titanium dioxide in drug delivery systems primarily relies on its distinctive physicochemical characteristics. These include its high surface area, exceptional biocompatibility, and capacity to react to external stimuli like light exposure, temperature shifts, pH fluctuations, etc. These properties effectively tackle the

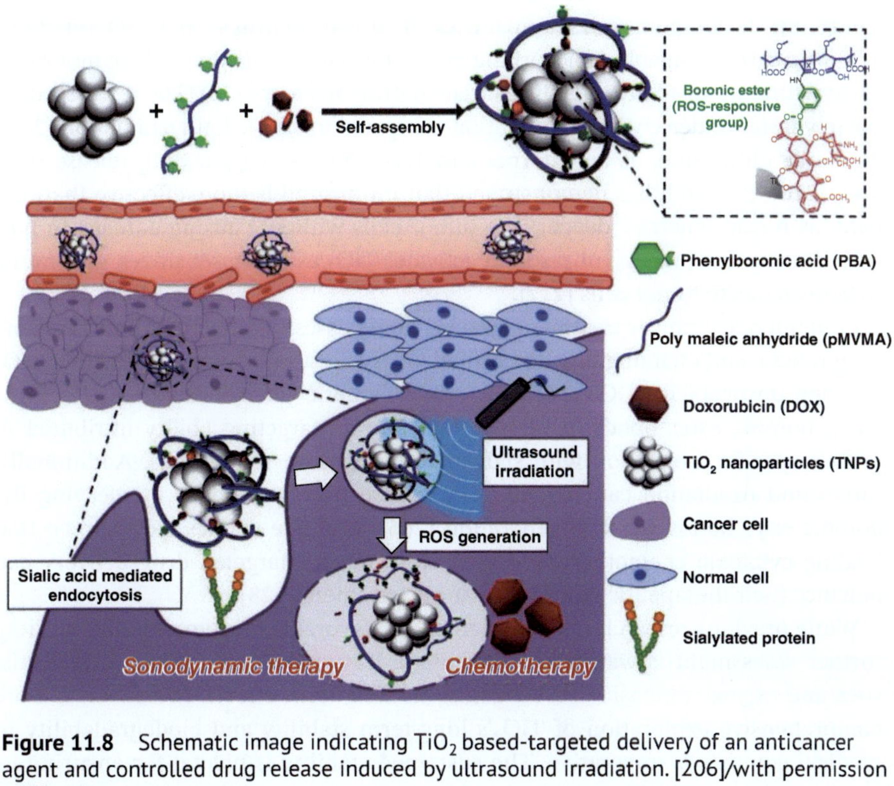

Figure 11.8 Schematic image indicating TiO_2 based-targeted delivery of an anticancer agent and controlled drug release induced by ultrasound irradiation. [206]/with permission of Elsevier.

aforementioned challenges [207]. Several studies have demonstrated that targeted drug delivery of cytotoxic chemo agents loaded onto TiO_2 could enhance the therapeutic efficiency of agents in cancer treatment [208, 209]. Furthermore, Faria and de Queiroz [210] have indicated the value of TiO_2/ZnS NTs as drug delivery systems. TiO_2 possesses unique physicochemical properties that enable precise drug delivery to diseased tissues, thereby enhancing local drug concentration and minimizing systemic side effects [211, 212].

nTiO_2 has the potential to serve as a carrier for controlled drug delivery systems. By encapsulating drugs on the surface or within TiO_2, a controlled and slow release of the drug can be achieved. This system of controlled release offers a more stable and extended drug delivery, leading to dose reduction and less frequent drug administration, thereby enhancing treatment efficacy and minimizing side effects [213]. Furthermore, precise control of the drug release rate can be attained by modifying the structure and surface properties of TiO_2. For instance, TiO_2 can be stimulated to release drugs in response to stimuli such as light exposure or changes in pH, facilitating timed and quantifiable drug delivery [214]. Owing to its distinctive physicochemical properties, TiO_2 can accurately target drug delivery to diseased tissues, thereby increasing local drug concentration and decreasing systemic side effects [211, 212].

The use of TiO_2 in cancer treatment has garnered significant attention. TiO_2 nanostructures are also recognized as promising photoactive drug delivery carriers. In innovative cancer treatment approaches, light serves as an external stimulus for controlling the temporal and spatial release of anticancer drugs [207]. TiO_2 functions as a drug carrier, capable of adsorbing or encapsulating anticancer drug molecules on or within its structure, enabling targeted drug delivery [215]. The release rate is notably faster under UV light irradiation compared to visible light irradiation [207]. Similarly, ultrasound, as an external stimulus, can also trigger drug release from TiO_2 [216]. Research has demonstrated that ultrasound is more effective than UV light, as it can penetrate deeper into tumor cells without causing damage to host cells. During this process, ultrasound activates TiO_2 as a sonosensitizer, generating ROS to eradicate tumor cells [217].

Contemporary cancer treatment systems now possess a dual function, managing drug release and enabling targeted drug delivery. Research has indicated that TiO_2 carrying doxorubicin (DOX), coated with polymeric phenylboronic acid (pPBA) using boronic ester bonds, exhibit notable tumor-targeting ability attributed to pPBA's specific interaction with sialylated epitopes on tumor cells. Additionally, ultrasound irradiation can trigger DOX release from these NPs by cleaving the boronic ester bonds via ROS generation [206]. Multiple studies have shown that loading cytotoxic chemotherapy drugs onto TiO_2 for targeted drug delivery can enhance their therapeutic efficacy in cancer treatment [128].

While previous research has highlighted the favorable biocompatibility of TiO_2, further assessment is warranted to evaluate its compatibility across diverse tissues and organs, alongside investigating its long-term effects. Moreover, a more comprehensive exploration of TiO_2's long-term stability and biodegradability in medical contexts is imperative. The sustained stability is pivotal for ensuring its persistence and reliability within drug delivery systems, while its biodegradability relates to its metabolism and elimination within the human body. Additionally, a thorough evaluation and validation of TiO_2's efficacy in medical imaging, drug delivery, and therapeutic applications are essential. Despite indications of potential benefits in these domains, rigorous clinical experiments and observations are necessary to ascertain its real-world application effects and safety. Delving deeper into these aspects will unlock the full potential of TiO_2 in medical settings, offering a more scientifically grounded approach and guidance for its safe and effective utilization.

11.5.3 The Application of TiO_2 in Tissue Engineering

Bone is one of the most commonly transplanted tissues. There are three primary approaches to address bone defects. The first approach is autologous bone grafting, which entails harvesting bone from the patient's own body. However, this method has limitations including short-term instability, limited supply, and an inability to address small defects. The second approach, allograft bone transplantation, involves transplanting bone from a donor and partially mitigates the drawbacks of autologous bone grafting. Nonetheless, this method exhibits a high clinical failure rate of 30–60% within a 10-year period. The third approach involves the use of synthetic implants and engineered bone tissue to replace damaged tissue, but it encounters challenges such as inadequate mechanical strength and load-bearing capacity [218, 219]. TiO_2 has been utilized as a scaffold material in bone tissue engineering due to its outstanding biocompatibility and bioactivity, which promote the adhesion, proliferation, and differentiation of bone cells, thus facilitating bone tissue regeneration and repair [219]. Additionally, TiO_2 scaffolds offer excellent structural support and mechanical properties, contributing to the reconstruction of damaged bone tissue. Hou et al. [220] conducted a study on Ag-doped TiO_2 NTs, demonstrating that the surface modification of Ti-based implants confers antibacterial properties to TiO_2, thereby reducing bacterial adhesion on the surface of titanium dioxide NTs.

The nanoporous morphology features of TiO_2 nanomaterials have enabled their successful integration as implants in medical applications, addressing specific limitations of titanium implants. Titanium and its various alloys have served as the primary materials in bone implants due to their high specific strength, satisfactory stability, low elastic modulus that aligns well with the tissue's elastic modulus, and their capability to form a thin and stable oxide (TiO_2) layer on the surface, providing corrosion resistance [221].

The skin acts as the body's primary defense against constant exposure to the external environment, playing a crucial role in preventing injury and damage [222]. Over the years, wound healing has become a major clinical concern globally due to the prolonged time required for the repair and regeneration of injured tissues, as well as the risks of bleeding, infection, and sepsis associated with wounds [223]. TiO_2 has garnered significant attention as a typical inorganic antibacterial biomaterial, attributed to its potent and broad-spectrum antibacterial activity [224]. The high specific surface area and active sites of TiO_2 facilitate the adsorption of proteins and the adhesion of cells, making it suitable for preparing wound dressings with antibacterial and wound healing properties. This can aid in preventing wound infections and promoting the wound healing process [128]. TiO_2 can inhibit bacterial growth and contribute to creating a hygienic treatment environment, thereby promoting wound healing through cell adhesion and proliferation [149], as well as enhancing cell proliferation and differentiation [225]. Additionally, TiO_2 can serve as a drug delivery carrier for promoting wound healing through the controlled release of bioactive substances, thereby facilitating tissue regeneration [226]. Furthermore, TiO_2 demonstrates excellent biocompatibility, making it an ideal candidate for enhancing fillers in polymer matrices [227].

While there has been some progress in utilizing TiO_2 for tissue engineering, significant challenges and unresolved issues still exist. It is imperative to conduct further comprehensive research to address these concerns. Specifically, a more thorough assessment of TiO_2's biocompatibility with human tissues is required, encompassing its impact on various tissue types and potential adverse reactions resulting from prolonged exposure. Additionally, investigating the long-term stability of TiO_2 within tissue engineering contexts is crucial to ensure its continued efficacy and reliability. Furthermore, gaining a deeper insight into the interaction mechanisms between TiO_2 and living organisms, as well as assessing its potential ecological impacts, is essential. By delving into these aspects, a more comprehensive understanding of TiO_2's true potential in tissue engineering can be achieved, thereby providing a scientific foundation and guiding its further advancement in this field.

11.6 The Use of Environment Friendly TiO_2

With the widespread utilization of nanomaterials, there is a growing concern regarding their potential implications for environmental and biological systems. TiO_2, being a form of nanomaterial, can endure in the environment for prolonged periods, exerting various effects on ecosystems. These effects may include influencing the growth of aquatic organisms, disrupting food chain dynamics, and contributing to soil pollution. Moreover, nanomaterials can pose adverse impacts on human health, potentially leading to health risks through inhalation or dermal exposure. Therefore, advocating for the adoption of environmentally friendly TiO_2 is imperative. Environmental-friendly TiO_2 denotes nanotitanium materials characterized by reduced environmental and biological toxicity, with the objective of mitigating potential adverse effects on the environment and human health. The utilization of ecologically responsible nano titanium dioxide entails the utilization of green synthesis methodologies and sustainable production practices to minimize environmental footprints and resource depletion.

Green Synthesis Methods: Implementing environmentally friendly synthesis methods is crucial for ensuring eco-responsibility. Various natural biological resources, including living plants [228], plant products, plant extracts, algae, fungi, yeast [229], bacteria [230], viruses, and more, are utilized for synthesizing NPs. Among these resources, plant-based materials are considered the most suitable for large-scale green NP synthesis due to their simplicity and ecotoxicology safety. In comparison to microorganisms, plant extracts facilitate a faster rate of metal ion reduction and yield stable particles [231]. Plants contain biomolecules such as phenolic compounds, nitrogen compounds, terpenoids, and other metabolites. The hydroxyl and carboxyl groups in these bioactive compounds act as stabilizers and reducing agents due to their high antioxidant activity [30]. Hence, plant extracts are regarded as one of the best green alternatives for metal oxide NP synthesis [232].

Principles of Circular Economy: The concept of circular economy aims to minimize waste and resource consumption, based on the principles of reduction, reuse, and recycling (commonly known as the 3R principle). When producing TiO_2, it is important to embrace the principles of the circular economy, encompassing waste reduction, efficient resource utilization, and waste recycling. Optimization of the production process should prioritize minimizing waste generation and utilizing by-products as raw materials for other products.

Energy and Water Resource Management: Efficient management of energy and water resources is crucial in TiO_2 production. This can be achieved through the adoption of energy-saving techniques and water recycling technologies to minimize the utilization of finite resources.

Environmental Monitoring and Assessment: Thorough environmental monitoring and assessment are essential throughout the production and utilization of TiO_2, encompassing the monitoring of pollutant emissions, assessment of by-product generation, and the conduction of a life cycle analysis of the product. Subsequent to the assessment results, essential measures need to be implemented to enhance the production process and minimize environmental impact.

Compliance With National and International Standards: Adherence to relevant national and international standards and regulations, including green chemistry and sustainable development standards, is essential to ensure that the production and usage of TiO_2 meet environmental and health safety requirements.

11.7 Conclusion

Due to its unique physicochemical properties, TiO_2 holds significant promise for applications in the medical field. Its exceptional antibacterial properties, photocatalytic activity, and potential as a drug carrier make it suitable for use in disinfection, photodynamic therapy, targeted drug delivery, and biomedical imaging. However, before these potential applications can be realized, several challenges must be addressed. First, critical considerations include assessing the biocompatibility and biodegradability of TiO_2, ensuring the stability and nontoxic nature of NPs in biological systems, as well as their post-functional safe degradation. This is essential to avoid potential long-term accumulation and associated toxicity issues. Second, a thorough evaluation of TiO_2's cytotoxicity and potential inflammatory reactions is necessary to guarantee its safety in medical contexts. In the realm of photodynamic therapy, while TiO_2 shows promise in therapeutic effects, further research is needed to elucidate its mechanisms in generating ROS within the body, understand ROS's impact on surrounding cells and tissues, and optimize photocatalytic efficiency. Challenges also persist in utilizing TiO_2 as a drug carrier, requiring improvements in drug loading efficiency and precise control over drug release kinetics. Meeting the demands of medical-grade applications requires establishing cost-effective, efficient, and eco-friendly production methods while ensuring consistent product quality. Moreover, tailored regulations and standardized systems are urgently needed for the medical use of TiO_2 to oversee research, manufacturing, and clinical implementation, thus safeguarding patient well-being. While TiO_2 showcases exceptional antibacterial properties, it is crucial not to overlook potential ecological toxicity and health hazards. Proactive management measures must be adopted to mitigate potential risks to ecosystems and human health during TiO_2 utilization. Although, TiO_2 holds substantial potential in medicine, additional research and development efforts are necessary to address the aforementioned challenges and ensure its safe and effective integration into medical settings. Furthermore, conducting in-depth studies and assessments on the toxicity and ecological risks of nanoscale titanium dioxide across various environments and organisms will provide valuable insights and guide its responsible use and disposal. Embracing circular economy principles, such as efficient waste resource utilization and developing eco-friendly synthesis methods for nanoscale titanium dioxide, can significantly contribute to reducing energy consumption, waste production, and the release of harmful substances.

References

1. Huang, Y., Zhang, G., and Ren, Q. (2019). Molecular characterization of two toll receptors in *Hyriopsis cumingii* and their potential roles in antibacterial response. *Frontiers in Physiology* 10: 952.
2. Lv, X., Tao, J., Chen, B. et al. (2016). Roles of temperature and flow velocity on the mobility of nano-sized titanium dioxide in natural waters. *Science of the Total Environment* 565: 849–856.
3. Mueller, N.C. and Nowack, B. (2008). Exposure modeling of engineered nanoparticles in the environment. *Environmental Science & Technology* 42 (12): 4447–4453.
4. Weir, A., Westerhoff, P., Fabricius, L. et al. (2012). Titanium dioxide nanoparticles in food and personal care products. *Environmental Science & Technology* 46 (4): 2242–2250.
5. Zhang, X., Yin, H., Li, Z. et al. (2016). Nano-TiO_2 induces autophagy to protect against cell death through antioxidative mechanism in podocytes. *Cell Biology and Toxicology* 32: 513–527.
6. Gao, J., Bonzongo, J.C.J., Bitton, G. et al. (2008). Nanowastes and the environment: using mercury as an example pollutant to assess the environmental fate of chemicals adsorbed onto manufactured nanomaterials. *Environmental Toxicology and Chemistry: An International Journal* 27 (4): 808–810.
7. Handy, R.D., Von der Kammer, F., Lead, J.R. et al. (2008). The ecotoxicology and chemistry of manufactured nanoparticles. *Ecotoxicology* 17: 287–314.
8. Klaine, S.J., Alvarez, P.J., Batley, G.E. et al. (2008). Nanomaterials in the environment: behavior, fate, bioavailability, and effects. *Environmental Toxicology and Chemistry: An International Journal* 27 (9): 1825–1851.

9 Nel, A., Xia, T., Madler, L. et al. (2006). Toxic potential of materials at the nanolevel. *Science* 311 (5761): 622–627.

10 Hagendorfer, H., Lorenz, C., Kaegi, R. et al. (2010). Size-fractionated characterization and quantification of nanoparticle release rates from a consumer spray product containing engineered nanoparticles. *Journal of Nanoparticle Research* 12: 2481–2494.

11 Parrino, F. and Palmisano, L. (2020). *Titanium Dioxide (TiO_2) and Its Applications.* Amsterdam: Elsevier.

12 Mandal, A. and Ray Banerjee, E. (2020). Introduction to nanoscience, nanotechnology and nanoparticles. In: *Nanomaterials and Biomedicine: Therapeutic and Diagnostic Approach* (ed. E. Ray Banerjee), 1–39.Singapore: Springer.

13 Bourikas, K., Kordulis, C., and Lycourghiotis, A. (2014). Titanium dioxide (anatase and rutile): surface chemistry, liquid–solid interface chemistry, and scientific synthesis of supported catalysts. *Chemical Reviews* 114 (19): 9754–9823.

14 Baranowska-Wójcik, E., Szwajgier, D., Oleszczuk, P. et al. (2020). Effects of titanium dioxide nanoparticles exposure on human health—a review. *Biological Trace Element Research* 193: 118–129.

15 Bai, S., Zhang, N., Gao, C. et al. (2018). Defect engineering in photocatalytic materials. *Nano Energy* 53: 296–336.

16 Fujishima, A. and Honda, K. (1972). Electrochemical photolysis of water at a semiconductor electrode. *Nature* 238: 37–38.

17 Maruska, H.P. and Ghosh, A.K. (1978). Photocatalytic decomposition of water at semiconductor electrodes. *Solar Energy* 20 (6): 443–458.

18 Hashimoto, K., Irie, H., and Fujishima, A. (2005). TiO_2 photocatalysis: a historical overview and future prospects. *Japanese Journal of Applied Physics* 44: 8269.

19 Gomes, J., Lincho, J., Domingues, E. et al. (2019). N–TiO_2 photocatalysts: a review of their characteristics and capacity for emerging contaminants removal. *Water* 11 (2): 373.

20 Carey, J.H., Lawrence, J., and Tosine, H.M. (1976). Photodechlorination of PCB's in the presence of titanium dioxide in aqueous suspensions. *Bulletin of Environmental Contamination and Toxicology* 16: 697–701.

21 Tennakone, K., Tilakaratne, C., and Kottegoda, I. (1995). Photocatalytic degradation of organic contaminants in water with TiO_2 supported on polythene films. *Journal of Photochemistry and Photobiology A: Chemistry* 87 (2): 177–179.

22 Malato, S., Blanco, J., Vidal, A. et al. (2003). Applied studies in solar photocatalytic detoxification: an overview. *Solar Energy* 75 (4): 329–336.

23 Nakajima, A., Hashimoto, K., Watanabe, T. et al. (2000). Transparent superhydrophobic thin films with self-cleaning properties. *Langmuir* 16 (17): 7044–7047.

24 Pelaez, M., Nolan, N.T., Pillai, S.C. et al. (2012). A review on the visible light active titanium dioxide photocatalysts for environmental applications. *Applied Catalysis B: Environmental* 125: 331–349.

25 Gilbertson, L.M., Albalghiti, E.M., Fishman, Z.S. et al. (2016). Shape-dependent surface reactivity and antimicrobial activity of nano-cupric oxide. *Environmental Science & Technology* 50 (7): 3975–3984.

26 Pal, S., Tak, Y.K., and Song, J.M. (2007). Does the antibacterial activity of silver nanoparticles depend on the shape of the nanoparticle? A study of the gram-negative bacterium *Escherichia coli*. *Applied and Environmental Microbiology* 73 (6): 1712–1720.

27 Hajipour, M.J., Fromm, K.M., Ashkarran, A.A. et al. (2012). Antibacterial properties of nanoparticles. *Trends in Biotechnology* 30 (10): 499–511.

28 Whitesides, G.M. (2005). Nanoscience, nanotechnology, and chemistry. *Small* 1 (2): 172–179.

29 Dasari, T.P. and Hwang, H.-M. (2013). Effect of humic acids and sunlight on the cytotoxicity of engineered zinc oxide and titanium dioxide nanoparticles to a river bacterial assemblage. *Journal of Environmental Sciences* 25 (9): 1925–1935.

30 He, Z., Cai, Q., Fang, H. et al. (2013). Photocatalytic activity of TiO_2 containing anatase nanoparticles and rutile nanoflower structure consisting of nanorods. *Journal of Environmental Sciences* 25 (12): 2460–2468.

31 Vimbela, G.V., Ngo, S.M., Fraze, C. et al. (2017). Antibacterial properties and toxicity from metallic nanomaterials. *International Journal of Nanomedicine* 12: 3941–3965.

32 Zhang, Q., Yan, X., Shao, R. et al. (2014). Preparation of nano-TiO_2 by liquid hydrolysis and characterization of its antibacterial activity. *Journal of Wuhan University of Technology-Material Science Edition* 29 (2): 407–409.

33 Yao, S., Ying, W., Yao, L. et al. (2022). Titanium dioxide nanoparticles in personal care products and titanium health risk assessment to adult consumers. *Asian Journal of Ecotoxicology* (4): 533–544.

34 Mamaghani, A.H., Haghighat, F., and Lee, C.-S. (2021). Effect of titanium dioxide properties and support material on photocatalytic oxidation of indoor air pollutants. *Building and Environment* 189: 107518.

35 Vijayaraj, V., Liné, C., Cadarsi, S. et al. (2018). Transfer and ecotoxicity of titanium dioxide nanoparticles in terrestrial and aquatic ecosystems: a microcosm study. *Environmental Science & Technology* 52 (21): 12757–12764.

36 Du, J., Xu, S., Zhou, Q. et al. (2019). The ecotoxicology of titanium dioxide nanoparticles, an important engineering nanomaterial. *Toxicological & Environmental Chemistry* 101 (3–6): 165–189.

37 Khanna, P., Ong, C., Bay, B.H. et al. (2015). Nanotoxicity: an interplay of oxidative stress, inflammation and cell death. *Nanomaterials* 5 (3): 1163–1180.

38 Sarkar, A., Ghosh, M., and Sil, P.C. (2014). Nanotoxicity: oxidative stress mediated toxicity of metal and metal oxide nanoparticles. *Journal of Nanoscience and Nanotechnology* 14 (1): 730–743.

39 Chen, Q., Xue, Y., and Sun, J. (2013). Kupffer cell-mediated hepatic injury induced by silica nanoparticles in vitro and in vivo. *International Journal of Nanomedicine* 8: 1129–1140.

40 Monteiro-Riviere, N.A. and Inman, A.O. (2006). Challenges for assessing carbon nanomaterial toxicity to the skin. *Carbon* 44 (6): 1070–1078.

41 Filippi, C., Pryde, A., Cowan, P. et al. (2015). Toxicology of ZnO and TiO_2 nanoparticles on hepatocytes: impact on metabolism and bioenergetics. *Nanotoxicology* 9 (1): 126–134.

42 Hussain, S., Hess, K., Gearhart, J. et al. (2005). In vitro toxicity of nanoparticles in BRL 3A rat liver cells. *Toxicology in Vitro* 19 (7): 975–983.

43 Hattori, K., Nakadate, K., Morii, A. et al. (2017). Exposure to nano-size titanium dioxide causes oxidative damages in human mesothelial cells: the crystal form rather than size of particle contributes to cytotoxicity. *Biochemical and Biophysical Research Communications* 492 (2): 218–223.

44 Wang, J., Zhu, X., Zhang, X. et al. (2011). Disruption of zebrafish (*Danio rerio*) reproduction upon chronic exposure to TiO_2 nanoparticles. *Chemosphere* 83 (4): 461–467.

45 Alidadi, H., Khorsandi, L., and Shirani, M. (2018). Effects of quercetin on tubular cell apoptosis and kidney damage in rats induced by titanium dioxide nanoparticles. *The Malaysian Journal of Medical Sciences: MJMS* 25 (2): 72.

46 Shirdare, M., Jabbari, F., Salehzadeh, M. et al. (2022). Curcuma reduces kidney and liver damage induced by titanium dioxide nanoparticles in male Wistar rats. *Avicenna Journal of Phytomedicine* 12 (5): 537.

47 Long, T.C., Tajuba, J., Sama, P. et al. (2007). Nanosize titanium dioxide stimulates reactive oxygen species in brain microglia and damages neurons in vitro. *Environmental Health Perspectives* 115 (11): 1631–1637.

48 Guo, Y., Chen, L., Wu, J. et al. (2019). Parental co-exposure to bisphenol A and nano-TiO_2 causes thyroid endocrine disruption and developmental neurotoxicity in zebrafish offspring. *Science of the Total Environment* 650: 557–565.

49 Shi, W., Han, Y., Guo, C. et al. (2017). Immunotoxicity of nanoparticle TiO_2 to a commercial marine bivalve species, *Tegillarca granosa*. *Fish & Shellfish Immunology* 66: 300–306.

50 Tsuda, H., Xu, J., Sakai, Y. et al. (2009). Toxicology of engineered nanomaterials—a review of carcinogenic potential. *Asian Pacific Journal of Cancer Prevention* 10 (6): 975–980.

51 Bacchetta, R., Santo, N., Fascio, U. et al. (2012). Nano-sized CuO, TiO_2 and ZnO affect *Xenopus laevis* development. *Nanotoxicology* 6 (4): 381–398.

52 Wang, L., Luanpitpong, S., Castranova, V. et al. (2011b). Carbon nanotubes induce malignant transformation and tumorigenesis of human lung epithelial cells. *Nano Letters* 11 (7): 2796–2803.

53 Fekete-Kertész, I., Piszmán, D., and Molnár, M. (2017). Particle size and concentration dependent ecotoxicity of nano-and microscale TiO_2—comparative study by different aquatic test organisms of different trophic levels. *Water, Air, & Soil Pollution* 228: 1–17.

54 Hou, J., Hu, C., Wang, Y. et al. (2022). Nano-bio interfacial interactions determined the contact toxicity of TiO_2 to nematodes in various soils. *Science of the Total Environment* 835: 155456.

55 Jośko, I. and Oleszczuk, P. (2013). Influence of soil type and environmental conditions on ZnO, TiO_2 and Ni nanoparticles phytotoxicity. *Chemosphere* 92 (1): 91–99.

56 Liu, W., Su, P., Chen, S. et al. (2015). Antibacterial and osteogenic stem cell differentiation properties of photoinduced TiO_2 nanoparticle-decorated TiO_2 nanotubes. *Nanomedicine* 10 (5): 713–723.

57 Sun, H., Zhang, X., Niu, Q. et al. (2007). Enhanced accumulation of arsenate in carp in the presence of titanium dioxide nanoparticles. *Water, Air, and Soil Pollution* 178: 245–254.

58 Wang, X., Liu, Y., Wang, J. et al. (2017). Amplification of arsenic genotoxicity by TiO_2 nanoparticles in mammalian cells: new insights from physicochemical interactions and mitochondria. *Nanotoxicology* 11 (8): 978–995.

59 Ahamed, M., Akhtar, M.J., and Alhadlaq, H.A. (2019). Preventive effect of TiO_2 nanoparticles on heavy metal Pb-induced toxicity in human lung epithelial (A549) cells. *Toxicology in Vitro* 57: 18–27.

60 Kongseng, S., Yoovathaworn, K., Wongprasert, K. et al. (2016). Cytotoxic and inflammatory responses of TiO_2 nanoparticles on human peripheral blood mononuclear cells. *Journal of Applied Toxicology* 36 (10): 1364–1373.

61 Zdravkovic, B., Zdravkovic, T.P., Zdravkovic, M. et al. (2019). The influence of nano-TiO_2 on metabolic activity, cytotoxicity and ABCB5 mRNA expression in WM-266-4 human metastatic melanoma cell line. *Journal of B.U.O.N.* 24 (1): 338–346.

62 Hou, J., Wang, L., Wang, C. et al. (2019). Toxicity and mechanisms of action of titanium dioxide nanoparticles in living organisms. *Journal of Environmental Sciences* 75: 40–53.

63 Li, S., Chen, H., Liu, C. et al. (2023). Dietary exposure to TiO_2 reduces byssus performance of mussels under ocean warming. *Science of the Total Environment* 881: 163499.

64 Trevisan, R., Delapedra, G., Mello, D.F. et al. (2014). Gills are an initial target of zinc oxide nanoparticles in oysters *Crassostrea gigas*, leading to mitochondrial disruption and oxidative stress. *Aquatic Toxicology* 153: 27–38.

65 Gao, X., Deng, R., and Lin, D. (2020). Insights into the regulation mechanisms of algal extracellular polymeric substances secretion upon the exposures to anatase and rutile TiO_2 nanoparticles. *Environmental Pollution* 263: 114608.

66 Shukla, R.K., Kumar, A., Gurbani, D. et al. (2013). TiO_2 nanoparticles induce oxidative DNA damage and apoptosis in human liver cells. *Nanotoxicology* 7 (1): 48–60.

67 Ma, Y., Guo, Y., Wu, S. et al. (2017). Titanium dioxide nanoparticles induce size-dependent cytotoxicity and genomic DNA hypomethylation in human respiratory cells. *RSC Advances* 7 (38): 23560–23572.

68 Luo, Z., Li, Z., Xie, Z. et al. (2020). Rethinking nano-TiO_2 safety: overview of toxic effects in humans and aquatic animals. *Small* 16 (36): 2002019.

69 Ahamed, A., Liang, L., Lee, M.Y. et al. (2021). Too small to matter? Physicochemical transformation and toxicity of engineered TiO_2, $nSiO_2$, nZnO, carbon nanotubes, and nAg. *Journal of Hazardous Materials* 404: 124107.

70 Aitken, R.J., Chaudhry, M., Boxall, A. et al. (2006). Manufacture and use of nanomaterials: current status in the UK and global trends. *Occupational Medicine* 56 (5): 300–306.

71 Chen, T.-H., Lin, C.-Y., and Tseng, M.-C. (2011). Behavioral effects of titanium dioxide nanoparticles on larval zebrafish (*Danio rerio*). *Marine Pollution Bulletin* 63 (5–12): 303–308.

72 Clemente, Z., Castro, V., Moura, M. et al. (2014). Toxicity assessment of TiO_2 nanoparticles in zebrafish embryos under different exposure conditions. *Aquatic Toxicology* 147: 129–139.

73 Hao, L., Wang, Z., and Xing, B. (2009). Effect of sub-acute exposure to TiO_2 nanoparticles on oxidative stress and histopathological changes in Juvenile Carp (*Cyprinus carpio*). *Journal of Environmental Sciences* 21 (10): 1459–1466.

74 Yeo, M.-K. and Kang, M. (2009). Effects of $Cu_x TiO_y$ nanometer particles on biological toxicity during zebrafish embryogenesis. *Korean Journal of Chemical Engineering* 26: 711–718.

75 Zhu, X., Teng, J., Xu, E.G. et al. (2022). Toxicokinetics and toxicodynamics of plastic and metallic nanoparticles: a comparative study in shrimp. *Environmental Pollution* 312: 120069.

76 Marisa, I., Matozzo, V., Martucci, A. et al. (2018). Bioaccumulation and effects of titanium dioxide nanoparticles and bulk in the clam *Ruditapes philippinarum*. *Marine Environmental Research* 136: 179–189.

77 Zha, S., Liu, H., Bao, Y. et al. (2024). Toxicological impacts of nanomaterials on marine bivalves. In: *Marine Bivalve Mollusks and Emerging Pollutants* (ed. G. Liu), 83–101. Amsterdam: Elsevier.

78 Shang, Y., Wu, F., Wei, S. et al. (2020). Specific dynamic action of mussels exposed to TiO_2 nanoparticles and seawater acidification. *Chemosphere* 241: 125104.

79 Hu, M., Lin, D., Shang, Y. et al. (2017). CO_2-induced pH reduction increases physiological toxicity of nano-TiO_2 in the mussel *Mytilus coruscus*. *Scientific Reports* 7 (1): 40015.

80 Han, Y., Shi, W., Rong, J. et al. (2019). Exposure to waterborne TiO_2 reduces fertilization success and increases polyspermy in a bivalve mollusc: a threat to population recruitment. *Environmental Science & Technology* 53 (21): 12754–12763.

81 Canesi, L., Ciacci, C., Vallotto, D. et al. (2010). In vitro effects of suspensions of selected nanoparticles (C60 fullerene, TiO_2, SiO_2) on *Mytilus* hemocytes. *Aquatic Toxicology* 96 (2): 151–158.

82 Middepogu, A., Hou, J., Gao, X. et al. (2018). Effect and mechanism of TiO_2 nanoparticles on the photosynthesis of *Chlorella pyrenoidosa*. *Ecotoxicology and Environmental Safety* 161: 497–506.

83 Cardinale, B.J., Bier, R., and Kwan, C. (2012). Effects of TiO_2 nanoparticles on the growth and metabolism of three species of freshwater algae. *Journal of Nanoparticle Research* 14: 1–8.

84 Yang, L., Xu, J., Gao, H. et al. (2023). Toxicity enhancement of nano titanium dioxide to *Brachionus calyciflorus* (Rotifera) under simulated sunlight and the underlying mechanisms. *Ecotoxicology and Environmental Safety* 251: 114556.

85 Chen, L., Zhou, L., Liu, Y. et al. (2012). Toxicological effects of nanometer titanium dioxide (nano-TiO_2) on *Chlamydomonas reinhardtii*. *Ecotoxicology and Environmental Safety* 84: 155–162.

86 Xia, B., Chen, B., Sun, X. et al. (2015). Interaction of TiO_2 nanoparticles with the marine microalga *Nitzschia closterium*: growth inhibition, oxidative stress and internalization. *Science of the Total Environment* 508: 525–533.

87 Matouke, M.M., Elewa, D.T., and Abdullahi, K. (2018). Binary effect of titanium dioxide nanoparticles ($nTiO_2$) and phosphorus on microalgae (*Chlorella 'Ellipsoides Gerneck*, 1907). *Aquatic Toxicology* 198: 40–48.

88 Wang, Y., Zhu, X., Lao, Y. et al. (2016). TiO_2 nanoparticles in the marine environment: physical effects responsible for the toxicity on algae *Phaeodactylum tricornutum*. *Science of the Total Environment* 565: 818–826.

89 Liu, W., Li, M., Li, W. et al. (2022). Metabolic alterations in alga *Chlamydomonas reinhardtii* exposed to TiO_2 materials. *Environmental Science: Nano* 9 (8): 2922–2938.

90 Cherchi, C., Miljkovic, M., Diem, M. et al. (2015). TiO_2 induced changes in intracellular composition and nutrient stoichiometry in primary producer—cyanobacteria. *Science of the Total Environment* 512: 345–352.

91 Ghosh, M., Bandyopadhyay, M., and Mukherjee, A. (2010). Genotoxicity of titanium dioxide (TiO_2) nanoparticles at two trophic levels: plant and human lymphocytes. *Chemosphere* 81 (10): 1253–1262.

92 Metzler, D.M., Li, M., Erdem, A. et al. (2011). Responses of algae to photocatalytic nano-TiO_2 particles with an emphasis on the effect of particle size. *Chemical Engineering Journal* 170 (2–3): 538–546.

93 Erdem, A., Metzler, D., Cha, D.K. et al. (2015). The short-term toxic effects of TiO_2 nanoparticles toward bacteria through viability, cellular respiration, and lipid peroxidation. *Environmental Science and Pollution Research* 22: 17917–17924.

94 Kim, B., Kim, D., Cho, D. et al. (2003). Bactericidal effect of TiO_2 photocatalyst on selected food-borne pathogenic bacteria. *Chemosphere* 52 (1): 277–281.

95 Bavanilatha, M., Yoshitha, L., Nivedhitha, S. et al. (2019). Bioactive studies of TiO_2 nanoparticles synthesized using *Glycyrrhiza glabra*. *Biocatalysis and Agricultural Biotechnology* 19: 101131.

96 Baysal, A., Saygin, H., and Ustabasi, G.S. (2018). Interaction of PM2.5 airborne particulates with ZnO and TiO_2 nanoparticles and their effect on bacteria. *Environmental Monitoring and Assessment* 190: 1–15.

97 Kasemets, K., Ivask, A., Dubourguier, H.-C. et al. (2009). Toxicity of nanoparticles of ZnO, CuO and TiO_2 to yeast *Saccharomyces cerevisiae*. *Toxicology in Vitro* 23 (6): 1116–1122.

98 Mattiello, A., Lizzi, D., and Marchiol, L. (2018). Influence of titanium dioxide nanoparticles ($nTiO_2$) on crop plants: a systematic overview. In: *Nanomaterials in Plants, Algae, and Microorganisms* (eds. D.K. Tripathi, P. Ahmad, S. Sharma, D.K. Chauhan, and N.K. Dubey), 277–296. Academic Press.

99 Mishra, V., Baranwal, V., Mishra, R.K. et al. (2016). Titanium dioxide nanoparticles augment allergic airway inflammation and Socs3 expression via NF-κB pathway in murine model of asthma. *Biomaterials* 92: 90–102.

100 Karimi, S., Khorsandi, L., and Nejaddehbashi, F. (2019). Protective effects of curcumin on testicular toxicity induced by titanium dioxide nanoparticles in mice. *JBRA Assisted Reproduction* 23 (4): 344.

101 Orazizadeh, M., Daneshi, E., Hashemitmar, M. et al. (2015). Protective effect of beta-carotene against titanium dioxide nanoparticles induced apoptosis in mouse testicular tissue. *Andrologia* 47 (7): 816–825.

102 Naserzadeh, P., Ghanbary, F., Ashtari, P. et al. (2018). Biocompatibility assessment of titanium dioxide nanoparticles in mice fetoplacental unit. *Journal of Biomedical Materials Research Part A* 106 (2): 580–589.

103 Chen, F., Wang, C., Yue, L. et al. (2022). Nano-TiO_2 retarded fetal development by inhibiting transplacental transfer of thyroid hormones in rats. *Environmental Science: Nano* 9 (7): 2351–2362.

104 Shukla, R.K., Kumar, A., Vallabani, N.V.S. et al. (2014). Titanium dioxide nanoparticle-induced oxidative stress triggers DNA damage and hepatic injury in mice. *Nanomedicine* 9 (9): 1423–1434.

105 Heringa, M., Peters, R., Bleys, R. et al. (2018). Detection of titanium particles in human liver and spleen and possible health implications. *Particle and Fibre Toxicology* 15: 1–9.

106 Rebleanu, D., Gaidau, C., Voicu, G. et al. (2019). The impact of photocatalytic Ag/TiO_2 and Ag/N-TiO_2 nanoparticles on human keratinocytes and epithelial lung cells. *Toxicology* 416: 30–43.

107 Petković, J., Žegura, B., Stevanović, M. et al. (2011). DNA damage and alterations in expression of DNA damage responsive genes induced by TiO_2 nanoparticles in human hepatoma HepG2 cells. *Nanotoxicology* 5 (3): 341–353.

108 Simon-Deckers, A., Loo, S., Mayne-L'hermite, M. et al. (2009). Size-, composition- and shape-dependent toxicological impact of metal oxide nanoparticles and carbon nanotubes toward bacteria. *Environmental Science & Technology* 43 (21): 8423–8429.

109 Santonastaso, M., Mottola, F., Colacurci, N. et al. (2019). In vitro genotoxic effects of titanium dioxide nanoparticles (n-TiO_2) in human sperm cells. *Molecular Reproduction and Development* 86 (10): 1369–1377.

110 Hartmann, N.B., Von der Kammer, F., Hofmann, T. et al. (2010). Algal testing of titanium dioxide nanoparticles—testing considerations, inhibitory effects and modification of cadmium bioavailability. *Toxicology* 269 (2–3): 190–197.

111 Deng, F., Liu, J., Xie, Z. et al. (2022). TiO_2 alleviates the toxic effects of TCPP on mussels by adjusting respiratory metabolism and gut microbiota. *Science of the Total Environment* 851: 158176.

112 Gao, F., Hong, F., Liu, C. et al. (2006). Mechanism of nano-anatase TiO_2 on promoting photosynthetic carbon reaction of spinach: inducing complex of rubisco-rubisco activase. *Biological Trace Element Research* 111: 239–253.

113 Shyla, K.K. and Natarajan, N. (2014). Customizing zinc oxide, silver and titanium dioxide nanoparticles for enhancing groundnut seed quality. *Indian Journal of Science and Technology* 7 (9): 1376–1381.

114 Doyle, J.J., Ward, J.E., and Mason, R. (2015). An examination of the ingestion, bioaccumulation, and depuration of titanium dioxide nanoparticles by the blue mussel (*Mytilus edulis*) and the eastern oyster (*Crassostrea virginica*). *Marine Environmental Research* 110: 45–52.

115 Yan, C., Yang, F., Wang, Z. et al. (2017). Changes in arsenate bioaccumulation, subcellular distribution, depuration, and toxicity in *Artemia salina* nauplii in the presence of titanium dioxide nanoparticles. *Environmental Science: Nano* 4 (6): 1365–1376.

116 Xin, X., Huang, G., Zhang, B. et al. (2021). Trophic transfer potential of TiO_2, nZnO, and triclosan in an algae-algae eating fish food chain. *Aquatic Toxicology* 235: 105824.

117 Zhu, X., Wang, J., Zhang, X. et al. (2010). Trophic transfer of TiO_2 nanoparticles from daphnia to zebrafish in a simplified freshwater food chain. *Chemosphere* 79 (9): 928–933.

118 Wang, Z., Yin, L., Zhao, J. et al. (2016). Trophic transfer and accumulation of TiO_2 nanoparticles from clamworm (*Perinereis aibuhitensis*) to juvenile turbot (*Scophthalmus maximus*) along a marine benthic food chain. *Water Research* 95: 250–259.

119 Kabir, E., Kumar, V., Kim, K.-H. et al. (2018). Environmental impacts of nanomaterials. *Journal of Environmental Management* 225: 261–271.

120 Maneerat, C. and Hayata, Y. (2006). Antifungal activity of TiO_2 photocatalysis against *Penicillium expansum* in vitro and in fruit tests. *International Journal of Food Microbiology* 107 (2): 99–103.

121 Wang, L., Ding, Y., Shen, Y. et al. (2014). Study on properties of modified nano-TiO_2 and its application on antibacterial finishing of textiles. *Journal of Industrial Textiles* 44 (3): 351–372.

122 Kim, H., Lee, S., Han, Y. et al. (2006). Preparation of dip-coated TiO_2 photocatalyst on ceramic foam pellets. *Journal of Materials Science* 41: 6150–6153.

123 Cho, M., Chung, H., Choi, W. et al. (2004). Linear correlation between inactivation of *E. coli* and OH radical concentration in TiO_2 photocatalytic disinfection. *Water Research* 38 (4), 1069–1077.

124 Li, Q., Mahendra, S., Lyon, D.Y. et al. (2008). Antimicrobial nanomaterials for water disinfection and microbial control: potential applications and implications. *Water Research* 42 (18): 4591–4602.

125 Kang, I.-C., Zhang, Q., Yin, S. et al. (2008). Improvement in photocatalytic activity of TiO2 under visible irradiation through addition of N-TiO_2. *Environmental Science & Technology* 42 (10): 3622–3626.

126 Wang, Y., Rao, L., Wang, P. et al. (2020). Photocatalytic activity of N-TiO_2/O-doped N vacancy g-C3N4 and the intermediates toxicity evaluation under tetracycline hydrochloride and Cr (VI) coexistence environment. *Applied Catalysis B: Environmental* 262: 118308.

127 Jayaseelan, C., Rahuman, A.A., Roopan, S.M. et al. (2013). Biological approach to synthesize TiO_2 nanoparticles using *Aeromonas hydrophila* and its antibacterial activity. *Spectrochimica Acta Part A: Molecular and Biomolecular Spectroscopy* 107: 82–89.

128 Jafari, S., Mahyad, B., Hashemzadeh, H. et al. (2020). Biomedical applications of TiO_2 nanostructures: recent advances. *International Journal of Nanomedicine* 15: 3447–3470.

129 Caratto, V., Ball, L., Sanguineti, E. et al. (2017). Antibacterial activity of standard and N-doped titanium dioxide-coated endotracheal tubes: an in vitro study. *Revista Brasileira de Terapia Intensiva* 29: 55–62.

130 Yuan, Y., Ding, L., Chen, Y. et al. (2022). Nano-silver functionalized polysaccharides as a platform for wound dressings: a review. *International Journal of Biological Macromolecules* 194: 644–653.

131 Chong, M.N., Jin, B., Chow, C.W. et al. (2010). Recent developments in photocatalytic water treatment technology: a review. *Water Research* 44 (10): 2997–3027.

132 Umpo, M., Lollen, T., Jini, M. et al. (2021). Antimicrobial efficacy of titanium dioxide coating in operating theaters at a tertiary care hospital in Arunachal Pradesh. *Journal of Pure and Applied Microbiology* 15 (4): 2177–2182.

133 Sun, R.-D., Nakajima, A., Fujishima, A. et al. (2001). Photoinduced surface wettability conversion of ZnO and TiO_2 thin films. *The Journal of Physical Chemistry B* 105 (10): 1984–1990.

134 Wang, R., Hashimoto, K., Fujishima, A. et al. (1997). Light-induced amphiphilic surfaces. *Nature* 388 (6641): 431–432.

135 Kalčíková, G., Roy, R., Klun, B. et al. (2023). Interactions between titanium dioxide nanoparticles and polyethylene microplastics: adsorption kinetics, photocatalytic properties, and ecotoxicity. *Chemosphere* 329: 138628.

136 Nabi, G., Raza, W., and Tahir, M. (2020). Green synthesis of TiO_2 nanoparticle using cinnamon powder extract and the study of optical properties. *Journal of Inorganic and Organometallic Polymers and Materials* 30 (4): 1425–1429.

137 Seitz, F., Bundschuh, M., Dabrunz, A. et al. (2012). Titanium dioxide nanoparticles detoxify pirimicarb under UV irradiation at ambient intensities. *Environmental Toxicology and Chemistry* 31 (3): 518–523.

138 Abdullah, H., Kuo, D.-H., and Gultom, N.S. (2019). N=N bond cleavage of azobenzene via photocatalytic hydrogenation with Dy-doped Zn(O,S): the progress from hydrogen evolution to green chemical conversion. *Catalysis Science & Technology* 9: 2651–2663.

139 Bandara, J., Mielczarski, J.A., and Kiwi, J. (1999). Molecular mechanism of surface recognition. Azo dyes degradation on Fe, Ti, and Al oxides through metal sulfonate complexes. *Langmuir* 15: 7670–7679.

140 Abel, S., Jule, L.T., Belay, F. et al. (2021). Application of titanium dioxide nanoparticles synthesized by sol-gel methods in wastewater treatment. *Journal of Nanomaterials* 2021: 1–6.

141 Grzechulska, J., Hamerski, M., and Morawski, A.W. (2000). Photocatalytic decomposition of oil in water. *Water Research* 34 (5): 1638–1644.

142 Mironyuk, I., Soltys, L., Tatarchuk, T. et al. (2020). Ways to improve the efficiency of TiO_2-based photocatalysts. *Physics and Chemistry of Solid State* 21 (2): 300–311.

143 Yang, Y.L. and Xie, Z.Y. (2013). Design of indoor harmful gas monitor system. *Applied Mechanics and Materials* 423: 2376–2379.

144 Takita, Y., Yamada, H.A., Hashida, M. et al. (1990). Conversion of 1,1,2-trichloro-1,2,2-trifluoroethane (CFC113) over TiO_2-supported metal and metal oxide catalysts. *Chemistry Letters* 19: 715–718.

145 Nguyen, H.H., Gyawali, G., Martinez-Oviedo, A. et al. (2020). Modified blue TiO_2 nanostructures for efficient photo-oxidative removal of harmful NO_x gases. *Korean Journal of Chemical Engineering* 37: 1507–1514.

146 Dibble, L.A. and Raupp, G.B. (1992). Fluidized-bed photocatalytic oxidation of trichloroethylene in contaminated air streams. *Environmental Science & Technology* 26: 492–495.

147 Stauff, I., Derman, S., Barbe, A. et al. (2018). Efficacy and acceptance of a high-velocity microdroplet device for interdental cleaning in gingivitis patients—a monitored, randomized controlled trial. *International Journal of Dental Hygiene* 16 (2): e31–e37.

148 Chung, I.-M., Park, I., Seung-Hyun, K. et al. (2016). Plant-mediated synthesis of silver nanoparticles: their characteristic properties and therapeutic applications. *Nanoscale Research Letters* 11: 1–14.

149 Wang, X., Li, Z., Shi, J. et al. (2014). One-dimensional titanium dioxide nanomaterials: nanowires, nanorods, and nanobelts. *Chemical Reviews* 114 (19): 9346–9384.

150 Han, G. and Ceilley, R. (2017). Chronic wound healing: a review of current management and treatments. *Advances in Therapy* 34: 599–610.

151 Gelmetti, C. (2008). Local antibiotics in dermatology. *Dermatologic Therapy* 21 (3): 187–195.

152 Siddiqui, A.R. and Bernstein, J.M. (2010). Chronic wound infection: facts and controversies. *Clinics in Dermatology* 28 (5): 519–526.

153 Shukla, K.K., Shah, S., Tyagi, C.K. et al. (2021). Dextran-chitosan-nTiO_2 tailored hydrogels (DEXCH-nTiO_2-HGS) as versatile tool in exploring the potential of kaempferol for effective diabetic wound healing. *Journal of Advanced Scientific Research* 12 (03 Suppl 2): 210–223.

154 Archana, D., Dutta, J., and Dutta, P. (2013). Evaluation of chitosan nano dressing for wound healing: characterization, in vitro and in vivo studies. *International Journal of Biological Macromolecules* 57: 193–203.

155 Mohandas, A., Deepthi, S., Biswas, R. et al. (2018). Chitosan based metallic nanocomposite scaffolds as antimicrobial wound dressings. *Bioactive Materials* 3 (3): 267–277.

156 Yadav, H.M., Kim, J.-S., and Pawar, S.H. (2016). Developments in photocatalytic antibacterial activity of nano TiO_2: a review. *Korean Journal of Chemical Engineering* 33: 1989–1998.

157 Ding, J. (2014). Preparation of TiO_2 photocatalysis antibacterial ceramics. *Key Engineering Materials* 575: 302–305.

158 Grande, F. and Tucci, P. (2016). Titanium dioxide nanoparticles: a risk for human health? *Mini Reviews in Medicinal Chemistry* 16 (9): 762–769.

159 Rashid, M.M., Forte Tavčer, P., and Tomšič, B. (2021). Influence of titanium dioxide nanoparticles on human health and the environment. *Nanomaterials* 11 (9): 2354.

160 Sadrieh, N., Wokovich, A.M., Gopee, N.V. et al. (2010). Lack of significant dermal penetration of titanium dioxide from sunscreen formulations containing nano-and submicron-size TiO_2 particles. *Toxicological Sciences* 115 (1): 156–166.

161 Dar, G.I., Saeed, M., and Wu, A. (2020). Toxicity of TiO_2 nanoparticles. In: Wu, A. & Ren, W. (eds.) *TiO_2 Nanoparticles: Applications in Nanobiotechnology and Nanomedicine*, pp. 67–103.Weinheim: Germany: Wiley-VCH (John Wiley & Sons, Inc.).

162 Madhubala, V., Pugazhendhi, A., and Thirunavukarasu, K. (2019). Cytotoxic and immunomodulatory effects of the low concentration of titanium dioxide nanoparticles (TiO_2 NPs) on human cell lines—an in vitro study. *Process Biochemistry* 86: 186–195.

163 Warheit, D.B. and Donner, E.M. (2015). Risk assessment strategies for nanoscale and fine-sized titanium dioxide particles: recognizing hazard and exposure issues. *Food and Chemical Toxicology* 85: 138–147.

164 Gamer, A., Leibold, E., and Van Ravenzwaay, B. (2006). The in vitro absorption of microfine zinc oxide and titanium dioxide through porcine skin. *Toxicology in Vitro* 20 (3): 301–307.

165 Schulz, J., Hohenberg, H., Pflücker, F. et al. (2002). Distribution of sunscreens on skin. *Advanced Drug Delivery Reviews* 54: S157–S163.

166 Pflücker, F., Wendel, V., Hohenberg, H. et al. (2001). The human stratum corneum layer: an effective barrier against dermal uptake of different forms of topically applied micronised titanium dioxide. *Skin Pharmacology and Physiology* 14 (Suppl 1): 92–97.

167 Monteiro-Riviere, N.A. (2014). Safety implications of nanomaterial exposure to skin. In: *Nanotoxicology: Progress Toward Nanomedicine* (eds. N.A. Monteiro-Riviere and C. Lang Tran), 247–272. Boca Raton: CRC Press.

168 Dreno, B., Alexis, A., Chuberre, B. et al. (2019). Safety of titanium dioxide nanoparticles in cosmetics. *Journal of the European Academy of Dermatology and Venereology* 33: 34–46.

169 Yang, W., Peters, J.I., and Williams III, R.O. (2008). Inhaled nanoparticles—a current review. *International Journal of Pharmaceutics* 356 (1–2): 239–247.

170 Merkel, O.M., Zheng, M., Debus, H. et al. (2012). Pulmonary gene delivery using polymeric nonviral vectors. *Bioconjugate Chemistry* 23 (1): 3–20.

171 Aruoja, V., Dubourguier, H.-C., Kasemets, K. et al. (2009). Toxicity of nanoparticles of CuO, ZnO and TiO_2 to microalgae *Pseudokirchneriella subcapitata*. *Science of the Total Environment* 407 (4): 1461–1468.

172 Sonwani, S., Madaan, S., Arora, J. et al. (2021). Inhalation exposure to atmospheric nanoparticles and its associated impacts on human health: a review. *Frontiers in Sustainable Cities* 3: 690444.

173 Kreyling, W.G. and Scheuch, G. (2000). Clearance of particles deposited in the lungs. *Lung Biology in Health and Disease* 143: 323–366.

174 Marshall, C.D., Hu, M.S., Leavitt, T. et al. (2018). Cutaneous scarring: basic science, current treatments, and future directions. *Advances in Wound Care* 7 (2): 29–45.

175 McGuinnes, C., Duffin, R., Brown, S. et al. (2011). Surface derivatization state of polystyrene latex nanoparticles determines both their potency and their mechanism of causing human platelet aggregation in vitro. *Toxicological Sciences* 119 (2): 359–368.

176 Warheit, D.B., Webb, T.R., Reed, K.L. et al. (2007). Pulmonary toxicity study in rats with three forms of ultrafine-TiO_2 particles: differential responses related to surface properties. *Toxicology* 230 (1): 90–104.

177 Roursgaard, M., Jensen, K.A., Poulsen, S.S. et al. (2011). Acute and subchronic airway inflammation after intratracheal instillation of quartz and titanium dioxide agglomerates in mice. *The Scientific World Journal* 11: 801–825.

178 Afshari-Kaveh, M., Abbasalipourkabir, R., Nourian, A. et al. (2021). The protective effects of vitamins A and E on titanium dioxide nanoparticles ($nTiO_2$)-induced oxidative stress in the spleen tissues of male Wistar rats. *Biological Trace Element Research* 199: 3677–3687.

179 Petrarca, C., Mangifesta, R., and Di Giampaolo, L. (2020). Immunotoxicity of nanoparticles. In T. Otsuki, M. Di Gioacchino & C. Petrarca (Eds.), *Allergy and Immunotoxicology in Occupational Health—The Next Step* (pp. 75–94). Singapore: Springer.

180 IARC Working Group on the Evaluation of Carcinogenic Risks to Humans. (2010). *Carbon Black, Titanium Dioxide, and Talc. IARC Monographs on the Evaluation of Carcinogenic Risks to Humans* 93: Lyon: International Agency for Research on Cancer.

181 Hu, H., Fan, X., Yin, Y. et al. (2019). Mechanisms of titanium dioxide nanoparticle-induced oxidative stress and modulation of plasma glucose in mice. *Environmental Toxicology* 34 (11): 1221–1235.

182 Deng, Z.J., Mortimer, G., Schiller, T. et al. (2009). Differential plasma protein binding to metal oxide nanoparticles. *Nanotechnology* 20 (45): 455101.

183 Hagens, W.I., Oomen, A.G., de Jong, W.H. et al. (2007). What do we (need to) know about the kinetic properties of nanoparticles in the body? *Regulatory Toxicology and Pharmacology* 49 (3): 217–229.

184 Wang, J., Zhou, G., Chen, C. et al. (2007). Acute toxicity and biodistribution of different sized titanium dioxide particles in mice after oral administration. *Toxicology Letters* 168 (2): 176–185.

185 Chen, J., Dong, X., Zhao, J. et al. (2009). In vivo acute toxicity of titanium dioxide nanoparticles to mice after intraperitioneal injection. *Journal of Applied Toxicology* 29 (4): 330–337.

186 Kreyling, W.G., Holzwarth, U., Schleh, C. et al. (2017). Quantitative biokinetics of titanium dioxide nanoparticles after oral application in rats: part 2. *Nanotoxicology* 11 (4): 443–453.

187 Teubl, B.J., Leitinger, G., Schneider, M. et al. (2015). The buccal mucosa as a route for TiO_2 nanoparticle uptake. *Nanotoxicology* 9 (2): 253–261.

188 MacNicoll, A., Kelly, M., Aksoy, H. et al. (2015). A study of the uptake and biodistribution of nano-titanium dioxide using in vitro and in vivo models of oral intake. *Journal of Nanoparticle Research* 17: 1–20.

189 Jones, K., Morton, J., Smith, I. et al. (2015). Human in vivo and in vitro studies on gastrointestinal absorption of titanium dioxide nanoparticles. *Toxicology Letters* 233 (2): 95–101.

190 Velasco, E., Baldovino-Medrano, V.G., Gaigneaux, E.M. et al. (2016). Development of an efficient strategy for coating TiO_2 on polyester-cotton fabrics for bactericidal applications. *Topics in Catalysis* 59 (2–4): 378–386.

191 Bianchi, M., Carnevale, G., Pisciotta, A. et al. (2017). Osteogenic differentiation of human dental pulp stem cells on bone-apatite like coatings obtained by pulsed electron deposition. *ECerS2017* 8: 516.

192 Bonetta, S., Bonetta, S., Motta, F. et al. (2013). Photocatalytic bacterial inactivation by TiO_2-coated surfaces. *Amb Express* 3: 1–8.

193 Sekiguchi, Y., Yao, Y., Ohko, Y. et al. (2007). Self-sterilizing catheters with titanium dioxide photocatalyst thin films for clean intermittent catheterization: basis and study of clinical use. *International Journal of Urology* 14 (5): 426–430.

194 Foster, H.A., Ditta, I.B., Varghese, S. et al. (2011). Photocatalytic disinfection using titanium dioxide: spectrum and mechanism of antimicrobial activity. *Applied Microbiology and Biotechnology* 90: 1847–1868.

195 Li, H., Cui, Q., Feng, B. et al. (2013). Antibacterial activity of TiO_2 nanotubes: influence of crystal phase, morphology and Ag deposition. *Applied Surface Science* 284: 179–183.

196 Gao, A., Hang, R., Huang, X. et al. (2014). The effects of titania nanotubes with embedded silver oxide nanoparticles on bacteria and osteoblasts. *Biomaterials* 35 (13): 4223–4235.

197 Hang, R., Gao, A., Huang, X. et al. (2014). Antibacterial activity and cytocompatibility of Cu–Ti–O nanotubes. *Journal of Biomedical Materials Research Part A* 102 (6): 1850–1858.

198 Giavaresi, G., Ambrosio, L., Battiston, G.A. et al. (2004). Histomorphometric, ultrastructural and microhardness evaluation of the osseointegration of a nanostructured titanium oxide coating by metal-organic chemical vapour deposition: an in vivo study. *Biomaterials* 25 (25): 5583–5591.

199 Cui, C., Liu, H., Li, Y. et al. (2005). Fabrication and biocompatibility of nano-TiO_2/titanium alloys biomaterials. *Materials Letters* 59 (24–25): 3144–3148.

200 Azzawi, Z.G., Hamad, T.I., Kadhim, S.A. et al. (2018). Osseointegration evaluation of laser-deposited titanium dioxide nanoparticles on commercially pure titanium dental implants. *Journal of Materials Science: Materials in Medicine* 29: 1–11.

201 Souza, J.C., Sordi, M.B., Kanazawa, M. et al. (2019). Nano-scale modification of titanium implant surfaces to enhance osseointegration. *Acta Biomaterialia* 94: 112–131.

202 Sodagar, A., Khalil, S., Kassaee, M.Z. et al. (2016). Antimicrobial properties of poly (methyl methacrylate) acrylic resins incorporated with silicon dioxide and titanium dioxide nanoparticles on cariogenic bacteria. *Journal of Orthodontic Science* 5 (1): 7.

203 Choi, S.-H., Jang, Y.-S., Jang, J.-H. et al. (2019). Enhanced antibacterial activity of titanium by surface modification with polydopamine and silver for dental implant application. *Journal of Applied Biomaterials & Functional Materials* 17 (3): 2280800019847067.

204 Prorokova, N., Kumeeva, T.Y., and Kuznetsov, O.Y. (2018). Antimicrobial properties of polyester fabric modified by nanosized titanium dioxide. *Inorganic Materials: Applied Research* 9: 250–256.

205 Li, S., Zhu, T., Huang, J. et al. (2017). Durable antibacterial and UV-protective Ag/TiO_2@ fabrics for sustainable biomedical application. *International Journal of Nanomedicine* 12: 2593–2606.

206 Kim, S., Im, S., Park, E.-Y. et al. (2020). Drug-loaded titanium dioxide nanoparticle coated with tumor targeting polymer as a sonodynamic chemotherapeutic agent for anti-cancer therapy. *Nanomedicine: Nanotechnology, Biology and Medicine* 24: 102110.

207 Nešić, M., Žakula, J., Korićanac, L. et al. (2017). Light controlled metallo-drug delivery system based on the TiO_2-nanoparticles and Ru-complex. *Journal of Photochemistry and Photobiology A: Chemistry* 347: 55–66.

208 Akram, M.W., Raziq, F., Fakhar-e-Alam, M. et al. (2019). Tailoring of Au-TiO_2 nanoparticles conjugated with doxorubicin for their synergistic response and photodynamic therapy applications. *Journal of Photochemistry and Photobiology A: Chemistry* 384: 112040.

209 Li, Q., Wang, X., Lu, X. et al. (2009). The incorporation of daunorubicin in cancer cells through the use of titanium dioxide whiskers. *Biomaterials* 30 (27): 4708–4715.

210 Faria, H.A.M. and de Queiroz, A.A.A. (2015). A novel drug delivery of 5-fluorouracil device based on TiO_2/ZnS nanotubes. *Materials Science and Engineering: C* 56: 260–268.

211 Darvishi, M.H., Nomani, A., Hashemzadeh, H. et al. (2017). Targeted DNA delivery to cancer cells using a biotinylated chitosan carrier. *Biotechnology and Applied Biochemistry* 64 (3): 423–432.

212 Hashemzadeh, H., Allahverdi, A., Ertl, P. et al. (2019). Comparison between three-dimensional spheroid and two-dimensional monolayer in A549 lung cancer and PC9 normal cell lines under treatment of silver Nanoparticles. *Modares Journal of Biotechnology* 10 (4): 573–580.

213 Esfandyari, J., Shojaedin-Givi, B., Mozafari-Nia, M. et al. (2019). *Diatom* biosilica shell manipulation with gold, spion nanoparticles and trastuzumab with aims of diagnostics of her2 cells. *Modares Journal of Biotechnology* 10 (4): 581–588.

214 Venkatasubbu, G.D., Ramasamy, S., Reddy, G.P. et al. (2013). In vitro and in vivo anticancer activity of surface modified paclitaxel attached hydroxyapatite and titanium dioxide nanoparticles. *Biomedical Microdevices* 15: 711–726.

215 Wang, T., Jiang, H., Wan, L. et al. (2015). Potential application of functional porous TiO_2 nanoparticles in light-controlled drug release and targeted drug delivery. *Acta Biomaterialia* 13: 354–363.

216 Kashkooli, F.M., Jakhmola, A., Hornsby, T.K. et al. (2023). Ultrasound-mediated nano drug delivery for treating cancer: fundamental physics to future directions. *Journal of Controlled Release* 355: 552–578.

217 Li, R., Li, T., and Zhou, Q. (2020). Impact of titanium dioxide (TiO_2) modification on its application to pollution treatment—a review. *Catalysts* 10 (7): 804.

218 Haugen, H.J., Monjo, M., Rubert, M. et al. (2013). Porous ceramic titanium dioxide scaffolds promote bone formation in rabbit peri-implant cortical defect model. *Acta Biomaterialia* 9 (2): 5390–5399.

219 Saiz, E., Zimmermann, E.A., Lee, J.S. et al. (2013). Perspectives on the role of nanotechnology in bone tissue engineering. *Dental Materials* 29 (1): 103–115.

220 Hou, X., Mao, D., Ma, H. et al. (2015). Antibacterial ability of Ag–TiO_2 nanotubes prepared by ion implantation and anodic oxidation. *Materials Letters* 161: 309–312.

221 Bosco, R., Van Den Beucken, J., Leeuwenburgh, S. et al. (2012). Surface engineering for bone implants: a trend from passive to active surfaces. *Coatings* 2 (3): 95–119.

222 Lu, C. and Fuchs, E. (2014). Sweat gland progenitors in development, homeostasis, and wound repair. *Cold Spring Harbor Perspectives in Medicine* 4 (2): a015222.

223 Sen, C.K., Gordillo, G.M., Roy, S. et al. (2009). Human skin wounds: a major and snowballing threat to public health and the economy. *Wound Repair and Regeneration* 17 (6): 763–771.

224 Yan, Y., Zhang, X., Huang, Y. et al. (2014). Antibacterial and bioactivity of silver substituted hydroxyapatite/TiO_2 nanotube composite coatings on titanium. *Applied Surface Science* 314: 348–357.

225 Shuai, C., Shuai, C., Feng, P. et al. (2018). Antibacterial capability, physicochemical properties, and biocompatibility of TiO_2 incorporated polymeric scaffolds. *Polymers* 10 (3): 328.

226 Sivaranjani, V. and Philominathan, P. (2016). Synthesize of titanium dioxide nanoparticles using *Moringa oleifera* leaves and evaluation of wound healing activity. *Wound Medicine* 12: 1–5.

227 Bajsić, E.G., Bulatović, V.O., Slouf, M. et al. (2014). Characterization of biodegradable polycaprolactone containing titanium dioxide micro and nanoparticles. *International Journal of Materials and Metallurgical Engineering* 8 (7): 611–615.

228 Bali, R., Razak, N., Lumb, A. et al. (2006). The synthesis of metallic nanoparticles inside live plants. *2006 International Conference on Nanoscience and Nanotechnology*, Brisbane, Australia (03–07 July 2006). IEEE.

229 Boroumand Moghaddam, A., Moniri, M., Azizi, S. et al. (2017). Biosynthesis of ZnO nanoparticles by a new *Pichia kudriavzevii* yeast strain and evaluation of their antimicrobial and antioxidant activities. *Molecules* 22 (6): 872.

230 Jha, Z., Behar, N., Sharma, S.N. et al. (2011). Nanotechnology: prospects of agricultural advancement. *Nano Vision* 1 (2): 88–100.

231 Nasrollahzadeh, M., Maham, M., and Sajadi, S.M. (2015). Green synthesis of CuO nanoparticles by aqueous extract of *Gundelia tournefortii* and evaluation of their catalytic activity for the synthesis of N-monosubstituted ureas and reduction of 4-nitrophenol. *Journal of Colloid and Interface Science* 455: 245–253.

232 Nadeem, M., Tungmunnithum, D., Hano, C. et al. (2018). The current trends in the green syntheses of titanium oxide nanoparticles and their applications. *Green Chemistry Letters and Reviews* 11 (4): 492–502.

Part III

Applications

12

Applications of TiO_2 for Photocatalytic Water Treatment

Muhammed A. Ballim[1,], Xinzhu Pang[2,*], Nosipho Moloto[1], Nathan Skillen[3], Peter K.J. Robertson[4], Kathryn Ralphs[4,*]*

[1] *School of Chemistry (Humphrey Raikes Building), University of the Witwatersrand, Braamfontein, Johannesburg, South Africa*
[2] *School of Forensic Medicine, China Medical University, Shenyang, Liaoning, People's Republic of China*
[3] *International Centre for Brewing and Distilling, Institute for Biological Chemistry, Biophysics and Bioengineering, School of Engineering and Physical Sciences, Heriot-Watt University, Riccarton, Edinburgh, UK*
[4] *School of Chemistry and Chemical Engineering, Queen's University Belfast, Belfast, UK*

*Corresponding authors: k.ralphs@qub.ac.uk; muhammed.ballim@yahoo.com; xzpang@cmu.edu.cn

12.1 Introduction

Clean water provision for an ever-rising global population is a serious challenge for the modern world. In 2021, over 2 billion people lived in water stressed countries and in 2022, and at least 1.7 billion people used drinking water contaminated with feces [1]. Drinking water contaminated with pathogenic micro-organisms can transmit diseases such as diarrhea, cholera, dysentery typhoid and polio and it is estimated to cause over half a million deaths each year [2]. With a continual rise in global population coupled with climate change, these statistics are only expected to worsen.

As a consequence of climate change, marine and freshwater ecosystems are warming, acidifying and deoxygenating in combination with the extensive use of phosphates in agriculture. Parallel to this, the impacts of harmful algal blooms on these ecosystems are intensifying [3]. Cyanotoxins are extremely toxic substances produced by bloom-forming cyanobacteria and display hepatotoxic and neurotoxic effects [4]. Heterogeneous photocatalysis, an advanced oxidation process has been demonstrated to be a potential method for degrading a broad range of contaminants in water [5]. TiO_2 has been shown to be a highly effective photocatalyst in the destruction of these harmful cyanobacterial toxins [4, 6].

The excess use of antibiotics, herbicides and pesticides has been identified to have a serious detrimental effect on human health and the environment by contaminating

Nanostructured TiO_2: Synthesis, Photocatalytic Properties and Applications, First Edition.
Edited by Jia Hong Pan, Wan In Lee and Detlef W. Bahnemann

water bodies [7]. In particular, the use of herbicides such as 2-methyl-4-chlorophenoxyacetic acid (MCPA) which is used to selectively control broad leaf weeds in grasslands and cereals crop fields. Kelly et al. have demonstrated the photocatalytic degradation of MCPA using commercial TiO_2 [8].

Intensive livestock farming has also led to an excessive use of antibiotics for prevention of infection and growth promotion in animals which has contributed to the formation of antibiotic resistant bacterial strains [9]. Antibiotics end up in water bodies through agricultural effluents. Again, TiO_2 have proved to be effective for antibiotic degradation in water [10].

As mentioned above, heterogeneous catalysis utilizing TiO_2 as a photocatalyst is an advanced oxidation process (AOP) and has the potential to completely degrade the parent component to safer products and intermediates. It is a low energy process that can treat a variety of contaminants, temperatures, and pressure conditions [11]. The different types of AOPs can include hydroxyl radicals, UV light, ozone, Fenton reagents and sulfate radicals.

Other typical contaminated water treatment options include coagulation, filtration, ozonation, adsorption, biological treatment, and plasma treatment [12]. These methods are, however, expensive, and can often yield harmful by-products. This can include potentially unsafe particulate matter in the treated water resulting from deficient filtration [13]. Additionally, disposal issues can arise with coagulation and the ineffective and low throughput biological treatment in non-biodegradable pollutants [14]. The disposal of consumed treatment media and ultimate fate of the contaminants that are removed needs to be considered in modern day water treatment solutions.

TiO_2 nanoparticles for water pollution remediation have been investigated extensively over the past few decades. TiO_2 has various excellent properties such as high photocatalytic activity, low cost, photo, chemical and biological stability [15]. TiO_2 however, has a large bandgap of 3.2 eV and requires ultraviolet (UV) light to promote electrons between the valence and conduction bands within the semiconductor particles. Following photoexcitation TiO_2 generates reactive oxygen species (ROS) following reaction of surface adsorbed species the photogenerated electron–hole pair in the semiconductor particle. These highly reactive species can completely degrade contaminants in water very rapidly. TiO_2 photocatalysis has been utilized for the treatment of a broad spectrum of water borne contaminants, including chlorinated organic compounds [16, 17], polycyclic aromatic hydrocarbons [18], dyes [19], phenols [20], pesticides [21], arsenic [22], cyanide [23], and heavy metals [24].

TiO_2 has some drawbacks, such as the large bandgap, which limits its photocatalytic activity to a narrow range within the light spectrum—primarily in the ultraviolet (UV) region. To improve their photocatalytic properties, metal doping has been considered. The dopants improve visible light absorbance and photocatalytic activity under UV irradiation [25, 26]. This will be discussed later in the chapter in more detail.

This chapter is not to provide an exhaustive review of TiO_2 as a photocatalyst for water treatment, material synthesis or mechanistic studies as this has already been done in an extensive range of excellent review articles to date [27–32]. This chapter aims to provide a broad overview of the potential of this simple metal oxide photocatalyst for the treatment of contaminated water, and more importantly explores the necessary steps to translate research into practical, large-scale solutions.

12.2 Photodegradation Mechanism

TiO_2 has been the most widely used photocatalyst for water treatment over the past several decades and photocatalytic mechanisms have been extensively discussed and reviewed in the literature [14, 33–38].

For a photocatalytic reaction to occur, thermodynamically, the conduction band and valence band edges are required to match the redox potential of the substrates. Due to the broad range of both photocatalyst materials and waterborne pollutants, the photocatalytic degradation mechanisms are typically very complicated. Hence this section will focus on a basic overview of this mechanism using unmodified TiO_2 and a simple organic pollutant, denoted as R.

The photocatalytic properties of titania were demonstrated effectively by Fujishima and Honda in 1972 [39]. Using TiO_2 they illustrated that when the energy of the incident light is larger than the bandgap it can be absorbed by the semiconductor [39].

The energy excites electrons in the valance band, causing them to migrate to the conduction band. This process generates photoelectrons (e^-) and electronic holes (h^+) as shown in Eq. (12.1), with all the reactions shown in Figure 12.1:

$$TiO_2 + hv \rightarrow e^- + h^+ \tag{12.1}$$

Recombination of the electron-hole (e^-/h^+) might occur on the surface of the catalyst, which is known as an oxy-reduction reaction [40]. When the electron-hole (h^+) in the valence band reacts with water or surface adsorbed hydroxide ions highly reactive hydroxyl radicals (·OH) can be generated (Eqs. 12.2 and 12.3).

$$h^+ + H_2O_{ads} \rightarrow H^+_{ads} + OH \cdot_{ads} \tag{12.2}$$

$$h^+ + OH^-_{ads} \rightarrow OH \cdot_{ads} \tag{12.3}$$

The valence band hole can also react with species directly adsorbed on the TiO_2 surface such as organic chemical compounds (R) including chlorinated organic compounds [16, 17], dyes [19], phenols [20], or agrichemicals [21], as detailed above (Eq. 12.4).

$$h^+ + R_{ads} \rightarrow R \cdot_{ads} \tag{12.4}$$

The photogenerated electrons (e^-) in the conduction band can also react with surface adsorbed oxygen generating superoxide radical anions (Eq. 12.5)

$$e^- + O_{2ads} \rightarrow O_2 \cdot -_{ads} \tag{12.5}$$

Electron/hole recombination can occur particularly if there are no appropriate acceptors or donors of electrons:

$$TiO_2 + h^+ + e^- \rightarrow TiO_2 \tag{12.6}$$

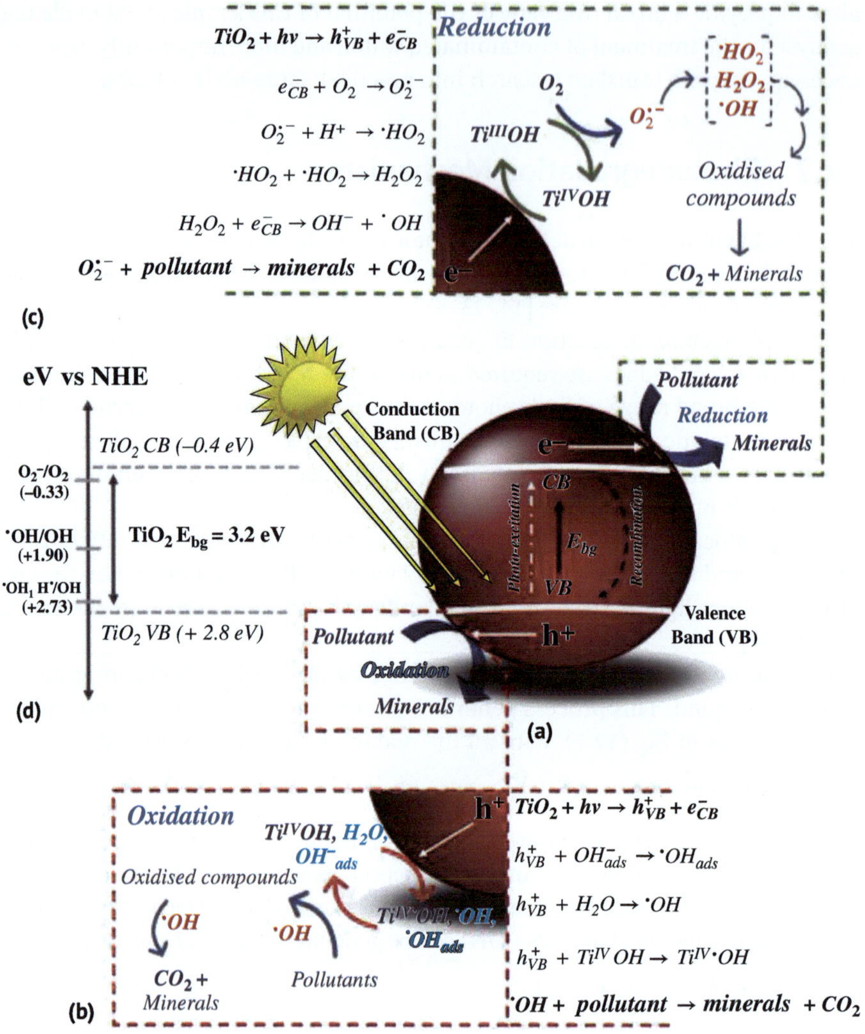

Figure 12.1 An overview of TiO_2 photocatalysis illustrating (a) photon absorption, photo-excitation and the generation of an electron–hole pair, (b) the oxidative process occurring at the VB which generates ˙OH, (c) the reduction process occurring at the CB which can generate $O_2^{\cdot -}$, $^{\cdot}HO_2$, H_2O_2, and ˙OH, and (d) the redox scale showing the potential required for radical formation (vs. NHE) concerning the electronic structure of TiO_2 [42]/with permission of Elsevier.

It is possible to protonate the superoxide radical to form a hydroperoxyl radical ($HO_2\cdot$) which then forms hydrogen peroxide (H_2O_2). Hydrogen peroxide can then dissociate into two highly reactive hydroxyl radicals:

$$O_2 \cdot_{ads} + H^+ \rightarrow HO_2 \cdot_{ads} \tag{12.7}$$

$$2HO_2 \cdot abs \rightarrow H_2O_{2ads} + O_2 \tag{12.8}$$

$$H_2O_{2abs} \rightarrow 2OH \cdot_{ads} \tag{12.9}$$

The hydroxy radicals may also degrade pollutants (R), ultimately resulting in mineralization to carbon dioxide, water, and mineral acids:

$$R + OH\cdot_{ads} \rightarrow \text{intermediate of reaction} \rightarrow CO_2 + H_2O \quad (12.10)$$

The overall photocatalytic degradation mechanism is as follows and the whole mechanism is displayed in Figure 12.1 [13]:

- Diffusion of the contaminant from the bulk of the solution towards the photocatalyst and subsequent adsorption to its surface.
- Photoexcitation of the catalyst when irradiated with light of energy greater than the bandgap. This enables electrons in the valence band to be excited to the conduction band leaving a "hole" in the valance band, creating an electron–hole pair.
- Migration of the electron–hole pairs from the bulk of the photocatalyst to the surface of hence enabling oxidation–reduction reactions can occur [41].
- Electrons can then react with H^+ to form hydrogen or oxygen to produce superoxide radical anions and holes can react with H_2O or hydroxide ions produce hydroxyl radicals [13].

12.3 Research on Modified TiO_2 Catalysts for Water Purification

Improvements to the light-absorbing properties of TiO_2 have been made by increasing its photosensitivity and identifying the correlation between its surface, interfacial and microstructural characteristics, and the corresponding mechanism [43]. Strategies to enhance the properties of TiO_2 include element doping [44] and, formation of a heterojunction [45], synthesis of a nanostructure [46], and enhancement of surface optical characteristics using metal nanostructures, metal complexes or organic dyes [47].

To enhance the catalytic activity of TiO_2 particularly under visible light, doping with metal-/non-metal-doped TiO_2 has been extensively investigated. Doping with metal/non-metals such as Pd [48], Pt [49], Au [50], Ag [50], Ce [51], Sr [52], and V [53] has been used to influence the light absorption characteristics in the visible region of the spectrum for TiO_2 [13]. Addition of such elements can create a new energy level or interband in the bandgap close to the conduction band (CB). Doping using metals modify the electronic structure and hence can lower the bandgap [29]. Additionally, the metal ions can also reduce the overall activity of TiO_2 by acting as recombination centers for electrons and holes [54]. More recently, non-metals such as N, C, S, and P have been used to narrow the bandgap of TiO_2 [55] thus significantly enhancing catalytic activity under visible light irradiation [55]. The presence of nonmetal elements such as N can interact with 2p electrons of oxygen (in TiO_2) and hence influence the valance band energy lowering the bandgap [29].

In addition to doping, the introduction of oxygen vacancies to the catalyst has also been used to enhance the photocatalytic performance of materials. Oxygen vacancies are intrinsic defects where the oxygen from the crystal lattice has detached. They are typically found in metal oxides [56]. Oxygen vacancies can be classified as surface, subsurface and bulk phase. They have found significant attention in the field of photocatalysis as an effective means to enhance the performance of the catalysts. For example, introducing oxygen vacancies by changing the coordination number of Ti from 6 to 4 has been shown to increase the selectivity of the NO photoreduction reaction to preferentially produce N_2 rather than nitrates [57].

Coupling TiO_2 with metal oxides such as ZnO [58], SiO_2 [59], Cu_2O [60], Bi_2O_3 and $ZnMn_2O_4$ [61], and graphene [62] has been employed in many applications. The presence of metal oxides in the catalyst has been found to enhance its lifetime by improving charge separation. When TiO_2 mixed with another metal oxide semiconductor is irradiated, both materials will have an electron excited from their valence bands to their conduction bands. However, the TiO_2 conduction band should be more favorable than the conduction band of the other semiconductor material and the valence band of the other material should be more cathodic than that of TiO_2. The electrons excited to the conduction band of the semiconductor can migrate to the conduction band of the TiO_2 thus increasing the concentration of electrons in the TiO_2 conduction band. Simultaneously, the hole generated at the valence band of the TiO_2 can be transferred to the valence band of the semiconductor material thus increasing the concentration of holes [63].

As well as doping and coupling TiO_2 with various elements, TiO_2 composites have been prepared. For example, Xie et al. have recently prepared TiO_2–biochar composites from the pyrolysis of inexpensive cornstalk powder followed by heating with TiO_2 [64]. Biochar has a large specific surface area and has been reported to have the potential to immobilize micro/nanoparticles for effective water treatment. Immobilization of TiO_2 onto porous solid matrices such as biochar enables facile separation, which is particularly useful following photocatalytic reactions.

As mentioned before, TiO_2 and TiO_2-based materials were employed in photocatalytic degradation processes for various pollutants, including organic compounds organisms and heavy metals. Key examples of these applications are summarized and shown in Table 12.1.

Table 12.1 Photocatalytic degradation efficiencies of TiO_2 and doped TiO_2 materials for various pollutants.

Type of compound	Classification	Pollutant	Efficiency	References
Organic compounds	Dyes	Resazurin redox dye	1.6%	[65]
		Methyl blue	99.6%	[66]
		Rhodamine B	99.2%	
		Methyl orange	99.4%	
	Petroleum hydrocarbon	Naphthalene	80%	[67]
	Phenolic compounds	Phenols	100%	[68]

Type of compound	Classification	Pollutant	Efficiency	References
Heavy metals	Cr	Cr(IV)	100%	[69]
	Pb	Pb(II)	99.6%	[70]
	Hg	Hg(II)	100%	[71]
Pesticides	-	Pestanal	90%	[72]
	-	Imazapyr	100%	[73]
	-	Phenoxyacetic acid	87%	[74]
Herbicides	-	Glyphosate	100%	[75]
	-			
Micro organisms	-	*E. coli*	99.99%	[76]
	-	*E. coli*	99%	[77]
Pharmaceutical	Antibiotic	Tetracycline	70.5%	[78]
	Anti-inflammatory	Ibuprofen	87%	[79]
	Lipid regulators	Metformin	50%	[80]
Inorganic		Nitrogen oxides (NO)	71%	[81]

12.4 Photocatalytic Reactor Design

While the development of novel and efficient TiO_2-based photocatalysts for water treatment is an important area of research, for practical application of this technology for water treatment, the development of photocatalytic reactors capable of treating cubic meter volumes of water is critical. In engineering, for a photocatalytic process that is viable from both a technical and economical basis, several key issues need to be taken into consideration:

- How the photocatalyst is distributed within the reactor unit.
- How effectively light is distributed within the reactor and the photocatalyst is being effectively irradiated.
- What are the mass transfer characteristics of the reactor and can mass transfer limitations be minimized.
- What are the reaction kinetics of the redox processes on the photocatalyst surface and how can these be maximized.

When it comes to reactor design, two major issues are limitations of photon transfer to the photocatalyst surface and mass transfer limitations within the reactor. Photocatalytic reactors are classified into different categories, based on size (e.g., microreactor), light source, geometry, agitation methodology, and bed dynamics [13, 82, 83]. The most common types of photocatalytic reactors are batch and continuous flow. Photocatalysts may be deployed as slurries (batch) and the reactions are run in individual batches or the photocatalyst is immobilized, and the reactor has a continuous flow of substrates often with a recirculation loop.

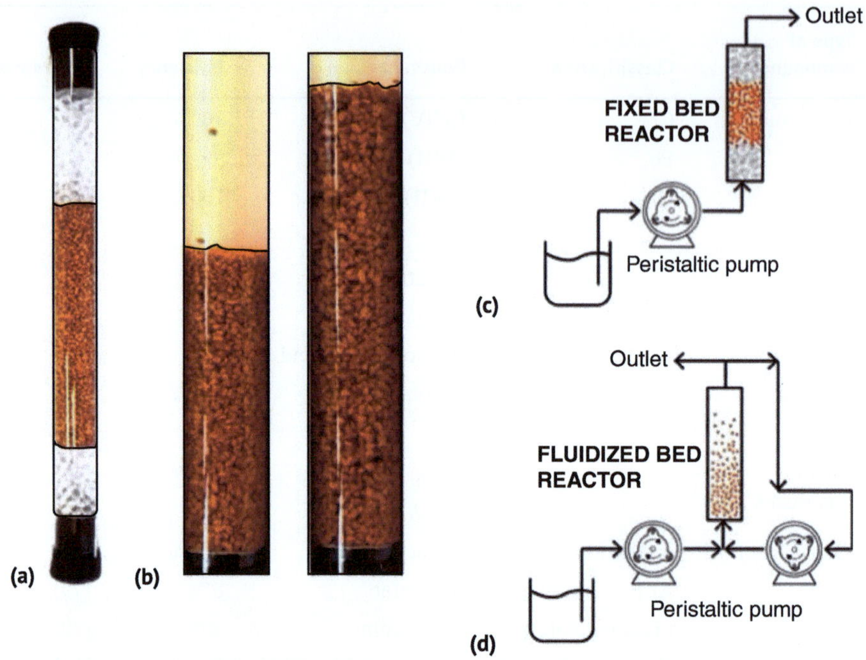

Figure 12.2 (a) Fixed bed reactor, D_p = (1.00–1.41) mm, (b) fluidized bed reactor D_p = (1.00–1.41) mm, left: no flow; right: flow rate of 313 mL min^{-1}. Experimental setup of (c) the fixed bed reactor and (d) the fluidized bed reactor [87]/with permission of Elsevier.

Figure 12.2 shows a fluidized bed reactor where the catalyst is directly loaded on the granular carrier and placed in the solution. The advantages of this type of reactor include small mass transfer restrictions, fast reaction rate, and a large surface area of catalyst. However, these reactors frequently suffer from blockages and difficulties in recovering the catalyst, making industrialization of these reactors more complicated [84, 85].

The fixed bed reactor is superior in the case that the retrieval of the catalyst following reaction is facile. However, by immobilization of the catalyst on the reactor, the surface area is significantly reduced. The largest challenges to scale these types of reactors, is the choice of catalyst and the fixing technology. Scaling up a photoreactor is no easy feat. Overcoming mass transfer issues and light penetration are the major obstacles. Additionally, significant reactor modeling is required [86]. Photocatalytic reactors have been modeled for both liquid and gas phases using computational fluid dynamics (CFD). A limitation in CFD is that the light source emission is represented by a linear source model, which only suits some reactors. Some examples of photocatalytic reactors that have been reported in the literature are detailed below.

In a paper by Adams et al., two reactor concepts were considered for the removal of pollutants from waste effluent sources [88]. The first reactor was based on a thin film photo reactor, which was scaled up to create a stack of thin films plates, shown in Figure 12.3. Polymethylmethacrylate (PMMA) and Ti metal were investigated as materials for the reactor.

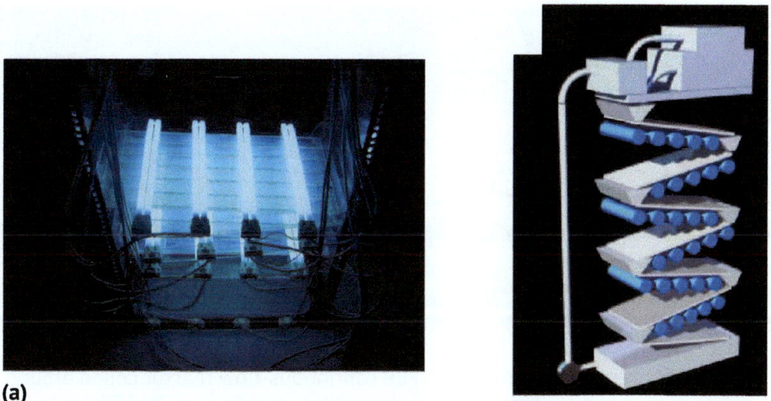

Figure 12.3 Multiple thin film plate reactor stack (a) lab-based unit and (b) concept design for scaled-up unit [88]/John Wiley & Sons/CC BY-3.0.

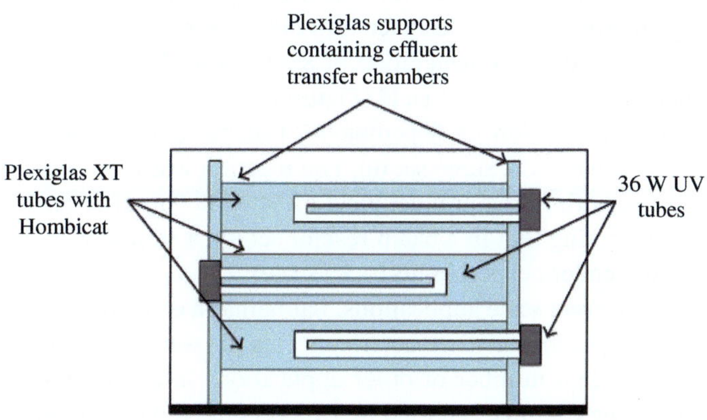

Figure 12.4 Drum reactor configuration in UV box [86]/with permission of Elsevier.

The mounting angle of the plates was investigated. It was found that by altering the coated plate angle with a 15 degrees tilt resulted in 60% destruction of the hydrocarbon after just 15 min irradiation. Using a lower angle of 15 degrees, increased the retention time of the catalyst and substrate, allowing for successful photocatalytic destruction. Additionally, it was found that adding H_2O_2 to the setup increased the degradation of the pollutant to 80% after 135 min.

The second reactor concept utilized a drum reactor (Figure 12.4). After 10 min, 90% of VOCs present in the wastewater effluent were destroyed using a photocatalyst loading of 600 g. The wastewater effluent was passed through three consecutive reactor drums with one third of the total photocatalyst.

A microreactor developed by Brooker-Milburn was utilized to facilitate the [2 + 2] photocycloaddition of maleimide and *n*-hexyne [89]. The first reactor was a reactor that consisted of three layers of fluorinated ethylene-propylene (FEP) tubing with an internal diameter of 2.7 mm ID. The FEP tubing was coiled around a Pyrex

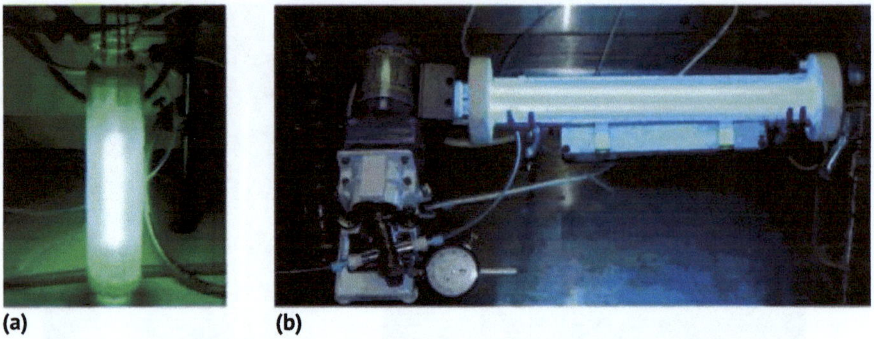

Figure 12.5 UV photomicroreactors: (a) FEP continuous-flow reactor coiled around a 400 W medium-pressure Hg lamp [90]/American Chemical Society. (b) FEP continuous-flow reactor with a 36 W PL-L lamp (254 nm) [91]/John Wiley & Sons.

tube which contained a 400 W Hg lamp. To avoid reactor clogging and reduce the build-up of back pressure at higher flow rates, tubing with a diameter of 2.7 mm was chosen. Reagents were continuously fed into the photoreactor over a 24-hr reaction period which resulted in the generation of 85 g of the [2 + 2] photocycloaddition product which equated to an 82% yield (Figure 12.5a).

The [5 + 2] cycloaddition of *N*-pentenyl-3,4-dichloromaleimide (Figure 12.5b) was also investigated in the same set up. The reaction was complete after a 6 min residence time resulting in a product yield of 64% (38.5 g) after an 11-hr reaction time. It is worth noting that in a batch reactor set up, a similar productivity would require 120 independent experiments (0.5 g scale).

With respect to larger scale applications, Bahnemann et al. investigated a number of solar activated photocatalytic reactor concepts for water treatment and considered these alongside a number of other applications in a very interesting review in 2004 [92]. In one example reported by Dillert et al., the development and application of the double sheet skin photocatalytic reactor which used a 5 g L^{-1} slurry of P25 TiO_2 photocatalyst deployed in an enclosed and channeled polymer methyl methacrylate rectangular box structure was reported (Figures 12.6 and 12.7) [93]. Arrays of up to 12 of these units were connected together providing a total (irradiated area of 27.6 m^2). The reactors were deployed for treatment of industrial effluents in both Germany and Brazil [93]. The reactor suspension had a total treatment volume of 1 m^3, and the effluent was recycled through reactor sheets at a rate of 800–1300 L h^{-1}. After about 11 h of solar irradiation, the organic contamination of the effluent reduced by over 50% as monitored using TOC and COD.

A comprehensive review by Spasiano et al. considered applications of solar photocatalysis in detail including materials and reactor concepts at prototype and commercial scale [94]. An interesting industrial application utilized a compound parabolic collector where the photocatalyst is immobilized on the inside to a glass or Perspex tube and the contaminated water flows through these tubes. Sun light is directed onto the tube coated with the photocatalysts using parabolic mirrors (Figure 12.8).

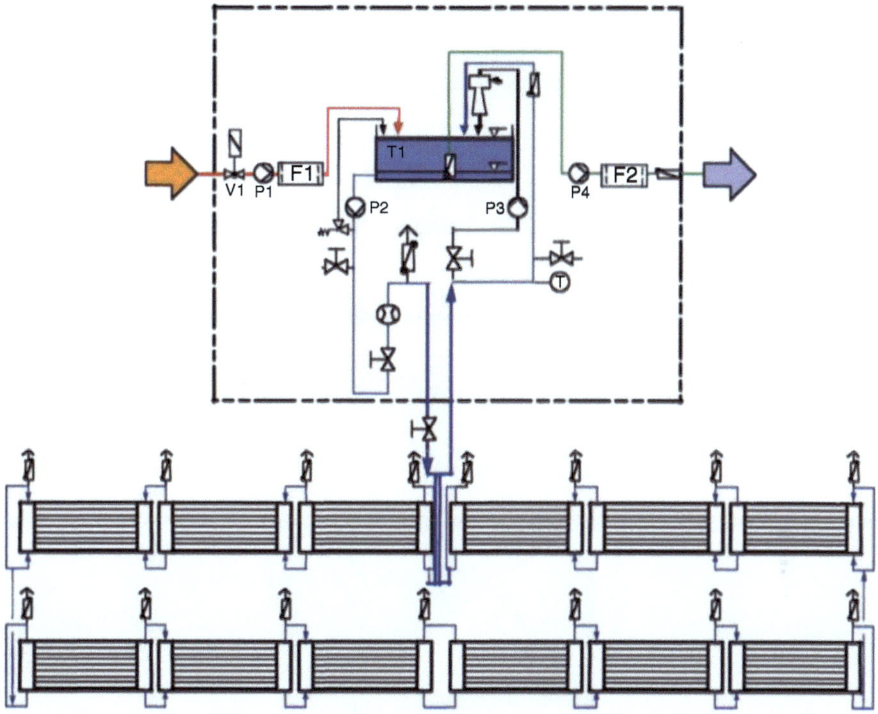

Figure 12.6 Flow chart of a DSSR pilot plant installed in the Wolfsburg factory of the Volkswagen AG. Reproduced from [86]/with permission of Elsevier.

Figure 12.7 Photograph of DSSR Pilot Plant installed in the Wolfsburg Factory of Volkswagen AG [93]. Reproduced from *Solar Energy*, 2005, 77, 445–459 with permission from Elsevier.

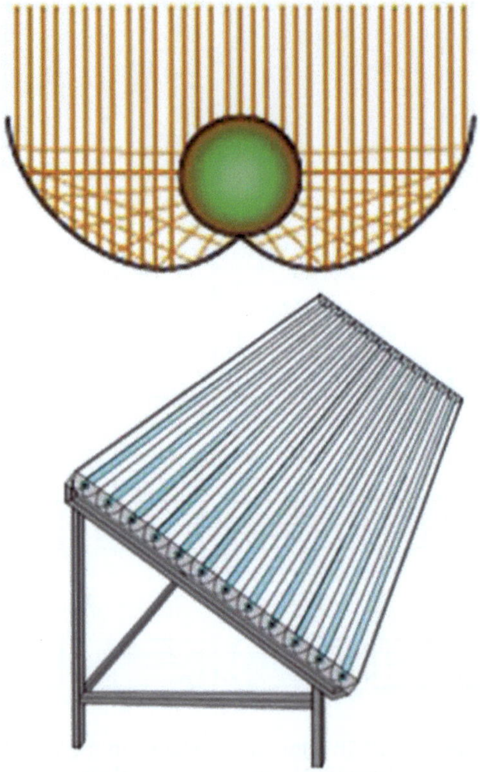

Figure 12.8 CPC solar photocatalytic reactor. Reproduced from [94]/with permission of Elsevier.

Martinz-Garcia reported the use of a pilot scale (up to 54 L per batch) V trough reactor for the solar disinfection of rainwater, which was an evolution of the CPC concept (Figure 12.9) [95]. This unit was demonstrated to be effective for the treatment of water contaminated with four strains of pathogenic bacteria including *E. faecalis, E. coli, S. enteriditis*, and *P. aeruginosa* [95].

In 2022, Pestana et al. reported a different approach to the application of photocatalysis for water treatment [96]. Rather than developing a water treatment reactor, a system that was deployed directly into a lake contaminated with a blue-green algal bloom was developed. In this system, the photocatalysts were distributed into 1 m length steel mesh pods which had UV-LED strips lining them. These pods were deployed in mesocosms within a eutrophic drinking water reservoir for a week (Figure 12.10). Following this treatment period, the abundance of cyanobacteria was reduced by 85% while other water quality parameters such as turbidity and transparency improved by 40% and 81%, respectively.

Transitioning photocatalysis toward pilot scale and industrial deployment is a challenging process. The technology readiness level (TRL) method was developed by NASA for measuring the maturity of a technology or process with levels 1–3 cover basic fundamental research at laboratory scale and demonstrating proof of principle, levels 4–6 covered the development of the technology to a demonstration or proof of concept scale, while level 7 corresponded to large-scale industrial and commercial applications. The majority of academic research on TiO_2

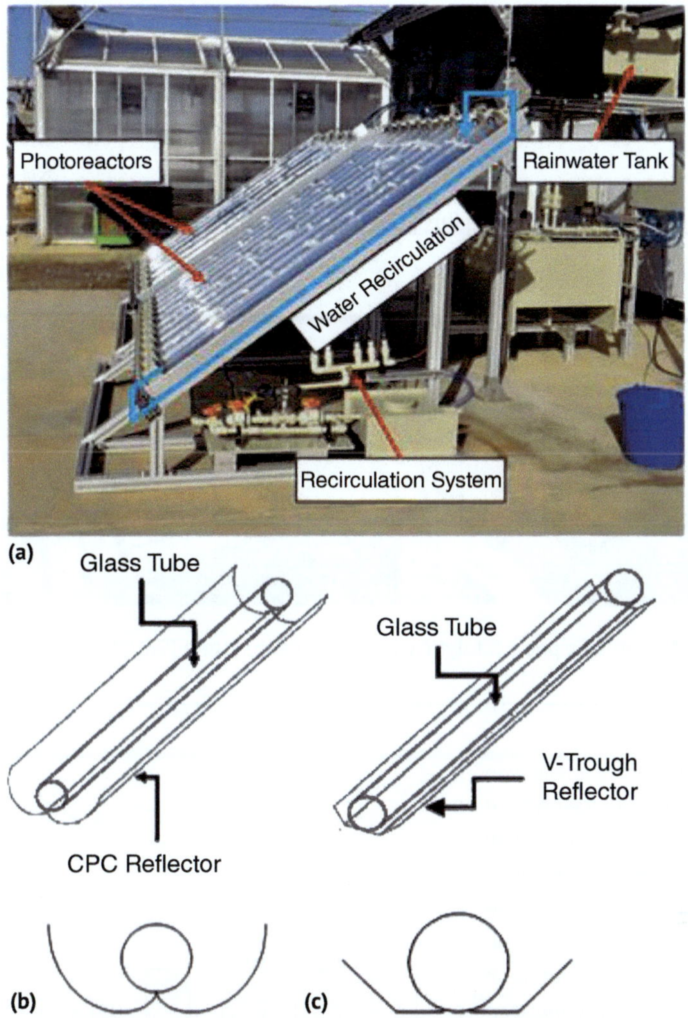

Figure 12.9 Photograph of the V-trough photo-reactor (a) diagram of the solar reflectors used, CPC-mirror (b), and V-trough mirror (c) [95]/Elsevier/CC BY-4.0.

photocatalysis for water treatment focuses on improving fundamental understanding and material synthesis and hence is carried out in controlled lab-based environments indicating a TRL of 1–3 with a few examples at 4–5. Transitioning beyond this, a TRL of 4–5 represents a photocatalytic system operating in a "relevant environment," which requires significant industrial involvement. To date, there have been few examples of photocatalytic water treatment systems achieving TRLs of 7–9. It could be argued that the focus of research on fundamentals and material design rather than on reactor design and scale up has hindered increasing the TRL of photocatalytic processes hence industrial implementation remains limited. As discussed above, the two significant challenges to scale up are photon transfer and utilization in the photocatalytic reactor and mass transfer limitations in units capable of operating at cubic meter quantities. These issues are discussed in more detail in the following section.

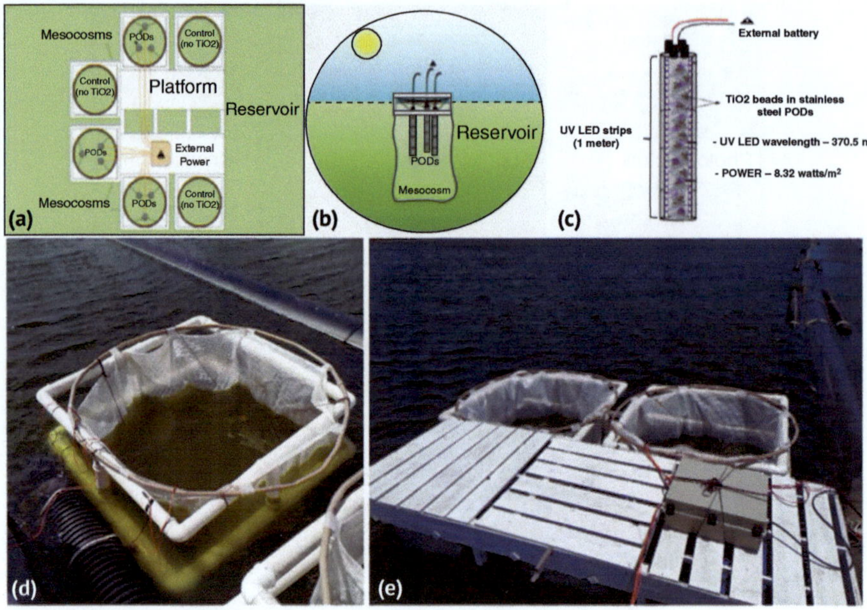

Figure 12.10 In lake photocatalytic water treatment system [96]/Elsevier/CC BY-4.0.

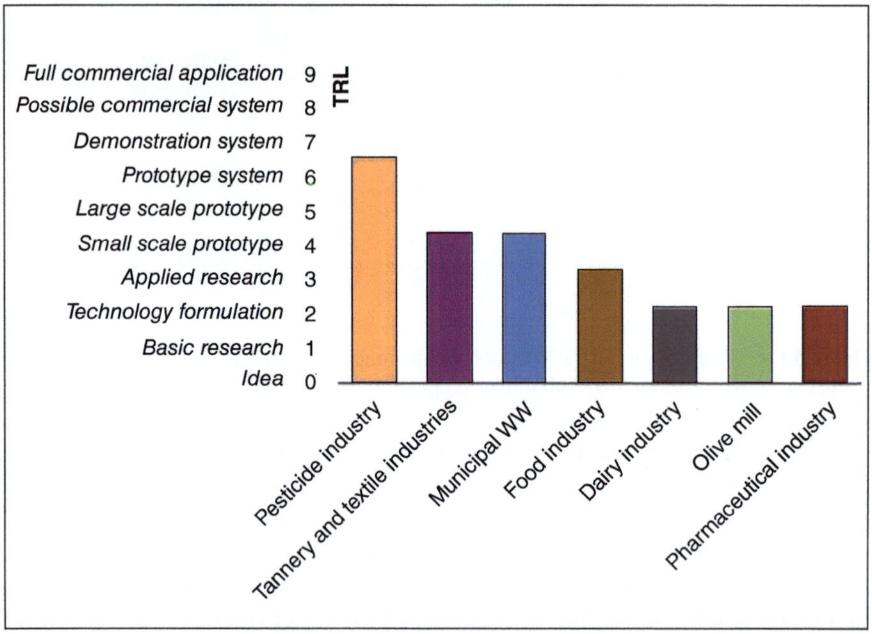

Figure 12.11 TRL of photocatalytic processes applied to different industrial effluents and municipal wastewater [97]/Elsevier/CC BY 4.0.

In their review on solar photocatalysis for water treatment, Muscetta et al. highlighted various TRLs of photocatalytic water treatment applications as shown in Figure 12.11 [97]. Overall, they considered that treatment of pesticide-contaminated water is probably at the highest TRL level to date, but this was still only at a TRL of 6.

Figure 12.12 Reactor technology from Purifics. Photo-Cat. Purifics, https://www.purifics.com/photo-cat, last accessed January 2025.

Despite the level of research activity in the field of photocatalysis for water treatment over the past five decades, it has been somewhat difficult to find examples of commercialized processes. A paper published by Mills et al. in 2020 detailed seven companies involved in the commercialization of photocatalytic technology for water treatments [98]. Some of the companies listed in the publications, such as Eco Metal were described as having no obvious products. A publication in 2007 listed a similar quantity of companies [99]. When searched online, information about these companies and their technologies was extremely limited. After significant research, some examples were found:

An industrial scale photocatalytic water purification technology by "Purifics Photo-Cat," which is described as the largest photocatalytic AOP system in the world (Figure 12.12) [100]. The technology uses a TiO_2 slurry-based photocatalytic system for water purification [100]. Another commercialized example is the solar bag from Paralytics which employs a patented photochemical water purification process to purify up to 3 L of contaminated water in 2–3 h (sunny conditions).

12.5 Kinetic Analysis

When it comes to reactor design, the most important challenge is the determination of the influence of the photon absorption rate on the reaction rate due to the process being light activated [101]. Early studies routinely utilized Langmuir–Hinshelwood rate equations to estimate kinetic parameters (see Eqs. 12.11 and 12.12)

$$\text{Rate} = \frac{k_{\text{cat}}\, K_{\text{ads}}\, \text{C}}{[1 + K_{\text{ads}}\text{C}]} \tag{12.11}$$

$$K_{\text{ads}} = \frac{k_1}{k_{-1}} \tag{12.12}$$

where k_{cat} represents the catalyst activity and K_{ads} represents the adsorption equilibrium of the Langmuir isotherm.

The rate constants, k_1 and k_{-1}, were not believed to be influenced by light intensity, while experimentally, it has been shown that both k_{cat} and K_{ads} depended on light intensity; thus, the Langmuir–Hinshelwood rate equations were not sufficient for photocatalytic reactions [102–104].

To correct this situation, a pseudo-steady state (PSS) assumption, involving extremely unstable intermediates such as free radicals, has been invoked. This approach enabled k_{cat} to have any value reactive to k_{-1}. The PSS analysis resulted in an apparent adsorption constant, K_{ads}, which was dependent on k_{cat}, and hence was influenced by light intensity [105].

$$K_{ads}(\text{apparent}) = \frac{k_1}{k_{-1} + k_{cat}} \tag{12.13}$$

Ollis et al. applied a PSS assumption to a study on the oxidation of acetophenone by Xu and Langford [105, 106]. Ollis extrapolated the experimental data from Xu and Langford [106] and plotted against $I^{0.5}$, which is shown in Figure 12.13. It was determined that first-order dependence passing through origin provides a better fit to data for k_{LH}; K plots are ambiguous.

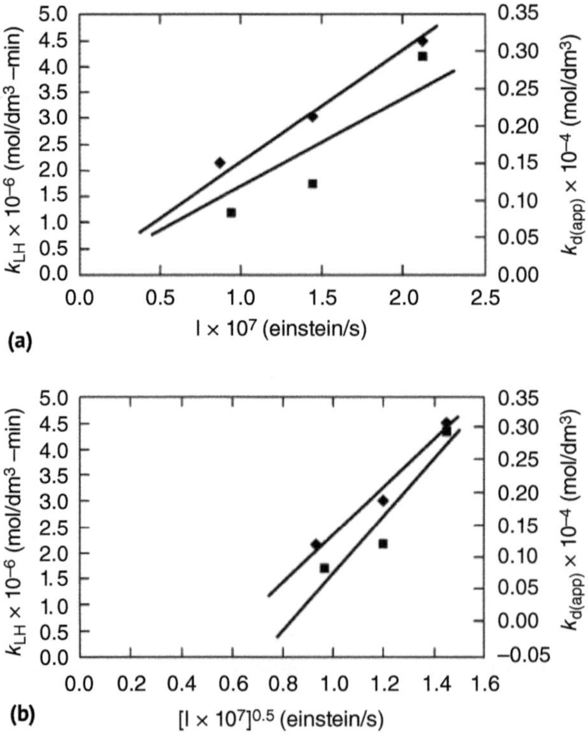

Figure 12.13 Plot of calculated $K_{d(app)}$ and k_{LH} for acetophenone degradation vs intensity, I (top) or square root of intensity (bottom). Reproduced from [105]/with permission of American Chemical Society.

It has been found that an air gap affects the efficiency of substrate degradation in continuous flow photocatalytic reactors. At low concentrations, the rate limiting step is the transfer of unhindered molecules (due to low concentration) from the bulk to the surface of the catalyst. Reactors that incorporate a recycling/recirculation loop have been shown to improve efficiency with the number of recycling runs having a significant impact on the overall efficiency of the process [107]. Reduction of efficiency of a photocatalytic process can also result from catalyst poisoning, which inactivates the photocatalyst surface. Additionally sintering and aging of the catalyst can also reduce the efficiency of the catalyst by resulting in a reduced adsorption capacity on the surface of the photocatalyst. Another cause of the reduction in efficiency is due to fouling, when an impurities are deposited on active sites. A critical parameter for maximizing the number of redox reactions is to ensure oxygen levels are maximized through effective aeration by stirring or air sparging. Airlift reactors are typically preferred in chemical and engineering fields due to the ease of construction, high removal efficiency, low sheer force, and better light availability ratios.

12.6 Challenges and Perspectives

12.6.1 Technical Challenges

A vast number of scientific reports in the literature that focus on photocatalytic water treatment, including the elucidation of complex mechanisms, material synthesis, and the enhancement of kinetics and selectivity of the processes. Despite significant academic advancement, there remains a considerable gap between the progress achieved at the academic lab scale and what is required for industrial application. Figure 12.14 shows publication trends in photocatalysis research and the percentage which focuses on industrial application. It can be seen that increasing attention is being placed on the transition from lab to industrial scale.

Despite various progress being made, transitioning of photocatalysis toward the pilot scale TRL method was developed by NASA as a method for measuring the maturity of a technology or process with levels 1–3 covering basic fundamental research at laboratory scale and demonstrating proof of principle, levels 4–6 covered the development of the technology to a demonstrated or proof of concept scale, while level 7 corresponded to large-scale industrial and commercial applications. The majority of academic research on TiO_2 photocatalysis for water treatment is typically focused on improving fundamental understanding and material synthesis and hence is carried out in controlled lab-based environments indicating a TRL of 1–3 with a few examples at 4–5. Transitioning from this, a TRL of 4–5 represents a photocatalytic system operating in a "relevant environment," which requires significant industrial involvement. To date, there have been few examples of photocatalytic water treatment systems achieving TRLs of 7–9.

It could be argued that the focus of research on fundamentals and material design rather than on reactor design and scale up has hindered increasing the TRL of

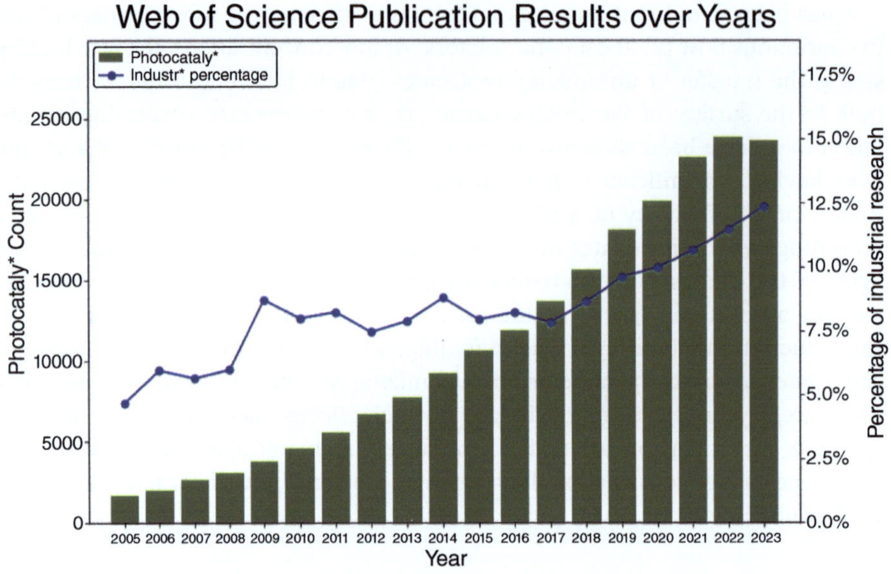

Figure 12.14 Publication trends in photocatalysis research. Web of Science results for the number of yearly publications from 2005 to 2023. The blue line represents the percentage of papers on industrial research about photocatalysis relative to the total publications.

photocatalytic processes; hence, industrial implementation remains limited. The two significant challenges to scale up remain: photon effective transfer and utilization in the photocatalytic reactor and mass transfer limitations in units capable of operating at cubic meter quantities.

12.6.1.1 Environmental and Health Concerns

Some materials used as catalysts for photodegradation or their degradation products may be toxic, posing risks on humans and the wider environment. Therefore, a careful and comprehensive environmental assessment about a reaction and its scale-up application is needed. On the other hand, safe handling, recycling, and disposal of photocatalysts are necessary to minimize environmental impact and lower the cost.

12.6.1.2 Integration with Existing Processes

Starting a factory from scratch is very difficult. Integrating photocatalysis with existing industrial processes is more realistic but still challenging, requiring significant modifications and investments. Maintaining consistent performance and control over the photocatalytic process in an industrial setting is complex due to varying light conditions and reactant properties.

Addressing these challenges requires interdisciplinary research and development efforts, including advances in material science, chemical engineering, and reactor design. New photocatalysts with improved efficiency, stability, and selectivity, coupled with innovative reactor designs and optimized reaction conditions, are essential for the successful industrial implementation of photocatalytic processes.

12.6.2 Future Perspectives

12.6.2.1 Innovative Reactor Designs

Future reactor designs are expected to focus on scalability and efficiency, ensuring uniform light distribution and maximizing photocatalyst exposure. Advanced reactor configurations, such as fluidized beds, microreactors, and photoreactors with optimized light sources, are expected to play a significant role. On the other hand, integrating photocatalytic processes with renewable energy sources, such as solar photovoltaics, is able to enhance sustainability. Hybrid systems that combine photocatalysis with other green technologies will be developed to leverage synergies and improve overall process efficiency.

12.6.2.2 Digitalization and Process Optimization

The integration of digital tools and data analytics is of great importance to virtualize and optimize photocatalytic processes. These will be employed to optimize reaction conditions and scale up relations efficiently. In a recent review, Muscetta et al. provided an excellent consideration of issues and challenges in the development of solar photocatalytic water treatment processes [97]. In their paper, they proposed an interesting and robust methodological approach for the initial modeling stages that need to be considered in developing a photocatalytic process (Figure 12.15). Advanced sensors and real-time monitoring systems will enable precise control over photocatalytic reactions, ensuring consistent performance and rapid response to process variations.

12.6.2.3 Economic Viability

Achieving sustainability and economic viability in photocatalysis for industrial applications involves a multifaceted approach. Accurate cost assessments enable stakeholders to make informed decisions regarding the implementation and scaling of photocatalytic processes. It may vary but all of them include the total installation cost (facility, project contingency, engineering project, replacement cost) and the operating expenses, personal cost, (maintenance, electricity, materials, and services) [97, 108]. In the study by Vela et al., an assumption that the average annual sunshine exposure in their treatment area was ~3000 h and that the useful time (UT) of their pilot plant was 480 min per day [109]. The treatment cost

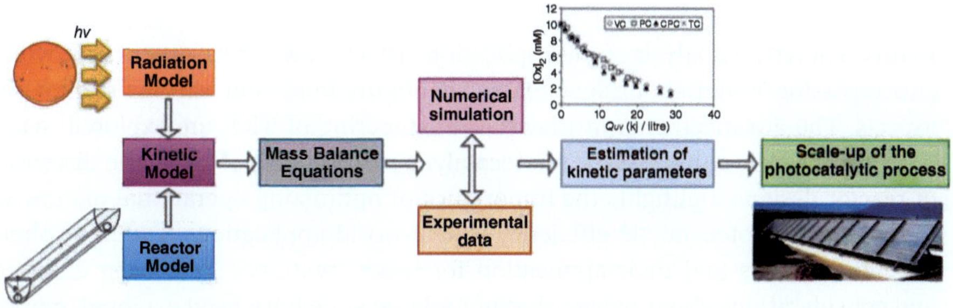

Figure 12.15 Methodological procedure for modeling a photocatalytic process [97]/ Elsevier/CC BY 4.0.

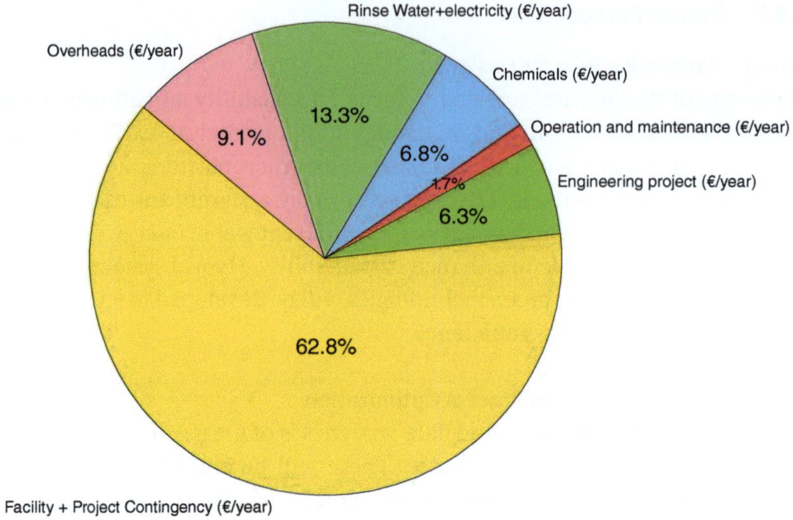

Figure 12.16 Annual expenditures breakdown with TiO_2 photocatalysis. Adapted from S. Dutta [110].

was calculated in the range 103–258 € m^{-3} for photocatalytic treatment using various doped TiO_2 materials based on an annual work cycle of 365 days and 8 h per day. With bad weather considerations, the price increased to ~144–406 € m^{-3} in 240 days of use, with the annual expenditures breakdown shown in Figure 12.16 [109]. Additionally, conducting comprehensive economic viability analyses, including return on investment, market analysis, and exploring economic incentives, ensures that photocatalytic technologies are both sustainable and economically attractive in the long term.

In summary, the future perspectives of photocatalysis in industry are centered around technological advancements, sustainability, and integration with digital tools and renewable energy. Continued innovation and collaboration will be key to overcoming current challenges and unlocking the full potential of photocatalysis for industrial applications.

12.7 Conclusion

In this chapter, an analysis of the application of TiO_2 in water treatment is provided, encompassing materials design, reactor configurations, and various engineering aspects. The advancements in materials engineering of TiO_2 are explored, which have significantly enhanced the photocatalytic performance of TiO_2. The discussion of reactor designs highlights the importance of optimizing operational parameters to maximize photocatalytic efficiency in real-world applications. Different photocatalytic reactors and their application for water treatment have been discussed and considerations about reactor design for large scale have been explored, emphasizing the need for tailored solutions depending on specific application requirements. Engineering aspects such as kinetic analysis and cost evaluation are critical for understanding the practical feasibility of TiO_2-based water treatment technologies. TRL scale of photocatalytic water treatment was examined, which emphasize the key considerations to advance this technique. These analyses are all essential for guiding the scale-up and commercialization of TiO_2 technologies. Despite the promising advancements, several challenges remain in the widespread adoption of TiO_2 for water treatment. These challenges include the limited light absorption range of TiO_2 and the recombination of electron–hole pairs. Moreover, environmental and regulatory considerations are crucial for the large-scale implementation of TiO_2 technologies. Moving forward, continued research and development efforts are necessary to overcome these challenges. Innovations in materials design, reactor engineering, and system integration, along with interdisciplinary collaborations, will be key to unlocking the full potential of TiO_2 in water treatment. By addressing these issues, TiO_2-based technologies hold significant promise for providing sustainable and efficient solutions for water purification.

References

1. UN-Water (2023). WHO/UNICEF Joint Monitoring Program for Water Supply, Sanitation and Hygiene (JMP) – Progress on household drinking water, sanitation and hygiene 2000–2022: Special focus on gender. https://www.unwater.org/publications/who/unicef-joint-monitoring-program-update-report-2023 (accessed 28 October 2024).
2. WHO (2023). Drinking-water. https://www.who.int/news-room/fact-sheets/detail/drinking-water (accessed 29 July 2024).
3. Griffith, A.W. and Gobler, C.J. (2020). Harmful algal blooms: a climate change co-stressor in marine and freshwater ecosystems. *Harmful Algae* 91: 101590. https://doi.org/10.1016/j.hal.2019.03.008.
4. Robertson, P.K.J., Lawton, L.A., Münch, B., and Rouzade, J. (1997). Destruction of cyanobacterial toxins by semiconductor photocatalysis. *Chem. Commun.* 4: 393–394. https://doi.org/10.1039/A607965B.
5. Wang, J.L. and Xu, L.J. (2012). Advanced oxidation processes for wastewater treatment: formation of hydroxyl radical and application. *Crit. Rev. Environ. Sci. Technol.* 42 (3): 251–325. https://doi.org/10.1080/10643389.2010.507698.
6. Pestana, C.J., Hui, J., Camacho-Muñoz, D. et al. (2023). Solar-driven semiconductor photocatalytic water treatment (TiO_2, g-C_3N_4, and TiO_2+g-C_3N_4) of cyanotoxins: proof-of-concept study with microcystin-LR. *Chemosphere* 310: 136828. https://doi.org/10.1016/j.chemosphere.2022.136828.
7. Du, C. (2022). Fe-based metal organic frameworks (Fe-MOFs) for organic pollutants removal via photo-Fenton: a review. *Chem. Eng. J.* 431: 133932.
8. Kelly, J., Morrison, G., Skillen, N. et al. (2019). An investigation of the role of pH in the rapid photocatalytic degradation of MCPA and its primary intermediate by low-power UV led irradiation. *Chem. Eng. J.* 359: 112–118.
9. Manyi-Loh, C., Mamphweli, S., Meyer, E., and Okoh, A. (2018). Antibiotic use in agriculture and its consequential resistance in environmental sources: potential public health implications. *Mol. J. Synth. Chem. Nat. Prod. Chem.* 23 (4): 795. https://doi.org/10.3390/molecules23040795.

10 Asih, A.K., Yetti, R.D., and Chandra, B. (2021). Photodegradation of antibiotic using TiO_2 as a catalyst: a review. *Int. J. Pharm. Sci. Med.* 6 (2): 37–43. https://doi.org/10.47760/ijpsm.2021.v06i02.004.

11 Zheng, X., Shen, Z.-P., Shi, L. et al. (2017). Photocatalytic membrane reactors (PMRs) in water treatment: configurations and influencing factors. *Catalysts* 7 (8): 224. https://doi.org/10.3390/catal7080224.

12 Mahmoodi, N.M., Arami, M., Limaee, N.Y., and Tabrizi, N.S. (2006). Kinetics of heterogeneous photocatalytic degradation of reactive dyes in an immobilized TiO_2 photocatalytic reactor. *J. Colloid Interface Sci.* 295 (1): 159–164. https://doi.org/10.1016/j.jcis.2005.08.007.

13 Binjhade, R., Mondal, R., and Mondal, S. (2022). Continuous photocatalytic reactor: critical review on the design and performance. *J. Environ. Chem. Eng.* 10 (3): 107746. https://doi.org/10.1016/j.jece.2022.107746.

14 Konstantinou, I.K. and Albanis, T.A. (2004). TiO_2-assisted photocatalytic degradation of azo dyes in aqueous solution: kinetic and mechanistic investigations. *Appl. Catal. B Environ.* 49 (1): 1–14. https://doi.org/10.1016/j.apcatb.2003.11.010.

15 Rawal, S.B., Bera, S., Lee, D. et al. (2013). Design of visible-light photocatalysts by coupling of narrow bandgap semiconductors and TiO_2: effect of their relative energy band positions on the photocatalytic efficiency. *Catal. Sci. Technol.* 3 (7): 1822–1830. https://doi.org/10.1039/C3CY00004D.

16 Kalantary, R.R., Dadban Shahamat, Y., Farzadkia, M. et al. (2015). Photocatalytic degradation and mineralization of diazinon in aqueous solution using nano-TiO_2(Degussa, P25): kinetic and statistical analysis. *Desalination Water Treat.* 55 (2): 555–563. https://doi.org/10.1080/19443994.2014.928795.

17 Ohsaka, T., Shinozaki, K., Tsuruta, K., and Hirano, K. (2008). Photo-electrochemical degradation of some chlorinated organic compounds on n-TiO_2 electrode. *Chemosphere* 73 (8): 1279–1283. https://doi.org/10.1016/j.chemosphere.2008.07.016.

18 Guo, H. and Barnard, A.S. (2013). Naturally occurring iron oxide nanoparticles: morphology, surface chemistry and environmental stability. *J. Mater. Chem. A* 1 (1): 27–42. https://doi.org/10.1039/C2TA00523A.

19 Lee, Y.-S., Kim, S.J., Jang, J.-S. et al. (2008). Anion Co-doped titania for solar photocatalytic degradation of dyes. *Carbon Lett.* 9 (2): 131–136. https://doi.org/10.5714/CL.2008.9.2.131.

20 Nguyen, A.T., Hsieh, C.-T., and Juang, R.-S. (2016). Substituent effects on photodegradation of phenols in binary mixtures by hybrid H_2O_2 and TiO_2 suspensions under UV irradiation. *J. Taiwan Inst. Chem. Eng.* 62: 68–75. https://doi.org/10.1016/j.jtice.2016.01.012.

21 Gar Alalm, M., Tawfik, A., and Ookawara, S. (2015). Comparison of solar TiO_2 photocatalysis and solar photo-Fenton for treatment of pesticides industry wastewater: operational conditions, kinetics, and costs. *J. Water Process Eng.* 8: 55–63. https://doi.org/10.1016/j.jwpe.2015.09.007.

22 Moon, G., Kim, D., Kim, H. et al. (2014). Platinum-like behavior of reduced graphene oxide as a cocatalyst on TiO_2 for the efficient photocatalytic oxidation of arsenite. *Environ. Sci. Technol. Lett.* 1 (2): 185–190. https://doi.org/10.1021/ez5000012.

23 Kim, S.H., Lee, S.W., Lee, G.M. et al. (2016). Monitoring of TiO_2-catalytic UV-LED photo-oxidation of cyanide contained in mine wastewater and leachate. *Chemosphere* 143: 106–114. https://doi.org/10.1016/j.chemosphere.2015.07.006.

24 Chen, Z., Li, Y., Guo, M. et al. (2016). One-pot synthesis of Mn-doped TiO_2 grown on graphene and the mechanism for removal of Cr(VI) and Cr(III). *J. Hazard. Mater.* 310: 188–198. https://doi.org/10.1016/j.jhazmat.2016.02.034.

25 Mu, W., Herrmann, J.-M., and Pichat, P. (1989). Room temperature photocatalytic oxidation of liquid cyclohexane into cyclohexanone over neat and modified TiO_2. *Catal. Lett.* 3 (1): 73–84. https://doi.org/10.1007/BF00765057.

26 Anpo, M., Kishiguchi, S., Ichihashi, Y. et al. (2001). The design and development of second-generation titanium oxide photocatalysts able to operate under visible light irradiation by applying a metal ion-implantation method. *Res. Chem. Intermed.* 27 (4): 459–467. https://doi.org/10.1163/156856701104202101.

27 Friedmann, D., Mendive, C., and Bahnemann, D. (2010). TiO_2 for water treatment: parameters affecting the kinetics and mechanisms of photocatalysis. *Appl. Catal. B Environ.* 99 (3): 398–406. https://doi.org/10.1016/j.apcatb.2010.05.014.

28 Lee, S.-Y. and Park, S.-J. (2013). TiO_2 photocatalyst for water treatment applications. *J. Ind. Eng. Chem.* 19 (6): 1761–1769. https://doi.org/10.1016/j.jiec.2013.07.012.

29 Armaković, S.J., Savanović, M.M., and Armaković, S. (2023). Titanium dioxide as the most used photocatalyst for water purification: an overview. *Catalysts* 13 (1): 26. https://doi.org/10.3390/catal13010026.

30 Schneider, J., Matsuoka, M., Takeuchi, M. et al. (2014). Understanding TiO_2 photocatalysis: mechanisms and materials. *Chem. Rev.* 114 (19): 9919–9986. https://doi.org/10.1021/cr5001892.

31 Tanos, F., Razzouk, A., Lesage, G. et al. (2024). A comprehensive review on modification of titanium dioxide-based catalysts in advanced oxidation processes for water treatment. *ChemSusChem* 17 (6): e202301139. https://doi.org/10.1002/cssc.202301139.

32 Al-Madanat, O., AlSalka, Y., Ramadan, W., and Bahnemann, D.W. (2021). TiO_2 photocatalysis for the transformation of aromatic water pollutants into fuels. *Catalysts* 11 (3): 317. https://doi.org/10.3390/catal11030317.

33 Anpo, M., Shima, T., and Kubokawa, Y. (1985). Electron-spin-resonance and photo-luminescence evidence for the photocatalytic formation of hydroxyl radiclas on small TiO_2 particles. *Chem. Lett.* 14 (11), 1799–1802, https://doi.org/10.1246/cl.1985.799

34 Bhatkhande, D.S., Pangarkar, V.G., and Beenackers, A.A.C.M. (2002). Photocatalytic degradation for environmental applications—a review. *J. Chem. Technol. Biotechnol.* 77: 102–116.

35 Chen, Y., Yang, S., Wang, K., and Lou, L. (2005). Role of primary active species and TiO_2 surface characteristic in UV-illuminated photodegradation of acid orange 7. *J. Photochem. Photobiol. Chem.* 172: 47–54. https://doi.org/10.1016/j.jphotochem.2004.11.006.

36 Hoffmann, M.R., Martin, S.T., Choi, W.Y., and Bahnemann, D.W. (1995). Environmental applications of semiconductor photocatalysis. *Chem. Rev.* 95: 69–96.

37 Jaeger, C.D. and Bard, A.J. (1979). Spin trapping and electron-spin resonance detection of radical intermediates in the photo-decomposition of water at TiO_2 particulate systems. *J. Phys, No. em.* 83: 3146–3152.

38 Yang, L., Yu, L.E., and Ray, M.B. (2008). Degradation of paracetamol in aqueous solutions by TiO_2 photocatalysis. *Water Res.* 42: 3480–3488.

39 Fujishima, A. and Honda, K. (1972). Electrochemical photolysis of water at a semiconductor electrode. *Nature* 238 (5358): 37–38. https://doi.org/10.1038/238037a0.

40 Lebeau, B., Jonas, F., Gaudin, P. et al. (2020). Dyes depollution of water using porous TiO_2-based photocatalysts. In: *Environmental Nanotechnology Volume 4* (eds. N. Dasgupta, S. Ranjan, and E. Lichtfouse), 35–92. Environmental Chemistry for a Sustainable World, Cham: Springer International Publishing. https://doi.org/10.1007/978-3-030-26668-4_2.

41 Singh, R. and Dutta, S. (2018). Synthesis and characterization of solar photoactive TiO_2 nanoparticles with enhanced structural and optical properties. *Adv. Powder Technol.* 29 (2): 211–219. https://doi.org/10.1016/j.apt.2017.11.005.

42 Skillen, N., Rice, C., Pang, X. et al. (2021). Photocatalytic radical species: an overview of how they are generated, detected, and measured. In: *Nanostructured Photocatalysts* (eds. V.-H. Nguyen, D.-V.N. Vo, and S. Nanda), 85–118. Amsterdam: Elsevier. https://doi.org/10.1016/B978-0-12-823007-7.00008-0.

43 Armaković, S.J., Mary, Y.S., Mary, Y.S. et al. (2021). Optoelectronic properties of the newly designed 1,3,5-triazine derivatives with isatin, chalcone and acridone moieties. *Comput. Theor. Chem.* 1197: 113160. https://doi.org/10.1016/j.comptc.2021.113160.

44 Huang, F., Yan, A., Zhao, H. et al. (2016). Influences of doping on photocatalytic properties of TiO_2 photocatalyst. In: *Semiconductor Photocatalysis - Materials, Mechanisms and Applications* (ed. W. Cao). IntechOpen. https://doi.org/10.5772/63234.

45 Kumar, A., Khan, M., He, J., and Lo, I.M.C. (2020). Recent developments and challenges in practical application of visible–light–driven TiO_2–based heterojunctions for PPCP degradation: a critical review. *Water Res.* 170: 115356. https://doi.org/10.1016/j.watres.2019.115356.

46 Hua, X., Liu, Z., Bruce, P.G., and Grey, C.P. (2015). The morphology of TiO_2 (B) nanoparticles. *J. Am. Chem. Soc.* 137 (42): 13612–13623. https://doi.org/10.1021/jacs.5b08434.

47 Goulart, S., Jaramillo Nieves, L.J., Dal Bó, A.G., and Bernardin, A.M. (2020). Sensitization of TiO_2 nanoparticles with natural dyes extracts for photocatalytic activity under visible light. *Dyes Pigments* 182: 108654. https://doi.org/10.1016/j.dyepig.2020.108654.

48 Wu, J., Lu, S., Ge, D. et al. (2016). Photocatalytic properties of Pd/TiO_2 nanosheets for hydrogen evolution from water splitting. *RSC Adv.* 6 (72): 67502–67508. https://doi.org/10.1039/C6RA10408H.

49 Yu, F., Wang, C., Ma, H. et al. (2020). Revisiting Pt/TiO_2 photocatalysts for thermally assisted photocatalytic reduction of CO_2. *Nanoscale* 12 (13): 7000–7010. https://doi.org/10.1039/C9NR09743K.

50 Sadrieyeh, S. and Malekfar, R. (2018). Photocatalytic performance of plasmonic Au/Ag-TiO_2 aerogel nanocomposites. *J. Non-Cryst. Solids* 489: 33–39. https://doi.org/10.1016/j.jnoncrysol.2018.03.020.

51 Cao, X., Yang, X., Li, H. et al. (2017). Investigation of Ce-TiO_2 photocatalyst and its application in asphalt-based specimens for NO degradation. *Constr. Build. Mater.* 148: 824–832. https://doi.org/10.1016/j.conbuildmat.2017.05.095.

52 Sood, S., Umar, A., Kumar Mehta, S. et al. (2015). Efficient photocatalytic degradation of brilliant green using Sr-doped TiO_2 nanoparticles. *Ceram. Int.* 41 (3, Part A): 3533–3540. https://doi.org/10.1016/j.ceramint.2014.11.010.

53 Rossi, L., Palacio, M., Villabrille, P.I., and Rosso, J.A. (2021). V-doped TiO_2 photocatalysts and their application to pollutant degradation. *Environ. Sci. Pollut. Res.* 28 (19): 24112–24123. https://doi.org/10.1007/s11356-021-12339-5.

54 Zheng, S.K., Wang, T.M., Hao, W.C., and Shen, R. (2002). Improvement of photocatalytic activity of TiO_2 thin film by Sn ion implantation. *Vacuum* 65 (2): 155–159. https://doi.org/10.1016/S0042-207X(01)00424-9.

55 Choudhury, B., Bayan, S., Choudhury, A., and Chakraborty, P. (2016). Narrowing of band gap and effective charge carrier separation in oxygen deficient TiO_2 nanotubes with improved visible light photocatalytic activity. *J. Colloid Interface Sci.* 465: 1–10. https://doi.org/10.1016/j.jcis.2015.11.050.

56 Zhan, F., Wen, G., Li, R. et al. (2024). A comprehensive review of oxygen vacancy modified photocatalysts: synthesis, characterization, and applications. *Phys. Chem. Chem. Phys.* 26 (15): 11182–11207. https://doi.org/10.1039/D3CP06126D.

57 Pan, X., Yang, M.-Q., Fu, X. et al. (2013). Defective TiO_2 with oxygen vacancies: synthesis, properties and photocatalytic applications. *Nanoscale* 5 (9): 3601–3614. https://doi.org/10.1039/C3NR00476G.

58 Siwińska-Stefańska, K., Kubiak, A., Piasecki, A. et al. (2018). TiO_2-ZnO binary oxide systems: comprehensive characterization and tests of photocatalytic activity. *Materials* 11 (5): 841. https://doi.org/10.3390/ma11050841.

59 Beyers, E., Biermans, E., Ribbens, S. et al. (2009). Combined TiO_2/SiO_2 mesoporous photocatalysts with location and phase controllable TiO_2 nanoparticles. *Appl. Catal. B Environ.* 88 (3): 515–524. https://doi.org/10.1016/j.apcatb.2008.10.009.

60 Xiong, L., Yang, F., Yan, L. et al. (2011). Bifunctional photocatalysis of TiO_2/Cu_2O composite under visible light: Ti^{3+} in organic pollutant degradation and water splitting. *J. Phys. Chem. Solids* 72 (9): 1104–1109. https://doi.org/10.1016/j.jpcs.2011.06.016.

61 Bessekhouad, Y., Robert, D., Weber, J.-V. (2005). Photocatalytic activity of Cu_2O/TiO_2, Bi_2O_3/TiO_2 and $ZnMn_2O_4/TiO_2$ heterojunctions. *Catal. Today* 101 (3): 315–321. https://doi.org/10.1016/j.cattod.2005.03.038.

62 Padmanabhan, N.T., Thomas, N., Louis, J. et al. (2021). Graphene coupled TiO_2 photocatalysts for environmental applications: a review. *Chemosphere* 271: 129506. https://doi.org/10.1016/j.chemosphere.2020.129506.

63 Tian, Y.; Yang, H.; Wu, S.; Xu, C.; Gong, B.; Yan, J.; Cen, K.; Bo, Z.; Ostrikov, K.; High-performance water purification and desalination by solar-driven interfacial evaporation and photocatalytic VOC decomposition enabled by hierarchical TiO_2-CuO nanoarchitecture – (2022) - International Journal of Energy Research - 46 (6), 7132–7145

64 Xie, Y., Liu, A., Bandala, E.R., and Goonetilleke, A. (2022). TiO_2-biochar composites as alternative photocatalyst for stormwater disinfection. *J. Water Process Eng.* 48: 102913. https://doi.org/10.1016/j.jwpe.2022.102913.

65 Sathasivam, S., Bhachu, D.S., Lu, Y. et al. (2015). Tungsten doped TiO_2 with enhanced photocatalytic and optoelectrical properties via aerosol assisted chemical vapor deposition. *Sci. Rep.* 5 (1): 10952. https://doi.org/10.1038/srep10952.

66 Nguyen, C.H., Tran, M.L., Tran, T.T.V., and Juang, R.-S. (2020). Enhanced removal of various dyes from aqueous solutions by UV and simulated solar photocatalysis over TiO_2/ZnO/rGO composites. *Sep. Purif. Technol.* 232: 115962. https://doi.org/10.1016/j.seppur.2019.115962.

67 Chen, Y., Hong, C., Xu, Q. et al. (2024). Visible light enhancement of biocarbon quantum-dot-decorated TiO_2 for naphthalene removal. *Molecules* 29 (11): 2708. https://doi.org/10.3390/molecules29112708.

68 Çinar, B., Kerimoğlu, I., Tönbül, B. et al. (2020). Hydrothermal/electrospinning synthesis of CuO plate-like particles/TiO_2 fibers heterostructures for high-efficiency photocatalytic degradation of organic dyes and phenolic pollutants. *Mater. Sci. Semicond. Process.* 109: 104919. https://doi.org/10.1016/j.mssp.2020.104919.

69 Yan, R., Luo, D., Fu, C. et al. (2020). Simultaneous removal of Cu(II) and Cr(VI) ions from wastewater by photoreduction with TiO_2–ZrO_2. *J. Water Process Eng.* 33: 101052. https://doi.org/10.1016/j.jwpe.2019.101052.

70 Bi, J., Huang, X., Wang, J. et al. (2019). Oil-phase cyclic magnetic adsorption to synthesize Fe_3O_4@C@TiO_2-nanotube composites for simultaneous removal of Pb(II) and rhodamine B. *Chem. Eng. J.* 366: 50–61. https://doi.org/10.1016/j.cej.2019.02.017.

71 Wang, X., Pehkonen, S.O., and Ray, A.K. (2004). Photocatalytic reduction of Hg(II) on two commercial TiO_2 catalysts. *Electrochimica Acta* 49 (9): 1435–1444. https://doi.org/10.1016/j.electacta.2003.10.030.

72 Kane, A., Assadi, A.A., and El Jery, A. (2022). Advanced photocatalytic treatment of wastewater using immobilized titanium dioxide as a photocatalyst in a pilot-scale reactor: process intensification. *Materials* 15 (13): 4547.

73 Mkhalid, I.A., Fierro, J.L.G., Mohamed, R.M., and Alshahri, A.A. (2020). Visible light driven photooxidation of imazapyr herbicide over highly efficient mesoporous Ag/Ag_2O–TiO_2 p-n heterojunction photocatalysts. *Ceram. Int.* 46 (16, Part A): 25822–25832. https://doi.org/10.1016/j.ceramint.2020.07.064.

74 Lannoy, A., Bleta, R., Machut-Binkowski, C. et al. (2017). Cyclodextrin-directed synthesis of gold-modified TiO_2 materials and evaluation of their photocatalytic activity in the removal of a pesticide from water: effect of porosity and particle size. *ACS Sustain. Chem. Eng.* 5 (5): 3623–3630. https://doi.org/10.1021/acssuschemeng.6b03059.

75 Verdi, I.R., Maroli Neto, A.J., Garcia, I.S. et al. (2024). Magnetic $NiFe_2O_4$/TiO_2 heterostructures for the photocatalytic decontamination of glyphosate in water. *Mater. Sci. Semicond. Process.* 174: 108205. https://doi.org/10.1016/j.mssp.2024.108205.

76 Mathew, S., Ganguly, P., Rhatigan, S. et al. (2018). Cu-doped TiO_2: visible light assisted photocatalytic antimicrobial activity. *Appl. Sci.* 8 (11): 2067. https://doi.org/10.3390/app8112067.

77 Song, J., Wu, X., Zhang, M. et al. (2020). Highly flexible, core-shell heterostructured, and visible-light-driven titania-based nanofibrous membranes for antibiotic removal and *E. coil* inactivation. *Chem. Eng. J.* 379: 122269. https://doi.org/10.1016/j.cej.2019.122269.

78 Wang, X., Jiang, L., Li, K. et al. (2020). Fabrication of novel Z-scheme $SrTiO_3$/$MnFe_2O_4$ system with double-response activity for simultaneous microwave-induced and photocatalytic degradation of tetracycline and mechanism insight. *Chem. Eng. J.* 400: 125981. https://doi.org/10.1016/j.cej.2020.125981.

79 Khalaf, S., Shoqeir, J.H., Lelario, F. et al. (2020). TiO_2 and active coated glass photodegradation of ibuprofen. *Catalysts* 10 (5): 560. https://doi.org/10.3390/catal10050560.

80 Carbuloni, C.F., Savoia, J.E., Santos, J.S.P. et al. (2020). Degradation of metformin in water by TiO_2–ZrO_2 photocatalysis. *J. Environ. Manage.* 262: 110347. https://doi.org/10.1016/j.jenvman.2020.110347.

81 Li, G., Huang, R., Zhu, C. et al. (2021). Effect of oxygen vacancies and its quantity on photocatalytic oxidation performance of titanium dioxide for NO removal. *Colloids Surf. Physicochem. Eng. Asp.* 614: 126156. https://doi.org/10.1016/j.colsurfa.2021.126156.

82 Tokode, O., Prabhu, R., Lawton, L.A., and Robertson, P.K.J. (2016). Controlled periodic illumination in semiconductor photocatalysis. *J. Photochem. Photobiol. Chem.* 319–320: 96–106. https://doi.org/10.1016/j.jphotochem.2015.12.002.

83 Tapia-Tlatelpa, T., Trull, J., and Romeral, L. (2019). In situ decolorization monitoring of textile dyes for an optimized UV-LED/TiO_2 reactor. *Catalysts* 9 (8): 669. https://doi.org/10.3390/catal9080669.

84 Bello, M.M., Abdul Raman, A.A., and Purushothaman, M. (2017). Applications of fluidized bed reactors in wastewater treatment—a review of the major design and operational parameters. *J. Clean. Prod.* 141: 1492–1514. https://doi.org/10.1016/j.jclepro.2016.09.148.

85 Ren, G., Han, H., Wang, Y. et al. (2021). Recent advances of photocatalytic application in water treatment: a review. *Nanomaterials* 11 (7): 1804. https://doi.org/10.3390/nano11071804.

86 Joven-Quintero, S.A., Castilla-Acevedo, S.F., Betancourt-Buitrago, L.A. et al. (2020). Photocatalytic degradation of cobalt cyanocomplexes in a novel LED photoreactor using TiO_2 supported on borosilicate sheets: a new perspective for mining wastewater treatment. *Mater. Sci. Semicond. Process.* 110: 104972. https://doi.org/10.1016/j.mssp.2020.104972.

87 Ulloa-Ovares, D., Rodríguez-Rodríguez, C.E., Masís-Mora, M., and Durán, J.E. (2021). Simultaneous degradation of pharmaceuticals in fixed and fluidized bed reactors using iron-modified diatomite as heterogeneous Fenton catalyst. *Process Saf. Environ. Prot.* 152: 97–107. https://doi.org/10.1016/j.psep.2021.05.032.

88 Adams, M., Campbell, I., and Robertson, P.K.J. (2008). Novel photocatalytic reactor development for removal of hydrocarbons from water. *Int. J. Photoenergy* 2008 (1): 674537. https://doi.org/10.1155/2008/674537.

89 Su, Y., Straathof, N.J.W., Hessel, V., and Noël, T. (2014). Photochemical transformations accelerated in continuous-flow reactors: basic concepts and applications. *Chem. Eur. J.* 20 (34): 10562–10589. https://doi.org/10.1002/chem.201400283.

90 Hook, B.D.A., Dohle, W., Hirst, P.R. et al. (2005). A practical flow reactor for continuous organic photochemistry. *J. Org. Chem.* 70 (19): 7558–7564. https://doi.org/10.1021/jo050705p.

91 Maskill, K.G., Knowles, J.P., Elliott, L.D. et al. (2013). Complexity from simplicity: tricyclic aziridines from the rearrangement of pyrroles by batch and flow photochemistry. *Angew. Chem.* 125: 1539–1542.

92 Bahnemann, D. (2004). Photocatalytic water treatment: solar energy applications. *Sol. Energy* 77: 445–459. https://doi.org/10.1016/j.solener.2004.03.031.

93 Dillert, R., Vollmer, S., Gross, E. et al. (1999). Solar-catalytic treatment of an industrial wastewater. *Z. Für Phys. Chem.* 213 (2): 141–147. https://doi.org/10.1524/zpch.1999.213.Part_2.141.

94 Spasiano, D., Marotta, R., Malato, S. et al. (2015). ChemInform abstract: solar photocatalysis: materials, reactors, some commercial, and pre-industrialized applications. A comprehensive approach. *Appl. Catal. B Environ.* 170–171: 90–123. https://doi.org/10.1016/j.apcatb.2014.12.050.

95 Martinez-Garcia, A., Vincent, M., Rubiolo, V. et al. (2020). Assessment of a pilot solar V-trough reactor for solar water disinfection. *Chem. Eng. J.* 399 (125719): 1. https://doi.org/10.1016/j.cej.2020.125719.

96 Pestana, C.J., Santos, A.A., Capelo-Neto, J. et al. (2022). Suppressing cyanobacterial dominance by UV-LED TiO_2-photocatalysis in a drinking water reservoir: a mesocosm study. *Water Res.* 226: 119299. https://doi.org/10.1016/j.watres.2022.119299.

97 Muscetta, M., Ganguly, P., and Clarizia, L. (2024). Solar-powered photocatalysis in water purification: applications and commercialization challenges. *J. Environ. Chem. Eng.* 12: 113073. https://doi.org/10.1016/j.jece.2024.113073.

98 Mills, A. and Lee, S.-K. (2002). A web-based overview of semiconductor photochemistry-based current commercial applications. *J. Photochem. Photobiol. Chem.* 152 (1): 233–247. https://doi.org/10.1016/S1010-6030(02)00243-5.

99 Gerven, T. V.; Mul, G.; Moulijn, G.; Stankiewicz, A.; (2007). A Review of Intensification of Photocatalytic Processes', Chem. Eng. Process.; 46; 781–789. 10.1016/j.cep.2007.05.012.'

100 Purifics. Photo-Cat. https://www.purifics.com/photo-cat (accessed 30 July 2024).

101 Schneider, J., Bahnemann, D., Ye, J. et al. (2016). *Photocatalysis: Fundamentals and Perspectives*. Royal Society of Chemistry.

102 Ollis, D.F. (2018). Kinetics of photocatalyzed reactions: five lessons learned. *Front. Chem.* 6: 378. https://doi.org/10.3389/fchem.2018.00378.

103 Mills, A., O'Rourke, C., and Moore, K. (2015). Powder semiconductor photocatalysis in aqueous solution: an overview of kinetics-based reaction mechanisms. *J Photochem Photobiol Chem* 310: 66–105. https://doi.org/10.1016/j.jphotochem.2015.04.011.

104 Emeline, A., Ryabchuk, V., and Serpone, N. (2000). Factors affecting the efficiency of a photocatalyzed process in aqueous metal-oxide dispersions: prospect of distinguishing between two kinetic models. *J Photochem Photobiol Chem* 133: 89–97. https://doi.org/10.1016/S1010-6030(00)00225-2.

105 Ollis, D.F. (2005). Kinetics of liquid phase photocatalyzed reactions: an illuminating approach. *J. Phys. Chem. B* 109 (6): 2439–2444. https://doi.org/10.1021/jp040236f.

106 Xu, Y. and Langford, C.H. (2000). Variation of Langmuir adsorption constant determined for TiO_2-photocatalyzed degradation of acetophenone under different light intensity. *J. Photochem. Photobiol. Chem.* 133 (1): 67–71. https://doi.org/10.1016/S1010-6030(00)00220-3.

107 EL-Mekkawi, D.M., Nady, N., Abdelwahab, N.A. et al. (2016). Flexible bench-scale recirculating flow CPC photoreactor for solar photocatalytic degradation of methylene blue using removable TiO_2 immobilized on PET sheets. *Int. J. Photoenergy* 2016: e9270492. https://doi.org/10.1155/2016/9270492.

108 Suhan, M.B.K., Al-Mamun, M.R., Farzana, N. et al. (2023). Sustainable pollutant removal and wastewater remediation using TiO_2-based nanocomposites: a critical review. *Nano-Struct. Nano-Objects* 36: 101050. https://doi.org/10.1016/j.nanoso.2023.101050.

109 Vela, N., Calín, M., Yáñez-Gascón, M.J. et al. (2018). Solar reclamation of wastewater effluent polluted with bisphenols, phthalates and parabens by photocatalytic treatment with $TiO_2/Na_2S_2O_8$ at pilot plant scale. *Chemosphere* 212: 95–104. https://doi.org/10.1016/j.chemosphere.2018.08.069.

110 Dutta, S. (2020). Wastewater treatment using TiO_2-based photocatalysts. In: *Handbook of Smart Photocatalytic Materials* (eds. C. Mustansar Hussain and A.K. Mishra), 303–323. Amsterdam: Elsevier. https://doi.org/10.1016/B978-0-12-81905 1-7.00010-5.

13

TiO_2 Photocatalytic Degradation of Antibiotics

Shiping Xu, Wanting Sun, Tianhao Song, Bo Wang*

School of Environmental Science and Engineering, Shandong University, Jimo, Qingdao, P.R. China

*Corresponding author: shiping.xu@sdu.edu.cn

13.1 Introduction

Antibiotics, a vital class of drugs used to treat infections from pathogenic microorganisms, are widely employed in both human and veterinary medicine. These drugs are often partially metabolized within organisms and can be excreted into the environment, either in their original form or as metabolites, through urine and feces. The overuse and misuse of antibiotics have significantly increased the prevalence of antibiotic-resistant bacteria (ARB) and antibiotic-resistant genes (ARGs), posing a significant threat to ecosystems and human health [1]. The World Health Organization (WHO) has identified antibiotic resistance as one of the major threats to global public health, leading to prolonged treatments, higher medical expenses, and even life-threatening situations. It is estimated that in 2019, antibiotic resistance was responsible for 1.27 million deaths globally, exceeding the number of deaths caused by AIDS and malaria [2]. The WHO further warns that without intervention, by 2050, the death toll from resistant bacterial infections could reach 10 million [3]. Moreover, by 2030, antibiotic resistance could plunge 240 million people into extreme poverty [2]. Therefore, the efficient and safe removal of antibiotics from water has become an urgent global priority. However, conventional wastewater treatment processes face considerable challenges in removing these drugs effectively and may inadvertently amplify the potential risk of spreading resistance.

Advanced oxidation processes (AOPs), recognized for their strong oxidative capacity, have been extensively explored for the remediation of refractory pollutants. Within the realm of AOPs, photocatalytic processes, leveraging the renewable resource of solar energy, are deemed the most environmentally benign and promising approach for the complete decomposition of antibiotics. A spectrum of photocatalytic materials, such as TiO_2, ZnO, g-C_3N_4, CdS, CuO, $BiVO_4$, WO_3, etc., have been employed for environmental remediation due to their efficient light absorption

Nanostructured TiO_2: Synthesis, Photocatalytic Properties and Applications, First Edition.
Edited by Jia Hong Pan, Wan In Lee and Detlef W. Bahnemann

and charge separation properties. Among these, TiO_2 is distinguished for its high efficiency, cost-effectiveness, physical and chemical stability, widespread availability, and noncorrosive nature, establishing it as a leading photocatalyst for practical environmental applications. Many studies have documented the application of TiO_2 for the degradation of a broad range of antibiotics. In this chapter, we will delve into the performance and mechanisms of TiO_2 photocatalytic degradation of these antibiotics in detail.

13.2 Performance and Mechanism of TiO_2 Photocatalytic Degradation of Antibiotics

Research has validated the potent degradative capacity of TiO_2 against a broad spectrum of antibiotics, encompassing tetracyclines (TCs), sulfonamides, quinolones, macrolides, β-lactams, aminoglycosides, peptide antibiotics, lincosamides, pleuromutilins, rifamycins, and nitroimidazoles. In this section, we will provide a detailed exploration of these antibiotic categories.

13.2.1 Tetracycline Antibiotics

The TC class of antibiotics, characterized by a distinctive tetracyclic naphthacene carboxamide ring system (depicted in Figure 13.1a), include prominent members, such as TC, oxytetracycline (OTC), chlortetracycline (CTC), and doxycycline. These compounds are widely utilized for the treatment and prevention of a broad spectrum of bacterial infections across both human and veterinary medicine. Despite their therapeutic benefits, their pronounced water solubility, enhanced stability in aquatic environments, and intrinsic resistance to natural degradation processes present significant ecological concerns.

Table 13.1 presents a compilation of recent studies utilizing TiO_2-based photocatalysts for the degradation of TC antibiotics. It is evident that the most extensively studied antibiotic is TC, followed by OTC and CTC, and the degradation kinetics of these antibiotics are relatively rapid. For instance, under the excitation of 350-nm UV light, P25 photocatalyst can degrade 94.8% of a 10 mg L^{-1} TC within 2 h, concurrent with a substantial degree of mineralization (35.7%) [4]. However, the efficiency of

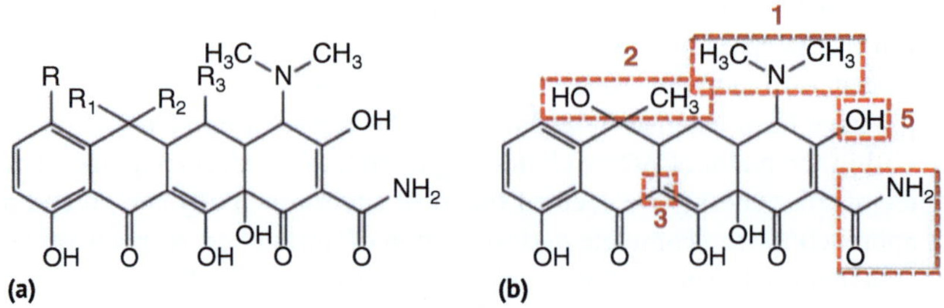

Figure 13.1 (a) The general chemical structure of tetracycline antibiotics and (b) the molecular structure of TC with the key degradation sites (encircled).

Table 13.1 Summary of recent studies employing TiO_2-based photocatalysts for the degradation of tetracycline antibiotics.

Photocatalyst	Antibiotic	Initial concentration of antibiotic (mg L^{-1})	Degradation efficiency	Degradation rate constant (min^{-1})	Degree of mineralization	Light condition	Photocatalyst dosage (g L^{-1})	Main active species	Ref.
P25	TC	10	94.8% (120 min)	0.038 (0–30 min) 0.02 (30–120 min)	35.7% (120 min)	UV light at 350 nm, 300 W xenon lamp	0.2	$\cdot O_2^-$, h^+	[4]
			76.6% (120 min)	0.036 (0–30 min) 0.0039 (30–120 min)	≈0	Visible light at 420 nm, 300 W xenon lamp		$\cdot O_2^-$	
P25	TC	0.5		0.11		UV light at 365 nm, 3 W LED lamp	0.002		[11]
N-doped TiO_2/ graphene	OTC	30	36% (160 min)	0.0063		UV light, 250 W mercury lamp			[12]
	CTC	30	45% (160 min)	0.0053					
Ni-P25	TC	10	82.5% (60 min)	0.0232		Visible light, 300 W xenon lamp		$\cdot O_2^-$, h^+	[13]
TiO_2 heterophase junction (anatase: rutile= 0.3:0.7)	TC	50	79.9% (120 min)	0.01069		Visible light > 400 nm, 300 W xenon lamp	0.2	·OH, $\cdot O_2^-$	[14]
TiO_2 (rutile)			38.4% (120 min)	0.00368					
P25			72.9% (120 min)	0.00802					
TiO_2@g-C_3N_4 core–shell quantum heterojunction	TC	20	100% (8 min)			300 W xenon lamp	1.0	$\cdot O_2^-$, h^+	[5]

(Continued)

Table 13.1 (Continued)

Photocatalyst	Antibiotic	Initial concentration of antibiotic (mg L^{-1})	Degradation efficiency	Degradation rate constant (min^{-1})	Degree of mineralization	Light condition	Photocatalyst dosage (g L^{-1})	Main active species	Ref.
Au-CuS-TiO_2 nanobelts	OTC	5	96% (60 min)	0.055		Visible light, 35 W xenon lamp		$\cdot O_2^-$	[6]
MIL-101(Fe)/TiO_2	TC	20	71.21% (visible light, 30 min)			Xenon lamp	0.8		[7]
			98.01% (sunlight, 30 min)			Sunlight			
			99.19% (UV light, 30 min)			300 W mercury lamp			
BiOCl/TiO_2/ sepiolite	TC	50	91% (180 min)	0.00437		Visible light > 400 nm, 400 W xenon lamp	0.6		[8]
Z-scheme CdTe/TiO_2	TC	20	78% (60 min)			Visible light > 400 nm, 300 W xenon lamp	0.6	$\cdot O_2^-$, $\cdot OH$	[9]
Black TiO_2	TC	10	66.2% (240 min)	0.0045	29.6% (240 min)	Visible light > 400 nm, 1000 W xenon lamp	0.2	$\cdot O_2^-$, h^+	[10]

TC degradation by P25 under visible light is markedly diminished. For example, under illumination of 420-nm visible light, TC degradation efficiency could only reach 76.6% within the same 2-h period, with negligible mineralization observed.

To enhance the photocatalytic activity of TiO_2 under visible light, researchers have integrated TiO_2 with a spectrum of other visible light-responsive photocatalytic materials, creating many robust composites. For example, the TiO_2@g-C_3N_4 core–shell quantum heterojunction, Au-CuS-TiO_2 nanobelts, MIL-101(Fe)/TiO_2, BiOCl/TiO_2/sepiolite composite, and Z-scheme CdTe/TiO_2 heterostructure all demonstrate excellent visible light degradation performance [5–9]. For TC concentrations ranging from 5 to 50 mg L^{-1}, their degradation efficiencies can reach 100% (within 8 min), 96% (within 60 min), 71.21% (within 30 min), 91% (within 180 min), and 78% (within 60 min), respectively. Additionally, black TiO_2 can also degrade TC rapidly under visible light, achieving a 66.2% degradation rate of a 10 mg L^{-1} TC solution within 4 h [10].

For the TC degradation mechanism, the literature reports that $\cdot O_2^-$ is the primary active species in the TiO_2-mediated photocatalytic degradation, as outlined in Table 13.1. According to Wu et al., $\cdot O_2^-$ is likely to first target the C—N—C moiety (designated as Region 1 in Figure 13.1b) of the TC molecule, resulting in the release of the *N*-methyl group [10]. Subsequent attacks by $\cdot O_2^-$ caused the detachment of the methyl and hydroxyl groups (Region 2) and ring-opening. The obtained intermediate then underwent a series of degradation reactions including hydroxylation, dealkylation, dehydration, and further ring-opening, eventually mineralizing into CO_2, H_2O and inorganic ions. Gong et al. also reported the early detachment of the methyl and hydroxyl groups at Region 2 during their investigation for TC degradation using the Z-scheme CdTe/TiO_2 heterojunction under visible light [9].

Beyond the primary reactive species $\cdot O_2^-$, both $\cdot OH$ and protons h^+ have also been observed to contribute to the TC degradation in certain systems. For example, Zhang et al., in their study using anatase/rutile TiO_2 with a heterophase junction, observed that the initial photodegradation stage involved the attack of $\cdot OH$ on the C=C of TC at Region 3 in Figure 13.1b, resulting in the formation of a hydroxylated intermediate [14]. The subsequent degradation steps continued with demethylation and deamination at Region 1, deamidation at Region 4 and the cleavage of the benzene ring. Concurrently, in the investigation by Wu et al. on the visible light-induced decomposition of TC using black TiO_2, h^+ could induce a dehydroxylation reaction by attacking the hydroxyl group at Region 5 [10].

The degradation pathway of OTC is quite similar to that of TC. In the study by Chen et al., which utilized Au-CuS-TiO_2 nanobelts for the degradation of OTC, similar processes were identified, including the removal of one *N*-methyl group (Region 1), deamidation (Region 4), and dehydroxylation (Region 5) [6].

13.2.2 Sulfonamide Antibiotics

Sulfonamide antibiotics (SAs), distinguished by their p-aminobenzene sulfonamide chemical structure, have become one of the most globally prevalent classes of antibiotics, and widely employed in the medical, veterinary, and aquaculture sectors due to their chemical stability, broad-spectrum antimicrobial properties, and affordability. As depicted in Figure 13.2a, the general structural formula of these antibiotics

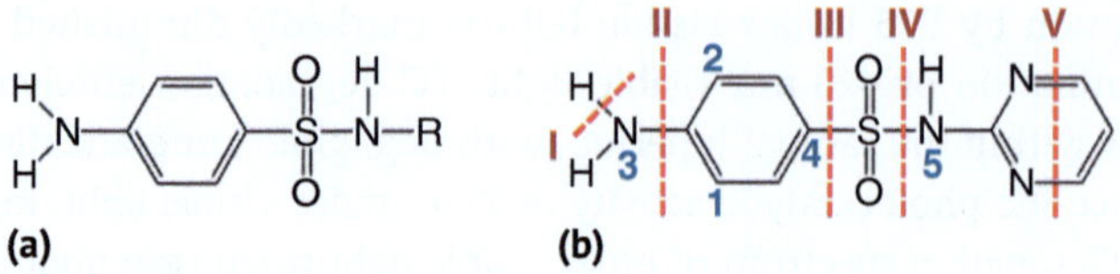

Figure 13.2 (a) The general chemical structure of SAs and (b) the molecular structure of SDZ with the key degradation sites.

features a free amino group ($-NH_2$) at one end and a diverse R-substituent group at the opposite end. The spectrum of commonly employed SAs includes, sulfadiazine (SDZ), sulfamethazine (SMT), sulfamethoxazole (SMX), sulfamerazine (SMZ), sulfisoxazole (SIZ), and sulfamethoxydiazine (SMD), among others. These antibiotics exhibit high solubility and chemical stability in water, which results in high residue levels in environment and poses serious threats to the ecosystem and human health. In TiO_2 photocatalytic degradation studies, SDZ and SMX have attracted the most extensive research attention as representatives for SAs. Additionally, some studies have also focused on the degradation of SMT and STZ.

Table 13.2 presents a compilation of recent studies utilizing TiO_2-based photocatalysts for the degradation of SAs. Upon comparison with Table 13.1, the degradation rates of SAs notably exceed those of TC antibiotics, with SAs exhibiting degradation rate constants approximately tenfold higher under both UV and visible light irradiation. Specifically, under UV irradiation from a 9 W LED lamp, P25 photocatalyzes the degradation of SMX with a rate constant of 0.1003 min^{-1} [15], contrasting with the 0.038 min^{-1} for TC degradation under 300 W xenon lamp UV irradiation [4]. In the visible light spectrum, Bi_2O_3-TiO_2/PAC achieves a SMX degradation rate constant of 0.159 min^{-1} [16] when excited by a 300 W xenon lamp, whereas the BiOCl/TiO_2/sepiolite composite yields a significantly lower rate constant of 0.00437 min^{-1} for TC under similar conditions [8]. This underscores the enhanced susceptibility of SAs to photocatalytic degradation compared to TCs. Furthermore, Table 13.2 indicates that the degradation rates of various SAs are distinct, with a degradation rate hierarchy of SMX > SDZ > SMT [16].

It is noteworthy that the molecularly imprinted TiO_2 (MIP-TiO_2) has demonstrated a robust photodegradation capacity for the specific target (e.g., SMX with a degradation rate constant of 0.426 min^{-1}), attributed to the selective enrichment [17]. Concurrently, the MIP-TiO_2 system has been observed to produce fewer intermediate products. These findings offer a promising strategy to augment the degradation rates of antibiotics, and manage the production of intermediates, thereby mitigate the associated environmental hazards.

The photocatalytic degradation mechanism of SAs exhibits clear regularities. As shown in Table 13.2, under UV irradiation, ·OH is the principal reactive species mediating SAs degradation, whereas $\cdot O_2^-$ fulfills this function under visible light. Furthermore, a comprehensive review of the literature indicates that the degradation of SAs is predominantly marked by processes of hydroxylation, bond cleavage, desulfonation, and oxidation, particularly under UV irradiation [18]. Taking SDZ as an example, the electrophilic attacks by ·OH on the aromatic ring at the C1, C2, or N3 atoms (designated in Figure 13.2b) lead to the production of mono- and di-hydroxylated products. These hydroxylated intermediates are then further targeted by reactive oxygen species (ROS), including h^+, $\cdot O_2^-$, and ·OH, with a notable example being the hydrogen abstraction from

Table 13.2 Summary of recent studies employing TiO_2-based photocatalysts for the degradation of sulfonamide antibiotics.

Photocatalyst	Antibiotic	Initial concentration of antibiotic (mg L^{-1})	Degradation efficiency	Degradation rate constant (min^{-1})	Degree of mineralization	Light condition	Photocatalyst dosage (g L^{-1})	Main active species	Ref.
P25	SMX	0.1	100% (60 min)	0.1003		UV light, 9 W LED	1.0	$\cdot OH$	[15]
TiO_2 nanosheets	SDZ		82% (60 min)	0.035	90% (240 min)	UV light		$\cdot OH$, h^+, $\cdot O_2^-$	[18]
N-doped TiO_2/graphene	SMT	30	43% (160 min)	0.0032		UV light, 250 W mercury lamp		$\cdot OH$, $\cdot O_2^-$	[12]
TiO_2/zeolite	SDZ	10	93.37% (120 min)	0.0212		UV light at 365 nm, 20 W lamp		$\cdot OH$, h^+	[1]
PVDF-PVP-TiO_2-dopamine membrane	SDZ	0.1	91.4% (120 min)	0.0216		UV light at 365 nm		$\cdot OH$, h^+	[19]
Molecularly imprinted TiO_2	SMX	10	99.9% (15 min)	0.426		UV light, 300 W mercury lamp	0.5	h^+, e^-, $\cdot O_2^-$	[17]
Bi_2O_3-TiO_2/PAC	SDZ	20	100% (60 min)	0.034		Visible light, 300 W xenon lamp		$\cdot O_2^-$, h^+, $\cdot OH$	[16]
	SMT		100% (120 min)	0.029					
	SMX		100% (30 min)	0.159				h^+, $\cdot O_2^-$, $\cdot OH$	
Bi_2WO_6/TiO_2/hydrochar	STZ	10	99.59% (60 min)	0.096		Visible light, 50 W LED	1.0	$\cdot O_2^-$, h^+	[20]
g-C_3N_4/Cu, N-TiO_2	SMZ	10	95.8% (240 min)		44% (240 min)	Visible light, 350 W xenon lamp	0.6	$\cdot O_2^-$, h^+, $\cdot OH$	[21]

Abbreviation: PAC, powdered activated carbon.

the N3 amino group by ·OH. This process is considered as a vital photocatalytic reaction for SDZ degradation within a UV-irradiated TiO_2 system [21]. Moreover, further attacks on the N3 amino group lead to the sequential formation of —NO and —NO_2 [21]. Bond cleavage is another prevalent photodegradation pathway for SAs, which can occur directly at sites II, III, or IV in Figure 13.2b. The cleavage at S—N and C—S bonds is particularly common, leading to the formation of intermediates such as 2-aminopyrimidine, aniline and other intermediates. In these products, the sulfur site is often further targeted by ·OH radicals, yielding a variety of hydroxylated derivatives [1, 18, 21]. Furthermore, nucleophilic addition and rearrangement at N5 and C4 position of the aniline ring facilitate the elimination of —SO_2— moiety from the SDZ molecule [18]. It is noteworthy that, for SDZ molecule, the pyrimidine ring is also susceptible to ·OH attack, leading to the cleavage of C—N and C═N bonds. Specifically, this results in the ring-opening at position V (Figure 13.2b), and is followed by subsequent degradation processes [1].

13.2.3 Quinolone Antibiotics

Quinolone antibiotics, a class of synthetic antibacterial agents, are widely employed for their broad-spectrum antibacterial activity and good bactericidal properties. Their general chemical structure is illustrated in Figure 13.3a. All quinolone antibiotics share a common core structure known as the quinolone ring, which is a fused heterocyclic ring system that includes a 4-oxo-1,4-dihydroquinoline nucleus featuring a pyridone ring fused to a benzene ring. Quinolone antibiotics have been categorized into four generations based on their discovery history and antibacterial properties. The first generation, exemplified by nalidixic acid (NA), and the second generation, typified by pipemidic acid (PA), are now rarely utilized in clinical practice due to their limited efficacy. The fourth generation, such as moxifloxacin (MOX), is noted for its enhanced antibacterial activity, but its application is often limited by its higher cost. The third-generation, commonly known as fluoroquinolones, are the most commonly used quinolones in clinical practice today. This group comprises ciprofloxacin (CIP), ofloxacin (OFL), enrofloxacin (ENR), norfloxacin (NOR), etc. The global presence of quinolone antibiotics, detected in major rivers, lakes, and offshore waters, has significantly raised environmental concerns. To mitigate this pollution, TiO_2-based photocatalysts have been extensively studied for their potential in removing CIP, OFL, ENR, MOX, NOR, and NA from the environment.

Table 13.3 presents a compilation of recent studies utilizing TiO_2-based photocatalysts for the degradation of quinolone antibiotics. Upon comparing Tables 13.1–13.3,

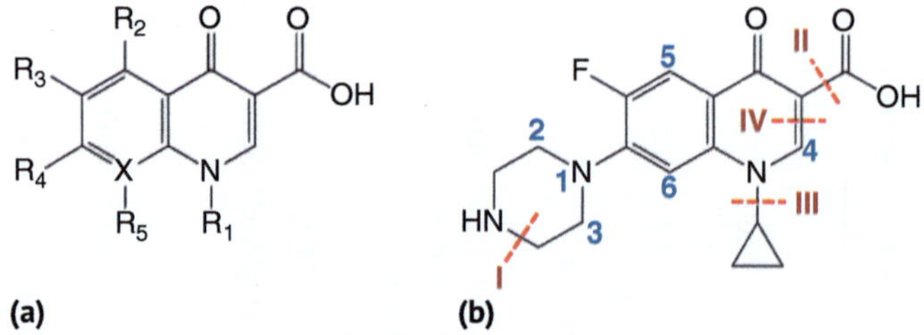

Figure 13.3 (a) The general chemical structure of quinolone antibiotics and (b) the molecular structure of CIP with the key degradation sites.

Table 13.3 Summary of recent studies employing TiO_2-based photocatalysts for the degradation of quinolone antibiotics.

Photocatalyst	Antibiotic	Initial concentration of antibiotic (mg L^{-1})	Degradation efficiency	Degradation rate constant (min^{-1})	Light condition	Photocatalyst dosage (g L^{-1})	Main active species	Ref.
P25	CIP	33.1	99% (60 min)	0.38	UV light, 125 W high-pressure mercury lamp	1.5	$\cdot OH$, h^+	[22]
TiO_2/Fe^0	CIP	30	94.6% (60 min)		UV light, 10 W mercury lamp	1.0	$\cdot OH$, $\cdot O_2^-$	[23]
N-doped TiO_2/graphene	NOR	30	65% (160 min)	0.0052	UV light, 250 W mercury lamp		$\cdot OH$, $\cdot O_2^-$	[12]
Molecularly imprinted TiO_2	CIP	15	93.75% (90 min)	0.015	UV light, 14 W lamp	0.2	h^+, $\cdot OH$, $\cdot O_2^-$	[24]
P-doped TiO_2	CIP	0.05	99% (60 min)	0.065	Visible light	0.5	h^+, $\cdot OH$, $\cdot O_2^-$	[25]
Z-scheme TiO_2 nanorod/g-C_3N_4 nanosheet	CIP	0.005	93.4% (60 min)	0.0366	Visible light, 500 W xenon lamp	0.2	h^+, $\cdot OH$, $\cdot O_2^-$	[26]
TiO_2/Au nanorods	NA	0.02	30% (30 min)	0.013	Visible light, 150 W Quartz tungsten halogen lamp	1.0		[27]
Z-scheme Co_3O_4/TiO_2	ENR	20	95% (100 min)	0.0269	Visible light, 500 W xenon lamp	0.4	$\cdot OH$, h^+, $\cdot O_2^-$	[28]
Bi_2O_3/TiO_2	OFL	25	92.4% (120 min)		Visible light	0.5	h^+, $\cdot OH$	[29]
Fe/TiO_2 biochar	CIP	40	88% (6 h)		Visible light		$\cdot OH$, h^+	[30]
	NOR		88.4% (6 h)					
Z-scheme CdS/Au/TiO_2 nanobelt	NOR	5	65% (60 min)	0.018	Visible light, 35 W xenon lamp		$\cdot O_2^-$, h^+, $\cdot OH$	[31]

it is evident that the degradation rate of quinolone antibiotics by P25 under UV light is more rapid than TCs and sulfonamides. Specifically, the degradation rate constant for CIP is notably high at 0.38 min^{-1} [22], in contrast to the lower constants of 0.038 min^{-1} for TC [4] and 0.1003 min^{-1} for SMX [15]. This underscores the rapidity of the photocatalytic process in mitigating quinolone antibiotic contamination. Additionally, under visible light, P-doped TiO_2 demonstrates an efficient degradation capability for CIP, with a degradation rate constant of 0.065 min^{-1} [25]. Notably, several Z-scheme heterojunctions have shown excellent performance in photocatalytic degradation of quinolone antibiotics, including the TiO_2 nanorod/g-C_3N_4 nanosheet, Co_3O_4/TiO_2, and CdS/Au/TiO_2 nanobelt [26, 28, 31]. Their degradation rate constants for CIP, ENR and NOR are 0.0366, 0.0269, and 0.018 min^{-1}, respectively. These findings confirm that photocatalytic technology is a promising strategy for the elimination of quinolone antibiotic pollutants from aquatic environments.

As shown in Table 13.3, under UV irradiation, ·OH is the most significant reactive species in degradation of quinolone antibiotics, followed by h^+ and $\cdot O_2^-$. However, under visible light irradiation, h^+ has been identified as the predominant reactive species in a majority of studies, with the impact of ·OH being relatively reduced. At present, third-generation fluoroquinolones are the predominant quinolone antibiotics utilized in clinical applications. Based on existing research, the photodegradation of these drugs is likely to occur at the piperazine ring, the fluorine atom, the carboxyl group and the cyclopropyl group [32]. Herein, we take CIP as an example to elucidate the photocatalytic degradation mechanism of fluoroquinolones.

One potential degradation pathway of CIP involves the oxidative attack of the piperazine ring at the C—N bond (Position I in Figure 13.3b), since the C—N bond within the piperazine ring is notably susceptible to direct oxidation by h^+ [30]. Through stepwise oxidative attacks, the piperazine ring is cleaved, leading to the formation of a dialdehyde derivative. Following this, the sequential loss of two aldehyde moieties and the secondary amine nitrogen from the amine side chain leads to the formation of aniline, which completely degrades the piperazinyl substituent of CIP [30]. Further oxidation of CIP leads to the decarboxylation of the quinolone at Position II and the removal of the cyclopropyl group from the quinolone ring at Position III as depicted in Figure 13.3b [30]. In another degradation pathway for CIP, the quinolone moiety is subjected to the attack by $h^+/\cdot OH/\cdot O_2^-$ at the C=C bond adjacent to the carboxyl group as indicated at Position IV in Figure 13.3b [30]. This reaction triggers the ring opening and the formation of aldehyde and ketone moieties. Subsequently, the further loss of these groups results in the decarboxylation of the quinolone, and the resulting intermediate is then susceptible to further oxidation by reactive species such as ·OH. In addition, hydroxylation is very common during CIP photodegradation. The N1 atom in piperazine ring (as marked in Figure 13.3b) is a tertiary amine which possesses strong electron absorption ability. Hence, its α-carbon (C2 and C3) is unstable and is preferred to be attacked by ·OH radical for hydroxyl substitution during photocatalytic reaction [26]. Meanwhile, the quinolone ring is susceptible to hydroxyl radical addition at the C4, C5, and C6 positions [32]. Additionally, the fluorine atom in CIP molecule is also a potential site for hydroxyl substitution. But beyond hydroxyl substitution, the fluorine atom may also undergo defluorination, representing another significant pathway during the CIP degradation process [32].

13.2.4 Macrolide Antibiotics

Since the discovery of erythromycin (ERY) in 1949, macrolide antibiotics have been utilized in the medical field for seven decades. The spectrum of commonly used clinical macrolides encompasses ERY, clarithromycin (CLA), azithromycin (AZM), roxithromycin (RXM), spiramycin (SP), acetylspiramycin, and tylosin. These antibiotics are characterized by a macrocyclic lactone ring with 14–16 atoms (Figure 13.4), to which one or more sugar moieties are affixed via glycosidic linkages. Specifically, ERY and CLA are characterized by their 14-membered lactone rings, whereas AZM features a 15-membered ring. Tylosin, with a 16-membered ring, represents an extended macrolactone structure and is widely used in veterinary medicine and occasionally in human therapeutics. In photocatalysis research, research predominantly focuses on ERY, AZM, tylosin, RXM, and SP.

Table 13.4 presents a compilation of recent studies utilizing TiO_2-based photocatalysts for the degradation of macrolide antibiotics. It can be found that the photocatalytic degradation rate of macrolide antibiotics is comparable to that of sulfonamide and quinolone antibiotics, and is faster than TC antibiotics. As shown in Table 13.4, among several macrolide antibiotics, SP exhibits the fastest degradation rate, with a rate constant approximately an order of magnitude faster than other macrolide antibiotics, such as AZM and ERY. Moreover, TiO_2-based photocatalysts, under both UV and visible light irradiation, demonstrate a robust capacity for the rapid degradation of macrolide antibiotics. For instance, under visible light irradiation, N-doped TiO_2 can mineralize 74% of a 40 mg L^{-1} solution of SP within 4 h [33]; ZrO_2/Ag@TiO_2 nanorods achieve a 92% degradation rate for a 2.5 mg L^{-1} solution of AZM within 2 h [34]. These findings underscore the efficacy of TiO_2-based photocatalysts in the degradation of macrolide antibiotics.

During the photocatalytic degradation of macrolide antibiotics, the ·OH radical is identified as the principal reactive species that drives the process, with the potential sites of attack highlighted in blue squares in Figure 13.4 [40]. This degradation

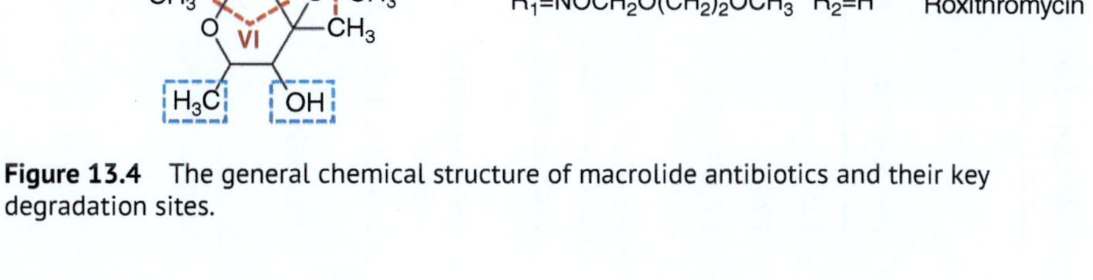

Figure 13.4 The general chemical structure of macrolide antibiotics and their key degradation sites.

Table 13.4 Summary of recent studies employing TiO_2-based photocatalysts for the degradation of macrolide antibiotics.

Photocatalyst	Antibiotic	Initial concentration of antibiotic (mg L^{-1})	Degradation efficiency	Degradation rate constant (min^{-1})	Degree of mineralization	Light condition	Photocatalyst dosage (g L^{-1})	Main active species	Ref
P25	AZM	0.1		0.0363		UVA, LEDs	1.0		[15]
P25	SP	10	100% (40 min)	0.282		UV light at 365 nm	2.0	·OH	[35]
Cu_2O-TiO_2 nanotubes	AZM	0.05	92.12% (90 min)	0.02436		UV light, mercury lamp	1.5	$·O_2^-$, ·OH	[36]
$Ti_{1-x}Sn_xO_2$ (Sn/Ti = 0.025)	ERY	50	63% (120 min)	0.04653	≈45%	UVA			[37]
N-doped TiO_2	SP	40			74% (240 min)	400 nm< visible light < 550 nm	3.0		[33]
ZrO_2/Ag@TiO_2 nanorods	AZM	2.5	92% (120 min)			Visible light, 250 W xenon lamp	0.1	·OH, $·O_2^-$	[34]
Fe-HQLC/TiO_2	ERY	10	96.83% (20 min)	0.048	28%	Visible light, 300 W xenon lamp	1.0		[38]
TiO_2-RGO	ERY	0.1	86% (60 min)			Visible light, 1000 W xenon lamp	0.1		[39]
	CLA	0.1	84% (60 min)						

Abbreviation: HQLC, 8-hydroxyquinoline-7-carboxylic.

process involves several key steps, including the hydroxylation of the aromatic ring through an electrophilic ·OH radical attack, the cleavage of C—O and C—N bonds, and the α-cleavage adjacent to the aromatic moiety, followed by ring opening. AZM degradation exemplifies this process, where the parent compound undergoes O-demethylation and *N*-demethylation at Positions I and II (Figure 13.4) [34]. The resulting products are prone to further lose the methylamino group at Position III. In addition, the degradation fragmentation pattern of AZM also reveals other significant degradation pathways, such as the cleavage of the amino sugar molecules D-desosamine at Position IV and L-cladinose at Position V. Interestingly, these amino sugars are subject to further degradation; for instance, the dehydrated product of D-desosamine along with the subsequent loss of the methylamino fragment can be detected. Additionally, the cleavage of the C—O bond and ring opening in L-cladinose at Position VI results in the formation of smaller degradation products.

13.2.5 β-lactam Antibiotics

β-Lactam antibiotics constitute a class of antimicrobial agents characterized by the presence of a β-lactam ring in their molecular structure (Figure 13.5). These antibiotics exhibit an extensive spectrum of antibacterial activity coupled with minimal side effects, making them the most prevalently utilized commercial antimicrobials for the treatment of infectious diseases. There is a wide variety of β-lactam antibiotics. Common classes of β-lactam antibiotics include penicillins (e.g., penicillin G, amoxicillin, ampicillin); cephalosporins (e.g., cefalexin, cefuroxime, cefotaxime, ceftriaxone); and atypical β-lactams, such as carbapenems (e.g., imipenem); monobactams (e.g., aztreonam); oxacephems (e.g., latamoxef); and cephamycins (e.g., cefotetan). In the field of TiO_2 photocatalysis, research primarily focuses on the degradation of penicillins, such as amoxicillin (AMX), ampicillin (AMP), penicillin G (PG), cloxacillin (CLX), and oxacillin (OXA), as well as cephalosporins, like cefalexin (CF), cefradine (CPD), and cefadroxil (CDX).

Table 13.5 presents a compilation of recent studies utilizing TiO_2-based photocatalysts for the degradation of β-lactam antibiotics. It can be found that the photocatalytic degradation of β-lactam antibiotics is highly efficient. Specifically, P25 can completely degrade 10 mg L^{-1} of AMX within just 20 min, achieving a mineralization rate of up to 90% [41]. Comparative studies indicate that penicillins

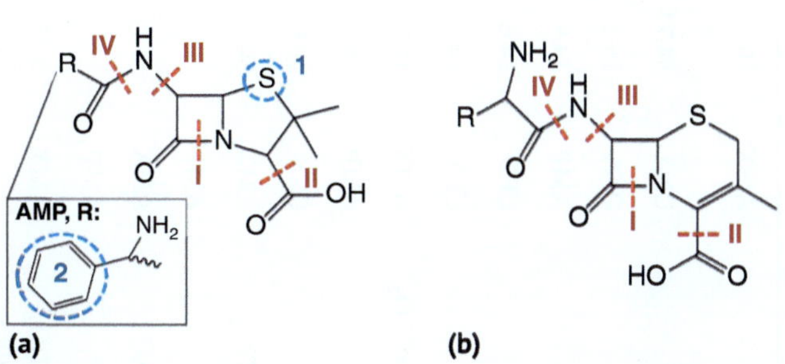

Figure 13.5 The general chemical structure of (a) penicillins and (b) cephalosporins with their key degradation sites.

Table 13.5 Summary of recent studies employing TiO_2-based photocatalysts for the degradation of β-lactam antibiotics.

Photocatalyst	Antibiotic	Initial concentration of antibiotic (mg L^{-1})	Degradation efficiency	Degradation rate constant (min^{-1})	Degree of mineralization	Light condition	Photocatalyst dosage (g L^{-1})	Main active species	Ref.
P25	AMX	10	100% (20 min)		>90%	350 nm < UV light < 400 nm, 9 W lamp	0.25	·OH	[41]
P25	OXA	94.74	84.58 (60 min)			UV light at 365nm	2.0	·OH	[43]
P25	OXA	16.1	93% (60 min)		65% (240 min)	UV light at 365 nm, 4 W lamp	0.5	·OH	[44]
	AMP	14.9	74% (60 min)						
	CPD	14.0	91% (60 min)						
	CDX	14.5	58% (60 min)		57% (240 min)				
Anatase TiO_2	AMX	104	35.4% (120 min)	0.004		UV light at 365 nm	1.0		[45]
	AMP	105	82.2% (120 min)	0.007					
	CLX	103	96.1% (120 min)	0.029					
Nano TiO_2	AMX	50	89% (180 min)	0.013		Solar irradiation	0.4	·OH	[42]
	AMP		83% (180 min)	0.011					
TiO_2/AC	AMX		100% (120 min)	0.037					
	AMP		100% (180 min)	0.022					
Cu_2O-TiO_2 nanotubes	AMX	0.05	94.31% (90 min)	0.02748		UV light, 5×11W mercury lamp	1.0	$\cdot O_2^-$	[36]
Fe-HQLC/TiO_2	AMX	10	97.65% (20 min)	0.044	22%	Visible light, 300 W xenon lamp	1.0		[38]
Ag-AgBr/TiO_2 / RGO	PG	10	99% (120 min)	0.0316		White light at 450 nm, LED lamp	0.5	$\cdot O_2^-$, ·OH	[46]
Ni-P25	CF	10	93.6% (60 min)	0.0336		Visible light, 300 W xenon lamp	0.2	$\cdot O_2^-$, h^+	[13]

Abbreviations: AC, activated carbon; HQLC, 8-hydroxyquinoline-7-carboxylic.

and cephalosporins have similar degradation rates. In addition, the photocatalytic efficiency is significantly improved by immobilizing TiO_2 on activated carbon (TiO_2/AC). With this photocatalyst, both AMX and AMP can be completely degraded within 2 h by TiO_2/AC [42], whereas the bare TiO_2 can only remove 89% of AMX and 83% of AMP over a longer period of 3 h. Moreover, modified TiO_2-based photocatalysts under visible light display remarkable degradation capabilities. For instance, Fe-HQLC/TiO_2 can degrade 97.65% of amoxicillin (10 mg L^{-1}) within just 20 min [38], demonstrating its robust photocatalytic activity.

As shown in Table 13.5, under UV light irradiation, ·OH is the primary reactive species responsible for the degradation of β-lactam antibiotics by TiO_2. This is attributed to the β-lactam ring's bond angle of 90°, which deviates from the typical 109.5° bond angle for sp3 hybridized carbon atoms. This deviation increases the ring's reactivity and susceptibility to ·OH attack, resulting in ring cleavage. However, in certain composite photocatalytic systems, such as those utilizing Cu_2O–TiO_2 or Ag–AgBr/TiO_2/RGO as photocatalysts, $\cdot O_2^-$ emerges as the primary reactive species, emphasizing the influence of photocatalyst composition on the degradation mechanism [36, 46].

Current research highlight the comparable degradation mechanisms of penicillins and cephalosporins, both of which commonly include the cleavage of the β-lactam ring and decarboxylation [44]. Herein, we delineate these pathways in detail. Figure 13.5a illustrates the key reactions involved in penicillin degradation [43, 44]: (i) the β-lactam ring, known for its high reactivity, is susceptible to attack by ·OH, which triggers ring opening at Position I and the formation of a carboxylic acid group; (ii) the thioether group (Region 1 depicted by blue circle in Figure 13.5a), acting as a nucleophile, is prone to oxidation by ·OH, resulting in the formation of a sulfoxide; (iii) the free carboxylic acid group within penicillin molecules engages in the photo-Kolbe reaction with h^+, leading to the decarboxylation at Position II; and (iv) the secondary amide moiety is vulnerable to ·OH attack, which causes the C—N bond linked to the β-lactam ring to cleave at Position III, producing a primary amide. Concurrently, the opposing C—N bond of the secondary amide may also be cleaved at Position IV by ·OH attack. Additionally, ·OH are documented to target the aromatic rings within the R group of penicillins, exemplified by Region 2 in Figure 13.5a, leading to the hydroxylation of the benzene ring. Similarly, the degradation pathway of cephalosporins, as depicted in Figure 13.5b, includes analogous key processes [44]: (a) cleavage of the β-lactam ring at Position I, (b) decarboxylation at Position II, and (c) fragmentation of the C—N bonds adjacent to the secondary amide at Positions III and IV [13]. This alignment underscores the structural and functional parallels between these two classes of antibiotics.

13.2.6 Other Classes of Antibiotics

Beyond the aforementioned classes of antibiotics, photocatalytic technology has shown strong efficacy in degrading other antibiotic categories, including aminoglycosides, peptides, lincosamides, amphenicols, rifamycins and nitroimidazoles.

Table 13.6 provides a compilation of recent studies employing TiO_2-based photocatalysts for the degradation of these antibiotics. It can be found that under UV light irradiation, TiO_2 can typically degrade more than 70% of antibiotics at

Table 13.6 Summary of recent studies employing TiO_2-based photocatalysts for the degradation of other classes of antibiotics.

Class of antibiotics	Antibiotic	Photocatalyst	Initial concentration of antibiotic (mg L^{-1})	Degradation efficiency	Degradation rate constant (min^{-1})	Degree of mineralization	Light condition	Photocatalyst dosage (g L^{-1})	Main active species	Ref.
Aminoglycosides	Amikacin	TiO_2 nanobelts	5	70% (150 min)	0.0083		Visible light, 35 W xenon lamp		$\cdot OH$	[47]
	Gentamicin	TiO_2 nanoparticle	50	95% (80 min)	0.948		UV light < 400 nm, 300 W Osram Lamp		$\cdot OH$, $\cdot O_2^-$	[48]
Peptides	Vancomycin	TiO_2-clinoptilolite	30	97% (50.9 min)	0.0187	56% (50.9 min)	UV light at 254 nm	0.2		[49]
Lincosamides	Lincomycin	Polycrystalline TiO_2	30	97.78% (60min)			Visible light	0.20	$\cdot OH$	[50]
Amphenicols	Chloramphenicol	P25	50	90% (90 min)		88% (120 min)	UV light at 365 nm	1.0		[51]
		TiO_2 (anatase)		37% (90 min)		50% (240 min)				
		Nano-TiO_2 (anatase)	19.97	85.97% (60 min)			UV light 365 nm, 300 W medium pressure mercury lamp	0.94	$\cdot OH$, $\cdot O_2^-$	[52]

Class of antibiotics	Antibiotic	Photocatalyst	Initial concentration of antibiotic (mg L^{-1})	Degradation efficiency	Degradation rate constant (min^{-1})	Degree of mineralization	Light condition	Photocatalyst dosage (g L^{-1})	Main active species	Ref.
Rifamycins	Rifampicin	Ag@IP-TiO_2	25	70% (180 min)	0.00592		Visible light, 30 W LED lamp	0.125		[53]
Nitroimidazoles	Metronidazole	TiO_2/ Ag_3PO_4/g-C_3N_4	8.2	95.6% (80 min)	0.0472		Visible light, 24 W LED lamp	1.248	·OH	[54]
Others	Trimethoprim	P25	0.1	93.87% (60 min)	0.0354		UV light at 381 nm, 9W LEDs	1.0	·OH	[15]
				97.72% (60 min)	0.0668		UV light at 381 nm, 18 W LEDs			
				100% (60 min)	0.1171		UV light at 381 nm, 36 W LEDs			
			20	100% (29 min)	0.22	64% (107 min)	UV light < 400 nm	0.2	·OH	[55]
	Berberine hydrochloride	TiO_2/ ZrO_2/g-C_3N_4	10	86.93% (60min)	0.02909		Visible light, 300 W xenon lamp	0.2	$\cdot O_2^-$, ·OH	[56]

Note: Ag@IP-TiO2: TiO2 inverse opal nanostructure modified by Ag nanoparticles concentrations below 50 mg L^{-1} within 1–3 h, while also achieving effective mineralization. Furthermore, with modifications such as doping, these photocatalysts can also efficiently degrade antibiotics under visible light, underscoring the strong practical potential of photocatalytic antibiotic degradation.

13.3 Modification of TiO_2 to Enhance the Efficiency of Antibiotic Photocatalytic Degradation

Despite the extensive research on TiO_2 due to its excellent photocatalytic activity, its low solar energy conversion efficiency and the low utilization rate of photogenerated charge carriers significantly limit its practical applications. Many studies have confirmed that modification of TiO_2 can effectively enhance its photocatalytic performance. In this section, we will succinctly outline the efficacious modification strategies and elucidate the underlying mechanisms for augmenting the photocatalytic degradation efficacy of antibiotics.

13.3.1 Construction of Heterojunctions

Heterojunctions, which typically arise between two semiconductors – referred to as photocatalyst I (PC I) and photocatalyst II (PC II) – with differing energy band structures, can enhance solar energy utilization and mitigate e^-–h^+ recombination, thus boosting photocatalytic performance. Classified by their energy band alignment, heterojunctions include Type I (straddling gap), Type II (staggered gap), and Type III (broken gap) configurations. Types I and III heterojunctions are less efficient due to the lack of staggered band structures. In contrast, Type II heterojunctions, which promote the directional transfer of photogenerated electrons and holes, are commonly employed. This transfer leads to spatial charge separation, for example, with electrons and holes respectively accumulating on PC II for reduction and PC I for oxidation. Despite improving charge separation, Type II heterojunctions may face a trade-off between separation efficiency and redox capacity, potentially hindering photocatalytic reactions. To overcome this limitation, derivative structures, such as Z-scheme and S-scheme heterojunctions, have been developed, leveraging the advantages of Type II while aiming to optimize both charge separation and redox potential.

Z-scheme heterojunction photocatalysts are particularly effective, combining two semiconductors with a well-matched interfacial coupling and an interleaved energy band structure. In this system, photogenerated holes in the valence band (VB) of PC I (with lower oxidation potential) react with an electron donor (D) to form an electron acceptor (A), while electrons in the conduction band (CB) of PC II (with lower reduction potential) reduce A back to D. This process results in spatially separated redox reactions, with electrons in CB of PC I (with higher reduction potential) participating in reduction and holes in VB of PC II (with higher oxidation potential) in oxidation. Thus, the charge transfer pattern of the Z-scheme ensures a robust redox capacity across distinct reaction sites. Gong et al. successfully synthesized a Z-scheme CdTe/TiO_2 heterostructure using a hydrothermal method, demonstrating its effectiveness in degrading TC hydrochloride under visible light. The heterostructure achieved a decomposition rate of 78% within 30 min, surpassing the 62% achieved by pure TiO_2 [9].

Furthermore, heterophase junctions, a unique class of heterojunctions formed by different crystallographic phases of the same semiconductor material, also drive the migration of photogenerated charge carriers due to differences in CB and VB positions of the different crystallographic phases. Unlike conventional heterojunctions, heterophase junctions can provide continuous band bending that accelerate electron transfer across the interface. For example, a built-in electric field of approximately 1 kV cm^{-1} has been identified at the interface of rutile/anatase heterophase junctions [57], which aids in the separation and transfer of e^-–h^+ pairs. In comparative studies, Chatzitakis et al. found that P25, with an anatase/rutile ratio of 3.6/1, achieved 90% degradation of chloramphenicol after 90 min, significantly outperforming commercial anatase form of TiO_2 (Tronox McGee), which only achieved 37% degradation [51]. Additionally, Zhang et al. synthesized the anatase/rutile TiO_2 (M-O-600) photocatalyst with heterophase junctions by calcining MIL-125(Ti) with 30% O_2/Ar at 600°C [14]. The weight fractions of rutile and anatase in M-O-600 are approximately 0.7 and 0.3, respectively. Under visible light, M-O-600 exhibited an apparent reaction rate constant of 0.01069 min^{-1} for the degradation of TC, outperforming both M-O-700 (rutile TiO_2, 0.00368 min^{-1}) and P25 (0.00802 min^{-1}).

The aforementioned examples demonstrate that through the strategic design of heterostructures, we can efficiently promote the separation and migration of photogenerated charges, which in turn notably boosts the photocatalytic performance of the materials.

13.3.2 Elemental Doping

Elemental doping is a widely recognized strategy for adjusting the bandgap of semiconductors and enhancing their light absorption capabilities. This strategy involves the incorporation of metallic elements, like Fe, Cd, Co, Mo, V, Ce, and Y, as well as nonmetallic elements, such as B, C, N, F, P, S, Cl, and Br. Advanced doping methods like co-doping and tri-doping are also employed to further refine these materials' properties.

Metallic doping typically introduces new energy levels into the band structure of semiconductors, which can effectively reduce the bandgap and improve the absorption of visible light. Cu and Ni are particularly favored for TiO_2 modification due to their high reactivity, wide availability, and cost-effectiveness compared to other metals. Nguyen et al. demonstrated the effectiveness of Ni-doped TiO_2 in degrading cefalexin and TC in wastewaters [13]. They observed that the Ni doping not only narrowed the bandgap of TiO_2 from 3.1 to 2.8 eV but also extended the light absorption spectrum beyond 550 nm. Notably, the Ni-TiO_2 achieved the photodegradation efficiencies of 93.6% for cefalexin and 82.5% for TC, significantly outperforming P25, which achieved only 50.2% and 40.9%, respectively.

Nonmetallic doping, characterized by high ionization energy and electronegativity, facilitates the formation of covalent bonds by capturing electrons during reactions with other compounds. This can lead to a reduction in the bandgap and an expansion of the light absorption spectrum. Recent research on nonmetallic doping of TiO_2 has focused on elements such as N, C, F, P, and S. Notably, P doping has been shown to significantly enhance the photocatalytic efficiency of TiO_2 by reducing the recombination of photogenerated charges. Feng et al. developed an efficient P doping method that can also simultaneously generate oxygen vacancies within TiO_2 in a single step [25]. The resulting P-doped TiO_2 with surface oxygen vacancies demonstrated a rate constant of 0.065 min^{-1} for CIP degradation under visible light, which is 16 times higher than that of the undoped TiO_2 (0.004 min^{-1}).

Co-doping or tri-doping approaches leverage the synergistic effects of multiple dopants to further amplify the photocatalytic activity. The presence of one dopant atom can disturb the charge and defect energy states of another, allowing co-doping to significantly reduce the bandgap without creating charge-carrier recombination centers. Jahdi et al. successfully synthesized the F-Pd co-doped TiO_2 nanocomposites by combining the hydrothermal synthesis and microwave heating [58]. The incorporation of F and Pd into TiO_2 lattice introduced new energy levels, thereby extending the light absorption spectrum to the visible and near-infrared regions, and narrowing the bandgap from 3.26 to 0.54 eV. This nanocomposite showed exceptional performance in the degradation of SMX (30 ppm), achieving 98.4% elimination within 40 min under direct sunlight irradiation.

These advancements in elemental doping – whether through single, co-, or tri-doping – substantially improve the photocatalytic performance of semiconductors, underscoring their value for environmental remediation and other applications.

13.3.3 Composite Photocatalysts

To enhance the photocatalytic performance of photocatalysts, researchers have strategically combined semiconductors with well-matched bandgaps to construct multifunctional composite photocatalysts. This approach is intended to alter the transport pathways of photogenerated electrons and holes, which significantly reduces the recombination probability of the photogenerated carriers. Concurrently, it improves the optical properties of the materials, thereby yielding composite photocatalysts with higher photocatalytic activity.

Wang et al. have successfully synthesized a highly efficient photocatalytic nanocomposite, Ag-AgBr/TiO_2/RGO, which has demonstrated superior photocatalytic activity in the degradation of penicillin G (PG) under white LED light [46]. This nanocomposite is capable of being excited by visible light with wavelengths up to approximately 600 nm, and achieves a remarkable 99% removal efficiency of PG within 2 h. The photocatalytic activity of the Ag-AgBr/TiO_2/RGO nanocomposite is superior to that of its individual components, with the rate constant (expressed in 10^{-3} min^{-1}) following the order of significance: Ag-AgBr/TiO_2/RGO (31.6) > Ag/TiO_2/RGO (26.9) > Ag-AgBr/TiO_2 (19.6) > Ag-AgBr/RGO (10.1) > Ag/TiO_2 (6.08) > Ag-AgBr (5.52) > TiO_2/RGO (4.75) > TiO_2 (2.84). The improved performance of the Ag-AgBr/TiO_2/RGO nanocomposite is ascribed to the synergistic effects of the surface plasmon resonance of Ag nanoparticles, the enhanced utilization of visible light by AgBr, efficient electron transfer at the AgBr-Ag-TiO_2 nanojunctions, and interfacial electron transfer within the RGO sheets. Moreover, the Ag-AgBr/TiO_2/RGO nanocomposite exhibits good photochemical stability and reusability, marking it as a promising candidate for advanced photocatalytic applications.

Hu et al. have developed a $BiOCl/TiO_2$/sepiolite ternary composite that exhibits a significant enhancement in photocatalytic performance under visible light for the degradation of TC [8]. The reaction rate constant of this composite is notably higher than that of its individual components, being approximately 9.20, 4.28, 2.41, 2.08, and 1.73 times that of BiOCl, BiOCl/sepiolite, P25, TiO_2, and TiO_2/sepiolite, respectively. The superior photocatalytic activity of the BiOCl/TiO_2/sepiolite composite is attributed to several key factors: (i) the introduction of sepiolite with superior adsorption capabilities, which serves to capture and concentrate TC molecules; (ii) an enhanced response to visible light; and (iii) a robust separation efficiency of photogenerated carriers due to the formation of a Type II heterojunctions between TiO_2 and BiOCl. Furthermore, by precisely controlling the mass ratio of BiOCl to TiO_2, the composite achieves optimal grain sizes, ensuring good dispersion and chemically bonded interfacial contact within the ternary system.

The aforementioned examples demonstrate that by strategically combining materials with unique properties in composite photocatalysts, we can capitalize on their individual advantages. This approach is a promising avenue for the design of advanced photocatalytic materials that offer increased practical applicability and superior performance characteristics.

13.3.4 Defect Engineering

Defect engineering plays a pivotal role in enhancing the photocatalytic activity of photocatalysts. TiO_2, a typical metal oxide photocatalyst, experiences significant performance enhancement through the strategic introduction of oxygen vacancy and Ti^{3+} defects.

Li et al. have successfully synthesized oxygen vacancy/Ti^{3+} engineered TiO_2 nanotube arrays (TiO_2 NTs) via an electrochemical strategy at a high potential of −6 V [59]. This material demonstrated exceptional efficiency in chloramphenicol removal, achieving a 91% removal rate within 120 min under visible light – nearly 9.1 times more effective than P25, which only managed 10%. This superior performance is attributed to the engineered defects (oxygen vacancy/Ti^{3+}) within the TiO_2 NTs, which not only augment the material's ability to absorb visible light, but also serve as active sites for activating O_2 to form $\cdot O_2^-$. Moreover, these defects also promote the separation of e^-–h^+ pairs, which in turn increases the production of ·OH compared to pristine TiO_2. Additionally, the oxygen vacancies improve the adsorption of chloramphenicol molecules onto the TiO_2 surface, facilitating their efficient degradation under visible light. It is important to recognize that the photocatalyst's activity is maximized with a balanced presence of defects, where an optimal number of defects can greatly enhance activity, but an excess of Ti^{3+} defects can act as recombination centers for electrons and holes, thus reducing the photocatalytic effectiveness.

Besides the aforementioned modification strategies, the efficiency of TiO_2 in degrading antibiotic pollutants can also be enhanced through approaches such as noble metal deposition, exposure of specific crystal facets and dye sensitization etc. Moreover, with the continuous development of nanotechnology and materials science, the methods for modifying TiO_2 will become more diverse and efficient, enabling TiO_2 to play an increasingly important role in addressing environmental issues.

13.4 Factors Influencing the Antibiotic Degradation Efficiency and the Underlying Mechanisms

Beyond the intrinsic properties of photocatalyst, the efficiency of photocatalytic antibiotic degradation is also influenced by a variety of factors, such as the solution pH, coexisting substances, illumination conditions, photocatalysts dosage, the initial concentration of pollutants, etc. In this section, we comprehensively analyze the effects of these parameters on antibiotic degradation efficiency and explore their potential mechanisms.

13.4.1 Effect of Solution pH

The photocatalytic degradation efficiency is greatly influenced by the solution pH, which determines the surface charge of TiO_2 and the ionization state of pollutants in aqueous solutions. The point of zero charge (pH_{PZC}) for Degussa P25 is approximately 6.3. Consequently, below this pH, P25 exhibits a positive surface charge, while above it, the surface charge becomes negative. This charge variation significantly influences the electrostatic interactions between the photocatalyst and the pollutants, further affecting the adsorption and subsequent degradation processes. In addition, pH variations can alter the concentration of H^+ and OH^- ions, which in turn directly affect the generation of reactive species, such as photogenerated h^+, $\cdot OH$ and $\cdot O_2^-$. Therefore, optimizing the pH is essential for enhancing the photocatalytic performance of TiO_2 in the treatment of aqueous pollutants.

Take CIP as an example, many research have reported the profound impact of pH on its degradation efficiency and the associated pathways. CIP is characterized by five protonation states at different pH levels: H_4CIP^{3+}, H_3CIP^{2+}, H_2CIP^+, $HCIP^0$, and CIP^-. Under acidic/alkaline pH conditions, both TiO_2 and CIP surface bear the same positive/negative charges, respectively, leading to electrostatic repulsion that reduces the adsorption of CIP on TiO_2 surface, and consequently inhibits the degradation of CIP. Conversely, within the pH range of 5–7, TiO_2 maintains a positive surface charge while CIP is in its neutral form, thus the CIP adsorption is most efficient due to van der Waals interactions, and concurrently, the degradation efficiency is also at its most favorable [60]. Additionally, the degradation pathways of CIP are pH-dependent as well. Under acidic conditions, CIP predominantly undergoes stepwise cleavage of the piperazine ring, whereas in alkaline conditions, defluorination becomes the main pathway. In a neutral pH environment, the main degradation routes involve oxidation of both the piperazine ring and the quinolone structure [32]. These findings highlight the critical role of pH in governing the chemical transformations and the overall efficiency of photocatalytic degradation process of antibiotics.

13.4.2 Effect of Coexisting Organic Substances

Dissolved organic matter (DOM), a common component in aquatic environments, significantly influences the photocatalytic degradation of antibiotics. Humic acid (HA), a key constituent of DOM, is known for its photochemical properties and its ability to alter the environmental chemistry of antibiotics. HA, containing chromophores like benzene rings and functional groups such as carbonyl and carboxyl groups, can both inhibit and enhance the photocatalytic process through a dualistic mechanism [61]. On the one hand, it acts as a light filter or quencher of reactive species, suppressing the degradation process. Studies have found that HA's absorption of UV light can impede the degradation of sulfonamides; meanwhile, HA can also suppress the photocatalytic conversion of water into ·OH and h^+, and hinder the generation of h^+ and e^- on the TiO_2 surface. On the other hand, HA can function as a photosensitizer, generating ROS that enhance the indirect photolysis of organic compounds. This dual role of HA underscores the complex interaction between DOM components and the photocatalytic system, highlighting the importance of understanding these interactions to optimize the degradation processes.

In a study by Li et al., the impact of HA on the photolytic and TiO_2 photocatalytic degradation of TC was compared [11]. They found that the presence of HA slightly enhanced the TC degradation in photolytic process, attributed to an augmented photosensitization pathway, supported by the presence of ·OH and the excited-state of HA $(^3HA^*)$. In contrast, in the photocatalytic process, HA significantly impeded the TC degradation, ascribed to the deactivation of TiO_2 surfaces and the quenching of ·OH. The addition of HA did not alter the intermediate species formed during TC degradation in either photolysis or photocatalysis, but did influence the quantity of identified degradation products, with an increase in photolysis and a decrease during photocatalysis. This study concluded that HA not only affects the reaction kinetics but also modulates the generation of ROS and formation of transformation products during both photolytic and photocatalytic processes. This highlights the complex role of DOM components in the environmental fate of pharmaceutical pollutants and underscores the importance of considering their presence in aquatic systems when assessing photocatalytic treatment efficacy.

13.4.3 Effect of Coexisting Ions

The common ions in water significantly influence the photocatalytic degradation process, often leading to reduced efficiency in practical water treatment. Cations, which can stabilize the antibiotic molecules through complexation, exert a relatively modest influence on the photocatalytic activity. Yang's study indicated that the inhibitory effect of cations follows the order of $Ca^{2+} > Mg^{2+} > K^+ > Na^+$ during the degradation of SMX by P25 [62]. Conversely, anions have a more significant and complex effect on the process. Common anions, such as SO_4^{2-}, CO_3^{2-} and Cl^-, etc., generally impede photocatalytic activity, though to different degrees. Yuan's study ranked the inhibitory effects as $HPO_4^{2-} > HCO_3^- > SO_4^{2-} > Cl^- > H_2PO_4^-$ during the photocatalytic decomposition of SMX [63]. However, Wang et al. reported an alternate order of anionic inhibition ($NO_3^- > HCO_3^- > Cl^- > SO_4^{2-}$) when using graphene/$WO_3$ co-decorated TiO_2 nanotube array photoelectrodes for the degradation of dimethyl phthalate (DMP) [64]. This variation underscores the inconsistency in the inhibitory effects across different photocatalytic systems.

The impact of SO_4^{2-}, Cl^-, and CO_3^{2-} on the photocatalytic degradation process is mediated through two primary mechanisms: (i) competing with pollutants for the active sites on TiO_2 surface and (ii) quenching the photocatalytic active species. Although these ions can produce corresponding free radicals, such as $\cdot SO_4^-$, $\cdot CO_3^-$, and ·Cl upon quenching the active species, the overall photocatalytic efficiency is frequently diminished due to the lower reactivity of these radicals. However, there are exceptions. For example, ·Cl radicals preferentially react with electron-rich functional moieties, such as olefins, phenols, anilines and alkyl-/alkoxybenzenes. This preferential reactivity can enhance the degradation of pharmaceuticals like carbamazepine, naproxen, clenbuterol, salbutamol, SMX, and trimethoprim [65]. Additionally, Yuan et al. reported a dual effect of Cl^- during the degradation of acid orange 7 (AO7) [66]. At concentrations ranging from 0 to 50 mM, Cl^- significantly improved the decolorization of AO7. This is attributed to the higher affinity of Cl^- for h^+ on the TiO_2 surface compared to ·OH, which in turn suppresses the recombination of e^-–h^+ pairs. However, this promotional effect is markedly reduced with increased Cl^- concentrations from 50 to 200 mM. This decrease is primarily due to the deactivation of photoactive species, such as h^+ and ·OH, in the presence of elevated Cl^- levels.

The influence of NO_3^- and $H_2PO_4^-$ on photocatalytic degradation varies across different systems. In a study by Wang et al., where a graphene/WO_3 co-decorated TiO_2 nanotube array photoelectrode was utilized to degrade DMP, NO_3^- exhibited a significant inhibitory effect [64]. This was because that NO_3^- could absorb photons to produce NO_2^-, and NO_2^- was believed to be the scavengers of ·OH, thus significantly reducing the photocatalytic degradation of DMP. In contrast, in the photocatalytic degradation of acetaminophen, Bai et al. observed that NO_3^- promoted the photocatalytic reaction [67]. This enhancement was attributed to the increased production of ·OH resulting from the photolysis of NO_3^-. $H_2PO_4^-$, despite its role as a scavenger for h^+ and ·OH, was found to slightly enhance the SMX photodegradation by P25 [63]. This enhancement is associated with the phosphorus-containing anion's strong hydrogen bonding interactions with water molecules, which establish itself as an effective anchor group for efficient electron transfer between TiO_2 and the chromophore. This facilitated charge transfer between h^+ and H_2O, leading to the formation of ·OH, which can offset the scavenging effect of $H_2PO_4^-$ on h^+ and ·OH. These findings illustrate that the role of anions within a photocatalytic system is highly system-dependent, emphasizing the importance of the specific degradation context in determining their influence.

It is worth noting that most ions primarily influence the rate of degradation, without influencing the degradation pathways. However, Cl^- represent an exception to this general behavior. Numerous studies have demonstrated that the quenching of ·OH by Cl^- results in the generation of chlorine radicals (e.g., ·Cl), which can further engage in the degradation process of pollutants, thereby leading to the formation of chlorinated byproducts. It is well known that chlorinated products are more stable and difficult to degrade; moreover, certain chlorinated products have been observed to induce the expression of resistance genes or genetic mutations via oxidative stress mechanisms, thereby exacerbating the issue of antibiotic resistance [68]. Chi et al. found that Cl^- quench $\cdot O_2^-$ through a cyclic indirect quenching model with h^+ and ·OH quenching as "initiators" [69]. Thus, scavenging h^+ and ·OH could successfully block the chain reactions between Cl^- and $\cdot O_2^-$. In the photocatalytic degradation of methyl orange, where $\cdot O_2^-$ is the dominant attacking species, inhibiting the quenching of $\cdot O_2^-$ by Cl^- resulted in remarkable enhancement in degradation efficiency by nearly 50 times, even with the presence of 10 wt% Cl^-. More importantly, both HPLC-MS analyses and DFT calculation validated that, by blocking its quenching effect, Cl^- could be effectively excluded from the pollutant degradation processes, thus preventing the generation of toxic chlorinated by-products. However, this strategy has a limited scope of application, and currently, there is no broad-spectrum strategy to counteract the interference of Cl^- or other coexisting ions in photocatalytic systems. To develop a universal mitigation strategy remains an area of ongoing research and innovation.

13.4.4 Effect of Light Conditions

The efficiency of photocatalytic process is profoundly influenced by the light conditions, which are governed by two critical parameters: wavelength and light intensity. The relationship between light intensity and photocatalytic activity is quite direct: higher light intensity leads to the excitation of more e^-–h^+ pairs, thereby boosting the photocatalytic activity [70]. Nevertheless, once the light intensity reaches a specific threshold, the photocatalyst is unable to produce more e^-–h^+ pairs, resulting in a plateau in photocatalytic performance. Moreover, excessively intense light can elevate the temperature of the reaction system, which may induce photocatalyst aggregation, thereby reducing the photocatalytic efficiency.

The influence of light wavelength on the photocatalytic process is complex, affecting not only the efficiency but also the underlying mechanisms and the behavior of coexisting organic matter. In Li et al.'s study on the degradation of SAs with P25, UV irradiation was much more effective than visible light, achieving over 99% degradation within 90 min versus less than 70% in 120 min under visible light [61]. This difference is mainly attributed to the high sensitivity of TiO_2 to UV light, which has photons with more energy than visible light. Wu et al.'s study on the photodegradation of TC using P25 also showed a higher mineralization rate of 35.7% within 120 min under UV irradiation, compared to no mineralization under visible light [4]. This further emphasizes the higher energy of UV photons and their effectiveness in degrading pollutants.

It is worth noting that the aforementioned studies by Li et al. and Wu et al. also clearly demonstrates the pivotal influence of light wavelength on the mechanisms of photocatalytic degradation. In Li et al.' study, the photocatalytic degradation of SAs was mainly triggered by TiO_2 excitation, and thus, variations in the light wavelength did not alter the primary reactive species and degradation mechanism [61]. In contrast, Wu's study revealed a different scenario under 700 nm visible light, where the TC degradation was initiated by its complexation with TiO_2, which introduced a distinct degradation mechanism compared to that under UV light at 350 nm [4].

Under visible light, $\cdot O_2^-$ generated from the photoexcitation of the surface complex played a pivotal role in TC degradation. In contrast, both $\cdot O_2^-$ and h^+ were identified as the primary reactive species under UV irradiation. The shift in the type of reactive species between visible and UV light led to the formation of a distinct set of intermediate products during TC degradation. This variation underscores the impact of light wavelength on the degradation mechanisms and the formation of intermediates.

The impact of DOM on antibiotic degradation rates also varies with light wavelength. Li et al. found that, under UV light, the introduction of HA significant inhibited SAs degradation by P25, with the inhibitory effect being proportional to the concentration of HA added [61]. This is due to HA's role as an active oxygen quencher, its capacity to absorb UV light, and its competition for the limited adsorption sites. Conversely, under visible light, a high concentration of HA was found to boost the degradation of SAs, with the enhancement being proportional to the HA concentration. This enhancement is attributed to HA's capacity to act as a photosensitizer, which facilitates the photodegradation process. These findings emphasize the complexity of the influence of light wavelength on degradation mechanisms.

13.4.5 Effect of Photocatalyst Dosage

The photocatalyst dosage is an important parameter in photocatalytic processes, significantly impacting the degradation efficacy of antibiotics. Chatzitakis et al. illustrated a direct correlation between the P25 dosage and the initial reaction rate (r_0) of Chloramphenicol degradation [51]. As the TiO_2 concentration increased from 0.25 to 4 g L^{-1}, the r_0 exhibited a nearly twofold increase. Interestingly, the correlation between TiO_2 dosage and r_0 was very similar to the Langmuir adsorption isotherm, suggesting that the initial reaction rate reaches a plateau at elevated TiO_2 concentrations.

It is worth noting that an optimal dosage of photocatalyst often exists, with increments beyond this dosage potentially leading to a decline in pollutant degradation efficiency. In a study by Wu et al. on the degradation of TC using black anatase TiO_2 (B-TiO_2), a notable escalation in the TC degradation rate was observed upon increasing the B-TiO_2 dosage from 10 to 20 mg L^{-1}, due to the increased number of available photocatalytic sites [10]. However, elevating the dosage further to 40 mg L^{-1} only resulted in a slight additional increase in the degradation rate, and further increasing the dosage to 80 mg L^{-1} resulted in an obvious decrease in the TC degradation rate. This decline is attributed to the worsened solution turbidity caused by the excess of photocatalyst particles, which in turn diminished the light transmittance and light intensity available to the photocatalyst.

Achieving an optimal balance in photocatalyst dosage is essential for maximizing the photocatalytic efficiency, managing costs, and maintaining operational feasibility. It is noteworthy that the optimal dosage is dependent upon the characteristics of the organic pollutant and the geometric configuration of the photoreactor. Tailoring the dosage to the specific degradation scenario is indispensable for realizing the most efficacious degradation outcomes.

13.4.6 Effect of Initial Antibiotic Concentration

The initial concentration of antibiotics significantly influences the photocatalytic degradation process, presenting numerous challenges at elevated level. First, the high concentration of antibiotics competes for the limited active sites on the photocatalyst surface, thereby hindering the formation of vital reactive species, such as $\cdot OH$ and $\cdot O_2^-$. Second, the elevated antibiotic concentration leads to an increased yield of intermediates and by-products, which further compete with the parent antibiotic molecules for the available active sites and reactive species. Thirdly, a high antibiotic concentration reduces the light transmittance of the solution; for instance, CIP itself can absorb photons, which further limits the number of photons absorbed by the photocatalyst, consequently decreasing the generation of reactive species [71].

Hu et al. investigated the degradation efficiency of CIP across a concentration range of 5–25 μM [72]. Below 15 μM, a positive correlation was observed between the initial CIP concentration and the degradation rate constant, which increased from 0.0219 to 0.0381 min^{-1} as the concentration rose from 5 to 15 μM. This enhancement is probably due to the increased adsorption of CIP molecules onto the active sites of TiO_2 surface. Conversely, beyond 15 μM, there was a notable decrease in the degradation rate, with the rate constant dropping to 0.0259 min^{-1} at 25 μM. This downturn can be ascribed to two primary factors. First, the fixed amount of TiO_2 photocatalyst may not offer adequate active sites to handle the elevated CIP concentrations. Second, the saturation of active sites by CIP molecules could obstruct the adsorption of O_2, which serves a critical role as an electron acceptor on photocatalyst surface. Additionally, the high concentration of CIP is capable to absorb the incident light, thereby reducing the available photons to activate the photocatalyst and further diminishing the photocatalytic efficiency.

He et al. also analyzed the impact of initial TC concentration on its degradation efficiency when using MIL-101(Fe)/TiO_2 composite as the photocatalyst [7]. Their research documented a progressive decrease in degradation efficiency, from 99.19% to 63.30% within 80 min, as the TC concentration elevating from 20 to 80 mg L^{-1}. They found that, at lower concentrations, the production rate of $\cdot OH$ and $\cdot O_2^-$ was significantly faster than their consumption rate, thereby resulting in faster TC degradation.

The analysis above indicates that the efficiency of photocatalytic degradation of pollutants is not only influenced by the intrinsic properties of the photocatalyst itself but also closely related to the actual operating conditions. To achieve efficient degradation of antibiotics, it is essential to optimize these operating conditions through systematic experiments, ensuring the efficiency and practicality of the photocatalytic process.

13.5 Photocatalysis Coupled with Other Technologies to Enhance the Efficiency of Antibiotic Degradation

Although photocatalytic antibiotic degradation has been extensively investigated so far, its practical applicability is still restricted because of the rapid recombination of photogenerated e^-–h^+ and the relatively poor solar energy conversion efficiency.

To improve the photocatalytic efficiency, researchers have made significant efforts to screen, synthesize and modify the current photocatalyst, but the efficiency gains were not very significant. Recently, researchers have discovered that combining photocatalysis with other technologies could stimulate their synergistic effect which could offer an additional avenue for practical implementation. To date, researchers have studied photocatalysis in conjunction with other AOPs, external electric field, biodegradation, membrane separation and photo-thermal technology etc. Their coupling mechanisms are briefly described below.

13.5.1 Coupled with Other AOPs

The integration of photocatalysis with other AOPs emerges as a robust strategy for enhancing the oxidation and mineralization of pollutants. This research field has encompassed both well-established methods, such as photo-Fenton, photocatalytic ozonation and persulfate activation, as well as innovative approaches involving the photocatalytic activation of chlorine, chlorite, peracetic acid, etc. Within these hybrid systems, the primary reactive species generated by photocatalysis, such as e^- and h^+, are capable of activating the coupled oxidants, leading to the production of secondary reactive species like $\cdot SO_4^-$ and $^1O_2$. This activation can not only significantly improve the separation of photogenerated e^-–h^+ pairs, but also substantially increases the concentration and diversity of oxidative species present within the photocatalytic systems. Moreover, this coupling system can introduce radicals with extended lifetimes and enhanced selectivity, which are more resistant to the quenching effects posed by complex background constituents in real water environment. In this section, we explore the coupling photocatalysis with Fenton process, persulfate oxidation, and ozone treatment, from the perspectives of performance and mechanism in environmental remediation.

13.5.1.1 Coupled with Fenton Process

The traditional Fenton process is limited by the slow reduction of Fe^{3+} to Fe^{2+}, which impedes the activation of H_2O_2 and leads to higher costs of iron sludge disposal. However, integrating photocatalysis with the Fenton process, known as photo-Fenton, can enhance the Fe^{3+}/Fe^{2+} cycle by utilizing conduction band electrons (CB-e^-) to reduce iron sludge accumulation. Additionally, the activation efficiency of H_2O_2 is increased, since both CB-e^- and valence band holes (VB-h^+) can directly activate H_2O_2 to produce $\cdot OH$ without the need for iron, thereby improving the mineralization of pollutants.

Hernandez-Coronado et al. evaluated the degradation of cefuroxime (CFX) using a heterogeneous photo-Fenton system with α-Fe_2O_3-TiO_2 as a catalyst under simulated solar radiation [73]. Compared to the photocatalytic system (TiO_2/UV–vis) and the Fenton system (Fe^{2+}/H_2O_2), the degradation of CFX was significantly enhanced in the heterogeneous photo-Fenton system, with degradation rates of 80.0%, 50.5%, and 94.8% for photocatalysis, Fenton, and photo-Fenton, respectively, and mineralization rates of 36.7%, 23.9%, and 50.4%, respectively. It is particularly noteworthy that the production of iron sludge is greatly reduced in this heterogeneous photo-Fenton system, thereby not only improving the degradation efficiency of the antibiotics, but also significantly enhancing the applicability and reducing the treatment costs.

13.5.1.2 Photocatalytic Persulfate Oxidation

$\cdot SO_4^-$-based AOPs have gained significant popularity in recent years due to their high redox potential ($E_0 = 2.6–3.1$ V vs. NHE), extended lifetime (30–40 μs), broad pH tolerance ranging from 2 to 10, and high selectivity for pollutant degradation. Numerous studies have demonstrated a synergistic effect between photocatalysis and persulfate (PS) oxidation in the removal of antibiotics. In such hybrid systems, PS can capture photogenerated electrons to generate $\cdot SO_4^-$ through self-activation, meanwhile, the photocatalytic activity is also enhanced due to the improved separation of photogenerated $e^-–h^+$ pairs [74].

In a study by Zhang et al., a TiO_2 composite doped with acetylene black (TiO_2/AB) was integrated with PS for the degradation of TC hydrochloride (TCH) [75]. Following 2 h of LED irradiation, the removal rates of TCH were as follows: 5.3% for PS alone, 18.2% for TiO_2 alone, 24.0% for TiO_2/PS, 54.8% for TiO_2/AB, and 93.3% for TiO_2/AB/PS. It was evident that the degradation efficiency of both TiO_2 and TiO_2/AB was significantly enhanced by the presence of PS. Meanwhile, the $\cdot SO_4^-$ was found to play a more crucial role in the degradation of TCH than $\cdot OH$. However, the degradation efficiency was dependent on PS concentration. When the PS concentration increased from 0 to 3.0 mM, the TCH removal rate after 2 h rose from 54.8% to 93.3%, due to a higher yield of $\cdot SO_4^-$. Conversely, when the PS concentration exceeded 3.0 mM, the removal rate of TCH declined, which can be attributed to the quenching of $\cdot SO_4^-$ by excess PS, as in the reaction $S_2O_8^{2-} + \cdot SO_4^- = SO_4^{2-} + \cdot S_2O_8^-$. Additionally, at higher PS concentrations, radical–radical reactions, such as $\cdot SO_4^- + \cdot SO_4^- = 2SO_4^{2-}$, become more pronounced, reducing the availability of these radicals for the degradation process. pH is another significant factor affecting the efficiency of this system. Results showed that the TCH degradation under acidic conditions was higher than under alkaline or neutral conditions. Therefore, it is essential to optimize operational parameters to maximize the degradation efficiency of antibiotics in such systems.

13.5.1.3 Photocatalytic Ozonation

O_3-based AOPs are effective in disinfection, oxidation, and elimination of taste and odor issues due to O_3's strong redox potential ($E_0 = 2.01$ V vs. NHE). However, ozonation faces challenges such as low mineralization efficiency, limited mass transfer, and a short lifetime, leading to the low utilization efficiency of O_3. Photocatalysis offers the benefit of complete pollutant mineralization, thus only a small amount of O_3 is required to attain high levels of mineralization when combined with ozonation. Meanwhile, the heterogeneous photocatalyst surface provides favorable active sites for ozonation and pollutant oxidation, effectively addressing the mass transfer challenges inherent in homogeneous ozonation. Moreover, in heterogeneous photocatalytic ozonation, O_3 acts as a superior electron acceptor compared to O_2, due to its higher electrophilicity, thereby significantly enhancing the separation of $e^-–h^+$ pairs and then boosting the photocatalytic efficiency.

Moreira et al. compared the degradation of amoxicillin (AMX) across different AOPs: ozonation (O_3), photolytic ozonation (O_3 + UV–vis), photocatalysis (UV/Vis + TiO_2) and photocatalytic ozonation (O_3 + UV–vis + TiO_2) system [76]. Although AMX was readily oxidized in all systems, the initial reaction rate for photocatalytic ozonation was more than 2 times higher than the sum of the ozonation and photocatalysis alone. Additionally, the mineralization behavior of AMX also varied significantly among the processes. Specifically, the single ozonation system achieved less than 41% mineralization in 180 min, while photocatalysis yielded 85%. Photolytic ozonation, despite achieving approximately 95% TOC removal, was notably slower and required 300 min. In contrast, the photocatalytic ozonation process achieved complete TOC reduction for AMX within just 30 min. These results clearly indicate a synergistic effect when combining photocatalytic and ozonation treatments. Beltran et al. evaluated the removal rates of SMX and TOC through simple photolysis (UVA), ozonation (O_3), catalytic ozonation (O_3/TiO_2), ozone photolysis (O_3/UVA), photocatalytic oxidation (O_2/TiO_2/UVA) and photocatalytic ozonation (O_3/UVA/TiO_2) [77]. Photocatalytic ozonation was found to be the most effective, achieving complete antibiotic removal in less than 5 min and a substantial 93% TOC removal. Additionally, this process demonstrated the lowest ozone consumption per TOC removed and the least toxicity. These findings underscore the practical potential of photocatalytic ozonation as an efficient treatment method.

It is worthy to note that photocatalytic ozonation technology has reached a relatively mature stage and has begun attempts at commercialization. It has been applied in the treatment of primary and secondary effluents from industries, such as pulp, pharmaceutical and winery [78, 79].

13.5.2 Coupled with External Electric Field

External electric field has been widely adopted to enhance photocatalytic performance by accelerating the separation of photoinduced carriers. This strategy has been developed into photoelectrocatalysis (PEC) and extensively studied in antibiotic degradation research. In PEC system, antibiotics degradation is commonly initiated by light activation of a semiconductor anode, followed by the separation of photogenerated e^-–h^+ pairs with the assistance of an electric field. The residual holes on the anode, along with the oxidative species (·OH and $\cdot O_2^-$) generated at the cathode, then contribute to the degradation of antibiotics that are either adsorbed on the electrodes surface or situated nearby [80].

Among all reported photoelectrode materials, TiO_2-based photoelectrodes outperform other materials in degrading a broader spectrum of antibiotics, exhibiting superior degradation and mineralization efficiency. This is attributed to their strong oxidative properties under UV illumination and remarkable stability. In a study conducted by Liu et al., highly ordered TiO_2 nanotubes were utilized as photoanode for the PEC degradation of ofloxacin [81]. This system achieved a remarkable removal efficiency of 90.1% within 120 min via PEC, surpassing the rates achieved through photocatalysis (75.6%), electrochemistry (54.6%) and direct photolysis (65.6%). This enhanced performance is predominantly due to the improved separation of photogenerated electrons and holes during the PEC process.

In PEC processes, various electrodes, including TiO_2/Ti, TiO_2/Au/TiO_2 multilayer thin films, and ZnO/TiO_2/Ag_2Se, have demonstrated high degradation efficiency for TCs (e.g., TC and OTC) and quinolones (e.g., ofloxacin, CIP, and norfloxacin) [81–83]. This high efficiency is attributed to the susceptibility of their functional groups, like —OH, $—CH_3$, $—NH_2$, and $—CONH_2$, as well as structural features of the antibiotics, including the benzene ring and C—C bonds in TCs, and the piperazinyl and carboxyl groups in quinolones, to the degradation by PEC electrodes. However, β-lactams (e.g., penicillin G), sulfonamides (e.g., SMX), macrolides (e.g., erythromycin), and berberine can only be degraded by a few types of photoelectrodes [84], indicating the weak degradability.

13.5.3 Intimately Coupled Photocatalysis and Biodegradation

Intimately coupled photocatalysis and biodegradation (ICPB) technology shows great potential for enhancing the removal and mineralization of refractory pollutants. In a typical ICPB system, the photocatalyst is immobilized on the surface of millimeter-sized macro-porous carriers, with the biofilm established within the pores of these carriers [85]. This configuration protects the bacteria from potential damage caused by free radicals, while ensuring their proximity to the biodegradable products of photocatalysis. Notably, the presence of biofilm enhances the degradation of pollutants by minimizing the interaction between photocatalytic intermediates and reactive species. Thus, the ICPB system adeptly overcomes the limitations of incomplete mineralization that may occur with photocatalysis alone, and mitigates the risk of toxic compound accumulation inherent in bioreactor-dependent processes.

To date, ICPB technology has been effectively utilized in the treatment of various recalcitrant pollutants, including phenols, dinitrotoluene, and dyes. Recently, TCH degradation and mineralization were proved to be significantly improved by a visible-light-driven ICPB reactor with Ag-doped TiO_2 as the photocatalyst [86]. This system achieved an 11% increase in TCH removal and a 20% improvement in COD reduction compared to photocatalytic processes alone. This study represents the inaugural instance of employing ICPB technology for the treatment of antibiotic compounds.

Despite its good efficacy in treating wastewater contaminated with antibiotics, ICPB technology still possesses some limitations. One notable concern pertains to the poor optical properties of the carriers. Additionally, another critical issue arises from the competitive consumption of electron acceptors, particularly O_2, between the photocatalytic reactions and the metabolic processes of aerobic bacteria. This competition can potentially diminish the availability of electron acceptors for photocatalysis, consequently impairing its effectiveness. In addressing the aforementioned limitations, Xiao et al. incorporated algal-bacterial biofilms that could generate O_2 essential for photocatalysis through photosynthesis process [87]. Additionally, a light-conducting nuclear pore filter membrane was employed to serve as carrier for the algal-bacterial biofilm, facilitating the transportation of photocatalytic intermediates and biofilm metabolites (O_2). This innovative design of ICPB system has yielded remarkable results, with the removal and mineralization rates of TC achieving 98.1% and 95.0% within 5 and 10 h, respectively. During a 120-day experiment, the TC degradation rate and COD removal rate were consistently maintained at 8.83 μmol/h and 16.83 mg/h, respectively. Concurrently, the biomass concentration within the reactor's biofilm increased by an impressive 11-fold. This research presents a novel strategy to mitigate the limitations of ICPB technology.

To augment the biotransformation and mineralization of antibiotics in the ICPB system, a readily biodegradable co-substrate, acetate, could be introduced as both an energy source and an electron donor. Xiong et al. demonstrated that the acetate addition significantly boosted the removal and mineralization of TCH by about 5% and 20%, respectively, and almost all photocatalytic products were consumed by the bacterial biofilm [88]. The improvement in TCH removal efficiency was attributed to the enhanced biomass activity, which in turn promoted the development of microbial community. Consequently, the evolved microbial community accelerated the photocatalytic and biotransformation of the intermediates.

13.5.4 Coupled with Mechanical Stress

Piezoelectric materials, characterized by their non-centrosymmetric crystalline structure, can harness mechanical stress to generate a piezoelectric potential, resulting in charge polarization. In piezoelectric semiconductors or composites integrating piezoelectric materials with semiconductors, this polarization facilitates the directional separation of photogenerated electrons and holes, and effectively modulates the barrier height at the heterojunction interface, thereby optimizing carrier migration [89]. Additionally, it promotes the spatial separation of redox-active sites on the photocatalyst surface, which efficiently inhibits the reverse reaction between reduction and oxidation products, thereby further enhancing the photocatalytic efficiency.

Ultrasonic irradiation is the most common approach to exert the periodic external strain on piezoelectric materials via the intense acoustic pressure of ultrasonic waves and the localized high pressure resulting from the implosion of cavitation bubbles. The ultrasonic vibration induces the deformation of piezoelectric semiconductors, which leads to the deviation of positive and negative charge centers in the unit cell. Thus, a polarization-induced internal electric field is generated to promote the migration of photoexcited electrons and holes along opposite directions, which significantly enhance the photocatalytic performance. In addition to the efficient charge separation, the distribution of surface active sites can also be effectively manipulated by ultrasonic-induced piezoelectric polarization. Tang et al. developed a twisty C-TiO_2/PCN S-scheme heterojunction for the degradation of CIP with the assistance of ultrasonic vibration and visible light irradiation [90]. They found that the nitrogen-rich twisty polymeric carbon nitride (PCN) induced a non-centrosymmetric structure with enhanced polarity for an excellent piezoelectric effect. Consequently, the degradation rate constant of CIP was elevated to 1.86 times that of photocatalysis alone, underscoring the potency of integrating mechanical stimulation with photocatalytic processes to achieve more effective environmental remediation.

13.5.5 Coupled with Other Processes

13.5.5.1 Adsorption

Adsorption is a highly effective method to remove antibiotics from water. Studies reveal that integrating photocatalysts with adsorbents significantly boosts photocatalytic efficiency, primarily due to the enhanced interaction between target organic molecules and active species. Carbon materials (such as activated carbon, carbon nanotubes, graphene, biochar, etc.) are favored for their excellent biocompatibility and electronic conductivity. They are widely utilized as carriers for photocatalysts, which augment surface area of photocatalyst for adsorption and facilitate charge separation for photodegradation. However, traditional activated carbon, which is rich in micropores, is less effective for adsorbing macromolecular antibiotics due to size exclusion effects. In contrast, mesoporous carbon, characterized by its tunable porous structure, strong thermal stability, and high surface area, has attracted considerable attention for antibiotics adsorption.

Ji et al. successfully synthesized ordered mesoporous carbons that exhibited superior adsorption capacities for TC and tylosin, surpassing those of commercial microporous activated carbons [91]. Based on these findings, Zheng et al. synthesized a graphitized mesoporous carbon–TiO_2 nanocomposite (GMC-TiO_2) via an extended resorcinol-formaldehyde approach [92]. This composite demonstrated a strong synergy between adsorption and photocatalytic process. They found that the lamellar GMC nanosheets, with large specific surface area and mesoporous structure, could effectively adsorb and concentrate antibiotics. This not only rapidly reduce the antibiotic concentration in water, but also greatly amplifies the probability of their interaction with photocatalytic active species. Interestingly, GMC can also enhance the transportation of photogenerated electrons to further boost the photocatalytic efficiency of TiO_2. As a result, a 15 mg L^{-1} solution of CIP could be completely mineralized in 1.5 h, and the biological toxicity of the reaction solution on luminescence bacteria diminished remarkably with antibiotics degradation, ultimately reaching nontoxic levels. This outcome underscores the significant potential of combining photocatalysts with adsorbents for the purification of antibiotic-contaminated wastewater.

13.5.5.2 Molecular Imprinting Technology

Photocatalysts encounter considerable challenges in practical water treatment due to the nonselective nature of the photogenerated active species. For example, within natural water bodies, natural organic matter (NOM) is abundant and usually present at concentrations higher than those of antibiotics. The competition for active species between NOM and antibiotics definitely diminish the efficiency of antibiotic degradation. Moreover, if the concentration of active species near antibiotics is insufficient for their complete degradation, intermediate products might react with other organic substances or ions, potentially leading to the formation of harmful by-products, such as chlorinated products. This not only impedes the mineralization process of antibiotics but could also result in the production of toxins. To address these issues, it is essential to selectively enrich antibiotic molecules on/near the surface of photocatalysts.

Molecular imprinting technology (MIT) is an advanced molecular recognition technique that has emerged as a versatile tool in fields such as artificial antibodies, sensors, chromatographic separation, catalysis, and drug delivery. MIT involves the creation of molecular imprinting polymers containing specific molecular recognition sites that closely match the chemical structure and spatial configuration of the target molecules, thus enabling selective binding and recognition. This technology has been validated for its effectiveness in the photocatalytic degradation of specific antibiotics in complex aquatic environments. Li et al. developed a molecularly imprinted photocatalyst (CT-MI) based on carbon nanosheets supported TiO_2, utilizing CIP as the template [24]. The obtained CT-MI demonstrated a synergistic effect between adsorption and photocatalytic activity, along with remarkable selectivity in CIP degradation. Compared to the competing pollutants, the selectivity coefficients of CT-MI for CIP were 7.2 for adsorption and 3.2 for photocatalysis. This superior performance is attributed to the specific molecular recognition provided by the imprinted layer, as well as the strong adsorption capability and improved electroconductivity of the carbon nanosheets. This study underscores the potential of integrating MIT with photocatalytic technology for the precise identification and selective degradation of highly toxic contaminants in diverse and complex systems.

13.5.5.3 Membrane Technology

The photocatalytic membrane reactor (PMR) is an innovative integration of photocatalytic degradation with membrane filtration, designed to enhance pollutant removal, mitigate membrane fouling, and ensure efficient photocatalyst recovery. There are two primary configurations of PMRs: (i) Slurry System: in this configuration, the photocatalyst is suspended in solution, which maximizes the exposure of active sites to enhance photocatalytic efficiency. However, membrane fouling from photocatalyst nanoparticles deposition potentially leads to a decrease in flux. Moreover, managing the photocatalyst – through separation, recovery, and reuse – can also be costly and time-consuming. Additionally, high concentrations of photocatalyst may impede light penetration due to increased light scattering [93]. (ii) Immobilization System: Here, the photocatalyst is entrapped or deposited in/on the membrane. Although this may slightly limit the accessibility of active sites, it facilitates straightforward the photocatalyst recovery and reuse, thereby minimizing operational costs. Additionally, the membrane fouling is greatly mitigated due to the photocatalytic membrane's self-cleaning properties, as well as the enhanced hydrophilicity owing to the photocatalyst decoration. Besides, immobilized PMRs effectively minimize light scattering [94]. Therefore, the immobilized PMR has attracted considerable research attention in recent years.

Ren et al. fabricated a photocatalytic reactive membrane by functionalizing polyvinylidene fluoride (PVDF) ultrafiltration membrane with TiO_2 nanoparticles for the removal of ARB and ARGs from a secondary wastewater effluent [95]. They found that, the TiO_2-modified PVDF membrane provided complete retention of ARB and effective photocatalytic degradation of ARGs and integrons. Specifically, the total removal efficiency of ARGs (i.e., plasmid-mediated *floR*, *sul1*, and *sul2*) reached ~98% after exposure to UV irradiation. Excellent removal of integrons (i.e., *intI1*, *intI2*, and *intI3*) after UV treatment indicated that the horizontal transfer potential of ARGs was effectively controlled by the TiO_2 photocatalytic reaction. Furthermore, long-term UF operation with several filtration cycles indicated that the TiO_2-modified PVDF membrane remained high water flux, and showed great membrane stability and antifouling performance. This demonstrates the great potential of photocatalytic reactive membranes in treating wastewater contaminated by ARB and ARGs.

In addition to the aforementioned technologies, photocatalytic process can be further enhanced by integrating with thermal catalysis, microwave radiation, plasma technique, magnetic fields, etc. These synergistic combinations can significantly increase light utilization efficiency, improve the separation of charge carriers, enhance pollutant adsorption, and facilitate the recovery of photocatalysts. These integrated approaches substantially promote the practical implementation and efficacy of photocatalytic technology.

13.6 Summary

TiO_2, an efficient photocatalyst, has demonstrated remarkable effectiveness in degrading antibiotic pollutants in water bodies. Extensive research has validated its capacity to degrade a broad spectrum of antibiotics, including but not limited to TCs, sulfonamides, quinolones, macrolides, β-lactams, aminoglycosides, peptides, lincosamides, pleuromutilins, rifamycins, and nitroimidazoles. The degradation process is influenced by many factors, such as pH levels, the presence of organic and inorganic substances, light exposure, photocatalyst concentration, and the initial concentration of the antibiotics. To maximize TiO_2's efficacy, it is crucial to optimize environmental conditions and operational parameters. Innovations like heterojunction construction, element doping, composite material synthesis, and defect engineering significantly enhance TiO_2's photocatalytic performance. Moreover, integrating TiO_2 photocatalysis with complementary technologies – such as oxidant activation, electric fields, biodegradation, mechanical stress, adsorption, molecular imprinting, and membrane processes – further amplifies its purifying capabilities. This chapter highlights the substantial potential of TiO_2 photocatalytic technology in addressing the contamination of water bodies by antibiotics. With ongoing technological progress, the modification and application of TiO_2 will become more diverse and efficient. It is anticipated that in the future, TiO_2 will make a greater contribution to environmental protection and human health.

References

1 Liu, X., Liu, Y., Lu, S. et al. (2018). Performance and mechanism into TiO_2/zeolite composites for sulfadiazine adsorption and photodegradation. *Chemical Engineering Journal* 350: 131–147.

2 World Health Organization. (2019). No time to wait: securing the future from drug-resistant infections. https://www.who.int/publications/i/item/no-time-to-wait-securing-the-future-from-drug-resistant-infections.

3 Murray, C.J.L., Ikuta, K.S., Sharara, F. et al. (2022). Global burden of bacterial antimicrobial resistance in 2019: a systematic analysis. *The Lancet* 399 (10325): 629–655.

4 Wu, S., Hu, H., Lin, Y. et al. (2020). Visible light photocatalytic degradation of tetracycline over TiO_2. *Chemical Engineering Journal* 382: 122842.

5 Wang, W., Fang, J., Shao, S. et al. (2017). Compact and uniform TiO_2@g-C_3N_4 core-shell quantum heterojunction for photocatalytic degradation of tetracycline antibiotics. *Applied Catalysis B: Environmental* 217: 57–64.

6 Chen, Q., Wu, S., and Xin, Y. (2016). Synthesis of Au–CuS–TiO_2 nanobelts photocatalyst for efficient photocatalytic degradation of antibiotic oxytetracycline. *Chemical Engineering Journal* 302: 377–387.

7 He, L., Dong, Y., Zheng, Y. et al. (2019). A novel magnetic MIL-101(Fe)/TiO_2 composite for photo degradation of tetracycline under solar light. *Journal of Hazardous Materials* 361: 85–94.

8 Hu, X., Sun, Z., Song, J. et al. (2019). Synthesis of novel ternary heterogeneous BiOCl/TiO_2/sepiolite composite with enhanced visible-light-induced photocatalytic activity towards tetracycline. *Journal of Colloid and Interface Science* 533: 238–250.

9 Gong, Y., Wu, Y., Xu, Y. et al. (2018). All-solid-state Z-scheme CdTe/TiO_2 heterostructure photocatalysts with enhanced visible-light photocatalytic degradation of antibiotic waste water. *Chemical Engineering Journal* 350: 257–267.

10 Wu, S., Li, X., Tian, Y. et al. (2021). Excellent photocatalytic degradation of tetracycline over black anatase-TiO_2 under visible light. *Chemical Engineering Journal* 406: 126747.

11 Li, S. and Hu, J. (2016). Photolytic and photocatalytic degradation of tetracycline: effect of humic acid on degradation kinetics and mechanisms. *Journal of Hazardous Materials* 318: 134–144.

12 Zhao, W., Duan, J., Ji, B. et al. (2020). Novel formation of large area N-TiO_2/graphene layered materials and enhanced photocatalytic degradation of antibiotics. *Journal of Environmental Chemical Engineering* 8 (1): 102206.

13 Nguyen, T.L., Pham, T.H., Viet, N.M. et al. (2022). Improved photodegradation of antibiotics pollutants in wastewaters by advanced oxidation process based on Ni-doped TiO_2. *Chemosphere* 302: 134837.

14 Zhang, X., Chen, J., Jiang, S. et al. (2021). Enhanced photocatalytic degradation of gaseous toluene and liquidus tetracycline by anatase/rutile titanium dioxide with heterophase junction derived from materials of Institut Lavoisier-125(Ti): degradation pathway and mechanism studies. *Journal of Colloid and Interface Science* 588: 122–137.

15 Biancullo, F., Moreira, N.F.F., Ribeiro, A.R. et al. (2019). Heterogeneous photocatalysis using UVA-LEDs for the removal of antibiotics and antibiotic resistant bacteria from urban wastewater treatment plant effluents. *Chemical Engineering Journal* 367: 304–313.

16 Wang, N., Li, X., Yang, Y. et al. (2020). Photocatalytic degradation of sulfonamides by Bi_2O_3-TiO_2/PAC ternary composite: mechanism, degradation pathway. *Journal of Water Process Engineering* 36: 101335.

17 Li, D., Yuan, R., Zhou, B., and Chen, H. (2023). Selective photocatalytic removal of sulfonamide antibiotics: the performance differences in molecularly imprinted TiO_2 synthesized using four template molecules. *Journal of Cleaner Production* 383: 135470.

18 Xiang, X., Wu, L., Zhu, J. et al. (2021). Photocatalytic degradation of sulfadiazine in suspensions of TiO_2 nanosheets with exposed (001) facets. *Chinese Chemical Letters* 32 (10): 3215–3220.

19 Zhou, A., Jia, R., Wang, Y. et al. (2020). Abatement of sulfadiazine in water under a modified ultrafiltration membrane (PVDF-PVP-TiO_2-dopamine) filtration-photocatalysis system. *Separation and Purification Technology* 234: 116099.

20 Qi, K., Wang, Z., Xie, X., and Wang, Z. (2023). Photocatalytic performance of pyrochar and hydrochar in heterojunction photocatalyst for organic pollutants degradation: activity comparison and mechanism insight. *Chemical Engineering Journal* 467: 143424.

21 Zeng, X., Shu, S., Meng, Y. et al. (2023). Enhanced photocatalytic degradation of sulfamethazine by g-C_3N_4/Cu,N-TiO_2 composites under simulated sunlight irradiation. *Chemical Engineering Journal* 456: 141105.

22 An, T., Yang, H., Li, G. et al. (2010). Kinetics and mechanism of advanced oxidation processes (AOPs) in degradation of ciprofloxacin in water. *Applied Catalysis B: Environmental* 94 (3–4): 288–294.

23 Diao, Z.-H., Xu, X.-R., Jiang, D. et al. (2017). Simultaneous photocatalytic Cr(VI) reduction and ciprofloxacin oxidation over TiO_2/Fe^0 composite under aerobic conditions: performance, durability, pathway and mechanism. *Chemical Engineering Journal* 315: 167–176.

24 Li, L., Zheng, X., Chi, Y. et al. (2020). Molecularly imprinted carbon nanosheets supported TiO_2: strong selectivity and synergic adsorption-photocatalysis for antibiotics removal. *Journal of Hazardous Materials* 383: 121211.

25 Feng, X., Wang, P., Hou, J. et al. (2018). Significantly enhanced visible light photocatalytic efficiency of phosphorus doped TiO_2 with surface oxygen vacancies for ciprofloxacin degradation: synergistic effect and intermediates analysis. *Journal of Hazardous Materials* 351: 196–205.

26 Hu, K., Li, R., Ye, C. et al. (2020). Facile synthesis of Z-scheme composite of TiO_2 nanorod/g-C_3N_4 nanosheet efficient for photocatalytic degradation of ciprofloxacin. *Journal of Cleaner Production* 253: 120055.

27 Truppi, A., Petronella, F., Placido, T. et al. (2019). Gram-scale synthesis of UV-vis light active plasmonic photocatalytic nanocomposite based on TiO_2/Au nanorods for degradation of pollutants in water. *Applied Catalysis B: Environmental* 243: 604–613.

28 Wang, Y., Zhu, C., Zuo, G. et al. (2020). 0D/2D Co_3O_4/TiO_2 Z-scheme heterojunction for boosted photocatalytic degradation and mechanism investigation. *Applied Catalysis B: Environmental* 278: 119298.

29 Sood, S., Mehta, S.K., Sinha, A.S.K., and Kansal, S.K. (2016). Bi_2O_3/TiO_2 heterostructures: synthesis, characterization and their application in solar light mediated photocatalyzed degradation of an antibiotic, ofloxacin. *Chemical Engineering Journal* 290: 45–52.

30 Cheng, N., Wang, B., Chen, M. et al. (2023). Adsorption and photocatalytic degradation of quinolone antibiotics from wastewater using functionalized biochar. *Environmental Pollution* 336: 122409.

31 Li, J., Xia, Z., Ma, D. et al. (2021). Improving photocatalytic activity by construction of immobilized Z-scheme CdS/Au/TiO_2 nanobelt photocatalyst for eliminating norfloxacin from water. *Journal of Colloid and Interface Science* 586: 243–256.

32 Hu, X., Hu, X., Peng, Q. et al. (2020). Mechanisms underlying the photocatalytic degradation pathway of ciprofloxacin with heterogeneous TiO_2. *Chemical Engineering Journal* 380: 122366.

33 Vaiano, V., Sacco, O., Sannino, D., and Ciambelli, P. (2015). Photocatalytic removal of spiramycin from wastewater under visible light with N-doped TiO_2 photocatalysts. *Chemical Engineering Journal* 261: 3–8.

34 Naraginti, S., Yu, Y.-Y., Fang, Z., and Yong, Y.-C. (2019). Visible light degradation of macrolide antibiotic azithromycin by novel ZrO_2/Ag@TiO_2 nanorod composite: transformation pathways and toxicity evaluation. *Process Safety and Environmental Protection* 125: 39–49.

35 Lofrano, G., Libralato, G., Casaburi, A. et al. (2018). Municipal wastewater spiramycin removal by conventional treatments and heterogeneous photocatalysis. *Science of the Total Environment* 624: 461–469.

36 Sharma, M., Rajput, D., Kumar, V. et al. (2023). Photocatalytic degradation of four emerging antibiotic contaminants and toxicity assessment in wastewater: a comprehensive study. *Environmental Research* 231: 116132.

37 Albornoz, L.L., da Silva, S.W., Bortolozzi, J.P. et al. (2021). Degradation and mineralization of erythromycin by heterogeneous photocatalysis using SnO_2-doped TiO_2 structured catalysts: activity and stability. *Chemosphere* 268: 128858.

38 Chen, Y., Zhang, X., Wang, L. et al. (2020). Rapid removal of phenol/antibiotics in water by Fe-(8-hydroxyquinoline-7-carboxylic)/TiO_2 flower composite: adsorption combined with photocatalysis. *Chemical Engineering Journal* 402: 126260.

39 Karaolia, P., Michael-Kordatou, I., Hapeshi, E. et al. (2018). Removal of antibiotics, antibiotic-resistant bacteria and their associated genes by graphene-based TiO_2 composite photocatalysts under solar radiation in urban wastewaters. *Applied Catalysis B: Environmental* 224: 810–824.

40 Babić, S., Ćurković, L., Ljubas, D., and Čizmić, M. (2017). TiO_2 assisted photocatalytic degradation of macrolide antibiotics. *Current Opinion in Green and Sustainable Chemistry* 6: 34–41.

41 Dimitrakopoulou, D., Rethemiotaki, I., Frontistis, Z. et al. (2012). Degradation, mineralization and antibiotic inactivation of amoxicillin by UV-A/TiO_2 photocatalysis. *Journal of Environmental Management* 98: 168–174.

42 Gar Alalm, M., Tawfik, A., and Ookawara, S. (2016). Enhancement of photocatalytic activity of TiO_2 by immobilization on activated carbon for degradation of pharmaceuticals. *Journal of Environmental Chemical Engineering* 4(2): 1929–1937.

43 Giraldo-Aguirre, A.L., Erazo-Erazo, E.D., Flórez-Acosta, O.A. et al. (2015). TiO_2 photocatalysis applied to the degradation and antimicrobial activity removal of oxacillin: evaluation of matrix components, experimental parameters, degradation pathways and identification of organics by-products. *Journal of Photochemistry and Photobiology A: Chemistry* 311: 95–103.

44 Estrada-Flórez, S.E., Serna-Galvis, E.A., and Torres-Palma, R.A. (2020). Photocatalytic vs. sonochemical removal of antibiotics in water: structure-degradability relationship, mineralization, antimicrobial activity, and matrix effects. *Journal of Environmental Chemical Engineering* 8 (5): 104359.

45 Elmolla, E.S. and Chaudhuri, M. (2010). Photocatalytic degradation of amoxicillin, ampicillin and cloxacillin antibiotics in aqueous solution using UV/TiO_2 and UV/H_2O_2/TiO_2 photocatalysis. *Desalination* 252 (1–3): 46–52.

46 Wang, P., Tang, Y., Dong, Z. et al. (2013). Ag–AgBr/TiO_2/RGO nanocomposite for visible-light photocatalytic degradation of penicillin G. *Journal of Materials Chemistry A* 1 (15): 4718–4727.

47 Chen, Q., Liu, H., Xin, Y., and Cheng, X. (2013). TiO_2 nanobelts – effect of calcination temperature on optical, photoelectrochemical and photocatalytic properties. *Electrochimica Acta* 111: 284–291.

48 Singh, H., Ahlawat, A., Dhiman, T.K., and Solanki, P.R. (2023). Photocatalytic degradation of gentamicin using TiO_2 nanoparticle driven by UV light irradiation. *Materials Letters* 346: 134504.

49 Dehghani, F., Yousefinejad, S., Dehghani, M. et al. (2022). Vancomycin removal using TiO_2-clinoptilolite/UV in aqueous media and optimisation using response surface methodology. *International Journal of Environmental Analytical Chemistry* 104: 1–19.

50 Do, T.C.M.V., Nguyen, D.Q., Nguyen, K.T., and Le, P.H. (2019). TiO_2 and Au-TiO_2 nanomaterials for rapid photocatalytic degradation of antibiotic residues in aquaculture wastewater. *Materials* 12 (15): 2434.

51 Chatzitakis, A., Berberidou, C., Paspaltsis, I. et al. (2008). Photocatalytic degradation and drug activity reduction of chloramphenicol. *Water Research* 42 (1–2): 386–394.

52 Zhang, J., Fu, D., Xu, Y., and Liu, C. (2010). Optimization of parameters on photocatalytic degradation of chloramphenicol using TiO_2 as photocatalyist by response surface methodology. *Journal of Environmental Sciences* 22 (8): 1281–1289.

53 Le, T.-H.T., Bui, T.-T., Van Bui, H. et al. (2021). TiO_2 inverse opals modified by Ag nanoparticles: a synergic effect of enhanced visible-light absorption and efficient charge separation for visible-light photocatalysis. *Catalysts* 11 (7): 761.

54 Abbasi-Asl, H., Sabzehmeidani, M.M., and Ghaedi, M. (2021). Efficient degradation of metronidazole antibiotic by TiO_2/Ag_3PO_4/g-C_3N_4 ternary composite photocatalyst in a continuous flow-loop photoreactor. *Journal of Environmental Chemical Engineering* 9 (5): 105963.

55 Sirtori, C., Agüera, A., Gernjak, W., and Malato, S. (2010). Effect of water-matrix composition on Trimethoprim solar photodegradation kinetics and pathways. *Water Research* 44 (9): 2735–2744.

56 Ding, P., Ji, H., Li, P. et al. (2022). Visible-light degradation of antibiotics catalyzed by titania/zirconia/graphitic carbon nitride ternary nanocomposites: a combined experimental and theoretical study. *Applied Catalysis B: Environmental* 300: 120633.

57 Zhang, J., Chen, X., Bai, Y. et al. (2019). Boosting photocatalytic water splitting by tuning built-in electric field at phase junction. *Journal of Materials Chemistry A* 7 (17): 10264–10272.

58 Jahdi, M., Mishra, S.B., Nxumalo, E.N. et al. (2020). Smart pathways for the photocatalytic degradation of sulfamethoxazole drug using F-Pd co-doped TiO_2 nanocomposites. *Applied Catalysis B: Environmental* 267: 118716.

59 Li, J., Cheng, D., Chen, Z. et al. (2021). Oxygen vacancy/Ti^{3+} engineered TiO_2 nanotube arrays prepared by in-situ exfoliation with H_2 bubbles: a visible-light-driven self-supporting photocatalyst for detoxfication of chloraphenicol. *Journal of Environmental Chemical Engineering* 9 (6): 106670.

60 Bayan, E.M., Pustovaya, L.E., and Volkova, M.G. (2021). Recent advances in TiO_2-based materials for photocatalytic degradation of antibiotics in aqueous systems. *Environmental Technology & Innovation* 24: 101822.

61 Li, D., Zhang, N., Yuan, R. et al. (2021). Effect of wavelengths on photocatalytic oxidation mechanism of sulfadiazine and sulfamethoxazole in the presence of TiO_2. *Journal of Environmental Chemical Engineering* 9 (5): 106243.

62 Yang, C.-C., Huang, C.-L., Cheng, T.C., and Lai, H.T. (2015). Inhibitory effect of salinity on the photocatalytic degradation of three sulfonamide antibiotics. *International Biodeterioration & Biodegradation* 102: 116–125.

63 Yuan, R., Zhu, Y., Zhou, B., and Hu, J. (2019). Photocatalytic oxidation of sulfamethoxazole in the presence of TiO_2: effect of matrix in aqueous solution on decomposition mechanisms. *Chemical Engineering Journal* 359: 1527–1536.

64 Wang, G., Chen, Q., Liu, Y. et al. (2018). In situ synthesis of graphene/WO_3 co-decorated TiO_2 nanotube array photoelectrodes with enhanced photocatalytic activity and degradation mechanism for dimethyl phthalate. *Chemical Engineering Journal* 337: 322–332.

65 Guo, K., Wu, Z., Yan, S. et al. (2018). Comparison of the UV/chlorine and UV/H_2O_2 processes in the degradation of PPCPs in simulated drinking water and wastewater: kinetics, radical mechanism and energy requirements. *Water Research* 147: 184–194.

66 Yuan, R., Ramjaun, S.N., Wang, Z., and Liu, J. (2012). Photocatalytic degradation and chlorination of azo dye in saline wastewater: kinetics and AOX formation. *Chemical Engineering Journal* 192: 171–178.

67 Bai, Y., Cui, Z., Su, R., and Qu, K. (2018). Influence of DOM components, salinity, pH, nitrate, and bicarbonate on the indirect photodegradation of acetaminophen in simulated coastal waters. *Chemosphere* 205: 108–117.

68 Zammit, I., Marano, R.B.M., Vaiano, V. et al. (2020). Changes in antibiotic resistance gene levels in soil after irrigation with treated wastewater: a comparison between heterogeneous photocatalysis and chlorination. *Environmental Science & Technology* 54 (12): 7677–7686.

69 Chi, Y., Xu, S., Li, M. et al. (2020). Effective blockage of chloride ion quenching and chlorinated by-product generation in photocatalytic wastewater treatment. *Journal of Hazardous Materials* 396: 122670.

70 Sheydaei, M., Shiadeh, H.R.K., Ayoubi-Feiz, B., and Ezzati, R. (2018). Preparation of nano N-TiO_2/graphene oxide/titan grid sheets for visible light assisted photocatalytic ozonation of cefixime. *Chemical Engineering Journal* 353: 138–146.

71 Kutuzova, A., Dontsova, T., and Kwapinski, W. (2021). Application of TiO_2-based photocatalysts to antibiotics degradation: cases of sulfamethoxazole, trimethoprim and ciprofloxacin. *Catalysts* 11 (6): 728.

72 Hu, K., Li, R., Ye, C. et al. (2020). Facile synthesis of Z-scheme composite of TiO_2 nanorodig-C_3N_4 nanosheet efficient for photocatalytic degradation of ciprofloxacin. *Journal of Cleaner Production* 253: 120055.

73 Hernandez-Coronado, E.E., Ruiz-Ruiz, E.J., Hinojosa-Reyes, L. et al. (2021). Effective degradation of cefuroxime by heterogeneous photo-Fenton under simulated solar radiation using α-Fe_2O_3-TiO_2. *Journal of Environmental Chemical Engineering* 9 (6): 106822.

74 Chen, G., Yu, Y., Liang, L. et al. (2021). Remediation of antibiotic wastewater by coupled photocatalytic and persulfate oxidation system: a critical review. *Journal of Hazardous Materials* 408: 124461.

75 Zhang, T., Liu, Y., Rao, Y. et al. (2020). Enhanced photocatalytic activity of TiO_2 with acetylene black and persulfate for degradation of tetracycline hydrochloride under visible light. *Chemical Engineering Journal* 384: 123350.

76 Moreira, N.F.F., Orge, C.A., Ribeiro, A.R. et al. (2015). Fast mineralization and detoxification of amoxicillin and diclofenac by photocatalytic ozonation and application to an urban wastewater. *Water Research* 87: 87–96.

77 Beltran, F.J., Aguinaco, A., Garcia-Araya, J.F., and Oropesa, A.L. (2008). Ozone and photocatalytic processes to remove the antibiotic sulfamethoxazole from water. *Water Research* 42 (14): 3799–3808.

78 Biglari, H., Afsharnia, M., Alipour, V. et al. (2017). A review and investigation of the effect of nanophotocatalytic ozonation process for phenolic compound removal from real effluent of pulp and paper industry. *Environmental Science and Pollution Research* 24 (4): 4105–4116.

79 Chavez, A.M., Quinones, D.H., Rey, A. et al. (2020). Simulated solar photocatalytic ozonation of contaminants of emerging concern and effluent organic matter in secondary effluents by a reusable magnetic catalyst. *Chemical Engineering Journal* 398: 125642.

80 Wu, S. and Hu, Y.H. (2021). A comprehensive review on catalysts for electrocatalytic and photoelectrocatalytic degradation of antibiotics. *Chemical Engineering Journal* 409: 127739.

81 Liu, L., Li, R., Liu, Y., and Zhang, J. (2016). Simultaneous degradation of ofloxacin and recovery of Cu(II) by photoelectrocatalysis with highly ordered TiO_2 nanotubes. *Journal of Hazardous Materials* 308: 264–275.

82 Olvera-Rodriguez, I., Hernandez, R., Medel, A. et al. (2019). TiO_2/Au/TiO_2 multilayer thin-film photoanodes synthesized by pulsed laser deposition for photoelectrochemical degradation of organic pollutants. *Separation and Purification Technology* 224: 189–198.

83 Su, Y.F., Wang, G.B., Kuo, D.T.F. et al. (2016). Photoelectrocatalytic degradation of the antibiotic sulfamethoxazole using TiO_2/Ti photoanode. *Applied Catalysis B-Environmental* 186: 184–192.

84 Ma, Q., Zhang, H., Guo, R. et al. (2018). Construction of CuS/TiO_2 nano-tube arrays photoelectrode and its enhanced visible light photoelectrocatalytic decomposition and mechanism of penicillin G. *Electrochimica Acta* 283: 1154–1162.

85 Zhou, D., Xu, Z., Dong, S. et al. (2015). Intimate coupling of photocatalysis and biodegradation for degrading phenol using different light types: visible light vs UV light. *Environmental Science & Technology* 49 (13): 7776–7783.

86 Xiong, H., Zou, D., Zhou, D. et al. (2017). Enhancing degradation and mineralization of tetracycline using intimately coupled photocatalysis and biodegradation (ICPB). *Chemical Engineering Journal* 316: 7–14.

87 Xiao, C., Yuan, J., Li, L. et al. (2023). Photocatalytic synergistic biofilms enhance tetracycline degradation and conversion. *Environmental Science and Ecotechnology* 14: 100234.

88 Xiong, H., Dong, S., Zhang, J. et al. (2018). Roles of an easily biodegradable co-substrate in enhancing tetracycline treatment in an intimately coupled photocatalytic-biological reactor. *Water Research* 136: 75–83.

89 Yang, Q., Guo, X., Wang, W. et al. (2010). Enhancing sensitivity of a single ZnO micro-/nanowire photodetector by piezo-phototronic effect. *ACS Nano* 4 (10): 6285–6291.

90 Tang, R., Zeng, H., Feng, C. et al. (2023). Twisty C-TiO_2/PCN S-scheme heterojunction with enhanced $n \rightarrow \pi^*$ electronic excitation for promoted piezo-photocatalytic effect. *Small* 19 (18): 2207636.

91 Ji, L., Liu, F., Xu, Z. et al. (2010). Adsorption of pharmaceutical antibiotics on template-synthesized ordered micro- and mesoporous carbons. *Environmental Science & Technology* 44 (8): 3116–3122.

92 Zheng, X., Xu, S., Wang, Y. et al. (2018). Enhanced degradation of ciprofloxacin by graphitized mesoporous carbon (GMC)-TiO_2 nanocomposite: strong synergy of adsorption-photocatalysis and antibiotics degradation mechanism. *Journal of Colloid and Interface Science* 527: 202–213.

93 Homocianu, M. and Pascariu, P. (2022). High-performance photocatalytic membranes for water purification in relation to environmental and operational parameters. *Journal of Environmental Management* 311: 114817.

94 Ashley, A., Thrope, B., Choudhury, M.R., and Pinto, A.H. (2021). Emerging investigator series: photocatalytic membrane reactors: fundamentals and advances in preparation and application in wastewater treatment. *Environmental Science-Water Research & Technology* 8 (1): 22–46.

95 Ren, S., Boo, C., Guo, N. et al. (2018). Photocatalytic reactive ultrafiltration membrane for removal of antibiotic resistant bacteria and antibiotic resistance genes from wastewater effluent. *Environmental Science & Technology* 52 (15): 8666–8673.

14

Ecotoxicological and Antibacterial Properties of TiO_2 and Processes Based on TiO_2

*Eryk Fernandes, João Lincho, Rui C. Martins, João Gomes**

University of Coimbra, CERES – Chemical Engineering and Renewable Resources for Sustainability, Department of Chemical Engineering, Faculty of Sciences and Technology, Coimbra, Portugal

*Corresponding author: jgomes@eq.uc.pt

14.1 Introduction

The photocatalytic process mediated by TiO_2 as active material is widely investigated for the removal of contaminants of emerging concern (CECs) from water. CECs are a big group of contaminants present in water, encompassing pharmaceutical and personal care products and pesticides. During photocatalysis, CECs can lead to by-products or intermediates that can be more toxic than the parent compounds. Several studies in the literature indicate the potential of TiO_2 under different radiation sources, mainly UV radiation for the abatement of such CECs [1–3]. Due to its bandgap energy, TiO_2 (3.2 eV) activation must occur under UV radiation which increases the operation cost. To overcome this drawback, TiO_2 doping can be an alternative route to improve its activity. The selection of dopants is usually made based on the textural and photocatalytic properties of TiO_2. However, some dopants and their precursors can introduce toxicity and may act as antimicrobial agents. Therefore, it is important to consider different issues during the photocatalytic application of TiO_2 for the CECs abatement:

- Understand the impact of TiO_2 nanoparticles and doping-based precursors on the different aquatic species in terms of ecotoxicity.
- Understand the benefit of TiO_2 nanoparticles and their precursors in doping cases for microbiological disinfection.
- During the CECs abatement, the resulting by-products toxicity must be assessed and compared with the original CECs.

These issues are typically found during the TiO_2 photocatalytic application, and in this chapter, these points were addressed and deeply evaluated.

Nanostructured TiO_2: Synthesis, Photocatalytic Properties and Applications, First Edition.
Edited by Jia Hong Pan, Wan In Lee and Detlef W. Bahnemann

14.2 Species Commonly Used in Ecotoxicity Tests

The literature studies for photocatalytic oxidation treatment normally follow the efficacy of the process in terms of pollutant abatement [4–6]. When water reuse is the perspective, the efficacy of the treatment should be evaluated in terms of the ecotoxicity impact over the different trophic levels. A wide range of living organisms can be used for ecotoxicity experiments. Table 14.1 presents the organisms already used for toxicological bioassays [7–9]. Some of these organisms can be considered also as indicators of microbial water quality, such as *Escherichia coli*. Another relevant information is that some organisms can suffer modification of species names. For example, *Raphidocelis subcapitata*, formerly known as *Selenastrum capricornutum* and *Pseudokirchneriella subcapitata* is the same microalgae.

Table 14.1 Species commonly used for toxicological bioassays.

Type of organism	Organism species
Bacteria	*Aliivibrio fischeri*
	Vibrio Harvey
	Photobacterium phosphoreum
	Bacillus subtilis
	Staphylococcus epidermidis
	Escherichia coli
	Pseudomonas fluorescens
	Pseudomonas liognathi
	Pseudomonas putida
Invertebrates	*Paracentrotus lividius (sea urchin)*
	Daphnia magna
	Artemia salina
Algae	*Raphidocelis subcapitata*
	Scenedesmus subspicatus
	Dunaliella tertiolecta
	Scenedesmus obliquus
	Nitellopsis obtuse
	Chlorella vulgaris
	Synechococcus leopoliensis
	Phaeodactylum tricornutum
Plants	*Avena sativa*
	Brassica campestres
	Lepidium sativum
Fish	*Poecilia vivipara (guppy)*
	Oncorhynchus mykiss (rainbow trout)
	Danio rerio (zebrafish)

14.3 TiO_2 Particles Ecotoxic Impact

Nanoparticles can accumulate in the environment and may have different types of impacts on the environment as changes in microbial communities, nutrient availability, pH, and turbidity [10]. Metal oxide nanoparticles can act as carriers of contaminants (as the pharmaceuticals and personal care products – PPCPs) promoting their bioaccumulation, and the interaction between both may disrupt cell membranes increasing the permeability which allows to other contaminants accumulate easily [10]. Impacts were already reported for TiO_2, ZnO, CuO, and carbon nanotubes [10]. For example, Li et al. [11] showed that CuO and ZnO nanoparticles injured cell membranes of earthworms which increased the penetration of bifenthrin, while TiO_2 nanoparticles served as carriers of benzophenone-3 and triclosan [12], and polybrominated diphenyl ethers [13]. TiO_2 also contributed to the enrichment of bisphenol A in the brain and gonads of zebrafish [14], and a study involving algae showed that TiO_2 and ZnO nanoparticles can accumulate more in algae due to their interaction with triclosan, which should prevent its aggregation and increase the penetration into the algae cells due to the cell damage caused by the triclosan [10, 15]. Carbon nanotubes were also responsible for carrying indoxacarb, fluoxetine, roxithromycin, or ciprofloxacin [10, 16–19]. However, the toxicity between PPCPs and nanoparticles should result more in synergetic and antagonistic effects and hardly result in additive effects, with a strong correlation between bioaccumulation and co-toxicity. It may also depend on the target species and their development stage [10]. For example, Gram-positive bacteria (such as *B. subtilis*) may be more sensitive than Gram-negative (such as *E. coli*) regarding the toxicity of substances, due to an outer membrane in the cells of Gram-negative bacteria [18]. In another example, graphene oxide decreased the bisphenol A absorption of zebrafish embryos but increased the endocrine disruptive effects of BPA in zebrafish adults, since the adults can have other different exposure pathways [20]. In water environments, there are several factors (pH, ionic strength, humic acid concentration, or temperature) that may influence the effect of the PPCPs and nanoparticles presence, which also changes its final impact on the biological effects [10]. For example, humic acids can decrease the toxicity of TiO_2 nanoparticles when irradiated by the Sun [10, 21], but it is possible to conclude that, in general, natural organic matter contributes to the toxicity decrease of nanoparticles in aquatic organisms [22]. With this, ecotoxicity tests that discard the presence of natural organic matter will result in conservative safety assessments [22].

The TiO_2 nanoparticles are the second most used after nanosized silver for several applications [9]. TiO_2 has been considered for sunscreens composition and other personal care products. Moreover, it is applied in advanced wastewater treatment technologies, such as photocatalysis or even in other applications [2, 9, 23, 24]. Different studies report toxicity or harmful effects of TiO_2 nanoparticles or the formation of more toxic compounds when TiO_2 is used as material in the photocatalyzed reactions for the generation of ROS used in the degradation of organic contaminants.

For aquatic organisms (fish, mollusks, algae, or marine invertebrates), the TiO_2 bioaccumulation was related to different problems in fish and mollusks, also increasing the absorption of other contaminants, and metals (such as As or Cd) in fish [9, 25–27]. Other toxic effects on algae, marine invertebrates, and fish are also reported [9, 28–32]. Due to these associated problems for aquatic and other types of organisms, it is important to ensure that TiO_2 nanoparticles are not released into the water and accumulate in nature.

For example, Soler de la Vega et al. [9] obtained a half maximal effective concentration, EC_{50} (72 h) for *D. magna* of 0.05, 8.89, 2.34, and 0.57 mg L^{-1} for TiO_2 nanoparticles, methylparaben, propylparaben, and benzylparaben, respectively, which shows a very danger toxic effect of TiO_2. Also, for the *Phaeodactylum tricornutum* microalgae, the TiO_2 and the benzylparaben revealed an EC_{50} (72 h) of 2.27 and 10.61 mg L^{-1}, respectively. Interestingly, when these two compounds were used in the mixture, the toxicity toward *D. magna* increased. This increase can be explained by the formation of more toxic by-products due to the interaction between the benzylparaben and the reactive oxidative species (ROS) formed by the TiO_2 nanoparticles. Therefore, it is evident that the presence of TiO_2 nanoparticles in the environment can be a source of contamination and toxicity for different trophic-level species. Moreover, the toxicity may also increase due to the generation of toxic by-products that are a consequence of the reaction between organic contaminants and the ROS generated during the TiO_2 irradiation by UV radiation. The generation and action of ROS may also occur in the environment if TiO_2 NPs and different organic compounds are present in water resources, in which the Sun acts as the radiation source. With this, it is important to ensure a solution that avoids the use and release of TiO_2 nanoparticles and its posterior occurrence, accumulation, and action in the ecosystems. For this, the use of supported/immobilized TiO_2 can be a possible solution, but there is the possibility of material leaching during the photocatalytic reaction [33]. Another option that does not face this problem is the synthesis of TiO_2 nanopores or nanotubes by anodic oxidation of a Ti foil. In this case, the TiO_2 is produced and remains attached to the Ti foil, being easy to recover from water and reused without existing loss of mass. Figure 14.1 resumes some of the problems caused by the use and release of TiO_2 nanoparticles and possible solutions for this problem.

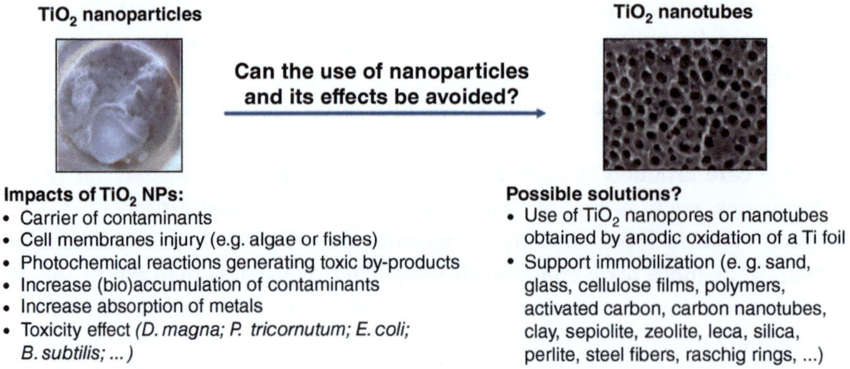

Figure 14.1 Problematic of TiO_2 NPs and possible solutions.

14.4 Ecotoxicity from Wastewater Treated by Photocatalytic Processes with TiO_2

During the photocatalytic processes, the incomplete mineralization of wastewater contaminants may generate reaction intermediaries that can be more toxic than the parent compound [8]. So treatment technologies should be optimized to reach total mineralization, or at least, to achieve the highest efficiency as possible, to avoid the release of more toxic compounds to aquatic resources. Anyway, toxicological effects of the treated water must be considered while evaluating treatment technologies efficiency. Figure 14.2 presents some ecotoxicity tests that can be made for this assessment.

For example, Lincho et al. [34] treated a mixture of parabens using TiO_2 nanotubes and conducted toxicity studies using *Lepidium sativum* seeds and *Aliivibrio fischeri* bacteria. Results showed different sensitivities for both species, with the initial parabens mixture revealing to have strong toxicity for *L. sativum* but not being toxic to *A. fischeri*. This shows the importance of evaluating the toxicity of different organisms (animals, plants, algae, etc.). The toxicity toward both species was evaluated for photocatalytic ozonation treated samples along the reaction time. Strong toxicity

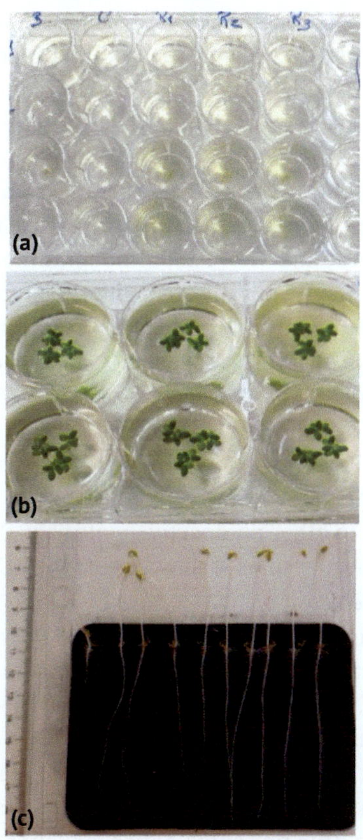

Figure 14.2 Example of ecotoxicity tests made for (a) *R. subcapitata*, (b) *L. minor*, and (c) *L. sativum*.

was detected for *L. sativum* seeds for the initial samples (low treatment time), decreasing to "medium" and "nontoxicity" for 5 and 10 min, respectively, and then at 15 and 20 min, the toxicity increased to "strong inhibition." This was revealed to occur due to the existence of more toxic intermediates, since other by-products (nonidentified) were detected by HPLC analysis, even after total parabens depletion. Interestingly, for *A. fischeri*, the contrary behavior occurred. The initial mixture was not toxic for this species, and at 10 and 15 min, the toxicity increased to "slightly toxic" decreasing again to "non-toxic" at 20 min. This study shows that the same compounds may present different sensitivities for different trophic-level species and should motivate toxicological studies for more than one species in a row. A similar change in the *A. fischeri* toxicity over time was observed by Vela et al. [35], while Tichonovas et al. [36] presented different sensitivities for different compounds considering the *D. magna* invertebrate.

In the case of wastewater treatment, the use of the photocatalyst as powder particles difficult its separation from the water after the treatment. This is the major limitation of implementing the industrial process using nanopowdered TiO_2. For the use of powders, the correct separation must be mandatory, but using supported TiO_2 (as in the study of Lincho et al. [34]) can overcome this drawback, and make the industrial application closer. However, there is no doubt that TiO_2 nanoparticles are probably the most studied material for environmental applications and that the use of TiO_2 nanoparticles in the removal of organic contaminants from water allows for decreasing the water toxicity, although all the problematics and interactions that nanoparticles may have in the ecosystems.

For example, Arslan-Alaton et al. [37] completely removed toxicity for *D. magna* using TiO_2 P25 UVA photocatalysis, in the degradation of a commercial biocide used in textile industries (removal of 95–99%). In fact, the most commonly used TiO_2 nanoparticles must be the TiO_2 P25, although other commercial TiO_2 can be available. The TiO_2 obtained from different manufacturers may present different performances due to different properties (purity, crystallite phase, BET surface area, particle size, etc.) [35]. For example, Vela et al. [35] showed that the P25 can be more efficient in removing toxicity than the TiO_2 Kronos vlp 7000. The materials showed different performances (dissolved organic carbon (DOC) removal was 84% and 76% in 240 min reaction for P25 and Kronos vlp 700), which may be explained by different properties of materials.

Also, when compared to other photoactive materials, the TiO_2 P25 can be advantageous. For example, Souza et al. [38] showed that the LC_{50} of Artemia salina invertebrate for a textile wastewater was 27.6%, and by using 0.25 g L^{-1} of photocatalyst (300 min, pH = 3), this increased to 90.9%, 61.6%, 66.6%, and 77.5% for P25, Kronos, ZnO, and Nb_2O_5, respectively. The chemical oxygen demand (COD) reduction was 70%, 71%, 50%, and 66%, and the absorbance removal (660 nm) was 97%, 85%, 68%, and 86% for P25, Kronos, ZnO, and Nb_2O_5, respectively. Contrary to what happened in the work of Vela et al. [35], in this case, higher degradations (measured by absorbance or COD reduction) did not lead to a lower toxicity (see the case of Kronos and ZnO). For this work, the ZnO presented better toxicity removal but lower contaminants degradation, which may suggest different degradation pathways, probably caused by the presence and action of different ROS, which contributed to the existence of different by-products (more/less toxic) [39].

In another work, Adams et al. [40] showed that the TiO_2, ZnO, and SiO_2 nanoparticles presented antibacterial activity which increased with the particle concentration. In this case, the highest antibacterial activity was obtained from ZnO followed by TiO_2 and SiO_2, respectively. For *B. subtilis*, using a concentration of 50 ppm, the growth inhibition was 0%, 0%, and 98% for TiO_2, SiO_2, and ZnO, respectively. Only by increasing this concentration to 1000 and 2000 ppm, the TiO_2 presented 75% and 99% inhibition, while the SiO_2 presented 7% and 84%, respectively. Evaluating the toxicity for *E. coli*, at 100 ppm, the growth inhibition was 0%, 0%, and 28% for TiO_2, SiO_2, and ZnO, while increasing to 1000 ppm, the inhibition was 44%, 19%, and 48% for the same nanoparticles order. These results were made under solar irradiation, and in general, the *B. subtilis* was more sensitive to the nanoparticles than the *E. coli*. Even in the dark, antimicrobial activity existed. For *B. subtilis*, the growth inhibition decreased from 75% to 30% for TiO_2 using light and at dark, respectively. For SiO_2, there is a decrease of about 25%, although a statistically analysis at 95% confidence suggests that the cell inhibition is similar for light and dark conditions. For the ZnO, the antimicrobial activity was even higher at dark conditions. This suggests that the light is only beneficial for the TiO_2 and that its antimicrobial activity is mostly related to the generation of ROS. In fact, Maness et al. [41] reported that the *E. coli* deactivation by TiO_2 photocatalysis is related to the peroxidation of the polyunsaturated phospholipid component of the lipid membrane by the generated ROS, which caused a disorder in the cell membrane and led to the cell death due to the loss of essential functions, as respiration, that rely on the intact cell membrane architecture. Regarding the *E. coli*, the TiO_2 had highest toxicity decrease in the presence of light due to the ROS action [41], the SiO_2 was almost the same, and the ZnO was way better with light, contrary to what happened for the *B. subtilis* [40].

TiO_2 is commonly used in photocatalyzed technologies, but it is pointed with two drawbacks. The first one is the high recombination rate which is traduced in low yield, and the second is the UV dependence due to its high bandgap energy (3–3.2 eV) [2]. Modifying the TiO_2 by doping of metals or nonmetals or by decorating the surface with (noble) metals may decrease these drawbacks [2, 24]. With this, the presence of metals may also influence the final toxicity for organisms, and it is known that some metals (as silver) present antibacterial properties [42]. As example, Gomes et al. [43] compared pure TiO_2 and TiO_2 decorated with 0.5% Au, Ag, or Pd for the elimination of *E. coli*. The pure TiO_2 only removed the bacteria after 30 min of irradiation, while the 0.5% Au, TiO_2 required 120 min for the same objective. Interestingly, the TiO_2 decorated with Ag or Pd removed all the bacteria in 20 min, without the radiation. In another work considering TiO_2 nanotubes, the doping of Ag led to higher antibacterial capacity for *E. coli* than bare TiO_2 nanotubes, in the presence or absence of radiation. Also, the presence of the catalyst was important to improve the *E. coli* inactivation, when compared to UVC radiation [44].

Even when real wastewater is considered, the use of TiO_2 can present efficiency in removing toxicity. As example, Jallouli et al. [45] showed that the use of TiO_2 (2.5 g L^{-1}) was able to reduce the toxicity (considering inhibition of *A. fischeri*) from 78.3% to 40.8% for municipal wastewater spiked with ibuprofen, and from 73.9% to 30.3% for pharmaceutical wastewater. Using immobilized TiO_2 on sand, He et al. [46] spiked a real wastewater with propranolol, diclofenac, carbamazepine, and ibuprofen (5 mg L^{-1} each), and treated the solution by photolysis and photocatalysis. The removal of propranolol, diclofenac, carbamazepine, and ibuprofen in 96 h irradiation by solar simulated light (300–800 nm) was 100%, 100%, 67%, and 68% for photolysis and 100%, 100%, 76%, and 74% for photocatalysis, respectively. Interestingly, the photolysis of these contaminants in deionized water led to the removal of 11–43%, which the authors point out that the presence of dissolved organic matter in the real wastewater is responsible for enhancing the degradation due to the generation of $^{\cdot}OH$ radicals. For *P. subcapitata* and *A. flos-aquae* algae, the toxicity decreased with the treatment, with better toxicity removal when photocatalysis was applied. However, in the case of *A. fisheri*, the inhibition maintained almost the same, due to the generation of toxic by-products. Moreover, the application of photocatalysis allowed to decrease the chemical oxygen demand of the treated solution in about 30%, while the photolysis led to about an 8% decrease.

14.5 Antibacterial Properties of TiO_2-based Photocatalytic Processes

Currently, chlorination is the most commonly applied treatment for water disinfection, due to its simplicity, low costs, and good antimicrobial characteristics, although the reaction between the chlorinated disinfectants and organic matter generates highly toxic disinfection by-products [47, 48]. Ozonation has also been growingly applied for water disinfection proposes, as part of the AOPs groups, producing various oxidative species, added to the already highly oxidative ozone molecules, but this treatment may also produce toxic by-products, especially due to reaction with bromine species, and the gas production has a high energy demand [49, 50]. Photocatalytic disinfection rises as a more prone to mineralization alternative, with lesser by-products formed, simple to be applied, and with high efficiencies, due to the great production of oxidative species [51].

TiO_2 for photocatalytic disinfection is a highly studied subject and has been employed for the elimination of a variety of species, such as bacteria, fungi, viruses, and other pathogenic microorganisms [50–54]. The highly reactive species produced through photon absorption, such as $^{\cdot}OH$, h^+, $^{\cdot}O_2^-$, and H_2O_2, are able to effectively damage the cellular membranes or capsid, promoting the leakage of cellular organelles and genetic material, and lastly their mineralization (Figure 14.3) [51, 53]. However, due to some drawbacks of TiO_2, such as its typical high energy bandgap, which makes its activation dependent on ultraviolet radiation, and the material separation post-treatment, modifications of the catalyst are generally employed and studied, more commonly, elemental doping and the combination with other materials to form composite catalysts with better properties [51, 55].

The *E. coli* bacteria is one of the most studied microorganisms in disinfection treatments, as it is a highly used indicator of water quality, possessing a good correlation with other pathogenic species [56]. Carré et al. [57] indicated an up to 3-log

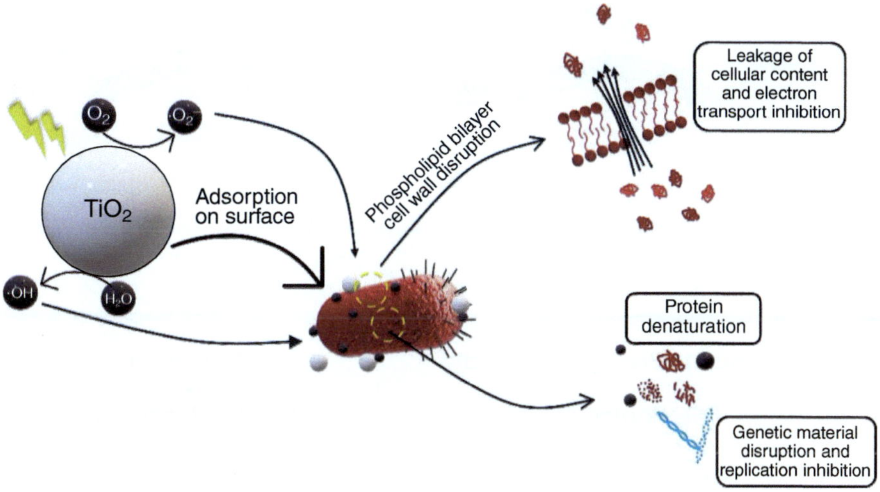

Figure 14.3 General mechanism of photocatalytic disinfection of bacteria.

reduction of *E. coli* after 60 min of UVA-assisted photocatalysis using P25 TiO_2 and indicated that the superoxide radical ($^{\cdot}O_2^-$) was particularly responsible for the lipid peroxidation, which is the main constituent of the bacteria cellular membrane, but also the alteration of multiple proteins, responsible for transport and metabolism within the cell, was appointed.

The complexity of real wastewater regarding the variety of bacteria and other microorganisms represents a challenge for the selected disinfection technology, as the different species may possess distinct responses to the treatment [58]. He et al. [58] compared the disinfection of different bacteria, both Gram-positive and Gram-negative, by a photocatalytic treatment using a TiO_2-based composite catalyst, combined with graphene oxide and iron oxide (Fe_3O_4). The authors indicated a higher leakage of nucleic acid, used as an indicator of damage to the cellular membrane, of the *Enterococcus faecalis*, a Gram-positive bacterium, than the other two analyzed organisms, *E. coli* and *Salmonella typhimurium*, both Gram-negative, due to single phospholipid bilayer cellular membrane of the former bacteria. Moreover, when tested the disinfection of all three bacteria in the same mixture, *E. coli* and *E. faecalis* had a higher inhibition of their elimination, while *S. typhimurium* presented no significant alteration, a result of the higher difference between the electrostatic charges of the catalyst surface and the *S. typhimurium*, which indicates a high influence of this parameter in the evaluation of photocatalytic disinfection in complex water matrices.

Generally, TiO_2 possesses no significant cytotoxicity in the absence of light, capable of only inactivating microorganisms after the production of radicals through its photoactivation [57]. Nonetheless, the incorporation of foreign elements in the catalyst structure may provide an antimicrobial characteristic for the material. Gomes et al. [43] indicated that the doping of silver and palladium over the TiO_2 catalyst resulted in a complete elimination of *E. coli*, with an initial concentration around 3-log, after 20 min of contact without any light source. Moreover, Zhang et al. [59]

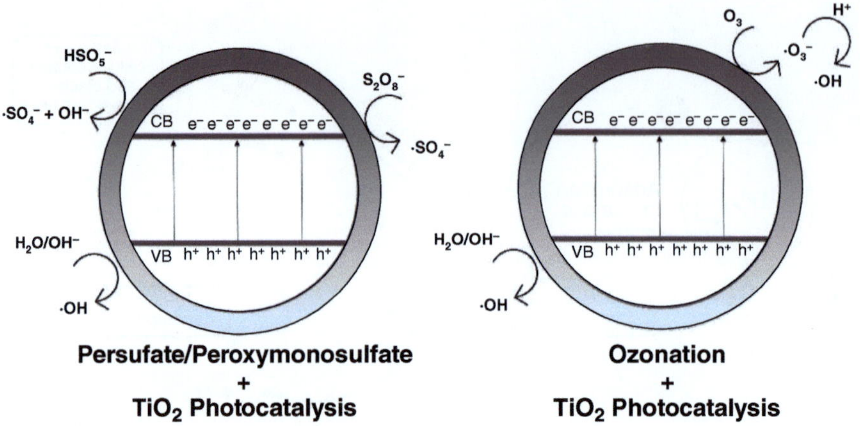

Figure 14.4 Enhancing the formation of radicals by combination of TiO_2 photocatalysis with other oxidant-based processes.

have also indicated that Pd doping improved the overall performance of the photocatalytic performance, with a 5% Pd–TiO_2 catalyst under visible light resulting in around 7-log and 4-log decrease of cellular density within 150 min of, respectively, *E. coli* and *Fusarium oxysporum f. sp. Vasinfectum* spores.

The presence of TiO_2 may also contribute to obtaining higher efficiencies when combined with other technologies, especially with other AOPs (Figure 14.4). John et al. [60] indicated that the incorporation of a TiO_2–GO composite in a persulfate-based oxidation under visible light for the disinfection of both *E. coli* and *Staphylococcus aureus* resulted in a great increase in the number of ROS produced, and consequently the more than a 4-log decrease in the population of both bacteria under 140 min, compared to a no significant alteration in the absence of the catalyst.

The combination of TiO_2-based photocatalysis with ozonation has also been vastly explored. Using real secondary wastewater (SWW), Mecha et al. [61] investigated the disinfection of four bacteria, *E. coli, Salmonella* species, *Shigella* species, and *Vibrio cholerae*, and found an increase from 0.4–1.0 to 1.0–2.2-log reductions when TiO_2 was added to the ozonation process, after 15 min of solar radiation. Moreover, the combined process resulted in no regrowth of all microorganisms evaluated after 48 h, while this was observed for the single ozonation-treated samples. The combination of these two AOPs provides a reduction of the necessary amount of ozone, decreasing its overall cost, lower reaction times, higher mineralization degree, and radicals' production [62].

To better highlight the benefits of incorporating TiO_2 catalysts in ozone-based disinfection treatments, a brief study using a sol–gel synthesized nitrogen-doped TiO_2 (described elsewhere [63]) was used for the elimination of *E. coli* present in fresh SWW. The SWW was collected in a wastewater treatment plant in the central region of Portugal, which possesses a primary settler, activated sludge treatment, and a secondary settler [64]. Both photocatalytic ozonation (UVA + TiO_2 + O_3) and photolytic ozonation (UVA + O_3) disinfection tests are present in Figure 14.5. The SWW was collected

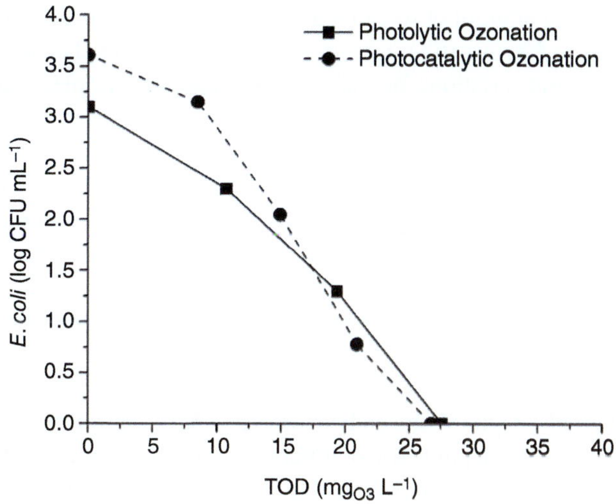

Figure 14.5 N-TiO_2 photocatalytic and photolytic ozonation disinfection treatments for the removal of *E. coli* in secondary wastewater.

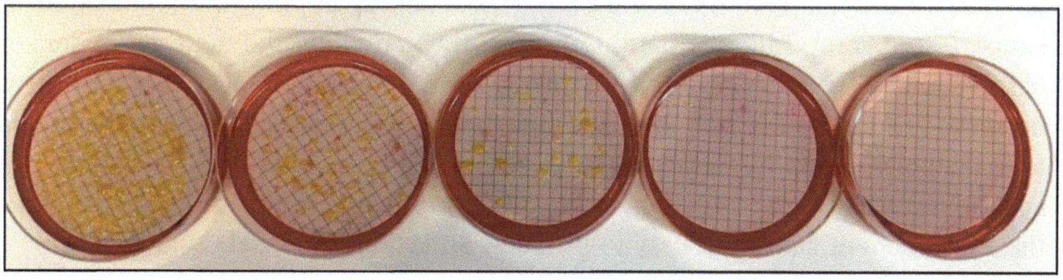

Figure 14.6 Membrane filters containing microorganism's colonies for different times of photocatalytic treatment.

at different moments to prevent microorganisms' decay during storage. The disinfection using only UVA light was also tested, and no significant result was found. The bacteria were quantified using a membrane filtration method, according to ISO 9308-1:2014. In Figure 14.6, the filters used in a first step of bacteria quantification are present, where the yellow colonies are possibly of *E. coli* and are verified after.

Even with a higher initial concentration of bacteria, the transferred ozone dose necessary to achieve a full removal was similar for both disinfection treatments, 26.7 and 27.6 mg L^{-1} for photocatalytic and photolytic ozonation, respectively. The higher production of ROS due to the combination of both AOPs, as well as a faster decomposition of ozone in the medium, which possesses a very low solubility in water, results in a faster elimination of the microorganisms. This higher number of available radicals also implicates in a lower overall consumption of ozone, thus reducing the process cost. A high concentration of microorganisms in the treated effluent of WWTPs has been reported in different studies, which would be released in the environment, emphasizing the importance of the development and application of efficient disinfection treatment, such as the advanced oxidation processes [64–67].

Despite the great amount of research development for TiO_2-based disinfection treatments, a number of factors still inhibit its further and larger applications [68]. The multiple parameters and variations for real water matrices, such as pH, organic matter, and inorganic ions, deeply affect the efficiency of the photocatalytic reactions [51]. The different microorganisms and how they interact with each other, the medium, and the catalyst also require a more specific evaluation for each application. Furthermore, the catalyst recuperation and activity after treatment microorganism elimination needs to be optimized to reduce losses and maximize the reuse of the materials.

14.6 Conclusion

The application of TiO_2 in photocatalytic processes has been successfully studied over the last decades for the degradation of different chemical contaminants, the reduction of contaminated water toxicities, and the removal of multiple pathogenic microorganisms. Despite its high efficiency, large-scale applications still face difficulties in being implemented for water treatment, due to crucial drawbacks of the material and the overall processes, such as artificial radiation dependency and post-treatment recuperation of the photocatalyst. The scientific community's current research is more directed to the adaptations of the base TiO_2, such as doping, immobilization into supports and composites formation, to boost the activity of the material under solar light, facilitate their separation from the medium, reduce possible deactivations, and decrease in the photoactivity over multiple uses, among other necessary parameters to make possible the implementation in water treatment facilities. The future of the application of TiO_2-based processes is highly promising, but it is also dependent on the attainment of a feasible, easy-to-produce, and stable material.

References

1 Khaki, M.R.D., Shafeeyan, M.S., Raman, A.A.A., and Daud, W.M.A.W. (2017). Application of doped photocatalysts for organic pollutant degradation—a review. *Journal of Environmental Management* 198: 78–94.

2 Pelaez, M., Nolan, N.T., Pillai, S.C. et al. (2012). A review on the visible light active titanium dioxide photocatalysts for environmental applications. *Applied Catalysis B: Environment and Energy* 125: 331–349.

3 Gomes, J., Lincho, J., Domingues, E. et al. (2019). N-TiO_2 photocatalysts: a review of their characteristics and capacity for emerging contaminants removal. *Water (Switzerland)* 11 (2): 373.

4 Murgolo, S., Franz, S., Arab, H. et al. (2019). Degradation of emerging organic pollutants in wastewater effluents by electrochemical photocatalysis on nanostructured TiO_2 meshes. *Water Research* 164: 114920.

5 Rueda-Márquez, J.J., Palacios-Villarreal, C., Manzano, M. et al. (2020). Photocatalytic degradation of pharmaceutically active compounds (PhACs) in urban wastewater treatment plants effluents under controlled and natural solar irradiation using immobilized TiO_2. *Solar Energy* 208 (August): 480–492.

6 Candia-Onfray, C., Irikura, K., Calzadilla, W. et al. (2023). Degradation of contaminants of emerging concern in a secondary effluent using synthesized MOF-derived photoanodes: a comparative study between photo-, electro- and photoelectrocatalysis. *Chemosphere* 315 (October 2022): 137683.

7 Girotti, S., Ferri, E.N., Fumo, M.G., and Maiolini, E. (2008). Monitoring of environmental pollutants by bioluminescent bacteria. *Analytica Chimica Acta* 608 (1): 2–29.

8 Babu, D.S., Srivastava, V., Nidheesh, P.V., and Kumar, M.S. (2019). Detoxification of water and wastewater by advanced oxidation processes. *Science of the Total Environment* 696: 133961.

9 Soler de la Vega, A.C., Molins-Delgado, D., Barceló, D., and Díaz-Cruz, M.S. (2019). Nanosized titanium dioxide UV filter increases mixture toxicity when combined with parabens. *Ecotoxicology and Environmental Safety* 184 (February): 109565.

10 Li, X., Li, L., Tang, L. et al. (2024). Unveiling combined ecotoxicity: interactions and impacts of engineered nanoparticles and PPCPs. *Science of the Total Environment* 921 (October 2023): 170746.

11 Li, M., Xu, G., Yang, X. et al. (2020). Metal oxide nanoparticles facilitate the accumulation of bifenthrin in earthworms by causing damage to body cavity. *Environmental Pollution* 263: 114629.

12 Sendra, M., Pintado-Herrera, M.G., Aguirre-Martínez, G.V. et al. (2017). Are the TiO_2 NPs a "Trojan horse" for personal care products (PCPs) in the clam Ruditapes philippinarum? *Chemosphere* 185: 192–204.

13 Tian, S., Zhang, Y., Song, C. et al. (2015). Bioaccumulation and biotransformation of polybrominated diphenyl ethers in the marine bivalve (Scapharca subcrenata): influence of titanium dioxide nanoparticles. *Marine Pollution Bulletin* 90 (1–2): 48–53.

14 Guo, Y., Chen, L., Wu, J. et al. (2019). Parental co-exposure to bisphenol A and nano-TiO_2 causes thyroid endocrine disruption and developmental neurotoxicity in zebrafish offspring. *Science of the Total Environment* 650: 557–565.

15 Xin, X., Huang, G., Zhang, B., and Zhou, Y. (2021). Trophic transfer potential of $nTiO_2$, nZnO, and triclosan in an algae-algae eating fish food chain. *Aquatic Toxicology* 235 (March): 105824.

16 Yan, Z., Lu, G., Sun, H., and Ma, B. (2017). Influence of multi-walled carbon nanotubes on the effects of roxithromycin in crucian carp (*Carassius auratus*) in the presence of natural organic matter. *Chemosphere* 178: 165–172.

17 Yan, Z., Lu, G., Sun, H. et al. (2019). Comparison of the accumulation and metabolite of fluoxetine in zebrafish larva under different environmental conditions with or without carbon nanotubes. *Ecotoxicology and Environmental Safety* 172 (January): 240–245.

18 Deng, R., Yang, K., and Lin, D. (2021). Pentachlorophenol and ciprofloxacin present dissimilar joint toxicities with carbon nanotubes to *Bacillus subtilis*. *Environmental Pollution* 270: 116071.

19 Ren, B., Jia, B., Zhang, X. et al. (2021). Influence of multi-walled carbon nanotubes on enantioselective bioaccumulation and oxidative stress toxicity of indoxacarb in zebrafish (Danio rerio). *Chemosphere* 267 (235): 128872.

20 Yang, J., Zhong, W., Chen, P. et al. (2019). Graphene oxide mitigates endocrine disruption effects of bisphenol A on zebrafish at an early development stage. *Science of the Total Environment* 697: 134158.

21 He, X., McAlliser, D., Aker, W.G., and Hwang, H.M. (2016). Assessing the effect of different natural dissolved organic matters on the cytotoxicity of titanium dioxide nanoparticles with bacteria. *Journal of Environmental Sciences (China)* 48: 230–236.

22 Arvidsson, R., Hansen, S.F., and Baun, A. (2020). Influence of natural organic matter on the aquatic ecotoxicity of engineered nanoparticles: recommendations for environmental risk assessment. *NanoImpact* 20 (June): 100263.

23 Carp, O., Huisman, C.L., and Reller, A. (2004). Photoinduced reactivity of titanium dioxide. *Progress in Solid State Chemistry* 32 (1–2): 33–177.

24 Gaya, U.I. and Abdullah, A.H. (2008). Heterogeneous photocatalytic degradation of organic contaminants over titanium dioxide: a review of fundamentals, progress and problems. *Journal of Photochemistry and Photobiology C: Photochemistry Reviews* 9 (1): 1–12.

25 Federici, G., Shaw, B.J., and Handy, R.D. (2007). Toxicity of titanium dioxide nanoparticles to rainbow trout (Oncorhynchus mykiss): gill injury, oxidative stress, and other physiological effects. *Aquatic Toxicology* 84 (4): 415–430.

26 Hao, L., Wang, Z., and Xing, B. (2009). Effect of sub-acute exposure to TiO_2 nanoparticles on oxidative stress and histopathological changes in Juvenile Carp (Cyprinus carpio). *Journal of Environmental Sciences* 21 (10): 1459–1466.

27 Zhang, X., Sun, H., Zhang, Z. et al. (2007). Enhanced bioaccumulation of cadmium in carp in the presence of titanium dioxide nanoparticles. *Chemosphere* 67 (1): 160–166.

28 Blaise, C., Gagné, F., Férard, J.F., and Eullaffroy, P. (2008). Ecotoxicity of selected nano-materials to aquatic organisms. *Environmental Toxicology* 23 (5): 591–598.

29 Johnson, A.C., Bowes, M.J., Crossley, A. et al. (2011). An assessment of the fate, behaviour and environmental risk associated with sunscreen TiO_2 nanoparticles in UK field scenarios. *Science of the Total Environment* 409 (13): 2503–2510.

30 Clemente, Z., Castro, V.L.S.S., Moura, M.A.M. et al. (2014). Toxicity assessment of TiO_2 nanoparticles in zebrafish embryos under different exposure conditions. *Aquatic Toxicology* 147: 129–139.

31 Minetto, D., Libralato, G., and Volpi Ghirardini, A. (2014). Ecotoxicity of engineered TiO_2 nanoparticles to saltwater organisms: an overview. *Environment International* 66: 18–27.

32 Della Torre, C., Buonocore, F., Frenzilli, G. et al. (2015). Influence of titanium dioxide nanoparticles on 2,3,7,8-tetrachlorodibenzo-p-dioxin bioconcentration and toxicity in the marine fish European sea bass (Dicentrarchus labrax). *Environmental Pollution* 196: 185–193.

33 Silva, M.J., Alves, R., Alves, P. et al. (2023). Photocatalytic degradation of parabens by P25 support in PET sheets from water bottles. In: *WASTES: Solutions, Treatments and Opportunities IV*, 1–6. CRC Press.

34 Lincho, J., Gomes, J., Kobylanski, M. et al. (2021). TiO_2 nanotube catalysts for parabens mixture degradation by photocatalysis and ozone-based technologies. *Process Safety and Environmental Protection* 152: 601–613.

35 Vela, N., Calín, M., Yáñez-Gascón, M.J. et al. (2018). Photocatalytic oxidation of six pesticides listed as endocrine disruptor chemicals from wastewater using two different TiO_2 samples at pilot plant scale under sunlight irradiation. *Journal of Photochemistry and Photobiology A: Chemistry* 353: 271–278.

36 Tichonovas, M., Krugly, E., Jankunaite, D. et al. (2017). Ozone-UV-catalysis based advanced oxidation process for wastewater treatment. *Environmental Science and Pollution Research* 24 (21): 17584–17597.

37 Arslan-Alaton, I. (2007). Degradation of a commercial textile biocide with advanced oxidation processes and ozone. *Journal of Environmental Management* 82 (2): 145–154.

38 Souza, R.P., Freitas, T.K.F.S., Domingues, F.S. et al. (2016). Photocatalytic activity of TiO_2, ZnO and Nb_2O_5 applied to degradation of textile wastewater. *Journal of Photochemistry and Photobiology A: Chemistry* 329: 9–17.

39 Rueda-Marquez, J.J., Levchuk, I., Fernández Ibañez, P., and Sillanpää, M. (2020). A critical review on application of photocatalysis for toxicity reduction of real wastewaters. *Journal of Cleaner Production* 258: 120694.

40 Adams, L.K., Lyon, D.Y., and Alvarez, P.J.J. (2006). Comparative eco-toxicity of nanoscale TiO_2, SiO_2, and ZnO water suspensions. *Water Research* 40 (19): 3527–3532.

41 Maness, P.C., Smolinski, S., Blake, D.M. et al. (1999). Bactericidal activity of photocatalytic TiO_2 reaction: toward an understanding of its killing mechanism. *Applied and Environmental Microbiology* 65 (9): 4094–4098.

42 Talapko, J., Matijević, T., Juzbašić, M. et al. (2020). Antibacterial activity of silver and its application in dentistry, cardiology and dermatology. *Microorganisms* 8 (9): 1–13.

43 Gomes, J.F., Lopes, A., Gonçalves, D. et al. (2018). Biofiltration using *C. fluminea* for *E. coli* removal from water: comparison with ozonation and photocatalytic oxidation. *Chemosphere* 208: 674–681.

44 Mao, K., Li, Y., Zhang, H. et al. (2013). Photocatalytic degradation of 17α-ethinylestradiol and inactivation of *Escherichia coli* using Ag-modified TiO_2 nanotube arrays. *Clean - Soil, Air, Water* 41 (5): 455–462.

45 Jallouli, N., Pastrana-Martínez, L.M., Ribeiro, A.R. et al. (2018). Heterogeneous photocatalytic degradation of ibuprofen in ultrapure water, municipal and pharmaceutical industry wastewaters using a TiO_2/UV-LED system. *Chemical Engineering Journal* 334 (June 2017): 976–984.

46 He, Y., Sutton, N.B., Rijnaarts, H.H.H., and Langenhoff, A.A.M. (2016). Degradation of pharmaceuticals in wastewater using immobilized TiO_2 photocatalysis under simulated solar irradiation. *Applied Catalysis B: Environment and Energy* 182: 132–141.

47 He, H., Sun, N., Li, L. et al. (2023). Effects of dissolved organic matter removal and molecular transformation in different water treatment processes on formation of disinfection byproducts. *Water Research* 245 (August): 120626.

48 Zhang, X., Feng, X., Ma, Y. et al. (2024). Comparison of chlorination resistance of biodegradable microplastics and conventional microplastics during the disinfection process in water treatments. *Science of the Total Environment* 908 (October 2023): 168229.

49 Varouqa, I.F. (2021). Detecting the concentration of aldehydes disinfection by-products formed due to the application of the ozonation process in water treatment plants. *Materials Today: Proceedings* 45: 5631–5634.

50 Gomes, J.F., Leal, I., Bednarczyk, K. et al. (2017). Photocatalytic ozonation using doped TiO_2 catalysts for the removal of parabens in water. *Science of the Total Environment* 609: 329–340.

51 He, J., Kumar, A., Khan, M., and Lo, I.M.C. (2021). Critical review of photocatalytic disinfection of bacteria: from noble metals- and carbon nanomaterials-TiO_2 composites to challenges of water characteristics and strategic solutions. *Science of the Total Environment* 758: 143953.

52 Zhang, J., Liu, S., Wang, X. et al. (2021). Highly efficient Ti^{3+} self-doped TiO_2 co-modified with carbon dots and palladium nanocomposites for disinfection of bacterial and fungi. *Journal of Hazardous Materials* 413 (February): 125318.

53 De Pasquale, I., Porto, C.L., Dell'Edera, M. et al. (2021). TiO_2-based nanomaterials assisted photocatalytic treatment for virus inactivation: perspectives and applications. *Current Opinion in Chemical Engineering* 34: 100716.

54 Liu, L., Laghari, A.A., Meng, G. et al. (2022). Photocatalytic disinfection of different airborne microorganisms by TiO_2/MXene filler: inactivation efficiency, energy consumption and self-repair phenomenon. *Journal of Environmental Chemical Engineering* 10 (3): 107641.

55 Fernandes, E., Gomes, J., and Martins, R.C. (2022). Semiconductors application forms and doping benefits to wastewater treatment: a comparison of TiO_2, WO_3, and g-C_3N_4. *Catalysts* 12 (10): 1218.

56 Moinet, M., Collis, R.M., Rogers, L. et al. (2024). Development of a multiplex droplet digital PCR assay for simultaneous detection and quantification of *Escherichia coli, E. marmotae*, and *E. Ruysiae* in water samples. *Journal of Microbiological Methods* 220: 106909.

57 Carré, G., Hamon, E., Ennahar, S. et al. (2014). TiO_2 photocatalysis damages lipids and proteins in *Escherichia coli*. *Applied and Environmental Microbiology* 80 (8): 2573–2581.

58 He, J., Zheng, Z., and Lo, I.M.C. (2021). Different responses of gram-negative and gram-positive bacteria to photocatalytic disinfection using solar-light-driven magnetic TiO_2-based material, and disinfection of real sewage. *Water Research* 207 (February): 117816.

59 Zhang, J., Zhai, M., Wang, X. et al. (2023). Pd nanoparticle-modified TiO_{2-x} composites as efficient visible-light active photocatalysts for tetracycline degradation and microbial disinfection. *Optical Materials (Amsterdam)* 136 (December 2022): 113396.

60 John, D., Jose, J., Bhat, S.G., and Achari, V.S. (2021). Integration of heterogeneous photocatalysis and persulfate based oxidation using TiO_2-reduced graphene oxide for water decontamination and disinfection. *Heliyon* 7 (7): e07451.

61 Mecha, A.C., Onyango, M.S., Ochieng, A., and Momba, M.N.B. (2017). Evaluation of synergy and bacterial regrowth in photocatalytic ozonation disinfection of municipal wastewater. *Science of the Total Environment* 601–602: 626–635.

62 Fernandes, E., Contreras, S., Medina, F. et al. (2020). N-doped titanium dioxide for mixture of parabens degradation based on ozone action and toxicity evaluation:

precursor of nitrogen and titanium effect. *Process Safety and Environmental Protection* 138: 80–89.

63 Fernandes, E., Martins, R.C., and Gomes, J. (2020). Photocatalytic ozonation of parabens mixture using 10% N-TiO_2 and the effect of water matrix. *Science of the Total Environment* **718**: 137321.

64 Fernandes, E., Mazierski, P., Miodyńska, M. et al. (2024). Emerging contaminants and pathogenic microorganisms elimination in secondary effluent by graphitic carbon nitride photocatalytic ozonation processes. *Catalysis Today* 432: 114624.

65 Chahal, C., van den Akker, B., Young, F. et al. (2016). Pathogen and particle associations in wastewater: significance and implications for treatment and disinfection processes. *Advances in Applied Microbiology* 97: 63–119.

66 Echeverry-Gallego, R.A., Espinosa-Barrera, P.A., Delgado-Vargas, C.A. et al. (2023). The application of the photo-electro-Fenton process in the treatment of wastewater reduces the abundance of genes associated with pathogenicity factors, antibiotic resistance, and metabolism: a metagenomic analysis. *Journal of Environmental Chemical Engineering* 11 (3): 109937.

67 Wen, Q., Tutuka, C., Keegan, A., and Jin, B. (2009). Fate of pathogenic microorganisms and indicators in secondary activated sludge wastewater treatment plants. *Journal of Environmental Management* 90 (3): 1442–1447.

68 Rengifo-Herrera, J.A. and Pulgarin, C. (2023). Why five decades of massive research on heterogeneous photocatalysis, especially on TiO_2, has not yet driven to water disinfection and detoxification applications? Critical review of drawbacks and challenges. *Chemical Engineering Journal* 477 (November 2023): 146875.

15

The Application of TiO_2 for the Photocatalytic Conversion of the Nitrogen Oxides – NO and NO_2

*Andreas Kafizas**

Department of Chemistry, Molecular Science Research Hub, Imperial College London, White City, London, UK

*Corresponding author: a.kafizas@imperial.ac.uk

15.1 Introduction

15.1.1 The Nitrogen Oxides: NO and NO_2

The most ubiquitous gas in the Earth's atmosphere is dinitrogen (N_2), making up ~78% of the air. Due to the strong triple bond in this diatomic, N_2 is often considered a stable molecule; however, its chemistry is diverse [1]. N_2 and its related compounds are essential to all life on the Earth, forming part of the nitrogen cycle [2]. In this cycle, nitrogen adopts a wide range of formal oxidation states, from 0 in N_2, to as low as −3, upon conversion to ammonia (NH_3) primarily through fixation by Diazotroph bacteria, to as high as +5, upon conversion to nitrates (NO_3^-) primarily by soil-living Nitrosomonas and Nitrobacter bacteria. However, since the industrial era, anthropogenic activities have altered the concentrations of nitrogen species present in land, sea, and air [3, 4]. With regards to nitric oxide (NO) and nitrogen dioxide (NO_2) gases, which together are often referred to as NO_x ($NO_x = NO + NO_2$), these gases were once predominantly produced naturally by lightning [5]. Today, anthropogenic activities dominate the formation of NO_x, with ~70% produced as a by-product of combusting fossil fuels and ~20% produced as a by-product of burning biomass [6]. Increases in the concentrations of NO_x gases are having profound consequences on humans and the environment. NO_x gases are highly toxic to humans when inhaled; even at parts per million concentrations [7]. NO_x gases also contribute to the depletion of the Earth's ozone layer and acid rain [8]. Present anthropogenic emissions of NO_x stand at ~100 Tg yr^{-1} [3], and to mitigate their negative effects on human health and the environment, targets have been set at national and international levels to reduce emissions.

To reduce NO_x, emissions will require wholesale changes in how people travel, produce energy, heat their homes and work, and will likely take decades to enact this transition [9]. Part of the strategy to reduce emissions of NO_x is the development of

Nanostructured TiO_2: Synthesis, Photocatalytic Properties and Applications, First Edition.
Edited by Jia Hong Pan, Wan In Lee and Detlef W. Bahnemann

disruptive technologies. Currently, the most used technology to reduce the emission of NO_x is the catalytic converter, which is most frequently applied on the exhausts of petrol-based automobiles [10]. NO_x emissions produced in the engines of vehicles – through the unwarranted reaction of N_2 and O_2 at the high temperatures found in engines – are reduced by thermal catalysis back to N_2 and O_2. Unfortunately, catalytic converters do not prevent all NO_x emissions from escaping the vehicle, and in situations of high congestion, this can result in unsafe levels of NO_x emissions into the surrounding air. As a consequence, in many highly populated areas the globe, NO_x levels surpass the guidelines set by the World Health Organization (WHO), and causes ~38 000 deaths per annum [11].

Once dispersed, NO_x emissions are challenging to treat, as dilutions in the atmosphere are typically lower than a part per million. However, one promising technology that can overcome this challenge of high dilution are photocatalysts [12]. Photocatalysts have been shown to be capable of treating dilute levels of NO_x, and function passively using ambient light (e.g., from the sun) [13]. The most commonly studied photocatalyst for treating these gases is TiO_2, and is now seeing use in a range of commercially available products for remediating NO_x [14].

In this chapter, we will present and discuss the recent key developments in the use of TiO_2-based photocatalysts for treating NO_x. It should also be noted that that this chapter will not be exhaustive in its presentation and discussion of studies on this topic, but rather, will present notable pieces of work to support a more general discussion on the current state and future directions of this growing area of research.

15.1.2 Photocatalytic NO_x Conversion Using TiO_2-based Materials

The first demonstration of photocatalytic NO_x conversion was by Pichat et al. in 1982, who showed that the commercially available composite material, TiO_2 Degussa P25 (now known as Evonik P25), was able to decompose NO under the action of UV light [15]. However, it was not until the early 2000s that that research on this topic really started to grow. Figure 15.1a shows the cumulative number of research articles published in this field since the year 2000; determined by our search of Web of Science.[1] We saw ~2100 research articles published on the topic, with ~1400 of these pertaining to the use of TiO_2 (~67% of the literature). Interestingly, a more detailed review of the literature shows that the development of TiO_2-based photocatalysts for NO_x conversion has focused more on product development and its implementation in the field than in other studied photocatalytic reactions, such as water splitting. An analysis of articles published by category shows that there is a significantly larger contribution of engineers and environmental scientists in the development of photocatalysts for NO_x conversion than the most popularly studied photocatalytic reaction water splitting, shown in Figure 15.1b. For example, the percentage of published articles on the topic of photocatalysts for NO_x conversion in the categories

1 Data were acquired from Web of Science using the search terms ((“TiO_2” OR “titania” OR “titanium dioxide”) AND (“photocat*”) AND (“NO_x” OR “nitrogen oxides” OR “nitric oxide” OR “nitrogen dioxide” OR “NO decomposition”)).

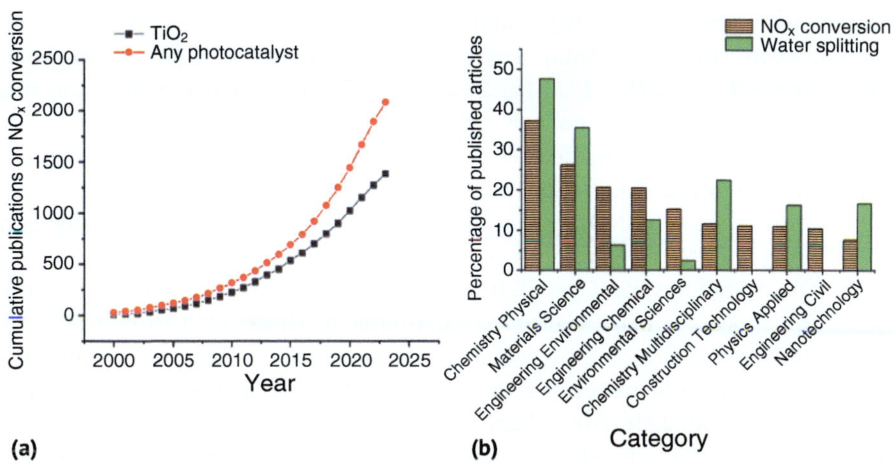

Figure 15.1 (a) Cumulative publications on the topic of the photocatalytic conversion of NO_x since the year 2000 for any photocatalyst or TiO_2 exclusively. (b) The percentage of published articles by category in the development of TiO_2-based photocatalysts for NO_x conversion and water splitting. The data used in these figures was acquired from Web of Science.

Engineering Environmental, Construction Technology, and Engineering Civil are 20.6%, 11.1%, and 10.4%, respectively, whereas on the topic of water splitting they are 6.2%, 0%, and 0%, respectively.[2] This, alongside the significant number of field trials that have been carried out to date, shows that the focus from the outset on TiO_2-based photocatalysts for NO_x conversion was on product development for commercial implementation, and less so on the rational design of photocatalysts for this purpose [16]. This was likely due to the high efficacy shown by various commercially available TiO_2-based photocatalysts, such as TiO_2 Degussa P25 and other materials, to indiscriminately treat NO_x pollution, even at high dilutions of the gas (i.e., from the parts per million to billion regime).

15.2 Mechanism and ISO Protocol

15.2.1 The Photocatalytic Oxidation Mechanism

The mechanism by which TiO_2 oxidizes NO_x gases through the action of photocatalysis has been outlined in various review articles [17–19]. The process starts with the photoexcitation of TiO_2 by ultra-bandgap energy (E_{bg}) light, promoting an electron from the valence band, into the conduction band, leaving behind a positive hole (Eq. 15.1a). Light in the ultraviolet (UV) region of the electromagnetic spectrum can excite the E_{bg} of TiO_2. The UV light can be provided from a UV lamp or from sunlight, which contains ~4% of photons in the UV region. Most have argued

2 Data were acquired from Web of Science using the search terms ((“TiO_2” OR “titania” OR “titanium dioxide”) AND (“photocat*”) AND (“water splitting”)).

that the conversion of NO_x is predominantly mediated by reactive oxygen species (ROS), where conduction band electrons reduce dioxygen (O_2) present in the air into superoxide ($O_2^{\bullet-}$) (Eq. 15.1b) and valence band holes oxidize surface absorbed water (H_2O) into hydroxyl radicals ($OH^\bullet$) (Eq. 15.1c) [20]. Surface-absorbed NO_x gases are subsequently oxidized by these ROS, with $O_2^{\bullet-}$ capable of oxidizing NO gas directly to nitrate (NO_3^-) (Eq. 15.1d), and $OH^\bullet$ capable of sequentially oxidizing NO to nitrous acid (HONO) to NO_2 to NO_3^- (Eq. 15.1e). Overall, the reaction converts nitrogen in NO_x gas from the +2 (NO) and +4 (NO_2) oxidation state into the +5 (NO_3^-) oxidation state. Overall, if the NO_x present in the air is fully oxidized to NO_3^-, this process is beneficial to the environment. However, it should be noted that if the photocatalyst is only able to partially oxidize the NO_x present in the air, then this can sometimes result in a situation where the treated air is more toxic than the non-treated air. This is because NO_2 – an intermediate on the pathway of oxidizing NO to NO_3^- – is between 5 and 25 times more toxic than NO [17]. As such, care should be taken when developing any photocatalyst for this process, with the de-NO_x index recently developed to ascertain if the overall toxicity level is reduced when treating NO_x gas [21].

$$TiO_2 \xrightarrow{hv \geq E_{bg}} TiO_2 + e^+ + h^+ \tag{15.1a}$$

$$e^- + O_2 \rightarrow O_2^{\bullet-} \tag{15.1b}$$

$$h^+ + H_2O \rightarrow OH^\bullet + H^+ \tag{15.1c}$$

$$NO \xrightarrow{O_2^{\bullet-}} NO_3^- \tag{15.1d}$$

$$NO \xrightarrow{OH^\bullet} HONO \xrightarrow{OH^\bullet\ (-H_2O)} NO_2 \xrightarrow{OH^\bullet(-H^+)} NO_3^- \tag{15.1e}$$

15.2.2 ISO Protocol for Measuring Photocatalytic NO_x Conversion Activity

An aspect that is quite unique to this photocatalytic reaction is the development of standardized protocol for measuring NO_x conversion activity by the International Organization for Standardization (ISO); previously BS ISO 22197-1:2007, and now superseded by BS ISO 22197-1:2016[3] [22]. In brief, as shown in Figure 15.2, a photocatalyst sample ($5.0 \times 10.0\ cm = 50\ cm^2$) is placed inside a reactor of defined dimensions and exposed to a gas stream of NO pollution in air (1.0 ppmv) under flow ($3.0\ L\ min^{-1}$) of constant humidity (50%). The sample is examined for 30 min in the dark, 5 h under ultraviolet light (315–400 nm, $1.0\ mW\ cm^{-2}$) and 30 min thereafter in the dark, with changes in the NO concentration measured continuously using a

3 Full ISO title: Fine ceramics (advanced ceramics, advanced technical ceramics) – Test method for air-purification performance of semiconducting photocatalytic materials; Part 1: Removal of nitric oxide.

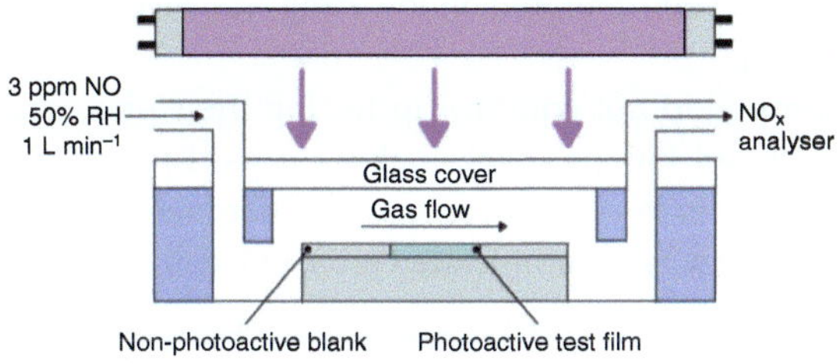

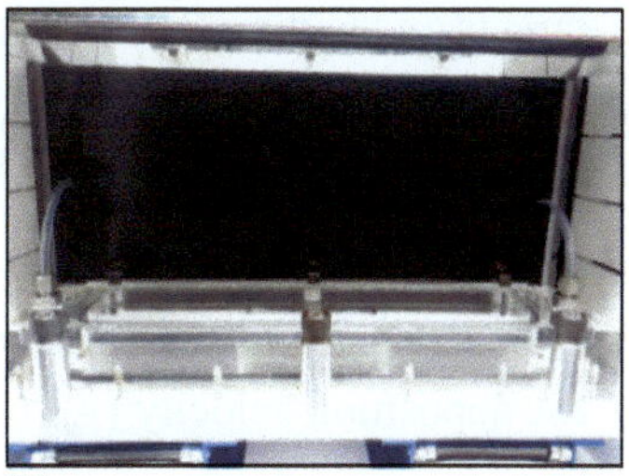

Figure 15.2 (a) Schematic of the reactor and test conditions used in the ISO (BS ISO 22197-1:2016) for measuring photocatalytic NO_x conversion activity. (b) A photograph of such a reactor under operation.

chemiluminescence analyzer.[4] With knowledge of: (i) the amount of NO absorbed during the initial dark period, (ii) the amount of NO converted into NO_2 and NO_3^- in the light, and (iii) the amount of NO_x desorbed during the final dark period, the photocatalytic NO_x conversion activity can be calculated. These results are typically represented as percentages of NO removed, NO_2 formed and total NO_x removed. Despite it being stated in the ISO protocol that washings from the photocatalyst sample should be analyzed to definitively determine the amount of NO_3^- produced, this aspect is rarely carried out in the literature [23].

Although it is highly beneficial that ISO protocol for measuring photocatalytic NO_x conversion activity has been established – enabling direct comparisons of activity between studies – the current measurement standard does have its limitations. Perhaps the biggest drawback is the test gas used, as the ISO prescribes the use of NO gas, that contains no NO_2. This means that without quantifying the amount of NO_3^- produced (through washings of the photocatalyst sample), one cannot quantify any potential photocatalytic reduction of NO gas into N_2 and O_2. This is important, as the reduction pathway (as opposed to the traditional oxidation pathway) for treating NO_x is increasingly being pursued [24–29], and will be discussed in more detail below. Further criticism of using a test gas that contains only NO stems from the fact that this is not typical of real-world conditions, where mixtures of NO and NO_2 are often present in polluted air [30]. For photocatalysts that oxidize NO, which as mentioned above, is typical of TiO_2-based photocatalysts, this often results in some steady-state formation of NO_2 as NO is oxidized to NO_3^-, and makes the assessment of the overall toxicity change of the treated air quite challenging. For future work in this field, it would be more prudent to test photocatalysts against both NO and NO_2 gases (either individually or together), to ascertain if the photocatalyst can treat the polluted gas in a manner that will reduce the overall toxicity level [31]. Such a change will likely be of benefit to those wishing to influence policy makers, as most

4 A device that can measure the NO and NO_2 concentrations in a gas stream through the process of chemiluminescence. However, such a device cannot measure the NO_3^- that is produced from the complete oxidation of NO_x.

targets on tackling NO_x pollution have been centered on lowering NO_2 levels, as it is significantly more toxic than NO [16].

Unfortunately, many studies simply do not conform to testing their photocatalysts in accordance with the prescribed ISO protocol, making it hard to compare the results produced from such work. However, to account for differences test conditions (e.g., NO_x concentration, gas flow rate, sample area etc.), one can determine the deposition velocity (v_d, cm s^{-1}):

$$v_d = log_e(c_0/c_i) \times F/A \tag{15.2}$$

where c_0 is the inlet gas concentration, c_i is the outlet gas concentration, F is the flow rate of the gas ($cm^3\ s^{-1}$) and A is the area of the photocatalytic coating (cm^2) [32]. This measure is useful, and can be used to predict the level of NO_x treatment that might be seen when these photocatalytic materials are applied at scale [33].

We therefore recommend that researchers continue to use the well-established ISO protocol for measuring the activity of their photocatalysts towards NO. Nevertheless, we believe that there will be great merit to assessing the activity towards NO_2 gas also (and/or mixtures of NO and NO_2), as polluted air in the field contains NO_2 as well. In addition, the determinations of the deposition velocity can also assist researchers compare the activity of their catalysts, especially if nonconventional testing conditions were used (e.g., a slower gas flow rate, a higher NO concentration, etc.), as it can account for these changes. Another benefit of the deposition velocity is that it enables researchers to make predictions on the level of NO_x removal their catalysts may show in the field, and guide the rational design and evaluation of new products.

15.3 Field Trials

15.3.1 Product Development

As discussed above, most studies on TiO_2-based photocatalysts for NO_x conversion have focused on developing methods to incorporate commercially available TiO_2 photocatalysts into a range of building materials, including concretes, paints and asphalts [34]. For samples measured in accordance with the ISO, activities for removing NO_x have been wide-ranging, with concretes showing reductions in NO ranging from ~7% to 45% (~3–20 mg $m^{-2}\ h^{-1}$), paints showing reductions in NO ranging from ~14% to 70% (~6–31 mg $m^{-2}\ h^{-1}$), and asphalts showing reductions in NO ranging from ~14% to 67% (~6–30 mg $m^{-2}\ h^{-1}$) [35].

One of the highest performing concretes to date was produced by Yang et al. [36]. They achieved high activity by producing nanocrystalline anatase TiO_2 photocatalysts on the surface of quartz sand, which was then incorporated into an aggregate layer on the surface of the concrete. Samples were examined in accordance with ISO protocol, where a typical sample showed ~45% removal of NO, ~25% formation of NO_2, and ~20% NO_x removal overall. Yang et al. also determined the quantum efficiency of NO reduction (activity per incident photon), which increased from 0.4%

to 0.8% as their TiO_2 loading was increased. One of the highest performing paints to date was produced by Ângelo et al., who studied their performance in both a laboratory and outdoor setting [37]. Two types of photocatalytic paint were examined: one composed of TiO_2 Degussa P25 and the other of PC500 CristalACTiV^{TM}. The paint contained pigmentary TiO_2 (~9 wt.%), photocatalytic TiO_2 (~9 wt.%), water (~30 wt.%), extenders (~18 wt.%: $CaCO_3$ and silicates), a polymer extender slurry (~8 wt.%), a binder slurry (~20 wt.%), and an additives slurry (~6 wt.%). The photocatalytic NO_x conversion activity was measured in accordance with ISO protocol, where the most active paint, which contained PC500 CristalACTiV^{TM}, showed ~70% NO removal, ~42% NO_2 formation, and ~28% overall NO_x removal. Under outdoor tests, the most active paint showed >95% removal of NO. The higher activity of the PC500 CristalACTiV^{TM} was attributed to the high surface area (~345 $m^2\ g^{-1}$) and nanocrystalline nature (~5–10 nm) of the photocatalyst. And one of the highest performing asphalts to date was developed by Wang et al. [38]. The asphalt was produced by bonding a pulverized TiO_2-cement mortar onto a pavement surface with epoxy resin. Various commercial TiO_2 photocatalysts were examined (4 wt.%), and their photocatalytic NO_x conversion activity measured in accordance with ISO protocol. Using nano-sized anatase, with a grain size of ~15 nm and surface area of ~90 $m^2\ g^{-1}$, an NO removal of up to 67% was observed. A test section of the photocatalytic asphalt was constructed at the Institute of Highway Engineering, Aachen, Germany. The activity of a drilled core from the test track was assessed in the laboratory and showed a NO reduction of ~25%. In a follow up study, Wang et al. investigated the use of surface coatings and pore-filling methods [39]. Photocatalytic NO removal rates were similar in both materials (~42%); however, pore-filled asphalt showed a more prolonged activity.

The wide range of studies that have been carried out, incorporating TiO_2-based photocatalysts into building materials, has resulted in a greater understanding of how the photocatalyst should be processed and incorporated into a given building material to achieve high activity and durability. For example, Maggos et al. found that photocatalytic paints that use water-based styrene acrylic binders show higher photocatalytic activity (12% NO removal, v_d of 0.13 cm s^{-1}; 12% NO_2 removal, v_d of 0.13 cm s^{-1}) than photocatalytic paints that use mineral silicate binders (4.9% NO removal, v_d of 0.05 cm s^{-1}; 2.0% NO_2 removal, v_d of 0.02 cm s^{-1}) [40]. In Table 15.1, we provide an overview of the performances of select examples of TiO_2-based concrete, paint, and asphalt products that have been developed and tested in a controlled laboratory setting.

15.3.2 Outdoor Testing

Many products have been piloted and trialed in the field; with field trials on photocatalytic concretes and paints carried out in Belgium, France, England, Italy, the Netherlands, the Philippines, the United States, and Denmark [35].

At least three trials have been carried out in road tunnels, including Rome [47], Brussels [48], and the Hague [49]. In the Rome trial, a cement-based paint, supplied by Italcementi, was coated over an area spanning ~9000 m^2 inside the

Table 15.1 A summary of the performances of select examples of TiO_2-based concrete, paint, and asphalt products that have been developed and tested in a controlled laboratory setting.

Product	TiO_2 type	ISO*	NO reduction	NO_2 formed	NO_x reduction	References
Concrete	P25 (1%)	yes	~25%, ~11 $mg\ m^{-2}\ h^{-1}$†	~5%	~20%	[41]
	P25 spray coating (60 $g\ m^{-2}$)	yes	~16%†, ~7 $mg\ m^{-2}\ h^{-1}$	*ns*	~60%	[42]
	Anatase hydrosol (0.1 g)	yes	~45%, ~20 $mg\ m^{-2}\ h^{-1}$†	~25%	~20%	[43]
	Anatase – nano-sized	no	~78%	*ns*	*ns*	[44]
	Rutile – nanorods (3%)	no	*Ns*	*ns*	~24%	[45]
Paint	VLP7101 Kronos™ C-doped anatase (17%)	yes	~27%, ~12 $mg\ m^{-2}\ h^{-1}$†	~16%	~11%	[46]
	PC500 CristalACTiV™ anatase (18%)	yes	~70%, ~31 $mg\ m^{-2}\ h^{-1}$†	~42%	~28%	[38]
	StoClimasan	yes	~14%, ~6 $mg\ m^{-2}\ h^{-1}$†	10.4%	3.8%	[22]
Asphalt	nano-sized anatase	yes	~67%, ~30 $mg\ m^{-2}\ h^{-1}$†	*ns*	*ns*	[39]
	micron-sized anatase: rutile	yes	~14%, ~6 $mg\ m^{-2}\ h^{-1}$†	*ns*	*ns*	[39]
	nano-sized anatase	yes	~42%, ~19 $mg\ m^{-2}\ h^{-1}$†	*ns*	*ns*	[40]

ns = not stated, *nm* = not measured.
* Measured in accordance with ISO protocol (either 22197-1:2007 or ISO 22197-1:2016)
† Calculated herein if sufficient information was present in the article.

Umberto I tunnel. In the Brussels trial, a 160 m strip of the Leopold II tunnel was coated with a cement-based paint (TX-Active® Skim Coat Boosted), again supplied by Italcementi (Figure 15.3a,b). And in The Hague trial, a 150 m trial section of the Koningstunnel was coated with a photocatalytic paint (KNOxOUT™ Clear), supplied by Boysen. Contrasting results were observed. In the Rome trial, significant reductions in both NO and NO_2 levels were seen. In the Hague trial, only reductions in NO were seen. And in the Brussels trial, no significant reductions in NO or NO_2 were seen. In such tunnel trials, where sunlight cannot reach the photocatalytic coatings, UVA lamps are installed to activate the TiO_2-based photocatalysts. As such, the observed differences may be explained, in part, by the differences in UVA light power used across the three trials, with the Rome trial receiving more than double the light power (~2.0 $mW\ cm^{-2}$) than the Brussels (~0.16 $mW\ cm^{-2}$) and The Hague trials (~0.1 $mW\ cm^{-2}$). Also worthy of note was the accumulation

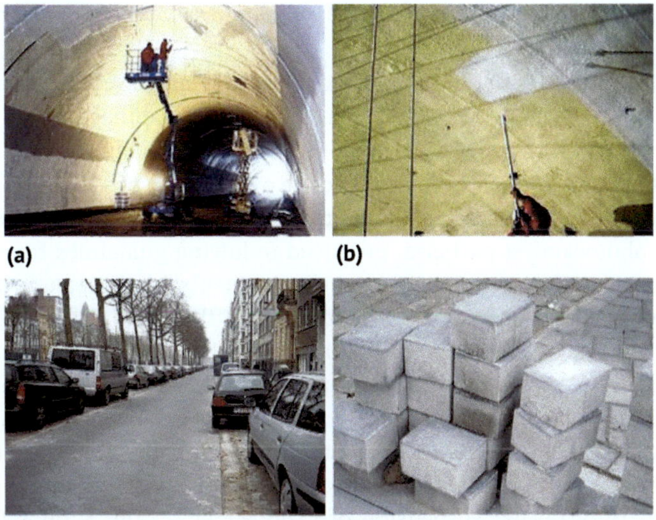

(a) (b) (c) (d)

Figure 15.3 Photographs of the photocatalytic coatings and building materials for NO_x conversion that have been trialed in various locations. (a, b) The Umberto I tunnel in Rome coated with TX-Active® Skim Coat Boosted, by Italcementi. Reproduced from [47]/with permission of Elsevier. (c, d) A parking lane in the Leien of Antwerp with concrete paving blocks coated with a top layer (~8 mm thick) of a hydraulic binder intermixed with anatase TiO_2 photocatalyst particles. Reproduced from [54]/Springer Nature/CC BY-2.0.

of soot, dust and other particulates had accumulated on the surface of these coatings during the trial, which resulted in a significant loss of function [33]. Their activity could be restored with extensive UV treatment or washing, highlighting the importance for the regular cleaning of such tunnel coatings so that their effectiveness can be retained.

Field trials of photocatalytic paints and coatings have been carried out in open areas of many cities, including London [50], Paris, and Manila. In London, only relatively small areas were trialed (up to 135 m^2) using a developmental water-based product (PCX-S7), supplied by Cristal. No significant change in NO_x was observed, with modelling showed that the benefits of such a small-scale trial would be limited to a small volume of air in close proximity to the coating. It was therefore recommended that larger, more sheltered areas be coated in future trials for significant improvements in air quality to be seen. Maggos et al. conducted a small-scale (~200 m^2) field trial in Guerville, near Paris, where they used a mortar with a mineral binder containing a 3 wt.% TiO_2 photocatalyst and sand, provided by Italcementi [51]. They created artificial street canyons, spaced ~2 m apart, and found that NO_x concentrations in the treated canyon were between ~40% and ~80% lower than reference canyons. Maggos et al. also conducted a small-scale field trial in a car park in La Défense, in the west of Paris, this time using an acrylic paint containing a TiO_2 photocatalyst, and saw a 19% reduction in NO and 20% reduction in NO_2 gas [52]. And Wahlstedt et al. conducted a large-scale (~6000 m^2) field trial at Guadalupe MRT, in Manila [53]. They applied a styrene acrylic water-based paint that contained ultra-fine TiO_2 (7.5 wt.%), supplied by Boysen, and saw reductions in ambient NO_2 levels of up to 23%.

And field trials of photocatalytic concrete paving have been carried out in Antwerp [54], Hengelo [55], Wijnegem [56], and Copenhagen [57]. In the field trial in Antwerp, Boonen and Beeldens examined the change in efficacy of after a five-year field trial of photocatalytic concrete over an area of ~10 m^2 (Figure 15.3c,d) [54]. The concrete contained an ~8 mm thick hydraulic binder intermixed with anatase TiO_2 photocatalyst particles, produced following guidelines from the patents of Italcementi (EP-patent 1600430 A1) and Mitsubishi (EP-patent 0786283 A1). Before the field trial, lab-scale tests showed that their samples showed ~40% NO removal, ~10% NO_2 formation, and ~30% overall NO_x removal. After the five-year field trial, their samples showed a lower activity. However, after washing, this was restored. In the Hengelo trial, Ballari and Brouwers examined the efficacy of photocatalytic concrete paving tiles along a 150 m footpath [55]. The photocatalytic concrete blocks were supplied by Struyk Verwo Infra, with the surface spray coated using a 4% (w/w) water suspension of C-doped TiO_2 from Kronos International (~2.67 g m^{-2}). Comparing data from a control street, the NO_x concentration was found to decrease, on average, by 19% during the day, and 28% during the afternoon. Under ideal weather conditions (i.e., high radiation and low relative humidity) a relative decrease of up to 45% was observed. In the Wijnegem trial, in Den Hoek 3, Boonen and Beeldens examined the activity of an enclosed area of concrete paving using a portable testing unit [56]. The concrete contained an active top layer (~50 mm thick) made of white cement with 4% anatase TiO_2 by weight, provided by the Heidelberg Cement Group. On site measurements showed reductions of up to ~60% NO removal, ~5% NO_2 formation, and ~55% overall NO_x removal, which corresponded to a deposition velocity for NO_x of ~0.02 cm s^{-1}. The Copenhagen trials were carried out by Folli et al., who compared differences in NO_x concentrations along a 200 m × 2 m sidewalk, in Gasværksvej, close to the Central Railway Station, where one half was coated with photocatalytic concrete paving [57]. The photocatalytic paving, provided by Starka Betongindustrier, contained a 10 mm thick active layer, with a TiO_2 photocatalyst intermixed into the concrete (0.04 g cm^{-3}). They found that under ideal weather conditions, the monthly average NO concentration in proximity to the photocatalytic coating was ~22% lower than the control. Moreover, data collected at noon, when sunlight was most intense, showed NO removal higher than 45%, and a total NO_x removal higher than 30%.

From a review of the literature, it is clear that field trials have shown varying levels of success. As such, for field trials to be more reliable and comparable, we recommend the use of a more standardized format. For example, the use of split sites to produce control data may not be as reliable as monitoring a single site before and after the application of the photocatalytic material. Other meteorological factors should also be considered, such as the wind speed, direction, solar intensity, and humidity, which can all influence photocatalyst performance [58]. To therefore ascertain is the photocatalytic material will function under the imposed conditions, we recommend that the material is examined in the laboratory under simulated conditions that are similar to those seen in the field. Subsequent testing should only be carried out if the photocatalytic material shows significant NO_x removal under simulated conditions. Such preliminary work could be supported by technoeconomic analysis to produce a cost–benefit analysis. As mentioned earlier, such an analysis could include consideration of the overall toxicity change in the air after photocatalytic treatment. And lastly, and perhaps most importantly, a reasonable consideration of the scale of the treatment applied and its potential impact should be made. This should include considerations of the surface area being treated, the expected volume of air that can be treated, the expected change in surrounding NO_x level, and expected level of NO_x removed from the air.

The product development and field trials of photocatalytic materials for NO_x conversion has resulted in a number of products entering the market, which include tiles from TOTO (Hydrotect), concrete from Italcementi (TX), paints from Keim (Ecosil-ME), and Boysen (KNOxOUT) and transparent sprays from Cristal Tronox (CristalACTiV-PC-S7). This has coincided with a booming market, with the global photocatalyst market valued at ~\$3.1 billion in 2021 and forecast to grow from to ~\$7.6 billion by 2030, with a compound annual growth rate of ~10.5% [59]. However, despite rapid growth in this market, few countries insist on providing independent certification of photocatalytic activity. One of the exceptions in this regard is Japan, which is the largest market for photocatalyst sales, and requires the independent certification of all products by the Photocatalysis Industry Association of Japan (PIAJ).

15.4 Materials Design

As mentioned above, the focus in the field of photocatalytic NO_x conversion has largely been on product development, where known, and commercially available TiO_2 photocatalyst media, have been incorporated into a wide range of building materials and examined in the laboratory and field. Nevertheless, there has been a significant body of research carried out on the "bottom-up" approach of improving the performance of TiO_2-based photocatalysts for NO_x conversion through materials design strategies, with several reviews on this topic published [60–62]. The most common strategies for improving the activity of any photocatalyst are to:

I. Increase the light harvesting capability of the photocatalyst, which can be achieved by impurity doping that causes a narrowing of the bandgap in the material [63].

II. Increase the lifetime of photogenerated electrons and holes, and therefore, increase the probability for surface redox reactions to occur, which is typically achieved through the formation of a composite with another semiconductor photocatalyst that results in staggered band energy levels that promote charge carrier separation [64].

III. Increase its catalytically active surface area, which is typically achieved by tailoring the nanostructure [65].

IV. Increase the reaction kinetics of surface redox reactions, which is typically achieved by affixing co-catalysts to the surface of the photocatalyst [66].

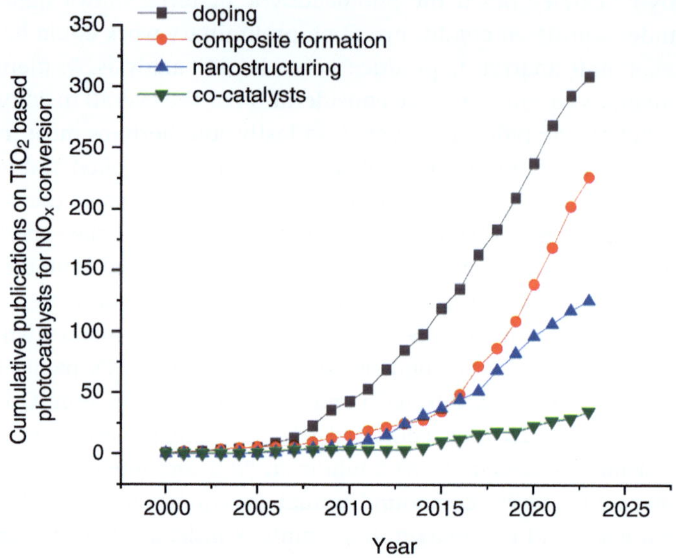

Figure 15.4 Cumulative publications on the topic of the TiO_2-based photocatalytic NO_x conversion since the year 2000 by modification strategy: doping, composite formation, nanostructuring, and co-catalysts. The data used in this figure was acquired from Web of Science.

From a Web of Science search, we can see that in the field of TiO_2-based photocatalysts for NO_x conversion,[5] articles that have included the strategies of doping, composite formation, nanostructuring, and use of co-catalysts have seen 311, 228, 135, and 35 publications, respectively, since the year 2000 (Figure 15.4). The earliest strategy to be adopted was doping, which gained traction in ~2008. This was followed by nanostructuring and composite formation, which both gained traction in ~2016, and then the use of co-catalysts, which is recently seeing attention. In the subsequent section, we will provide an overview of the various types of dopants, composite materials, nanostructures and co-catalysts that have been explored to date, going in-depth for some key examples from the literature to illustrate the merits of each strategy.

15.4.1 Doping Strategies

With regards to doping strategies, a wide range of dopants have been explored, including, carbon [67], nitrogen [68], fluorine [69], sulfur [70], vanadium [71],

5 Data were acquired from Web of Science using the search terms ((“TiO_2” OR “titania” OR “titanium dioxide”) AND (“photocat*”) AND (“NO_x” OR “nitrogen oxides” OR “nitric oxide” OR “nitrogen dioxide” OR “NO decomposition”)), where for the strategy of doping the additional search term of (“dope*”) was added, for the strategy of composite formation the additional search terms of (“composite” OR “heterojunction” OR “Z-scheme”) were added, for the strategy of nanostructuring the additional search terms of (“nanostruct*”or “nanotube*” or “nanomaterial”) were added, and for the strategy of use of a co-catalyst the additional search terms of (“co-catalyst” OR “cocatalyst” OR “surface mod*”) were added.

manganese [72], cobalt [73], iron [74], tin [75], lanthanum [76], cerium [77], tungsten [78], and co-doping with nitrogen and sulfur [79], and other pairs of elements [80–84]. As mentioned above, doping strategies have predominantly been explored in an effort to increase the light harvesting capability of TiO_2. This occurs when there are interactions between the orbitals of dopant and the orbitals in TiO_2 that form the valence band maximum (primarily O 2p) and conduction band minimum (primarily Ti 3d) energy levels, and results in a narrowing of the bandgap.

For example, Zhang et al. studied the effect of nitrogen doping in TiO_2 synthesised using a microwave-assisted hydrothermal method [85]. Their materials adopted the anatase crystal structure, contained substitutional nitrogen (for lattice oxygen) and interstitial nitrogen states, and were of high surface area (~300 $m^2\,g^{-1}$). In their photocatalytic tests using a visible light source (>510 nm), up to 30% NO_x oxidation was seen, which was more than three times higher than the commercial TiO_2 Degussa P25 benchmark material. Also, Katsanakia et al. investigated nitrogen and fluorine co-doping in TiO_2 using a sol–gel synthesis method [86]. They found that the samples were active photocatalysts under fluorescent lighting for NO_x oxidation, showing NO removal of up to ~24%. And Cheng et al. studied the effect of iron doping in TiO_2 using a sol–gel synthesis method [87]. Their materials were of the anatase crystal structure, and showed a reduced bandgap energy from ~3.20 to 2.85 eV. They found that 5% iron doping was optimal for photocatalytic NO conversion under visible light irradiation (>420 nm), showing up to 90% NO_x conversion and highly suppressed NO_2 formation (~10%). They attributed the increase in activity to an increase propensity for NO to bind to surface iron sites, and an increase in photo-induced oxygen vacancy formation, which they purported enhanced the formation of ROS that oxidized NO_x species.

15.4.2 Composite Formation Strategies

A wide range of TiO_2-containing composites have been explored in an effort to improve the performance of NO_x conversion photocatalysts [88, 89]. Composites, when possessing the correct band alignment, can improve the separation of photogenerated charge carriers, increasing their lifetime and probability to take part in redox reactions that result in the conversion of gaseous NO_x species. Typically, a type II band alignment is targeted, where the conduction and valence bands in both materials stagger in a manner that provides a thermodynamic driving force for electrons and holes to spatially separate across the two materials in opposite directions. Composites can also possess complementary bandgaps that absorb light in different regions of the electromagnetic spectrum. This can increase the light harvesting capability of the composite photocatalyst, extending it into the visible region of the electromagnetic spectrum. The range of semiconductor composites with TiO_2 that have been explored for photocatalytic NO_x conversion include carbon nitrides [90], silicon oxide [91], iron oxide [92], zinc ferrite [93], zinc oxide [94], tin oxide [95], sodium tantalite [96], tungsten oxide [97], bismuth iodide [98], and bismuth tungstate [99].

For example, Balbuena et al. explored hematite iron oxide (Fe_2O_3) composites with TiO_2 prepared using a plasma-assisted chemical vapor deposition method

[100]. Various iron to titanium ratios were explored, resulting in plate-like nanocomposites. The activity of their samples against NO gas was examined using simulated solar irradiation, where synergistic improvements in photocatalytic NO_x removal activity were seen in their composites (up to ~12%) compared with the parent materials (~2% and ~4% seen in Fe_2O_3 and TiO_2 parent materials, respectively). Moreover, improved selectivity for the formation of NO_3^- was seen in their composites (up to ~60%) compared with the parent materials (~10% and ~20% seen in Fe_2O_3 and TiO_2 parent materials, respectively). And Ma et al. investigated graphitic carbon nitride (g-C_3N_4) composites with TiO_2 produced using a simple solid state synthesis route using melamine and TiO_2 Degussa P25 as precursors [101]. They found that an intimate electronic contact between the two materials was essential for improvements in photocatalytic activity to be seen, where for an optimal loading of g-C_3N_4 of ~15 wt.% they saw a ~27% photocatalytic NO_x removal under simulated solar irradiation (>420 nm), which was synergistically higher than pure TiO_2 Degussa P25 (~17% NO_x removal) and g-C_3N_4 (~7% NO_x removal).

Worthy of note are alternative other types of composite that have been explored with TiO_2. One type is to couple TiO_2 to a metallic material to increase charge carrier separation, which includes the use of graphene [102] and MXenes [103]. And another type is to couple TiO_2 to a robust, high surface area support material, which includes the use of clays [104], hydroxyapatite [105], saponite [106], and zeolites [107].

15.4.3 Nanostructuring Strategies

A range of nanostructures have been investigated to improve the activity of TiO_2-based photocatalysts for NO_x conversion [108], which include the use of nanotubes [109], nanorods [110], nanoparticlcs [111], nanostructured coatings [112], and hierarchical structures [113].

Huang et al. investigated highly nanostructured TiO_2 nanotube structures prepared using a microwave-assisted method [114]. The structures possessed both anatase (E_{bg} ~3.2 eV) and rutile (E_{bg} ~3.0 eV) crystal phases, enhancing light-harvesting efficiency in the near visible. Their photocatalytic activity was examined under visible light and showed significantly higher NO conversion (~95%) than a TiO_2 Degussa P25 standard (~54%). Bahşi et al. explored the effect of TiO_2 particle size in photocatalytic cements for NO_x conversion [111]. They explored nano-sized, submicron-sized, and micron-sized anatase-phase TiO_2 powders, each at 5 wt.% in the cement mixture. The most active cements contained nano-sized TiO_2 powders, showing up to ~50% NO_x removal and ~60% selectivity for NO_3^- formation. They concluded that a key parameter for improving the photocatalytic performance of photocatalytic cements is to increase the loading of TiO_2 particles at/near the surface.

In general, it should be noted that any increase in TiO_2 nanostructure will likely come at the expense of mechanical strength, and that this strategy lends itself to applications where the photocatalyst will not be subjected to regular abrasion (e.g., not for applications as paving stones, but rather, for applications in cladding, paints or on filters within encapsulated air purification systems).

15.4.4 Co-catalyst Strategies

One of the least pursued strategies to date for the photocatalytic conversion of NO_x is the use of surface co-catalysts. However, this strategy has been shown to be critical to facilitating certain photocatalytic reactions, such as the most popularly studied water splitting reaction [115]. The use of catalysts can not only improve the spatial separation of charge carriers [116], but can improve the kinetics of photocatalytic reactions [117].

Various noble and platinum group metals have been examined for use as co-catalysts to drive the photocatalytic conversion of NO_x, including silver [118], gold [119], palladium [120], platinum [121, 122], and platinum oxide [123]. It should be noted that catalytic converters, which drive the thermal catalytic conversion of NO_x, contain the platinum group metals rhodium, palladium, and platinum, attesting to the affinity of their interactions and activation of NO_x gases [124].

For example, Fujiwara et al. explored the use of single-atom palladium (Pd) co-catalysts on TiO_2, produced using a flame spray pyrolysis method [125]. Isolated Pd atoms were identified by diffuse reflectance infrared Fourier transform spectroscopy (DRIFTS) and high-resolution transmission electron microscopy (HR-TEM) imaging. They were able to achieve between 40% and 70% surface coverage of TiO_2 with single atom Pd co-catalyst for loadings between 0.025 and 1.0 wt.%. The photocatalytic activity toward NO gas was examined under solar simulated light, with Pd single atom loaded TiO_2 showing NO_x conversion of up to ~45%, whereas a commercial TiO_2 Degussa P25 standard showed significantly lower NO_x conversion of ~5%. The superior performance of isolated Pd atoms was attributed to their high resistance to NO_3^- poisoning and high selectivity for the conversion of NO to NO_3^-. Wang et al. studied the effect of platinum (Pt)-loaded TiO_2 on photocatalytic NO_x conversion activity [126]. They found that a neutralization method, which produced platinum oxide (PtO_x) particles on the surface of TiO_2 was most beneficial to activity; forming new electronic states above the valence band of TiO_2 that increased the separation and lifetime of photogenerated charge carriers. The photocatalytic activity of a 0.05 at.% PtO_x co-catalyst loaded on TiO_2 (~70% NO_x removal) was more than double that of a commercial TiO_2 Degussa P25 standard (~30% NO_x removal).

15.5 Challenges and Perspectives

15.5.1 An Emerging Strategy; Targeting the Reduction of NO_x to N_2 and O_2

Before closing this review, we would like to highlight one emerging area of research in the field of photocatalytic NO_x conversion that could have significant implications on the field if proven to be successful. This strategy is to develop TiO_2-based photocatalysts that show a preference for the photocatalytic reduction of NO_x gases to N_2 and O_2 (i.e., the decomposition pathway), and has been the subject of discussion in recent review papers [18, 19]. The reduction pathway is highly advantageous over the oxidation pathway as it: (i) avoids the formation of surface NO_3^-, which not only inhibits the photocatalyst from functioning but can also drive deleterious back reactions, where NO_3^- reacts with photogenerated holes and NO to form more toxic

NO_2 [22] and (ii) avoids the formation of NO_2, as NO is sequentially oxidized by hydroxyl radicals on its pathway to forming NO_3^-.

As early as 2006, Bowering et al. showed that silver-loaded TiO_2 dramatically increased the selectivity for the reduction pathway in photocatalytic NO conversion [24]. They prepared their materials by annealing silver nitrate salt that was loaded onto the surface of commercial TiO_2 Degussa P25 powders, studying a range of mass loadings and annealing temperatures. They carried out their photocatalytic reactions under anaerobic reactions using argon carrier gas, and using mass spectrometry (MS), saw up to 100% selectivity for the reduction reaction in samples that had been annealed at higher temperatures (between 450 °C and 600 °C). X-ray photoelectron spectroscopy (XPS) analysis showed that their silver particles contained both metallic and singly oxidized states. Interestingly, catalysts that possessed a higher proportion of metallic silver showed a higher selectivity for the reduction pathway. Nevertheless, high selectivities were at the expense of photocatalytic activity, with lower NO conversion rates seen in samples annealed at high temperature. Supporting this work, in 2016, Duan et al. also studied silver-loaded TiO_2 [28]. They loaded their silver onto commercial TiO_2 Degussa P25 powders using a photochemical reduction process followed by annealing. Using gas chromatography (GC), they confirmed the reduction of NO gas to N_2 and O_2, purporting that this pathway was mediated by oxygen vacancy (V_O) states. However, the selectivity of their catalysts for the reduction pathway was not quantified in their work.

One of the most recognized examples of the photocatalytic reduction of NO was provided by Wu et al. in 2012 in their study of iron-doped TiO_2 [27]. They found that iron dopants (Fe^{3+}) in TiO_2 resulted in the co-formation of V_O states, with a solubility limit of ~10 at.%. Extended X-ray absorption fine structure (EXAFS) and X-ray absorption near edge structure (XANES) revealed the presence of tetrahedrally coordinated titanium clusters (TiO_4) at the surface of the material. Interestingly, they found that during the photocatalytic conversion of NO, their catalysts showed some preference for NO to be decomposed into N_2 and O_2 [127]. As this preference was found to be proportional to the degree of Fe^{3+} doping and V_O concentration, they concluded that the reduction pathway was dependent on surface V_O states acting as catalytic centers. Their proposed mechanism entailed the trapping of photogenerated electrons and gaseous NO (Eq. 15.3a), with neighboring trapping sites reacting and releasing N_2 (Eq. 15.3b), and the oxidation of surface trapped oxygen by photogenerated holes to release O_2 and regenerate the V_O state (Eq. 15.3c):

$$V_O^{\bullet\bullet} + 2e^- + NO_{(gas)} \rightarrow NO_{(surface)} \tag{15.3a}$$

$$2NO_{(surface)} \rightarrow N_{2(gas)} + 2O_{(surface)} \tag{15.3b}$$

$$2O_{(surface)} + 2h^+ \rightarrow O_{2(gas)} + V_O^{\bullet\bullet} \tag{15.3c}$$

Wu et al. saw only ~4.5% selectivity for the reduction pathway in their catalyst (and therefore ~95.5% selectivity for the oxidation pathway), but proposed that the selectivity for this pathway would increase in catalysts with a higher population of V_O states.

Anpo and Takeuchi showed in 2001 that chromium (Cr)- or vanadium (V)-doped TiO_2 coatings, produced using a high-voltage acceleration method, similarly showed the ability to decompose NO under solar lighting conditions to N_2 and O_2 [29]. The reduction products were measured using GC and MS; however, any potential oxidation products were not measured; therefore, the selectivity for the reduction pathway was not quantified. Nevertheless, they found that pristine TiO_2 coatings were not able to decompose NO. As the Cr- or V-doping of TiO_2 is known to form V_O states (through the substitution of Ti^{4+} states with Cr^{3+} or V^{3+} states and co-creation of V_O), their observation lends credence to the argument of Wu et al. that V_O states are integral to facilitating the reduction pathway [27].

Although not a TiO_2-based catalyst, in 2017, Dong et al. showed that carbon vacancy (V_C) states in g-C_3N_4 nanosheet structured photocatalysts showed a preference for the reduction of NO [26]. Their materials showed a high selectivity for the reduction pathway, estimated to be as high as ~66% (Figure 15.5). They attributed this to the high concentration of V_C states and ultrathin nanostructure. Computational calculations suggested an increased preference for absorption of NO at V_C defect sites, and by comparing the performance of their defective catalyst with a pristine g-C_3N_4 catalyst, they concluded that V_C states were essential in driving the reduction pathway with high selectivity.

15.5.2 Final Thoughts

Herein, the state of the field of TiO_2-based photocatalysts for NO_x conversion has been reviewed. It is a rich field of research, that has resulted in a wide range of commercial products entering the market to remedy this air pollutant; including photocatalytic tiles, concretes, paints, and transparent sprays. Current limitations of the International Organization for Standardization (ISO) protocol for measuring the activity of photocatalytic materials for NO_x conversion were discussed, as well as the need for national

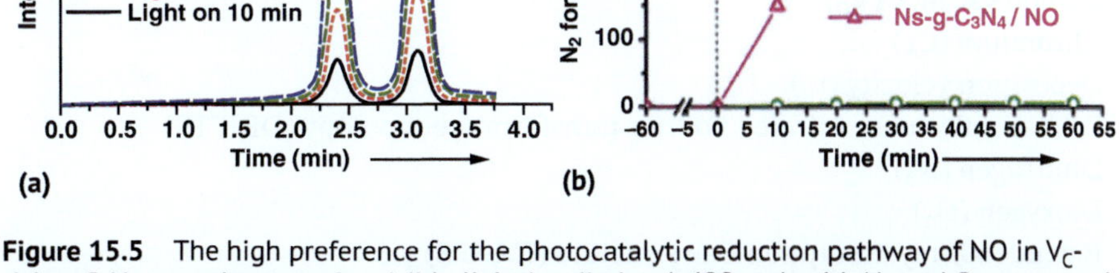

Figure 15.5 The high preference for the photocatalytic reduction pathway of NO in V_C-rich g-C_3N_4 nanosheets under visible light irradiation (>420 nm), with N_2 and O_2 measured by GC. (a) The production of N_2 and O_2 from the photocatalytic reduction of NO under anaerobic conditions for a V_C-rich g-C_3N_4 catalyst. (b) A comparison of the activity of V_C-rich g-C_3N_4 and g-C_3N_4 nanosheets under anaerobic conditions in the production of N_2, where the photocatalytic reduction pathway of NO was only seen in the V_C-rich material. Reproduced from Ref. [27]/with permission of Elsevier.

and/or internationally recognized bodies that can independently certify photocatalytic performance and increase consumer confidence in the products they purchase. Although laboratory testing of such products under controlled conditions often yields high efficacy for the photocatalytic conversion of NO_x gas, the same level of performance is not often translated when products are examined in the field. Reasons for these discrepancies were discussed, with suggestions provided on how future field trials could be improved. Improved standardization of the use of ISO testing, measures of activity, field trial testing, analysis of any hazardous by-products that may form, considerations of cost–benefit, and the certification of newly developed consumer products that meet strict regulatory requirements are all required to guide this rapidly developing field forward.

Alongside product design and development, there is a substantial strand of research being carried out on the development of next-generation TiO_2-based photocatalytic materials for NO_x conversion. In this review, we summarized ongoing work in this area, which included the use of dopants, composites, nanostructuring, and surface co-catalysts. We ended the review by discussing a highly promising emerging strategy, where TiO_2-based photocatalysts are being developed that can drive the photocatalytic reduction of NO with high selectivity (as opposed to the traditional pathway, where NO is oxidized). The reduction pathway promotes the decomposition of NO_x into N_2 and O_2, and avoids the accumulation of NO_3^- on the surface of the photocatalyst, which inhibits the photocatalyst from carrying out further oxidation reactions. The development of photocatalysts that promote reduction reactions will not only be of benefit to those working in the field of NO_x conversion, but may also result in development of transferable technologies, for example, in the field of greenhouse gas (GHG) removal, where the reduction of carbon dioxide (CO_2) and nitrous oxide (N_2O) gases are currently being targeted; a topic increasing in importance due to the emerging climate crisis.

List of Abbreviations

Ammonia (NH_3)
Area of photocatalytic coating (A)
Bandgap energy (E_{bg})
Carbon dioxide (CO_2)
Carbon vacancy (V_C)
Chromium (Cr)
Deposition velocity (v_d)
Diffuse reflectance infrared Fourier transform spectroscopy (DRIFTS)
Dinitrogen (N_2)
Dioxygen (O_2)
Extended X-ray absorption fine structure (EXAFS)
Flow rate of gas (F)
Gas chromatography (GC)
Graphitic carbon nitride (g-C_3N_4)
Greenhouse gas (GHG)
Hematite iron oxide (Fe_2O_3)

High-resolution transmission electron microscopy (HR-TEM)
Hydroxyl radicals ($OH^{\bullet}$)
Inlet gas concentration (c_0)
International Organization for Standardization (ISO)
Mass spectrometry (MS)
Nitrate (NO_3^-)
Nitric oxide (NO)
Nitrogen dioxide (NO_2)
Nitrous acid (HONO)
Nitrous oxide (N_2O)
$NO_x = NO + NO_2$
Outlet gas concentration (c_i)
Oxygen vacancy (V_O)
Palladium (Pd)
Photocatalysis Industry Association of Japan (PIAJ).
Platinum oxide (PtO_x)
Reactive oxygen species (ROS)
Scanning electron microscopy (SEM)
Superoxide ($O_2^{\bullet -}$)
Ultraviolet (UV)
Vanadium (V)
Water (H_2O)
X-ray absorption near edge structure (XANES)
X-ray photoelectron spectroscopy (XPS)

References

1 Wang, S., Ichihara, F., Pang, H. et al. (2018). Nitrogen fixation reaction derived from nanostructured catalytic materials. *Advanced Functional Materials* 28: 1803309.

2 Takai, K. (2019). The nitrogen cycle: a large, fast, and mystifying cycle. *Microbes and Environments* 34: 223–225.

3 Fowler, D., Coyle, M., Skiba, U. et al. (2013). The global nitrogen cycle in the twenty-first century. *Philosophical Transactions of the Royal Society B: Biological Sciences* 368: 20130164.

4 Schlesinger, W.H. (2009). On the fate of anthropogenic nitrogen. *Proceedings of the National Academy of Sciences of the United States of America* 106: 203–208.

5 Tie, X., Zhang, R., Brasseur, G., and Lei, W. (2002). Global NO_x production by lightning. *Journal of Atmospheric Chemistry* 43: 61–74.

6 Olivier, J.G.J., Bouwman, A.F., Van der Hoek, K.W., and Berdowski, J.J.M. (1998). Global air emission inventories for anthropogenic sources of NO_x, NH_3 and N_2O in 1990. *Environmental Pollution* 102: 135–148.

7 Gaston, B., Drazen, J.M., Loscalzo, J., and Stamler, J.S. (1994). The biology of nitrogen oxides in the airways. *American Journal of Respiratory and Critical Care Medicine* 149: 538–551.

8 Boningari, T. and Smirniotis, P.G. (2016). Impact of nitrogen oxides on the environment and human health: Mn-based materials for the NO_x abatement. *Current Opinion in Chemical Engineering* 13: 133–141.
9 Winiwarter, W. and Klimont, Z. (2011). The role of N-gases (N_2O, NO_x, NH_3) in cost-effective strategies to reduce greenhouse gas emissions and air pollution in Europe. *Current Opinion in Environmental Sustainability* 3: 438–445.
10 Dey, S. and Mehta, N.S. (2020). Automobile pollution control using catalysis. *Resources, Environment and Sustainability* 2: 100006.
11 Anenberg, S.C., Miller, J., Minjares, R. et al. (2017). Impacts and mitigation of excess diesel-related NO_x emissions in 11 major vehicle markets. *Nature* 545: 467–471.
12 de Richter, R. and Caillol, S. (2011). Fighting global warming: the potential of photocatalysis against CO_2, CH_4, N_2O, CFCs, tropospheric O_3, BC and other major contributors to climate change. *Journal of Photochemistry and Photobiology C: Photochemistry Reviews* 12: 1–19.
13 Ovcharov, M.L. and Granchak, V.M. (2021). Photocatalytic conversion of nitrogen oxides: current state and perspectives: a review. *Theoretical and Experimental Chemistry* 57: 30–63.
14 Wei, Y., Meng, H., Wu, Q. et al. (2023). TiO_2-based photocatalytic building material for air purification in sustainable and low-carbon cities: a review. *Catalysts* 13: 1466.
15 Pichat, P., Herrmann, J.-M., Courbon, H. et al. (1982). Photocatalytic oxidation of various compounds over TiO_2 and other semiconductor oxides; mechanistic considerations. *The Canadian Journal of Chemical Engineering* 60: 27–32.
16 Russell, H.S., Hertel, O., Ellerman, T., and Jensen, S.S. (2021). A review of photocatalytic materials for urban NO_x remediation. *Catalysts* 11: 1–45.
17 Balbuena, J., Cruz-Yusta, M., and Sánchez, L. (2015). Nanomaterials to combat NO_x pollution. *Journal of Nanoscience and Nanotechnology* 15: 6373–6385.
18 Li, N., Wang, C., Zhang, K. et al. (2022). Progress and prospects of photocatalytic conversion of low-concentration NO. *Chinese Journal of Catalysis* 43: 2363–2387.
19 Nguyen, V.-H., Nguyen, B.-S., Huang, C.-W. et al. (2020). Photocatalytic NO_x abatement: recent advances and emerging trends in the development of photocatalysts. *Journal of Cleaner Production* 270: 121912.
20 Dillert, R., Engel, A., Große, J. et al. (2013). Light intensity dependence of the kinetics of the photocatalytic oxidation of nitrogen(ii) oxide at the surface of TiO_2. *Physical Chemistry Chemical Physics* 15: 20876.
21 Bloh, J.Z., Folli, A., and Macphee, D.E. (2014). Photocatalytic NO_x abatement: why the selectivity matters. *RSC Advances* 4: 45726–45734.
22 Mills, A. and Elouali, S. (2015). The nitric oxide ISO photocatalytic reactor system: measurement of NO_x removal activity and capacity. *Journal of Photochemistry and Photobiology A: Chemistry* 305: 29–36.
23 Mills, A., Hill, C., and Robertson, P.K.J. (2012). Overview of the current ISO tests for photocatalytic materials. *Journal of Photochemistry and Photobiology A: Chemistry* 237: 7–23.
24 Bowering, N., Croston, D., Harrison, P.G., and Walker, G.S. (2007). Silver modified Degussa P25 for the photocatalytic removal of nitric oxide. *International Journal of Photoenergy* 2007: 1–8.

25 Iwamoto, M., Furukawa, H., Mine, Y. et al. (1986). Copper(II) ion-exchanged ZSM-5 zeolites as highly active catalysts for direct and continuous decomposition of nitrogen monoxide. *Journal of the Chemical Society, Chemical Communications* 1986: 1272. https://doi.org/10.1039/c39860001272.

26 Dong, G., Jacobs, D.L., Zang, L., and Wang, C. (2017). Carbon vacancy regulated photoreduction of NO to N_2 over ultrathin g-C_3N_4 nanosheets. *Applied Catalysis B: Environment and Energy* 218: 515–524.

27 Wu, Q., Zheng, Q., and van de Krol, R. (2012). Creating oxygen vacancies as a novel strategy to form tetrahedrally coordinated Ti^{4+} in Fe/TiO_2 nanoparticles. *The Journal of Physical Chemistry C* 116: 7219–7226.

28 Duan, Y., Zhang, M., Wang, L. et al. (2017). Plasmonic Ag-TiO_{2-x} nanocomposites for the photocatalytic removal of NO under visible light with high selectivity: the role of oxygen vacancies. *Applied Catalysis B: Environment and Energy* 204: 67–77.

29 Anpo, M. and Takeuchi, M. (2001). Design and development of second-generation titanium oxide photocatalysts to better our environment—approaches in realizing the use of visible light. *International Journal of Photoenergy* 3: 89–94.

30 Clapp, L. (2001). Analysis of the relationship between ambient levels of O_3, NO_2 and NO as a function of NO_x in the UK. *Atmospheric Environment* 35: 6391–6405.

31 Engel, A., Glyk, A., Hülsewig, A. et al. (2015). Determination of the photocatalytic deposition velocity. *Chemical Engineering Journal* 261: 88–94.

32 Boonen, E., Akylas, V., Barmpas, F. et al. (2015). Construction of a photocatalytic de-polluting field site in the Leopold II tunnel in Brussels. *Journal of Environmental Management* 155: 136–144.

33 Bloh, J.Z. (2021). Photocatalytic NO_x abatement. In: *Heterogeneous Photocatalysis*, 303–329. Wiley. https://doi.org/10.1002/9783527815296.ch13.

34 Pedersen, P.D., Lock, N., and Jensen, H. (2021). Removing NO_x pollution by photocatalytic building materials in real- life: evaluation of existing field studies. *Journal of Photocatalysis* 2: 84–96.

35 The Environmental Industries Commission. (2018). Towards purer air: a review of the latest evidence of the effectiveness of photocatalytic materials and treatments in tackling local air pollution. https://eic-uk.co.uk/media/baecbnd4/towards-purer-air.pdf.

36 Yang, L., Hakki, A., Wang, F., and Macphee, D.E. (2018). Photocatalyst efficiencies in concrete technology: the effect of photocatalyst placement. *Applied Catalysis B: Environment and Energy* 222: 200–208.

37 Ângelo, J., Andrade, L., and Mendes, A. (2014). Highly active photocatalytic paint for NO_x abatement under real-outdoor conditions. *Applied Catalysis A: General* 484: 17–25.

38 Wang, D., Leng, Z., Hüben, M., Oeser, M., and Steinauer, B. (2016). Photocatalytic pavements with epoxy-bonded TiO_2-containing spreading material. *Construction and Building Materials* 107: 44–51.

39 Wang, D., Leng, Z., Yu, H. et al. (2017). Durability of epoxy-bonded TiO_2-modified aggregate as a photocatalytic coating layer for asphalt pavement under vehicle tire polishing. *Wear* 382–383: 1–7.

40 Maggos, T., Bartzis, J.G., Leva, P., and Kotzias, D. (2007). Application of photocatalytic technology for NO_x removal. *Applied Physics A: Materials Science and Processing* 89: 81–84.

41 Sugrañez, R., Álvarez, J. I., Cruz-Yusta, M. et al. (2013). Enhanced photocatalytic degradation of NO_x gases by regulating the microstructure of mortar cement modified with titanium dioxide. *Building and Environment* 69: 55–63.

42 Guo, M.Z., Ling, T.C., and Poon, C.S. (2017). Photocatalytic NO_x degradation of concrete surface layers intermixed and spray-coated with nano-TiO_2: influence of experimental factors. *Cement and Concrete Composites* 83: 279–289.

43 Yang, L., Hakki, A., Wang, F., and Macphee, D.E. (2017). Photocatalyst efficiencies in concrete technology: the effect of photocatalyst placement. *Applied Catalysis B: Environment and Energy* 222: 200–208.

44 Chen, M. and Chu, J.W. (2011). NO_x photocatalytic degradation on active concrete road surface – from experiment to real-scale application. *Journal of Cleaner Production* 19: 1266–1272.

45 De Melo, J.V.S., Trichês, G., Gleize, P.J.P., and Villena, J. (2012). Development and evaluation of the efficiency of photocatalytic pavement blocks in the laboratory and after one year in the field. *Construction and Building Materials* 37: 310–319.

46 Águia, C., Ângelo, J., Madeira, L.M., and Mendes, A. (2011). Photo-oxidation of NO using an exterior paint – screening of various commercial titania in powder pressed and paint films. *Journal of Environmental Management* 92: 1724–1732.

47 Guerrini, G.L. (2012). Photocatalytic performances in a city tunnel in Rome: NOx monitoring results. Construction and Building Materials 27: 165–175.

48 Gallus, M., Akylas, V., Barmpas, F. et al. (2015). Photocatalytic de-pollution in the Leopold II tunnel in Brussels: NO_x abatement results. *Building and Environment* 84: 125–133.

49 Kerrod, J. and McIntyre, R. (2013). The effectiveness of CristalACTiV^{TM} for depollution in tunnels with low levels of light. https://www.cristalactiv.com/wp-content/uploads/2019/04/Konningstunnel-Trial.pdf.

50 Barratt, B. (2007). Statistical analysis of monitoring results from the city of London's NO_x – reducing paint study. https://www.londonair.org.uk/london/reports/CoL_de-nox_paint_statistical_report.pdf.

51 Maggos, T., Plassais, A., Bartzis, J. G. et al. (2007). Photocatalytic degradation of NO_x in a pilot street canyon configuration using TiO_2-mortar panels. *Environmental Monitoring and Assessment* 136: 35–44.

52 Maggos, T., Bartzis, J.G., Liakou, M., and Gobin, C. (2007). Photocatalytic degradation of NO_x gases using TiO_2-containing paint: a real scale study. *Journal of Hazardous Materials* 146: 668–673.

53 Wahlstedt, B. (2010). Impact of photocatalytic paint on NO_2 levels in a heavily polluted area in Metro Manila. https://ecoevolution.org/wp-content/uploads/2020/08/Manila-Observatory-Ateneo-University-of-Manila.pdf.

54 Boonen, E. and Beeldens, A. (2013). Photocatalytic roads: from lab tests to real scale applications. *European Transport Research Review* 5: 79–89.

55 Ballari, M.M. and Brouwers, H.J.H. (2013). Full scale demonstration of air-purifying pavement. *Journal of Hazardous Materials* 254–255: 406–414.

56 Boonen, E. and Beeldens, A. (2014). Recent photocatalytic applications for air purification in Belgium. *Coatings* 4: 553–573.

57 Folli, A., Strøm, M., Madsen, T. P. et al. (2015). Field study of air purifying paving elements containing TiO_2. *Atmospheric Environment* 107: 44–51.

58 Kafizas, A. and Rhys-Tyler, G. (2023). M1 SmogStop® air quality barrier trial. https://nationalhighways.co.uk/media/qp3fr5mg/smogstop-report-final-002.pdf.

59 SkyQuest. (2024). Global photocatalyst market. https://www.skyquestt.com/report/photocatalyst-market.

60 Roy, S. and Madras, G. (2015). Photocatalytic NO_x abatement: a short review. *Current Organic Chemistry* 19: 2122–2131.

61 Ângelo, J., Andrade, L., Madeira, L.M., and Mendes, A. (2013). An overview of photocatalysis phenomena applied to NO_x abatement. *Journal of Environmental Management* 129: 522–539.

62 Lasek, J., Yu, Y.-H., and Wu, J.C.S. (2013). Removal of NO_x by photocatalytic processes. *Journal of Photochemistry and Photobiology C: Photochemistry Reviews* 14: 29–52.

63 Le, P.T., Cao, T.M., Doan, T.C.D., and Van Pham, V. (2024). Treating NO_x gas pollution by visible light photocatalytic reaction of S-doped TiO_2 nanotubes. *Topics in Catalysis* 67: 1129–1140.

64 Van Pham, V., Bui, T. K., Nguyen, T. et al. (2024). Bifunctional photocatalysis toward efficient NO_x removal performance and water splitting activity: a case of TiO_2/g-C_3N_4. *Topics in Catalysis* 67: 1055–1065.

65 Slapničar, Š., Žerjav, G., Zavašnik, J. et al. (2024). Novel TiO_2-supported gold nanoflowers for efficient photocatalytic NO_x abatement. *Molecules* 29: 3333.

66 Ma, H., Li, D., Li, Y. et al. (2024). Atomically dispersed Mn sites on TiO_2(B) microspheres enables efficient photocatalytic abatement of NO_x. *Separation and Purification Technology* 337: 126357.

67 Ghezzo, F. (2021). Visible light performance of carbon doped nanoporous titania photocatalysts. In: *2021 IEEE International Conference on Manipulation, Manufacturing and Measurement on the Nanoscale (3M-NANO)*, 254–257. IEEE. https://doi.org/10.1109/3M-NANO49087.2021.9599797.

68 Tobaldi, D.M., Pullar, R.C., Gualtieri, A.F. et al. (2015). Nitrogen-modified nano-titania: true phase composition, microstructure and visible-light induced photocatalytic NO abatement. *Journal of Solid State Chemistry* 231: 87–100.

69 Huang, J., Ho, W., and Lee, F.S.C. (2012). Facile synthesis of visible-light-activated F-doped TiO_2 hollow spheres by ultrasonic spray pyrolysis. *Science of Advanced Materials* 4: 863–868.

70 Kalampaliki, T., Makri, S. P., Papadaki, E. et al. (2021). Visible-light active sulfur-doped titania nanoparticles immobilized on a silica matrix: synthesis, characterization and photocatalytic degradation of pollutants. *Nanomaterials* 11: 2543.

71 Wang, T., Shen, D., Xu, T., and Jiang, R. (2017). Photocatalytic degradation properties of V-doped TiO_2 to automobile exhaust. *Science of the Total Environment* 586: 347–354.

72 Binas, V.D., Sambani, K., Maggos, T. et al. (2012). Synthesis and photocatalytic activity of Mn-doped TiO_2 nanostructured powders under UV and visible light. *Applied Catalysis B: Environment and Energy* 113–114: 79–86.

73 Bengtsson, N., Castellote, M., López-Muñoz, M.J., and Cerro, L. (2009). Preparation of Co-doped TiO_2 for photocatalytic degradation of NO_x in air under visible light. *Journal of Advanced Oxidation Technologies* 12: 55–64.

74 Ma, J., He, H., and Liu, F. (2015). Effect of Fe on the photocatalytic removal of NO over visible light responsive Fe/TiO_2 catalysts. *Applied Catalysis B: Environment and Energy* 179: 21–28.

75 Martinez-Oviedo, A., Ray, S.K., Nguyen, H.P., and Lee, S.W. (2019). Efficient photo-oxidation of NO_x by Sn doped blue TiO_2 nanoparticles. *Journal of Photochemistry and Photobiology A: Chemistry* 370: 18–25.

76 Ho, C.-C., Kang, F., Chang, G.-M. et al. (2019). Application of recycled lanthanum-doped TiO_2 immobilized on commercial air filter for visible-light photocatalytic degradation of acetone and NO. *Applied Surface Science* 465: 31–40.

77 Khan, H., Gar Alalm, M., Lalonde-Lavoie, M. et al. (2021). Photocatalytic degradation of NO_x and ethanol in the gas phase by spray dried Ce-TiO_2. *Journal of Environmental Chemical Engineering* 9: 106813.

78 Wu, X., Yin, S., Dong, Q. et al. (2014). Photocatalytic performance and near infrared absorption property of tungsten and tungsten–carbon doped titania. *Materials Technology* 29: A20–A27.

79 Todorova, N., Vaimakis, T., Petrakis, D. et al. (2013). N and N,S-doped TiO_2 photocatalysts and their activity in NO_x oxidation. *Catalysis Today* 209: 41–46.

80 Zhou, L., Tan, X., Zhao, L., and Sun, M. (2007). Photocatalytic degradation of NO_x over platinum and nitrogen codoped titanium dioxide under visible light irradiation. *Collection of Czechoslovak Chemical Communications* 72: 379–391.

81 Pérez-Nicolás, M., Navarro-Blasco, I., Fernández, J.M., and Alvarez, J.I. (2017). Atmospheric NO_x removal: study of cement mortars with iron- and vanadium-doped TiO_2 as visible light–sensitive photocatalysts. *Construction and Building Materials* 149: 257–271.

82 Zollo, A., Livraghi, S., Giamello, E. et al. (2023). C-N co-doped titanium dioxide. Key aspects in the assessment of the air pollutants abatement capability. *Journal of Environmental Chemical Engineering* 11: 109451.

83 Komatsuda, S., Asakura, Y., Vequizo, J.J.M. et al. (2018). Enhanced photocatalytic NO decomposition of visible-light responsive F-TiO_2/(N,C)-TiO_2 by charge transfer between F-TiO_2 and (N,C)-TiO_2 through their doping levels. *Applied Catalysis B: Environment and Energy* 238: 358–364.

84 Zhang, P. Yu, Y., Wang, E. et al. (2014). Structure of nitrogen and zirconium co-doped titania with enhanced visible-light photocatalytic activity. *ACS Applied Materials & Interfaces* 6: 4622–4629.

85 Zhang, P., Liu, B., Yin, S. et al. (2009). Rapid synthesis of nitrogen doped titania with mixed crystal lattice via microwave-assisted hydrothermal method. *Materials Chemistry and Physics* 116: 269–272.

86 Katsanaki, A.V., Kontos, A. G., Maggos, T. et al. (2013). Photocatalytic oxidation of nitrogen oxides on N-F-doped titania thin films. *Applied Catalysis B: Environment and Energy* 140–141: 619–625.

87 Cheng, G., Liu, X., Song, X. et al. (2020). Visible-light-driven deep oxidation of NO over Fe doped TiO_2 catalyst: synergic effect of Fe and oxygen vacancies. *Applied Catalysis B: Environment and Energy* 277: 119196.

88 Boyjoo, Y., Sun, H., Liu, J., Pareek, V.K., and Wang, S. (2017). A review on photocatalysis for air treatment: from catalyst development to reactor design. *Chemical Engineering Journal* 310: 537–559.

89 Gopalan, A.-I., Lee, J.-C., Saianand, G. et al. (2020). Recent progress in the abatement of hazardous pollutants using photocatalytic TiO_2-based building materials. *Nanomaterials* 10: 1854.
90 Giannakopoulou, T., Papailias, I., Todorova, N. et al. (2017). Tailoring the energy band gap and edges' potentials of g-C_3N_4/TiO_2 composite photocatalysts for NO_x removal. *Chemical Engineering Journal* 310: 571–580.
91 Kim, S.K., Chang, H., Cho, K. et al. (2011). Enhanced photocatalytic property of nanoporous TiO_2/SiO_2 micro-particles prepared by aerosol assisted co-assembly of nanoparticles. *Materials Letters* 65: 3330–3332.
92 Li, Y., Yi, H., Tang, X. et al. (2016). Study on the performance of simultaneous desulfurization and denitrification of Fe_3O_4-TiO_2 composites. *Chemical Engineering Journal* 304: 89–97.
93 Zhao, S., Fu, M., Li, Y. et al. (2021). Facile hydrothermal preparation of a $ZnFe_2O_4$/TiO_2 heterojunction for NO_x removal. *Molecular Catalysis* 507: 111570.
94 Todorova, N., Giannakopoulou, T., Pomoni, K. et al. (2015). Photocatalytic NO_x oxidation over modified ZnO/TiO_2 thin films. *Catalysis Today* 252: 41–46.
95 Huy, T.H., Bui, D.P., Kang, F. et al. (2019). SnO_2/TiO_2 nanotube heterojunction: the first investigation of NO degradation by visible light-driven photocatalysis. *Chemosphere* 215: 323–332.
96 Wu, X., Yin, S., Dong, Q., and Sato, T. (2013). Preparation and visible light induced photocatalytic activity of C-$NaTaO_3$ and C-$NaTaO_3$–Cl-TiO_2 composite. *Physical Chemistry Chemical Physics* 15: 20633.
97 Mendoza, J.A., Lee, D.H., and Kang, J.-H. (2017). Photocatalytic removal of gaseous nitrogen oxides using WO_3/TiO_2 particles under visible light irradiation: effect of surface modification. *Chemosphere* 182: 539–546.
98 Li, Q., Hu, J., Wang, H., and Wu, Z. (2021). In situ synthesis of an advanced Z-scheme Bi/BiOI/black TiO_2 heterojunction photocatalysts for efficient visible-light-driven NO purification. *Applied Surface Science* 562: 150250.
99 Yuan, R., Wang, M., Liao, L. et al. (2023). 100% N_2O inhibition in photocatalytic NO_x reduction by carbon particles over Bi_2WO_6/TiO_2 Z-scheme heterojunctions. *Chemical Engineering Journal* 453: 139892.
100 Balbuena, J., Carraro, G., Cruz, M. et al. (2016). Advances in photocatalytic NO_x abatement through the use of Fe_2O_3/TiO_2 nanocomposites. *RSC Advances* 6: 74878–74885.
101 Ma, J., Wang, C., and He, H. (2016). Enhanced photocatalytic oxidation of NO over g-C_3N_4-TiO_2 under UV and visible light. *Applied Catalysis B: Environment and Energy* 184: 28–34.
102 Trapalis, A., Todorova, N., Giannakopoulou, T. et al. (2016). TiO_2/graphene composite photocatalysts for NO_x removal: a comparison of surfactant-stabilized graphene and reduced graphene oxide. *Applied Catalysis B: Environment and Energy* 180: 637–647.
103 Hu, X., Wang, Y., Ling, Z. et al. (2021). Ternary g-C_3N_4/TiO_2/Ti_3C_2 MXene S-scheme heterojunction photocatalysts for NO_x removal under visible light. *Applied Surface Science* 556: 149817.
104 Todorova, N., Giannakopoulou, T., Karapati, S. et al. (2014). Composite TiO_2/clays materials for photocatalytic NO_x oxidation. *Applied Surface Science* 319: 113–120.

105 Mitsionis, A., Vaimakis, T., Trapalis, C. et al. (2011). Hydroxyapatite/titanium dioxide nanocomposites for controlled photocatalytic NO oxidation. *Applied Catalysis B: Environment and Energy* 106: 398–404.

106 Nikolopoulou, A., Papoulis, D., Komarneni, S. et al. (2009). Solvothermal preparation of TiO_2/saponite nanocomposites and photocatalytic activity. *Applied Clay Science* 46: 363–368.

107 Hashimoto, K., Wasada, K., Osaki, M. et al. (2001). Photocatalytic oxidation of nitrogen oxide over titania–zeolite composite catalyst to remove nitrogen oxides in the atmosphere. *Applied Catalysis B: Environment and Energy* 30: 429–436.

108 Truppi, A., Petronella, F., Placido, T. et al. (2017). Visible-light-active TiO_2-based hybrid nanocatalysts for environmental applications. *Catalysts* 7: 100.

109 Nguyen, N.H. and Bai, H. (2014). Photocatalytic removal of NO and NO_2 using titania nanotubes synthesized by hydrothermal method. *Journal of Environmental Sciences* 26: 1180–1187.

110 Li, F., Liu, G., Liu, F., and Yang, S. (2023). A WO_3–TiO_2 nanorod/$CaCO_3$ photocatalyst with degradation-regeneration double sites for NO_2-inhibited and durable photocatalytic NO. *Chemosphere* 324: 138277.

111 Bahşi, E., Şahin, O., İlcan, H. et al. (2022). Role of inclusion size distribution of titanium dioxide on the nitrogen oxides reduction capability and microstructural characteristics of cementitious systems. *Construction and Building Materials* 318: 125992.

112 Toma, F.-L., Bertrand, G., Chwa, S. et al. (2006). Microstructure and photocatalytic properties of nanostructured TiO_2 and TiO_2–Al coatings elaborated by HVOF spraying for the nitrogen oxides removal. *Materials Science and Engineering: A* 417: 56–62.

113 Erdogan, D.A. and Ozensoy, E. (2017). Hierarchical synthesis of corrugated photocatalytic TiO_2 microsphere architectures on natural pollen surfaces. *Applied Surface Science* 403: 159–167.

114 Huang, K.-C. and Chien, S.-H. (2013). Improved visible-light-driven photocatalytic activity of rutile/titania-nanotube composites prepared by microwave-assisted hydrothermal process. *Applied Catalysis B: Environment and Energy* 140–141: 283–288.

115 Takata, T., Jiang, J., Sakata, Y. et al. (2020). Photocatalytic water splitting with a quantum efficiency of almost unity. *Nature* 581: 411–414.

116 Ma, Y., Mesa, C.A., Pastor, E. et al. (2016). Rate law analysis of water oxidation and hole scavenging on a $BiVO_4$ photoanode. *ACS Energy Letters* 1: 618–623.

117 Corby, S., Rao, R.R., Steier, L., and Durrant, J.R. (2021). The kinetics of metal oxide photoanodes from charge generation to catalysis. *Nature Reviews Materials* 6: 1136–1155.

118 Kusiak-Nejman, E., Czyżewski, A., Wanag, A. et al. (2020). Photocatalytic oxidation of nitric oxide over AgNPs/TiO_2-loaded carbon fiber cloths. *Journal of Environmental Management* 262: 110343.

119 Luna, M., Gatica, J.M., Vidal, H., and Mosquera, M.J. (2019). Au-TiO_2/SiO_2 photocatalysts with NO_x depolluting activity: influence of gold particle size and loading. *Chemical Engineering Journal* 368: 417–427.

120 Sheng, Z., Song, S., Wang, H., Wu, Z., and Liu, Y. (2012). One-step hydrothermal synthesis of Pd-modified TiO_2 with high photocatalytic activity for nitric oxide oxidation in gas phase. *Environmental Engineering Science* 29: 972–978.

121 Shu, Y., Ji, J., Zhou, M. et al. (2022). Selective photocatalytic oxidation of gaseous ammonia at ppb level over Pt and F modified TiO_2. *Applied Catalysis B: Environment and Energy* 300: 120688.

122 Skalska, K., Malankowska, A., Balcerzak, J. et al. (2022). NO_x photooxidation over different noble metals modified TiO_2. *Catalysts* 12: 857.

123 Majrik, K., Pászti, Z., Korecz, L. et al. (2018). Study of PtO_x/TiO_2 photocatalysts in the photocatalytic reforming of glycerol: the role of co-catalyst formation. *Materials* 11: 1927.

124 Morcali, M.H. (2020). A new approach to recover platinum-group metals from spent catalytic converters via iron matte. *Resources, Conservation and Recycling* 159: 104891.

125 Fujiwara, K. and Pratsinis, S.E. (2018). Single Pd atoms on TiO_2 dominate photocatalytic NO_x removal. *Applied Catalysis B: Environment and Energy* 226: 127–134.

126 Wang, H., Wu, Z., Liu, Y., and Wang, Y. (2009). Influences of various Pt dopants over surface platinized TiO_2 on the photocatalytic oxidation of nitric oxide. *Chemosphere* 74: 773–778.

127 Wu, Q. and van de Krol, R. (2012). Selective photoreduction of nitric oxide to nitrogen by nanostructured TiO2 photocatalysts: role of oxygen vacancies and iron dopant. Journal of the American Chemical Society 134: 9369–9375.

16

Recycling of Heavy Metal Ions from Wastewater by TiO_2 Photocatalysis

Liqiang Yao[1], *Chongchong She*[1], *Kun Chen*[1,*], *Jia Hong Pan*[2,*]

[1] *School of Materials Science and Engineering, Beijing Institute of Technology, Beijing, China*
[2] *State Key Laboratory of Featured Metal Materials and Life-Cycle Safety for Composite Structures, School of Resources, Environments and Materials, Guangxi University, Nanning, China*

*Corresponding authors: kchen@bit.edu.cn; jhpan@gxu.edu.cn

16.1 Introduction

"Heavy metal" is a broad term of encompassing metals and metalloids with a density exceeding 5 g · cm^{-3} [1]. Common heavy metals include copper, cadmium, zinc, iron, manganese, chromium, arsenic, cobalt, titanium, and mercury, which are widely spread on the earth. Industrial activities, wastewater discharges, and improper waste disposal led to the release of heavy metal ions into soils, water bodies, and the atmosphere. Although some heavy metals are essential for metabolism, exceeding the tolerance limit of these metals in the human body can be harmful or even fatal [2]. Long-term exposure may also trigger chronic poisoning, affect the nerves, kidneys and respiratory system, or even induce cancer [3]. In the meanwhile, these heavy metal ions are also valuable natural resources. The reduction and recycling of heavy metals in a green way is beneficial to establish a sustainable social and economic system. However, heavy metals always exhibited the excellent stability and infinite lifetime, thus they do not only disrupt ecological and geological equilibrium but also incur substantial economic losses [4].

Adsorption, ion exchange, extraction, photocatalytic treatment and electrochemical methods [5–9] are widely adopted to remove and recover heavy metal in water. The adsorption process is a conventional method, which is applicable to a wide pH range, easy operating, and cost-effective. However, a potential drawback is the detachment of the adsorbed metal ions from the absorbents owing to the weak interaction [10]. The relatively higher operating costs and the need for more complex follow-up treatments [11] limit the application of the ion exchange, extraction, and electrochemical methods. In contrast, photocatalytic treatment based on the semiconductors offers an energy-saving and environmentally friendly

Nanostructured TiO_2: Synthesis, Photocatalytic Properties and Applications, First Edition.
Edited by Jia Hong Pan, Wan In Lee and Detlef W. Bahnemann

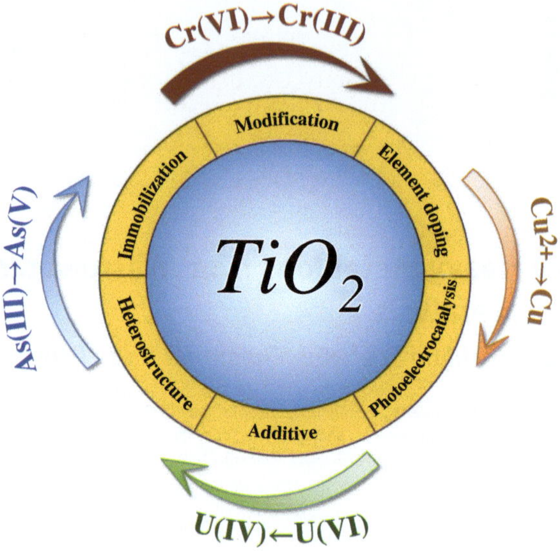

Figure 16.1 Photocatalytic treatment of heavy metals by TiO_2 photocatalysis.

approach to treat heavy metals in water with the minimal secondary pollution [12]. Undoubtedly, TiO_2 is the most popular photocatalysts for the photocatalytic treatment of heavy metals owing to its notable advantages of high chemical stability, affordability, and nontoxic nature [13]. Considerable efforts have been devoted to enhancing the photocatalytic performance of TiO_2 on the efficient reduction and recovery of heavy metals [14–16]. In this chapter, we will address the current advances in photocatalytic treatment of Cu(II), Cr(VI), U(VI), and As by TiO_2, as summarized in Figure 16.1.

16.2 Mechanism of Photocatalytic Reduction Recovery

Photocatalytic treatment of heavy metals is generally based on the reduction of the metal ions by the virtue of the photogenerated electrons [17]. During the photocatalysis, the photogenerated electrons reduced the metal ions and resulted into the simultaneous deposition of metal when the reduction potentials of the metal ions are lower than the semiconductor conduction band (CB) edges [18]. Meanwhile, the holes can oxidize the species on the surface. Thus, suppressing the recombination of the photogenerated electrons and holes is vital for the photocatalytic recovery of the heavy metals [19]. In the actual reaction process, the recovery of heavy metals can be summarized in the following four steps:

1. The adsorption of heavy metal ions on the surface of TiO_2. The hydroxy groups spread over all the surface of TiO_2, which shows the affinity to the heavy metal ions in water. Thus, adding more adsorbents with higher specific surface area is favorite for improving the adsorption performance of TiO_2 [20–22].

2. The separation of the photogenerated electron–hole pair under light irradiation. Under the light illumination, the electrons in the filled valence band (VB) are excited into the empty CB, leaving holes in the VB. The value of E_g and the positions of the band edge of the CB and the VB determine the light-absorption coefficiency, and thus the photocatalytic performance of TiO_2 [23]. Thermodynamically, the photocatalytic reduction occurred only when the reduction potential of the metal ions is lower than the band edges of the CB. In terms of kinetic, the photogenerated electron should reduce the metal ions rather than recombine with the hole. However, since the energy bands are non-continuous and the bandgap is wide, the photogenerated electron–hole pairs easily recombined, and thus these recombined pairs are nullified and the absorbed energy is released in the form of heat or light energy. To suppress the recombination process, element doping [24] and constructing the heterojunctions [25] are adopted to improve the photocatalytic performance of TiO_2.
3. The reduction of heavy metal ions by photogenerated electrons. The excited electrons may reduce the absorbed metal ions on the catalyst surface. In the meanwhile, it may also react with the dissolved oxygen or recombine with the holes. Thus, deoxygenation or introducing the hole scavenger should be beneficial for the reduction of the metal ions by promoting the separation of the electron–hole pair [26–28].
4. Elution of the adsorbed metals. To recover the heavy metal and regenerate the photocatalysts, it is necessary to desorb the heavy metals from the catalysts. Acid washing [29], alkali washing [30], and chelating agent were employed to recover the metal ions [31], and damage to the photocatalyst should be avoided to preserve the functionality during the elution process [29, 31].

16.3 Photocatalytic Treatment of Cu(II), Cr(VI), U(VI), and As(III/IV) by TiO_2

In photocatalytic treatment of heavy metal, numerous studies have highlighted TiO_2 as a promising photocatalyst [32–35]. However, the wide bandgap (3.0–3.2 eV) limits its utilization to UV radiation only, and the rapid recombination of photogenerated electron–hole pairs hampers the catalytic efficiency of TiO_2. To improve the photocatalytic performance of TiO_2, a series of measurement of doping elements [24, 36–39], constructing heterojunction [25, 40–42] and introducing functional additives [22, 43, 44] have been adopted to modify TiO_2 (Figure 16.2) [30, 34, 45]. In the following, the photocatalytic reduction/oxidation recovery strategies based on TiO_2 will be presented as an example for Cu(II), Cr(VI), U(VI), and As(III/IV), respectively.

16.3.1 Copper

Cu(II) is one of the most prevalent heavy metals mainly existing in the shallow natural waters. The contaminated water bodies by Cu(II) and its complexes has emerged as a significant global environmental concern. Ingesting excessive amounts

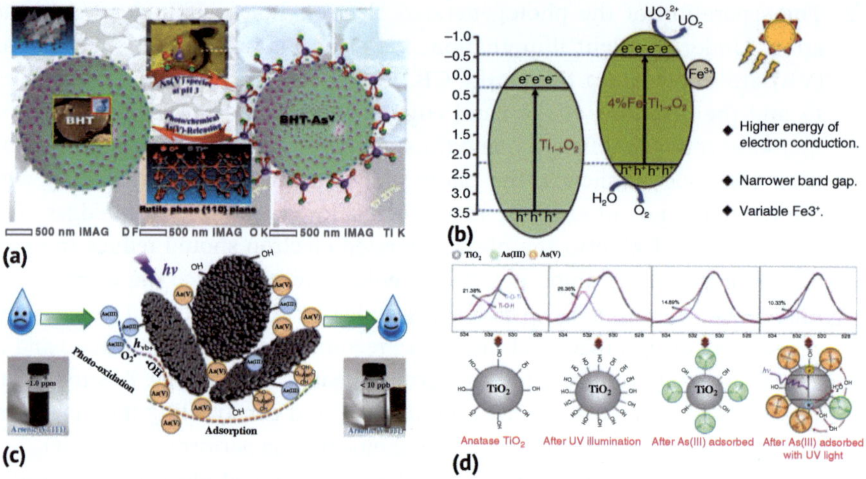

Figure 16.2 (a) Three-dimensional whole of modified mesoporous TiO_2 microspheres. [30]/with permission of American Chemical Society. (b) The Ti vacancies and doped Fe atoms were simultaneously introduced into TiO_2 nanosheet. [37]/with permission of Elsevier. (c) A novel and uniform sandwich-like magnetic $Fe_3O_4@TiO_2$ sheets heterojunction. [40]/with permission of Elsevier. (d) The number of hydroxyl groups on the anatase TiO_2 increases upon UV illumination. [22]/with permission of Elsevier.

of Cu(II) can lead to symptoms such as epigastric pain, nausea, vomiting, diarrhea, and in severe cases, even death [14, 46].

16.3.1.1 Adding Functional Additives

In order to reduce the harm to the environment and achieve the recycling of Cu resources, Satyro et al. [27] directly used TiO_2 to photocatalytically oxidize *N*,*N*-ethylenediamine-disuccinic acid (EEDS) and reduce Cu(II) in the acidic solution, and the conversion of EEDS and Cu(II) was close to 100% with a small degree of mineralization. Considering the presence of the dissolved oxygen in the system, photogenerated electrons may preferentially combine with oxygen to generate superoxide radicals, and thus the reaction needs to be carried out under anaerobic conditions (Figure 16.3). When the loading of TiO_2 exceeds 500 mg·L^{-1}, the increasing trend of reactivity tends to decrease, which should be attributed to the light scattering effect of TiO_2 particles [47]. Yeber et al. investigated the effect of the operating conditions on the photocatalytic reduction by using the response surface methodology. At the optimum pH = 4.3, the highest reduction efficiency of Cu(II) ions reached 80% after treating for 60–70 min [14]. The process was divided into two parts: the adsorption dominated in the early stage of the reaction, while the reduction dominated in the later stage. This facilitated the reduction of the Cu(II); however, the surplus Cu(II) might occupy the active sites on the catalyst, thereby reducing the reaction activity.

After reduction, copper was recovered by acid elution. As shown in Table 16.1, eluting with different types of acids did not showed significant difference, suggesting that the copper recovery process is associated with the presence of H^+ rather than the specific type of the anions [15].

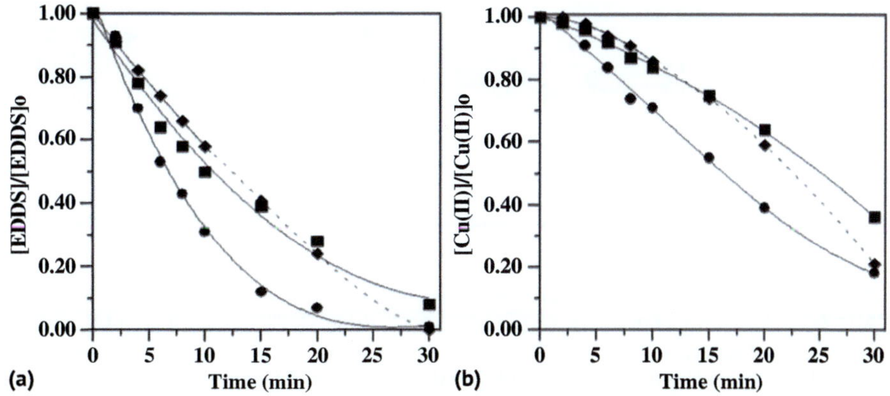

Figure 16.3 Effect of initial pH value on the photodegradation of EDDS (a) and the photoreduction of Cu(II) (b). [EDDS] = 0.2 mM, $[CuSO_4]$ = 1.0 mM, TiO_2 (pure anatase) 50 mg · L^{-1}, purging gas: N_2· Hg vapor lamp. (●) pH 2.0, (■) pH 3.0, (♦) pH 3.8. T: 25 °C. [27]/with permission of Elsevier.

Table 16.1 The recovery of copper by different types of acid and washing time. [15]/with permission of Springer Nature.

Acid used for regeneration	Time (h)	Initial concentration (mg · L^{-1})	Concentration after recovery (mg · L^{-1})	Recovery rate (R%)
1 mol · L^{-1} HNO_3	4	113.86	2029.00	89.10
1 mol · L^{-1} HCI		113.86	2075.00	91.12
0.5 mol · L^{-1} H_2SO_4		113.86	2017.00	88.57
1 mol · L^{-1} HNO_3	16	113.86	2247.00	98.67
1 mol · L^{-1} HCI		113.86	2240.00	98.37
0.5 mol · L^{-1} H_2SO_4		113.86	2266.00	99.51

16.3.1.2 Modifying TiO_2 with Metal Deposition

Lee et al. successfully developed a green method to recover Cu(II) by using the synergistic integration of Cu-EDTA photocatalytic reaction with one-dimensional ultra-long and ultrathin TiO_2 nanofibers [34]. Under the UV irradiation, electrons and holes are generated in the CB and VB of the TiO_2 nanofibers, respectively. The holes oxidized the hydroxyl groups on the surface of the photocatalysts, producing a strong oxidant in the form of reactive oxygen species $(\cdot O^{2-})$, which removed the Cu^{2+} from the Cu-EDTA complex. Subsequently, Cu^{2+} is adsorbed onto the TiO_2 nanofibers and reduced to Cu(0) by the photogenerated electrons (Figure 16.4a). The advantage of this method is that the TiO_2 nanofibers can be recycled into TiO_2–CuO composites after a simple heat treatment without additional chemicals,

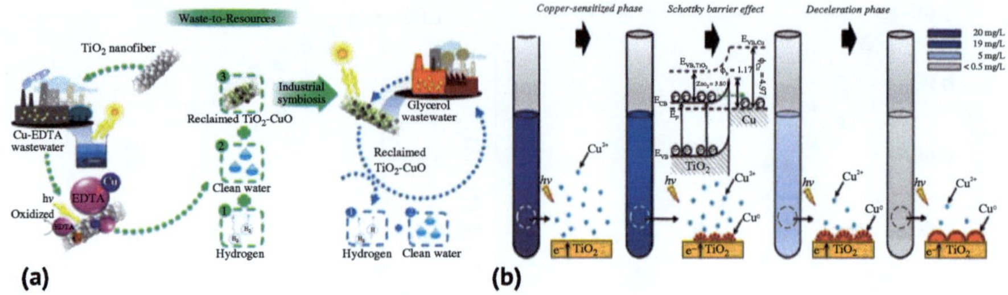

Figure 16.4 (a) Schematic diagram behind the cogeneration of H_2, clean water, and R-TC from the photocatalytic treatment of Cu-EDTA using 1D TiO_2 nanofibers. [34]/with permission of American Chemical Society. (b) Photodeposition process of Cu^{2+} with TiO_2 under UV light and acidic conditions. [48]/with permission of Elsevier.

which can subsequently be reused for photocatalytic treatment of other wastewaters. Similarly, Luo et al. utilized TiO_2 photocatalysts for photodeposition of Cu(II) from Cu-containing dilution [48]. The process of photodeposition of Cu(II) by TiO_2 under UV light can be divided into three stages: (i) The stage of the gradual enhancement of the photocatalytic performance. In this stage, the dissolved Cu(II) undergoes adsorption and is reduced to Cu(0), which is then establish Schottky junctions with TiO_2, promotes the separation of electron–hole pairs and thus enhances the photocatalytic activity. This enhancement results in an accelerating performance until nearly all active sites are covered by Cu(0) nuclei, ultimately leading to the photoreaction reaching its maximum rate [49, 50]. (ii) The stage of the growth of Cu(0) nuclei. The photogenerated electrons slightly accumulated on the surface of Cu, which in turn attracts Cu(II) ions and induces successive photoreduction. (iii) The deceleration rate of the epitaxial reaction, which is attributed to the slow mass transfer of the decreased concentration of Cu(II) (Figure 16.4b). The generated Cu(0)/TiO_2 composite exhibited the catalytic activity in the reverse water–gas shift [48].

In order to further improve the photocatalytic activity of TiO_2, Xu et al. successfully synthesized one-dimensional mesoporous TiO_2 nanotubes (TiNT) by hydrothermal calcination and applied them to photocatalytic removal of Cu(II) and reduced the production of H_2[45]. Once the reduction of Cu^{2+} begins, the Cu species deposited on TiNT act as electron reservoirs, preventing the recombination of electron holes in TiO_2 and providing reduction sites for the formation of H_2. Although a small amount of Cu is oxidized back to Cu^{2+} by holes again, TiNT photocatalytic reduction still has a great potential for Cu(II)-containing wastewater treatment due to the small concentration of Cu(II) remaining in the cycle and the fast rate of Cu^{2+} elimination.

16.3.1.3 Combing with Electrochemistry

Since heavy metal ions are mostly present as complexes in water, it is difficult to degrade them using single component TiO_2 catalysts [51]. The combination of photocatalysis and electrochemistry is more effective in removing pollutants than either photocatalytic or electrochemical process alone. As shown in Figure 16.5, in the photoelectrochemical cell, a photocatalyst immobilized on a conductive substrate serves as the anode, while an external voltage is applied to drive light-generated

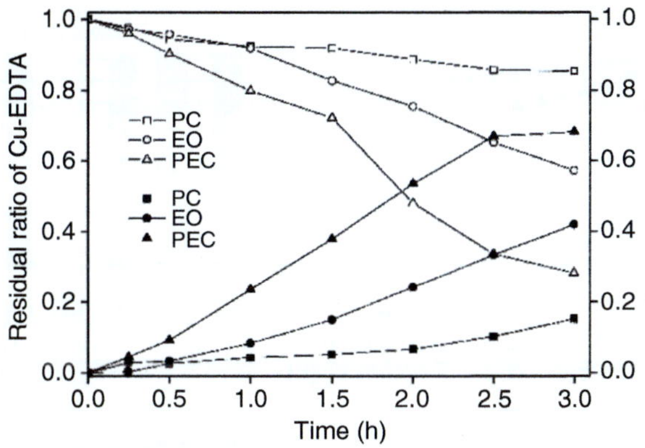

Figure 16.5 Removal of Cu–EDTA and recovery of Cu^{2+} ions via photocatalysis (PC), electrooxidation (EO), and PEC processes ($C_{(Cu-EDTA)}$, 0.05 mM; pH, 6.6; and current density, 0.5 mA cm^{-2}). [51]/with permission of American Chemical Society.

electrons toward the cathode. This setup minimizes the rate of electron–hole recombination. Moreover, photoelectrons on the cathode induce the reductive deposition of metal ions.

Byrne et al. placed TiO_2 electrodes in the anode chamber of a two-compartment photoelectrochemical cell and short-circuited them to a copper wire counter electrode in the cathode chamber, which was filled with oxalic acid solution and the cathode chamber was filled with copper sulfate solution [52]. The TiO_2 anode was irradiated with light, leading to the photo-oxidation of oxalates. Subsequently, the photogenerated electrons were directed through an external circuit to the counter electrode, where Cu^{2+} was reduced. Under nonoptimized conditions, 97% of the copper in the solution was recovered after illumination for 120 min. Such a high reduction rate depends largely on the constant current mode of the system, which not only promotes the movement of electrons, but also inhibits the reaction with dissolved oxygen to a certain extent [53, 54]. In addition, flushing the aqueous solution of metal ions with the inert gas to discharge the oxygen in the solution is also an effective method of improving the separation of electron–hole pairs. Wang et al. used platinum as the cathode, coated a layer of nano TiO_2 film on the photoanode of the cell, and the film was covered with a block copolymer (poly(ethylene glycol)-*b*-polypropylene glycol-*b*-poly(ethylene glycol)) cavity scavenger. After the deoxygenation using N_2, the Cu(II) was completely removed within 160 min. Finally, the reduced metal on the cathode was recovered by mechanical scraping (Figure 16.6) [55].

There are many factors of affecting the photoelectrocatalytic efficiency, such as cathode material, electrolyte, current density, and interfering ions [56–58]. The stability, electrolyte compatibility, catalytic activity, and conductivity of the cathodes should be considered when they are adopted. Other common cathode materials include graphite [59], stainless steel [57], and large-area Ti mesh [60]; the electrolyte indirectly affects the reduction effect by influencing the electron migration rate; the high current not only improves the electron migration rate in the solution, but also accelerates the reduction rate of Cu(II) at the cathode. Consequently, the overall performance of the photoelectrocatalytic system depends on the synergistic

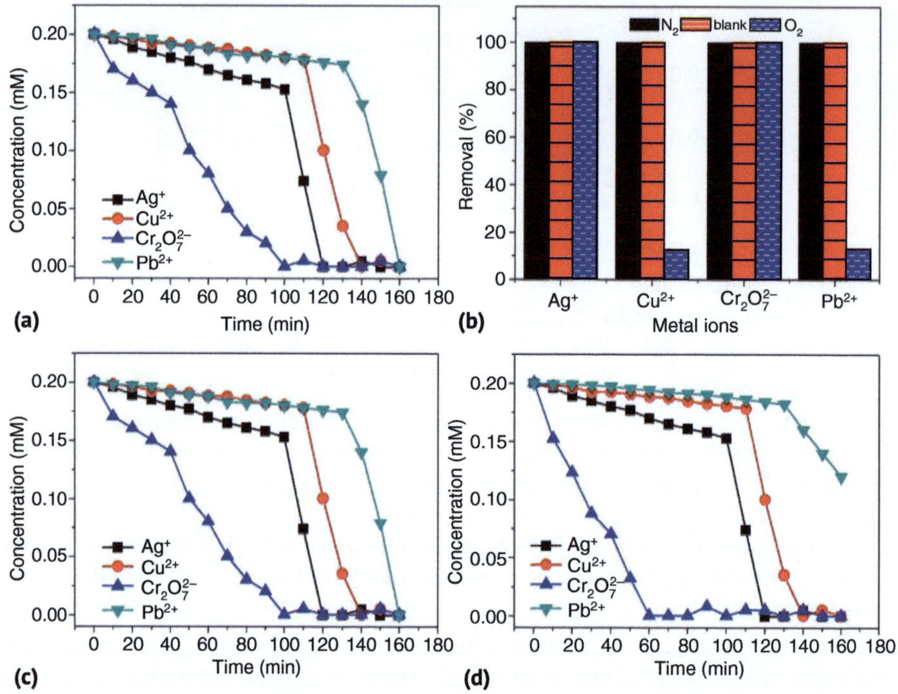

Figure 16.6 The removal of different ions under: (a) the blank experiment, Ag^+, Cu^{2+}, and Pb^{2+} were added as their nitrate salts while the experiment was conducted with no injection of N_2 or O_2 and no control of pH (3.0); (b) the effect of N_2 and O_2 on the removal of these four metal ions while other conditions were the same as in the blank experiment, the data was recorded at 160 min; (c) the pH was tuned to be 2.0 by adding HNO_3 while the other conditions were the same as in the blank experiment; (d) metal ions were added as sulfate salts, the pH was tuned to 2.0 by adding H_2SO_4, and other conditions were the same as in the blank experiment. [55]/with permission of Elsevier.

effect of the cathode material and other components, and the effects of each part on the experimental results need to be fully considered before the experiment.

16.3.2 Chromium

Chromium is widely used in pigments, catalysts, metallurgy, leather tanning industry, and electroplating process. Cr(VI) and Cr(III) are the two stable forms of chromium. Cr(VI) is a nondegradable and toxic metal ion. It exhibits high mobility in water and is carcinogenic, mutagenic, and teratogenic to organisms. In contrast, Cr(III) shows low mobility, is nontoxic, and plays a role in carbohydrate metabolism. Therefore, the reduction of Cr(VI) to Cr(III) is the predominant method for treating chromium-containing solutions [39, 61].

16.3.2.1 Adding Functional Additives

Chen et al. added diethylene glycol (DEG) as a hole scavenger to the reaction system, and diethylene glycol is chemically bonded to the TiO_2 surface, where its hydroxyl group on the surface can chelate Ti atoms and form chemical bonds. This results in an internal hole scavenging effect and a high electron-releasing capacity, rendering it superior to single TiO_2 photocatalytic materials [62]. The reduced Cr(III) can be easily removed by filtration, while Cr(III) deposited on the TiO_2 surface must also be removed in order to regenerate the TiO_2 and avoid to produce toxic wastes. Djellabi et al. used tartaric acid as a hole scavenger and citric acid/EDTA as leaching agents to extract Cr(III), and the process was mainly achieved by the formation of complexes between Cr(III) and the leaching agent (Figure 16.7a) [63]. Li et al. prepared hydroxyl-modified TiO_2 in alkaline solution using a simple immersion method [64]. The presence of the protonated hydroxyl groups ($—OH_2^+$) resulted in a more positively charged surface, which similarly improved the selective adsorption of Cr(VI) and the desorption of Cr(III) (Figure 16.7b). Cho et al. significantly improved the catalytic performance by complexing TiO_2 with a surfactant containing polyoxyethylene groups (Brij) [65]. The surfactants do not only form photoactive surface complexes with TiO_2 through hydrogen bonding to promote the excited electron transfer but also enhance the concentration of the substrate within the adsorbed surfactant layer.

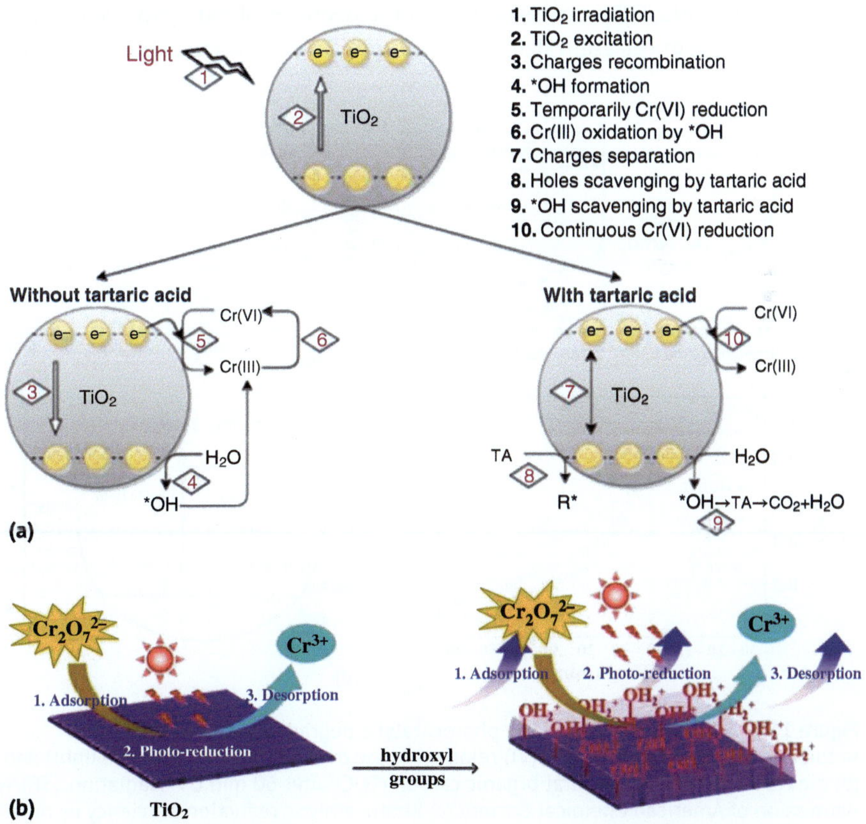

Figure 16.7 (a) Scheme reporting the main steps of tartaric acid effects on Cr(VI) photoreduction. [63]/with permission of Elsevier. (b) The process of photocatalytic reduction of Cr(VI) on TiO_2 surface. [64]/with permission of Elsevier.

16.3.2.2 Modifying TiO_2 with Metal Particles

Similar as photocatalytic treatment of Cu(II), metal particles deposition is also one effective strategy of removing Cr(VI) from the polluted water. By depositing Ag on TiO_2, Choi et al. enhanced the degradation of organic pollutants on Ag–TiO_2 under UV irradiation with the help of the successive dark reduction of Cr(VI) (Figure 16.8a, b) [38]. The degradation of pollutants was achieved through the action of holes, while photogenerated electrons were transferred to Ag for storage, and then utilized for the reduction of Cr(VI) during the subsequent dark period. Considering the great interests in utilizing the solar energy not only during the daytime but also the night time, this study is the cutting edge of the postirradiation removal of the heavy metal pollution. Pu et al. doped TiO_2 with ultra-trace Ag, and the photo-responsive range was distinctively enhanced along with the formation of the Schottky junctions at the interface after the Ag deposits [39]. Under light irradiation, photogenerated electrons tended to migrate and accumulate on Ag nanoparticles, which served as effective electron acceptors. These additional acceptors subsequently suppressed electron–hole pair recombination, ultimately enhancing the photocatalytic reduction of Cr(VI) (Figure 16.8c).

16.3.2.3 Elemental Doping TiO_2

Cu was doped into TiO_2 and resulted into the decrease of bandgap, well-dispersed TiO_2 particles, reduced grain size, increased specific surface area, and thus improved

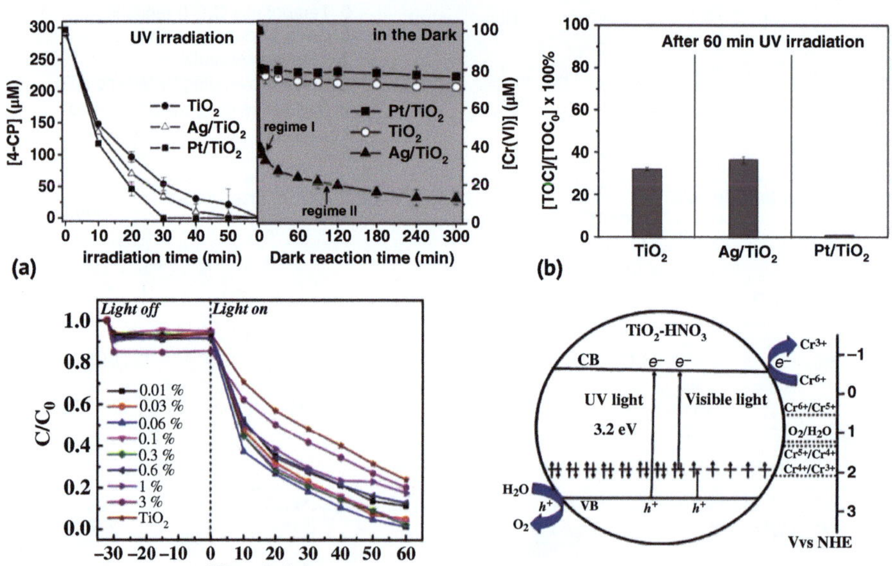

Figure 16.8 (a) Time profiles of the photocatalytic degradation of 4-CP under UV irradiation and the subsequent Cr(VI) removal in the dark (after turning off the light) and (b) the resulting removal of total organic carbon (TOC) after 60 min UV irradiation. [38]/with permission of American Chemical Society. (c) Photocatalytic reduction efficiency by pure TiO_2 and Ag@TiO_2 composites with various Ag loading ratio (effect of different loading on adsorption balance before irradiation, insert). [39]/MDPI/CC BY 4.0. (d) Schematic diagram for TiO_2-HNO_3-mediated photocatalytic reduction of Cr(VI) in water. [67]/with permission of Elsevier.

the visible light utilization. The first-order rate constant of Cr(VI) removal for Cu-TiO_2 is *ca.* 1.5 times of that for pristine TiO_2 [66]. For nonmetallic doping, Zhang et al. tried to introduce N into TiO_2, and similar to Ag doping, TiO_2 obtained smaller grain size and larger specific surface area [67]. Meanwhile, TiO_2 doped with N at interstitial positions produces bandgap states of about 0.7–0.8 eV on the VB edges (Figure 16.8d), which are responsible for the absorption of visible light [68, 69].

16.3.2.4 TiO_2-based Heterojunctions

To ameliorate the wide bandgap and suppress the recombination of photogenerated carriers, constructing heterojunctions is also a powerful strategy. Djellabi et al. synthesized TiO_2-montmorillonite (TiO_2-M) for removing Cr(VI) from water under sunlight [70]. Montmorillonite exhibits a high cation exchange capacity due to its ability of adsorbing a wide range of cations in water. After the reduction of Cr(VI) by TiO_2 particles immobilized on the montmorillonite carrier, the resulting Cr(III) is directly adsorbed by the negative charge of the montmorillonite itself. This process releases the sites on the TiO_2 and further reduces Cr(VI), while also preventing the infiltration of Cr(III) material into the solution. Zhao et al. modified the TiO_2 nanoparticles with the reduced graphene oxide (RGO). The introduced RGO increased the absorption coefficiency, extended the light absorption range and reduced the electron–hole pair recombination in TiO_2. Such enhancement significantly improved the removal efficiency of Cr(VI) (54.2% without RGO incorporation, and 86.5% with RGO incorporation) (Figure 16.9) [71]. In addition, TiO_2-based heterojunctions have performed equally well, such as TiO_2–ZrO_2 [72], TiO_2/BC/g–C_3N_4 [73] and HC/TiO_2 [74].

The main external factors of affecting the removal of Cr(VI) are the values of pH, the presence of the competing ions in solution, and the dissolved oxygen [39, 63, 75, 76]. It should be noted that the reduction of Cr(VI) favors low pH, while the neutral or alkaline conditions are beneficial for the precipitation and immobilization of

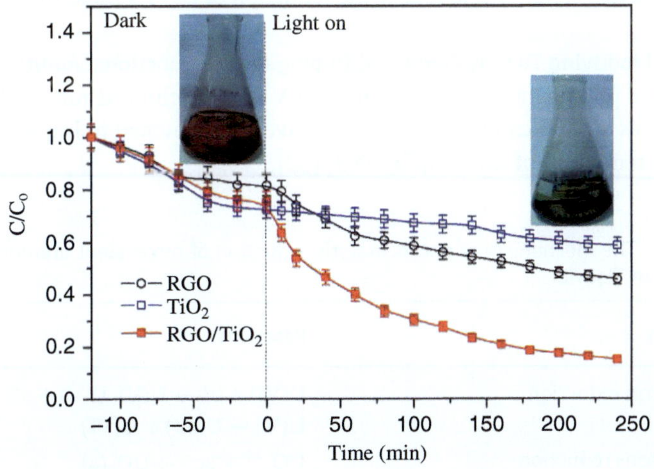

Figure 16.9 Variation of Cr(VI) concentration as a function of irradiation time (with the time of light on set as 0) under visible light irradiation with different samples. [71]/with permission of Elsevier.

Cr(III) as an oxide or hydroxide, promoting further separation; dissolved oxygen has an effect on the reduction of Cr(VI) only under alkaline conditions, and has no effect under acidic conditions.

16.3.3 Uranium

Uranium is a key element in the nuclear fuel. In one hand, proven uranium reserves on land are about 7.6 million tons, and conventional uranium reserves could be depleted within a century [77]. In the meanwhile, the total amount of uranium in the oceans is about 4.5 billion tons, which is more than 1000 times larger than the known terrestrial supply, and the extraction of uranium from seawater makes sense for the sustainable use of uranium resources [78]. On the other hand, the chemical and radiological toxicity of uranium leads to irreversible kidney damages, urinary tract disorders [79], DNA damages [80], and destruction of biomolecules [81]. Therefore, the extraction and recovery of uranium from seawater and radioactive uranium-containing wastewater is important for nuclear energy conservation and environmental sustainability.

In uranium-containing wastewater, the valence state of uranium ions is generally VI, and U(VI) usually exists in the form of uranyl ions (UO_2^{2+}), which are easy to be transported in the aquatic environment. U(IV) has a very small ionic radius of 1.05Å, rendering it insoluble in solution. It typically exists in the form of solid UO_2 or U_3O_8, which exhibits relatively high stability [82, 83]. Currently, there are three pathways for the reduction of U(VI): (i) the single-electron reduction mechanism; U(VI) is firstly reduced to U(V), and then U(V) is further reduced to U(IV). Meanwhile, under some specific condition, U(V) can automatically and disproportionately generate U(VI) and U(IV). (ii) The two-electron reduction mechanism; U(VI) is directly converted to U(IV) by a two-electron photoreduction reaction. (iii) The synergetic reduction of protons and electrons. The protons in the solution can participate in the two-electron reduction process under acidic conditions to realize the reduction of U(VI) (Table 16.2) [84].

16.3.3.1 Modifying TiO_2 by Elemental Doping and/or Functional Additives

In 1991, the photocatalytic reduction of U(VI) was achieved for the first time by using TiO_2 as a photocatalyst, and efforts have been devoted to improving the photocatalytic reduction of uranium by TiO_2 [32].

Table 16.2 The mechanism of photocatalytic reduction of hexavalent uranium. [84]/with permission of Elsevier.

Mechanism	Reactions
One-electron reduction	$U(VI) + e^- \rightarrow U(V)$, $U(V) + e^- \rightarrow U(IV)$ or $U(V) \rightarrow U(IV) + U(VI)$
Two-electrons reduction	$UO_2^{2+} + 2e^- \rightarrow UO_2(s)$
In the presence of acidity and two-electrons reduction	$UO_2^{2+} + 4H^+ + 2e^- \rightarrow U^{4+} + 2H_2O$

Gong et al. introduced Fe into the Ti vacancies of TiO_2 to fabricate a high active and stable catalyst for U(VI) photoreduction [37], and the introduction of heteroatoms helped to improve the carrier separation ratio and visible light absorption [85]. Wang et al. prepared Ti^{3+} self-doped black TiO_2 (BTn) by hydrogenation of TiO_2 using $NaBH_4$, which prolonged the light absorption, narrowed the bandgap, and reduced electron–hole combining through the introduction of Ti^{3+} and oxygen vacancies [86]. The results showed that U(VI) was photoreduced to UO_2 solid on the BTn surface.

Zhang et al. prepared TiO_2-polyacrylonitrile photocatalytic membranes for efficient uranium extraction using a phase conversion method [87]. The TPM photocatalytic membranes consisted of TiO_2, PAN, and a conductive carbon cloth. TiO_2 was the main active substance in the photoreduction process, and PAN was the film-forming substrate. The carbon cloth improves the electrical conductivity of the polymer substrate, enhances the electron transfer efficiency, and improves the photocatalytic reduction.

16.3.3.2 TiO_2-based Heterojunction

Jiang et al. prepared g-C_3N_4-TiO_2 heterojunctions using a combination of hydrothermal and calcination methods [88]. Under UV irradiation, electrons (e^-) in the VB of g-C_3N_4-TiO_2 were excited to the CB, leaving the same number of holes (h^+) in the VB, and the electrons in the CB of the former flowed to the latter because g-C_3N_4 has more negative charge than TiO_2. U(VI) was adsorbed on the surface of the heterojunction catalysts, which was then photogenerated electronically reduction to U(IV). In addition to the complexation with g-C_3N_4, Fe_3O_4 has been shown to be a good co-catalyst (Figure 16.10). Under UV irradiation, the photoactivity of TiO_2-Fe_3O_4 heterojunction was 19.3 times higher than that of pure TiO_2 [89]. This is mainly attributed to the reduction of Fe(II) and Fe(III) ions on the surface of Fe_3O_4, and produces the reduced Fe(0) and additional Fe(II), which furtherly participate in the reduction of U(VI) together with the surviving photoelectrons on TiO_2 and

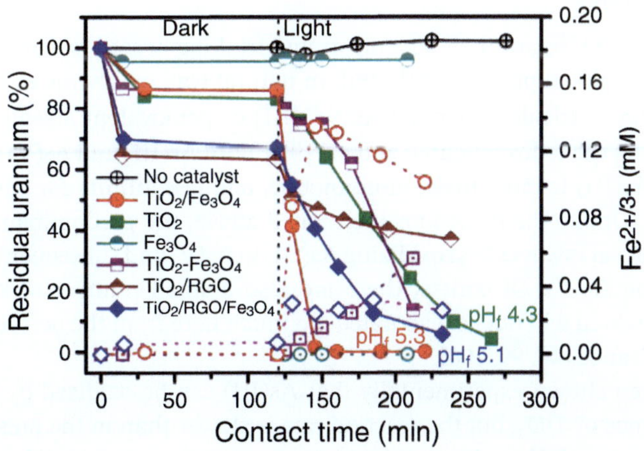

Figure 16.10 Comparison of U(VI) removal among different photocatalysts, pH values of the solutions after photocatalysis (pH$_f$) were indicated in the figure. [89]/with permission of American Chemical Society.

Fe_3O_4. On this basis, if graphene material is introduced, it may extract electron from TiO_2, thus inhibits the photodissolution of Fe_3O_4 nanoparticles and improve the stability of the catalyst.

16.3.3.3 Recovery and Elution of U(VI)

During the photocatalytic reduction of U(VI), the generated UO_2 will predominantly adsorb onto the TiO_2 surface. It is necessary to remove the adsorbed uranium from the catalyst to regenerate the TiO_2. In uranium-containing solutions, CO_3^{2-} has a strong inhibitory effect on the adsorption of U(VI) because $UO_2(CO_3)_2^{2-}$ and $UO_2(CO_3)_3^{4-}$ are the dominant substances in aqueous solutions, and the electrostatic repulsion between the anionic uranyl species and the negatively charged chalcogenides surface reduces the adsorption of U(VI) [16, 31, 90, 91]. Therefore, the deposited uranium can be recovered by adding Na_2CO_3 to the solution. In addition, U(VI) species can also be removed by adding acid [29] or ammonia [92], and electrochemical methods [77]. However, the introduced acid may cause some damage to the catalyst and electrochemical methods are relatively complicated to operate, and thus they are not recommended.

Similar to Cu(II) and Cr(VI), the factors of affecting the recovery of reduced U(IV) are none other than the property of the photocatalysts, the pH of the solution, the dissolved oxygen, and the interfering ions [28, 93, 94]. The dissolved oxygen may compete with U(VI) to receive the photogenerated electrons and also oxidize the reduced uranium species again. Moreover, the presence of proton competes with U ions for electrons and significantly decreases the rate of U(VI) reduction. Moreover, the number of the adsorption sites directly influences the probability of U(VI) contact with the photocatalyst, serving as another key factor in the photocatalytic reduction of U(VI) [95]. When the interfering ions exhibit the high binding affinity with the adsorption sites, the recovery efficiency declines.

16.3.4 Arsenic

Maximizing the removal of arsenic from drinking water is currently a major challenge for global public health [96]. In natural waters, arsenic exists as the oxyanionic form of trivalent arsenite AsO_3^{3-}(III) or pentavalent arsenate AsO_4^{3-}(V), depending on the redox potential and pH [97]. Both As(III) and As(V) are stable and toxic, but As(III) is more toxic, more mobile, and less affinity for adsorbents than As(V) [98]. Unlike the three ions mentioned above, the predominant approach to arsenic removal involves first oxidizing As(III) to As(V) and subsequently adsorbing/precipitating As(V). Of course, there are also a small number of studies on As reduction, which will not be elaborated too much here from the perspective of practical significance [44, 99, 100].

It has been shown experimentally that As(III) can be oxidized by UV radiation in the absence of TiO_2, but the reaction rate is slower than in the presence of TiO_2, limiting the practical application [101]. Yang et al. demonstrated for the first time that in the presence of UV light, TiO_2, and electron scavengers (O_2), the photogenerated hole/hydroxyl radicals in TiO_2 in a medium of $pH = 9$ would have sufficient

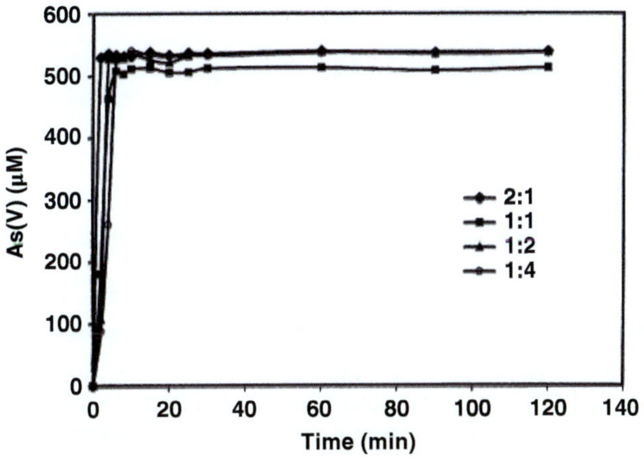

Figure 16.11 Evolution of As(V) "under UV-irradiation" with time in an air-saturated solution (pH = 9) initially containing sodium arsenite (525 µM) and dosed with H_2O_2 as shown. The numbers in the figure refer to the molar ratio of H_2O_2:As(III). [26]/with permission of Elsevier.

energy to oxidize As(III) to As(V), which laid the foundation for future studies on the catalytic oxidation of arsenic ions by TiO_2 (Figure 16.11) [26].

16.3.4.1 Oxidation of As(III) by TiO_2 Films on Inert Substates

Depositing TiO_2 films on inert substates is one strategy for the practical application of photocatalysts since it preserves the high activity of nano TiO_2 and avoids of the difficulty in separating and recycling the tiny TiO_2 particles. Efforts were devoted to immobilizing TiO_2 photocatalysts on porous carriers, such as glass, silica gel, metal, ceramics, polymers, films, fibers, zeolites, alumina clay, activated carbon, cellulose, and reactor walls [102]. Fostier et al. immobilized TiO_2 in PET bottles in the presence of natural light and iron salts, which allowed the oxidation of As(III) to As(V) [103]. As the local concentration of TiO_2 increased, the formation of Fe(III) hydroxide (oxide) increased, promoting a larger surface adsorption area, while the adsorption of As(V) on Fe(III) hydroxide (oxide) was much more efficient than the adsorption of As(V) on TiO_2, so that As(V) removal from solution could be achieved. Similarly, TiO_2 was dip-coated on a glass carrier and the synthetic nanoscale zero-valent Fe was added to the reaction system (Figure 16.12a) [104]. This study concluded that the oxidation of As(III) to As(V) was achieved by a combination of $\cdot HO$, $\cdot O_2^-$ and VB holes (h_{VB}^+), while the main role of zero-valent Fe was to provide adsorption sites for arsenic.

Vahala et al. prepared nano TiO_2/feldspar (FS) encapsulated chitosan (CS) composites by immobilizing TiO_2 nanoparticles and feldspar particles in a chitosan matrix [105]. The composites were effective in removing arsenic over a wide pH range, and the removal of arsenate decreased slightly with elevating pH (Figure 16.12b). Selective removal of arsenate was achieved in the presence of coexisting fluoride and sulfate ions. The possible oxidation mechanisms include the co-oxidation of $\cdot HO$, $\cdot O_2^-$ and h_{VB}^+, while UV irradiation can lead to polymer chain breakage in the

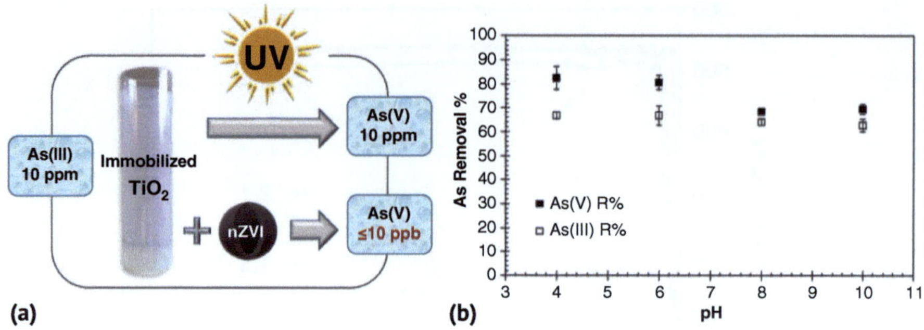

Figure 16.12 (a) The combination of immobilized TiO_2 and zero-valent iron for efficient removal of arsenic. [104]/with permission of Elsevier. (b) Effect of solution pH on As adsorption onto n.TiO_2.FS.CS. (Experimental conditions: As concentration: 1000 μg · L^{-1}; dosage: 1 g · L^{-1}; contact time: 1440 min). [105]/with permission of Elsevier.

CS matrix and introduce new carboxyl groups on the CS structure, which can be used as additional active sites for arsenic oxidation and adsorption; feldspars mainly act as adsorption of As(V).

The advantages of the immobilized photocatalysts are mainly in terms of high stability, homogeneous dispersion, and separable recyclability, which are essential to avoid the loss of catalyst particles and the introduction of new pollutants contaminated with TiO_2 in treated water [106].

16.3.4.2 Oxidation of As(III) by Doped TiO_2

Surface modification was carried out by depositing metals (*e.g.*, Fe, Pt, Au, and Ag) on TiO_2 particles. The deposited metals accelerated the transfer of photogenerated electrons, and improved the photocatalytic efficiency by trapping CB electrons and delaying the charge-pair recombination [107, 108]. Choi et al. found that the transfer rate of CB electrons to O_2 was much faster on Pt-coated TiO_2 than on bare TiO_2 surface by doping Pt on TiO_2 [36]. Iervolino et al. prepared Fe-doped TiO_2 photocatalysts, which reduced the bandgap value from 3.2 to 2.9 eV, and the photocatalysts are active under visible light [24]. For the undoped TiO_2 photocatalyst, the concentration of As(III) was basically unchanged after irradiating under UV light for 30 min. While in the presence of Fe-doped TiO_2 photocatalyst, As(III) was completely oxidized to As(V) after irradiating under visible light for 30 min. This result clearly indicates that iron doping induces photocatalytic activity under visible light irradiation.

16.3.4.3 TiO_2-based Heterojunction

Deng et al. anchored dispersed anatase TiO_2 nanoparticles on the nanocomposite Fe_3O_4 sheets with ferromagnetism, and the sandwich-shaped Fe_3O_4@TiO_2 sheets can be easily collected using an external magnetic field and rapidly dispersed when the magnetic field disappears. Thus, it can be separated directly in the reactive aqueous solution without secondary contamination [40]. The adsorption of arsenic on the catalyst experienced two stages. The initial fast adsorption stage was attributed to the large concentration gradient and a large number of available adsorption sites for arsenic species on the Fe_3O_4@TiO_2 sheet; the second slow adsorption stage was mainly ascribed to the accumulation of adsorbed arsenic on the surface of the adsorbent. Michelle et al. employed the sol–gel method to fabricate WO_3-TiO_2 heterojunction nanomaterials. These materials exhibit the capability to oxidize As(III) through two distinct pathways. First, oxidation occurs via superoxide radicals, which are produced by the interaction of photogenerated electrons with oxygen molecules. Second, oxidation can also proceed through hydroxyl radicals and photogenerated holes generated within the WO_3–TiO_2 structure [25].

Combining with the construction of the heterojunction, the catalytic oxidation performance can be further improved by doping elements. Boron doping in TiO_2/g-C_3N_4 nanocomposites can effectively promote the separation of photogenerated electron–hole pairs and reduce the undesirable complexation of e^- and h^+, and the B-OH bond on TiO_2 can enhance the adsorption of arsenic [41]. Carbon doping in TiO_2/Fe_3O_4 heterojunction can change the photocatalyst activity from a single UV wavelength to visible excitation, expanding the utilization of light energy, and DFT calculations further confirmed that the C—O—Ti bond can reduce the theoretical bandgap of TiO_2 [109].

The current discussion on the mechanism of arsenite oxidation focuses on whether the oxidation is mediated by OH radicals, VB holes, or superoxide radicals [110]. Choi et al. added an excess of *tert*-butanol (an OH radical scavenger) to a catalytic reaction system, but the oxidation rate of arsenite did not significantly decrease. Thus, it is not OH radicals but superoxide radicals that lead to the oxidation of As(III) [43]. By adding benzoquinone ($\cdot O_2^-$ trapping agent), EDTA-2Na$^+$ (h^+ trapping agent) and isopropanol (·HO trapping agent) to the TiO_2/Fe_3O_4 heterojunction system, it was found that the oxidation reaction was essentially unaffected only in the isopropanol group at the end of the reaction, indicating that $\cdot O_2^-$ and h^+ were the main substances for the catalytic oxidation (Figure 16.13a) [111]. In the case of the TiO_2/Fe_3O_4 heterojunction doped with carbon followed by the addition of the above three radical scavengers, the catalytic oxidation was inhibited in all three cases (Figure 16.13b), indicating that the arsenite oxidation process in this case is mainly a synergistic effect of ·HO, $\cdot O_2^-$, and h^+ reactive substances [109].

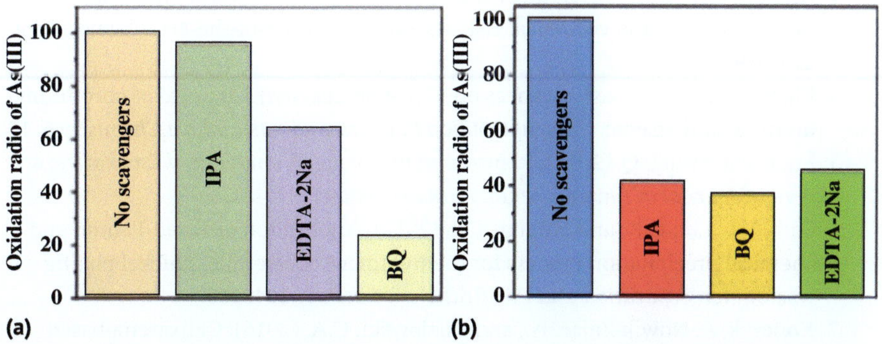

Figure 16.13 (a) Effect of different radical scavengers on oxidation of As(III) in the TF-0.10 system. [111]/with permission of Elsevier. (b) Effect of different radical scavengers on oxidation of As(III) in the C/TiO_2@Fe_3O_4 system. [109]/with permission of Elsevier.

16.4 Conclusions and Outlook

The photocatalytic treatment for removing Cu^{2+}, Cr(VI), U(VI), and As(III/IV) from water presents a promising and environmentally friendly approach. The three heavy metals of Cu(II), Cr(VI), and U(VI) can be processed by the reduction reaction; however, the toxicity and mobility of As(III) are stronger in comparison with As(V), and thus the oxidation treatment is generally done for the removal of As from water.

In order to improve the deficiency of TiO_2 in the photocatalytic treatment of heavy metals, the modification of TiO_2, the construction of heterojunctions and the introduction of scavenger aids not only address the kinetic challenges associated with photocatalytic reactions and optimizes the pathways for the transport of photogenerated electrons but also strengthen the interaction between catalysts and heavy metals, which effectively facilitated the reduction and oxidation processes to achieve the effective capture of the transformed heavy metal substances, so as to ultimately achieve the purpose of removing or sequential recovery.

Up to now, TiO_2 treatment of heavy metals in water mainly focused on single ion. From a practical point of view, future efforts may devote to improving the reduction and recovery efficiency of different heavy metal ions and exploring the synergistic treatment of multiple pollutants. With ongoing technological advancements and innovation, we anticipate witnessing TiO_2 photocatalytic technology achieving even more remarkable milestones in the realm of water treatment and resource recovery in the future.

References

1 Hawkes, S.J. (1997). What is a "heavy metal"? *Journal of Chemical Education* 74 (11): 1374.

2 Eid, M.H., Eissa, M., Mohamed, E.A. et al. (2024). New approach into human health risk assessment associated with heavy metals in surface water and groundwater using Monte Carlo Method. *Scientific Reports* 14 (1): 1008.

3 Kumar, V., Sharma, A., Kaur, P. et al. (2019). Pollution assessment of heavy metals in soils of India and ecological risk assessment: a state-of-the-art. *Chemosphere* 216: 449.

4 Litter, M.I. (2017). Last advances on TiO_2-photocatalytic removal of chromium, uranium and arsenic. *Current Opinion in Green and Sustainable Chemistry* 6: 150.

5 Fu, F. and Wang, Q. (2011). Removal of heavy metal ions from wastewaters: a review. *Journal of Environmental Management* 92 (3): 407.

6 Shih, Y.J., Lin, C.P., and Huang, Y.H. (2013). Application of Fered-Fenton and chemical precipitation process for the treatment of electroless nickel plating wastewater. *Separation and Purification Technology* 104: 100.

7 Konczyk, J., Nowik-Zajac, A., and Kozlowski, C.A. (2016). Calixarene-based extractants for heavy metal ions removal from aqueous solutions. *Separation Science and Technology (Philadelphia)* 51 (14): 2394.

8 Peng, H. and Guo, J. (2020). Removal of chromium from wastewater by membrane filtration, chemical precipitation, ion exchange, adsorption electrocoagulation,

electrochemical reduction, electrodialysis, electrodeionization, photocatalysis and nanotechnology: a review. *Environmental Chemistry Letters* 18 (6): 2055.

9 Kabra, K., Chaudhary, R., and Sawhney, R.L. (2008). Solar photocatalytic removal of Cu(II), Ni(II), Zn(II) and Pb(II): speciation modeling of metal-citric acid complexes. *Journal of Hazardous Materials* 155 (3): 424.

10 Dong, Y., Ma, T., Zhang, P., and Lin, H. (2024). Preparation of modified chitosan and its application to remove cadmium ions in wastewater: a review. *Journal of Environmental Chemical Engineering* 12 (1): 111626.

11 Wang, Y., Li, J., Xu, L. et al. (2024). EDTA functionalized Mg/Al hydroxides modified biochar for Pb(II) and Cd(II) removal: adsorption performance and mechanism. *Separation and Purification Technology* 335: 126199.

12 Zhao, H., Mao, Q., Jian, L. et al. (2022). Photodeposition of earth-abundant cocatalysts in photocatalytic water splitting: methods, functions, and mechanisms. *Chinese Journal of Catalysis* 43 (7): 1774.

13 Nakata, K. and Fujishima, A. (2012). TiO_2 photocatalysis: design and applications. *Journal of Photochemistry and Photobiology C: Photochemistry Reviews* 13 (3): 169.

14 Yeber, M.C., Soto, C., Riveros, R. et al. (2009). Optimization by factorial design of copper (II) and toxicity removal using a photocatalytic process with TiO_2 as semiconductor. *Chemical Engineering Journal* 152 (1): 14.

15 Zhang, Y., Luo, D., Wu, P. et al. (2023). Ultra-high selective recovery of Cu^{2+} and Ni^{2+} by a combination of photocatalysis and adsorption from electroless plating wastewater. *Journal of the Iranian Chemical Society* 20 (11): 2665.

16 Li, P., Wang, J., Wang, Y. et al. (2019). Photoconversion of U(VI) by TiO_2: an efficient strategy for seawater uranium extraction. *Chemical Engineering Journal* 365: 231.

17 Pomilla, F.R., García-López, E.I., Marcì, G. et al. (2021). Heterogeneous photocatalytic materials for sustainable formation of high-value chemicals in green solvents. *Materials Today Sustainability* 13: 100071.

18 Litter, M.I. (1999). Heterogeneous photocatalysis: transition metal ions in photocatalytic systems. *Applied Catalysis B: Environmental* 23 (2–3): 89.

19 Yuan, Z., Zhu, X., Gao, X. et al. (2024). Enhancing photocatalytic CO_2 reduction with TiO_2-based materials: strategies, mechanisms, challenges, and perspectives. *Environmental Science and Ecotechnology* 20: 100368.

20 Marinho, B.A., Cristóvão, R.O., Djellabi, R. et al. (2017). Photocatalytic reduction of Cr(VI) over TiO_2-coated cellulose acetate monolithic structures using solar light. *Applied Catalysis B: Environmental* 203: 18.

21 Zhao, X., Zhang, J., and Qu, J. (2015). Photoelectrocatalytic oxidation of Cu-cyanides and Cu-EDTA at TiO_2 nanotube electrode. *Electrochimica Acta* 180: 129.

22 Li, Y., Cai, X., Guo, J., and Na, P. (2014). UV-induced photoactive adsorption mechanism of arsenite by anatase TiO_2 with high surface hydroxyl group density. *Colloids and Surfaces A: Physicochemical and Engineering Aspects* 462: 202.

23 Guo, Q., Zhou, C., Ma, Z., and Yang, X. (2019). Fundamentals of TiO_2 photocatalysis: concepts, mechanisms, and challenges. *Advanced Materials* 31 (50): 1901997.

24 Iervolino, G., Vaiano, V., and Rizzo, L. (2018). Visible light active Fe-doped TiO_2 for the oxidation of arsenite to arsenate in drinking water. *Chemical Engineering Transactions* 70: 1573.

25 Navarrete-Magaña, M., Estrella-González, A., May-Ix, L. et al. (2021). Improved photocatalytic oxidation of arsenic (III) with WO_3/TiO_2 nanomaterials synthesized by the sol-gel method. *Journal of Environmental Management* 282: 111602.

26 Yang, H., Lin, W.Y., and Rajeshwar, K. (1999). Homogeneous and heterogeneous photocatalytic reactions involving As(III) and As(V) species in aqueous media. *Journal of Photochemistry and Photobiology A: Chemistry* 123 (1–3): 137.

27 Satyro, S., Marotta, R., Clarizia, L. et al. (2014). Removal of EDDS and copper from waters by TiO_2 photocatalysis under simulated UV-solar conditions. *Chemical Engineering Journal* 251: 257.

28 Chen, J., Ollis, D.F., Rulkens, W.H., and Bruning, H. (1999). Photocatalyzed deposition and concentration of soluble uranium (VI) from TiO_2 suspensions. *Colloids and Surfaces A: Physicochemical and Engineering Aspects* 151: 339.

29 Seko, N., Katakai, A., Tamada, M. et al. (2004). Fine fibrous amidoxime adsorbent synthesized by grafting and uranium adsorption–elution cyclic test with seawater. *Separation Science and Technology* 39 (16): 3753.

30 Gomaa, H., Khalifa, H., Selim, M.M. et al. (2017). Selective, photoenhanced trapping/detrapping of arsenate anions using mesoporous blobfish head TiO_2 monoliths. *ACS Sustainable Chemistry and Engineering* 5 (11): 10826.

31 Pan, H.B., Liao, W., Wai, C.M. et al. (2014). Carbonate-H_2O_2 leaching for sequestering uranium from seawater. *Dalton Transactions* 43 (28): 10713.

32 Amadelli, R., Maldotti, A., Sostero, S., and Carassiti, V. (1991). Photodeposition of uranium oxides onto TiO_2 from aqueous uranyl solutions. *Journal of the Chemical Society, Faraday Transactions* 87 (19): 3267.

33 Guan, X., Du, J., Meng, X. et al. (2012). Application of titanium dioxide in arsenic removal from water: a review. *Journal of Hazardous Materials* 215–216: 1.

34 Lee, S.S., Bai, H., Liu, Z., and Sun, D.D. (2015). Green approach for photocatalytic Cu(II)-EDTA degradation over TiO_2: toward environmental sustainability. *Environmental Science and Technology* 49 (4): 2541.

35 Kyung, H., Lee, J., and Choi, W. (2005). Simultaneous and synergistic conversion of dyes and heavy metal ions in aqueous TiO_2 suspensions under visible-light illumination. *Environmental Science and Technology* 39 (7): 2376.

36 Ryu, J. and Choi, W. (2004). Effects of TiO_2 surface modifications on photocatalytic oxidation of arsenite: the role of superoxides. *Environmental Science and Technology* 38 (10): 2928.

37 Gong, X., Tang, L., Zou, J. et al. (2022). Introduction of cation vacancies and iron doping into TiO_2 enabling efficient uranium photoreduction. *Journal of Hazardous Materials* 423: 126935.

38 Choi, Y., Koo, M.S., Bokare, A.D. et al. (2017). Sequential process combination of photocatalytic oxidation and dark reduction for the removal of organic pollutants and Cr(VI) using Ag/TiO_2. *Environmental Science and Technology* 51 (7): 3973.

39 Pu, S., Hou, Y., Chen, H. et al. (2018). An efficient photocatalyst for fast reduction of Cr(VI) by ultra-trace silver enhanced titania in aqueous solution. *Catalysts* 8 (6): 251.

40 Deng, M., Wu, X., Zhu, A. et al. (2019). Well-dispersed TiO_2 nanoparticles anchored on Fe_3O_4 magnetic nanosheets for efficient arsenic removal. *Journal of Environmental Management* 237: 63.

41 Ouyang, L., Zhang, Y., Wang, Y. et al. (2021). Insights into the adsorption and photocatalytic oxidation behaviors of boron-doped TiO_2/g-C_3N_4 nanocomposites toward As(III) in aqueous solution. *Industrial and Engineering Chemistry Research* 60 (19): 7003.

42 Qin, J., Ye, S., Yan, K., and Zhang, J. (2022). Visible light-driven photoelectrocatalysis for simultaneous removal of oxytetracycline and Cu (II) based on plasmonic Bi/Bi_2O_3/TiO_2 nanotubes. *Journal of Colloid and Interface Science* 607: 1936.

43 Lee, H. and Choi, W. (2002). Photocatalytic oxidation of arsenite in TiO_2 suspension: kinetics and mechanisms. *Environmental Science and Technology* 36 (17): 3872.

44 Levy, I.K., Mizrahi, M., Ruano, G. et al. (2012). TiO_2-photocatalytic reduction of pentavalent and trivalent arsenic: production of elemental arsenic and arsine. *Environmental Science and Technology* 46 (4): 2299.

45 Xu, S., Ng, J., Wang, Y. et al. (2012). Simultaneous copper ion removal and hydrogen production from water over a TiO_2 nanotube photocatalyst. *Water Science and Technology* 65 (3): 533.

46 Liu, F., Zhou, K., Chen, Q. et al. (2019). Application of magnetic ferrite nanoparticles for removal of Cu(II) from copper-ammonia wastewater. *Journal of Alloys and Compounds* 773: 140.

47 Herrmann, J.M. (2005). Heterogeneous photocatalysis: state of the art and present applications. *Topics in Catalysis* 34 (1–4): 49.

48 Luo, D., Yan, R., Fu, C. et al. (2019). Cu(0)/TiO_2 composite byproduct from photo-reduction of acidic Cu-containing wastewater and its reuse as a catalyst. *Journal of Water Process Engineering* 32: 100958.

49 Zhou, C., Wang, S., Zhao, Z. et al. (2018). A facet-dependent Schottky-junction electron shuttle in a $BiVO_4$\{010\}–Au–Cu_2O Z-scheme photocatalyst for efficient charge separation. *Advanced Functional Materials* 28 (31): 1801214.

50 Jiang, Z.Y. and Zhao, Z.Y. (2018). Comparison studies of interfacial energetic and electronic properties of bimetallic AuCu/TiO_2 hetero-structures from DFT calculations. *Inorganic Chemistry Frontiers* 5 (5): 1062.

51 Zhao, X., Guo, L., Zhang, B. et al. (2013). Photoelectrocatalytic oxidation of CuII-EDTA at the TiO_2 electrode and simultaneous recovery of CuII by electrodeposition. *Environmental Science and Technology* 47 (9): 4480.

52 Byrne, J.A., Eggins, B.R., Byers, W., and Brown, N.M.D. (1999). Photoelectrochemical cell for the combined photocatalytic oxidation of organic pollutants and the recovery of metals from waste waters. *Applied Catalysis B: Environmental* 20 (2): L85.

53 Liu, L., Li, R., Liu, Y., and Zhang, J. (2016). Simultaneous degradation of ofloxacin and recovery of Cu(II) by photoelectrocatalysis with highly ordered TiO_2 nanotubes. *Journal of Hazardous Materials* 308: 264.

54 Qi, J., Li, X., Zheng, H. et al. (2015). Simultaneous removal of methylene blue and copper(II) ions by photoelectron catalytic oxidation using stannic oxide modified iron(III) oxide composite electrodes. *Journal of Hazardous Materials* 293: 105.

55 Wang, D., Li, Y., Li Puma, G. et al. (2017). Photoelectrochemical cell for simultaneous electricity generation and heavy metals recovery from wastewater. *Journal of Hazardous Materials* 323: 681.

56 Doan, H.D. and Saidi, M. (2008). Simultaneous removal of metal ions and linear alkylbenzene sulfonate by combined electrochemical and photocatalytic process. *Journal of Hazardous Materials* 158 (2–3): 557.

57 Zeng, H., Liu, S., Chai, B. et al. (2016). Enhanced photoelectrocatalytic decomplexation of Cu-EDTA and Cu recovery by persulfate activated by UV and cathodic reduction. *Environmental Science and Technology* 50 (12): 6459.

58 Jiang, S., Fu, F., Qu, J., and Xiong, Y. (2008). A simple method for removing chelated copper from wastewaters: $Ca(OH)_2$-based replacement-precipitation. *Chemosphere* 73 (5): 785.

59 Wei, P., Zhang, Y., Huang, Y. et al. (2022). Effect and mechanism of cyanide degradation and Cu/Zn recovery by photoelectro-catalytic oxidation. *Separation and Purification Technology* 282: 120050.

60 Wang, Q., Shang, J., Zhu, T., and Zhao, F. (2011). Efficient photoelectrocatalytic reduction of Cr(VI) using TiO_2 nanotube arrays as the photoanode and a large-area titanium mesh as the photocathode. *Journal of Molecular Catalysis A: Chemical* 335 (1–2): 242.

61 Levina, A. and Lay, P.A. (2005). Mechanistic studies of relevance to the biological activities of chromium. *Coordination Chemistry Reviews* 249 (3–4 SPEC. ISS.): 281.

62 Chen, G., Feng, J., Wang, W. et al. (2017). Photocatalytic removal of hexavalent chromium by newly designed and highly reductive TiO_2 nanocrystals. *Water Research* 108: 383.

63 Djellabi, R., Ghorab, F.M., Nouacer, S. et al. (2016). Cr(VI) photocatalytic reduction under sunlight followed by Cr(III) extraction from TiO_2 surface. *Materials Letters* 176: 106.

64 Li, Y., Bian, Y., Qin, H. et al. (2017). Photocatalytic reduction behavior of hexavalent chromium on hydroxyl modified titanium dioxide. *Applied Catalysis B: Environmental* 206: 293.

65 Cho, Y., Kyung, H., and Choi, W. (2004). Visible light activity of TiO_2 for the photoreduction of CCl4 and Cr(VI) in the presence of nonionic surfactant (Brij). *Applied Catalysis B: Environmental* 52 (1): 23.

66 Zhang, J., Fu, D., Wang, S. et al. (2019). Photocatalytic removal of chromium(VI) and sulfite using transition metal (Cu, Fe, Zn) doped TiO_2 driven by visible light: feasibility, mechanism and kinetics. *Journal of Industrial and Engineering Chemistry* 80: 23.

67 Zhang, Y.C., Yang, M., Zhang, G., and Dionysiou, D.D. (2013). HNO_3-involved one-step low temperature solvothermal synthesis of N-doped TiO_2 nanocrystals for efficient photocatalytic reduction of Cr(VI) in water. *Applied Catalysis B: Environmental* 142–143: 249.

68 Barolo, G., Livraghi, S., Chiesa, M. et al. (2012). Mechanism of the photoactivity under visible light of N-doped titanium dioxide. Charge carriers migration in irradiated N-TiO_2 investigated by electron paramagnetic resonance. *Journal of Physical Chemistry C* 116 (39): 20887.

69 Di Valentin, C., Finazzi, E., Pacchioni, G. et al. (2007). N-doped TiO_2: theory and experiment. *Chemical Physics* 339 (1–3): 44.

70 Djellabi, R., Ghorab, M.F., Bianchi, C.L. et al. (2016). Recovery of hexavalent chromium from water using photoactive TiO_2-montmorillonite under sunlight. *Mediterranean Journal of Chemistry* 5 (3): 442.

71 Zhao, Y., Zhao, D., Chen, C., and Wang, X. (2013). Enhanced photo-reduction and removal of Cr(VI) on reduced graphene oxide decorated with TiO_2 nanoparticles. *Journal of Colloid and Interface Science* 405: 211.

72 Yan, R., Luo, D., Fu, C. et al. (2020). Harmless treatment and selective recovery of acidic Cu(II)-Cr(VI) hybrid wastewater via coupled photo-reduction and ion exchange. *Separation and Purification Technology* 234: 116130.

73 Chen, M., Wang, G., Dai, J. et al. (2024). Indirect Z-scheme TiO_2/BC/g-C_3N_4 for efficient photocatalytic reduction of Cr(VI) in aqueous solution. *Journal of Chemical Technology and Biotechnology* 99 (2): 415.

74 Zhang, Y., Liu, C., Nian, P. et al. (2024). Facile preparation of high-performance hydrochar/TiO_2 heterojunction visible light photocatalyst for treating Cr(VI)-polluted water. *Colloids and Surfaces A: Physicochemical and Engineering Aspects* 681: 132775.

75 Litter, M.I. (2015). Mechanisms of removal of heavy metals and arsenic from water by TiO_2-heterogeneousphotocatalysis. *Pure and Applied Chemistry* 87: 557.

76 Wang, X., Pehkonen, S.O., and Ray, A.K. (2004). Removal of aqueous Cr(VI) by a combination of photocatalytic reduction and coprecipitation. *Industrial and Engineering Chemistry Research* 43 (7): 1665.

77 Liu, C., Hsu, P.C., Xie, J. et al. (2017). A half-wave rectified alternating current electrochemical method for uranium extraction from seawater. *Nature Energy* 2 (4): 17007.

78 Endrizzi, F., Leggett, C.J., and Rao, L. (2016). Scientific basis for efficient extraction of uranium from seawater. I: understanding the chemical speciation of uranium under seawater conditions. *Industrial and Engineering Chemistry Research* 55 (15): 4249.

79 Sather, A.C., Berryman, O.B., and Rebek Jr, J. (2010). Selective recognition and extraction of the uranyl ion. *Journal of the American Chemical Society* 132 (39): 13572.

80 Cooper, K.L., Dashner, E.J., Tsosie, R. et al. (2016). Inhibition of poly(ADP-ribose) polymerase-1 and DNA repair by uranium. *Toxicology and Applied Pharmacology* 291: 13.

81 Mathews, G., Nagaiah, N., Karthik Kumar, M.B., and Ambika, M.R. (2015). Radiological and chemical toxicity due to ingestion of uranium through drinking water in the environment of Bangalore, India. *Journal of Radiological Protection* 35 (2): 447.

82 Abney, C.W., Mayes, R.T., Saito, T., and Dai, S. (2017). Materials for the recovery of uranium from seawater. *Chemical Reviews* 117 (23): 13935.

83 Yu, F., Zhu, Z., Wang, S. et al. (2021). Tunable perylene-based donor-acceptor conjugated microporous polymer to significantly enhance photocatalytic uranium extraction from seawater. *Chemical Engineering Journal* 412: 127558.

84 Chen, T., Yu, K., Dong, C. et al. (2022). Advanced photocatalysts for uranium extraction: elaborate design and future perspectives. *Coordination Chemistry Reviews* 467: 214615.

85 Bao, S., Liu, H., Liang, H. et al. (2021). Electrospinned silk-ribbon-like carbon-doped TiO_2 ultrathin nanosheets for enhanced visible-light photocatalytic activity. *Colloids and Surfaces A: Physicochemical and Engineering Aspects* 616: 126289.

86 Wang, J., Wang, Y., Wang, W. et al. (2020). Visible light driven Ti^{3+} self-doped TiO_2 for adsorption-photocatalysis of aqueous U(VI). *Environmental Pollution* 262: 114373.

87 Zhang, B., Shan, X., Yu, J. et al. (2024). Facile synthesis of TiO_2-PAN photocatalytic membrane with excellent photocatalytic performance for uranium extraction from seawater. *Separation and Purification Technology* 328: 125026.

88 Jiang, X.H., Xing, Q.J., Luo, X.B. et al. (2018). Simultaneous photoreduction of uranium(VI) and photooxidation of arsenic(III) in aqueous solution over g-C_3N_4/TiO_2 heterostructured catalysts under simulated sunlight irradiation. *Applied Catalysis B: Environmental* 228: 29.

89 Li, Z.J., Huang, Z.W., Guo, W.L. et al. (2017). Enhanced photocatalytic removal of uranium(VI) from aqueous solution by magnetic TiO_2/Fe_3O_4 and its graphene composite. *Environmental Science and Technology* 51 (10): 5666.

90 Zhang, P., Li, H., Wang, Y. et al. (2023). Highly efficient uranium (VI) remove from aqueous solution using nano-TiO_2-anchored polymerized dopamine-wrapped magnetic photocatalyst. *Journal of Cleaner Production* 425: 138796.

91 Wang, Y., Wang, J., Wang, J. et al. (2020). Efficient recovery of uranium from saline lake brine through photocatalytic reduction. *Journal of Molecular Liquids* 308: 113007.

92 Qin, Z., Ye, Y., Li, C. et al. (2023). Removal and recovery of aqueous uranium using photocatalytic reduction method: performance and implication. *Separation and Purification Technology* 306: 122670.

93 Salomone, V.N., Meichtry, J.M., and Litter, M.I. (2015). Heterogeneous photocatalytic removal of U (VI) in the presence of formic acid: U (III) formation. *Chemical Engineering Journal* 270: 28.

94 Li, P., Wang, J., Peng, T. et al. (2019). Heterostructure of anatase-rutile aggregates boosting the photoreduction of U(VI). *Applied Surface Science* 483: 670.

95 Liu, X., Du, P., Pan, W. et al. (2018). Immobilization of uranium(VI) by niobate/titanate nanoflakes heterojunction through combined adsorption and solar-light-driven photocatalytic reduction. *Applied Catalysis B: Environmental* 231: 11.

96 Kuroda, K., Lu, B., Hama, Y., and Yang, Y. (2023). Recent progress in photocatalysts for oxidation of As(III) and photocatalyst-impregnated adsorbents for removing aqueous arsenic. *Current Opinion in Environmental Science and Health* 35: 100498.

97 Smedley, P.L. and Kinniburgh, D.G. (2002). A review of the source, behaviour and distribution of arsenic in natural waters. *Applied Geochemistry* 17 (5): 517.

98 Liu, B., Kim, K.H., Kumar, V., and Kim, S. (2020). A review of functional sorbents for adsorptive removal of arsenic ions in aqueous systems. *Journal of Hazardous Materials* 388: 121815.

99 Tufano, K.J., Reyes, C., Saltikov, C.W., and Fendorf, S. (2008). Technology. reductive processes controlling arsenic retention: revealing the relative importance of iron and arsenic reduction. *Environmental Science & Technology* 42 (22): 8283.

100 Weber, F.-A., Hofacker, A.F., Voegelin, A., and Kretzschmar, R. (2010). Temperature dependence and coupling of iron and arsenic reduction and release during flooding of a contaminated soil. *Environmental Science & Technology* 44 (1): 116.

101 Bissen, M., Vieillard-Baron, M.M., Schindelin, A.J., and Frimmel, F.H. (2001). TiO_2-catalyzed photooxidation of arsenite to arsenate in aqueous samples. *Chemosphere* 44 (4): 751.

102 Chong, M.N., Jin, B., Chow, C.W.K., and Saint, C. (2010). Recent developments in photocatalytic water treatment technology: a review. *Water Research* 44 (10): 2997.

103 Fostier, A.H., Pereira, M.S.S., Rath, S., and Guimarães, J.R. (2008). Arsenic removal from water employing heterogeneous photocatalysis with TiO_2 immobilized in PET bottles. *Chemosphere* 72 (2): 319.

104 Raez, J.M., Arencibia, A., Segura, Y. et al. (2021). Combination of immobilized TiO_2 and zero valent iron for efficient arsenic removal in aqueous solutions. *Separation and Purification Technology* 258: 118016.

105 Yazdani, M.R., Bhatnagar, A., and Vahala, R. (2017). Synthesis, characterization and exploitation of nano-TiO_2/feldspar-embedded chitosan beads towards UV-assisted adsorptive abatement of aqueous arsenic (As). *Chemical Engineering Journal* 316: 370.

106 Yang, G.C.C. and Li, C.J. (2007). Electrofiltration of silica nanoparticle-containing wastewater using tubular ceramic membranes. *Separation and Purification Technology* 58 (1): 159.

107 Bae, E. and Choi, W. (2003). Highly enhanced photoreductive degradation of perchlorinated compounds on dye-sensitized metal/TiO_2 under visible light. *Environmental Science & Technology* 37 (1): 147.

108 Kamat, P.V. (2002). Photophysical, photochemical and photocatalytic aspects of metal nanoparticles. *The Journal of Physical Chemistry B* 106 (32): 7729–7744.

109 Xiao, M., Li, R., Hu, X. et al. (2022). Construction of in-situ carbon-doped TiO_2 decorated Fe_3O_4 heterojunction and their enhanced photocatalytic oxidation of As(III) under visible light. *Separation and Purification Technology* 300: 121836.

110 Monllor-Satoca, D., Gómez, R., and Choi, W. (2012) Concentration-dependent photoredox conversion of As(III)/As(V) on illuminated titanium dioxide electrodes. *Environmental Science and Technology* 46 (10): 5519.

111 Xiao, M., Li, R., Yin, J. et al. (2022). Enhanced photocatalytic oxidation of As(III) by TiO_2 modified with Fe_3O_4 through Ti-O-Fe interface bonds. *Colloids and Surfaces A: Physicochemical and Engineering Aspects* 651: 129678.

17

Solar Fuel Generation

Shuai Wang[1], Hairong Yue[1], Patrik Schmuki[2], Xuemei Zhou[1,]*

[1] *School of Chemical Engineering, Sichuan University, Chengdu, P. R. China*
[2] *Department of Materials Science and Engineering, University of Erlangen-Nuremberg, Erlangen, Germany*

*Corresponding author: xuemeizhou@scu.edu.cn

17.1 Introduction to the Solar Fuel Generation Using Nanostructured TiO_2

Inspired by the intricate process of plant photosynthesis, the concept of artificial photosynthesis first emerged in the early 1900s through the visionary ideas of Ciamician [1]. In the current era marked by the transition from traditional fossil fuels toward the development and utilization of renewable energy sources, significant research efforts have been directed toward harnessing abundant solar energy and converting it into chemical energy. In recent years, the utilization of photocatalysis has witnessed a surge in its applications across diverse domains including removal of liquid and gaseous pollutants, carbon dioxide (CO_2) reduction [2–4], N_2 fixation [5], hydrogen evolution from water splitting [6, 7], from hydrogen sulfide (H_2S) splitting [8], or from plastics and biomass [9], and artificial photosynthesis. Figure 17.1 illustrates three sections that will be discussed in this book chapter using TiO_2 as a photocatalyst.

17.2 Solar Energy: Hydrogen Production

17.2.1 Introduction

In the early 1970s, Honda and Fujishima has conducted pioneering experiments in solar-powered water splitting using TiO_2 [10]. Notably, hydrogen, produced through this process, offers an environmentally friendly alternative to traditional fossil fuels. The combustion of hydrogen does not yield pollutants or greenhouse gas emissions and therefore photocatalytic H_2 is regarded as a promising energy source for the future.

Nanostructured TiO_2: Synthesis, Photocatalytic Properties and Applications, First Edition.
Edited by Jia Hong Pan, Wan In Lee and Detlef W. Bahnemann

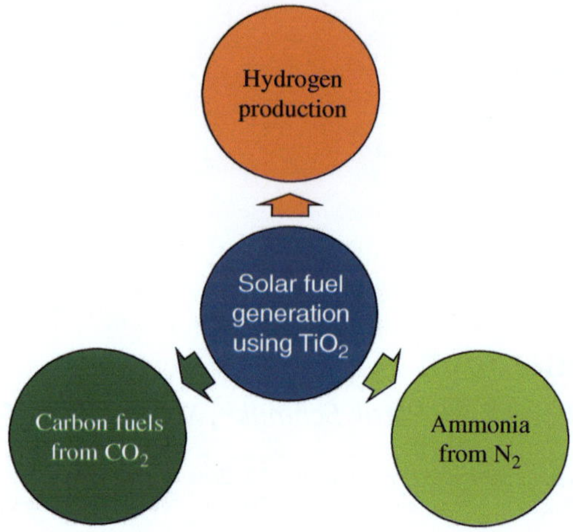

Figure 17.1 Three aspects that are included in the solar fuel generation using TiO_2 (this chapter).

For hydrogen production using TiO_2 by photocatalysis, non-recombined electrons participate in the reduction of protons (H^+) to yield H_2 molecules $\left(E_{CB} < E^0(H_2/H^+)\right)$ that are released as gas from the electrodes. The oxidation occurs between holes and H_2O or e.g., organic molecules, to produce O_2 molecules or, for example, oxidized hydrocarbons $\left(E_{VB} > E^0(O_2/H_2O)\right)$. The overall reaction is thermodynamically governed by (i) the Gibbs energy (water splitting requires a theoretical Gibbs energy > 1.23 eV to satisfy thermodynamic conditions, or if higher energetic intermediates are considered, > 2.0 eV [11] and (ii) the energetic positions of CB and VB (requiring sufficiently negative and positive potentials, respectively, for water reduction and oxidation half-reactions). The main problems of TiO_2 in photocatalysis are the slow charge transfer kinetics and the wide bandgap allows the utilization of UV light only. Modifications are made to effectively accelerate surface reaction kinetics and serve as charge trapping and transfer centers, facilitating the efficient separation of electron–hole pairs under illumination, thereby increasing the yield of H_2. This section primarily focuses on the application and replacement of noble metal cocatalysts, the fabrication of heterojunctions, as well as advanced photocatalytic systems to generate hydrogen from different substrates such as seawater, waste plastics, and biomass.

17.2.2 Development to Replace Noble Metal Cocatalysts for Hydrogen Evolution

Introduction of noble metal cocatalysts can significantly enhance the photocatalytic performance of TiO_2-based photocatalysts [12–14]. However, the high cost of precious metal cocatalysts increases the cost of photocatalysts and hinders the applications in large-scale manufacturing. A large range of non-noble metal catalysts (e.g., MoS_2, NiP, etc.) have been explored to replace expensive Pt metals [15–17]. And most recently, highly dispersed single-atom catalysts (SACs) have emerged as a solution to solve the problem [18]. Single atoms provide intrinsic energy levels and electronic structure with an unsaturated coordination environment and the ability to regulate metal–carrier interactions, thus to promote charge transfer and hence reaction activation. Moreover, the active site of SACs typically involves a single metal atom and adjacent atoms on the surface of the support or other functional structure, maximizing atom utilization efficiency (ideal 100% atom utilization) thus reducing the use of precious metals, for example, both Li and Schmuki's group has reported the interaction between Ti^{3+} and Pt_1 [19]. The H_2 evolution rate per Pt is 150 times higher than the H_2 generation rate of the traditional Pt NPs-modified TiO_2 layer. Generally, the greater the percentage of SACs, the higher the activity. However, with a loading above 0.5 wt%, single Pt distribution is challenging, particularly on particles with low surface areas. Obtaining highly dispersed SACs at high loading remains a major bottleneck for photocatalytic H_2 production. Nevertheless, please note that in photocatalysis only a very low loading is sufficient to reach a saturated activity [20].

Another approach is the use of noble-metal-free cocatalyst. Transition metals, such as Ni, Cu, and Co, have emerged as promising materials for enhancing the photocatalytic activity of TiO_2. Liu et al. [21] reported a remarkable loading (up to 1.5 wt%) of Cu SAC on TiO_2, with a surprisingly high H_2 evolution rate of $101.7\,mmol\,g^{-1}\,h^{-1}$ under simulated sunlight irradiation and an apparent quantum efficiency of 56% under 365 nm irradiation. To replace noble metals, the activation by introduction of intrinsic active sites is another approach. Schmuki's group has reported reduction or implantation of N-Ti^{3+} or protons for activation of TiO_2 for hydrogen evolution [22–24].

17.2.3 Novel Heterojunctions for Hydrogen Evolution

TiO_2-based heterojunctions are typically composed of TiO_2 and a secondary semiconductor, showing close contact interfaces, different band energy levels, close lattice fringes, and similar thermal expansion coefficients. Various heterojunctions can be classified as traditional heterojunctions (Figure 17.2) (Types I, II, and III), p–n junctions, Z-type junctions, and S-scheme heterojunction [25].

A special Type II heterojunction is the p–n junction, which is composed of p-type and n-type semiconductors. The electric field formed at the interface accelerates the transfer of electrons from the CB of the p-type semiconductor to the CB of the n-type semiconductor and the transfer of holes from the VB of the n-type semiconductor to the VB of the p-type semiconductor. For example, Kim et al. [26] constructed a p–n junction TiO_2-based (Pd/TiO_2) nanotube photocatalyst with a p-type at the bottom and n-type at the top, and exhibited a high photocatalytic hydrogen production rate of $25.2\,\mu L\,cm^{-2}\,h^{-1}$ under sunlight, that is an increase of over 250 times compared to n-type TiO_2 nanotubes.

Type II heterojunctions facilitate charge separation but may sacrifice strong redox potentials. To address this issue, Z-type heterojunctions have emerged. Based on the migration paths of electrons and holes, Z-type heterojunctions can be divided into direct and indirect Z-type junctions. The S-scheme heterojunction, composed of a reduction photocatalyst (RP) and an oxidation photocatalyst (OP), is similar to the direct Z-type

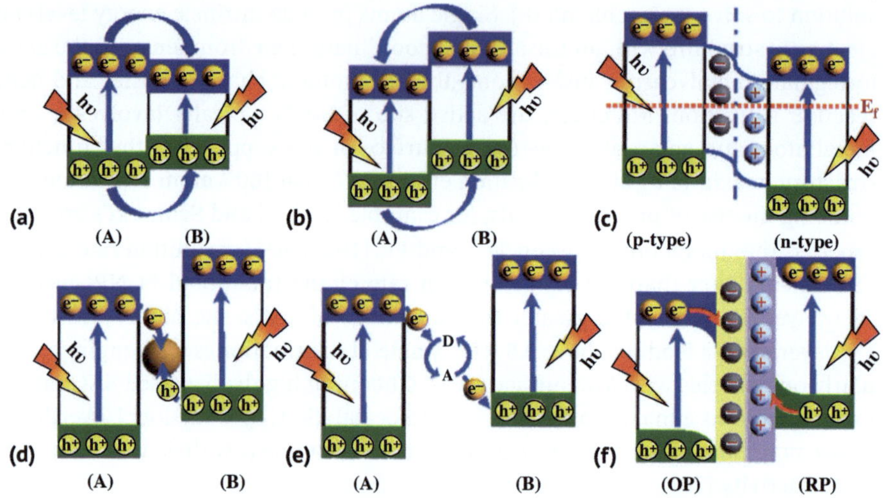

Figure 17.2 Five typical types of heterojunctions with TiO_2: (a) Type I; (b) Type II; (c) p–n heterojunction; (d) direct Z-scheme; (e) indirect Z-scheme; and (f) S-scheme. D: electron donor; A: electron acceptor. Reproduced from Ref. [27]/with permission of Elsevier.

heterojunction. Like the Z-type heterojunction, the S-scheme heterojunction also possesses a strong built-in electric field, which facilitates the efficient separation of electron–hole pairs while maintaining their high redox ability [28, 29]. Yu's group has made significant contributions to the field of S-scheme heterojunctions, designing various S-scheme heterojunction catalysts that offer new perspectives for enhancing photocatalytic performance in artificial photosynthesis systems. For instance, two-dimensional graphitic carbon nitride nanosheets can be synthesized with rutile TiO_2 NPs and H-doped anatase TiO_2 nanorods, which exhibits a dual S-scheme transport mechanism, overcoming the limitations of large contact potential barriers, lattice mismatch, and slow charge transfer typically associated with traditional heterojunctions [30]. The S-scheme of $TiO_2/FePS_3$ (FPS) nanosheets exhibited significantly enhanced photocatalytic activity for hydrogen evolution, with a rate increase of approximately 1700% [29]. Recently, Zhou et al. have shown the fabrication of S-scheme heterojunctions on thin TiO_2 nanosheets incorporating a few-layer MoO_3 structure, that enables efficient electron transfer at the interface, effectively mitigating charge transfer resistance [31]. Further, a hybrid S-scheme heterojunction is developed by incorporating COF films onto TiO_2 NTAs that shows a hydrogen evolution rate of $74.1\,\text{mmol}\,\text{h}^{-1}\,\text{g}_{\text{COF}}^{-1}$ with a 15-nm thick COF film [32].

17.2.4 Hydrogen from Seawater

Extensive research has been conducted on photocatalytic hydrogen production from freshwater resources; more challenging is to produce hydrogen from seawater, which constitutes approximately 97% of the total water volume. Seawater contains various inorganic salts that can act as sacrificial agents, and thus can beneficially improve the hydrogen production efficiency. However, inorganic salt ions in

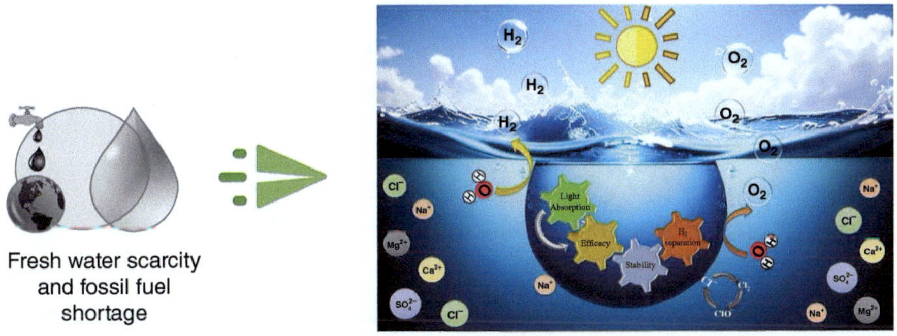

Figure 17.3 Green fuels derived from photocatalytic seawater splitting. Reproduced from Ref. [33]/with permission of Elsevier.

seawater pose greater challenges to the durability of photocatalysts. Therefore, the development of highly durable photocatalysts is crucial for the future development of seawater photocatalytic hydrogen production (Figure 17.3).

TiO_2 as a high chemical and photo resistive semiconductor is an ideal candidate to be developed. Chen et al. [34] synthesized Pt/brookite, Pt/anatase, and Pt/rutile TiO_2 NPs, with the highest hydrogen production rate of Pt/brookite, reaching $1476\,\mu mol\,g^{-1}\,h^{-1}$ from seawater. Unfortunately, the photocatalytic activity significantly declined after continuous reaction for 40 h. To address the poor stability in seawater, Cheng et al. [35] designed a superhydrophilic mesoporous anatase/rutile TiO_2 photocatalyst that achieved a hydrogen production rate of $6.49\,mmol\,g^{-1}\,h^{-1}$. Moreover, the catalyst exhibited high stability after being tested in seawater for 100 h, as evidenced by XRD spectrum. Obviously further efforts are required to enhance the activity and durability of photocatalyst to produce hydrogen from seawater.

17.2.5 Hydrogen from Organic Substrates

Photocatalytic water splitting for hydrogen production is challenging due to the thermodynamic stability of water and the high kinetic barrier for O_2 production, which requires a significant input of energy to generate H_2. In contrast, the photoreforming (PR) of waste polymers, such as biomass-derived natural polymer waste and synthetic polymer (plastics) waste, is a less explored approach. PR involves two independent half-reactions (Figure 17.4). One is hydrogen evolution reaction (HER) and the other is the oxidation of organic substrates (biomass, plastic), where the substrate directly participates in a complex multistep process. The oxidation of organics is thermodynamically (and kinetically) more favored than the oxidation of water.

In addition, PR may be particularly attractive when biomass and plastic waste are used as feedstocks, providing an opportunity for simultaneous production of H_2, reduction of organic environmental pollutants and generation of high-value-added organic compounds. Using waste polymers (natural polymer waste (biomass) and synthetic polymer waste (plastics) as a multifunctional renewable resource can achieve stable and scalable production of H_2 and convert waste into chemical feedstocks.

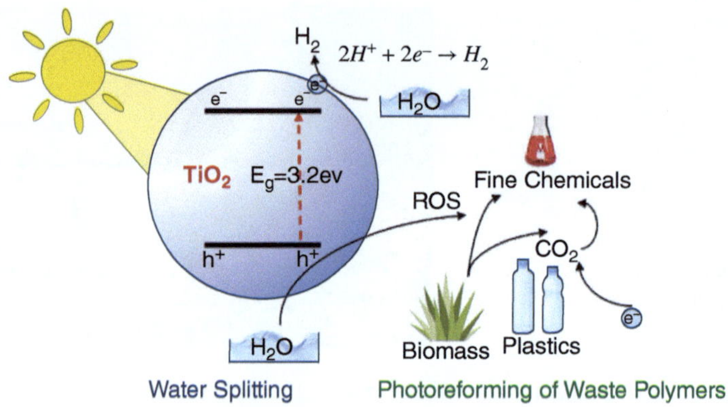

Figure 17.4 The proposed possible hydrogen evolution mechanism from organic substrates.

The utilization of biomass for photocatalytic reforming reactions offers a sustainable source of H_2 fuel. Renewable biomass includes agricultural waste, grass, and wood, which are typically composed of energy-rich and structurally stable lignocellulosic materials. The recalcitrant structure of lignocellulosic biomass prevents easy degradation, making the photocatalytic hydrogen production from lignocellulose challenging and requiring efficient photocatalysts that can absorb visible light. For example, Andrea et al. [36] conducted a systematic study using cellulose (CLS) and rice husks as sacrificial agents, in the presence of Pt/TiO_2, the H_2 evolution from CLS water suspension is 10-fold compared to pure water.

Low-energy-consuming PR processes at ambient conditions to convert plastic waste into H_2 and other high-value-added organic products [37]. For example, Reisner et al. [38] constructed a 5% Pt/TiO_2 catalyst for the photocatalytic reforming of PET, achieving a hydrogen production efficiency of up to 153 μmolgcat^{-1}h^{-1}, providing a pathway for the photocatalytic reforming of plastic waste (PET). Despite significant achievements in TiO_2-based photocatalysts for biomass and plastic photoreforming in the past decade, the current yields of target products (solar-to-hydrogen efficiency of at least 10%) are still not commercially viable. The use of large amounts of acids or bases in the pretreatment of plastics and biomass, as well as the subsequent processing, adds significant cost in the process. Future research can be directed to the synthesis of TiO_2-based photocatalysts with high selectivity, strong stability, and optimized spectral absorbance.

17.3 Solar Fuel: Carbonaceous Solar Fuels from CO_2

17.3.1 Introduction

The conversion of CO_2 to valuable chemicals in the form of CH_4, HCOOH, HCHO, or CH_3OH through solar energy is considered as a promising and attractive approach for sustainable development. In 1979, Inoue et al. [4] reported pioneering work on the photocatalytic conversion of CO_2 molecules into organic compounds using semiconductors. In heterogeneous photocatalysis, CO_2 reduction involves five consecutive steps [27]: (*i*) photoexcitation to generate electrons and holes, (*ii*) effective separation of charge carriers, (*iii*) CO_2 adsorption, (*iv*) surface redox reactions, and (*v*) desorption of products. The reduction of CO_2 is often competing with HER from environmental H_2O, that is, the photogenerated electrons are consumed in this side reaction. Therefore, various approaches have been explored to avoid the hydrogen production.

17.3.2 Gas Solid Photocatalytic CO_2 Reduction with Water Vapor

The use of semiconductor powder for photocatalytic CO_2 reduction has been studied with water vapor that can in a certain extent minimize the hydrogen evolution in the reduction reaction and often CO or methane gas can be produced. The oxidation of water vapor leads to the production of oxygen as well. Therefore, thermodynamically, the photocatalysts need to meet the oxidation potential of water vapor and reduction potential of CO_2.

SACs can be applied to improve the stability, selectivity, and photocatalytic activity [18]. Li et al. [39] reported that Cu_1 species supported on defective TiO_2 exhibited activity for CO_2 photoreduction to CO under open-circuit conditions. The presence of surface defects (Ti^{3+}) provided active sites for the adsorption of oxygen atoms from CO_2 molecules. Photogenerated electrons transferred from bulk TiO_2 to surface Ti^{3+} sites and onto the π orbitals of adsorbed CO_2. According to experimental characterization and quantum chemical calculations, Cai et al. [40] showed that oxygen vacancy defect-rich TiO_2 (V_O-TiO_2) can firmly anchor single Pt atoms on the TiO_2 surface. The enhancement in the interaction between Vo-TiO_2 and single Pt atoms can promote the accumulation of electrons on the surface of the catalyst for the photocatalytic reaction. The optimal V_O-TiO_2 calcined at 500 °C exhibited the highest CH_4 and CO yields, which were nearly 2.2 times higher than those of P25.

Also in gas vapor solid reaction, the overall photocatalytic efficiency can also be improved by constructing heterojunction structures [41, 42]. Yu's group [28] proposed the S-type heterojunction structure to enhance the redox ability of photogenerated electrons and holes. Wang et al. [43] used a hydrothermal method to construct a hollow sphere structure of CdS-TiO_2 with an S-type heterojunction. Because the work function of TiO_2 is 6.8 eV and the work function of cadmium sulfide is 5.7 eV, when they form a closely contacted heterojunction, electrons flow from the CdS conduction band to TiO_2 conduction band at the interface (Figure 17.2f). Under illumination, due to the effect of the internal electric field, the electrons in the conduction band of CdS participate in the reduction of carbon dioxide, and the holes in the valence band of TiO_2 participate in the water oxidation reaction. This S-scheme heterojunction structure of CdS-TiO_2 enables the catalyst to exhibit excellent catalytic activity in the carbon dioxide reaction.

17.3.3 Gas Solid Photothermal Catalytic CO_2 Reduction with Hydrogen Gas

To avoid HER competing with the CO_2 reduction reaction, another approach is to separate the reactions, where water splitting occurs first and hydrogen gas is introduced to the second step, instead of water vapor. However, to trigger the CO_2 methanation reaction at the second step, often a photothermal reaction is used.

Thermochemical catalytic reduction of CO_2 to solar fuels is one of the most promising technologies for industrial development. In this process, the ultraviolet–visible (UV–vis) portion of solar light is utilized to excite charge carriers, while the visible–infrared (Vis–IR) light is used for in situ surface temperature enhancement of the catalyst [44]. The thermochemical catalytic reduction of CO_2 not only improves the utilization of solar light but enhances catalytic efficiency. The thermal effect addresses the bottleneck problems of low catalytic efficiency and limited energy transfer in photocatalysis, providing energy for overcoming the activation energy barrier and breaking through the kinetic limitations.

Based on the understanding of the interaction between light and matter, the current understanding of the photothermal mechanism can be categorized into three different modes: activation of thermal vibration of molecules, plasmon-induced localized heating, and nonradiative relaxation. To achieve efficient photothermal conversion, it is necessary to select suitable photothermal catalytic materials, taking into account various important factors such as optical and thermal properties.

TiO_2 is considered as one of the most active and stable photocatalysts. To accelerate the activity of TiO_2, efforts have been made to explore the thermal response of modified TiO_2 (Figure 17.5). For example, the introduction of oxygen vacancies can alter the band structure, resulting in broad and strong light absorption from UV to near-infrared, thus enhancing the photothermal effect. It was found that photothermal catalysis outperforms photocatalysis in CO_2 reduction, and oxygen vacancies (O_v) were found to enhance the activity of TiO_2 in photothermal catalysis [45].

Depositing noble metals on TiO_2 can significantly enhance catalytic activity. Noble metals exhibit significant photothermal effects in energy conversion, providing direct local heating through the localized surface plasmon resonance effect or carrier relaxation [46]. For example, Pd NPs-loaded TiO_2 photocatalysts exhibit photothermal effects in the reduction of CO_2 to CO, resulting in an improved CO_2 reduction rate [47].

17.3.4 Liquid Solid Photocatalytic CO_2 Reduction

To compete with HER from water, in literature, reduction of CO_2 is performed using saturated CO_2 solution or alkaline solution, where carbonate species are adsorbed

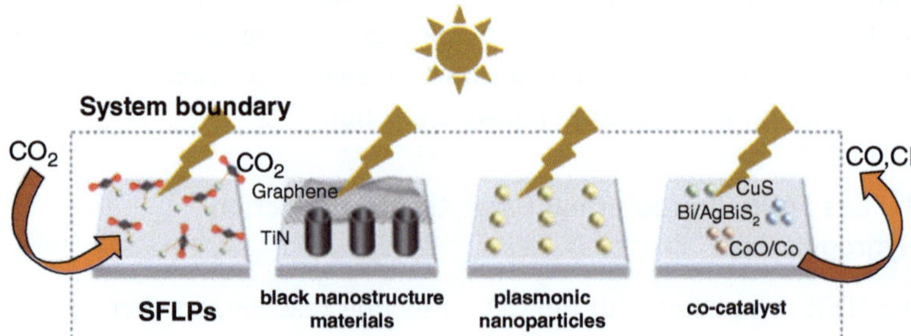

Figure 17.5 An overview of the methods to improve the photothermal activity. Reproduced from Ref. [27]/with permission of Elsevier.

and reacted on the surface of solid photocatalysts. The use of TiO_2 as powder photocatalyst in the suspension often leads to insufficient conversion of CO_2. Therefore, modifications on TiO_2 photocatalysts or development on the device can be adopted.

For example, Jin et al. [48] reported a method for light-driven conversion of bicarbonate (a model compound of CO_2) to formate on a Pd/TiO_2 catalyst. The localized surface plasmon resonance effect of Pd NPs promoted light absorption, while the local temperature increase enhanced the reaction rate and mass transfer. As a result, the synthesized Pd/TiO_2 catalyst exhibited excellent catalytic activity in the photothermal reduction of bicarbonate to formate, with a formate yield of 68.3%.

Additionally, PEC systems can promote charge separation through an externally applied bias-induced electric field, which often results in higher conversion efficiency than the PC system (Figure 17.6). Also the reduced and oxidized products can be spatially separated into two half-cells by an ion exchange membrane, thus effectively eliminating the possibility of product re-oxidation.

For example, Castro et al. [49] constructed a photoelectrochemical (PEC) system for CO_2 reduction using a TiO_2-based membrane electrode as the photoanode and Cu as the cathode. The observed maximum increase in current density of 4.3 mA cm^{-2} under UV irradiation, highlighting the potential advantages of the TiO_2-based photoanode-driven system in terms of productivity and efficiency. In the system with a copper plate as the cathode, products such as methanol, ethanol, hydrogen, carbon monoxide, and ethylene were detected. Under an external bias of −1.8 V (vs. Ag/AgCl), the maximum methanol production was 9.5 μmol m^{-2} s^{-1} with a Faradaic efficiency (FE) of 16.2%.

In the PEC process, maintaining long-term stability of the electrodes is challenging due to photocorrosion, electrocorrosion, and electrolyte degradation. In this regard, TiO_2, as a light- and chemically stable metal oxide, serves as an ideal electrode coating and has been extensively studied for PEC CO_2 reduction. Li et al. [50] demonstrated that dense ALD-TiO_2 protective film enhances the corrosion resistance of the photoelectrochemical cathode in near-neutral electrolytes. This film

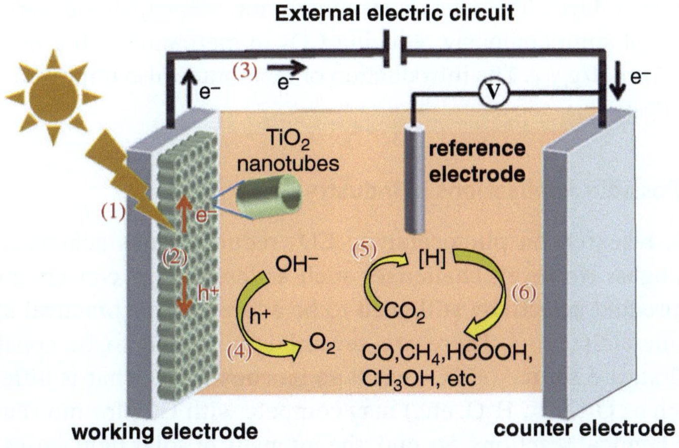

Figure 17.6 Schematic of a photoelectrochemical device for CO_2 reduction. Reproduced from Ref. [27]/with permission of Elsevier.

acts as a light absorber in the photoelectrode, while the catalytic layer consisting of grain-boundary-rich Au NPs promotes CO_2 reduction. Remarkably high synthesis gas production was achieved over an unprecedented wide range of PEC potentials, ranging from $-0.1\,V_{RHE}$ to as high as $0.4\,V_{RHE}$.

17.3.5 Selectivity Regulation for CO_2 Reduction

In general, the primary products in gas-phase CO_2 photocatalysis include CO or CH_4, and for CO_2 reduction reaction in liquid-phase, formates are the mostly reported product. In the case of photothermal systems, the main products are also CO, CH_4, and CH_3OH (with the assistance of H_2). Although the formation of C_2+ organic compounds has been occasionally observed in these studies, the yields and selectivity are relatively low. Moreover, there is still a gap in understanding the complex kinetics of CO_2 conversion to C_2+ products and the rate-determining steps that drive the selectivity of C_2+ products in the reaction mechanism.

Various strategies have been developed to synthesize high-performance photocatalysts that effectively reduce CO_2 by enhancing visible light utilization, promoting charge separation and transfer, improving CO_2 transport and adsorption, and increasing reaction kinetics. For instance, Wei et al. [51] selectively deposited Pt-Ru alloy NPs on the {101} facet of TiO_2, and the synergistic effect of Pt and Ru noble metals enhanced the efficiency of electron–hole pair separation, resulting in a CH_4 yield of $38.7\,\mu mol\,g^{-1}\,h^{-1}$ with a selectivity of 93.7%.

In addition, rational design of reactor, for example, photocatalytic membrane reactors integrate membrane separation and reaction to achieve chemical transformations, representing a process intensification strategy. This system allows for the continuous separation and recovery of reaction products while retaining the catalyst. For example, Baniamer et al. [52] synthesized a dual-layer photocatalytic membrane by immobilizing TiO_2 on porous polyethersulfone as the photocatalytic layer and using a nonporous thin layer of poly-ether-block-amide (PEBAX-1657) as the selective layer. This dual-layer membrane was capable of separating CO_2 from CH_4 and simultaneously reducing CO_2 to methanol with a high methanol yield of $697\,\mu mol/g_{cat\cdot h}$. The introduction of membrane also improved the stability of the catalyst.

17.3.6 Possible Applications in Industry

Currently, research on photocatalytic CO_2 reduction has achieved significant breakthroughs. However, challenges such as low energy conversion efficiency and low product selectivity still need to be addressed for practical applications and commercialization. There are several issues need to be considered and solved. First, the source of CO_2 needs to be considered, that is different impurities (such as O_2, SO_2, H_2O, etc.) may compete with CO_2 for molecular adsorption and chemical reactions. Second, the intensity of solar light varies depending on location, weather, and time. Third, the scale up of photocatalytic reactors that is the theory for lab scale batch reactors may not be suitable for large scale flow reactor.

17.4 Solar Fuel: Ammonia from N_2

Ammonia (NH_3) is a key component in the commercial synthesis of valuable chemicals, such as fertilizers and cleaning agents. Liquid ammonia, with its higher hydrogen content (17.6 wt%) and well-established production and transportation networks, is considered as a highly promising hydrogen carrier. The combustion products are nitrogen and water, making it an alternative to traditional fossil fuels and even hydrogen.

NH_3 is mainly synthesized through a Haber–Bosch process, which requires high temperatures of 400–500 °C and pressures of 15–25 MPa [53]. Many attempts have been taken to develop green and economic methods to synthesize ammonia. In nature, NH_3 fixation primarily occurs through biological nitrogen fixation inside organisms. Over 90% of the molecular nitrogen in the atmosphere is reduced to ammonia through the action of nitrogen-fixing microorganisms. Inspired by biological nitrogen fixation in nature, artificial solar-driven nitrogen fixation reaction is investigated. Photocatalytic nitrogen fixation utilizes abundant solar energy resources and water to provide the active electrons and protons for N_2 fixation directly (see reaction equation below), presents an attractive alternative to the Haber–Bosch process for NH_3 production.

$$N_2 + 3H_2O \rightarrow 2NH_3 + \frac{3}{2}O_2$$

The photocatalytic process for nitrogen reduction reaction (NRR) generally involves an electron and hole transfer process as illustrated in Figure 17.7. The holes oxidize water to produce O_2 and the electrons reduce N_2 to NH_3. Though the standard equilibrium potential of the overall reaction is small (N_2/NH_3 *vs* RHE $= -0.148$V) [54], the NRR process requires six electrons and the HER competes for electrons, making the reaction kinetically slow. In addition, N_2 is chemically inert, thus the yield of NH_3 remains generally very low.

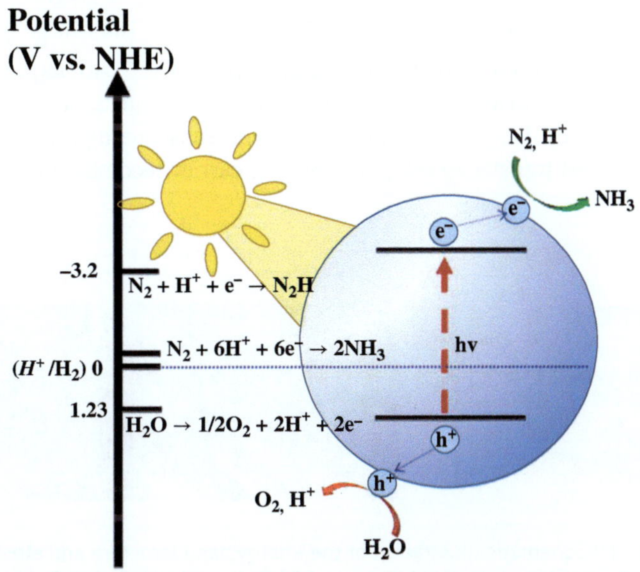

Figure 17.7 Schematic diagram of NRR process.

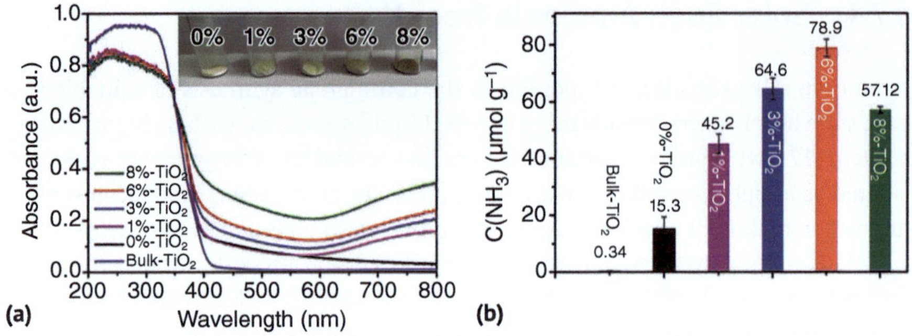

Figure 17.8 (a) UV-DRS of different TiO_2 nanosheets. (b) Yield of NH_3 under UV–vis illumination after 1 h. Reproduced from [5]/with permission of John Wiley & Sons.

Among the photocatalysts developed, TiO_2-based catalysts have attracted the widest attention [54], since in 1977, Schrauzer and Guth discovered that Fe-doped TiO_2 can catalyze the photocatalytic fixation of N_2 [55]. In order to enhance the NRR performance of TiO_2, efforts have been made include defect engineering, heterojunction engineering, doping, and loading noble metals on TiO_2.

For example, ultra-thin TiO_2 nanosheets with Cu as a dopant provide abundant oxygen vacancies and intrinsic compressive strain, thus enhance N_2 adsorption and provide a modified low-energy pathway to NH_3 [5]. As a result, a high NH_3 release rate is achieved $(78.9\,\mu mol\,g^{-1}\,h^{-1})$ under full sunlight irradiation and it reaches 1.54 and 0.72 $\mu mol\,g^{-1}h^{-1}$ under monochromatic irradiation at 600 and 700 nm (Figure 17.8).

Liu et al. prepared $TiO_2@C/g\text{-}C_3N_4$ heterostructures for NRR, showing excellent performance with an NH_3 yield of $250.6\,\mu mol\,g^{-1}\,h^{-1}$ [56]. The catalyst was synthesized by annealing a mixture of Ti_3C_2 and melamine. A unique arrangement of three Ti layers covering two C layers forms TiO_2 loaded on two-dimensional carbon nanosheets (denoted as $TiO_2@C$). Due to the alleviation of $g\text{-}C_3N_4$ aggregation by $TiO_2@C$, $TiO_2@C/g\text{-}C_3N_4$ exhibits a higher specific surface area $(49.5\,m^2g^{-1})$ compared to $g\text{-}C_3N_4$. Furthermore, the formation of a large amount of Ti^{3+} cations provide sufficient active sites. Under visible light irradiation, electrons transition from the valence band of $g\text{-}C_3N_4$ and TiO_2 to the conduction band. Subsequently, electrons transfer to TiO_2, while holes transfer to $g\text{-}C_3N_4$, reducing carrier recombination (Figure 17.9).

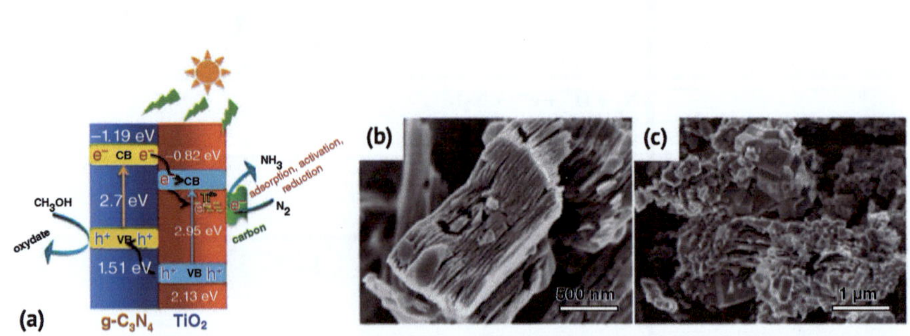

Figure 17.9 (a) Schematic illustration of the energy band structure and electron–hole separation of $TiO_2@C/g\text{-}C_3N_4$. SEM images of (b) Ti_3C_2 and (c)$TiO_2@C$. Reproduced from Ref. [56]/with permission of The Royal Society of Chemistry.

As a result, the NRR performance of the TiO_2@C/g-C_3N_4 heterostructure is approximately 18 and 10 times higher than that of g-C_3N_4 and TiO_2@C, respectively. Nevertheless, it has to be pointed out that a lot of experimental artifacts do interfere with many reports on successful N_2 splitting [57].

17.5 Summary

The use of TiO_2 has laid the foundation to scale up the photocatalytic reaction for solar fuel generation, such as hydrogen, CO/methane, methanol, and ammonia. The road ahead is challenging, but achieving the transition from laboratory to scalable practical applications is highly desirable in the future. This requires significant efforts in catalyst design, the construction of catalytic reactors, as well as supporting facilities. Solar fuel generation remains a most promising pathway for a sustainable development.

References

1 Ciamician, G. (1912). The photochemistry of the future. *Science* 36 (926): 385–394.
2 Li, X., Yu, J., Jaroniec, M. et al. (2019). Cocatalysts for selective photoreduction of CO_2 into solar fuels. *Chemical Reviews* 119 (6): 3962–4179.
3 Kong, T., Jiang, Y., and Xiong, Y. (2020). Photocatalytic CO_2 conversion: what can we learn from conventional CO_x hydrogenation? *Chemical Society Reviews* 49 (18): 6579–6591.
4 Inoue, T., Fujishima, A., Konishi, S. et al. (1979). Photoelectrocatalytic reduction of carbon dioxide in aqueous suspensions of semiconductor powders. *Nature* 277 (5698): 637–638.
5 Zhao, Y., Zhao, Y., Shi, R. et al. (2019). Tuning oxygen vacancies in ultrathin TiO_2 nanosheets to boost photocatalytic nitrogen fixation up to 700 nm. *Advanced Materials* 31 (16): 1806482.
6 Rahman, M.Z., Kibria, M.G., and Mullins, C.B. (2020). Metal-free photocatalysts for hydrogen evolution. *Chemical Society Reviews* 49 (6): 1887–1931.
7 Wang, Z., Li, C., and Domen, K. (2019). Recent developments in heterogeneous photocatalysts for solar-driven overall water splitting. *Chemical Society Reviews* 48 (7): 2109–2125.
8 Vikrant, K., Kim, K.-H., and Deep, A. (2019). Photocatalytic mineralization of hydrogen sulfide as a dual-phase technique for hydrogen production and environmental remediation. *Applied Catalysis B: Environmental* 259: 118025.
9 Ashraf, M., Ullah, N., Khan, I. et al. (2023). Photoreforming of waste polymers for sustainable hydrogen fuel and chemicals feedstock: waste to energy. *Chemical Reviews* 123 (8): 4443–4509.
10 Fujishima, A. and Honda, K. (1972). Electrochemical photolysis of water at a semiconductor electrode. *Nature* 238 (5358): 37–38.
11 Bolton, J.R., Strickler, S.J., and Connolly, J.S. (1985). Limiting and realizable efficiencies of solar photolysis of water. *Nature* 316 (6028): 495–500.

12 Zhou, X., Liu, G., Yu, J. et al. (2012). Surface plasmon resonance-mediated photocatalysis by noble metal-based composites under visible light. *Journal of Materials Chemistry* 22 (40): 21337–21354.

13 Cao, S., Chan, T.-S., Lu, Y.-R. et al. (2020). Photocatalytic pure water splitting with high efficiency and value by Pt/porous brookite TiO_2 nanoflutes. *Nano Energy* 67: 104287.

14 Wang, X., Long, R., Liu, D. et al. (2016). Enhanced full-spectrum water splitting by confining plasmonic Au nanoparticles in N-doped TiO_2 bowl nanoarrays. *Nano Energy* 24: 87–93.

15 Gupta, U. and Rao, C. (2017). Hydrogen generation by water splitting using MoS_2 and other transition metal dichalcogenides. *Nano Energy* 41: 49–65.

16 Zeng, H., Li, Z., Li, G. et al. (2022). Interfacial engineering of TiO_2/Ti_3C_2 MXene/ carbon nitride hybrids boosting charge transfer for efficient photocatalytic hydrogen evolution. *Advanced Energy Materials* 12 (1): 2102765.

17 Zhao, W., Chen, Z., Yang, X. et al. (2020). Recent advances in photocatalytic hydrogen evolution with high-performance catalysts without precious metals. *Renewable and Sustainable Energy Reviews* 132: 110040.

18 Zhou, X. (2021). TiO_2-supported single-atom catalysts for photocatalytic reactions. *Acta Physico-Chimica Sinica* 37 (6): 2008064.

19 Hejazi, S., Mohajernia, S., Osuagwu, B. et al. (2020). On the controlled loading of single platinum atoms as a co-catalyst on TiO_2 anatase for optimized photocatalytic H_2 generation. *Advanced Materials* 32 (16): 1908505.

20 Qin, S., Will, J., Kim, H. et al. (2023). Single atoms in photocatalysis: low loading is good enough! *ACS Energy Letters* 8 (2): 1209–1214.

21 Zhang, Y., Zhao, J., Wang, H. et al. (2022). Single-atom Cu anchored catalysts for photocatalytic renewable H_2 production with a quantum efficiency of 56%. *Nature Communications* 13 (1): 58.

22 Zhou, X., Häublein, V., Liu, N. et al. (2016). TiO_2 Nanotubes: nitrogen-ion implantation at low dose provides noble-metal-free photocatalytic H_2-evolution activity. *Angewandte Chemie International Edition* 55 (11): 3763–3767.

23 Zhou, X., Liu, N., Yokosawa, T. et al. (2018). Intrinsically activated $SrTiO_3$: photocatalytic H_2 evolution from neutral aqueous methanol solution in the absence of any noble metal cocatalyst. *ACS Applied Materials & Interfaces* 10 (35): 29532–29542.

24 Domaschke, M., Zhou, X., Wergen, L. et al. (2019). Magnéli-phases in anatase strongly promote cocatalyst-free photocatalytic hydrogen evolution. *ACS Catalysis* 9 (4): 3627–3632.

25 Liu, L., Zhang, Y., and Huang, H. (2021). Junction engineering for photocatalytic and photoelectrocatalytic CO_2 reduction. *Solar RRL* 5 (2): 2000430.

26 Kim, M., Lee, J., Je, M. et al. (2021). Electric field-driven one-step formation of vertical p–n junction TiO_2 nanotubes exhibiting strong photocatalytic hydrogen production. *Journal of Materials Chemistry A* 9 (4): 2239–2247.

27 Zhang, T., Han, X., Nguyen, N.T. et al. (2022). TiO_2-based photocatalysts for CO_2 reduction and solar fuel generation. *Chinese Journal of Catalysis* 43 (10): 2500–2529.

28 Xu, Q., Zhang, L., Cheng, B. et al. (2020). S-scheme heterojunction photocatalyst. *Chem* 6 (7): 1543–1559.

29 Xia, B., He, B., Zhang, J. et al. (2022). $TiO_2/FePS_3$ S-scheme heterojunction for greatly raised photocatalytic hydrogen evolution. *Advanced Energy Materials* 12 (46): 2201449.

30 Ruan, X., Huang, C., Cheng, H. et al. (2023). A twin S-scheme artificial photosynthetic system with self-assembled heterojunctions yields superior photocatalytic hydrogen evolution rate. *Advanced Materials* 35 (6): 2209141.

31 Yang, L., Zhang, T., Han, X. et al. (2023). Fabrication of step-scheme heterojunction between layered MoO_3/TiO_2 for photocatalytic H_2 evolution and study on the mechanism. *Advanced Sustainable Systems* 7 (1): 2200402.

32 Han, X., Dong, W., Li, L. et al. (2023). Improved photocatalytic activity of TiO_2 with a regulated covalent organic framework thin film. *Chemical Communications (Cambridge)* 59 (79): 11863–11866.

33 Dang, V.-H., Nguyen, T.-A., Le, M.-V. et al. (2024). Photocatalytic hydrogen production from seawater splitting: current status, challenges, strategies and prospective applications. *Chemical Engineering Journal* 484: 149213.

34 Zhang, J., Lei, Y., Cao, S. et al. (2022). Photocatalytic hydrogen production from seawater under full solar spectrum without sacrificial reagents using TiO_2 nanoparticles. *Nano Research* 15 (3): 2013–2022.

35 Cheng, Z., Zhang, X., Bo, C. et al. (2024). Precise design of TiO_2 photocatalyst for efficient photocatalytic H_2 production from seawater splitting. *International Journal of Hydrogen Energy* 55: 542–549.

36 Speltini, A., Sturini, M., Dondi, D. et al. (2014). Sunlight-promoted photocatalytic hydrogen gas evolution from water-suspended cellulose: a systematic study. *Photochemical & Photobiological Sciences* 13 (10): 1410–1419.

37 Tang, X., Han, X., Sulaiman, N.H.M. et al. (2023). Recent advances in the photoreforming of plastic waste: principles, challenges, and perspectives. *Industrial & Engineering Chemistry Research* 62 (23): 9032–9045.

38 Uekert, T., Kasap, H., and Reisner, E. (2019). Photoreforming of nonrecyclable plastic waste over a carbon nitride/nickel phosphide catalyst. *Journal of the American Chemical Society* 141 (38): 15201–15210.

39 Liu, L., Zhao, C., and Li, Y. (2012). Spontaneous dissociation of CO_2 to CO on defective surface of $Cu(I)/TiO_{2-x}$ nanoparticles at room temperature. *The Journal of Physical Chemistry C* 116 (14): 7904–7912.

40 Cai, S., Wang, L., Heng, S. et al. (2020). Interaction of single-atom platinum–oxygen vacancy defects for the boosted photosplitting water H_2 evolution and CO_2 photoreduction: experimental and theoretical study. *The Journal of Physical Chemistry C* 124 (45): 24566–24579.

41 Bai, S., Jiang, J., Zhang, Q. et al. (2015). Steering charge kinetics in photocatalysis: intersection of materials syntheses, characterization techniques and theoretical simulations. *Chemical Society Reviews* 44 (10): 2893–2939.

42 Yang, L., Du, J., Deng, J. et al. (2023). Defective Nb_2C MXene cocatalyst on TiO_2 microsphere for enhanced photocatalytic CO_2 conversion to methane. *Small* 20 (19): e2307007.

43 Wang, Z., Chen, Y., Zhang, L. et al. (2020). Step-scheme CdS/TiO_2 nanocomposite hollow microsphere with enhanced photocatalytic CO_2 reduction activity. *Journal of Materials Science & Technology* 56: 143–150.

44 Wang, X., Wang, F., Sang, Y. et al. (2017). Full-spectrum solar-light-activated photocatalysts for light–chemical energy conversion. *Advanced Energy Materials* 7 (23): 1700473.

45 Li, D., Huang, Y., Li, S. et al. (2020). Thermal coupled photoconductivity as a tool to understand the photothermal catalytic reduction of CO_2. *Chinese Journal of Catalysis* 41 (1): 154–160.

46 Gao, M., Peh, C.K., Zhu, L. et al. (2020). Photothermal catalytic gel featuring spectral and thermal management for parallel freshwater and hydrogen production. *Advanced Energy Materials* 10 (23): 2000925.

47 Xu, C., Huang, W., Li, Z. et al. (2018). Photothermal coupling factor achieving CO_2 reduction based on palladium-nanoparticle-loaded TiO_2. *ACS Catalysis* 8 (7): 6582–6593.

48 Jin, B., Ye, X., Zhong, H. et al. (2020). Light-driven hydrogenation of bicarbonate into formate over nano-Pd/TiO_2. *ACS Sustainable Chemistry & Engineering* 8 (17): 6798–6805.

49 Castro, S., Albo, J., and Irabien, A. (2020). Continuous conversion of CO_2 to alcohols in a TiO_2 photoanode-driven photoelectrochemical system. *Journal of Chemical Technology & Biotechnology* 95 (7): 1876–1882.

50 Li, C., Wang, T., Liu, B. et al. (2019). Photoelectrochemical CO_2 reduction to adjustable syngas on grain-boundary-mediated a-Si/TiO_2/Au photocathodes with low onset potentials. *Energy & Environmental Science* 12 (3): 923–928.

51 Wei, Y., Wu, X., Zhao, Y. et al. (2018). Efficient photocatalysts of TiO_2 nanocrystals-supported PtRu alloy nanoparticles for CO_2 reduction with H_2O: synergistic effect of Pt-Ru. *Applied Catalysis B: Environmental* 236: 445–457.

52 Baniamer, M., Aroujalian, A., and Sharifnia, S. (2021). Photocatalytic membrane reactor for simultaneous separation and photoreduction of CO_2 to methanol. *International Journal of Energy Research* 45 (2): 2353–2366.

53 Erisman, J.W., Sutton, M.A., Galloway, J. et al. (2008). How a century of ammonia synthesis changed the world. *Nature Geoscience* 1 (10): 636–639.

54 Chen, J., Zhang, W., Li, H. et al. (2021). Recent advances in TiO_2-based catalysts for N_2 reduction reaction. *SusMat* 1 (2): 174–193.

55 Schrauzer, G. and Guth, T. (2002). Photolysis of water and photoreduction of nitrogen on titanium dioxide. *Journal of the American Chemical Society* 99 (22): 7189–7193.

56 Liu, Q., Ai, L., and Jiang, J. (2018). MXene-derived TiO_2@C/g-C_3N_4 heterojunctions for highly efficient nitrogen photofixation. *Journal of Materials Chemistry A* 6 (9): 4102–4110.

57 Andersen, S.Z., Čolić, V., Yang, S. et al. (2019). A rigorous electrochemical ammonia synthesis protocol with quantitative isotope measurements. *Nature* 570 (7762): 504–508.

18

Visible-light-induced Selective Oxidation of Organic Molecules over Nanostructured TiO_2

*Yuexin Wang, Fengwei Huang, Xianjun Lang**

Hubei Key Lab on Organic and Polymeric Optoelectronic Materials, College of Chemistry and Molecular Sciences, Wuhan University, Wuhan, China

*Corresponding author: xianjunlang@whu.edu.cn

18.1 Introduction

Because of various functional properties and composition of two earth-abundant elements, TiO_2, as one of the most attractive materials, has been most widely exploited in solar cells, transparent conductive films, sensors, communication devices, and so on. Especially, the chemical and thermal stability, intrinsic reusability, nontoxicity, and strong redox ability endow TiO_2 with great potential as a vital photocatalyst, and TiO_2 provides an effective avenue to implement green and sustainable chemical transformations [1]. Notably, with anatase, rutile, brookite, and mixed-phase, TiO_2 exhibits various patterns and morphologies, leading to different physical and chemical properties [2, 3]. However, pristine TiO_2 without any modification only harvests light in the ultraviolet (UV) range caused by inherent large bandgap (E_g). Moreover, the high recombination rate of charge carriers significantly constrains the practical application of TiO_2. Thus, there have been many strategies to modify pristine TiO_2 for altering the surface and band structure to extend the absorption range into visible light.

Based on acid–base interaction, organic substrates can be adsorbed onto the surface of TiO_2, affording surface complexes, which have the ability to absorb visible light through electron transfers from the ground state of organic substrates into the conduction band (CB) of TiO_2 [4]. Apart from TiO_2 surface complexes with substrates, the other organic molecules and organic polymers are also used to extend the absorption in visible light [5]. In contrast to TiO_2 surface complexes, the adsorption of photoactive molecules, such as organic dyes, serves for surface sensitization. By utilizing organic dye, sensitized TiO_2 further extends the absorption range of visible light. The photoinduced electron (e^-) transfers from the excited state of the organic dye into the CB of TiO_2.

Moreover, doping with transition metal ions is also a viable strategy. The *d* and *f* orbitals of transition metal can alter the band structure of TiO_2, which also

Nanostructured TiO_2: Synthesis, Photocatalytic Properties and Applications, First Edition.
Edited by Jia Hong Pan, Wan In Lee and Detlef W. Bahnemann

contributes to visible light absorption. Introducing oxygen defects through lattice distortion is another merit of metal-doped TiO_2, which relies on the radius and concentration of transition metal ions [6]. Meanwhile, the Fermi energy level of supported noble metal is more adjacent potential of the CB of TiO_2. The combinations of pristine TiO_2 with other materials, such as CdS, by heterostructures, can prolong the recombination path of photoinduced charge carriers, resulting in a longer lifetime. Besides, the other materials with narrower E_g promote the absorption of visible light [7, 8]. Due to the small particle size of TiO_2, porous materials with sufficient Brunauer–Emmett–Teller (BET) specific surface areas are commonly loaded onto TiO_2. These heterostructures improve the interaction of TiO_2 and organic substrates, as well as increase the density of active sites [9].

Based on various merits, TiO_2 has been explored for photocatalytic applications, in which organic transformations often require more control of the reaction for desired products, contrasting to degradation of organic pollutants. Thus, there still remains a great challenge for modified TiO_2 photocatalysts. During the last decade, modified TiO_2 has attracted ample attention toward selective organic transformation. Therefore, it is highly valuable and desirable to meticulously understand the state-of-the-art development of photocatalytic organic transformations over TiO_2-based materials. In this chapter, visible-light-induced organic transformations over TiO_2 are summarized with a particular focus on selective oxidation. Here, these visible-light-induced selective oxidations over nanostructured TiO_2 are categorized into four functional group transformations: oxidation of alcohols, oxidation of amines, oxidation of sulfides, and coupling of C—C bonds.

18.2 Visible-light-induced Selective Oxidation of Organic Molecules

Visible light photocatalysis is of great significance for biomedicine, energy storage, materials synthesis, and pollutant degradation [10, 11]. In particular, selective oxidation of organic molecules is essential toward organic transformation. However, unlike photocatalytic selective oxidations, conventional oxidation methods of organic molecules generate substantial by-products, chemical waste, toxic compounds, and hazardous gases, ultimately disrupting ecological balance. In this case, due to the use of environment-friendly and energy-efficient solar energy, visible-light-induced selective oxidation of organic molecules over TiO_2 is burgeoning to address these issues [12, 13]. TiO_2 typically enables visible-light-induced selective oxidation of organic molecules for four major functional group transformations (Figure 18.1).

18.2.1 Visible-light-induced Selective Oxidation of Alcohols

The oxidation of alcohols is a significant method to produce aldehydes, which can be realized by TiO_2 photocatalysis. Since the first report on TiO_2 for selective oxidation of alcohols under UV irradiation, TiO_2-based materials have been widely adopted as photocatalysts for this transformation. It can be concluded that the oxidation of

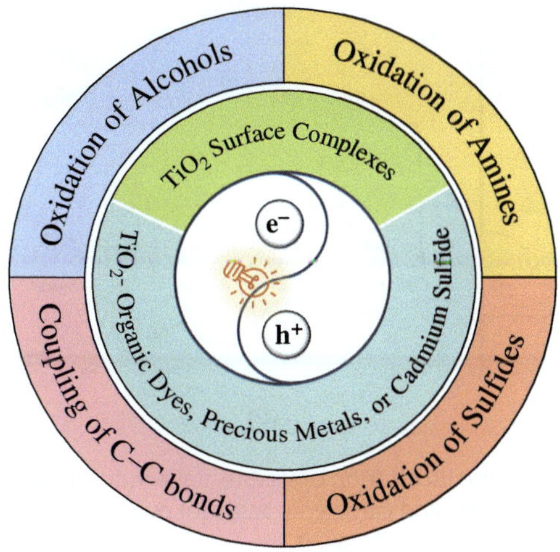

Figure 18.1 The visible-light-induced selective oxidation of organic molecules over nanostructured TiO_2.

alcohols with molecular oxygen (O_2) is a radical chain reaction. Thereinto, the photoinduced abstraction of α-H leads to a short-lived radical, and O_2 partakes in the radical chain reaction. Meanwhile, the participation of O_2 has been further verified through isotope labeling studies [14]. Without O_2, two electron transfer can also induce the oxidation of alcohols with inferior conversion. In contrast, with O_2 in benzotrifluoride, the oxygen atom of aldehyde is transferred from O_2 to the C_α of the alcohol rather than from the substrate alcohol (Eq. 18.1). In detail, the alcohols adsorb on the surface of TiO_2 with the production of C—O—Ti. The photoinduced hole (h^+) attacks the alcohols and induces the formation of the C-centered radical connected with Ti^{3+}. Meanwhile, the formed C-centered cationic radical, upon combining with O_2 and e^- produces the hyperoxide intermediate. The breakages of C—O and O—O lead to the formation of targeted aldehyde. At last, Ti/peroxide species is decomposed by H^+.

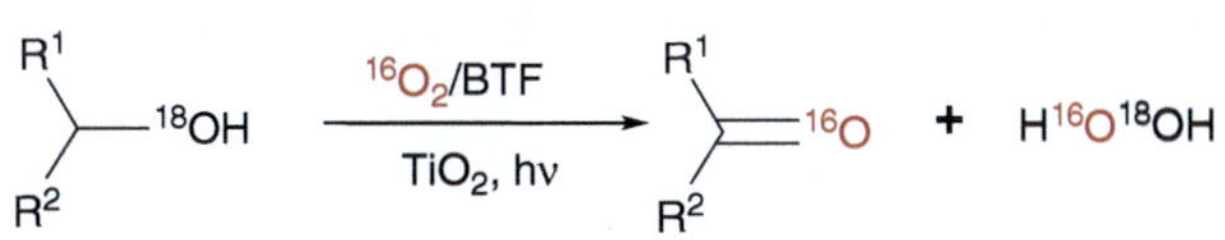

Eq. (18.1)

Notably, the decomposition of surface Ti/peroxide species caused by the oxidation restrains the reaction obviously. The regeneration of active Ti sites results in acceleration of the oxidation of alcohols. To improve the utilization of solar energy, many strategies are adopted to enhance the efficiency of charge carrier separation for visible-light-induced selective oxidation of alcohols. The acid–base interaction of alcohols with surface OH groups on TiO_2 leads to the formation of the alkoxide ([Ti]-O-CH_2-Ph) species (Figure 18.2a), which endows formed interfacial TiO_2 surface complex with visible-light absorptions [15]. Photocatalytic selective oxidation of CH_2—OH bonds was exploited to afford carbonyl acids (Figure 18.2b). In detail, it exposed that the complexing with alcohol species on TiO_2 surface led to absorption

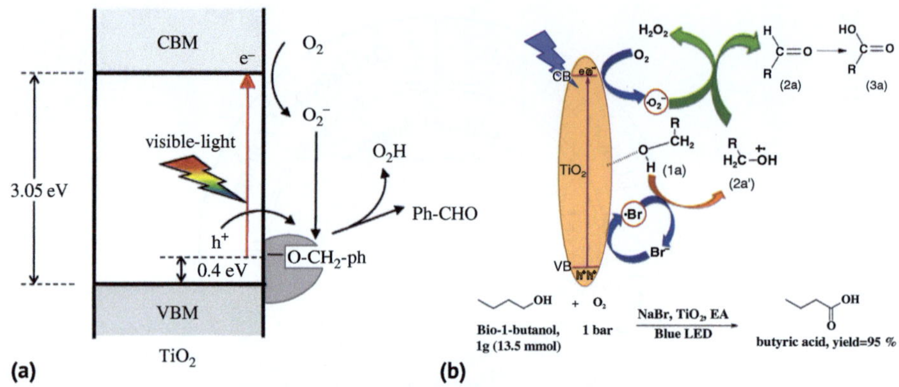

Figure 18.2 The schematic illustration of visible-light-induced selective oxidation of alcohols into aldehydes (a) and carboxylic acids (b) over TiO_2. Reproduced from Refs. [15, 16] with permission from Elsevier.

of visible light. Also, the bromine radicals, from the combination of NaBr and h^+, enhance the oxidation of alcohol [16]. On the basis of TiO_2 surface complexes, tetrahydroquinoline-linked covalent organic framework interacts with TiO_2, forming QH-COF-TiO_2. As a visible light photocatalyst, QH-COF-TiO_2 induces the oxidation of benzyl alcohols with high selectivities [17].

Besides, organic dyes, such as eosin Y [18], alizarin red S (ARS) [19], and 5(6)-carboxyfluorescein (5(6)-FAM) [20] sensitized TiO_2 for visible-light-induced selective oxidation of alcohols with air have been achieved more efficiently. The selective oxidation of benzyl alcohols with O_2 was directly implemented over ARS-TiO_2 under green light irradiation (Figure 18.3a). The visible-light-induced oxidation can be neatly merged with tetramethylpiperidine *N*-oxyl (TEMPO) as a redox mediator, overcoming the barrier of electron transfer from the dye cation radical to alcohols that induces the reaction through direct two electron transfer for selective oxidation of benzyl alcohols over 5(6)-FAM-TiO_2 with TEMPO under green light irradiation (Figure 18.3b). On the other hand, with suitable redox potential, TEMPO accelerates the electron transfer to afford the reaction product aldehydes with higher selectivities.

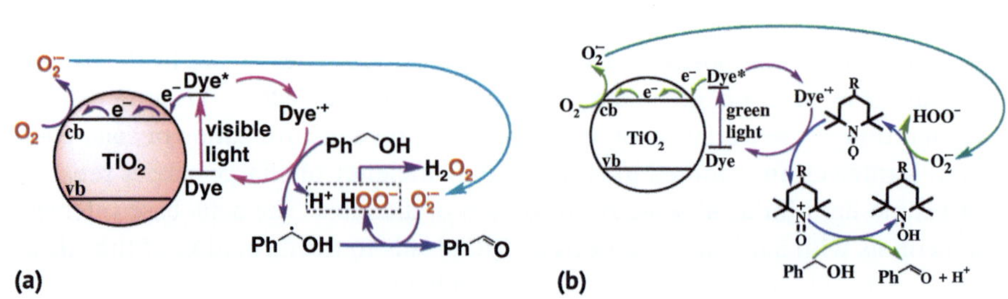

Figure 18.3 The schematic illustration of visible-light-induced selective oxidation of alcohols into aldehydes without (a) or with (b) TEMPO over dye-TiO_2. Reproduced from Refs. [19, 20] with permission from Elsevier/The Royal Society of Chemistry.

Intriguingly, N-doped TiO_2 possesses inferior photocatalytic activity to undoped TiO_2 nanofibers in the oxidation of 4-methoxybenzyl alcohol. Replacing the coordinated O with N atoms reduces the adsorption free energy of O_2 in the vicinity of N-dopant sites [21].

Based on the underlying photocatalytic mechanism, many strategies focus on improving selectivities under greener and milder reaction conditions. Because of the valence band (VB) potential of 2.7 V *vs.* normal hydrogen electrode (NHE), TiO_2 owns a moderate selectivity in the photocatalytic oxidation of alcohols in water [22]. Especially, the formed aldehyde is decomposed by further reaction on the TiO_2 surface. In various photocatalytic reactions, acetonitrile (CH_3CN) is used as the reaction medium to enhance the selectivity due to the shield effect [23]. Besides, acting as a weak base, CH_3CN stabilizes reaction products through solvation [24]. In continuation, maghemite-TiO_2 was prepared through ultrasound-assisted wet impregnation for highly effective visible-light-induced oxidation of benzyl alcohol in an organic or aqueous medium [25]. Therefore, using CH_3CN as the reaction medium inhibits the proton transfer and formation of radical species, thereby preventing the formation of undesired side reaction products. Besides, Fe^{3+}-modified TiO_2 improves the photocatalytic activity for visible-light-induced selective oxidation of benzyl alcohol, which is ascribed to the electronic transition from O 2*p* with occupied Fe 3*d* orbitals energy levels to CB of TiO_2 and subsequent to unoccupied Fe 3*d* orbitals (Figure 18.4a) [26].

Notably, CdS quantum dots modified TiO_2 with exposed (001) facets were synthesized to extend the absorption range of TiO_2 for visible-light-induced selective oxidation of alcohols (Figure 18.4b). CdS quantum dots enhance photo-response through heterojunction, as well, the exposed (001) facets provide oxygen vacancy (OV) and higher surface energy. Thus, it was realized with high activities for visible light photocatalytic aerobic oxidation of alcohols into aldehydes [27].

18.2.2 Visible-light-induced Selective Oxidation of Amines

In contrast to alcohols, amines are more active in nature. Thus, it is more difficult for selective oxidation of amines into corresponding imines without undesired

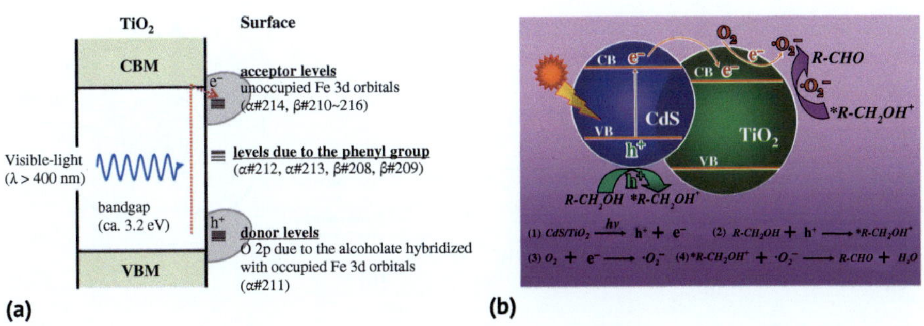

Figure 18.4 The schematic illustration of visible-light-induced selective oxidation of alcohol into aldehyde over Fe^{3+}-TiO_2 (a) and CdS-TiO_2 (b). Reproduced from Refs. [26, 27] with permission from Elsevier.

by-products. However, the oxidation of amines is commonly adopted to generate imines, which are important intermediates for organic transformations. Like the oxidation of alcohols, the oxidation of primary and secondary amines over TiO_2 with O_2 agrees well with that of alcohol at the first step, accompanied by the formation of the same aldehyde intermediate (Eq. 18.2) [28]. The presence of benzaldehyde as the intermediate was revealed through the photocatalytic oxidation of benzylamine. Next, the nucleophilic attack of the afforded aldehyde by benzylamine produces the corresponding imine *N*-benzylidenebenzylamine.

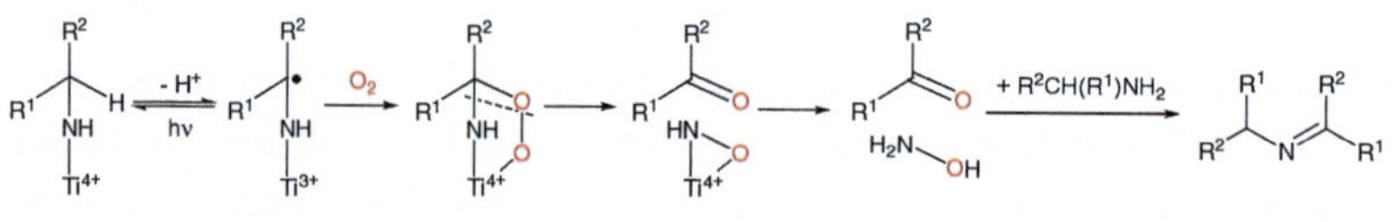

Eq. (18.2)

Simply by switching off all the UV irradiation, the reaction can be turned into a visible-light-induced one (Figure 18.5a) [29]. The visible light absorption is attributed to a TiO_2 surface complex with amines. In detail, the benzylamine is oxidized into *N*-centered cationic radical by an h^+. Furthermore, the *N*-centered cationic radical is converted into benzylideneamine by superoxide ($O_2^{\bullet-}$), resulting in the production of hydrogen peroxide (H_2O_2) rather than hydroxylamine (NH_2OH). In the end, by coupling benzylamine with benzylideneamine, the *N*-benzylidenebenzylamine as the desirable product is obtained, accompanied with the release of NH_3.

Except for the selective oxidation of amines into imines, the oxidative dehydrogenation of *N*-heterocyclic amines was induced by Rh-photodeposited TiO_2 nanoparticles, accompanied with the release of hydrogen (Figure 18.5b). *N*-heterocycle can be adsorbed onto TiO_2 to form a visible-light-responsive surface complex. In essence, Rh^0 is the most efficient in diverse noble metals as a cocatalyst. The oxidative dehydrogenation of various *N*-heterocycles, such as tetrahydroisoquinolines, tetrahydroquinolines, and indolines, was accomplished in good to excellent yields [30].

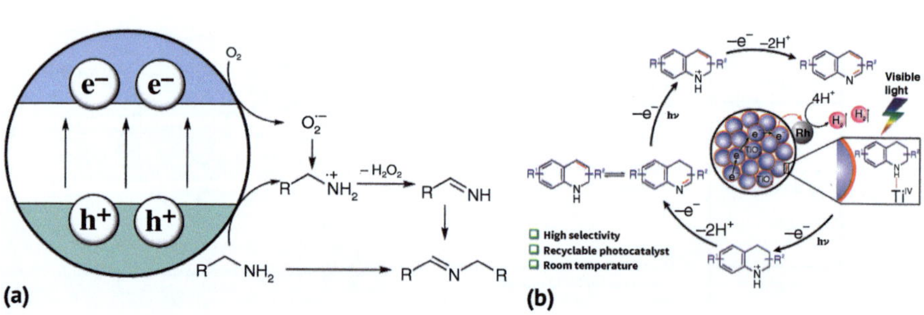

Figure 18.5 The schematic illustration of visible-light-induced selective oxidation of amines into imines over TiO_2 (a) and the oxidative dehydrogenation of *N*-heterocyclic amines over Rh/TiO_2 (b). Reproduced from Ref. [30] with permission from American Chemical Society.

The selectivity of visible-light-induced oxidation of amines into imines is more challenging to control. A redox mediator can be introduced to overcome these challenges. As such, the incorporation of TEMPO as a redox mediator can inhibit the direct involvement of h^+ and bypass benzaldehyde as an intermediate during the visible-light-induced oxidation of benzylamine, resulting in smooth conversions with satisfactory selectivities. Extending the TiO_2 surface with other organic molecules also induces the visible-light-induced selective oxidation of amines. The *ortho*-hydroxyl and sodium sulfonate groups endow the robust interaction of sodium 6,7-dihydroxynaphthalene-2-sulfonate (DHNS) with TiO_2 to form a surface complex (Figure 18.6a). On one hand, TEMPO mediates electron and proton transfers for the oxidation of amines. On the other hand, extending the π-conjugated system of organic molecules broadens visible light absorption. Polyimide, an organic polymer, is used for the construction of polyimide-TiO_2 as an effective visible light photocatalyst for selective oxidation of amines (Figure 18.6b). Because of the high oxidation potential of VB, various ROS are generated on the surface of TiO_2, such as hydroxyl radical ($^{\bullet}OH$) and H_2O_2. (Figure 18.6b) [31]. In detail, polyimide adsorbed on TiO_2 can inject e^- from the highest occupied molecular orbital (HOMO) of polyimide into CB of TiO_2, forming polyimide$^{\bullet+}$, which leads to the oxidation of TEMPO into $TEMPO^+$. The transformation of $TEMPO^+$ into TEMPOH with proton and two electron transfers in one step can cooperate with the oxidation of benzylamine, resulting in the direct product of benzylideneamine as the intermediate. Ultimately, *N*-benzylidenebenzylamine as the final product is obtained through coupling the unreacted benzylamine with benzylideneamine. Importantly, the return of TEMPOH to the state of TEMPO is realized through the hydrogen atom abstraction of $O_2^{\bullet-}$, closing the catalytic cycle. Besides, the derivatives of TEMPO, including 4-amino-TEMPO and 4-acetamido-TEMPO, have also been exploited for visible-light-induced selective oxidation of amines as a redox mediator [32].

The rise of activity over TiO_2 for visible-light-induced oxidation of amines is of great interest. A mechanistic study on Pd_3Pb/TiO_2 for visible-light-induced oxidation of amine to imine was implemented [33]. This study verified that Pb accelerates

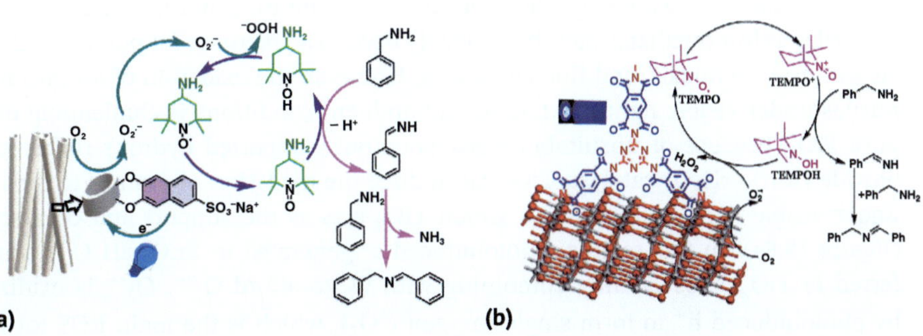

Figure 18.6 The schematic illustration of visible-light-induced oxidation of amines into imines over DHNS-TiO_2 (a) and polyimide-TiO_2 (b). Reproduced from Refs. [31, 32] with permission from John Wiley & Sons/Elsevier.

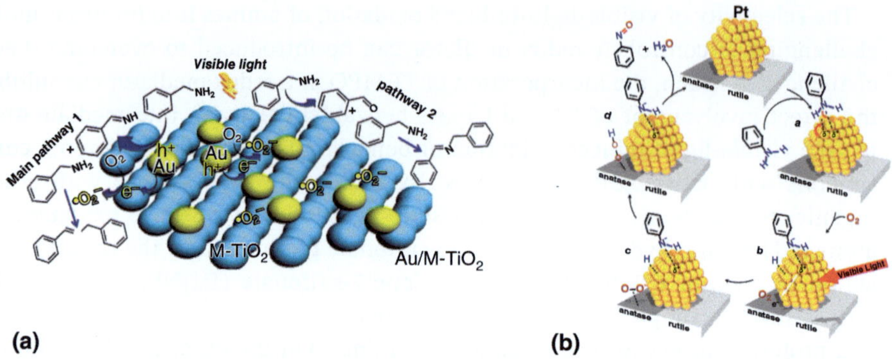

(a) (b)

Figure 18.7 The schematic illustration of visible-light-induced selective oxidation of amines into imines over Au/TiO_2 (a) and oxidation of aniline into nitrobenzene over Pt/TiO_2 (b). Reproduced from Refs. [35, 36] with permission from Elsevier/American Chemical Society.

the desorption effect of imines, while the basic sites on the support promote the adsorption of amines. Notably, Pd_3Pb is superior to Pb for the desorption of imines. Besides, it was found that mixed-phase TiO_2(B)-anatase TiO_2 shows higher photocatalytic activity than pure TiO_2(B) and anatase TiO_2, which is ascribed to the interfacial charge separation due to phase heterojunction and different adsorption of benzylamine due to surface property [34]. Notably, visible-light-induced oxidation of benzylamine over Au/TiO_2 also can be performed in CH_3CN [35]. However, the mechanism of photocatalytic plasmonic oxidation of amines is different. The key intermediate during the oxidation of benzylamine is benzylideneamine rather than benzaldehyde (Figure 18.7a). The rate-determining step did not involve the cleavage of the C_α—H bond.

Nitrosobenzene is a multifaceted intermediate in organic transformations. Pt/TiO_2 was used for visible-light-induced selective oxidation of aromatic amines into nitrosobenzene compounds (Figure 18.7b) [36]. Loading magnesium oxide (MgO) onto TiO_2 promoted the emergence of electron-rich lattice oxygen to facilitate the deprotonation of aniline. Similar to visible-light-induced selective oxidation of amines into imines, aromatic amines also were adsorbed onto Pt/TiO_2 to form a surface complex, in which $O_2^{\bullet-}$ occupies the dominant ROS in this reaction.

Nitriles as intermediates have been widely used as a precursor for polymers, dyes, agricultural chemicals, and fine chemicals. However, the oxidation of amines into nitriles under visible light irradiation and ambient conditions is challenging over pure TiO_2, because of unsuitable redox potential. Supported hydrous ruthenium dioxide ($RuO_2{\cdot}xH_2O$) catalyst, TiO_2 can induce the oxidation of amines to nitriles under visible light irradiation, in which TiO_2 acts as the support and cocatalyst (Figure 18.8a). In this regime, photoinduced e^- generated in $RuO_2{\cdot}xH_2O$ is transferred to TiO_2, subsequently combining with O_2 to afford $O_2^{\bullet-}$. $O_2^{\bullet-}$ is oxidized by photoinduced h^+ to form singlet oxygen ($^1O_2$), which is the main ROS for the oxidation of amines into corresponding nitriles [37]. The amount of Ru contributes greatly to the reaction. As the content of Ru increases, there is a tendency of increasing the ratio of benzonitrile in the presence of O_2 [38].

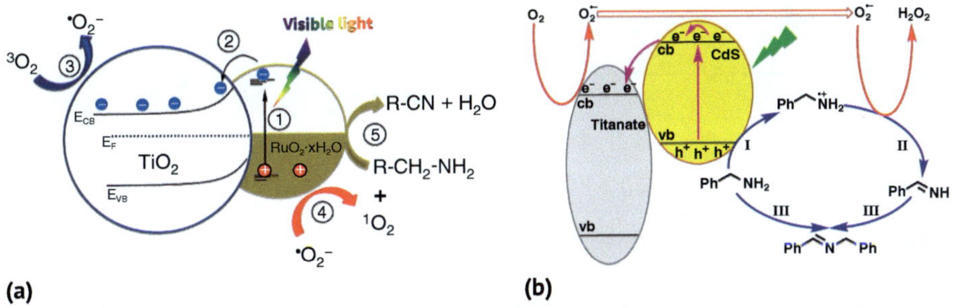

Figure 18.8 The schematic illustration of visible-light-induced selective oxidation of amines into nitriles over $RuO_2{\cdot}xH_2O/TiO_2$ (a) and into imines over CdS/TiO_2 (b). Reproduced from Refs. [37, 39] with permission from CC BY/Elsevier.

CdS/titanate induces the highly selective oxidation of amines under green light irradiation (Figure 18.8b) [39]. The optimized CdS/titanate hybrid can induce the highly selective oxidative homocoupling of benzyl amines in methanol under 520 nm green light irradiation. Meanwhile, the adsorption of amines on TiO_2 not only enhances the interaction between Ag-TiO_2 photocatalyst and amine substrates but also leads to reduced E_g of photocatalysts. Therefore, the activity was enhanced through facilitating efficient electron transfers and increasing visible light absorption [40].

Thermo-enhanced visible-light-induced selective oxidation of benzyl amines is achieved over Au/TiO_2 [41]. The benzylamine-TiO_2 complex, influenced by the OVs of TiO_2, plays a vital role in achieving good conversions of benzylamine and selectivity of imine. The influence of OV concentration of TiO_2 on their photocatalytic activity needs to be further clarified. $TiO_2/Ti_3C_2/CdS$ was applied for oxidation of benzylamine with O_2 [42]. The surface OVs are essential for the activation of the adsorbed O_2 on the TiO_2 photocatalyst surfaces to generate ROS. The surface OVs form shallow donor states to promote photoinduced charge carrier separation.

18.2.3 Visible-light-induced Selective Oxidation of Organic Sulfides

Due to various oxidation states of sulfur atom (−2, −1, 0, +2, +4, and +6), numerous methods have been developed to convert organic sulfides into more useful products. The high reactivity of organic sulfides enables their effective conversion over TiO_2. The reactions are induced by energy transfer, hydrogen atom transfer, or electron transfer, leading to the generation of sulfur-centered radical cations. The reaction then proceeds through C_α—H protonation, sulfoxidation, and C—S bond cleavage.

As oxidation products of thiols, disulfides are relatively stable and play important roles in various biological systems. Visible-light-induced oxidation of thiols into disulfides represents a green and sustainable process. Under 355 nm fluorescent lamp irradiation, thiols underwent oxidation into disulfides, albeit with low yields [43]. Notably, thiol complexation with TiO_2 results in an upshift in its HOMO energy level and a reduction in the oxidation potential of TiO_2. This offers a practical and mild approach for the synthesis of disulfides via visible-light-induced

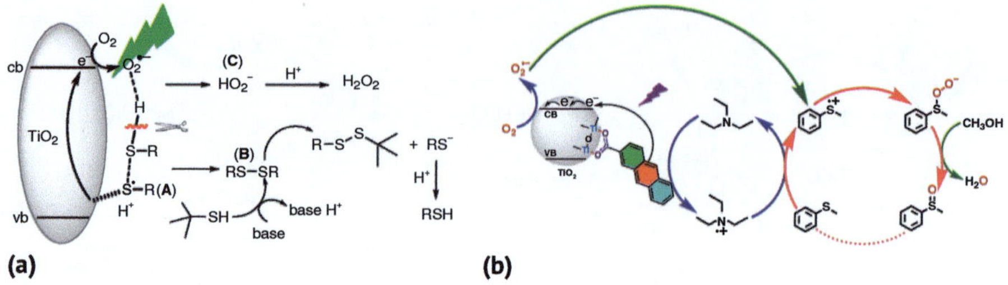

Figure 18.9 The schematic illustration of visible-light-induced selective oxidation of thiols into disulfides over TiO_2 (a) and oxidation of sulfides into sulfoxides over 2-AA-TiO_2 with TEA (b). Reproduced from Refs. [45, 47] with permission from Elsevier.

selective oxidation of thiols with O_2 [44]. Further, the conversion of thiol was up to 96% under 520 nm green light irradiation in 30 min with the presence of triethylamine (TEA) (Figure 18.9a) [45]. Notably, asymmetric disulfides were effectively synthesized with high selectivities through visible-light-induced selective oxidation of mixed thiols.

Under UV irradiation, phenyl methyl sulfides were oxidized into the corresponding sulfoxides over TiO_2, with H_2O enhancing the reaction [46]. In contrast, the incorporation of benzylamine gives rise to visible-light-induced selective oxidation of phenyl methyl sulfides over TiO_2 [28]. Visible light can only induce modest oxidation of benzylamine, with essentially no observable oxidation of phenyl methyl sulfide over TiO_2. Importantly, the adsorption of benzylamine onto the TiO_2 surface affords a complex that extends the absorption range of the system into visible light. Thus, sulfides are oxidized into sulfoxide under visible light irradiation over benzylamine-TiO_2. Based on this regime, trimethylamine (TMA) and TEA as organic molecules are accessible to extend the absorption range of TiO_2. Besides, the potential of single electron oxidation of sulfide and VB of Degussa P25 rutile are 1.75 and 2.31 V (*vs.* NHE), respectively. The E_{HOMO} of the TMA-TiO_2 complex is about 2.23 V (*vs.* NHE), suggesting that the incorporation of TEA as a redox mediator leads to a smooth electron transfer.

2-Anthracene acid (2-AA) as an organic molecule can be adsorbed onto TiO_2 surface, affording the 2-AA-TiO_2 surface complex for visible-light-induced oxidation of sulfides into sulfoxides (Figure 18.9b) [47]. There are predominantly two pathways to ameliorate the activity: extending the absorption range of TiO_2 through assembling surface complexes with adsorbed various small molecules; improving the stability and efficiency of the reaction by the incorporation of a more suitable redox mediator.

TiO_2 with an organic polymer polydopamine was constructed for blue-light-induced selective oxidation of organic sulfides (Figure 18.10a) [48]. Similarly, as an organic dye, ARS leads to the oxidation of sulfides over TiO_2 under $\lambda > 450$ nm visible light irradiation [49]. Both erythrosin B (ErB) [50] and anthraquinones (AQs) [51] induce the same reaction over TiO_2 under 520 nm green light irradiation. The adsorption model of ARS was explored using a series of AQs on TiO_2. The results indicate that the hydroxy groups of AQs serve as the binding site for the TiO_2 surface, in which the bidentate chelation shows enhanced stability.

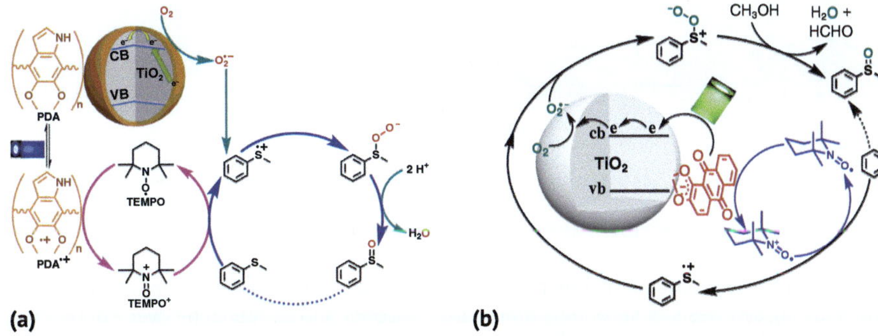

Figure 18.10 The schematic illustration of visible-light-induced selective oxidation of sulfides into sulfoxides over polymer-TiO_2 with TEMPO (a) and dye-TiO_2 with TEMPO (b). Reproduced from Refs. [48, 51] with permission from Elsevier.

Extending the π-conjugated molecules and altering the adsorption groups and their relative location obviously ameliorate the photocatalytic activity of TiO_2 surface complexes. For the latter case, the redox mediator facilitates rapid transfer the h^+ from the HOMO of the adsorbed molecule to the substrate, thereby preventing accumulation of oxidative pressure on the adsorbed molecules. In addition, the redox mediator creates a stepwise potential gradient between TiO_2 and substrates, effectively suppressing side reactions and enabling milder reaction conditions. Beyond TMA and TEA, TEMPO and its 4-substituted derivatives exhibit tunable electron transfer capabilities.

Compellingly a mechanism is concluded for visible-light-induced oxidation of phenyl methyl sulfide over dye-TiO_2 with TEMPO (Figure 18.10b). Under visible light irradiation, e^- are transferred from the adsorbed dye via the CB of TiO_2 to O_2, affording $O_2^{\bullet-}$. Meanwhile, h^+ oxidizes TEMPO into $TEMPO^+$, which oxidizes phenyl methyl sulfide to a sulfur-centered radical cation. Next, the combination of the generated sulfur-centered radical cation and $O_2^{\bullet-}$ leads to a persulfoxide intermediate. Ultimately, proton and electron transfer from the reaction medium methanol to the persulfoxide intermediate affords phenyl methyl sulfoxide.

The oxidation of sulfur compounds to sulfones enhances molecular polarity, offering significant value of practical desulfurization processes for fuel oil. Dibenzothiophene (DBT) and its derivatives were oxidized into sulfones under UV or xenon lamp irradiation over TiO_2 [52–54]. To enable visible-light-induced reactions, TiO_2 was modified with carbon or Pt-decorated carbon for the oxidation of DBT with O_2 [55]. The photocatalytic oxidation of diphenyl sulfide over TiO_2 polymorphs involves both energy transfer and electron transfer pathways for the generation of $^1O_2$ and $O_2^{\bullet-}$, leading to the selective formation of diphenyl sulfoxide and diphenyl sulfone [56]. Combining with other materials, such as SiO_2, realizes the same reaction under visible light irradiation [57].

18.2.4 C—C Coupling Reactions

In photocatalytic systems, h^+-induced radicals and radical cations are prone to rapid overoxidation by O_2 and ROS, favoring C—C bond scission and degradation over

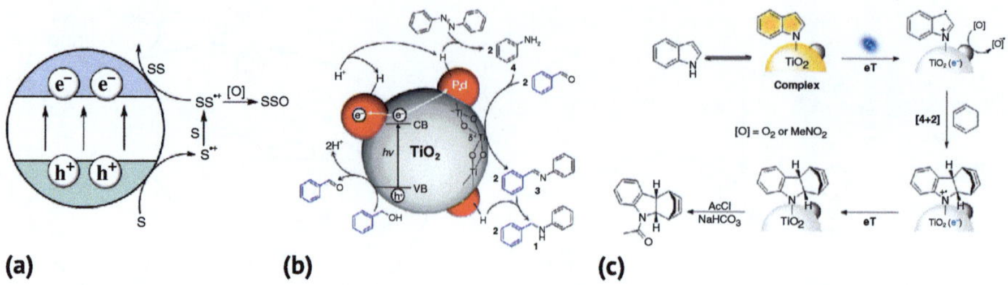

Figure 18.11 The schematic illustration of visible-light-induced C—C coupling reactions over TiO_2 (a), synthesis of secondary amine from alcohol and azobenzene over Pd/TiO_2 (b), and cycloaddition of indole with 1,3-cylohexadiene over Pt(0.2%)@TiO_2 (c). Reproduced from Refs. [66, 69] with permission from The Royal Society of Chemistry/American Chemical Society.

coupling reactions. Under UV irradiation, TiO_2 photocatalytic cycloaddition enables the C—C couplings by overcoming the electronic mismatching [58, 59]. However, heterogeneous photocatalytic cycloaddition is challenging under UV irradiation for organic transformation because of overoxidation over TiO_2. In 2016, styrenes were converted into aryltetralones intramolecular cyclization with O_2, under 395 nm visible LED irradiation over TiO_2 with excellent diastereoselectivity (Figure 18.11a) [60]. Besides, oxidative C_α—H arylation of heteroarenes provides methods for C—C bond construction. TiO_2 was applied in the arylation of electron-rich and electron-deficient heteroarenes for the formation of azoethers [61]. In this reaction, the formation of the azoether-TiO_2 surface complex extended absorption into the visible light region, which is similar to TiO_2 surface complex with organic molecules. The Ugi-type three-component reaction for α-functionalization of tertiary amines facilitates C—C bond construction, yielding α-amino amides [62]. The observed red shift in absorption is attributed to the amine-TiO_2 surface interaction.

As an effective cooperative system, Ru^{2+} catalysis for activation of C—H bond, and ErB-TiO_2 photocatalysis accomplished oxidative olefination/annulation reaction between olefins and arene carboxylic acid [63]. Carbonyl products were obtained through oxidative cleavage of C=C bonds [64]. In parallel, double-shelled TiO_2@ mesoporous organosilica served as an amphiphilic nanoreactor for visible-light-induced oxidation of styrenes into the corresponding aldehydes [65].

Under UV irradiation, Pd/TiO_2 as a photocatalyst converts alcohols and azobenzenes into secondary amines (Figure 18.11b). The process involves three steps: (1) aldehyde formation by oxidation of alcohols coupled with aniline formation by reduction of azobenzenes; (2) catalytic condensation of the resulting anilines and aldehydes on the TiO_2 surface; (3) photocatalytic hydrogenation of the formed imines [66, 67]. The coupling reaction can also proceed using nitrobenzene instead of azobenzene, affording asymmetric imines. Especially, benzaldehyde surface complexation generates OVs *in situ* on the TiO_2 surface, which facilitates interfacial electron transfer and extends the light absorption range [68]. The metal deposition-extended light absorption allows Pt(0.2%)@TiO_2 photocatalysis for Diels–Alder cycloaddition functionalize tetrahydrocarbazoles under 460 nm LED light irradiation (Figure 18.11c) [69].

18.3 Summary and Outlook

As a classic UV-induced photocatalyst TiO_2 is well-known for its great potential in generating a portfolio of ROS, making it suitable for selective oxidation of organic molecules with O_2. To this end, the generation of ROS from H_2O can be effectively suppressed by employing an inert organic solvent as the reaction medium. As such, there are several notable reports of selective aerobic oxidation by the conventional UV-induced TiO_2 systems. More broadly, visible light can be utilized to moderate the oxidative power of the h^+ over TiO_2 for selective oxidation of organic molecules. This is achievable through strategies such as forming TiO_2 surface complexes and modifying TiO_2 with organic dyes, precious metals, and CdS. In these systems, TiO_2 retains inherent remarkable properties for activating O_2 while reducing the oxidation potential of h^+, which is conducive to the formation of oxidation products in a selective manner.

Thus, the application of TiO_2 has been expanded to include photocatalytic selective oxidation of alcohols, amines, sulfides, and C—C couplings under visible light irradiation. Besides, the fundamentals of visible-light-induced selective oxidation of organic molecules over TiO_2 are explicitly explored to elucidate the underlying principles governing the photocatalytic activity. However, a critical challenge lies in maintaining the stability of these surface modifications, as they are typically less durable than TiO_2. Hence, the incorporation of redox mediators has proven effective to facilitate electron transfer. However, deeper fundamental understanding remains necessary to fully comprehend these complex systems of visible-light-induced selective organic transformations over TiO_2. This summary foretells the growing importance of visible-light-induced photocatalysis over nanostructured TiO_2 in addressing contemporary energy and environmental challenges.

Acknowledgments

This work was supported by the National Natural Science Foundation of China (Grants 22072108 and 22372124).

References

1 Schaub, R., Thostrup, P., Lopez, N. et al. (2001). Oxygen vacancies as active sites for water dissociation on rutile TiO_2(110). *Physical Review Letters* 87: 266104.
2 Zhao, T., Qian, R.F., Tang, Y. et al. (2020). Controllable synthesis and crystallization of nanoporous TiO_2 deep-submicrospheres and nanospheres via an organic acid-mediated sol–gel process. *Langmuir* 36: 7447–7455.
3 Ding, Y., Yang, I.S., Li, Z.Q. et al. (2020). Nanoporous TiO_2 spheres with tailored textural properties: controllable synthesis, formation mechanism, and photochemical applications. *Progress in Materials Science* 109: 100620.
4 Fujihira, M., Satoh, Y., and Osa, T. (1981). Heterogeneous photocatalytic oxidation of aromatic-compounds on TiO_2. *Nature* 293: 206–208.

5 Sheng, W.L., Shi, J.L., Hao, H.M. et al. (2020). Selective aerobic oxidation of sulfides by cooperative polyimide-titanium dioxide photocatalysis and triethylamine catalysis. *Journal of Colloid and Interface Science* 565: 614–622.

6 Xie, S.T., Zhao, X.Y., Wang, D.X. et al. (2024). MoO_x nanoclusters on Mo-doped TiO_2 nanosheets with enhanced singlet oxygen generation and sulfide conversion abilities for photocatalytic aerobic oxidative desulfurization. *Science China Chemistry* 67: 408–414.

7 Chen, F., Fu, H.T., Yang, X.H. et al. (2022). Fabrication of TaON/CdS heterostructures for enhanced photocatalytic hydrogen evolution under visible light irradiation. *Catalysts* 12: 1110.

8 Low, J.X., Dai, B.Z., Tong, T. et al. (2019). In situ irradiated X-ray photoelectron spectroscopy investigation on a direct Z-scheme TiO_2/CdS composite film photocatalyst. *Advanced Materials* 31: 1802981.

9 Lu, Y., Liu, X.L., He, L. et al. (2020). Spatial heterojunction in nanostructured TiO_2 and its cascade effect for efficient photocatalysis. *Nano Letters* 20: 3122–3129.

10 Prakash, J., Kumar, P., Saxena, N. et al. (2023). CdS based 3D nano/micro-architectures: formation mechanism, tailoring of visible light activities and emerging applications in photocatalytic H_2 production, CO_2 reduction and organic pollutant degradation. *Journal of Materials Chemistry A* 11: 10015–10064.

11 Pachfule, P., Acharjya, A., Roeser, J. et al. (2018). Diacetylene functionalized covalent organic framework (COF) for photocatalytic hydrogen generation. *Journal of the American Chemical Society* 140: 1423–1427.

12 Niu, Q., Dong, S.F., Tian, J.J. et al. (2022). Rational design of novel COF/MOF S-scheme heterojunction photocatalyst for boosting CO_2 reduction at gas–solid interface. *ACS Applied Materials & Interfaces* 14: 24299–24308.

13 Niu, J.N., Albero, J., Atienzar, P., and García, H. (2020). Porous single-crystal-based inorganic semiconductor photocatalysts for energy production and environmental remediation: preparation, modification, and applications. *Advanced Functional Materials* 30: 1908984.

14 Zhang, M., Wang, Q., Chen, C.C. et al. (2009). Oxygen atom transfer in the photocatalytic oxidation of alcohols by TiO_2: oxygen isotope studies. *Angewandte Chemie International Edition* 48: 6081–6084.

15 Kobayashi, H. and Higashimoto, S. (2015). DFT study on the reaction mechanisms behind the catalytic oxidation of benzyl alcohol into benzaldehyde by O_2 over anatase TiO_2 surfaces with hydroxyl groups: role of visible-light irradiation. *Applied Catalysis B: Environmental* 170: 135–143.

16 Wei, X.Z., Wang, H.Y., Xu, B. et al. (2023). Efficient visible light-driven oxidation of bio-1-butanol over a TiO_2-based photocatalyst system. *Colloids and Surfaces A: Physicochemical and Engineering Aspects* 665: 131216.

17 Li, H., Liu, H.R., Li, C.Z. et al. (2020). Micro-scale spatial location engineering of COF-TiO_2 heterojunctions for visible light driven photocatalytic alcohol oxidation. *Journal of Materials Chemistry A* 8: 18745–18754.

18 Zhang, Y.C., Wang, Z., and Lang, X.J. (2017). Merging visible light photocatalysis of dye-sensitized TiO_2 with TEMPO: the selective aerobic oxidation of alcohols. *Catalysis Science & Technology* 7: 4955–4963.

19 Li, X., Shi, J.L., Hao, H.M., and Lang, X.J. (2018). Visible light-induced selective oxidation of alcohols with air by dye-sensitized TiO_2 photocatalysis. *Applied Catalysis B: Environmental* 232: 260–267.

20 Wang, N.N., Shi, J.L., Hao, H.M. et al. (2019). Cooperative photocatalytic selective aerobic oxidation of alcohols on anatase TiO_2. *Sustainable Energy & Fuels* 3: 1701–1712.

21 Huang, L.Z., Jia, J.F., Liu, H.W. et al. (2015). Surface-mediated selective photocatalytic aerobic oxidation reactions on TiO_2 nanofibres. *RSC Advances* 5: 56820–56831.

22 Palmisano, G., Yurdakal, S., Augugliaro, V. et al. (2007). Photocatalytic selective oxidation of 4-methoxybenzyl alcohol to aldehyde in aqueous suspension of home-prepared titanium dioxide catalyst. *Advanced Synthesis & Catalysis* 349: 964–970.

23 Shiraishi, Y. and Hirai, T. (2008). Selective organic transformations on titanium oxide-based photocatalysts. *Journal of Photochemistry and Photobiology C: Photochemistry Reviews* 9: 157–170.

24 Chen, L., Tang, J., Song, L.N. et al. (2019). Heterogeneous photocatalysis for selective oxidation of alcohols and hydrocarbons. *Applied Catalysis B: Environmental* 242: 379–388.

25 Colmenares, J.C., Ouyang, W.Y., Ojeda, M. et al. (2016). Mild ultrasound-assisted synthesis of TiO_2 supported on magnetic nanocomposites for selective photo-oxidation of benzyl alcohol. *Applied Catalysis B: Environmental* 183: 107–112.

26 Higashimoto, S., Shirai, R., Osano, Y. et al. (2014). Influence of metal ions on the photocatalytic activity: selective oxidation of benzyl alcohol on iron (III) ion-modified TiO_2 using visible light. *Journal of Catalysis* 311: 137–143.

27 Li, X.R., Wang, J.G., Men, Y., and Bian, Z.F. (2016). TiO_2 mesocrystal with exposed (001) facets and CdS quantum dots as an active visible photocatalyst for selective oxidation reactions. *Applied Catalysis B: Environmental* 187: 115–121.

28 Lang, X.J., Ji, H.W., Chen, C.C. et al. (2011). Selective formation of imines by aerobic photocatalytic oxidation of amines on TiO_2. *Angewandte Chemie International Edition* 50: 3934–3937.

29 Lang, X.J., Ma, W. H., Zhao, Y.B. et al. (2012). Visible-light-induced selective photocatalytic aerobic oxidation of amines into imines on TiO_2. *Chemistry – A European Journal* 18: 2624–2631.

30 Balayeva, N.O., Mamiyev, Z., Dillert, R. et al. (2020). Rh/TiO_2-photocatalyzed acceptorless dehydrogenation of N-heterocycles upon visible-light illumination. *ACS Catalysis* 10: 5542–5553.

31 Sheng, W.L., Shi, J.L., Hao, H.M. et al. (2020). Polyimide-TiO_2 hybrid photocatalysis: visible light-promoted selective aerobic oxidation of amines. *Chemical Engineering Journal* 379: 122399.

32 Wang, Y.X., Zhou, J., Ma, X.M. et al. (2021). Cooperative photocatalysis with 4-amino-TEMPO for selective aerobic oxidation of amines over TiO_2 nanotubes. *Chemistry – An Asian Journal* 16: 2659–2668.

33 Furukawa, S., Suga, A., and Komatsu, T. (2015). Mechanistic study on aerobic oxidation of amine over intermetallic Pd_3Pb: concerted promotion effects by Pb and support basicity. *ACS Catalysis* 5: 1214–1222.

34 Dai, J., Yang, J., Wang, X.H. et al. (2015). Enhanced visible-light photocatalytic activity for selective oxidation of amines into imines over TiO_2(B)/anatase mixed-phase nanowires. *Applied Surface Science* 349: 343–352.

35 Yang, J.L. and Mou, C.Y. (2018). Ordered mesoporous Au/TiO_2 nanospheres for solvent-free visible-light-driven plasmonic oxidative coupling reactions of amines. *Applied Catalysis B: Environmental* 231: 283–291.

36 Shiraishi, Y., Sakamoto, H., Fujiwara, K. et al. (2014). Selective photocatalytic oxidation of aniline to nitrosobenzene by Pt nanoparticles supported on TiO_2 under visible light irradiation. *ACS Catalysis* 4: 2418–2425.

37 Ovoshchnikov, D.S., Donoeva, B.G., and Golovko, V.B. (2015). Visible-light-driven aerobic oxidation of amines to nitriles over hydrous ruthenium oxide supported on TiO_2. *ACS Catalysis* 5: 34–38.

38 Kim, Y., Kim, Y.J., Ko, D.H., and Kwon, K.Y. (2017). Oxidation of amine under aerobic and anaerobic condition catalyzed by ruthenium doped titanium oxides. *Bulletin of the Korean Chemical Society* 38: 993–994.

39 Dong, X.Y., Hao, H.M., Wang, N.N. et al. (2021). Cadmium sulfide/titanate hybrid green light photocatalysis for selective aerobic oxidative homocoupling of amines. *Journal of Colloid and Interface Science* 590: 387–395.

40 Mogal, S.I., Shah, D.O., Mukherjee, T. et al. (2018). Enhanced photocatalytic efficiency of a least active Ag-TiO_2 by amine adsorption. *ACS Omega* 3: 12802–12812.

41 Zhang, K.Y., Lu, G.L., Chu, F., and Huang, X.B. (2021). Au/TiO_2 nanobelts: thermal enhancement *vs.* plasmon enhancement for visible-light-driven photocatalytic selective oxidation of amines into imines. *Catalysis Science & Technology* 11: 7060–7071.

42 Ge, Y.F., Hu, Y.L., Ma, X.J. et al. (2024). Oxygen vacancy-modified TiO_2/Ti_3C_2/CdS nanocomposites for photocatalytic selective oxidation of amines. *ACS Applied Nano Materials* 7: 1233–1244.

43 Bettoni, M., Del Giacco, T., Elisei, F. et al. (2016). Substituent effects on the TiO_2 photosensitized oxidation reaction of benzyl thioethers and thiols in deaerated acetonitrile. *Journal of Photochemistry and Photobiology A: Chemistry* 324: 159–164.

44 Bottecchia, C., Erdmann, N., Tijssen, P.M.A. et al. (2016). Batch and flow synthesis of disulfides by visible-light-induced TiO_2 photocatalysis. *ChemSusChem* 9: 1781–1785.

45 Xu, H., Shi, J.L., Lyu, S.S., and Lang, X.J. (2020). Visible-light photocatalytic selective aerobic oxidation of thiols to disulfides on anatase TiO_2. *Chinese Journal of Catalysis* 41: 1468–1473.

46 Ding, L.Y., Zhao, Y.K., Li, Y.J. et al. (2023). Water-promoting effect on reaction selectivity in the photocatalytic oxidation of sulfides to sulfoxides on TiO_2. *Journal of Catalysis* 426: 328–335.

47 Li, H., Li, X., Zhou, J. et al. (2022). Extending aromatic acids on TiO_2 for cooperative photocatalysis with triethylamine: violet light-induced selective aerobic oxidation of sulfides. *Chinese Chemical Letters* 33: 3733–3738.

48 Shi, J.L. and Lang, X.J. (2020). Assembling polydopamine on TiO_2 for visible light photocatalytic selective oxidation of sulfides with aerial O_2. *Chemical Engineering Journal* 392: 123632.

49 Lang, X.J., Zhao, J.C., and Chen, X.D. (2016). Visible-light-induced photoredox catalysis of dye-sensitized titanium dioxide: selective aerobic oxidation of organic sulfides. *Angewandte Chemie International Edition* 55: 4697–4700.

50 Ma, X.M. and Lang, X.J. (2020). Titanate nanotube confined merger of organic photocatalysis and TEMPO catalysis for highly selective aerobic oxidation of sulfides. *Sustainable Energy & Fuels* 4: 1754–1763.

51 Hao, H.M., Li, X., and Lang, X.J. (2019). Anthraquinones as photoredox active ligands of TiO_2 for selective aerobic oxidation of organic sulfides. *Applied Catalysis B: Environmental* 259: 118038.

52 Tiple, A., Sinhmar, P.S., and Gogate, P.R. (2021). Improved direct synthesis of TiO_2 catalyst using sonication and its application for the desulfurization of thiophene. *Ultrasonics Sonochemistry* 73: 105547.

53 Wu, P.W., Liu, P.H., Chen, L.L. et al. (2022). Synergistic effect of Au–Cu alloy nanoparticles on TiO_2 for efficient aerobic catalytic oxidative desulfurization. *Industrial & Engineering Chemistry Research* 61: 6292–6300.

54 Molinari, A., Bonino, F., Magnacca, G. et al. (2015). Relation between phase composition and photocatalytic activity of TiO_2 in a sulfoxide deoxygenation reaction. *Materials Chemistry and Physics* 158: 60–66.

55 Abdollahi, M., Larimi, A., Jiang, Z. et al. (2022). Photocatalytic oxidative desulfurization of model fuel over visible light-active Cu-impregnated carbon-doped TiO_2. *Journal of Cleaner Production* 380: 134968.

56 Mikrut, P., Swies, A., Kobielusz, M. et al. (2022). Selective and efficient catalytic and photocatalytic oxidation of diphenyl sulphide to sulfoxide and sulfone: the role of hydrogen peroxide and TiO_2 polymorph. *RSC Advances* 12: 1862–1870.

57 Huang, F.W., Hao, H.M., Sheng, W.L. et al. (2022). Embedding an organic dye into Ti-MCM-48 for direct photocatalytic selective aerobic oxidation of sulfides driven by green light. *Chemical Engineering Journal* 432: 134285.

58 Nakayama, K., Maeta, N., Horiguchi, G. et al. (2019). Radical cation Diels–Alder reactions by TiO_2 photocatalysis. *Organic Letters* 21: 2246–2250.

59 Okada, Y., Maeta, N., Nakayama, K., and Kamiya, H. (2018). TiO_2 photocatalysis in aromatic "redox tag"-guided intermolecular formal [2+2] cycloadditions. *The Journal of Organic Chemistry* 83: 4948–4962.

60 Liu, Y.L., Zhang, M.F., Tung, C.H., and Wang, Y.F. (2016). TiO_2 photocatalytic cyclization reactions for the syntheses of aryltetralones. *ACS Catalysis* 6: 8389–8394.

61 Zoller, J., Fabry, D.C., and Rueping, M. (2015). Unexpected dual role of titanium dioxide in the visible light heterogeneous catalyzed C—H arylation of heteroarenes. *ACS Catalysis* 5: 3900–3904.

62 Vila, C. and Rueping, M. (2013). Visible-light mediated heterogeneous C—H functionalization: oxidative multi-component reactions using a recyclable titanium dioxide (TiO_2) catalyst. *Green Chemistry* 15: 2056–2059.

63 Dana, S., Dey, P., Patil, S.A., and Baidya, M. (2020). Enhancing Ru(II)-catalysis with visible-light-mediated dye-sensitized TiO_2 photocatalysis for oxidative C—H olefination of arene carboxylic acids at room temperature. *Chemistry – An Asian Journal* 15: 564–567.

64 Gottuso, A., De Pasquale, C., Livraghi, S. et al. (2023). Nitrate radicals generated by TiO_2 heterogeneous photocatalysis: application to the cleavage of C=C double bond to carbonyl compounds. *Molecular Catalysis* 550: 113607.

65 Qiao, X.Y., Xiong, Z.S., Wang, Y. et al. (2023). Double shelled titanium dioxide@ mesoporous organosilica nanotube as an amphiphilic photoactive nanoreactor for efficient photocatalytic oxidation of styrene. *Journal of Colloid and Interface Science* 651: 235–242.

66 Selvam, K., Sakamoto, H., Shiraishi, Y., and Hirai, T. (2015). Photocatalytic secondary amine synthesis from azobenzenes and alcohols on TiO_2 loaded with Pd nanoparticles. *New Journal of Chemistry* 39: 2856–2860.

67 Shiraishi, Y., Fujiwara, K., Sugano, Y. et al. (2013). N-monoalkylation of amines with alcohols by tandem photocatalytic and catalytic reactions on TiO_2 loaded with Pd nanoparticles. *ACS Catalysis* 3: 312–320.

68 Yang, X., Tao, H.L., Leow, W.R. et al. (2019). Oxygen-vacancies-engaged efficient carrier utilization for the photocatalytic coupling reaction. *Journal of Catalysis* 373: 116–125.

69 Pitre, S.P., Scaiano, J.C., and Yoon, T.P. (2017). Photocatalytic indole Diels–Alder cycloadditions mediated by heterogeneous platinum-modified titanium dioxide. *ACS Catalysis* 7: 6440–6444.

19

Photoelectrochemical Sensing Applications of Nanostructured TiO_2

*Miao Li, Xiaoqing Qiu**

College of Chemistry and Chemical Engineering, Central South University, Changsha, China

*Corresponding author: xq-qiu@csu.edu.cn
Co-author's email: 2567935624@qq.com (M. Li)

19.1 Introduction

Photoelectrochemical (PEC) sensing stands out as a groundbreaking technology for monitoring crucial chemical and biological processes in environmental protection, food safety, and medical diagnostics, garnering significant attention in analytical chemistry. PEC sensors incorporate photocatalytic and electrocatalytic features with the sensor to significantly enhance the sensing activity. The interaction of the target analyte with the electrode can change the type/rate of the reactions that occur at the electrode/electrolyte interface, which affects the charge separation/migration process. Detection can be realized by monitoring changes in photocurrent, as illustrated in Figure 19.1.

In contrast to traditional detection methods characterized by complex systems, prolonged processes, extensive spatial requirements, high costs, and stringent demands, high-performance PEC detection transforms input optical signals into electrical signals. This dual-energy channel effectively prevents interference, offering notable advantages such as rapid response, low noise, and high sensitivity [2].

Photocatalytic materials employed in constructing PEC sensors usually possess several desirable features, including swift generation and transfer of the photogenerated charge carriers for substantial signal intensity, excellent stability to prevent photocorrosion. Among numerous semiconductor photocatalytic materials, the n-type semiconductor TiO_2 has proven feasible for applications in dielectric materials, sensors, environment purification, coatings, catalysts. Notably, TiO_2 exhibits great application potential in the realm of PEC sensor due to its chemical stability, robust oxidizing ability, good biocompatibility, corrosion resistance, cost-effectiveness, and nontoxic nature [3].

The amplification of the sensor signal is a critical factor for PEC sensors, intricately tied to the light-harvesting capability, charge separation, and interfacial redox

Nanostructured TiO_2: Synthesis, Photocatalytic Properties and Applications, First Edition.
Edited by Jia Hong Pan, Wan In Lee and Detlef W. Bahnemann

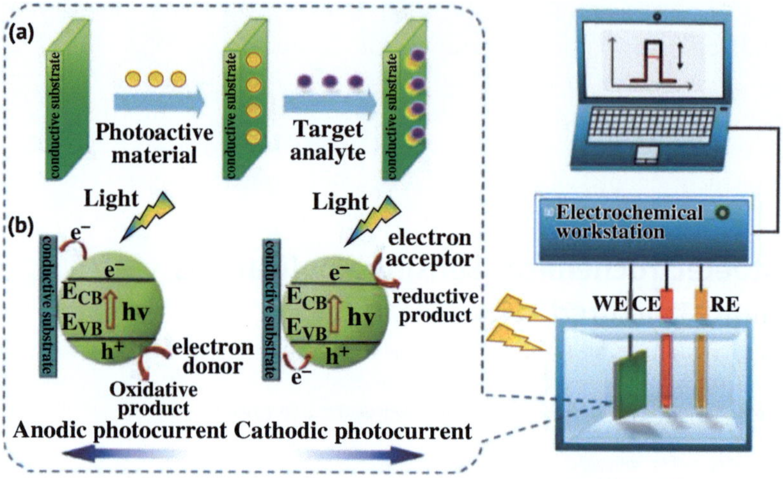

Figure 19.1 Schematic of PEC sensor: (a) working principle of PEC detection. (b) The photocurrent generation mechanism [1]/with permission from Elsevier.

reactions. Although TiO_2 possess good stability, limitations such as poor utilization efficiency of visible light and rapid electron–hole recombination hinder its application. Researches have carried out extensive studies on delaying the combination of electron–hole pairs, expanding the utilization of light, and regulating the interaction with target substances [4]. Various methods, such as the regulation of the energy bands, control of morphology and nanostructure, heterojunction construction, noble metal deposition, and surface modification, have been implemented to enhance the sensing activity of TiO_2 [5].

In recent years, nanostructured TiO_2-based PEC sensor technology has seen rapid development in life science research, water environment monitoring, and clinical diagnosis. Consequently, a concise review on the application of TiO_2-based materials in PEC detection becomes imperative. The following sections offer a versatile exploration of TiO_2-based sensors, including their applications in the PEC detection of biomolecules, harmful ions, and organic pollutants. Finally, the challenges and future development prospects of TiO_2-based materials in PEC sensing are briefly examined.

19.2 Fundamentals of PEC Sensing

A typical PEC sensing system comprises a three-electrode setup, a light source (e.g., xenon lamps, halogen lamps, light-emitting diodes), and an electrochemical workstation [6]. The standard three-electrode cell consists of a TiO_2-based photoanode as the working electrode, Ag/AgCl as the reference electrode, Pt as the counter electrode, and a PBS buffer electrolyte. Common substrate materials for TiO_2-based electrodes include FTO, ITO, or Ti foil, often assembled through methods like drop casting, dip coating, spin coating, or hydrothermal in situ growth.

During the detection process, valence band electrons in the TiO_2-based working electrode are excited by light into the conduction band, then transferred to the counter electrode through an external circuit to generate photocurrent. The oxidation reaction occurs at the working electrode, and the reduction reaction takes place at the counter electrode [7]. Applying an appropriate bias voltage facilitates photogenerated electron transfer, or the addition of electron donors/acceptors to the electrolyte effectively stabilizes the PEC signal. It is worth noting that the bias voltage generally selects the saturated photovoltage or not set to prevent the background noise from interfering with the sample test during the test process to achieve a high-sensitivity detection.

Key parameters for assessing the performance of PEC sensors include the detection limit, linear range, sensitivity, selectivity, reproducibility, and stability. Generally, the detection limit and sensitivity are determined by the linear relationship between the concentration of the detection target and the response photocurrent. Sensitivity measures the variation in the sensor's response signal due to changes in the unit concentration or mass of the detection substance, calculated from the ratio of the calibration curve slope to the working electrode area. Higher sensitivity implies greater precision in analysis results. The limit of detection refers to the minimum concentration of the target substance measured from the sample, typically estimated from the noise (*N*) and the slope of the calibration curve (*s*).

$$LOD = \left|3 \times \frac{N}{s}\right| \tag{19.1}$$

where the noise *N* represents the standard deviation of the peak current repeatedly measured in the blank sample [8]. The selectivity of the sensor refers to the ability to detect the target substance in the presence of interfering substances, where the structure and properties of the interfering substance are similar to the detection substance. Under the same experimental conditions, the relative standard deviation of the same sample was measured by multiple electrodes to evaluate the reproducibility. The stability of the sensor is evaluated by measuring the response of the sensor to the target material after repeated use for a period of time. High-performance PEC sensors ought to possess high sensitivity, low detection limit, wide linear range, good selectivity, reproducibility and stability [9]. Nanostructured TiO_2-based sensors are now widely used for the detection of biomolecules, harmful ions, and organic pollutants.

19.3 PEC Detection of Biomolecules

Biomolecules such as DNA, RNA, proteins, and enzymes play an important role in many physiological activities. The occurrence of growth, fertility, cancer, and most diseases is related to their level changes in organisms. For instance, the detection of extracellular β-amyloid peptide in blood and cerebrospinal fluid can diagnose Alzheimer's disease, and blood glucose levels provide important indicators for the diagnosis and treatment of diabetes [10]. Therefore, the efficient and sensitive detection of biomarkers is important for disease screening in medical diagnosis, which also helps us to understand the mechanism of many physiological phenomena. At present, the construction of efficient and simple TiO_2-based PEC biosensors for protein detection and pathogen recognition in medical diagnosis to achieve early intervention and management of diseases has attracted widespread attention.

The assembly and working principle of the specific recognition sensor in the preparation step of the biosensor are the two main issues that we will discuss. The most representative of PEC biosensors is their specific recognition elements, mainly including antibodies, aptamers, and peptides. The activity of biological molecules such as antibodies and enzymes will be destroyed due to the photochemical changes of proteins and nucleic acids caused by ultraviolet irradiation. Therefore, it is necessary to use visible light as the excitation light source to improve the service life of PEC immunoassay. The preparation of TiO_2-based photoactive materials with visible light absorption is the primary condition for constructing TiO_2-based biosensors. In other words, TiO_2 is modified to expand its light absorption range and optimize the photoelectric conversion performance in order to obtain a high sensitivity sensor. The methods of constructing vacancies, heterojunctions, doping, and cocatalysts are commonly used in modification, the principles of which is not explained in detail here [11].

The second step is to assemble the specific recognition element on the TiO_2-based photoactive material. The principles commonly used to construct TiO_2-based working electrodes include noble metal nanoparticle (NP) deposition, EDC/NHS coupling reaction, glutaraldehyde (GLD) cross-linking reaction, etc. Precious metal deposition is very useful for biological coupling and can provide anchor sites for biomolecules. Thio-biomolecules can be immobilized on precious metals through metal–sulfur bonds, which is a common method of antibody modification on the electrode surface. In addition, noble metal NPs (Au, Ag, etc.) generate strong local surface plasmon resonance (SPR) under shorter wavelength light, which enhances the photoelectric performance [12]. It is worth noting that the binding of metal–sulfur bonds is not specific. After incubation with the target molecule, blocking agents (bovine serum albumin solution, 6-aminocaproic acid, etc.) are required to block nonspecific binding sites. In addition, substances with abundant functional groups are modifiable on TiO_2-based surfaces through coupling interactions with the surface Ti sites or direct drop-coating, and then connected to the target molecule through the chemical reaction between the functional groups. In Figure 19.2a, the adjacent phenolic hydroxyl groups on the aromatic ring of catechol compounds can form bonds that are strongly adsorbed on the surface of metal oxides [13]. *N*-(3-dimethylaminopropyl)-*N*′-ethylcarbodiimide hydrochloride (EDC) and 1-hydroxy-2,5-pyrrolidinedione (NHS) are two coupling reagents for chemical coupling between biomolecules. EDC and NHS activate carboxyl groups and react with amino compounds to form aminoamide linkages, which can be used for electrode surface functionalization and binding to target molecules, as shown in Figure 19.2b [14]. Figure 19.2c shows the cross-linking reaction of GLD. GLD reacts with the amino groups on the electrode surface and in the biomolecules to form a highly stable cross-linking product with an imine structure for immobilization [15]. Abovementioned reactions are achieved by soaking the TiO_2-based electrode in a buffer solution containing the corresponding antibody or target molecule. The incubation time, concentration, and pH of the solution all need to be optimized to

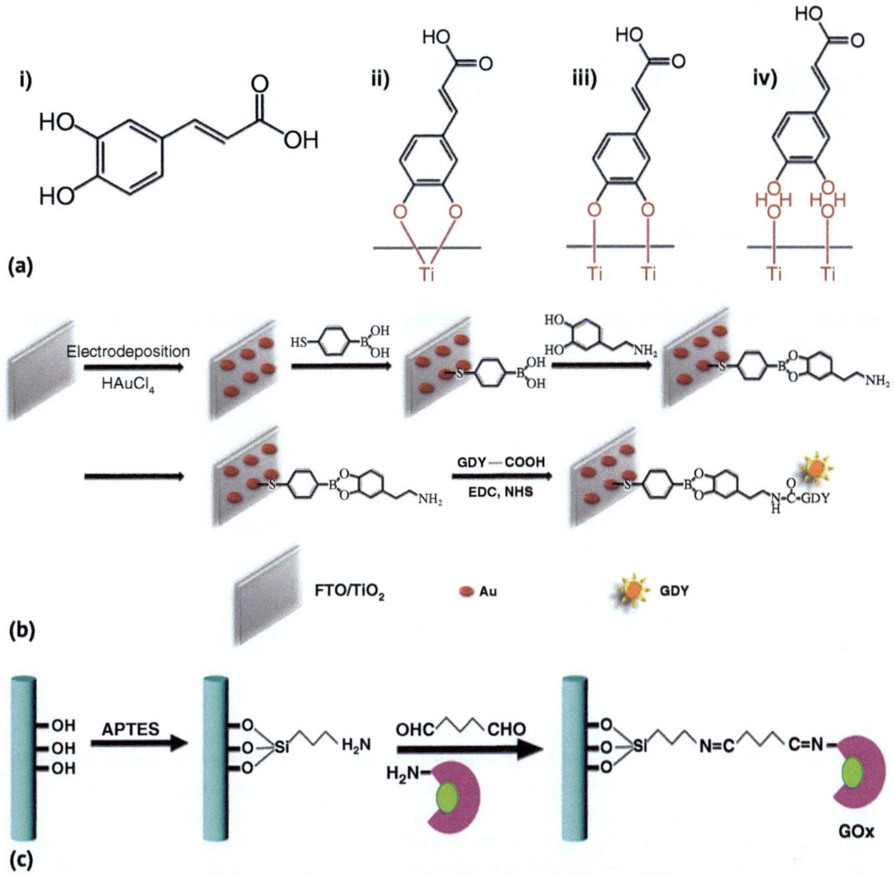

Figure 19.2 (a) Chemical structure of caffeic acid (i) and its bonding mechanism to TiO_2 surfaces: (ii) bidentate chelating, (iii) bidentate bridging (inner sphere), and (iv) bidentate bridging (outer sphere) [13b]/with permission from Elsevier. (b) The construction process of the dopamine sensor through EDC-NHS reaction [16]/with permission from Elsevier. Copyright 2022, Elsevier. (c) Schematic of GO_x functionalization on the TiO_2 NW surface via silane/GLD cross-linking reaction. [15b]/with permission from The Royal Society of Chemistry.

obtain the best results. Therefore, the preparation process of the sensor is complex and time-consuming.

The principle of specific recognition elements interacting with target molecules for detection will be described below, including steric hindrance effect, electron donor/acceptor interaction, photoactive/inert material modification, etc. [17]. The detected biomolecules have insulation and steric effects. After binding to the specific recognition element (such as antigen/antibody) on the surface of the modified TiO_2 photoelectrode, it will hinder the interfacial charge transfer of the electrode/electrolyte and reduce the light absorption, resulting in an increase in steric hindrance and a decrease in photocurrent as illustrated in Figure 19.3a [18]. Most biosensors operate based on the principle that the photocurrent decreases as the concentration of the target substance on the electrode surface increases is called

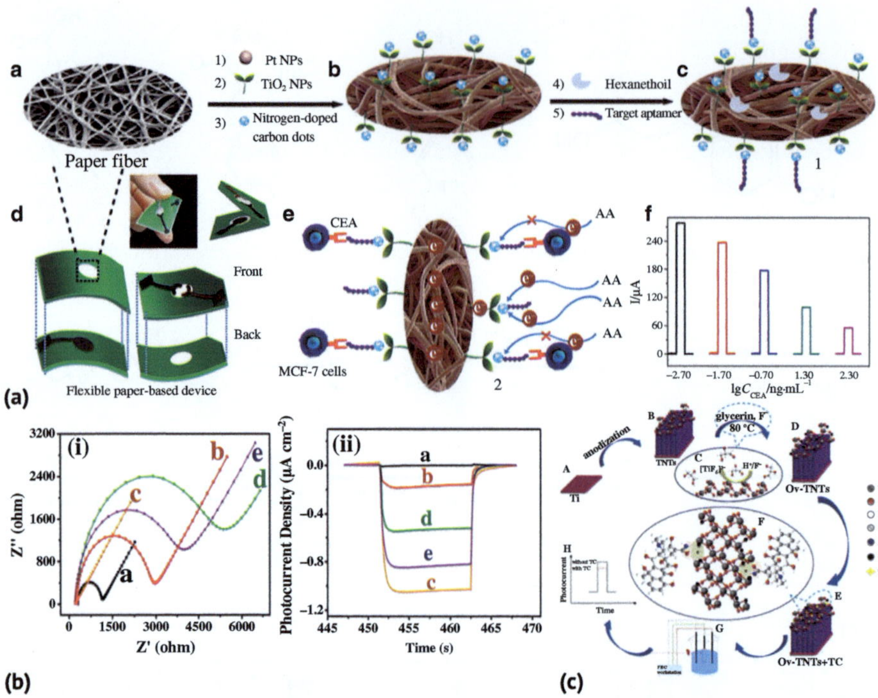

Figure 19.3 (a) Schematic illustration of (i–iii) the preparation procedure of the photoelectrochemical biosensor, (iv) the flexible paper-based platform; (v) detection mechanism of the biosensor for CEA on the surface of MCF-7 cells, (vi) photocurrent responses of the proposed paper-based sensing platform toward target antigen with different concentrations [15a]/with permission from American Chemical Society. (b) EIS spectra (i) and PEC response (ii) of different modified electrodes. a–e, GCE, GCE/OAM, GCE/OAM/PEI, GCE/OAM/PEI/Peptide@B-CQDs, electrode d after trypsin (0.1 mg/mL) incubation [20]/with permission from Elsevier. (c) Mechanisms of O_V-TNTs material synthesis and photoelectrochemical TC detection [21]/with permission from Elsevier.

a "signal-off" sensor. Notably, the steric hindrance effect caused by biomolecule modification on the electrode surface can also serve to detect the successful assembly of specific recognition elements on the electrode surface. Similarly, the reaction of a biomolecule with a specific recognition element (inducing conformational changes to reduce weak binding on the surface, or specific cleavage by enzymes, etc.) causes the dissociation from the electrode surface, where the spatial resistance effect lifts to achieve a "signal-on" for the target detection [19]. Dai et al. [20] used boron-doped carbon quantum dots (B-CQDs) labeled peptides as signal amplification markers, and used trypsin to specifically cleave the modified peptide chain (containing arginine) on the polyethyleneimine-sensitized TiO_2 mesocrystal substrate to achieve quantitative detection of trypsin. Figure 19.3b illustrates the EIS spectra and PEC responses for different modified electrodes. B-CQDs-peptide biocouplings acting as bursting agents are broken under the action of enzyme, resulting in a decrease in steric hindrance. At the same time, B-CQDs with competitive light absorption characteristics are stripped from the electrode allowing the double inhibition relieved to achieve the signal variations, which is suitable for monitoring various proteases and early diagnosis of different diseases.

The effective adsorption of analyte on the catalyst surface to change the steric hindrance is a valid way to achieve selective determination. Modulation of oxygen vacancies in TiO_2-based materials alters the positive electrostatic potential of the material, and its coordination-unsaturated sites can be used as electron traps to preferentially adsorb molecules with negative electrostatic potentials (or electron-donating groups). Cui et al. [21] used electrostatic adsorption to detect tetracycline hydrochloride (TC). The positively charged oxygen vacancy-regulated TiO_2 nanotube arrays (Ov-TNTs) induce the stable adsorption of negatively charged TC through electrostatic interaction without degradation. Then the electron-donating groups ($-NH_2$ and $-OH$) of TC are occupied in oxygen vacancies, which reduces the photoelectric performance of the sensor. The detection mechanism is shown in Figure 19.3c. When the adsorption reaches saturation, the excess TC allows the photocurrent to continuously decrease through the steric hindrance effect, but the decreasing trend is no longer linear. This specific detection is attributed to the adsorption of amide groups of the electron donor. The more electron-donor groups in the molecule, the more sensitive the photocurrent response.

Additionally, different properties of the target substance can be utilized for detection. Similar to the principle of the spatial potential resistance effect, the formation of an insoluble substance on the electrode surface is an efficient approach to the sensing process. The insoluble material covered on the surface of the photoelectrode not only blocked the interface contact between the electrode and the electrolyte, but also affected the light absorption of the TiO_2-based photoactive material. In Figure 19.4a, H_2O_2 accelerated the oxidation of 4-chloro-1-naphthol in the presence of horseradish peroxidase, and the insoluble product benzo-4-chlorohexanedione was produced on the electrode surface, resulting in a significant decrease in photocurrent [22]. The electron acceptor also quenches the photocurrent by trapping the photogenerated electrons during the transfer process. For example, dopamine, a neurotransmitter that plays an important role in the function of our central nervous system, can be oxidized by photogenerated holes into the electron acceptor dopamine quinine, thereby realizing the photoelectric detection of dopamine [23]. On the contrary, the generation of electron donors can provide electrons to enhance the photocurrent. Some detection substances can change the photocurrent signal act as electron acceptors/donors during PEC detection, thereby improving the high sensitivity and wide linear range of biosensors. For example, microcystins, lactic acid, and glucose can be oxidized by holes to hinder the recombination of photoelectron pairs [24]. The variation of photoresponse current with electron acceptor/donor concentration is shown in Figure 19.4b,c.

The recognition method described above is based on first fixing the specific recognition element on a biosensing electrode and then incubating in the target molecule solution. Alternatively, it is possible to exchange the position of them, which means the target substance is immobilized on the surface of the TiO_2-based photoelectrode and incubated in the solution with the signal amplification unit. Photoreactive substances/quenchers are commonly employed as signal amplification units, and the photocurrent signal resulting from their binding to the target molecule is

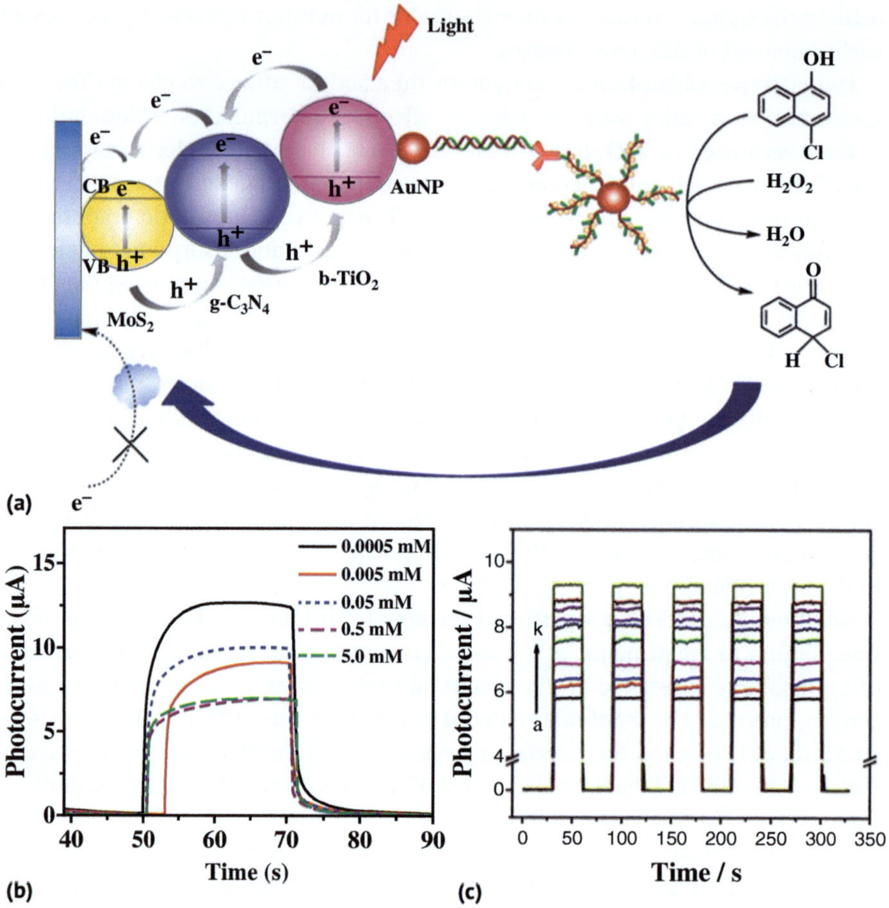

Figure 19.4 (a) Detection mechanism diagram of microRNA-396a by generating insoluble product benzo-4-chlorohexanedione on the electrode surface [22]/with permission from Elsevier. (b) Photocurrent response of the TiO_2/MAPTPA electrodes with different DA concentrations [23]/with permission from Elsevier. (c) Photocurrent signal of $BiFeO_3$@TiO_2-aptamer-BSA-MC-LR upon addition of MC-LR concentration (the concentration increases gradually from a to k) [24a]/with permission from Elsevier.

significantly altered. Most of the materials used as signal amplification units have photoelectric activity to achieve the change of photocurrent and the rich functional groups support their binding to the target molecules, such as carbon materials, metal–organic frameworks (MOFs), and so on [25]. Most of the carbon-based materials have excellent electronic conductivity, biocompatibility, and photoelectric properties. Modifying different functional groups on the surface of carbon materials and then reacting with biomolecules on the electrode surface to form chemical bonds can effectively enhance photoelectric properties. MOF materials are commonly used as signal amplification materials or carriers for signal probes due to their tunable structure and large number of pores, which is common in PEC detection of mRNA modifiers. It is captured by the strong binding ability with the phosphate group on the electrode surface, which can absorb visible light to generate

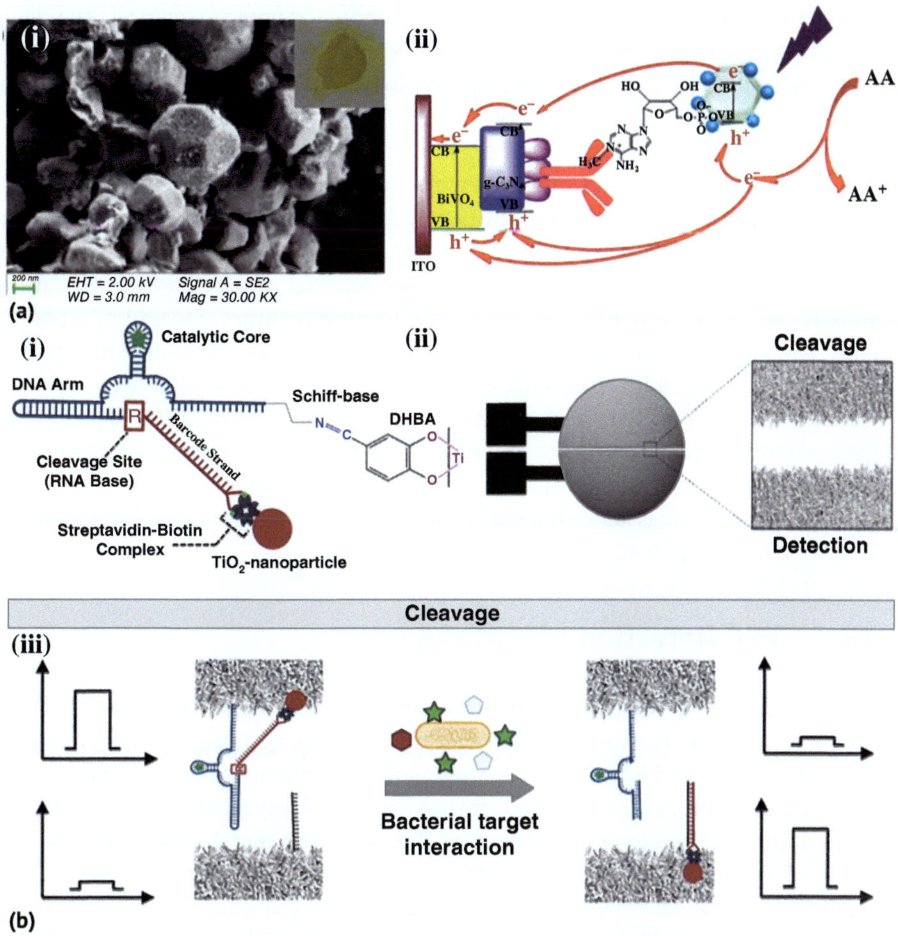

Figure 19.5 (a) SEM images of TiO_2@NH2-MIL-125(Ti) (i), (ii) schematic diagram of the photocurrent generation mechanism of the PEC immunosensor [27]/with permission from Elsevier. (b) The components of the photoactive DNAzyme strand (i), (ii) illustration of the TiO_2-assembly photoelectrode, (iii) operation of the PEC DNAzyme assay [13b]/with permission from Elsevier.

more photogenerated electrons to realize the enhancement of photocurrent, similar to the signal amplification principle of electron donor [16, 26]. In Figure 19.5a, Zhou et al. [27] achieved specific detection of N1-methyladenosine using TiO_2@NH_2-MIL-125(Ti) as a signal amplification unit and anchoring the MOF through the interaction of TiO_2 with phosphate groups to form Ti—O—P bonds, which provided a new alternative detection platform for biomarker detection to evaluate the ecotoxicological effects of environmental pollutants on food crops. Furthermore, the signal change caused by the transfer of signal amplification molecules between two closely spaced electrodes can also be used for detection. As shown in Figure 19.5b, using DNA-anatase TiO_2 NPs as signal amplification units, photoactive DNA barcodes were released from a rutile TiO_2 photoelectrode in the presence of bacteria and captured by a probe on the surface of another rutile TiO_2 photoelectrode. Using the fabrication and disruption of TiO_2 semiconductor heterostructures, the detection of *Escherichia coli* was realized using DNA as a switch [13b].

A large number of studies have shown that TiO_2-based biosensors have great potential in the detection of biomolecules. In the immobilization of specific recognition elements, a large number of organic reaction principles are worth exploring and applying to detection. Compared with TiO_2-based photoactive substrate materials, the specific recognition molecules modified on the surface of the electrode are more likely to be inactivated leading to the decrease of detection ability, and how to preserve the prepared TiO_2-based biosensing electrodes also deserves our attention. In practical assays, other interfering factors like temperature, PH and microorganisms may affect the activity, which requires the enhancement of the anti-interference ability. In addition, it is not enough to be satisfied with the detection of a single substance. How to simplify manufacturing steps and operation methods, develop flexible, portable, environmentally friendly and miniaturized equipment, distinguish and capture different signals of each analyte in a test sample for rapid diagnosis of multiple biomarkers is also what we need to pursue. TiO_2-based PEC biosensor may have a lot to be explored.

19.4 PEC Detection of Harmful Ions

The amount of agricultural, industrial, and urban sewage discharged increases annually with the development of society, while large quantities of ions and their compounds are discharged into the water source. Although ions are ubiquitous in biology and nature, excess ions are invariably harmful, especially nonbiodegradable and highly bio-toxic ions in wastewater. Taking heavy metal ions as an example, many heavy metals such as lead, cadmium, and cobalt are released into the environment and enter the ecosystem during the development of mineral resources. Heavy metals cannot be biodegraded and are bioaccumulative. They will interact strongly with proteins and various enzymes in organisms to inactivate them. Therefore, it is necessary to develop advanced sensing technology to detect the concentration of trace harmful ions in water [28]. In recent years, TiO_2-based PEC sensors have been widely used for the quantitative detection of harmful ions in water, including Cr^{6+},Hg^{2+},F^-, etc. The detection mechanisms include direct reduction of target metal ions, steric hindrance effect, photoactive/inert material modification, and energy transfer, which affect the separation and migration of photogenerated carriers on TiO_2-based photoelectrodes and cause significant changes in photocurrent response [1, 29].

The effect of electron donor/acceptor on photocurrents is pronounced. Cr^{6+} and Cr^{3+} are two common oxidation states of chromium. Cr^{3+} has a positive role in controlling blood glucose levels, cardiovascular protection, and weight control that is beneficial to the human body, whereas Cr^{6+} is toxic and carcinogenic with health risks even at trace levels. In TiO_2-based PEC sensing detection, Cr^{6+} with high oxidizing ability is easily combined with photogenerated electrons and reduced. Besides, it is easier to trigger the reduction reaction of Cr^{6+} by regulating the energy band structure of TiO_2 to make the conduction band negative shift or

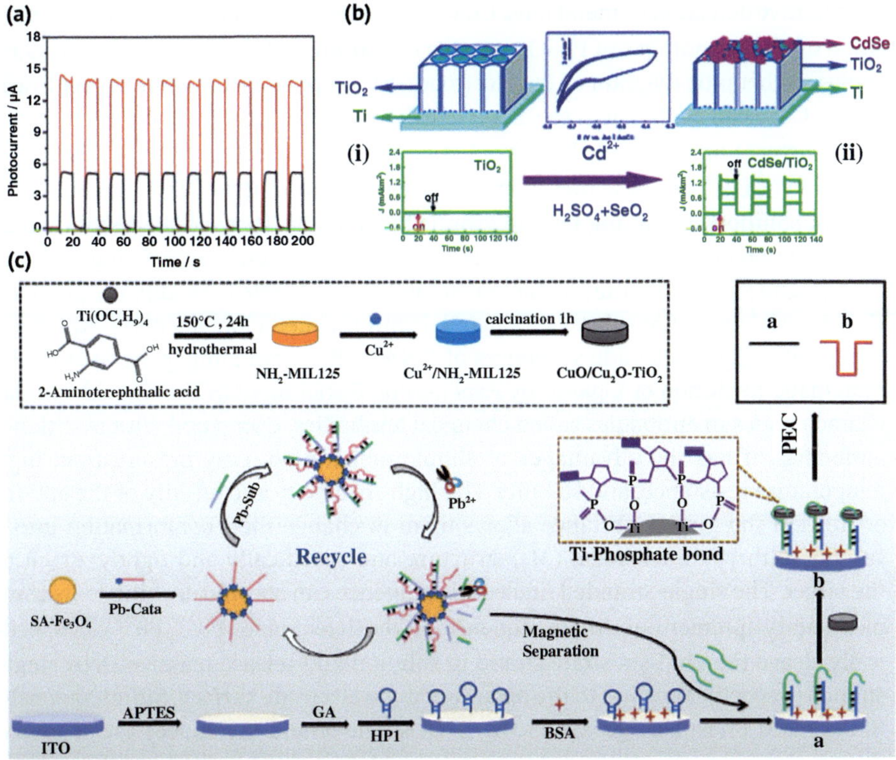

Figure 19.6 (a) Stability of the PEC system with and without 0.1 mM Cr(VI) at a NiCo-LDHs/TiO_2 NTAs/Ti Photoanode [32]/with permission from American Chemical Society. Copyright 2022, American Chemical Society. (b) Schematic illustration for photoelectrochemical detection of Cd^{2+} based on in situ electrodeposition of CdSe clusters on TiO_2 nanotubes. The lower figures: short-circuit photocurrents obtained at (i) TiO_2 nanotubes and (ii) CdSe/TiO_2 film in 1 M NaOH solution containing 1 M Na_2S and S under visible light illumination (l 4 420 nm) [33]/with permission from The Royal Society of Chemistry. Copyright 2012, The Royal Society of Chemistry. (c). Schematic illustration of the PEC biosensor for the detection of Pb^{2+} [34]/with permission from Elsevier.

modifying the reducing agent on the surface of TiO_2. As the Cr^{6+} concentration increases, more photogenerated electrons are consumed decreasing the photocurrent in Figure 19.6a. When the Cr^{6+} concentration is higher than a certain value, the photogenerated electrons are overconsumed and the electrons transferred to the external circuit will be "withdrawn" switching the photocurrent from the anode to the cathode. It is worth noting that the presence of other high-valent metal ions in the wastewater obviously interferes with the detection of Cr^{6+} [30]. Likewise, NO^{2-} and HS^- can be oxidized by photogenerated holes generated by TiO_2, which contributes to the separation of electron–hole pairs and enhances photocurrent. Meanwhile, some susceptible oxidized organics will also cause some interference to its detection [31].

The principle of photoactive material modification in biosensors is equally applicable to the detection of harmful ions. The in situ electrodeposition method can directly treat and convert harmful ions into useful PEC materials while achieving quantitative detection of metal ions. $CdSe/TiO_2$ was prepared by in situ electrodeposition of TiO_2 nanotubes in H_2SO_4 solution containing Cd^{2+} and SeO_2. Figure 19.6b shows a schematic diagram of the electrodeposition process and photocurrent variations. CdSe can be used as a sensitizer to extend light absorption and thus generate photocurrents under visible light irradiation for Cd^{2+} detection [33]. Similarly, the detection of Pb^{2+} can be achieved by in situ electrodeposition of TiO_2 nanotubes in Na_2SO_3 solution containing Pb^{2+} to form PbS/TiO_2 [35]. To avoid the generation of an initial photocurrent on the photoelectrode before detection interfering with the background signal and limiting the sensitivity, it is advisable to modify the photoactive material on a pure substrate in parallel with the metal ion detection. Aptamers are specific oligonucleotide sequences of DNA or RNA randomly generated by the Systematic Evolution of Ligands by Exponential Enrichment (SELEX), with similar characteristics to antibodies called chemical antibodies. Compared with traditional antibodies, it has the advantages of simple preparation, easy preservation, high temperature resistance, and stability. The high specificity and affinity of the aptamers toward the target substance allows them to change their conformation into a secondary three-dimensional (3D) structure and specifically and tightly attach to the target. The single-stranded nucleotide sequence can specifically bind to its complementary aptamer nucleotide sequence. In the detection of Pb^{2+}, Pb^{2+} can specifically cleave the Pb^{2+}-substrate strand in solution and release massive DNA single strands to specifically bind to the aptamer on the electrode surface. Simultaneously, the exposed phosphate group on the DNA single strand introduces the semiconductor material to the electrode by chelating with Ti^{3+} to achieve a "signal-on" state. The detection schematic is shown in Figure 19.6c [34].

Conversely, harmful ions generating spatial site-blocking effects or photo-inert substances on the electrode surface can lead to "signal-off." Luo et al. [36] used $TiO_2/AgInS_2/ZnS$ composite as a photoelectrically active anode substrate, NiCo-MOF@Au as a quenching element with aptamers SH-DNA and NH_2-DNA modified on both, respectively. As shown in Figure 19.7a, the selective detection of Hg^{2+} was achieved by using the specific binding of Hg^{2+} to thymine $(T\text{-}Hg^{2+}\text{-}T)$ to hinder electron transfer on the electrode surface. The detection of Cu^{2+}/S^{2-} can be achieved by capturing Cu^{2+} with TiO_2/CdS or capturing S^{2-} with TiO_2/Cu_2O to form Cu_xS on the surface of the photoelectrode makes the photocurrent decrease [37]. Nevertheless, the generated compounds do not fall off the electrode surface; therefore, the sensor cannot be reused.

The energy transfer between noble metal NPs and semiconductor materials can amplify the detection signal. When the emission spectrum of semiconductor QDs overlaps with the absorption spectrum of metal NPs, the incident light and the spontaneous emission generated by the radiation decay of QDs under illumination stimulate the local SPR of metal NPs. When the semiconductor closely contacts the metal NPs, the collective oscillation of the electrons in the metal NPs under excitation promotes the electron transfer from the metal NPs to the semiconductor leads to an increase in the photocurrent. When the QDs are close to the metal NPs (~10 nm), the SPR of the metal NPs will generate a local electric field to perturb the excited state of the QDs, thereby increasing the recombination of electron–hole

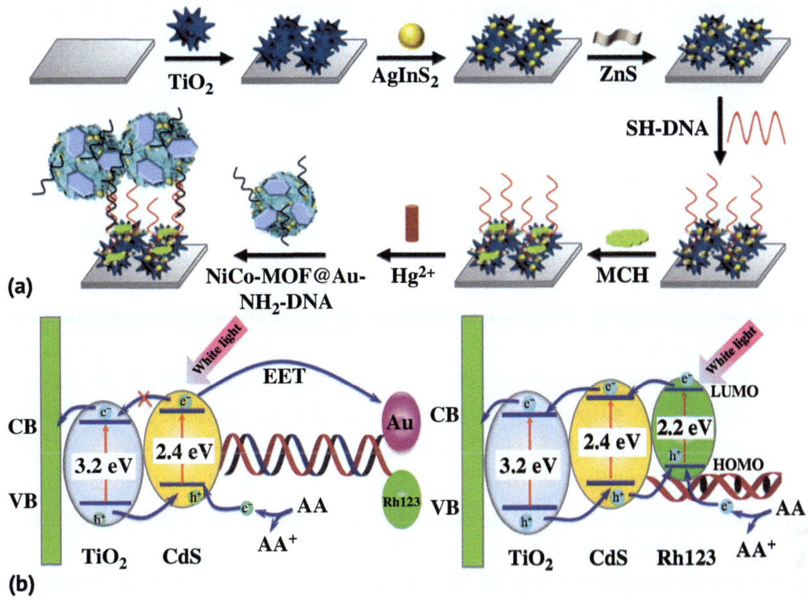

Figure 19.7 (a) The illustration of the construction PEC sensing platform of Hg^{2+} [36]/with permission from The Royal Society of Chemistry. (b) PEC Mechanism of the Biosensor (A) before and (B) after Incubation with Hg^{2+} [38]/with permission from American Chemical Society.

pairs [39]. Using rhodamine123 (Rh123) as a sensitizer, Zhao et al. [38] constructed a biosensor for Hg^{2+} detection based on the exciton energy transfer (EET) between CdS QDs and Au NPs as well as the distance tuning by DNA conformational changes. As shown in Figure 19.7b, the Au NPs-labeled target DNA hybridized with the Rh123-binding probe DNA (P-DNA) to form a rod-shaped double-helix structure to modify the TiO_2/CdS electrode before the detection. Rh123 is far away from the electrode surface and the EET between CdS QDs and Au NPs leads to a decrease in photocurrent. After Hg^{2+} incubation, the rod-like double helix structure and the EET were disrupted, while the folding of p-DNA made Rh123 closer to the electrode surface to achieve sensitization effect.

Different from heavy metal ions and NO^{2-}, F^- is generally beneficial to the human body at low concentrations, but high levels can cause related diseases. F^- with strong affinity can be easily adsorbed on the surface of TiO_2. Additionally, F^- can etch TiO_2 to break the Ti-O bond and form $[TiF_6]^{2-}$ ions under acidic conditions. Afterwards, Ti-F slowly dissolves in reducing glycerol due to affinity effects and eventually formed a defective TiO_2 layer rich in Ti^{3+} and O vacancies, which increased the surface active sites and enhanced the photocurrent [40].

The selectivity and repeatability of ion detection are the most important points of TiO_2-based PEC ion sensor. In the presence of multiple high valence metal ions, the reduction method of target metal ions is obviously not suitable for detection. Other harmful ions also consume electrons inevitably interfering with the detection. We also need to develop more mechanisms for ion selective detection. In addition, the sensors of Pb^{2+},Cd^{2+}, and F^- mentioned above are obviously not reusable. There is still considerable headroom for advances in selectivity, multiple detection, continuous monitoring, and commercialization in the detection of actual polluted water bodies.

19.5 PEC Detection of Organic Pollutants and Oxygen Demand

Polychlorinated biphenyls (PCBs), phthalic acid esters (PAEs), polycyclic aromatic hydrocarbons, phenols, ethers, and pesticides are representative synthetic organic pollutants detected in water, most of which are highly toxic or highly carcinogenic. They exist for a long time after entering the water, and can be absorbed by organisms and stored in adipose tissue, which is gradually amplified by the food chain to cause harm to the biota [41]. Therefore, it is very necessary to design TiO_2-based PEC equipment for selective detection of organic pollutant content in water.

PCBs are a class of industrially synthesized aromatic organochlorine chemicals with long half-life and high toxicity. They are classified as class I carcinogens by International Agency for Research on Cancer and have been banned from commercial production. The sensitive detection of these compounds can be carried out by using the steric hindrance sensing mechanism. Liao et al. [42] used CdTe@CdS QDs electrostatically adsorbed TiO_2 NTs as a photoactive material substrate, supplemented by aptamer probe modification for the detection of 2,3′,5,5′-tetrachlorobiphenyl (PCB72) in the environment. Figure 19.8a shows the process of preparing PEC aptasensor and the mechanism of detecting PCB72. The conductivity of the aptamer and the complex formed by the combination with PCB72 is poor, resulting in increased resistance to electron transfer on the electrode surface and a sharp decrease in photocurrent. Similarly, the pesticides widely used in agricultural pest control, such as diazinon and chlorpyrifos, can also be detected by this method [43]. The chelation between TiO_2-based surface-modified BiOI and chlorpyrifos was utilized to form complexes on the electrode surface and change the steric hindrance. The chelation only exists between Bi and the adjacent C═N and P═S in chlorpyrifos. Both the introduction of aptamers and specific chelation significantly increase the selectivity of the assay.

PAEs are widely used as plasticizers in the manufacture and processing of plastic products in the industrial, life, and medical fields. However, PAEs are physically dispersed in the polymer substrate and are easily leached and released into the environment. After absorbed by animals, it exerts hormone-like effects to disrupt endocrine. Dibutyl phthalate is a kind of PAE compound with high content in water environment, which can be specifically detected by using molecularly imprinted polymers (MIPs) to simulate the interaction between enzyme–substrate and antigen–antibody. MIPs have a certain spatial structure and strength, which can selectively adsorb and specifically recognize template molecules and their structural analogues, also known as artificial antibodies. First, the template molecule forms a complex with the functional monomer, followed by the addition of a cross-linking agent that is captured into the stereo structure of the polymer via a polymerization reaction. Then the template molecule is eluted, leaving a cavity with multiple action

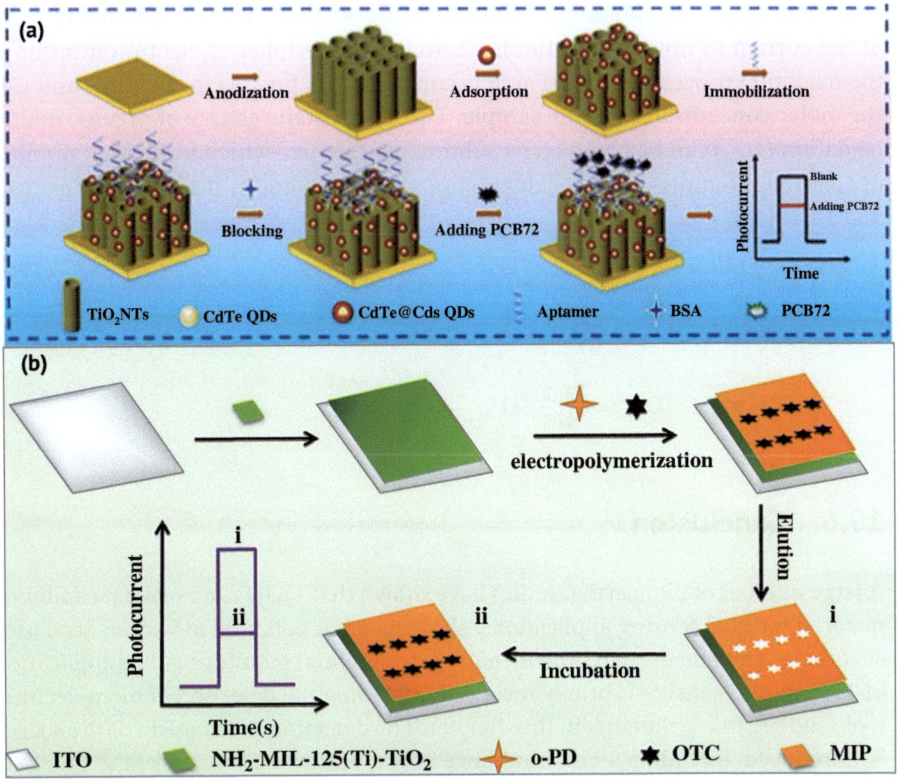

Figure 19.8 (a) The process of preparing PEC aptasensor and the mechanism of detecting PCB72 [42]/with permission from Elsevier. (b) Schematic illustration of the fabrication process of MIP–PEC sensor [44]/with permission from Elsevier.

points matching the spatial configuration of the template molecule for recognition [44, 45]. As shown in Figure 19.8b, the MIPs polymerized on the surface of the TiO_2-based electrode as a recognition element specifically binds and accumulates molecules, which is conducive to its oxidation by holes to produce a higher photocurrent [46].

Moreover, the PEC oxidation process provides a highly sensitive method for the detection of chemical oxygen demand COD of organic pollutants in wastewater. COD is a measure of the amount of oxygen (mg L^{-1}) required for the degradation of organic compounds contained in water by strong oxidants, which can be used as an important parameter for water quality and determination of oxygen demand of organic pollutants [47]. As a substitute for chemical oxidants, the inherent high stability and strong oxidizing ability of TiO_2 gives it a significant advantage in photoelectrocatalytic COD determination, which is capable of oxidizing almost any organic compound under UV irradiation. Organic compounds can be directly oxidized by photogenerated holes in the valence band of TiO_2, or oxidized by hydroxyl radical (·OH) active species generated by water oxidation. The photocurrent (i_{total}) generated by the TiO_2-based photoelectrode in the solution containing organic compounds is the total current of organic compounds (i_{net}) and water oxidation (i_{blank}). The amount of charge can be obtained according to the integral of the current to time in Eq. (19.2), where n is the number of electron transfer in the oxidation process, F is the Faraday constant, V is the reactor volume, and C is the molar concentration of the sample. The electrostatic charge of organic matter oxidation (Q_{net}) can be obtained by subtracting the two, which is the total number of electrons obtained by PEC oxidation of all compounds in the sample (Eqs. 19.3 and 19.4) [48].

$$Q = \int i\, dt = nFVC \tag{19.2}$$

$$O_2 + 4H^+ + 4e^- \rightarrow 2H_2O \tag{19.3}$$

$$\mathrm{COD}\left(\mathrm{mg\ L^{-1}O_2}\right) = \frac{32\,000}{4FV} Q_{net} \tag{19.4}$$

19.6 Conclusion

A large number of pioneering studies have shown that TiO_2 is an excellent candidate material for PEC sensing applications, showing great potential in various scenarios such as food and medical treatment, and environmental monitoring. The application of TiO_2 in PEC sensing is briefly reviewed, including the detection of biomolecules, ions, and organic pollutants in this chapter. The detection mechanism of the sensor is introduced, including steric hindrance effect, electron donor/acceptor, photoactive/inert material modification, energy transfer, etc. It is worth noting that in the design and implementation of the sensing system, its sensing behavior remains to be considered in terms of selectivity and multiple detection. In the process of PEC analysis, the photogenerated holes generated by TiO_2-based materials will oxidize most substances, so it is not accurate to analyze only by the signal changes generated by the oxidation process in practical applications. Although specific recognition substances such as antibodies, aptamers, enzymes, and MIPs can achieve selective detection, biological molecules such as antibodies and enzymes are easily inactivated, and the storage and test conditions of the electrode are strictly required. The aptamer has cross-reactivity, and it is difficult to detect multiple analytes. MIPs are suitable for small molecular substances and have poor preparation repeatability. The binding sites are easily buried in the three-dimensional structure of the polymer, resulting in low imprinting capacity. We have to explore more mechanisms for selective detection. At present, the detection of TiO_2-based sensors is generally only for one target substance, rarely for two or more substances, which is closely related to the selectivity of the electrode. On the basis of selective detection, how to distinguish and capture different signals of different substances in the test sample for multiple detection is a major problem in TiO_2-based PEC sensing. In addition, the advantages of PEC sensing over other detection methods are miniaturization and sensitive detection. Therefore, it is more necessary for us to explore the construction of TiO_2-based PEC small sensors, expand the detection range and improve the detection limit.

References

1 Li, L. Y., Chen, J. L., Xiao, C. B., et al. (2023). Recent advances in photoelectrochemical sensors for detection of ions in water. *Chinese Chemical Letters 34* (6): 107904.

2 Zhou, Y., Yin, H. and Ai, S. (2021). Applications of two-dimensional layered nanomaterials in photoelectrochemical sensors: A comprehensive review. *Coordination Chemistry Reviews 447*: 214156.

3 a) Fujishima, A. and Honda, K. (1972). Electrochemical Photolysis of water at a semiconductor electrode. *Nature 238* (5358), 37–38; b) Ullattil, S. G., Narendranath, S. B., Pillai, S. C., et al. (2018). Black TiO_2 nanomaterials: a review of recent advances. *Chemical Engineering Journal 343*: 708–736.

4 a) Kment, S., Riboni, F., Pausova, S., et al. (2017). Photoanodes based on TiO_2 and α-Fe_2O_3 for solar water splitting-superior role of 1D nanoarchitectures and of combined heterostructures. *Chemical Society Reviews 46* (12): 3716–3769; b) Qiu, J. X., Zhang, S. Q. and Zhao, H. J. (2011). Recent applications of TiO_2 nanomaterials in chemical sensing in aqueous media. *Sensors and Actuators B-Chemical 160* (1): 875–890.

5 a) Kurnaravel, V., Mathew, S., Bartlett, J., et al. (2019). Photocatalytic hydrogen production using metal doped TiO_2: A review of recent advances. *Applied Catalysis B-Environmental 244*: 1021–1064; b) Li, Z. Z., Wang, S. J., Wu, J. X., et al. (2022). Recent progress in defective TiO_2 photocatalysts for energy and environmental applications. *Renewable & Sustainable Energy Reviews 156*: 111980; c) Liang, D., Huang, Y. L., Wu, F., et al. (2019). In situ synthesis of g-C_3N_4/TiO_2 with {001} and {101} facets coexposed for water remediation. *Applied Surface Science 487*: 322–334; d) Qamar, O. A., Jamil, F., Hussain, M., et al. (2023). Advances in synthesis of TiO_2 nanoparticles and their application to biodiesel production: A review. *Chemical Engineering Journal 460*: 141734; e) Sun, X. Y., Wang, C. Y., Su, D. W., et al. (2020). Application of Photocatalytic Materials in Sensors. *Advanced Materials Technologies 5* (5): 1900993.

6 Wang, H. X., Xu, Y., Xu, D. F., et al. (2022). Graphitic Carbon nitride for photoelectrochemical detection of environmental pollutants. *ACS ES&T Engineering 2* (2): 140–157.

7 a) Li, M., Chen, L., Maigbay, M. A., et al. (2023). Applications of zinc indium sulfur-based photoelectrodes. *Materials Today Chemistry 31*: 101622; b) Liu, Q., Zhang, H., Jiang, H. H., et al. (2022). Photoactivities regulating of inorganic semiconductors and their applications in photoelectrochemical sensors for antibiotics analysis: A systematic review. *Biosensors & Bioelectronics 216*: 114634.

8 Moon, S., Charyulu, D. K., Lee, J., et al. (2022). Chemical-bath-deposited rutile TiO_2 film for electrochemical detection of 2, 4, 6-trinitrotoluene. *Thin Solid Films 748*: 139172.

9 Tembo, P. M., Dhabarde, N. and Subramanian, V. (2021). TiO_2 Nanotube-based sensor for the detection of cyanide in water. *Journal of the Electrochemical Society 168* (5): 057527.

10 Lu, Y. J., Purwidyantri, A., Liu, H. L., et al. (2020). Photoelectrochemical detection of β-amyloid peptides by a TiO_2 nanobrush biosensor. *IEEE Sensors Journal 20* (12): 6248–6255.

11 a) Cai, J. S., Shen, J. L., Zhang, X. N., et al. (2019). Light-Driven sustainable hydrogen production utilizing TiO_2 nanostructures: A Review. *Small Methods 3* (1): 1800184; b) Ge, M. Z., Cao, C. Y., Huang, J. Y., et al. (2016). A review of one-dimensional TiO_2 nanostructured materials for environmental and energy applications. *Journal of Materials Chemistry A 4* (18): 6772–6801; c) Hu, H. Y., Lin, Y.and Hu, Y. H. (2019). Synthesis, structures and applications of single component core-shell structured TiO_2: A review. *Chemical Engineering Journal 375*: 122029; d) Li, Z. L., Li, Z. Q., Zuo, C. L., et al. (2022). Application of nanostructured TiO_2 in UV photodetectors: A review. *Advanced Materials 34* (28): 2109083; e) Qiu, X. Q., Miyauchi, M., Sunada, K., et al. (2012). Hybrid Cu_xO/TiO_2 nanocomposites as risk-reduction materials in indoor environments. *ACS Nano 6* (2): 1609–1618; f) Rajaraman, T. S., Parikh, S. P.and Gandhi, V. G. (2020). Black TiO_2: A review of its properties and conflicting trends. *Chemical Engineering Journal 389*: 123918.

12 a) Sakib, S., Maclachlan, R., Gao, Y. X., et al. (2023). Catecholate dye-assisted decoration of Ag on TiO_2 nanoparticles for detection of cancer-related microRNA Biomarkers on a multiplexed photoelectrochemical platform. *ACS Applied Nano Materials 6* (21): 20375–20388; b) Wu, M., Jing, T., Tian, J. Z., et al. (2022). Synergistic effect of silver plasmon resonance and p-n heterojunction enhanced photoelectrochemical aptasensing platform for detecting Chloramphenicol. *Advanced Composites and Hybrid Materials 5* (3): 2247–2259.

13 a) Bakhshandeh, F., Saha, S., Sakib, S., et al. (2022). TiO_2 nanoparticles co-sensitized with graphene quantum dots and pyrocatechol violet for photoelectrochemical detection of Cr(VI). *Journal of the Electrochemical Society 169* (5): 057520; b) Sakib, S., Zhang, Z. J., Osman, E., et al. (2023). Programming synthetic DNA molecules for building and breaking heterostructures between TiO_2 nanomaterials for photoelectrochemical bacterial detection. *Nano Today 49*: 101783.

14 a) Li, Y. W., Xu, R., Wei, D., et al. (2020). A photoelectrochemical aptasensor for the detection of 17β-estradiol based on In_2S_3 and CdS co-sensitized cerium doped TiO_2. *New Journal of Chemistry 44* (2): 346–353; b) Liu, X., Fan, D. W., Duan, S. Q., et al. (2019). An ultrasensitive label-free photoelectrochemical sensor based on Ag_2O-sensitized WO_3/TiO_2 acicular composite for AFB1 detection. *Analytical Methods 11* (30): 3890–3897.

15 a) Li, L., Wang, T., Zhang, Y., et al. (2018). Editable TiO_2 nanomaterial-modified paper in situ for highly efficient detection of carcinoembryonic antigen by photoelectrochemical method. *ACS Applied Materials & Interfaces 10* (17): 14594–14601; b) Tang, J., Wang, Y. C., Li, J., et al. (2014). Sensitive enzymatic glucose detection by TiO_2 nanowire photoelectrochemical biosensors. *Journal of Materials Chemistry A 2* (17): 6153–6157.

16 Xie, Y., Wang, Y., Ma, Y., et al. (2022). Photoelectrochemical sensor based on carboxylated graphdiyne co-sensitized TiO_2 for sensitive detection of dopamine. *Materials Today Chemistry 26*: 101143.

17 Wang, H. X., Liang, D., Xu, Y., et al. (2021). A highly efficient photoelectrochemical sensor for detection of chlorpyrifos based on 2D/2D β-Bi_2O_3/g-C_3N_4 heterojunctions. *Environmental Science-Nano 8* (3): 773–783.

18 Ganganboina, A. B., Khoris, I. M., Konno, A., et al. (2023). CdSe-Co_3O_4@TiO_2 nanoflower-based photoelectrochemical platform probing visible light-driven virus detection. *Microchimica Acta 190* (1): 46.

19 a) Jiang, D., Du, X. J., Chen, D. Y., et al. (2016). One-pot hydrothermal route to fabricate nitrogen doped graphene/Ag-TiO_2: Efficient charge separation, and high-performance "on-off-on" switch system based photoelectrochemical biosensing. *Biosensors & Bioelectronics 83*: 149–155; b) Qin, X. F., Wang, Q. Q., Geng, L. P., et al. (2019). A "signal-on" photoelectrochemical aptasensor based on graphene quantum dots-sensitized TiO_2 nanotube arrays for sensitive detection of chloramphenicol. *Talanta 197*: 28–35.

20 Dai, H., Chen, S. H., Li, Y. L., et al. (2017). Photoelectrochemical biosensor constructed using TiO_2 mesocrystals based multipurpose matrix for trypsin detection. *Biosensors & Bioelectronics 92*: 687–694.

21 Cui, H., Yao, C. F., Cang, Y. G., et al. (2022). Oxygen vacancy-regulated TiO_2 nanotube photoelectrochemical sensor for highly sensitive and selective detection of tetracycline hydrochloride. *Sensors and Actuators B-Chemical 359*: 131564.

22 Wang, M. H., Yin, H. S., Zhou, Y. L., et al. (2019). Photoelectrochemical biosensor for microRNA detection based on a MoS_2/g-C_3N_4/black TiO_2 heterojunction with Histostar@AuNPs for signal amplification. *Biosensors & Bioelectronics 128*: 137–143.

23 Shang, X., Weng, Q. H., Wang, F., et al. (2021). Non-enzymatic photoelectrochemical sensors based on Schiff base and chitosan co-decorated TiO_2 nanosheets for dopamine detection. *Journal of Photochemistry and Photobiology A-Chemistry 418*: 113426.

24 a) Liao, X. L., Li, T. T., Ren, H. T., et al. (2022). Photoelectrochemical aptasensor driven by visible-light based on $BiFeO_3$@TiO_2 heterostructure for microcystin-LR detection. *Microchemical Journal 176*: 107201; b) Liu, W. Q., Duan, W., Jia, L. Q., et al. (2022). Surface Plasmon-enhanced photoelectrochemical sensor based on au modified TiO_2 nanotubes. *Nanomaterials 12* (12): 2058; c) Wang, M. Y., Zhan, Y., Wang, H. X., et al. (2022). A photoelectrochemical sensor for glutathione based on Bi_2S_3-modified TiO_2 nanotube arrays. *New Journal of Chemistry 46* (17): 8162–8170; d) Yang, Z. Y., Song, J., Jiang, R. L., et al. (2023). Construction of TiO_2@ZnO nanofibers with beads-on-a-string heterostructures for photoelectrochemical detection of lactic acid. *Journal of Alloys and Compounds 960*: 170659; e) Zhang, Y. J., Yu, J. R., Huang, W. J., et al. (2023). Detection of L-cysteine in urine samples based on CdS/TiO_2-modified extended-gate field-effect transistor photoelectrochemical sensor. *Microchimica Acta 190* (7): 280.

25 a) Ma, X. H., Kang, J. S., Wu, Y. W., et al. (2022). Recent advances in metal/covalent organic framework-based materials for photoelectrochemical sensing applications. *TrAC-Trends in Analytical Chemistry 157*: 116793; b) Shi, Y. T., Zou, Y. T., Khan, M. S., et al. (2023). Metal-organic framework-derived photoelectrochemical sensors: structural design and biosensing technology. *Journal of Materials Chemistry C 11* (11): 3692–3709.

26 a) Wang, Y., Yin, H. S., Li, X. H., et al. (2019). Photoelectrochemical immunosensor for N6-methyladenine detection based on Ru@UiO-66, Bi_2O_3 and Black TiO_2. *Biosensors & Bioelectronics 131*: 163–170; b) Wang, Z. H., Yan, Z. Y., Wang, F., et al. (2017). Highly sensitive photoelectrochemical biosensor for kinase activity detection and inhibition based on the surface defect recognition and multiple signal amplification of metal-organic frameworks. *Biosensors & Bioelectronics 97*: 107–114.

27 Zhou, Y. L., Wang, Y., Li, S. R., et al. (2020). A novel photoelectrochemical immunosensor for N1-methyladenine detection based on $BiVO_4$/g-C_3N_4 heterojunction with signal amplification of TiO_2@NH_2-MIL-125(Ti). *Sensors and Actuators B-Chemical 318*: 128310.

28 Shi, L., Yin, Y., Zhang, L. C., et al. (2019). Design and engineering heterojunctions for the photoelectrochemical monitoring of environmental pollutants: A review. *Applied Catalysis B-Environmental 248*: 405–422.

29 Zhao, W. W., Xu, J. J. and Chen, H. Y. (2016). Photoelectrochemical detection of metal ions. *Analyst 141* (14): 4262–4271.

30 a) Liu, Q. Y., Kim, J. and Cui, T. H. (2020). A highly sensitive photoelectrochemical sensor with polarity-switchable photocurrent for detection of trace hexavalent chromium. *Sensors and Actuators B-Chemical 317*: 128181; b) Meng, L. X., Zhang, Y., Wang, J. L., et al. (2023). Construction of MIL-125 derived 2D TiO_2@C nanocake/Au/3D peony-like BiOI Z-scheme heterojunction for sensitive and facile photoelectrochemical sensing assay of Cr (VI) based on competitive consumption of surface sacrificial agent. *Sensors and Actuators B-Chemical 396*: 134578; c) Moakhar, R. S., Goh, G. K. L., Dolati, A., et al. (2017). Sunlight-driven photoelectrochemical sensor for direct determination of hexavalent chromium based on Au decorated rutile TiO_2 nanorods. *Applied Catalysis B-Environmental 201*: 411–418.

31 a) Mokhtar, B., Kandiel, T. A., Ahmed, A. Y., et al. (2021). New application for TiO_2 P25 photocatalyst: A case study of photoelectrochemical sensing of nitrite ions. *Chemosphere 268*: 128847; b) Vázquez, C. I., Bruni, F. M., Baruzzi, A. M., et al. (2023). CdS|TiO_2 Non-Enzymatic Sulfide Photoelectrochemical Sensor. *IEEE Sensors Journal 23* (2): 856–863.

32 Qiao, J. Y., Wang, Y. H., Liang, Q., et al. (2022). A photoelectrochemical sensor based on TiO_2 nanotube arrays decorated with nickel-cobalt layered double hydroxides for the effective and sensitive detection of chromium(VI). *ACS Applied Nano Materials 5* (4): 5535–5543.

33 Liang, Y., Kong, B., Zhu, A., et al. (2012). A facile and efficient strategy for photoelectrochemical detection of cadmium ions based on in situelectrodeposition of CdSe clusters on TiO_2 nanotubes. *Chemical Communications 48* (2): 245–247.

34 Yu, Q., Fu, Y. M., Xiao, K., et al. (2022). A label-free photoelectrochemical biosensor with ultra-low-background noise for lead ion assay based on the Cu_2O-CuO-TiO_2 heterojunction. *Analytica Chimica Acta 1195*: 339456.

35 Luo, Y. P., Dong, C., Li, X. G., et al. (2015). A photoelectrochemical sensor for lead ion through electrodeposition of PbS nanoparticles onto TiO_2 nanotubes. *Journal of Electroanalytical Chemistry 759*: 51–54.

36 Luo, J., Ren, X., Wang, H., et al. (2023). Photoelectrochemical aptamer sensor based on $AgInS_2$ and ZnS co-sensitized TiO_2 for the detection of Hg^{2+}. *New Journal of Chemistry 47* (31): 14801–14807.

37 a) Feng, K. J., Yang, M. H., Xie, F., et al. (2017). TiO_2/CdS nanorod array-based photoelectrochemical sensing of Cu^{2+} in human serum samples. *Analytical Methods 9* (48): 6754–6759; b) Su, Y. Z., Yang, S. Y., Liu, W. P., et al. (2017). Visible light photoelectrochemical sulfide sensor based the use of TiO_2 nanotube arrays loaded with Cu_2O. *Microchimica Acta 184* (10): 4065–4072; c) Wang, Y. H., Ge, S. G., Zhang, L. N., et al. (2017). Visible photoelectrochemical sensing platform by in situ generated CdS quantum dots decorated branched-TiO_2 nanorods equipped with Prussian blue electrochromic display. *Biosensors & Bioelectronics 89*: 859–865.

38 Zhao, M., Fan, G. C., Chen, J. J., et al. (2015). Highly sensitive and selective photoelectrochemical biosensor for Hg^{2+} detection based on dual signal amplification by exciton energy transfer coupled with sensitization effect. *Analytical Chemistry 87* (24): 12340–12347.

39 a) Luo, C., Chen, X. X., Li, P., et al. (2023). A photoelectrochemical sensor based on DNA bio-dots-induced aggregation of AuNPs for methionine detection. *Molecules 28* (23): 7740; b) Yue, Z., Lisdat, F., Parak, W. J., et al. (2013). Quantum-dot-based photoelectrochemical sensors for chemical and biological detection. *ACS Applied Materials & Interfaces 5* (8): 2800–2814.

40 Su, Y. Z., Chen, D. T., Yang, S. Y., et al. (2019). Photoelectrochemical detection of ultra-trace fluorine ion using TiO_2 nanorod arrays as a probe. *RSC Advances 9* (46): 26712–26717.

41 Yan, K., Kannan, P. K., Doonyapisut, D., et al. (2021). Advanced functional electroactive and photoactive materials for monitoring the environmental pollutants. *Advanced Functional Materials 31* (12): 2008227.

42 Liao, D. Y., Liang, G. F., Liu, Y. Y., et al. (2023). Design an efficient photoelectrochemical aptasensor for PCB72 based on CdTe@CdS core@shell quantum dots-decorated TiO_2 nanotubes. *Journal of Hazardous Materials 441*: 129901.

43 a) Deiminiat, B. and Rounaghi, G. H. (2023). Fabrication of a Novel Photoelectrochemical Aptasensor Using Gold Nanoparticle-Sensitized TiO_2 Film for Quantitative Determination of Diazinon in Solutions. *Electrocatalysis 14* (3): 484–498; b) Lyu, R., Lei, Y. T., Zhang, C., et al. (2023). An ultra-sensitive photoelectrochemical sensor for chlorpyrifos detection based on a novel BiOI/TiO_2 n-n heterojunction. *Analytica Chimica Acta 1275*: 341579.

44 Yang, Y. K., Yan, W. Y., Wang, X. M., et al. (2021). Development of a molecularly imprinted photoelectrochemical sensing platform based on NH_2-MIL-125(Ti)-TiO_2 composite for the sensitive and selective determination of oxtetracycline. *Biosensors & Bioelectronics 177*: 113000.

45 Ma, X. G., Wang, C., Wu, F. X., et al. (2020). TiO_2 Nanomaterials in photoelectrochemical and electrochemiluminescent biosensing. *Topics in Current Chemistry 378* (2): 28.

46 Zeng, Y. A., Zhang, M., Peng, K. F., et al. (2023). Au/TiO_2-based molecularly imprinted photoelectrochemical sensor for dibutyl phthalate detection. *Microchimica Acta 190* (6): 244.

47 Azer, B. B., Gulsaran, A., Pennings, J. R., et al. (2022). A Review: TiO_2 based photoelectrocatalytic chemical oxygen demand sensors and their usage in industrial applications. *Journal of Electroanalytical Chemistry 918*: 116466.

48 a) Feng, Y. Y., Rijnaarts, H. H. M., Yntema, D., et al. (2020). Applications of anodized TiO_2 nanotube arrays on the removal of aqueous contaminants of emerging concern: A review. *Water Research 186*: 116327; b) Nurdin, M., Muzakkar, M. Z., Maulidiyah, M., et al. (2022). High-Performance COD Detection of Organic Compound Pollutants using Sulfurized-TiO_2/Ti Nanotube Array Photoelectrocatalyst. *Electrocatalysis 13* (5): 580–589; c) Wang, Y. Z., Zu, M., Li, S., et al. (2017). Dual modification of TiO_2 nanorods for selective photoelectrochemical detection of organic compounds. *Sensors and Actuators B-Chemical 250*: 307–314.

20

Reactor Engineering in Competitive Photocatalytic Performance: Advancements and Challenges

Jéssica Ariane de Oliveira[1], Daniel Raimundo Sevene[2], Osmando F. Lopes[2], Caue Ribeiro[1,]*

[1] *Nanotechnology National Laboratory for Agriculture (LNNA), Embrapa Instrumentation, São Carlos, São Paulo, Brazil*
[2] *Laboratory of Photochemistry and Materials Science, Institute of Chemistry, Federal University of Uberlandia, Uberlandia, Minas Gerais, Brazil*

*Corresponding author: caue.ribeiro@embrapa.br

20.1 Brief Overview

Photochemical reactors have undergone significant evolution driven by the demand for efficient and sustainable chemical processes, including organic synthesis, photodegradation of pollutants, and photopolymerization of materials. Historically, they have transitioned from simple glass vessels exposed to sunlight to sophisticated, controlled environments utilizing artificial light sources [1]. During the early 1990s, the parabolic trough reactor emerged as the pioneering solar reactor, a result of collaborative efforts between the Plataforma Solar de Almería in Spain and the National Renewable Energy Laboratory in the United States [2–4]. Subsequently, several photocatalytic solar reactors were designed, such as the film fixed bed reactor, parabolic collecting reactor, and double-skin sheet reactor [2]. In 1992, the patent for the photocatalytic slurry reactor was obtained [5], while the classical annular reactor (CAR) was developed in 1993 [6], leading to a predominance of reactor designs based on CAR variants with a lamp positioned within the inner glass/quartz tube and the reactant situated in the outer casing. The first LED photocatalytic reactor was created in 2005 [2, 7]. Figure 20.1 provides a comprehensive overview of the evolution of photochemical reactor engineering.

The design of a photochemical reactor requires careful consideration of several critical parameters, including (i) the selection of light source that can range from solar illumination, which experiences fluctuations in light intensity, to conventional options like ultraviolet (UV) and visible (Vis) lamps, as well as light emitting diodes (LEDs); (ii) the utilization of light-concentrating systems; (iii) the geometric configuration of the reactor; (iv) the choice of construction material considering that while quartz and glass remain widely used for their transparency, chemical, and

Nanostructured TiO_2: Synthesis, Photocatalytic Properties and Applications, First Edition.
Edited by Jia Hong Pan, Wan In Lee and Detlef W. Bahnemann

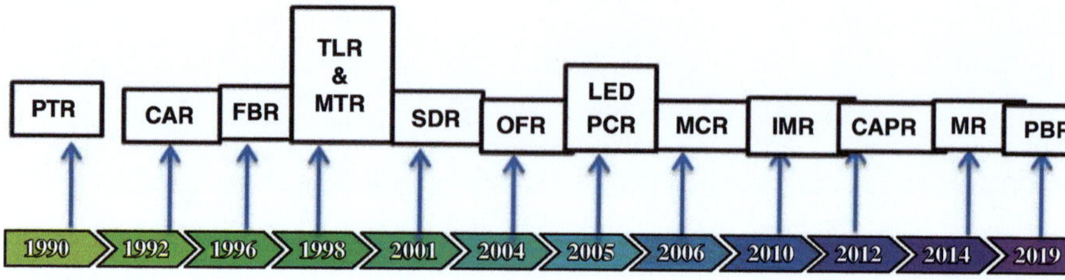

Figure 20.1 Evolution of photochemical reactors: parabolic trough reactor (PTR), CAR, fluidized bed reactor (FBR), tube light reactor (TLR), multitube reactor (MTR), spinning disk reactor (SDR), optical fiber reactor (OFR), light emitting diode photocatalytic reactor (LED PCR), microchannel reactor (MCR), integrated membrane reactor (IMR), capillary array photocatalytic reactor (CAPR), microreactor (MR), and packed bed reactor (PBR). Reproduced from Ref. [2] with permission from Elsevier.

thermal stability, more advanced materials like stainless steel coated with titanium oxide (TiO_2) can be used to improve light reflectivity and reaction efficiency; (v) mechanisms for heat exchange; (vi) the phases involved in the reaction (gas–solid, liquid–solid, or gas–liquid–solid); (vii) the operating mode (batch, semi-batch, or continuous); (viii) the type of catalyst employed, whether TiO_2-based or non-TiO_2-based; (ix) the size of the catalyst particles, whether in micron or nanometer scale; and (x) the form of the catalyst since suspended reactors are recognized for their simple design and efficiency, while immobilized reactors offer easy separation of the catalyst [2, 8, 9].

In industrial settings, the most common reactor designs include continuously stirred tank reactor (CSTR), plug flow reactor (PFR), and the packed bed reactor (PBR). The CSTR, idealized as perfectly mixed, is primarily used for liquid-phase reactions, maintaining steady-state conditions without time or position dependence on temperature, concentration, or reaction rate within the reactor. On the other hand, the PFR is typically employed for gas-phase reactions, with reactants constantly consumed along the reactor's length, resulting in a continuous variation in concentration along the axial direction. Like the PFR, the PBR is assumed to lack radial gradients in concentration, temperature, or reaction rate. However, in the PBR, the reaction occurs on the surface of the catalyst, with the reaction rate determined by the mass of the solid catalyst rather than the volume of the reactor [9, 10].

Advancements in photochemical reactor technology have resulted in the development of sophisticated photoreactors, often derived from the standard reactor types – CSTR, PFR, and PBR. These modern photoreactors offer precise control over reaction parameters, thereby enhancing process reproducibility and efficiency [1, 9]. Despite these advancements, several limitations persist, hindering their widespread adoption. These include challenges related to low efficiency in transferring light to the reaction, formation of gas bubbles that interfere with reaction effectiveness, availability and cost of suitable light sources, and scalability of photochemical processes [11]. Additionally, ensuring selectivity and product purity in photochemical reactions remains a critical issue, requiring careful optimization of reaction conditions and the development of selective photocatalysts. On the other hand, the involvement of sensitive or reactive intermediates in photochemistry necessitates specialized reactor materials and handling procedures to ensure safety and product stability. Furthermore, controlling temperature, pressure, and oxygen levels within the reactor environment becomes crucial to mitigate unwanted side effects or degradation pathways [12, 13].

Therefore, ensuring technical feasibility and economic viability is essential for sustainable processes. Achieving a balance between these two aspects of design while also considering the constraints associated with photon transfer and mass transfer is pivotal for successfully implementing the photocatalytic process in industrial applications [2]. The following section describes the technological advancements reported in recent literature over the past decade concerning reactor design for enhancing photoactivity and reaction rates.

20.2 Photoreactor Architecture

Most of the research has concentrated on optimizing the photocatalyst to achieve a high photonic yield, while there has been limited exploration into the configuration of the reactor. Several crucial aspects must be considered to meet the ideal requirements for the large-scale implementation of photocatalytic reactors: (i) the maximization of the specific surface area of the catalyst to enhance the interaction between photons and the catalyst, thus optimizing the efficiency of the photocatalytic process; (ii) the proper illumination that ensures a uniform and appropriate distribution of irradiation on the reaction surface, thus optimizing light absorption by the catalyst and boosting the effectiveness of photocatalytic reactions; and (iii) the efficient mass transfer to promote the effective transport of reactants and products in photocatalytic reactions, contributing to an overall more efficient reactor performance. In addressing these challenges and incorporating the main operating mode (batch and continuous flow), different reactor designs have been explored in the literature, such as packed bed, fluidized bed, cylindrical, and filter-press. Due to the intricate nature of photocatalytic reactions, effectively coordinating light irradiation, reactants, and catalysts poses a foremost challenge in designing photoreactors [8, 14–17].

20.2.1 Batch-mode Reactor

Batch-mode reactors operate intermittently, with reagents added in a single batch, allowing for greater control over reaction conditions. This system is suitable for small-scale operations and processes where initial and final conditions may vary. It is commonly used for testing new routes, manufacturing expensive products, and addressing processes that pose challenges for adaptation to continuous operations. Despite the advantage of achieving high conversion rates, batch-mode reactors typically incur elevated operating costs and exhibit product variations from one batch to another. Critical assumptions for batch-mode reactors include no material input or output, perfect mixing in the reactor, and negligible volume variation [10].

20.2.1.1 Slurry Photocatalytic Reactor (SPR)

The slurry photocatalytic reactor (SPR) is a type of continuous stirred tank reactor where the photocatalyst is suspended within the liquid phase. Usually, the effects

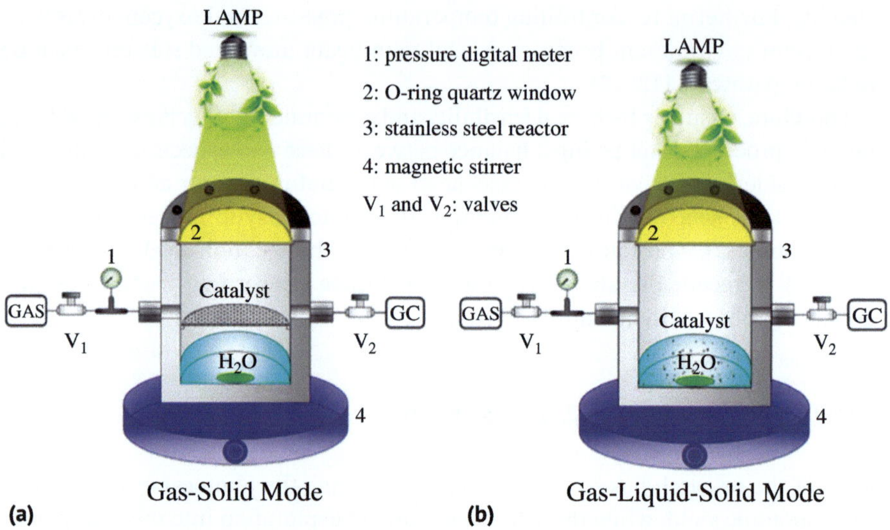

Figure 20.2 Slurry photoreactor operating in (a) gas–solid mode and (b) gas–liquid–solid mode. Reproduced from Ref. [21]/with permission from American Chemical Society.

of powder loading, particle size, adsorption of the target molecule, and shear forces determine the reaction rate [10]. The SPR operating in batch mode and containing suspended TiO_2 particles are widely used in research, leveraging the substantial surface area of the catalytic particles in suspension, high mass transfer rate, and simple system design. The catalyst concentration stands out as a crucial design parameter in this setup, influencing the dispersion of radiation and, consequently, the photoactivity. Beyond a specific catalyst amount, there is a decline in photon penetration within the reactor, resulting in decreased reaction rates [16, 18]. Achieving substantial photocatalytic yields frequently necessitates prolonged reaction durations. Additionally, SPR can lead to the leaching and settling of the catalysts, presenting challenges in the separation of the catalyst after the reaction process [19, 20]. Figure 20.2 illustrates a batch stainless-steel slurry reactor containing a quartz window on the top. This versatile system is suitable for both gas–solid or gas–liquid–solid modes. The solid catalyst is positioned on a Teflon catalyst holder in the reactor's upper section at the gas–solid interface (Figure 20.2a), while the gas can be preloaded at the bottom. For the reaction occurring in the gas–solid–liquid interface (Figure 20.2b), the catalyst is dispersed in a liquid solution precharged with gas at the bottom of the reactor [21].

Sauer et al. [22] exemplify an alternative slurry photoreactor configuration, with their findings resulting in the publication of the most extensively cited paper in the literature concerning the use of photochemical reactors employing TiO_2 as photocatalysts. As depicted in Figure 20.3, the lamp is immersed in the photochemical reactor without its glass cover. The solution inside the photoreactor is continuously magnetic stirred while the reactor's temperature remains constant through a thermostatic bath. Analytical control samples can be periodically extracted for monitoring purposes.

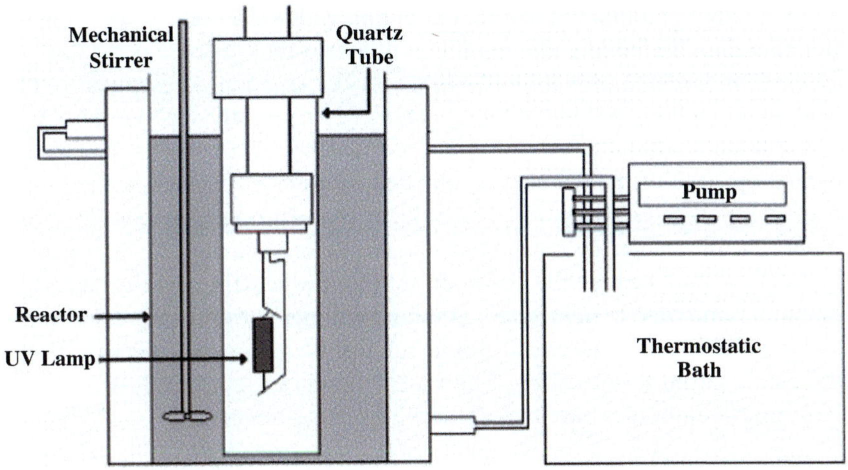

Figure 20.3 Scheme of the slurry photochemical reactor (SPR) used in the most cited paper regarding the degradation of the azo dye safira dye HEXL over TiO_2 P25. Reproduced from Ref. [22] with permission from Elsevier.

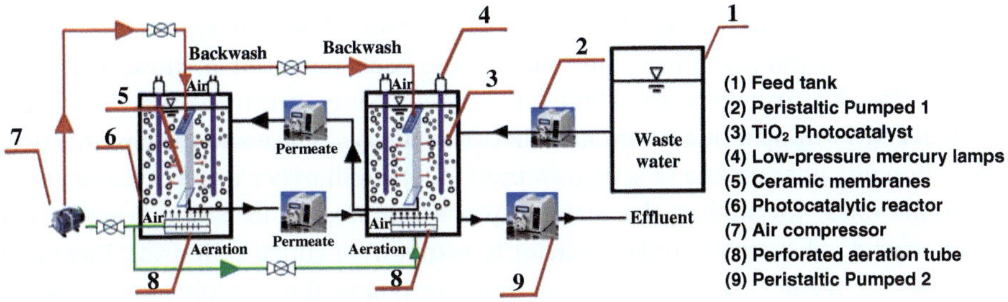

Figure 20.4 Scheme of the SCMPR employed for the degradation of amoxicillin. Reproduced from Ref. [23] with permission from Elsevier.

20.2.1.2 Submerged Ceramic Membrane Photocatalytic Reactor

Integrating TiO_2 with a submerged membrane configuration emerges as a viable solution for addressing the challenges associated with slurry-type photoreactors, particularly in the efficient separation of TiO_2 particles from the solution. Figure 20.4 depicts a schematic diagram illustrating the setup of the submerged ceramic membrane photocatalytic reactor (SCMPR). The two stainless steel rectangular reactors share identical dimensions and structures. An antifouling ceramic membrane is incorporated into the reactor, with a hollow structure and numerous rectangular holes upon cross-section examination. To facilitate backwashing, an air compressor is connected to the top of the ceramic membrane, while a peristaltic pump linked to the bottom ensures the separation of effluents from the catalyst slurry. Photocatalysis is enabled by lamps strategically arranged around the ceramic membrane. Moreover, perforated aeration pipes at the reactor's bottom generate microporous bubbles rich in dissolved oxygen. This fluidization of the catalysts effectively prevents the accumulation of a substantial number of catalysts on the membrane surface [23].

Prior to the reaction, the solution containing the chemical to be oxidized or reduced is directly fed into the reactor from a feed tank using a peristaltic pump. The ceramic membrane is fully immersed in the solution. Subsequently, TiO_2 is introduced into the solution, and the aeration system is activated to fluidize the catalysts. After a 30-min interval, the lamps are switched on to initiate photocatalysis. Upon completion of the reaction, the treated solution is removed from the system via another peristaltic pump. The TiO_2 utilized in the process remains in the reactor and can be reused for subsequent reactions. Before starting the next cycle, the solution from the feed tank is reintroduced into the reactor, and the backwashing procedure is initiated to minimize membrane fouling. The air compressor forces air into the ceramic membrane, creating enough turbulence on its surface to effectively stop catalysts from accumulating. The reactor operates as previously described after the ceramic membranes have had a backwashing treatment [23].

20.2.1.3 Capillary Photocatalytic Reactor

Capillary photocatalytic reactors (CPRs) offer advantages for photochemical applications over traditional batch reactors, boasting an increased lighted surface area, lower pressure drop, functionalized surfaces, and enhanced mass and heat transfer. Differential hydrodynamic properties lead to superior mass and heat transfer performance in capillary channels. Various flow patterns, including aerated, annular, bubbly, churn, and slug, have been observed in these channels. Taylor flow, featuring elongated bubbles and liquid slugs, minimizes barrier to mass transfer by forming a thin liquid layer between the gas phase and the channel wall. Additionally, phenomena observed in individual capillaries can be extrapolated to multiple ones, simplifying the scale-up process compared to conventional stirred tank reactors. A typical capillary reactor is a straight or coiled cylindrical form with an inner diameter ranging from 1 to 4 mm. For practical use in photocatalysis, the capillary must be constructed from a transparent material suitable for the applied radiation wavelength [17, 24].

Notably, the mass transfer coefficients and the photonic efficiency in capillary multiphase reactors, whether operating in slurry or fixed-bed mode, have been seen to surpass those in conventional setups by two to three orders of magnitude. A high surface-to-volume ratio, a short diffusion length, and quick reaction times are responsible for this improved performance. According to the liquid-to-gas velocity ratio inside the capillary, the kind of flow that CPR displays determines the mass transfer efficiency [17]. Figure 20.5 illustrates the CPR experimental arrangement, using carbon dioxide (CO_2) as an example of photoreaction. This setup consists of a quartz capillary reactor equipped with a T junction at the top, allowing simultaneous feeding of liquid and gas phases to generate Taylor flow. The peristaltic pump circulates the liquid solution through the system, while the gas flow is controlled using a mass flow controller [17].

20.2.2 Flow-mode Reactor

Continuous flow reactors (CFRs) are often preferred over batch-mode reactors due to several advantages. The continuous operation of CFR, with reagents constantly

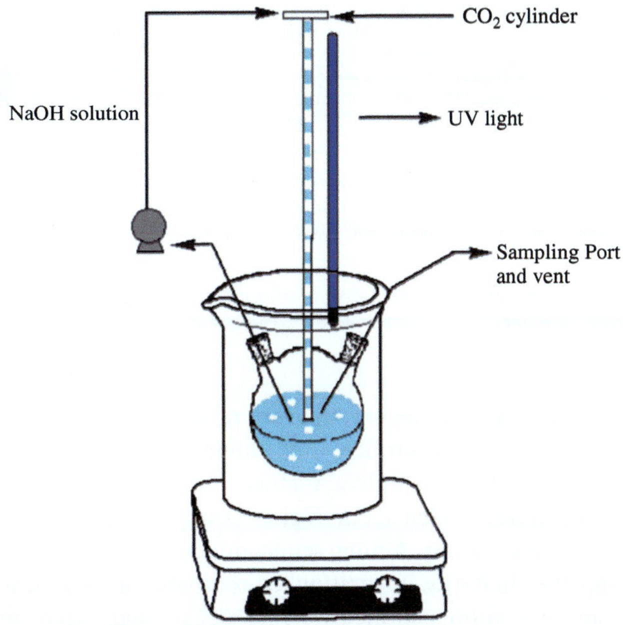

Figure 20.5 Experimental setup for the CPR used for the CO_2 reduction reaction. Reproduced with permission from Springer Nature [17].

fed over time, enables cost savings, increased efficiency, and higher productivity for extended processes. Additionally, CFRs provide easy scalability and superior control over temperature, pressure, product quality, and safety. A fundamental hypothesis in CFRs is their nearly constant operation at a steady state, meaning that conditions remain relatively constant over time. In these systems, the residence time rises proportionally with an increase in reactor volume. In other words, the larger or longer the reactor, the more time it takes for the chemicals to react. Consequently, the conversion is dependent on the reactor's volume. In this sense, recycling proves to be a valuable strategy to optimize the performance of chemical processes and address the challenge of conversion without the need for an excessively large reactor. The use of recycle reactors can (i) increase conversion and process efficiency by recycling unconsumed products back into the reactor, thereby boosting reagents conversion; (ii) control temperature in exothermic or endothermic reactions; (iii) enhance selectivity by recycling specific products or intermediates to guide the formation of desired products; and (iv) minimize waste by reintroducing byproducts or unused reagents [10].

20.2.2.1 Photocatalytic Membrane Reactor

Conventional photocatalyst application in fine powder form often leads to challenges such as aggregation and agglomeration in aqueous solutions, diminishing overall effectiveness and efficiency. Additionally, the photocatalyst recovery from the reaction medium becomes essential for applications on a large scale. Photocatalytic membrane reactors (PMRs) offer a promising solution by immobilizing nanoparticles

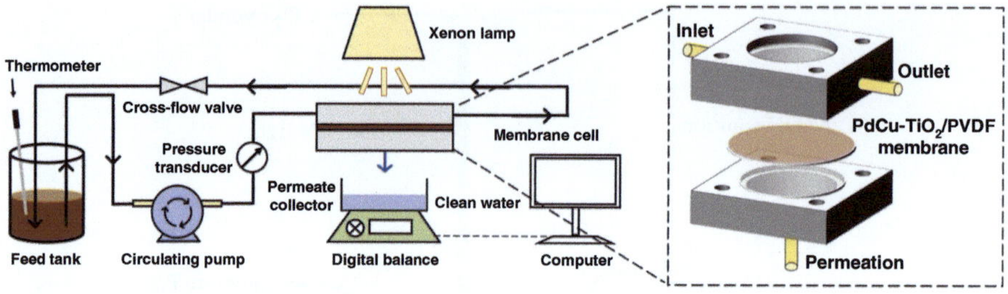

Figure 20.6 Diagram illustrating the PMR configuration. Reproduced from Ref. [26] with permission from Elsevier.

within a membrane, leading higher recovery rates and improved interaction between pollutants and catalyst surface. Embedding nanoparticles within the membrane enhances recyclability, unlike methods involving sequential filtering or coating onto substrates, which incur higher costs, longer processing times, and lower membrane permeability. Among semiconductors, TiO_2 stands out as the most frequently employed in PMRs, integrating redox capabilities, physical filtration, and antibacterial abilities within a unified system [20, 25]. Figure 20.6 illustrates a PMR equipped with quartz glass in its upper section to facilitate the transmission of light onto the membrane. The lamp is positioned above the quartz window, allowing adequate illumination. Collection and measurement of the permeation solution are conducted at regular intervals using a digital balance connected to a computer for gathering data acquisition. A temperature probe in the feed tank records real-time temperature changes in the feed solution.

20.2.2.2 Fixed Photocatalytic Reactor

Reactor systems with immobilized TiO_2 are favored for practical use. However, optimal efficiency requires a high ratio of active surface area to reactor volume, ensuring effective distribution of the incident illumination. The fixed bed photocatalytic reactor (FBPR) is a PBR-type reactor that eliminates catalyst particle separation, enabling catalytic support reuse. Yet, immobilized systems face limitations due to their area-to-volume ratio, leading to mass transfer constraints. Effective reactor design is crucial, focusing on optimizing catalyst exposure to light and addressing mass transfer limitations [16, 18].

Manassero et al. [18] developed a unique system operating as both FBPR and a modified fixed film photocatalytic reactor (FFPR), resembling a PFR-type reactor. Borosilicate glass rings support catalyst immobilization in FBPR, while in FFPR, the catalyst is on a flat borosilicate glass window. FFPR offers two illumination configurations: from the support side or the solution side. Optimal film thickness is crucial in the former, while in the latter, reaction rate increases with film thickness up to a specific value. This unique photocatalytic reactor is cylindrical, featuring two circular and flat windows. Illumination passes through one of these windows, with the lamp positioned along the focal axis of a parabolic reflector, as depicted in Figure 20.7. A container of $CoSO_4$ solution can be inserted between the reactor and the lamp to block visible radiation [18].

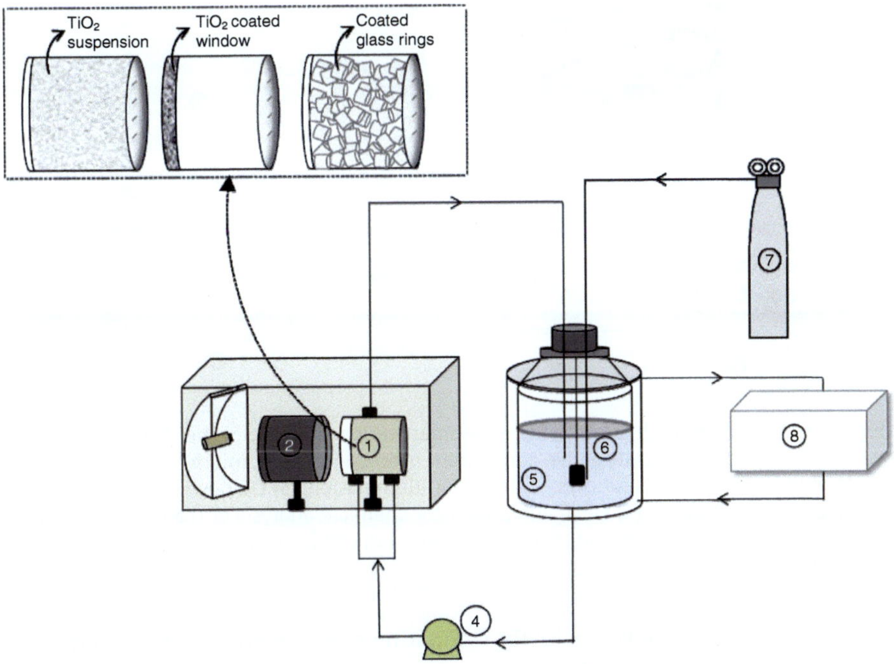

Figure 20.7 Diagram illustrating the experimental arrangement of the fixed photocatalytic reactor: (1) reactor, (2) filter solution, (3) light source and reflector, (4) pump, (5) reservoir for the aqueous medium, (6) thermometer, (7) O_2, and (8) thermostatic bath. Reproduced from Ref. [18] with permission from Elsevier.

20.2.2.3 Photocatalytic Microreactor

The microreactor's miniaturized dimensions provide advantages, including (i) a shortened light path that reduces photon transfer limitations, ensuring high luminous homogeneity throughout the reactor and (ii) a high surface/volume ratio, promoting enhanced heat and mass transfer [27]. Figure 20.8 illustrates a microreactor with a coil-type design enclosed in a UV-transparent Plexiglas housing with a mechanically engraved cylindrical channel to accommodate a catalytically active unit. The active unit consists of an anodized titanium-anode coil and two titanium-cathode coils, all made from tightly 250 μm titanium wire. The photoanode coil underwent anodization and subsequent annealing to achieve highly ordered, vertically oriented, and photocatalytically active TiO_2 nanotubes. The anode coil is centrally positioned within a silica glass rod, flanked on both sides by the cathode coils. Plastic spacers are employed to avoid any possible short circuits. The total length of the active unit in the microreactor is approximately 92 mm, resulting in a surface-to-volume ratio of around 5550 $m^2\ m^{-3}$. Two light sources illuminate the top and bottom of the anode coil [28].

The primary challenge in photocatalysis arises from the rapid recombination of photogenerated charges (e^-/h^+), typically occurring within the nanosecond range. This recombination rate is significantly faster, by two or three orders of magnitude, than the interfacial electron transport process. The transfer of the e^-/h^+ pairs to the

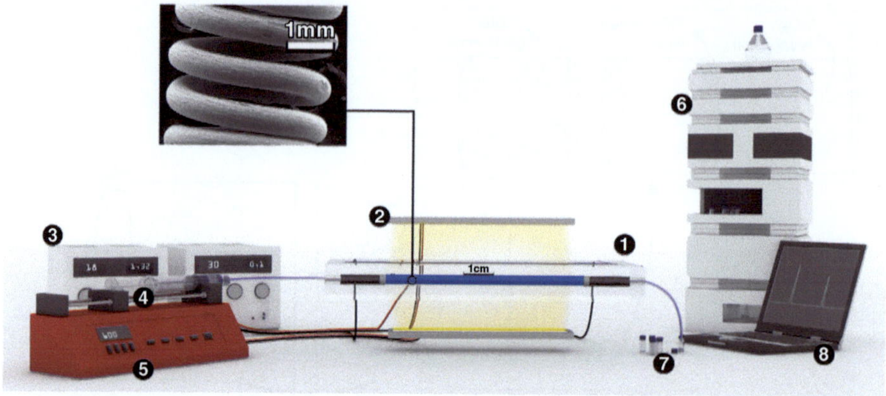

Figure 20.8 Diagram illustrating the experimental arrangement of the (1) microreactor, where the photocatalytically active component is an anodized titanium coil. Additional parts of the system are as follows: (2) lamp for illumination, (3) DC power supply, (4) glass syringe and its (5) pump, (6) HPLC equipment for chemical analysis of the liquid aliquots, (7) vials for sample storage, and (8) computer. Reproduced from [28]/with permission of Elsevier.

semiconductor surface is crucial for facilitating redox reactions. PEC emerges as a viable solution to address this issue since, under an applied potential, photogenerated carriers are effectively separated in two distinct physical sides – electrons (e^-) migrate to the cathode, while the holes (h^+) remain at the anode. This combination of processes enables the direct generation of electric current from light, offering a more efficient approach for various applications such as energy conversion, pollutant degradation, and fuel production [29]. The system illustrated in Figure 20.8 is versatile and can be utilized for both photo- and photoelectro-processes.

20.2.2.4 Staircase Photocatalytic Reactor (SPR)

Developed by Ahlstrom, the staircase photocatalytic reactor (SPR) concept comprises a reactor with a falling film that flows along a staircase, as depicted in Figure 20.9. The falling film is created as the solution containing the molecule to be reacted flows through the system while the unreacted components proceed horizontally. This reactor configuration offers some advantages, including (i) controlling a thin layer of the solution, approximately 1 mm thick, to enhance the light absorption; (ii) the step shape, improving the exchange surface between the catalyst and the liquid film; and (iii) increasing the exchange surface of the solution with the air, thereby enhancing the oxygen transfer into the solution and, consequently, enhancing oxidation for the O_2-dependent reactions. The reactor also includes lamps, a tank equipped with a magnetic agitator for solution stirring, and a peristaltic pump [30].

20.2.2.5 Solar Parabolic Photocatalytic Reactor

The solar parabolic photocatalytic reactor (SPPR) emerges as a favorable alternative for efficient photocatalytic reactions. By harnessing parabolic trough collectors (PTCs), the SPPR concentrates sunlight onto a reflective surface, significantly enhancing photocatalysis. PTCs not only absorb thermal energy but also offer

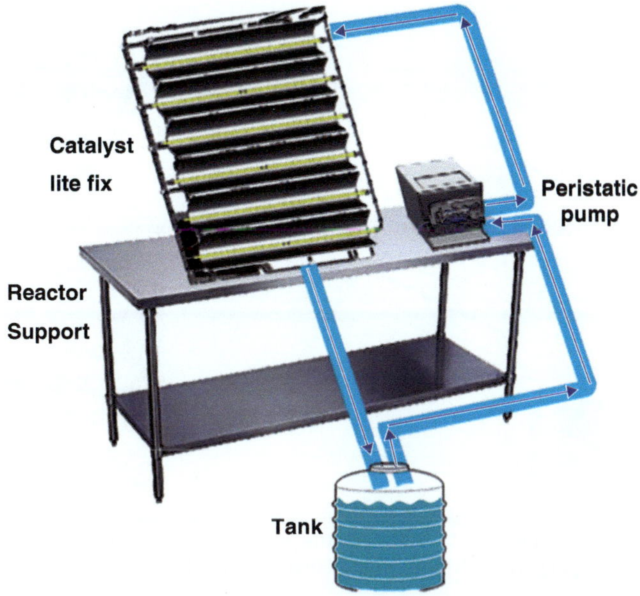

Figure 20.9 Diagram illustrating the experimental arrangement of the staircase photocatalytic reactor (SPR) [30]/MDPI/CC BY 4.0.

cost-effective and streamlined designs, widely used in solar power, heat generation, and wastewater treatment systems. By efficiently directing solar heat and intense UV light toward the photocatalyst, PTCs facilitate the redox process. The plant configuration includes a parabolic solar collector, a centrifugal pump, and a stirred tank reactor, as depicted in Figure 20.10. Oriented at a fixed 30° angle southward to optimize solar radiation collection, it operates continuously, concentrating irradiation in outer glass absorber tubes [31].

20.2.2.6 Filter-press Photocatalytic Reactor

As previously described, PEC combines electrocatalysis and photocatalysis by applying an external electrical potential that effectively separates the photogenerated charges. PEC processes offer additional benefits, including high energy efficiency, lower power consumption, versatility, accelerated reaction rates, potential for industrial-scale application, and prevention of oxidation of organic products formed at the cathode through the use of separate half-cells [32, 33]. In this context, the filter-press photocatalytic reactor (FPPR) is one of the most consistent architectural designs due to its capacity to diminish costs and generate a uniform current. The filter press-type geometry, particularly under the "zero-gap" condition, effectively minimizes ohmic losses [34], while also offering easy scalability and enhanced mass transfer rates. Consequently, FPPR holds significant potential to improve reaction kinetics and product selectivity. The system illustrated in Figure 20.11 facilitates process scalability as various filter press-type reactors can be connected either in parallel or in series to meet the specific requirements of different industry processing [35].

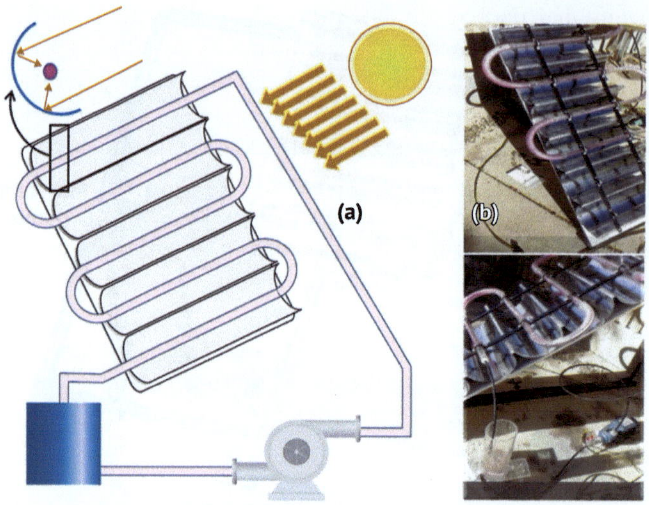

Figure 20.10 The experimental setup for the SPPR: (a) schematic and (b) photography. Reproduced from [31]/with permission of Elsevier/CC BY 4.0.

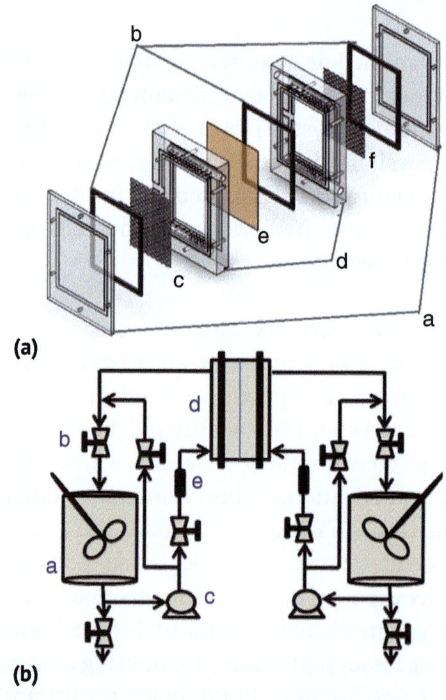

Figure 20.11 (A) Diagrammatic illustration of the photoelectrochemical filter-press reactor: (a) main plates; (b) nitrile seals; (c) photoanode; (d) flow distribution systems; (e) ion-exchange membrane; and (f) cathode. (B) Illustration depicting the experimental setup: (a) reservoir for the electrolyte; (b) valve; (c) pump; (d) photoelectrochemical reactor; and (e) flow meter. Reproduced from Ref. [36] with permission from Elsevier.

In a conventional photoelectrochemical setup, the photoanode typically comprises TiO_2 supported on a conductive substrate. The cathode is composed of an active metal or semiconductor surface tailored for the desired reactions. The system works by applying an electric potential to separate the electrons and holes generated by photoactivation. This allows the electrons to be transferred from the photoanode to the cathode and prevents them from recombining. The configuration of the flow distributors depicted in Figure 20.11 offers several advantages over a standard filter-press reactor: (i) the fluid elements enter the electroactive area at the same velocity as they leave, as required by the "closed-closed vessel" boundary conditions; (ii) the velocity discrepancies within the electroactive area are minimized, thereby reducing zones of low and high velocity; and (iii) a fully developed flow throughout the electroactive area is achieved. The reactor compartments are separated with an ion-exchange membrane, and each electrode is positioned in its respective compartment [36].

20.3 Main Applications of TiO_2 in Photocatalytic Reactors

TiO_2 is extensively used in different applications owing to its numerous merits, including cost-effectiveness, abundance, non-toxicity, chemical/thermal stability, and affordability [16, 17, 24, 31]. Noteworthy, many of these applications utilize commercial TiO_2 P25, indicating significant efforts in the literature to comprehensively study proposed reactor designs and their corresponding operational conditions to enhance process efficiency. It is worth noting that the majority of investigations concerning TiO_2 focus on oxidation reactions involving dyes [22, 23, 31], phenol [28], cyanide [19, 37], pharmaceuticals [18, 20, 23, 25, 30], and natural organic matter (NOM) [26], primarily due to the potent oxidative capacity of TiO_2. However, there are few reports exploring emerging applications of reduction reactions, such as carbon dioxide reduction (CO_2) [16, 17, 21].

Another important consideration is that photocatalysis still often faces challenges in achieving complete mineralization, leading to the formation of byproducts or intermediate compounds. While understanding the degradation efficiency of the target substance is crucial, it alone is insufficient to evaluate the quality of decontamination. Achieving mineralization of both the initial molecules and their byproducts is imperative to ensure the most effective decontamination [30]. In this sense, two significant challenges hinder the practical application of TiO_2: (i) its exclusive photoactivity under UV light owing to its wide bandgap energy ($\sim$ 3.2 eV for anatase) and (ii) the relatively short lifetime of photogenerated charges (approximately 30 ns), resulting in poorer photoactivity [16, 17, 24, 31]. Various strategies have been developed to overcome these issues and further enhance the efficiency of the photocatalytic process. Doping with metal or nonmetal elements narrows the bandgap of the semiconductor, enabling materials initially responsive only to ultraviolet radiation to become active under visible light [38, 39]. On the other hand, the formation of type II heterojunctions promotes the separation of photogenerated charges by exploiting the band edge positions of the isolated semiconductors [40–42]. The subsequent sections provide a deeper exploration of TiO_2's primary applications in photocatalytic reactors and the approaches employed to boost TiO_2's photoactivity.

20.3.1 Dyes

Dyes from the textile industry used to be a primary focus in photocatalysis research, mainly due to substantial pollutant losses during dyeing processes, ultimately being discharged into wastewater. The release of these colored effluents into the ecosystem poses significant threats, leading to aesthetic contamination, eutrophication, and disruptions in aquatic life. However, conventional treatment methods often involve mere adsorption onto sludge without actual degradation. The most extensively cited paper regarding the utilization of TiO_2 in photochemical reactors is related to the degradation of the azo dye, sapphire HEXL, emphasizing the dependence of the degradation process on the prior adsorption of the dye [22].

The degradation rate of the dye exhibits an increase with a growing catalyst quantity, eventually reaching a plateau. As the catalyst concentration rises, both photon absorption and dye molecule adsorption intensify due to a higher number of TiO_2 particles. This rise in particle density within the illuminated area further amplifies the reaction rate. Nevertheless, once a specific threshold is surpassed, the available substrate molecules become insufficient for adsorption by the increased TiO_2 particles, rendering additional catalyst powder inactive. This excess of TiO_2 can also result in light-scattering and light-blocking effects [22, 23]. Consequently, the reaction rate ceases to increase beyond a specific limit. Consideration of the potential aggregation of TiO_2 particles at higher concentrations is essential, as it can lead to a decrease in the number of active surface sites. Furthermore, the photodegradation rate is clearly influenced by the initial dye concentration. Given the brief lifetime of hydroxyl radicals (mere nanoseconds), their reactivity is confined to the vicinity where they are generated. Elevated dye concentrations naturally enhance the probability of collisions between organic matter and oxidizing species, thereby accelerating the decolorization rate [22].

On the other hand, the pH serves as a complex parameter intricately linked to the ionization state of the TiO_2 surface, as indicated by Eqs. (20.1) and (20.2). The impact of pH on the photocatalytic reaction primarily stems from the surface charge of TiO_2 (point of zero charge, pZc of $TiO_2 \approx 6.8$) and its correlation with the acid dissociation constants of S-dye. Below pH 6, as the pH decreases, there is pronounced adsorption of the dye onto TiO_2 particles due to the electrostatic attraction between the positively charged TiO_2 and the anionic S-dye. Conversely, the reaction rate decreases at pH 6, reaching a minimum at pH 11. This decrease illustrates how challenge is the adsorption of anionic dye molecules on the negatively charged TiO_2 surface as the solution pH rises [22].

$$TiOH_{(s)} + H^+_{(aq)} \leftrightarrow TiOH^+_{2\,(s)} \tag{20.1}$$

$$TiOH_{(s)} + OH^-_{(aq)} \leftrightarrow TiO^-{}_{(s)} \tag{20.2}$$

However, the study of dye degradation via photocatalysis is no longer a prevalent focus. This shift is primarily attributed to limitations in applicability, as photocatalysis may prove less efficient for dye degradation when compared to alternative methods. The implementation of large-scale photocatalytic systems is often economically unviable and impractical, leading to a preference for more straightforward and more cost-effective approaches in industrial settings. Moreover, advancements in other water and wastewater treatment technologies have yielded more effective and efficient means of dye removal, diminishing the emphasis on photocatalysis for this specific application. Additionally, the dynamic nature of scientific research responds to emerging needs, potentially redirecting attention toward the treatment of other pollutants deemed more critical or pressing than dyes. Thus, while photocatalysis remains valuable in various applications, research and practical application are shaped by factors such as effectiveness in comparison to alternative techniques, the complexity of target pollutants, and considerations related to practicality and economics.

20.3.2 Phenol

Phenol, a prominent organic pollutant with prospective environmental accumulation, poses health risks to living organisms. Suhadolnik et al. [28] utilized anodized TiO_2 nanotubes in a photoelectrochemical microreactor to mineralize phenol. The process involves the adsorption of phenol (Ph) onto TiO_2, where it reacts with hydroxyl radicals (Figure 20.12a), resulting in its decomposition, primarily yielding benzoquinone (Bq) along with the simultaneous generation of acids, CO_2, and H_2O (Figure 20.12b). Additionally, benzoquinone can further decompose to acids, CO_2, and H_2O. However, only the photocatalysis process did not lead to significant phenol decomposition (~11%) due to the decreasing levels of oxygen within the microchannel. The initial oxygen concentration corresponds to the saturated oxygen level at 25 °C. Molecular O_2 plays a crucial role in photodegradation by consuming photogenerated electrons, thus preventing their direct recombination with the holes. The presence of oxygen did not decisively impact phenol degradation in photoelectrocatalytic processes, as the electrical current effectively stabilized the positive holes. In this sense, PEC facilitated phenol mineralization, ensuring the complete phenol degradation [28]. Another example is the study conducted by Kane et al. [30] in

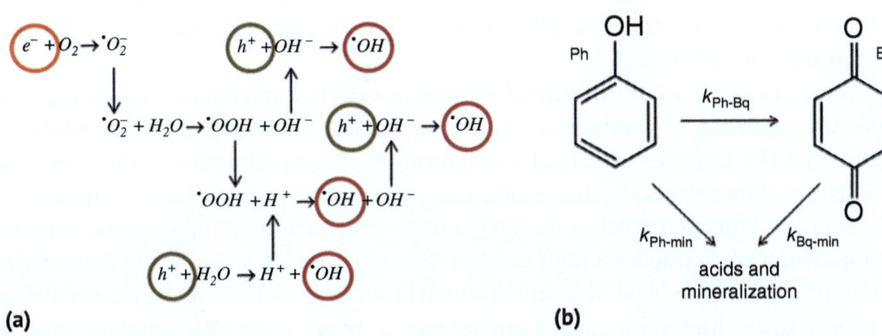

Figure 20.12 (a) Mechanism for the generation of •OH radicals over TiO_2 photocatalyst. (b) Mechanism involved in the degradation of phenol [28]/Elsevier.

a pilot-scale reactor, where nonylphenol, a synthetic organic compound classified as an alkylphenol, was decomposed over commercial TiO_2 P25. It highlights once more the importance of focusing not only on material synthesis but also on operational conditions for achieving optimal photoactivity.

Jaramillo-Gutiérrez et al. [36] decomposed phenol in a photoelectrochemical filter-press reactor, yielding simultaneous hydrogen (H_2) evolution. Fe-doped TiO_2 deposited on stainless steel under visible illumination was utilized. Four distinct zones were observed in the polarization curve: (i) 0.80–1.05 V vs. Ag/AgCl, indicating primarily charge transfer control; (ii) 1.05–1.19 V vs. Ag/AgCl, suggesting mixed control involving both charge transfer and mass transport; (iii) 1.19–1.30 V vs. Ag/AgCl, with current density reliant on flow rate, indicating mass transport control; and (iv) above 1.30 V vs. Ag/AgCl, where significant H_2 evolution occurred. Lower flow rates correlated with enhanced phenol decomposition and increased H_2 production, contrary to the expected behavior, possibly due to electrode passivation from deposition of a product from phenol electropolymerization during oxidation. Phenol rapidly reaching the anode's surface, at higher flow rates, led to quick polymer formation, consequently, resulting in decreased current density and process efficiency.

20.3.3 Pharmaceuticals

Pharmaceuticals, along with their resistant degradation byproducts, are frequently detected in urban wastewater channels, agricultural facilities, and pharmaceutical plants, indicating recalcitrance that impedes their natural degradation processes. Consequently, it is imperative to mitigate and convert these pollutants before they contaminate drinking water sources. Due to the increasing prevalence of antibiotic-resistant infections and their ecotoxic effects, antibiotics are extensively studied in photocatalysis due to their potential hazard to aquatic and land-based organisms. Taking a break from this continuous input into the environment becomes essential for safeguarding ecosystems and human health [23, 25]. The antibiotic sulfamethoxazole can be degraded over an ultrafiltration (UF) membrane by incorporating 1% *mpg*-C_3N_4/TiO_2 nanocomposite into a polysulfone (PSf) matrix. The PSf polymer, known for its high thermal, mechanical, and chemical resistance, is extensively used in UF membrane fabrication. This integrated system achieves a 69% mineralization of sulfamethoxazole after a continuous process for 30 h under solar light in a PMR [25].

Kane et al. [30] used commercially available cellulose paper coated with TiO_2 P25 for the oxidation of flumequine, a quinolone-class antibiotic molecule inhibiting bacterial DNA gyrases. The ideal concentration of the pollutant is critical because, at larger concentrations, the molecules serve as UV light filters, reducing the amount of light that reaches the TiO_2 surface and slowing photocatalytic degradation. Moreover, while both high and low flow rates lead to less efficient degradation kinetics, there is an ideal flow rate at which faster degradation can be accomplished. Mass transfer limitations play a crucial role in continuous photocatalytic process, especially in reactions with catalytic solids. At low flow rates, prolonged residence time allows interaction with radicals, while excessively low flow rates hinder solution reaching the TiO_2 surface. On the other hand, the highest flow rates result in lower photocatalytic performance due to decreased residence time. The addition of H_2O_2 as a sacrificial chemical also enhanced the photocatalytic degradation of flumequine by generating $^•OH$ through H_2O_2 photolysis, thereby improving the oxidative capability of the reaction medium. However, at elevated concentrations, H_2O_2 starts to act as an effective scavenger for $^•OH$, according to Eqs. (20.3) and (20.4), meaning that there is also an ideal H_2O_2 concentration [22].

$$H_2O_{2(aq)} + {}^•OH_{(aq)} \rightarrow HO_2{}^•{}_{(aq)} + H_2O_{(l)} \tag{20.3}$$

$$HO_2{}^•{}_{(aq)} + {}^•OH_{(aq)} \rightarrow H_2O_{(l)} + O_{2(g)} \tag{20.4}$$

Amoxicillin (AMX), the most often used antibiotic, can be oxidized in an SCMPR. Air injection into the porous ceramic membrane facilitates TiO_2 particle fluidization along the membrane surface, enhancing backwashing effectiveness in removing impurities, recovering TiO_2 particles, and improving permeate quality. Light intensity and pH significantly affect the photocatalytic oxidation process. For AMX decomposition, a pH of 6.5 is ideal. When the pH is less than 6.8, TiO_2 P25 displays a positively charged surface and a point of zero charge. On the other hand, AMX remains stable under acidic environments, and its adsorption is unfavorable at pH values lower than the zero charge. More hydroxyl (OH^-) ions are accessible on the surface of TiO_2 when pH rises. However, the holes created by photocatalysis interact with the excess of OH^- ions at pH values higher than 9.0, which results in a reduction in photocatalytic effectiveness. Regarding the light intensity, values below 300 W led to a notable rise in the reaction rate as the generation of the photogenerated charges became the predominant process. As the light intensity continued to increase beyond 300 W, the reaction rate showed a slight increase, attributed to the heightened recombination of photoinduced charges [23].

The two possible degradation pathways for amoxicillin are very complex, as depicted in Figure 20.13. One involves hydroxylation, resulting in the creation of intermediates B1 and B2. These intermediates undergo sequential oxidation processes, releasing NH_4^+, consequently, B1 and B2 are converted into 2-(4-hydroxyphenyl)-2-oxoacetic

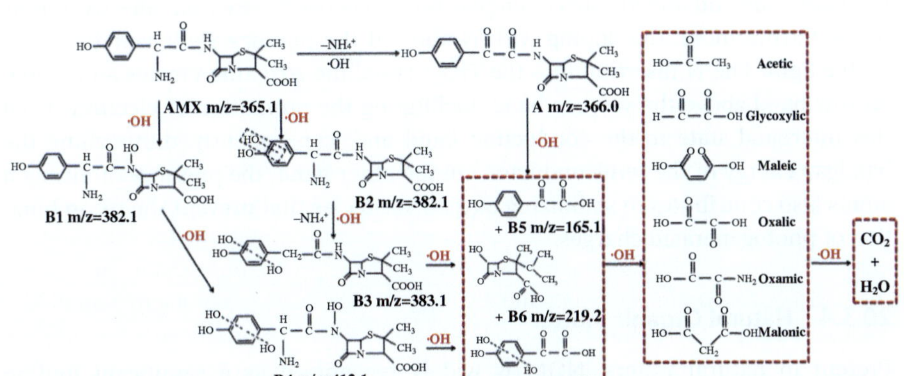

Figure 20.13 Mechanism reaction for the amoxicillin mineralization [23]/Elsevier.

acid and an open-chain structure containing a sulfonic group. Another degradation pathway initiates with the formation of intermediate A, resulting from hydroxyl radical attack on the primary amino group, leading to cleavage of carbonyl and secondary amine groups. Subsequent intermediates B5 and B6 undergo analogous oxidation, further reacting with additional hydroxyl radicals generated by TiO_2. This leads to the generation of diverse organic acids, ultimately mineralizing into CO_2, water, and inorganic ions [23].

The literature also evaluates clofibric acid, an active metabolite used to regulate blood lipid levels. Manassero et al. [18] conducted a comparative analysis of reactors with three distinct catalyst configurations, as depicted in Figure 20.7: (i) a system with continuously mixed TiO_2 particles (SPR), (ii) TiO_2 immobilized on the irradiated reactor window (FFPR), and (iii) TiO_2 supported on the surface of glass rings that fill the reactor (FBPR). Both suspended and immobilized systems utilized the same experimental setup. The study assessed photocatalytic performance using parameters such as photonic and quantum efficiencies. The authors demonstrated that photons absorbed in the SPR are more efficiently utilized for degrading clofibric acid. The lowered interfacial area in the FFPR contributed to decreased efficiency, while comparable rates and photonic efficiencies were achieved in both immobilized catalyst reactors. Despite FFPR having nearly twice the photon absorption rate, FBPR exhibited a quantum efficiency of approximately 1.5 times higher. This indicates that although the absorbed energy in the FBPR is lower, it is more effectively utilized for photocatalytic reactions. Immobilized systems, due to diminished interfacial area, are less efficient compared to suspended systems, with SPR being the most effective. However, besides FBPR exhibiting a more evenly distributed radiation in the reaction space, it also offers a higher catalytic area, making FBPR an extremely favorable configuration for photocatalytic reactors. Carbamazepine (CBZ) is another persistent pharmaceutical that is a commonly used antiepileptic medication. Horovitz et al. [20] evaluated its degradation over N-doped TiO_2/Al_2O_3 composite membranes in a PMR system. The membrane's high tortuosity promotes forced convection transport during permeation, improving reactant-catalyst interaction. This setup enhances the chances of molecules reaching active sites within the pores, reducing mass transfer resistance between the active surface and fluid bulk. Both diffusion and forced convection play roles in boosting reaction rates within the pores. Furthermore, the doping with N allowed the photocatalytic activity under visible light. The N insertion into the TiO_2 crystalline structure creates an isolated narrow band above the valence band, facilitating the promotion of electrons from this interband state to the conduction band and, consequently, diminishing the bandgap energy of the semiconductor. On the other hand, the presence of nitrogen atoms also contributes to stabilizing oxygen vacancies that prevent the recombination of photogenerated charges.

20.3.4 Natural Organic Matter

Present in natural waters, NOM is widely recognized as a significant fouling agent. Humic acid (HA), a predominant NOM component, is known for its reaction with common disinfectants in water treatment, resulting in the formation of diverse disinfection byproducts (DBPs). The removal of HA is important for adhering to rigorous DBP regulations and ensuring the provision of safe drinking water. Incorporating TiO_2-based photocatalysts into a membrane structure within nanoscale-constrained spaces in a PMR has the potential to create a more efficient catalytic system for foulant removal [26]. Yin et al. [26] prepared PdCu alloy-coupled TiO_2 (PdCu-TiO_2) due to the significant synergistic effect among metal atoms in the alloy nanoparticles, making them excellent cocatalysts for photocatalytic redox reactions.

The PdCu alloy can decrease the resistance of the final photocatalyst by facilitating charge transfer. A detailed analysis of the transfer of photogenerated charges and the electronic band structure also reveals that the pivotal role of the PdCu alloy is to diminish the bandgap of TiO_2 through the decrease of the valence band edge position. Consequently, this improvement of the modified TiO_2 contributes to an enhanced photocatalytic performance. The PdCu-TiO_2/PVDF membrane in a PMR achieved a total HA removal efficiency of ca. 46% at 25 °C. However, without temperature control, the degradation reaction can reach approximately 64 °C, significantly enhancing the decomposition rate to almost 100%. It suggests a more pronounced photothermal impact induced by the alloying effect during light irradiation, according to Figure 20.14.

Based on the assessment of electron spin resonance measurements, $\cdot O_2^-$ and $\cdot OH$ were detected. DFT calculations demonstrated that the coupled Pd-Cu sites within the PdCu alloy act as stepping bases to assist the transfer of photogenerated electrons (from TiO_2 to Pd-Cu sites and then to HA). The detection of CO_2 suggests that photocatalysis can mineralize HA, but it is still possible to observe some intermediate byproducts, including 2-furoic acid, oxydimethanol, phenols, 2-(propylamino) ethanol, acetoacetic acid, phthalic acid, and 2-hydroxy-4-methylpentanoic acid [26].

20.3.5 Carbon Dioxide Reduction

The extensive utilization of oil, natural gas, and coal has led to a significant rise in the release of CO_2 into the Earth's atmosphere. Given that CO_2 stands out

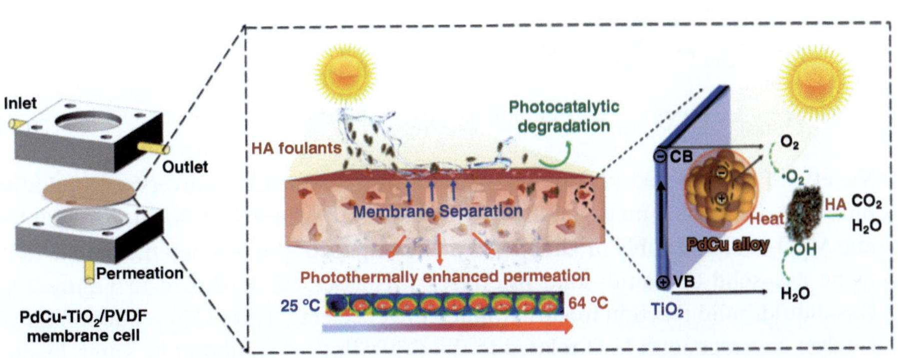

Figure 20.14 Diagrammatic representation of the PMR system utilizing the PdCu-TiO_2/PVDF photocatalytic membrane for the removal of humic acid. Reproduced from Ref. [26] with permission from Elsevier.

as one of the major greenhouse gases contributing to the climate crisis and the depletion of fossil resources has prompted research efforts to explore alternative carbon sources for fuel and chemical production, there has been substantial research attention in recent years towards capturing, storing, and utilizing CO_2 as an alternative carbon feedstock. The aim of converting CO_2 generated in the chemical and energy industries into fuels and valuable compounds is to establish a carbon-neutral cycle. However, due to the high stability of the CO_2 molecule and its dissociation energy of 750 kJ mol^{-1}, activating and converting CO_2 present formidable challenges [16, 17, 21].

CO_2 conversion follows two primary pathways: (i) co-feeding with a high-energy reactant like H_2, unsaturated compounds, epoxides, or organometallic compounds; and (ii) providing external energy, such as solar or electric energy. Inspired by natural photosynthesis, several research teams are diligently working on employing an artificial system to perform the CO_2 reduction reaction (CO2RR) into valuable compounds [21]. The main drawback of artificial photosynthesis is the low solubility of CO_2 in aqueous media. Typically, an alkaline environment is utilized to enhance CO_2 dissolvability in a liquid phase [21]. Na_2CO_3 and NaOH have been reported as suitable electrolytes by prioritizing prolonged CO2RR, given the depletion of H^+ in the solution [17, 43]. Equations (20.5)–(20.11) represent the reaction between CO_2 and the electrolytes that leads to the formation of carbonate (CO_3^{2-}) and bicarbonate (HCO_3^-), improving the availability of carbon sources in the aqueous media [17, 44].

$$NaOH_{(s)} \leftrightarrow Na^+{}_{(aq)} + OH^-{}_{(aq)} \tag{20.5}$$

$$CO_{2(g)} + OH^-{}_{(aq)} \leftrightarrow HCO^-_{3\,(aq)} \tag{20.6}$$

$$CO_{2(g)} + 2OH^-{}_{(aq)} \leftrightarrow CO_3^{2-}{}_{(aq)} + H_2O_{(l)} \tag{20.7}$$

$$Na_2CO_{3(s)} + H_2O_{(l)} + CO_{2(g)} \leftrightarrow 2NaHCO_{3(s)} \tag{20.8}$$

$$NaHCO_{3(s)} \leftrightarrow Na^+{}_{(aq)} + HCO^-_{3\,(aq)} \tag{20.9}$$

$$HCO^-_{3\,(aq)} \leftrightarrow H^+{}_{(aq)} + CO_3^{2-}{}_{(aq)} \tag{20.10}$$

$$H^+{}_{(aq)} + HCO^-_{3\,(aq)} \leftrightarrow CO_{2(g)} + H_2O_{(l)} \tag{20.11}$$

Xie et al. [21] explored various TiO_2-based catalysts for CO_2 conversion, including pure TiO_2 P25, noble metal-modified TiO_2 (M-TiO_2, where M = Pt, Pd, Au, Rh, Ag), and MgO- and Pt-doubly modified TiO_2. The impact of the reaction mode in a CSTR using gas–solid or liquid–solid interfaces was assessed, as shown in Figure 20.2. Gas–liquid–solid reaction mode exhibited lower selectivity for CO_2 reduction due to catalyst coverage with H_2O related to the CO_2's limited solubility in water, leading to higher H_2 formation rates but lower rates of carbon compound production. Essentially, a larger proportion of photogenerated electrons are utilized in reducing

H_2O rather than CO_2. Additionally, noble metal cocatalysts on TiO_2 were found to enhance photocatalytic efficiency, with electron utilization for reductive reactions increasing in the sequence TiO_2 < Ag-TiO_2 < Rh-TiO_2 < Au-TiO_2 < Pd-TiO_2 < Pt-TiO_2. This aligns closely with noble metals' electron-donating or accepting capabilities. Pt-TiO_2 exhibited a remarkable CH_4 selectivity of 83%, surpassing TiO_2's selectivity of 24%. Moreover, the inclusion of MgO enhanced TiO_2-based catalysts' photocatalytic capabilities in CO_2 reduction, potentially by improving CO_2 chemisorption or creating Ti^{3+} sites, resulting in >99% CH_4 selectivity with 1.0 wt% MgO. The presence of Ti^{3+} sites facilitates the formation of a Ti^{3+}—C bond, leading to a better selectivity for carbon products [45]. On the other hand, MgO offers alkaline sites that facilitate the selective chemisorption of CO_2 onto the catalyst surface. However, an optimal MgO concentration is crucial to prevent surface masking or impediment of photogenerated electron migration from TiO_2 to Pt particles. The CO2RR can occur through a mechanism involving proton-assisted multi-electron transfer, as described in Eqs. (20.12)–(20.16) [21].

$$CO_{2(g)} + 2H^+_{(aq)} + 2e^- \rightarrow CO_{(g)} + H_2O_{(aq)} \tag{20.12}$$

$$CO_{2(g)} + 8H^+_{(aq)} + 8e^- \rightarrow CH_{4(g)} + 2H_2O_{(aq)} \tag{20.13}$$

$$CO_{2(g)} + 2H^+_{(aq)} + 2e^- \rightarrow HCOOH_{(aq)} \tag{20.14}$$

$$2H^+_{(aq)} + 2e^- \rightarrow H_{2(g)} \tag{20.15}$$

$$H_2O_{(aq)} + 2h^+ \rightarrow \tfrac{1}{2}O_{2(g)} + 2H^+_{(aq)} \tag{20.16}$$

Tahir et al. [16] also investigate the CO_2 reduction to carbon dioxide (CO) and methane but over Cu/TiO_2 and N/TiO_2 supported on cordierite honeycomb monoliths. This photocatalytic system possesses a distinctive structure characterized by a high ratio of surface area to volume. Consequently, it has the potential to enhance the interaction between light irradiation and the catalyst. The authors performed a gas–solid reaction in a continuous flow, using CO_2/H_2 feed ratio of 1.0 at 100 °C and atmospheric pressure. The addition of Cu and N metals does not cause a notable impact on the surface area and bandgap of TiO_2, indicating that these elements' insertion was not a doping but a decoration. However, copper exhibits a higher capacity for trapping electrons, thereby impeding the recombination of charges at a more effective rate when compared to N/TiO_2 and pure TiO_2 samples. The primary product of CO_2 reduction was identified as CO, accompanied by a lesser quantity of CH_4. Using 3 wt.% Cu, it was possible to increase the CO yield by more than 14 times compared to the pure N/TiO_2 and 64-fold than pure TiO_2. More Cu or N can decrease the photoactivity through the shielding effect by an excess of the doping elements that can decrease the active site and increase the recombination centers available over the TiO_2 surface. The continuous flow mode allowed CO to be actively streamed outside the reactor throughout irradiation. In this gas-phase reaction, the protons for the methane generation, as previously shown in Eq. (20.13), were produced from the oxidation of H_2, according to Eq. (20.17).

$$H_{2(g)} + h^+ \rightarrow H^+_{(g)} + H^+_{(g)} \tag{20.17}$$

The increased photoactivity of Cu/TiO_2 may result from the elevated mobility of electrons observed in Cu-doped TiO_2 samples that can be easily reversible oxidized and reduced in the catalyst surface, as depicted in Eqs. (20.18) and (20.19). The introduction of copper (Cu^0, Cu^+, Cu^{2+}) into TiO_2 facilitates electron transfer at the metal/semiconductor junction, thereby enhancing its photoactivity. In the context of the CO_2 photoreduction reaction, the presence of an electron donor (reducing agent) serves to impede the recombination of the hole-electron pair [16, 17]. On the contrary of Tahir et al. [16], Peña et al. [17] could obtain a proper Cu-doping in TiO_2 crystalline structure since the diffusive reflectance spectroscopy showed a considerable shifting towards lower energy in the visible range compared to the undoped material. The best doping concentration was also around 3 wt.%, and the authors found that the presence of Ti^{3+} and Ti^{4+} could improve the overall efficiency of the system. The authors found mainly formic acid (HCOOH), according to Eq. (20.14), in a production rate of up to 4033 μmol g^{-1} h^{-1}.

$$Cu^0_{(s)} + h^+ \leftrightarrow Cu^+_{(s)} + e^- \tag{20.18}$$

$$Cu^+_{(s)} + h^+ \leftrightarrow Cu^{2+}_{(s)} + e^- \tag{20.19}$$

20.4 Conclusion

Photocatalysis research has predominantly focused on catalyst development, but optimizing reactor architecture and operational conditions is equally critical for enhancing performance and selectivity. The design of a photochemical reactor necessitates careful consideration of various parameters, including the selection of light source, geometric configuration, construction material, and operating mode. Recent advancements have resulted in the development of sophisticated reactors based on conventional types, offering precise control over reaction parameters to boost reaction yield. However, some challenges persist such as low light transfer efficiency, scalability issues, and obstacles with selectivity and product purity, necessitating further research and optimization. Achieving technical feasibility and economic viability is paramount for the sustainable implementation of photocatalytic processes in industrial settings, requiring a delicate balance between design aspects and constraints associated with photon and mass transfer. Furthermore, addressing challenges related to TiO_2's exclusive photoactivity under UV light and its short photogenerated charge lifetime is essential for enhancing photoactivity, with strategies like doping and heterojunction formation showing considerable potential. Overall, sustained research efforts are required to overcome these challenges and fully exploit the capabilities of photocatalytic reactors across various applications, including wastewater treatment and artificial photosynthesis.

Acknowledgments

The authors acknowledge FAPEMIG (Minas Gerais Research Foundation, grant: APQ-00282-21, APQ-00366-24, APQ-06609-24 and RED-00175-22) and CNPq (Brazilian National Council for Scientific and Technological Development, grants: 406026/2023-8, 405819/2022-6 and 440117/2022-4) FAPESP CINE (grant: 2017/11986-5) and FINEP (grant: 01.22.0274.00) for the financial support. J. A. de Oliveira is indebted to FAPESP (grants #2019/21496–0, #2018/01258–5) and Shell nº 24450-9 (Shell OP001I)- PROEX 23112.001274/2024-30.

References

1 Aguirre-Cortés, J.M., Moral-Rodriguéz, A.I., Bailón-García, E. et al. (2023). 3D printing in photocatalysis: methods and capabilities for the improved performance. *Applied Materials Today* 32: 101831. https://doi.org/10.1016/j.apmt.2023.101831.

2 Sundar, K.P. and Kanmani, S. (2020). Progression of photocatalytic reactors and it's comparison: a review. *Chemical Engineering Research & Design* 154: 135–150. https://doi.org/10.1016/j.cherd.2019.11.035.

3 Mehos, M., Turchi, C., Pacheco, J. et al. (1992). Pilot-scale study of the solar detoxification of VOC-contaminated groundwater. Technical Report, National Renewable Energy Lab., Golden, Colorado, United States. NREL/TP-432-4981, UC Category 233, DE92016405.

4 Blanco, J. (1991). Final configuration of PSA solar detoxification loop. Technical Report, Plataforma Solar de Almería, Almería, Andalusia, Spain, TR 06/91.

5 Cooper, G., Ratcliff, M.A., and Sczechowski, J.G. (1992). Photocatalytic slurry reactor having turbulence generating means. Patent No. 5116582, USA.

6 Assink, J.W., Koster, T.P.M., and Slaager, J.M. (1993). Fotokatalytische oxydatie voor afvalwaterbehandeling. Internal Report No. 93-137, TNO-Milieu en Energie, Apeldoorn, the Netherlands.

7 Chen, D.H., Ye, X.J., and Li, K.Y. (2005). Oxidation of PCE with a UV LED photocatalytic reactor. *Chemical Engineering & Technology* 28 (1): 95–97. https://doi.org/10.1002/ceat.200407012.

8 Castro, S., Albo, J., and Irabien, A. (2018). Photoelectrochemical reactors for CO_2 utilization. *ACS Sustainable Chemistry & Engineering* 6 (12): 15877–15894. https://doi.org/10.1021/acssuschemeng.8b03706.

9 Anaya-Rodríguez, F., Durán-Álvarez, J.C., Drisya, K.T., and Zanella, R. (2023). The challenges of integrating the principles of green chemistry and green engineering to heterogeneous photocatalysis to treat water and produce green H_2. *Catalysts* 13 (1): 154. https://doi.org/10.3390/catal13010154.

10 Fogler, H.S. (1987). *Elements of Chemical Reaction Engineering*, 2e. Prentice Hall.

11 Li, X., Chen, Y., Tao, Y. et al. (2022). Challenges of photocatalysis and their coping strategies. *Chem Catalysis* 2 (6): 1315–1345. https://doi.org/10.1016/j.checat.2022.04.007.

12 de Oliveira, J.A., da Cruz, J.C., Nascimento, O.R., and Ribeiro, C. (2022). Selective CH_4 reform to methanol through partial oxidation over Bi_2O_3 at room temperature and pressure. *Applied Catalysis B: Environmental* 318: 121827. https://doi.org/10.1016/j.apcatb.2022.121827.

13 Braham, R.J. and Harris, A.T. (2009). Review of major design and scale-up considerations for solar photocatalytic reactors. *Industrial & Engineering Chemistry Research* 48 (19): 8890–8905. https://doi.org/10.1021/ie900859z.

14 Meng, X.C., Zhang, Z.H., and Li, X.G. (2015). Synergetic photoelectrocatalytic reactors for environmental remediation: a review. *Journal of Photochemistry and Photobiology C: Photochemistry Reviews* 24: 83–101. https://doi.org/10.1016/j.jphotochemrev.2015.07.003.

15 Li, K.F., An, X., Park, K.H. et al. (2014). A critical review of CO_2 photoconversion: catalysts and reactors. *Catalysis Today* 224: 3–12. https://doi.org/10.1016/j.cattod.2013.12.006.

16 Tahir, M. and Tahir, B. (2016). Dynamic photocatalytic reduction of CO_2 to CO in a honeycomb monolith reactor loaded with Cu and N doped TiO_2 nanocatalysts. *Applied Surface Science* 377: 244–252. https://doi.org/10.1016/j.apsusc.2016.03.141.

17 Peña, R., Romero, R., Amado-Piña, D., and Natividad, R. (2023). Cu/TiO_2 photocatalyzed CO_2 chemical reduction in a multiphase capillary reactor. *Topics in Catalysis* 67: 377–393. https://doi.org/10.1007/s11244-023-01875-8.

18 Manassero, A., Satuf, M.L., and Alfano, O.M. (2017). Photocatalytic reactors with suspended and immobilized TiO_2: comparative efficiency evaluation. *Chemical Engineering Journal* 326: 29–36. https://doi.org/10.1016/j.cej.2017.05.087.

19 Coronel, S., Endara, D., Lozada, A.B. et al. (2021). Photocatalytic study of cyanide oxidation using titanium dioxide (TiO_2)-activated carbon composites in a continuous flow photo-reactor. *Catalysts* 11 (8): 924. https://doi.org/10.3390/catal11080924.

20 Horovitz, I., Avisar, D., Baker, M.A. et al. (2016). Carbamazepine degradation using a N-doped TiO_2 coated photocatalytic membrane reactor: influence of physical parameters. *Journal of Hazardous Materials* 310: 98–107. https://doi.org/10.1016/j.jhazmat.2016.02.008.

21 Xie, S.J., Wang, Y., Zhang, Q. et al. (2014). MgO- and Pt-promoted TiO_2 as an efficient photocatalyst for the preferential reduction of carbon dioxide in the presence of water. *ACS Catalysis* 4 (10): 3644–3653. https://doi.org/10.1021/cs500648p.

22 Sauer, T., Cesconeto Neto, G., José, H.J., and Moreira, R.F.P.M. (2002). Kinetics of photocatalytic degradation of reactive dyes in a TiO_2 slurry reactor. *Journal of Photochemistry and Photobiology A: Chemistry* 149 (1–3): 147–154. https://doi.org/10.1016/S1010-6030(02)00015-1.

23 Li, Q.L., Jia, R., Shao, J., and He, Y. (2019). Photocatalytic degradation of amoxicillin via TiO_2 nanoparticle coupling with a novel submerged porous ceramic membrane reactor. *Journal of Cleaner Production* 209: 755–761. https://doi.org/10.1016/j.jclepro.2018.10.183.

24 Hurtado, L., Solís-Casados, D., Escobar-Alarcón, L. et al. (2016). Multiphase photo-capillary reactors coated with TiO_2 films: preparation, characterization and photocatalytic performance. *Chemical Engineering Journal* 304: 39–47. https://doi.org/10.1016/j.cej.2016.06.003.

25 Yu, S., Wang, Y., Sun, F. et al. (2018). Novel *mpg*-C_3N_4/TiO_2 nanocomposite photocatalytic membrane reactor for sulfamethoxazole photodegradation. *Chemical Engineering Journal* 337: 183–192. https://doi.org/10.1016/j.cej.2017.12.093.

26 Yin, H.B., Zhang, J., Wang, Y. et al. (2023). Alloying effect of PdCu-TiO_2 in photocatalytic membrane reactor for efficient removal of humic acid foulant. *Applied Catalysis B: Environmental* 338: 123085. https://doi.org/10.1016/j.apcatb.2023.123085.

27 Zeghioud, H., Khellaf, N., Djelal, H. et al. (2016). Photocatalytic reactors dedicated to the degradation of hazardous organic pollutants: kinetics, mechanistic aspects, and design – a review. *Chemical Engineering Communications* 203 (11): 1415–1431. https://doi.org/10.1080/00986445.2016.1202243.

28 Suhadolnik, L., Pohar, A., Likozar, B., and Ceh, M. (2016). Mechanism and kinetics of phenol photocatalytic, electrocatalytic and photoelectrocatalytic degradation in a TiO_2-nanotube fixed-bed microreactor. *Chemical Engineering Journal* 303: 292–301. https://doi.org/10.1016/j.cej.2016.06.027.

29 White, J.L., Baruch, M.F., Pander III, J.E. et al. (2015). Light-driven heterogeneous reduction of carbon dioxide: photocatalysts and photoelectrodes. *Chemical Reviews* 115 (23): 12888–12935. https://doi.org/10.1021/acs.chemrev.5b00370.

30 Kane, A., Assadi, A.A., El Jery, A. et al. (2022). Advanced photocatalytic treatment of wastewater using immobilized titanium dioxide as a photocatalyst in a pilot-scale reactor: process intensification. *Materials* 15 (13): 4547. https://doi.org/10.3390/ma15134547.

31 Barzegar, M.H., Sabzehmeidani, M.M., Ghaedi, M. et al. (2021). S-scheme heterojunction g-C_3N_4/TiO_2 with enhanced photocatalytic activity for degradation of a binary mixture of cationic dyes using solar parabolic trough reactor. *Chemical Engineering Research & Design* 174: 307–318. https://doi.org/10.1016/j.cherd.2021.08.015.

32 Garcia-Segura, S. and Brillas, E. (2017). Applied photoelectrocatalysis on the degradation of organic pollutants in wastewaters. *Journal of Photochemistry and Photobiology C: Photochemistry Reviews* 31: 1–35. https://doi.org/10.1016/j.jphotochemrev.2017.01.005.

33 Xie, S.J., Zhang, Q., Liu, G., and Wang, Y. (2016). Photocatalytic and photoelectrocatalytic reduction of CO_2 using heterogeneous catalysts with controlled nanostructures. *Chemical Communications* 52 (1): 35–59. https://doi.org/10.1039/C5CC07613G.

34 Vennekoetter, J.B., Sengpiel, R., and Wessling, M. (2019). Beyond the catalyst: how electrode and reactor design determine the product spectrum during electrochemical CO_2 reduction. *Chemical Engineering Journal* 364: 89–101. https://doi.org/10.1016/j.cej.2019.01.045.

35 Deshmukh, S.R., Tonkovich, A.L.Y., Jarosch, K.T. et al. (2010). Scale-up of microchannel reactors for Fischer-Tropsch synthesis. *Industrial & Engineering Chemistry Research* 49 (21): 10883–10888. https://doi.org/10.1021/ie100518u.

36 Jaramillo-Gutiérrez, M.I., Rivero, E.P., Cruz-Díaz, M.R. et al. (2016). Photoelectrocatalytic hydrogen production from oilfield-produced wastewater in a filter-press reactor using TiO_2-based photoanodes. *Catalysis Today* 266: 17–26. https://doi.org/10.1016/j.cattod.2015.12.008.

37 Motegh, M., van Ommen, J.R., Appel, P.W., and Kreutzer, M.T. (2014). Scale-up study of a multiphase photocatalytic reactor: degradation of cyanide in water over TiO_2. *Environmental Science & Technology* 48 (3): 1574–1581. https://doi.org/10.1021/es403378e.

38 Kane, P.F., Larrabee, G.B., and Knight, S. (1970). Characterization of semiconductor materials. *Journal of the Electrochemical Society* 117 (11): 411C–412C. https://doi.org/10.1149/1.2407352.

39 Schneider, J., Matsuoka, M., Takeuchi, M. et al. (2014). Understanding TiO_2 photocatalysis: mechanisms and materials. *Chemical Reviews* 114 (19): 9919–9986. https://doi.org/10.1021/cr5001892.

40 Mourao, H., Avansi, W., and Ribeiro, C. (2012). Hydrothermal synthesis of Ti oxide nanostructures and TiO_2:SnO_2 heterostructures applied to the photodegradation of rhodamine B. *Materials Chemistry and Physics* 135 (2–3): 524–532. https://doi.org/10.1016/j.matchemphys.2012.05.019.

41 de Mendonca, V.R., Dalmaschio, C.J., Leite, E.R. et al. (2015). Heterostructure formation from hydrothermal annealing of preformed nanocrystals. *Journal of Materials Chemistry A* 3 (5): 2216–2225. https://doi.org/10.1039/C4TA05926C.

42 Lopes, O.F., Carvalho, K.T.G., Nogueira, A.E. et al. (2016). Controlled synthesis of $BiVO_4$ photocatalysts: evidence of the role of heterojunctions in their catalytic performance driven by visible-light. *Applied Catalysis B: Environmental* 188: 87–97. https://doi.org/10.1016/j.apcatb.2016.01.065.

43 Dias, E.H., da Silva, G.T.S.T., da Cruz, J.C., and Ribeiro, C. (2022). One-pot solvothermal synthesis of carbon black-supported CuO for catalysis of CO_2 electroreduction. *ChemElectroChem* 9 (11): e202200206. https://doi.org/10.1002/celc.202200206.

44 Hikita, H., Asai, S., and Takatsuka, T. (1976). Absorption of carbon dioxide into aqueous sodium hydroxide and sodium carbonate-bicarbonate solutions. *The Chemical Engineering Journal* 11 (2): 131–141. https://doi.org/10.1016/S0300-9467(76)80035-4.

45 Li, W., He, D., Hu, G. et al. (2018). Selective CO production by photoelectrochemical methane oxidation on TiO_2. *ACS Central Science* 4 (5): 631–637. https://doi.org/10.1021/acscentsci.8b00130.

21

TiO_2-based Protective Coatings for Buildings and Monuments

Yijian Cao[1], Cong Wang[2], Wanqian Feng[1], Mara Camaiti[3,]*

[1] *Key Laboratory of Archaeological Exploration and Cultural Heritage Conservation Technology (Northwestern Polytechnical University), Ministry of Education, Xi'an, China*
[2] *Key Laboratory of Cultural Heritage Research and Conservation (Northwest University), Ministry of Education, Xi'an, China*
[3] *CNR-Institute of Geosciences and Earth Resources, Florence, Italy*

*Corresponding author: mara.camaiti@igg.cnr.it

21.1 Introduction

The degradation of natural stones, bricks, mortars, concretes, tiles, and other materials used in the construction of contemporary and historic buildings has been a global concern since the past. Due to both intrinsic factors (mineralogical composition, textures, porosity, etc.) and extrinsic conditions (the climate, specific functions/locations, and other anthropogenic activities), building materials suffer from severe degradation phenomena. Disintegration, fissures, cracks, surface scaling/flaking/peeling/blistering, fractures, crack, and biological colonization are only a small fraction of the standard terminologies describing the degradation patterns [1], which consist of a series of chemical/physical reactions [2, 3]. Materials decay undermines the structural function, esthetic, and economic value, and also poses safety and health risks to humans. Therefore, it is of great significance to well conserve the building materials in order to preserve the entire structure and its intangible values.

The degradation of a given stony construction is closely related to the specific environmental/climate conditions in which it is situated. These include the temperature and humidity fluctuations, rainfall/snowfall/hail, solar irradiation and atmospheric pollution, etc. To prevent contact with extraneous deteriorating factors (e.g., water, air pollutants, microorganisms, human vandalism, graffiti) and to mitigate the decay process, the application of protective coatings on building materials is highly advisable.

Dating back to the Greek and Roman periods, waxes and oils were known as protectives of various artifacts [4, 5], and surface treatments were conducted on stone surfaces for both aesthetic and protective reasons, using these natural products [6].

Nanostructured TiO_2: Synthesis, Photocatalytic Properties and Applications, First Edition.
Edited by Jia Hong Pan, Wan In Lee and Detlef W. Bahnemann

In the Middle Ages, the practice of painting stone sculptures became widespread, especially in Northern Europe. During the Renaissance and the Baroque periods, shining and polishing marble with beeswax/olive oil was very popular. In 1567, the Public Palace of Louvain was painted with oil, which was formally recognized as a protection treatment [6]. Oil (walnut oil, linseed oil, etc.), wax (beeswax, ceresin wax, etc.), natural resins (rosin, dammar, mastic, etc.), and their mixtures were prevalent as water repellents until the advent of the petroleum industry revolution in the 19th century. Beeswax, paraffin wax, and its modern equivalent microcrystalline wax were widely applied as water repellents for sculptures in the 19th century and early 20th century. Since the second half of the 20th century, synthetic polymeric materials that were originally manufactured for industrial applications have almost completely replaced natural resins in stone protection. Acrylic resins, silicon-based resins, microcrystalline waxes, perfluoropolyethers (PFPEs), polyfluorourethanes, fluorinated polymers, and fluoroelastomers are representative in this era.

Recently, since the beginning of the 21st century, the study of protective coatings based on photocatalytic nanomaterials (e.g., TiO_2, ZnO) has been a hot research topic, and some of these coatings have already been practically applied for the maintenance of architectural surfaces [7–10], owing to the photoinduced multifunctionality, for example, self-cleaning, de-polluting, and antibacterial properties. As illustrated in Figure 21.1, TiO_2-based coatings facilitate the formation of an effective protective layer between the substrate surfaces and the deteriorating agents (e.g., liquid water, water vapor, dust/dirt deposits, atmospheric pollutants, etc.). Under UV irradiation, these coatings exhibit photocatalytic and hydrophilic/hydrophobic self-cleaning properties, keeping stone surfaces intact by easily removing or eliminating the undesignable deposits.

The following sections of this chapter explore the recent technological advancements in the preparation, application, and comparative evaluation of TiO_2-based protective coatings for buildings and monuments. The text emphasizes elucidating the fundamental working mechanisms that underpin the appealing properties, and the specific considerations for intervention on heritage objects.

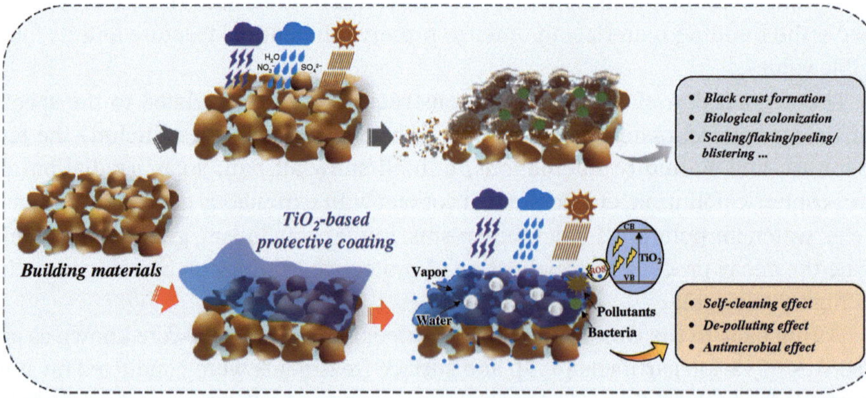

Figure 21.1 Schematic description of stone degradation and the multifunctionalities of TiO_2-based protective coatings. Reproduced from Ref. [11] with permission from Elsevier.

21.2 TiO_2-based Protective Coatings

With regard to the preparation method of TiO_2-based coatings, two general strategies may be employed [12], that is, the aqueous or solvent dispersion of nano-TiO_2, and the incorporation of nano-TiO_2 within a polymeric or inorganic matrix. Whereas concerning the wettability properties of TiO_2-based coatings, there are basically two broad types, namely hydrophilic coatings and hydrophobic coatings.

21.2.1 Photo-generated Superhydrophilic Self-cleaning Property

TiO_2-based hydrophilic coatings usually demonstrate UV-induced self-cleaning property. Upon UV irradiation, TiO_2 thin film exhibits photocatalytic activation and superhydrophilic self-cleaning property, simultaneously. The mechanism of photocatalysis during the period of charge carrier generation by photon irradiation has already been well understood and is discussed in other chapters of this book. The reactive oxygen species generated by the photoexcitation of electrons from the valence band to the conduction band of TiO_2 are responsible for its photocatalytic behavior, which accounts for the de-polluting and antibacterial properties. For the application in the field of building and monument maintenance, the self-cleaning property of TiO_2-based coatings is particularly desirable.

Beside the photocatalytic pathway, a different mechanism is proposed for the superhydrophilic self-cleaning effect. This process is generally explained through a two-step mechanism: the generation of oxygen vacancies and the reconstruction of Ti—OH bonds [13, 14]. In the first step, the photogenerated electrons reduce the Ti^{4+} cations to Ti^{3+}, while the holes oxidize the O^{2-} anions to produce O_2 ejection (Figure 21.2), thereby creating oxygen vacancies. The OH^- anions are produced during photocatalysis and water molecules are adsorbed at the vacancies, transforming

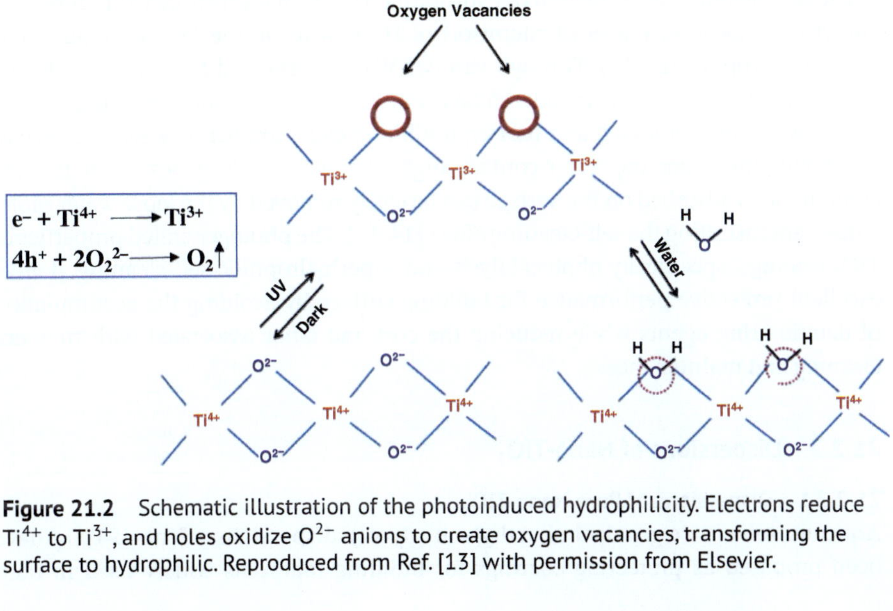

Figure 21.2 Schematic illustration of the photoinduced hydrophilicity. Electrons reduce Ti^{4+} to Ti^{3+}, and holes oxidize O^{2-} anions to create oxygen vacancies, transforming the surface to hydrophilic. Reproduced from Ref. [13] with permission from Elsevier.

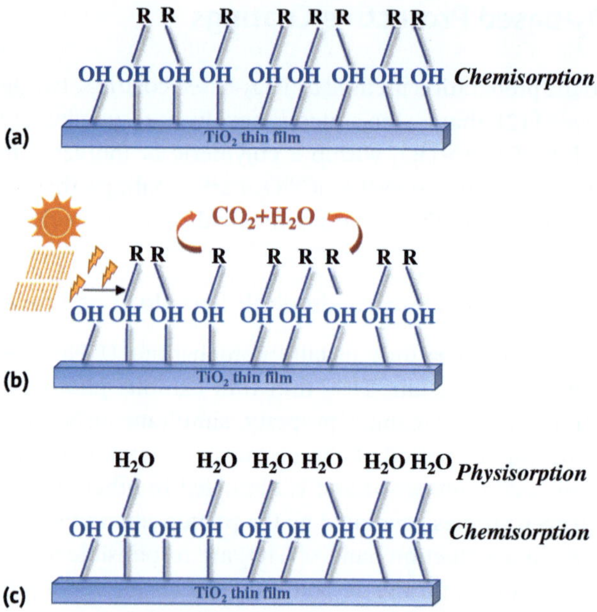

Figure 21.3 Schematic representation of self-cleaning mechanism by the adsorption of water. Reproduced from Ref. [16] with permission from Elsevier.

the surface to hydrophilic. In the second step, the photogenerated holes play a vital role. The holes can migrate to the surface and get trapped at the lattice oxygen sites to weaken Ti—O bond, leading to the formation of new hydroxyl bonds with the adsorbed water molecules [15]. Consequently, the charge is equalized by a proton separation and a new OH bond is formed. Thus, an OH group doubly coordinated to a Ti atom is transformed into two OH groups singly coordinated to the Ti atom [13].

The self-cleaning phenomenon of hydrophilic TiO_2 film is illustrated in Figure 21.3. Under UV exposure, a layer of chemisorbed H_2O forms on the TiO_2 film, due to its surface hydrophilicity [16]. Through van der Waals forces and hydrogen bonds, the chemisorbed H_2O can further adsorb layers of water by physisorption. These physisorbed water molecules act as a barrier, inhibiting close contact between the surface and pollutants. Since the water contact angle of the TiO_2 film is less than 10°, the contaminants adsorbed on the surface can be easily removed by the loose water molecules, demonstrating the self-cleaning effect [14, 16]. The photogenerated properties of TiO_2 coatings, specifically photocatalysis and superhydrophilic self-cleaning, deliver excellent protective performance for building surface by avoiding the accumulation of deteriorating agents while reducing the cost and labor associated with frequent cleaning and maintenance.

21.2.2 Dispersions of Nano-TiO_2

21.2.2.1 Dispersions of Pure Nano-TiO_2

Aqueous/solvent (e.g., alcohol, ethylene glycol) dispersions of nano-TiO_2 have been proposed as protecting coatings for building materials widely used in both historical monuments and contemporary architectures in Europe and around the Mediterranean region. In 1999, Poulios et al. pioneered the use of an aqueous dispersion of commercial TiO_2 (Degussa, P25) for marble protection [17]. Two TiO_2 coatings with different thicknesses were prepared by immersion. Under UV irradiation, their anti-sulfation and antibacterial properties were found to be much better than the pure organic coating (Paraloid B72). The positive findings of Poulios et al. have encouraged other researchers in the field. Potenza et al. studied the application of aqueous dispersions of nano-TiO_2 [18], for the protection of a highly porous limestone (i.e., Lecce stone) which has an open porosity 35–45% [19]. The nano-TiO_2 was prepared via sol–gel and hydrothermal process with tetrapropyl orthotitanate as the precursor, and products in both anatase/rutile phase were obtained. They primarily evaluated the compatibility of nano-TiO_2 coatings with stone substrate (i.e., the surface morphology and color), as well as the UV-induced self-cleaning and depolluting effects (NO_x degradation).

Bondioli et al., Luvidi et al., Licciulli et al., Quagliarini et al., Bergamonti et al., Gherardi et al., etc., later tested the protection performance of the lab-synthesized and commercial nano-TiO_2 suspensions/dispersions on different building stones [20–23], for example, limestone, marble, and ceramic tile. Franzoni et al., Graziani et al., etc., extended the application of aqueous dispersions of nano-TiO_2 to other building materials [24, 25], such as historic plasters, fired clay bricks, mortars, and so on.

Likewise, in these studies, the surface morphology, vapor permeability, water absorption by capillary, and surface color changes after coating application were evaluated, in order to assess the compatibility of the surface engineering on natural stones and heritage objects. They found that the UV-induced self-cleaning hydrophilicity did not lead to increased water absorption in the stones, thus favoring the use of TiO_2 coatings for architectural heritage.

For the photocatalytic property study, organic dyes (e.g., rhodamine B, methylene blue, methyl orange, etc.), NO_x decomposition and soiling removal tests were used. It is noticed, in the field of building and heritage conservation, instead of using common UV–vis spectrometry to measure the color intensity changes of the organic dyes upon photodecomposition [26], the chromatic difference of dye-stained coatings upon irradiation has been employed to estimate the photocatalytic activity (Figure 21.4) [27]. The former method is popular for industrial photocatalysts, since it requires a liquid phase which does not account for the possible interactions between the dye, the coating and the porous stone substrate [11]. In the case of porous building materials, the dye solution will penetrate the substrate, thereby hindering the photodecomposition process if the upper surface blocks the UV irradiation. To start the test, the water solution of dye was transferred onto the samples. After complete drying, colorimetric measurements by the spectrophotometer were carried out during UV irradiation at varied time intervals. Both the total color difference (ΔE^*) and the relative change of the characteristic chromatic coordinate (e.g., a^*, b^* in the CIE-$L^*a^*b^*$ standard color system) were used to assess the photocatalytic decoloration of the stain. The total color difference is calculated as:

$$\Delta E^* = \sqrt{(\Delta L^*)^2 + (\Delta a^*)^2 + (\Delta b^*)^2]} \tag{21.1}$$

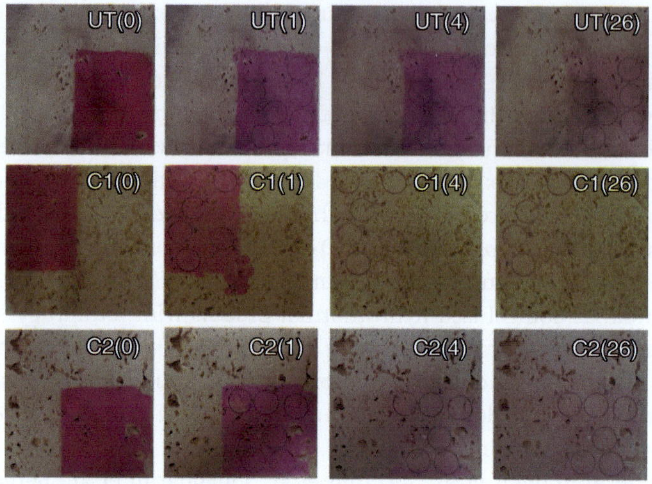

Figure 21.4 Rhodamine B test performed on travertine samples treated with varied nano-TiO_2 suspensions before UV exposure and after 1, 4, 26 h of UV illumination. Reproduced from [27] with permission of Elsevier.

Figure 21.5 Schematic illustration of coating adhesion due to mechanical interlocking.

where Δa^*, Δb^*, and ΔL^* were the differences of coordinates L^*, a^*, and b^* before irradiation and after certain time of irradiation. Our group also calculated the discoloration of the blue color (b^*) of methylene blue (MB) stained coating as [11]:

$$D_{b^*} = \frac{|b_t^* - b_{MB}^*|}{|b_{MB}^* - b_0^*|} * 100 \tag{21.2}$$

where b_0^* is the average value of color coordinate b^* measured before MB stain, whereas b_{MB}^* is the mean value of b^* just after MB application and drying. b_t^* indicates the mean value of b^* after t hour/s of irradiation.

21.2.2.2 Introduction of "Binder" for Nano-TiO_2

In the absence of effective chemical bonding, TiO_2 nanoparticles adhere to the porous stone substrate mainly through the bonding resulting from the diffusion, wetting, void/pore-filling of the porous microstructures, and the subsequent mechanical interlocking by the dispersions (Figure 21.5). For porous substrates as building stones, the mechanical bond force is dominant [28].

Consequently, the coatings can be easily washed away by rainfall or mechanical abrasion. To eliminate the penetration of nano-TiO_2 inside the stone pores and to avoid their leaching, the Si-based intermediate layer/binder was introduced. Bondioli et al. [29] prepared the TiO_2–SiO_2 binary film with varied contents of TiO_2 to enhance the adhesion between the polished ceramic tile surfaces, by using tetraethoxysilane (TEOS) and nano-TiO_2 as precursors. Scratch tests (Micro-Combi tester) were performed on as-prepared coatings, and results showed that the film scratch resistance increased with the increasing of TiO_2 content and annealing temperature. This is attributed to the high structural integrity of the coatings and the increased final E modulus with increasing TiO_2 content.

A silica gel intermediate layer was also employed to enhance the TiO_2-porous substrate adhesion [30]. A pretreatment with TEOS was done first on two building stones with different porosities, that is, Noto stone (porous calcarenite) and Carrara marble (low porous, very compact). Water dispersion of anatase nano-TiO_2 was brushed on the pretreated stone surfaces via "wet-on-wet" (on wet TEOS surface) and "wet-on-dry" (on dry/gelled TEOS surface) techniques. In the case of nano-TiO_2 applied on the wet TEOS surface, effective chemical bonding between nano-TiO_2 and the silica gel was evidenced by the stretching vibrations of Si—O—Ti bonds (around 910 cm^{-1}) in the FTIR spectrum. The chemical interaction resulted in a higher amount and more homogeneous embedment of nano-TiO_2 in the SiO_2 gel network, which led to a higher photocatalytic activity compared to the "wet-on-dry" coating. Additionally, the authors also applied the promising "wet-on-wet" coating on marble tiles in a pilot area of the façade of Monza Cathedral (Italy) [31] to assess its effectiveness in outdoor conditions.

21.2.2.3 Dispersions of Doped Nano-TiO_2

The main drawback of TiO_2-based coatings is their low light-harvesting efficiency (<5% of solar light) due to their large bandgap (3.2 eV) and the fast electron–hole pair recombination, which makes their photocatalytic performance highly dependent on light conditions. To improve the light utilization efficacy by shifting the maximum light absorption into the visible light region, core–shell-structured nanoparticles [32], impurity doping [33], plasmonic nanoparticles [34], and heterojunctions [35] strategies have been mainly explored. Since the doping of TiO_2 with metal ions offers a feasible and cost-effective approach to tune surface states, electronic properties, and charge transfer processes to enhance photocatalytic activity [36], nano-TiO_2 doped with transition metal ions were used for protecting building materials and stone artworks.

Transition metals are elements that have partially filled d-orbitals, and the incorporation of them into the lattice of TiO_2 induces the formation of new energy levels near the conduction band [33]. Essentially, the valence band maximum (O 2p) and conduction band minimum (Ti 3d) are formed when the orbitals of the impurity dopants interact with the orbitals in TiO_2, resulting in a narrower bandgap (Figure 21.6).

Either by insertion at interstices or lattice position occupation, Fonseca et al. doped nano-TiO_2 with Fe^{3+} which has a narrow bandgap (~2.6 eV) and a similar size to Ti^{4+} [38]. The antibiofouling performance of Fe^{3+}-doped nano-TiO_2 and conventional biocides were tested both in the laboratory on mortar slabs and in situ

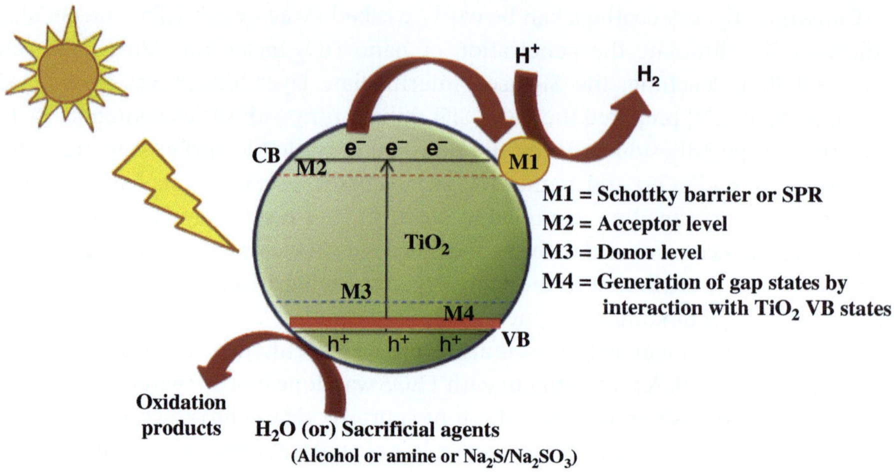

Figure 21.6 The mechanistic pathways of metal dopants for the enhancement of charge carrier separation and light absorption capacity of TiO_2. Reproduced from Ref. [37] with permission from Elsevier.

on walls of Palácio Nacional da Pena (Sintra, Portugal). Results showed that doped anatase nano-TiO_2 is a better biocide for prohibiting lichen and other phototrophic microorganisms, both in lab and in situ conditions.

There are several applications of Ag-doped nano-TiO_2 as photocatalyst, antibacterial, and protective coatings for building materials. The reasons for integrating Ag mainly lie in three aspects [39]. First, Ag can improve the trapping of electrons and inhibits the electron–hole pair recombination. Second, Ag can also shift the response wavelength of TiO_2 toward the visible light region. Third, silver ions can attack multiple cellular processes and trigger oxidative stress, protein dysfunction, membrane and DNA damage, leading to microbial cell damage.

La Russa et al. prepared metal-doped TiO_2 nanoparticles, dispersed in a binder (nanosilica, CTS, diluted 1/10 in water) at different powder/binder ratios, that is, 1/1, 1/10, and 1/100 [40]. The chromatic changes induced by coating application on stone surfaces, the photocatalytic activity, and antibacterial activity against selected Gram-positive and Gram-negative bacteria were evaluated. The color change decreased as the decreasing of TiO_2 content, but not in a proportional way. In the case with the molar ratio TiO_2/binder = 1/100, the color alteration of the surface due to Ag-TiO_2 and Sr-TiO_2 is negligible (Figure 21.7). All doping elements and mixtures (Ag-Fe and Ag-Sr) increased the visible light absorption capacity of TiO_2. However, the best antibacterial effect was observed for the Ag doped TiO_2, owing to the biocidal effects of Ag. Indeed, the major antibacterial effect was observed with Ag-TiO_2 and Ag-Sr-TiO_2 and a minor effect with Ag-Fe-TiO_2, explained with the different ionic radius of the three metals. They concluded that TiO_2 doped with Ag alone or in combination with Fe^{3+} or Sr has the best antibacterial activity, as well as good compatibility with building stones. In another study by Ruffolo et al. [41], the medium-term biocidal effect of Ag-doped

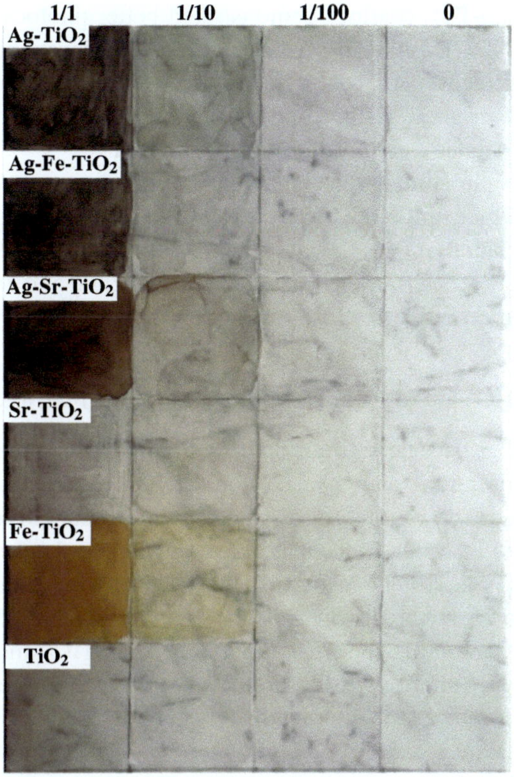

Figure 21.7 Surface chromatic change of marble treated with undoped/doped nano-TiO_2 coatings. Reproduced from [40] with permission of Elsevier.

nano-TiO_2 has been studied in-situ in the archaeological site of Ercolano (on the wall of Villa dei Papiri in Ercolano), in comparison with undoped nano-TiO_2.

Doping with nitrogen is also attractive because of the comparable atomic size of nitrogen with oxygen, small ionization energy, and stability. Two mechanisms have been identified for the incorporation of nitrogen into the TiO_2 structure: substitutional (replacement of oxygen) and interstitial (addition of nitrogen into the TiO_2 lattice). [42]. Nitrogen as a dopant can alter the bandgap of TiO_2 introducing filled N2p states above the TiO_2 valence band, which decreases the apparent gap of TiO_2 (around 2.5 eV) [43, 44]. Besides, N-doping can create oxygen vacancies that can shift the absorption edge into the visible region or enhance the photocatalytic efficiency by electron trapping, slowing down the recombination of photogenerated electrons and holes [43].

Bergomonti et al. [42] and Fornasini et al. [45] studied the protective performance of self-cleaning coatings based on N-doped TiO_2 nanoparticles, synthesized via the sol–gel process, for limestones and travertine. The red shift in the absorption spectra of N-doped TiO_2 nanoparticles was characterized by diffuse reflectance spectroscopy or XPS spectroscopy [45]. The possible doping mechanisms were summarized as: (i) nitrogen in substitutional positions for doping realized via an acidic process and (ii) nitrogen in interstitial positions for doping using urea as a nitrogen source.

The N-doped nano-TiO_2 coatings demonstrated better photocatalytic activity than the commercial TiO_2 (Evonik P25), tested by UV-induced wettability and dye decomposition measurements. All coatings showed good compatibility with building stones, since the original surface appearance and the capillary water absorption properties of the stones were not significantly altered.

However, as some studies have shown, strong color variations can be caused by the dopant, which shifts the absorption of nano-TiO_2 into the visible region of electromagnetic radiation. To alleviate the unacceptable color effect, proper tunning of the dopant type/concentration, the application method and amount and the coating is of great importance.

21.2.3 Nano-TiO_2 Embedded in Organic/Inorganic Matrices

Nano-TiO_2 coatings have poor adhesion to common building materials [12, 46], and they are susceptible to leaching in outdoor environments. Moreover, these coatings do not reduce the wettability of stone surfaces and, consequently, the negative effects caused by water. Indeed, as stated previously, the degradation of porous building materials is largely related to the water-driven damaging effects, such as freeze-thaw cycles, salt dissolution and recrystallization, thermo-hygrometric expansion, etc. Combining the multifunctional nano-TiO_2 and organic/inorganic matrices yields even more versatile protective materials for buildings and monuments, and also improves the interaction between the coating and the substrate.

21.2.3.1 TiO_2–SiO_2 Nanocomposites

Silica-based compounds are used as inorganic matrices. TiO_2–SiO_2 nanocomposites were synthesized by adding commercial/preformed/modified TiO_2 colloidal particles to the silica matrix (e.g., silica oligomer, tetraethyl orthosilicate, hydroxyl-terminated polydimethylsiloxane, etc.), aiming to achieve enhanced photocatalytic and protective/consolidation effects (hydrophobicity and/or surface mechanical strengthening) in addition to good adhesion to building materials. To apply, the starting sol can be facilely coated on building surfaces via an aerosol-assisted process (e.g., by a spray gun). Upon the hydrolysis-condensation of oligomeric ethoxysilane and the gel transition, TiO_2 nanoparticles were embedded in nano-SiO_2 matrices that formed via an aggregative growth model in the presence of the catalyst [47].

By tuning the dimensions (varying from micrometers to nanometers), the surface modifications (pure or Ag/Au/Cu-/N-modified TiO_2) and the loading amounts of TiO_2 particles, as well as the type/concentrations of the surfactant, a series of TiO_2–SiO_2, TiO_2–SiO_2–PDMS/TEOS nanocomposites were prepared [48–52]. Mosquera et al. have found out [47], that during the sol preparation, the nonionic surfactant (*n*-octylamine) plays a quadruple role, including (i) preventing crack formation by coarsening the pore structure of the gel network, (ii) enhancing photocatalytic activity by creating mesoporous nanocomposites, (iii) improving nano-TiO_2 dispersion by increasing pH, and (iv) serving as a basic catalyst of the sol-transition inside the porous stone substrate.

The TiO_2–SiO_2 nanocomposite coatings can effectively adhere to stone substrates, with the crack-free feature. Besides the decent self-cleaning and hydrophobic properties, it is worth noting that the nanocomposites also improved the mechanical strength of the substrate due to the silica gel network produced within the porous substrate. Under the equal TiO_2 load (4%, w/v), the integration of larger and sharper TiO_2 particles in silica matrices significantly improves self-cleaning effectiveness, due to the higher availability of surface photoactive sites. Yet, the photocatalytic activity decreases with higher TiO_2 load (10%, w/v), because of the drastically reduced porous volume, and hence more difficult access to the photoactive sites [47]. The performance of TiO_2–SiO_2 nanocomposites in outdoor environments was further evaluated in a four-year in situ monitoring study [53], by application on a historic mortar belonging to the bleachers of the Cadiz Roman Theatre in Spain (located in the south of the Andalusia region and the Atlantic coast, characterized by an oceanic Mediterranean climate). Results revealed that TiO_2–SiO_2 nanocomposites can reduce the deposit of organic substances and biological material for at least four years. Antifouling measurements also demonstrated that the treatment is very effective against biofouling (up to 98% inhibition efficacy) in the first two years. Re-application is suggested after two years, in order to maintain the high efficacy.

21.2.3.2 Nano-TiO_2 Embedded in Organic/Hybrid Matrices

To exploit the good wetting behavior, hydrophobicity, and good adhesive properties of organic polymers, the nanocomposite formulations were prepared by mixing dispersion/powder of nano-TiO_2 with polymeric solution/emulsion. The most commonly used organic matrices are acrylic polymers, alkyl silane/polysiloxanes, and fluorinated/partially fluorinated polymers. La Russa et al. comparatively studied the performance of nanocomposite formulations based on nano-TiO_2 and acrylic polymers/PFPE [54]. The polymer–TiO_2 formulation provides decent photocatalytic and water-repellent features, owing to the film-forming ability of the polymers, which act as an effective binder (partial or complete pore filling) between the porous substrate and the TiO_2 nanoparticles.

Unlike the dispersions of nano-TiO_2 mentioned above, these nanocomposite formulations usually alter the original wettability of the stone substrate from hydrophilic to hydrophobic, via the low-surface-tension groups (e.g., alkyl or fluorinated alkyl chains). Apart from the low surface tension materials, nano-TiO_2 dispersed in polymers can induce surface roughness at the micro-nanometer scales, thus fulfilling the two basic requirements for achieving superhydrophobicity [55]. TiO_2 nanoparticles have been widely used for fabricating superhydrophobic coatings in recent decades [56].

Inspired by natural plants and animals (e.g., lotus leaves, rice leaves, rose petals, cicada/butterfly wings, gecko feet, etc.), surfaces with water contact angle (CA) $> 150°$ while sliding angle (SA) $< 10°$ have attracted great interest in both fundamental research and functional applications [57]. For buildings and monuments protection, superhydrophobic coatings show particular advantages over traditional organic/inorganic coatings [58–61], which can be summarized as: (i) preventing liquid water adhesion/penetration or water vapor condensation by ultra-repellency; (ii) spontaneous surface maintenance by self-cleaning or "Lotus effect"; and (iii) inhibition of initial attachment of bacterial cells, facilitating biofilm removal through decreased interaction force/contact area with

Figure 21.8 Schematic representation of water contact angle on (a) a superhydrophobic surface and (b) a superhydrophilic surface.

cells. Therefore, superhydrophobic self-cleaning materials represent an optimal protective coating for outdoor natural and artificial building stones (e.g., natural stones, bricks, cement, ceramics, mortars, tiles, etc.).

Although it seems superhydrophobic coatings exhibit self-cleaning property as superhydrophilic coatings obtained from nano-TiO_2 dispersions, their underlying mechanisms are distinct. In general, there are two broad types of self-cleaning surfaces, being hydrophilic surfaces and hydrophobic surfaces, respectively. A major difference between these two types of surface is the surface roughness and/or micro-texture. The superhydrophobicity can be understood through the wettability theory. In order for a surface to exhibit superhydrophobic properties, water droplets must remain in the Cassie–Baxter state, which differs from the classical Young's model and Wenzel's model. The real contact occurs at the solid–liquid–air composite interface, as illustrated in Figure 21.8a. The presence of micro-grooves or protrusions on the surface can trap air to form air pockets, above which the real contact takes place. In such circumstances, overcoming the surface tension and passing through the air pocket to wet the surface is difficult for the droplet, resulting in ultra-repellency or superhydrophobicity, whereas nano-TiO_2 dispersion coated surface becomes superhydrophilic even if the surface is quite smooth (Figure 21.8b). The real contact occurs at the solid–liquid interface, creating a superficial water film. The thin water film has the advantage of protecting/cleaning the surface from contamination, yet the efficiency is compromised by its high surface adhesion compared with a superhydrophobic surface.

Figure 21.9 shows the schematic illustrations of the self-cleaning process on superhydrophilic and superhydrophobic surfaces. On the superhydrophilic surface (Figure 21.9a), the water drops form a superficial and continuous film, with the water contact angle below 10°. Once contaminants are deposited, the water film goes beneath and moves the contaminants out along spreading [62]. Nevertheless, the self-cleaning process observed is not as rapid as on superhydrophobic surfaces. Furthermore, the superhydrophilic surfaces remain in a state of prolonged wetness due to the presence of a superficial water film, which can contribute to the degradation of the stone. In fact, as discussed at the beginning of the chapter, the majority of the degradation phenomena observed in building materials can be attributed to water-related factors. On the contrary the superhydrophobic surfaces are in a dry state (Figure 21.9b), owing to their ultra-repellency and low adhesive property with water drops. The water drops in contact are spherical, and they can pick up the dust/dirt and roll over the surface instantly. As a consequence, the optimal self-cleaning efficiency is achieved with a superhydrophobic surface rather than a

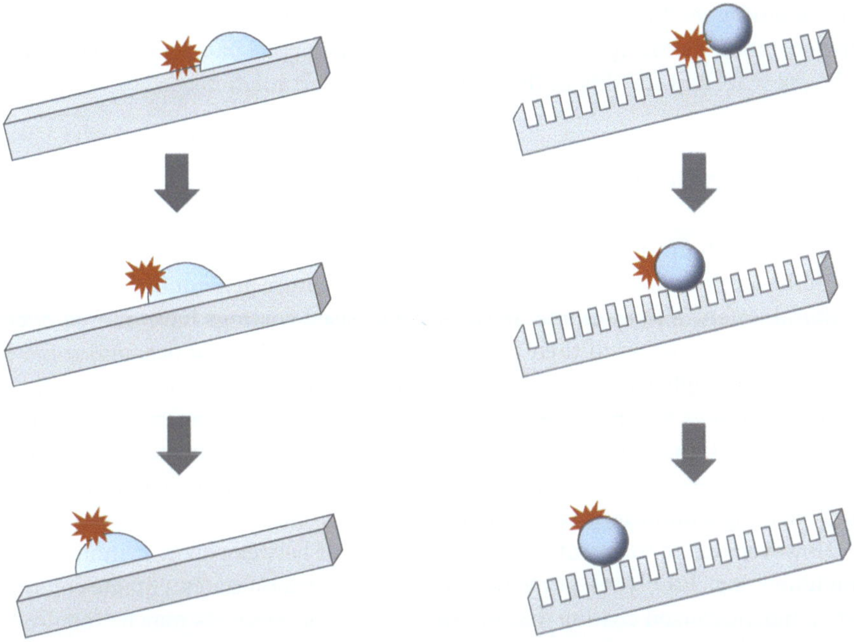

Figure 21.9 Schematic representation of self-cleaning processes on (a) a superhydrophilic and (b) a superhydrophobic surface.

superhydrophilic one, in addition to the prevention of water-related deteriorating agents.

By integrating photocatalytic nano-TiO_2, superhydrophobic protective coatings demonstrate more promising multifunctionalities, such as antimicrobial and de-polluting effects. By combining nano-TiO_2 with a partially fluorinated siloxane via Ti—O—Si hetero-linkages, our group has developed TiO_2-based nanocomposite coatings that exhibit efficient photoactivity, in addition to the superhydrophobicity [11]. The Ti—O—Si hetero-linkages can improve photon efficiency and stimulate the oxidation process, producing improved photocatalytic and antibacterial action. In comparison with pure anatase nano-TiO_2, the nanocomposites showed much greater antibacterial efficacy against both *Escherichia coli* and *Staphylococcus aureus* at the same concentrations. Bai et al. dispersed preformed TiO_2@SiO_2 core–shelled nanoparticles in acrylic resin (Paraloid B72) for marble protection [63]. The superhydrophobicity of the acrylic polymer was attributed to the enhanced micro-nano roughness induced by the nanoparticles. The adsorption rate and degradation efficiency of the organic dye in the nanocomposite coating were 2.19 and 1.45 times higher than that in the pure TiO_2 coating.

Despite the very appealing versatilities of TiO_2-based nanocomposite coatings, it should not be neglected, the introduction of photocatalytic nano-TiO_2 may also induce adverse effects. To activate photocatalysis, UV or UV–visible irradiation is a necessity, which limits the application in well-illuminated areas. Besides, the photocatalytic oxidation activity of TiO_2 nanoparticles will inevitably accelerate the photodegradation of organic polymers, thereby compromising the durability of nanocomposite coatings. It is essential to achieve a good balance between the high photoactivity and stability of the nanocomposite formulations.

21.3 Summary and Outlook

The preservation and maintenance of historical monuments and contemporary buildings are of great significance to present and future generations of humans, for cultural, safety, and economic reasons. TiO_2-based coatings have offered effective options mainly owing to their photoinduced multifunctionalities, and some have already been applied in conserving European churches and archaeological sites. This chapter outlines the main research directions in the development and application of TiO_2-based protective coatings, with emphasis on the basic mechanisms underlying their protective properties, for example, superhydrophilic/superhydrophobic self-cleaning, antibacterial and depolluting effects, etc.

However, due to the great individual differences between specific monuments or buildings (e.g., lithotypes, porosities, colors, conservation states, climate conditions, etc.), individualized coating designs and application methods may be required. No single coating or application method is universally applicable in the field. Recent advances have resulted in combinations/enhancements of protective properties via novel synthetic routes, but the physicochemical and artistic properties of building materials must be considered before real application. The feasibility and compatibility of most of the newly proposed approaches need further verification, with more fundamental research to better understand the nanoparticle–matrix/coating–substrate interactions. Additionally, in situ evaluation of the performance durability and stability of TiO_2-based coatings in outdoor conditions is also necessary, prior to scale-up of fabrication and application. Furthermore, for the promising nanocomposite coatings, achieving a good balance between the photoactivity and stability of themselves is practical for their sustainable implementation.

Acknowledgments

This work was supported by the National Natural Science Foundation of China (Grants 52202028), the “Belt and Road” Innovation Talent Exchange Program for Foreign Experts (Grant DL2021183001L), the Social Science Fund of Shaanxi Province (Grant N. 2021G011) and the Fundamental Research Funds for Central Universities.

References

1 ICOMOS-International Scientific Committee for Stone (ISCS) (2008). *Illustrated Glossary on Stone Deterioration Patterns*. Champigny/Marne, France: ICOMOS-ISCS.

2 Doehne, E. and Price, C.A. (2010). *Stone Conservation: An Overview of Current Research*, 2e. Los Angeles: The Getty Conservation Institute.

3 Steiger, M., Charola, A.E., and Sterflinger, K.(2014). Weathering and deterioration. In: *Stone in Architecture* (ed S. Siegesmund and R. Snethlage). Springer-Verlag Berlin Heidelberg.

4 Vitruvius (2014). *The ten books of Architecture*, translated by Morgan, M.H. London: Harvard University Press.

5 Plinio Secondo, G. (1561). *Historia Naturale*, translated for M. Lodovico Domenichi. Vinegia (Venezia): appresso Gabriel Giolito de' Ferrari.

6 Charola, A.E. (1995). Water-repellent treatments for building stones: a practical overview. *APT Bulletin* 26: 10–17.

7 Ruffolo, S.A. and La Russa, M.F. (2019). Nanostructured coatings for stone protection: an overview. *Frontiers in Materials* 6: 147.

8 Munafò, P., Goffredo, G.B., and Quagliarini, E. (2015). TiO_2-based nanocoatings for preserving architectural stone surfaces: an overview. *Construction and Building Materials* 84: 201–218.

9 Ben Chobba, M., Weththimuni, M.L., Messaoud, M. et al. (2024). Recent advances in the application of metal oxide nanomaterials for the conservation of stone artefacts, ecotoxicological impact and preventive measures. *Coatings* 14: 203.

10 Yang, Z., Xu, J., Wang, W. et al. (2023). Surface properties and de-polluting performance of a photocatalytic coating incorporating novel core@shell nanospheres for cementitious substrate. *Cement and Concrete Composites* 139: 105036.

11 Cao, Y., Salvini, A., and Camaiti, M. (2023). Multi-functional TiO_2-based nanocomposite coating with durable superhydrophobicity and enhanced photocatalytic and antimicrobial properties for the sustainable maintenance of building stones. *Construction and Building Materials* 404: 13339.

12 Gherardi, F. (2022). Current and future trends in protective treatments for stone heritage. In: *Conserving Stone Heritage* (ed. F. Gherardi and P. Noni Maravelaki). Cham, Switzerland: Springer Nature Switzerland.

13 Padmanabhana, N.T. and John, H. (2020). Titanium dioxide based self-cleaning smart surfaces: a short review. *Journal of Environmental Chemical Engineering* 8: 104211.

14 Banerjee, S., Dionysiou, D.D., and Pillai, S.C. (2015). Self-cleaning applications of TiO_2 by photo-induced hydrophilicity and photocatalysis. *Applied Catalysis B: Environmental* 176–177: 396–428.

15 Sakai, N., Fujishima, A., Watanabe, T., and Hashimoto, K. (2003). Quantitative evaluation of the photoinduced hydrophilic conversion properties of TiO_2 thin film surfaces by the reciprocal of contact angle. *The Journal of Physical Chemistry B* 107: 1028–1035.

16 Guan, K. (2005). Relationship between photocatalytic activity, hydrophilicity and self-cleaning effect of TiO_2/SiO_2 films. *Surface and Coatings Technology* 191: 155–160.

17 Poulios, I., Spathis, P., Grigoriadou, A. et al. (1999). Protection of marbles against corrosion and microbial corrosion with TiO_2 coatings. *Journal of Environmental Science and Health, Part A* 34 (4): 1455–1471.

18 Potenza, G., Licciulli, A., Diso, D. et al. (2007). Surface engineering on natural stone through TiO_2 photocatalytic coating. *Proceedings of International RILEM Symposium on Photocatalysis, Environment and Construction Materials* (ed. P. Baglioni and L. Cassar) (8–9 October 2007), pp. 315–322.

19 Cao, Y., Salvini, A., and Camaiti, M. (2021). One-step fabrication of superamphiphobic, self-cleaning surface for outdoor and in situ application on various building substrates. *Journal of Colloid and Interface Science* 591: 239–252.

20 Goffredo, G.B., Quagliarini, E., Bondioli, F., and Munafò, P. (2014). TiO_2 nanocoatings for architectural heritage: self-cleaning treatments on historical stone surfaces. *Proceedings of the Institution of Mechanical Engineers, Part N* 228 (1): 2–10.

21 Licciulli, A., Calia, A., Lettieri, M. et al. (2011). Photocatalytic TiO_2 coatings on limestone. *Journal of Sol-Gel Science and Technology* 60: 437–444.

22 Quagliarini, E., Bondioli, F., Goffredo, G.B. et al. (2012). Smart surfaces for architectural heritage: preliminary results about the application of TiO_2-based coatings on travertine. *Journal of Cultural Heritage* 13: 204–209.

23 Bergamonti, L., Alfieri, I., Lorenzi, A. et al. (2013). Nanocrystalline TiO_2 by sol-gel: characterisation and photocatalytic activity on Modica and Comiso stones. *Applied Surface Science* 282: 165–173.

24 Franzoni, E., Fregni, A., Gabrielli, R. et al. (2014). Compatibility of photocatalytic TiO_2-based finishing for renders in architectural restoration: a preliminary study. *Building and Environment* 80: 125–135.

25 Graziani, L., Quagliarini, E., Bondioli, F., and D'Orazio, M. (2014). Durability of self-cleaning TiO_2 coatings on fired clay brick façades: effects of UV exposure and wet & dry cycles. *Building and Environment* 71: 193–203.

26 Ko, W. and Jeon, S. (2014). An electrodeless quartz crystal resonator integrated with UV/Vis spectroscopy for the investigation of the photodecomposition of methylene blue. *Sensors Actuators B* 193: 774–777.

27 Quagliarini, E., Bondioli, F., Goffredo, G.B. et al. (2012). Self-cleaning and de-polluting stone surfaces: TiO_2 nanoparticles for limestone. *Construction and Building Materials* 37: 51–57.

28 Bruns, C.J. and Stoddart, J.F. (2016). An introduction to the mechanical bond. In: *The Nature of the Mechanical Bond: From Molecules to Machines* (ed. C.J. Bruns and J.F. Stoddart). Weinheim: Wiley-VCH.

29 Bondioli, F., Taurino, R., and Ferrari, A.M. (2009). Functionalization of ceramic tile surface by sol-gel technique. *Journal of Colloid and Interface Science* 334: 195–201.

30 Gherardi, F., Goidanich, S., Dal Santo, V., and Toniolo, L. (2018). Layered nano-TiO_2 based treatments for the maintenance of natural stones in historical architecture. *Angewandte Chemie International Edition* 130: 7360–7363.

31 Gherardi, F., Gulotta, D., Goidanich, S. et al. (2017). On-site monitoring of the performance of innovative treatments for marble conservation in architectural heritage. *Heritage Science* 5 (4): 4.

32 Li, Z., Xiong, J., Huang, Y. et al. (2024). Synergy mechanism of confined effect and Z-scheme electron transfer in core-shell structure photocatalyst for boosting photoreduction CO_2 activity. *Chemical Engineering Journal* 486: 150304.

33 Khlyustov, A., Sirotkin, N., Kusova, T. et al. (2020). Doped TiO_2: the effect of doping elements on photocatalytic activity. *Materials Advances* 1: 1193

34 Kang, S., Choi, J., Park, G.Y. et al. (2022). A novel and facile synthesis of Ag-doped TiO_2 nanofiber for airborne virus/bacteria inactivation and VOC elimination under visible light. *Applied Surface Science* 599: 153930.

35 Zhao, Y., Shu, Y., Linghu, X. et al. (2024). Modification engineering of TiO_2-based nanoheterojunction photocatalysts. *Chemosphere* 346: 140595.

36 Huang, L., He, G., Yuan, Y. et al. (2024). Trivalent metal ions (Al, Ga, In)-Doped TiO_2 for enhanced photocatalytic desulfurization of H_2S: band structure regulation, performance, and mechanism. *Industrial & Engineering Chemistry Research* 63: 7154–7165.

37 Kumaravel, V., Mathew, S., Bartlett, J., and Pillai, S.C. (2019). Photocatalytic hydrogen production using metal doped TiO_2: a review of recent advances. *Applied Catalysis B: Environmental* 244: 1021–1064.

38 Fonseca, A.J., Pina, F., Macedo, M.F. et al. (2010). Anatase as an alternative application for preventing biodeterioration of mortars: evaluation and comparison with other biocides. *International Biodeterioration & Biodegradation* 64: 388–396.

39 Ahmad, S.A., Das, S.S., Khatoon, A. et al. (2020). Bactericidal activity of silver nanoparticles: a mechanistic review. *Materials Science for Energy Technologies* 3: 756–769.

40 La Russa, M.F., Macchia, A., Ruffolo, S.A. et al. (2014). Testing the antibacterial activity of doped TiO_2 for preventing biodeterioration of cultural heritage building materials. *International Biodeterioration & Biodegradation* 96: 87–96.

41 Ruffolo, S.A., De Leo, F., Ricca, M. et al. (2017). Medium-term in situ experiment by using organic biocides and titanium dioxide for the mitigation of microbial colonization on stone surfaces. *International Biodeterioration & Biodegradation* 123: 17–26.

42 Bergamonti, L., Predieri, G., Paz, Y. et al. (2017). Enhanced self-cleaning properties of N-doped TiO_2 coating for cultural heritage. *Microchemical Journal* 133: 1–12.

43 Ansari, S.A., Khan, M.M., Ansari, M.O., and Cho, M.H. (2016). Nitrogen-doped titanium dioxide (N-doped TiO_2) for visible light photocatalysis. *New Journal of Chemistry* 40: 3000–3009.

44 Basavarajappa, P.S., Patil, S.B., Ganganagappa, N. et al. (2020). Recent progress in metal-doped TiO_2, non-metal doped/codoped TiO_2 and TiO_2 nanostructured hybrids for enhanced photocatalysis. *International Journal of Hydrogen Energy* 45: 7764–7778.

45 Fornasini, L., Bergamonti, L., Bondioli, F. et al. (2019). Photocatalytic N-doped TiO_2 for self-cleaning of limestones. *The European Physical Journal Plus* 134: 539.

46 Calia, A., Lettieri, M., and Masieri, M. (2016). Durability assessment of nanostructured TiO_2 coatings applied on limestones to enhance building surface with self-cleaning ability. *Building and Environment* 110: 1–10.

47 Pinho, L. and Mosquera, M.J. (2013). Photocatalytic activity of TiO_2-SiO_2 nanocomposites applied to buildings: influence of particle size and loading. *Applied Catalysis B: Environmental* 134: 205–221.

48 Khannyra, S., Mosquera, M., Addou, M., and Gil, M. (2021). Cu-TiO_2/SiO_2 photocatalysts for concrete-based building materials: self-cleaning and air de-pollution performance. *Construction and Building Materials* 313: 125419.

49 Khannyra, S., Luna, M., Gil, M.L.A. et al. (2022). Self-cleaning durability assessment of TiO_2/SiO_2 photocatalysts coated concrete: effect of indoor and outdoor conditions on the photocatalytic activity. *Building and Environment* 211: 108743.

50 Crupi, V., Fazio, B., Gessini, A. et al. (2018). TiO_2-SiO_2-PDMS nanocomposite coating with self-cleaning effect for stone material: finding the optimal amount of TiO_2. *Construction and Building Materials* 166: 464–471.

51 Kapridaki, C., Verganelaki, A., Dimitriadou, P., and Maravelaki-Kalaitzaki, P. (2018). Conservation of monuments by a three-layered compatible treatment of TEOS-nano-calcium oxalate consolidant and TEOS-PDMS-TiO_2 hydrophobic/photoactive hybrid nanomaterials. *Materials* 11: 684.

52 Fahmy, A., Gołąbiewska, A., Wojnicz, W. et al. (2023). Multi-functional monodispersed SiO_2-TiO_2 core-shell nanostructure and TEOS in the consolidation of archaeological lime mortars surfaces. *Journal of Building Engineering* 79: 107809.

53 Gemelli, G.M.C., Luna, M., Zarzuela, R. et al. (2022). 4-Year in-situ assessment of a photocatalytic TiO_2/SiO_2 antifouling treatment for historic mortar in a coastal city. *Building and Environment* 225: 109627.

54 La Russa, M.F., Rovella, N., Alvarez de Buergo, M. et al. (2016). Nano-TiO_2 coatings for cultural heritage protection: the role of the binder on hydrophobic and self-cleaning efficacy. *Progress in Organic Coatings* 91: 1–8.

55 Feng, X., Zhang, X., and Tian, G. (2022). Recent advances in bioinspired superhydrophobic ice-proof surfaces: challenges and prospects. *Nanoscale* 14: 5960–5993.

56 Zhang, H., Wang, F., and Guo, Z. (2024). The antifouling mechanism and application of bio-inspired superwetting surfaces with effective antifouling performance. *Advances in Colloid and Interface Science* 325: 103097.

57 Liu, M., Wang, S., and Jiang, L. (2017). Nature-inspired superwettability systems. *Nature Reviews Materials* 2: 17036.

58 Cao, Y., Salvini, A., and Camaiti, M. (2020). Current status and future prospects of applying bioinspired superhydrophobic materials for conservation of stone artworks. *Coatings* 10: 353.

59 Karapanagiotis, I. and Manoudis, P.N. (2020). Superhydrophobic and superamphiphobic materials for the conservation of natural stone: an overview. *Construction and Building Materials* 320: 126175.

60 Liu, G., Zhao, T., Fei, H. et al. (2023). A review of various self-cleaning surfaces, durability and functional applications on building exteriors. *Construction and Building Materials* 409: 134084.

61 Chen, H., Cao, Y., Wang, C., et al. (2025). Superhydrophobic surfaces for the sustainable maintenance of building materials and stone-built heritage: the challenges, opportunities and perspectives, *Advances in Colloid and Interface Science*, 335: 103343

62 Nishimoto, S. and Bhushan, B. (2013). Bioinspired self-cleaning surfaces with superhydrophobicity, superoleophobicity, and superhydrophilicity. *RSC Advances* 3: 671–690.

63 Bai, X., Yang, S., Tan, C. et al. (2022). Synthesis of TiO_2 based superhydrophobic coatings for efficient anti-corrosion and self-cleaning on stone building surface. *Journal of Cleaner Production* 380: 134975.

Index

Note: Page numbers in *italics* and **bold** refers to figures and tables respectively.

a

accumulation of TiO_2 in organisms 238
acetic acid (HAc) 79
acetylspiramycin 311
acid-hydrothermal process 44
acidic condition, TiO_2 NPs synthesis by sol–gel reaction under 3–12
activated carbon 96–97
adsorption 332–333
advanced oxidation processes (AOPs) 272, 328–330, 352
Ag-doped nano-TiO_2 504
alcohols, visible-light-induced selective oxidation 432–435, *434*, *435*
aldehydes 432–435
algae, toxic impact of TiO_2 on 235–236, *235*
Aliivibrio fischeri 347–348, *347*
alizarin red S (ARS) 434
alkali-hydrothermal process 42–44, *44*
allograft bone transplantation 249
all-solid-state Z-scheme 188–189, *188*
alumina (Al_2O_3) 87, 94
amines, visible-light-induced selective oxidation 435–439, *436–439*
amino acids 34–35, *35*
aminoglycosides 315
3-aminopropyltrimethoxysilane (APTMS) 212
ammonia from N_2 425–427, *425–427*
amorphous 2D TiO_2 NSs 59
amorphous HTCS 72
amoxicillin (AMX) 313, 315, 329–330, 487
amphenicols 315
ampicillin (AMP) 313
anatase [100] direction 29
  ester elimination reaction 30–31, *31*, *32*
  fundamental growth mechanism 29–30, *29*
  growth mechanism in presence of TMAH 31–33
anatase nanocrystallites 79
anatase NPs 10
anatase TiO_2 NRs, preparation of 33–36
  effect of amino acids 34–35, *35*
  effect of DEA and TMAH 35–36, *35*, *36*
  effect of lauric acid and TOPO 33–34, *34*
2-anthracene acid (2-AA) 440
anthraquinones (AQs) 440
antibacterial properties of TiO_2 350–354, *351–353*
antibiotic-resistant bacteria (ARB) 301
antibiotic-resistant genes (ARGs) 301
antibiotics 301–302
  classes of 315–318, **316–317**
  degradation 272
  resistance 301
antidotal effects of TiO_2 238
antipode heterojunctions 185
apparent quantum efficiency (AQE) 126
apparent quantum yield (AQY) 219
aquatic organisms, toxicity impact of TiO_2 on 233–235, *234*
arsenic 402–403, *403*
  oxidation of As(III), doped TiO_2 404
  oxidation of As(III), TiO_2 films on inert substates 403–404, *404*
  TiO_2-based heterojunction 404–405, *405*
artificial antibodies 462
artificial photosynthesis 415
aspect ratio 27
atomic layer deposition (ALD) 57, 60, *60*
  procedure 94

Nanostructured TiO_2: Synthesis, Photocatalytic Properties and Applications, First Edition.
Edited by Jia Hong Pan, Wan In Lee and Detlef W. Bahnemann

autologous bone grafting 249
azithromycin (AZM) 311, 313

b

Bacillus subtilis 236, 349
batch-mode reactor 473
  capillary photocatalytic reactors (CPRs) 476, *477*
  slurry photocatalytic reactor (SPR) 473–475, *474*, *475*
  submerged ceramic membrane photocatalytic reactor (SCMPR) 475–476
β-lactam antibiotics 313–315, *313*, **314**
binder for nano-TiO_2 502–503
biodegradable polymer 99, 101–110
biomolecules, PEC detection of 451–458
black anatase TiO_2 (B-TiO_2) 326
block copolymer 70
bone 249
boron-doped carbon quantum dots (B-CQDs) 454
bottom-up methods for 2D material synthesis 57
  atomic layer deposition 60, *60*
  topochemical method 60–61, *61*
  wet chemical methods 57–60
brookite NR, formation of 37–39, *38*
Brunauer–Emmett–Teller (BET) 432

c

capacitance–voltage (C–V) 195
capillary photocatalytic reactors (CPRs) 476, *477*
capping agents
  preparation of anatase TiO_2 NRs with various 33–36
  shape control with 20–22, *21*, *22*
carbamazepine (CBZ) 488
carbonaceous solar fuels
  from CO_2 reduction 420–421
  gas solid photocatalytic reduction with water vapor 421
  gas solid photothermal catalytic reduction with hydrogen gas 421–422, *422*
  industry applications 424
  liquid solid photocatalytic reduction 422–424, *423*
  selectivity regulation for 424
carbon-based substrates 96
  chemical vapor deposition 96, *96*
  hydrothermal synthesis 97
carbon dioxide reduction 489–492
carbon nanotubes (CNTs) 96–97, 345
carboxylic acid 20
carrier transport in sensitized TiO_2 163–177
catalyst dispersion, challenges in 106
C—C coupling reactions 441–442
cefadroxil (CDX) 313
cefalexin (CF) 313
cefradine (CPD) 313
cefuroxime (CFX) 328
cellulose (CLS) 104, 420
cellulose-based TiO_2 photocatalysts
  applications and future directions 110
  challenges and recommendations 109
  challenges in catalyst dispersion 106
  characterization of 105
  coating process for photocatalytic microreactors 108–109
  film characterization and mechanical stability 109
  forward osmosis applications 105
  FTIR spectroscopy 104–105
  future recommendations 107
  impact of sol–gel parameters on photocatalytic performance 107
  optimization of sol–gel parameters 106–107
  photocatalytic efficiency of 106
  rationale for material selection 108
  SEM 105
  synthesis techniques for 104
  TiO_2 gel formation 108
cellulose phosphate (CP) 105
cephalosporins 314–315
cephamycins 313
ceramic substrates 97
  spray pyrolysis 97
charge transfer ligand-to metal (CTLM) 118
charge transfer mechanism 191
charge transfer process 192
chemical diffusion coefficient 167, 169
chemical oxygen demand (COD) 348, 463
chemical vapor deposition (CVD) 88, 96, *96*
  procedure 94
chemiluminescence analyzer 365
chitosan (CS) 101–102
chitosan–TiO_2 (CS/TiO_2) composite
  advantages of 103
  photocatalytic applications of 102–103
  photocatalytic mechanism of 103
  synthesis techniques 102
Chlamydomonas reinhardtii 238
chlorination 350
4-chlorophenol (4-CP) 118
chlortetracycline (CTC) 302
chromium 396
  adding functional additives 396–397, *397*
  elemental doping TiO_2 398–399
  modifying TiO_2 with metal particles 398, *398*
  TiO_2-based heterojunctions 399–400, *399*

ciprofloxacin (CIP) 308–311, 322, 327
circular economy 251
clarithromycin (CLA) 311
classical annular reactor (CAR) 471
cloxacillin (CLX) 313
CO_2 reduction reaction (CO2RR) 490, 491
coarsening process 6–9, *8*
coating process for photocatalytic microreactors 108–109
co-catalyst strategies 375
co-doping 320
coexisting ions effects 323–325
coexisting organic substances effects 323
colloidal SiO_2 spheres 73
composite formation strategies 373–374
composite photocatalysts 320–321
computational fluid dynamics (CFD) 278
conduction band (CB) 118, 186, 318–319, 390–391, 431
contaminants of emerging concern (CECs) 343
continuous flow reactors (CFRs) 476–477, *477*
continuously stirred tank reactor (CSTR) 472
copper 391–392
  adding functional additives 392–393, *393*, **393**
  combing with electrochemistry 394–396, *395, 396*
  modifying TiO_2 with metal deposition 393–394, *394*
crystalline silicon (c-Si) 94
crystal violet (CV) 105
cyanotoxins 271

d

Daphnia magna 346, 348
defect engineering 321–322
degradation patterns 497
dendrimer, synthesis of extremely small TiO_2 NPs using 23–24, *23*
density functional theory 194–195, *195*
density of states (DOS) 168
deposition metal NPs on TiO_2 by reduction 212–214, *213, 214, 215*
dibenzothiophene (DBT) 441
diethylamine (DEA) 14–15
  effect of 35–36, *35, 36*
diethylene glycol (DEG) 396
diffuse reflectance infrared Fourier transform spectroscopy (DRIFTS) 375
diffusion model 163–166, *166*
6,7-dihydroxynaphthalene-2-sulfonate (DHNS) 437
dimethyl sulfoxide (DMSO) 106
5,5-dimethyl-1-pyrroline N-oxide (DMSO) 194
dinitrogen (N_2) 361
dip-coating process for TiO_2 immobilization 92–93, *93*
direct calcination of HTCS 75
direct electron transfer (DET) mechanism 207–209, *209, 210*
direct Z-scheme 189–190
dispersions of nano-TiO_2
  binder for nano-TiO_2 502–503
  doped nano-TiO_2 503–506, *504*
  pure nano-TiO_2 500–502, *502*
dissolved organic matter (DOM) 323, 326
diverse disinfection byproducts (DBPs) 489
dopamine 455
doped nano-TiO_2 503–506, *504*
doping strategies 372–373
double sheet skin photocatalytic reactor 280, *281*
doxorubicin (DOX) 248
doxycycline 302
drinking water contamination 271–272
drug delivery systems and therapy, application of TiO_2 in 247–249, *247*
drum reactor 279, *279*
dyes 484–485
dye-sensitized solar cells (DSSCs) 159–160

e

ecotoxicity
  particles impact 345–346
  species used for tests 344, **344**
  from wastewater 347–350, *347*
ecotoxicological effects of TiO_2 232
  accumulation of TiO_2 in organisms 238
  algae and plants, toxicity impact 235–236, *235*
  antidotal effects 238
  aquatic organisms, toxicity impact 233–235, *234*
  mice and humans, toxicity impact 237
  microorganisms, toxicity impact 236–237, *236*
  toxicity mechanisms and modes of action 232–233, *232*
ecotoxicology 247
edible risk of TiO_2 244–245
electrochemical exfoliation 56–57, *57*
electron continuity equation 164, 171
electron diffusion length 165
electron lifetime 170
electron–redox mediator recombination 170–176, *172, 173, 175*
electrons 149–150, *150*
electron–sensitizer recombination 176–177
electron spin resonance (ESR) 121, 194
electron transfer path 208

electrospinning 100
  TiO_2 NFs by 45, *45*
electrospraying 100
electrospun nanofibrous membrane 93–94
elemental doping 319–320
elongated TiO_2 NTs 40–41, *41*
enhanced photocatalytic mechanism based on LSPR 207
  DET mechanism 207–209, *209, 210*
  plasmon-induced resonance energy transfer mechanism 209–211, *210*
  resonant photon scattering mechanisms 211
enrofloxacin (ENR) 308
Enterococcus faecalis 351
erythromycin (ERY) 311
erythrosin B (ErB) 440
Escherichia coli 236, 344, 348, 350–353
ester elimination reaction 30–31, *31, 32*
ethanol solution impregnation method 129
evaporation-induced self-assembly (EISA) process 70–71
exciton energy transfer (EET) 461
extended X-ray absorption fine structure (EXAFS) 376
external electric field 330–331
external field enhancements
  crystallization of HTCS self-template by 75–76, *75*

f
falling film 480
Fenton process 328
Fickian diffusion 169
filter-press photocatalytic reactor (FPPR) 481–483, *482*
fixed bed photocatalytic reactor (FBPR) 478, 488
fixed bed reactor 278, *278*
fixed film photocatalytic reactor (FFPR) 478, 488
fixed photocatalytic reactor 478, *479*
flow-mode reactor 476–477
  filter-press photocatalytic reactor (FPPR) 481–483, *482*
  fixed photocatalytic reactor 478, *479*
  photocatalytic membrane reactor (PMRs) 477–478, *478*
  photocatalytic microreactor 479–480, *480*
  solar parabolic photocatalytic reactor (SPPR) 480–481, *482*
  staircase photocatalytic reactor (SPR) 480, *481*
fluidized bed reactor 278, *278*
fluorinated ethylene-propylene (FEP) 279, *280*
fluorine-doped tin oxide (FTO) 161
fluoroquinolones 308
foreign atom doping concept 120
forward osmosis (FO) applications 105
FTIR spectroscopy 104–105
functionalizing polyvinylidene fluoride (PVDF) 334

g
gas chromatography (GC) 376
gas-phase deposition 90–91
glass, immobilization of TiO_2 on 91–92, *92*
glutaraldehyde (GLD) 452
Gold (Au) nanocrystals 220
Gold nanoparticles (Au NPs) 221
graphene 51, 124
graphene oxide (GO) 58
graphitized mesoporous carbon (GMC) 333
green synthesis methods 250–251
gross quantum yield (GQY) 149

h
Haber–Bosch process 425
hard template method 69–70
harmful ions, PEC detection of 458–462
health, TiO_2 impact on 242–245, *242*
heavy metal 389–390
hematite iron oxide (Fe_2O_3) 373–374
heterogeneous photocatalysis 88, 271–272
heterojunctions 185
  classification of 186
  construction of 318–319
  P–N heterojunction 186
  roles of 197–201, *198–201*
  types I, II, and III 186–187, *187*, 318, 417–418, *418*
  Z-scheme family 187–192
hexadecylamine (HDA) 30
hierarchical anatase TiO_2 NW arrays 46–47
highly crystallized TiO_2 NBs 43
high-resolution transmission electron microscopy (HR-TEM) 121, 375
homogeneous ligand-free metal NPs 222
homotypic heterojunctions 185
hot electron injection. *see* direct electron transfer (DET) mechanism
humans, toxicity impact of TiO_2 on 237
humic acid (HA) 323, 326, 488–489
hydrofluoric acid (HF) 58
hydrogen evolution (HE) 141
hydrogen evolution reaction (HER) 419
hydrogen peroxide (H_2O_2) 60
hydrogen production 415–416
  development to replace noble metal cocatalysts 416–417
  heterojunctions 417–418, *418*
  from organic substrates 419–420, *420*
  from seawater 418–419

hydro/solvothermal methods 57–58
hydrothermal process 77–80, *78*
hydrothermal synthesis 97
hydrothermal treatment 100
hydrous TiO_2 colloidal spheres (HTCS) 68–69, 72
  crystallization of HTCS self-template 75–76, *75*
  self-organization theory from 72–73
  structure-directing agents for 73–74, *74*
hydroxyl radicals (OH) 147
1-hydroxy-2,5-pyrrolidinedione (NHS) 452

i
immobilization of TiO_2
  carbon-based substrates 96–97
  ceramic substrates 97
  critical parameters for 99
  dip-coating process for 92–93, *93*
  electrospun nanofibrous membrane 93–94
  future perspective 110
  future research directions 101
  gas-phase deposition 90–91
  on glass or quartz surfaces, sol–gel method 91–92, *92*
  impact on photocatalytic efficiency 100–101
  importance of 99
  liquid-phase deposition 89–90
  metal oxide substrates 94
  microplasma printer for TiO_2 deposition 94–95, *95*
  organically modified montmorillonite nanofibrous membrane 93–94
  photocatalytic performance evaluation 100
  polymeric substrates 95–96
  polymers 101–110
  substrates, techniques to 91
  techniques 89, 100
  textile substrates 97–99
  TiO_2-immobilized biodegradable polymers 99
immobilization system 334
immobilized Pt/TiO_2 photocatalyst 147, *147*
  for efficient PFBR 148
  estimate utilization of electrons, reduction cocatalyst 149–150, *150*
  light absorption 150–152, *151, 152*
  long-distance electron transport 153–154, *154*
  spatial separation of oxidation and reduction half-reactions 148–149, *148*
inhalation risk of TiO_2 243–244
initial antibiotic concentration effects 327
in-situ FTIR analyses 130
in situ-irradiated X-ray photoelectron spectroscopy (ISIXPS) 192, 193, *193*
interleukin (IL) 6, 237
International Organization for Standardization (ISO) 364, 377
International Union of Pure and Applied Chemistry (IUPAC) 68
intimately coupled photocatalysis and biodegradation (ICPB) 331–332

k
Kasuka method 39–40, *39*
Kelvin probe force microscopy (KPFM) 195

l
laccase (Lac) 93
lake water treatment system 283, *284*
LaMer model 73
lauric acid, effect of 33–34, *34*
layer-by-layer deposition 97
layered double hydroxides (LDHs) 124
Lepidium sativum seeds 347–348, *347*
Lifshitz–Slyozov–Wagner model 7
light absorption 150–152, *151, 152*
light conditions effects 325–326
light intensity 325
  dependent electronic process 142–144, *143*
light wavelength 325–326
lincosamides 315
liquid-phase deposition 89–90
liquid-phase exfoliation 53–56, *55, 56*
liquid-phase Z-scheme 187–188
lithium–sulfur (Li—S) batteries 54
localized surface plasmon resonance (LSPR) 207
  enhanced photocatalytic mechanism 207–211
long-distance electron transport 153–154, *154*
low lethal concentration (LC50) 236

m
macrolide antibiotics 311–313, *311*, **312**
Marcus–Gerischer theory 173–174
mass spectrometry (MS) 376
mechanical exfoliation 53
mechanical stress 332
medical devices and products, application of TiO_2 in 245–247, *246*
membrane technology 334–335
3-mercaptopropionic acid (MPA) 215
mesoporous carbon 333
metal-doped TiO_2 nanoparticles 504
metallic doping 319
metal nanoparticles 207, 211
metal–organic frameworks (MOFs) 72, 456
metal oxide clusters (MO_x) 118
  cocatalysts 121
  modification with 118–120, *119*
metal oxide nanoparticles 20, 345

metal oxide substrates 94
  atomic layer deposition 94
  chemical vapor deposition 94
metal–TiO2 (M— TiO_2) nanocomposites 211
2-methyl-4-chlorophenoxyacetic acid (MCPA) 272
methylene blue (MB) 89
mice, toxicity impact of TiO_2 on 237
microorganisms, toxicity impact of TiO_2 on 236–237, *236*
microplasma printer for TiO_2 deposition 94–95, *95*
microplasma system 94
modes of action of TiO_2 232–233, *232*
molecular imprinting technology (MIT) 333–334
molecularly imprinted photocatalyst (CT-MI) 333
molecularly imprinted polymers (MIPs) 462–463
molecularly imprinted TiO_2 (MIP-TiO_2) 306
monobactams 313
moxifloxacin (MOX) 308
multiple trapping model 166–169, *167*

n

nalidixic acid (NA) 308
nanobelts (NBs) 27
nanobuilding blocks (NBBs) 67
nanocrystals (NCs) 10, 28
nanofibers (NFs) 27
nanomaterials 250
nanoparticles (NPs) 207, 230, 345–346
nanorods (NRs) 27
  arrays 45
  hierarchical anatase TiO_2 NW arrays 46–47
  rutile NR arrays 45–46, *46*
nanosheet (NS) 51
nano-sized TiO_2 particles 3, 67
nanostructured systems 247
nanostructured TiO_2 159, 229, *433*
  electrodes 160–161
  photoelectrochemical (PEC) sensing 449–464
  visible-light-induced selective oxidation of organic molecules 431–443
nanostructuring strategies 374
nano-TiO_2 embedded 506
  organic/hybrid matrices 507–510
  TiO_2–SiO_2 nanocomposites 506–507
nanotube arrays (NTAs) 213
nanotubes (NTs) 27
nanowires (NWs) 27
  arrays 45
  hierarchical anatase TiO_2 NW arrays 46–47
  rutile NR arrays 45–46, *46*
National Renewable Energy Laboratory, USA 471
natural organic matter (NOM) 333, 488–489, *489*
natural polymer 101–110
natural resins 498
N-benzylidenebenzylamine 436, 437
N-doped TiO_2 nanoparticles 505–506
neutral pH, shape control with capping agent under 20–22, *21, 22*
N-heterocyclic amines 436
nickel nanoparticles (Ni NPs) 221
nitric oxide (NO) 5, 361–378
nitriles 438
nitrogen dioxide (NO_2) 361–378
nitrogen oxides (NO_x) 361–362
nitrogen reduction reaction (NRR) 425–427, *425*
nitroimidazoles 315
nitrosobenzene 438
noble metal cocatalysts 116, 416–417
  modification with 116–117, **117**
noble-metal-free cocatalyst 417
non-centrosymmetric structure 332
nonhydrolytic sol–gel process 12, *12*
nonmetallic doping 319–320
norfloxacin (NOR) 308
nuclear factor (NF) κB 237

o

ofloxacin (OFL) 308
oil 497–498
oleate 18–19, *18*
oleic acid (OA) 20–21
oleic amine (OM) 21
1D→3D Pathway 79–80
one-dimensional TiO_2 nanomaterials synthesis
  nanorods arrays 45–47
  nanowires arrays 45–47
  TiO_2 NB 42–44
  TiO_2 NFs by electrospinning 45, *45*
  TiO_2 NRs 27–39
  TiO_2 NT structure 39–42
open-circuit photovoltage 171
open-circuit potential (OCP) 144
optimization of sol–gel parameters 106–107
organically modified montmorillonite (O-MMT) nanofibrous membrane 93–94
organic pollutants
  PEC detection of 462–464
  photodegradation of 217–219, *218*
organic solar cells (OSCs) 159
organic sulfides, visible-light-induced selective oxidation 439–441, *440, 441*
organisms, accumulation of TiO_2 in 238
oriented aggregation mechanism 28–29, *29*
Ostwald ripening. *see* coarsening
oxacephems 313
oxacillin (OXA) 313
oxidation, spatial separation of 148–149, *148*

oxidation photocatalysts (OPs) 191
oxide semiconductors 153
oxygen demand, PEC detection of 462–464
oxygen-excessive surface 120
oxygen vacancy (OV) 435, 439
oxy-reduction reaction 273
oxytetracycline (OTC) 302, 305
ozonation 350, 352–353

p
packed bed reactor (PBR) 472
pad-dry-cure method 97
palladium (Pd) 116
parabolic trough collectors (PTCs) 480–481
particles ecotoxic impact 345–346, *346*
PC500 CristalACTiV™ 367
penicillin degradation 315
penicillin G (PG) 313, 320
penicillins 314–315
peptides 315
peptization process 5–6, *6*
perovskite solar cells (PSCs) 159
peroxotitanium (PTA) 59
persulfate (PS) oxidation 329
pharmaceuticals 486–488
pharmaceuticals and personal care products (PPCPs) 345
phenol 485–486, *485*
phosphomolybdic acid (HPMo) 123
phosphotungstic acid (HPW) 123
photocatalysis 87
photocatalysis coupled technologies 327–328
  adsorption 332–333
  AOPs 328–330
  external electric field 330–331
  Fenton process 328
  intimately coupled photocatalysis and biodegradation 331–332
  mechanical stress 332
  membrane technology 334–335
  molecular imprinting technology 333–334
  photocatalytic ozonation 329–330
  photocatalytic persulfate oxidation 329
Photocatalysis Industry Association of Japan (PIAJ) 171
photocatalyst dosage effects 326
photocatalyst I (PC I) 318
photocatalyst I (PC II) 318
photocatalyst immobilization 141
photocatalysts 197–198
photocatalytic cement-based paint 367–368
photocatalytic CO_2 reduction 221–224, *222, 223*
photocatalytic concretes 370
photocatalytic conversion of nitrogen oxides 362–363, *363*
  challenges and perspectives 375–378
  co-catalyst strategies 375
  composite formation strategies 373–374
  doping strategies 372–373
  emerging strategy, target reduction of NO*x* to N_2 and O_2 375–377
  field trials 366–371
  future of 377–378
  ISO protocol for measuring conversion activity 364–366
  materials design 371–372, *372*
  mechanism and ISO protocol 363–366
  nanostructuring strategies 374
  nitrogen oxides, NO and NO_2 361–362
  outdoor testing 367–371, *369*, **369**
  photocatalytic oxidation mechanism 363–364
  product development 366–367, **368**
photocatalytic degradation of antibiotics 301–302
  factors influencing efficiency and underlying mechanisms 322–327
  modification to enhance efficiency 318–322
  performance and mechanism of 302–318
  photocatalysis coupled technologies, enhance efficiency of 327–335
photocatalytic disinfection 350–351, *351*
photocatalytic efficiency
  of cellulose-based TiO_2 composites 106
  impact of immobilization on 100–101
photocatalytic fixed bed reactions (PFBRs) 141, *142*
  applications of 154–155
  cognition of time consumption for photoinduced electronic processes 142–146
  configuration features of immobilized TiO_2/Pt photocatalyst for efficient 148–154
  immobilized Pt/TiO_2 photocatalyst 147, *147*
photocatalytic hydrogen evolution 219–221, *219, 220, 221*
photocatalytic materials 301–302
photocatalytic mechanism of CS–TiO_2 composites 103
photocatalytic membrane reactor (PMR) 334, 477–478, *478*
photocatalytic microreactors 479–480, *480*
  coating process for 108–109
photocatalytic NO*x* conversion 362–363
photocatalytic oxidation mechanism 363–364
photocatalytic oxidation treatment 344
photocatalytic ozonation 329–330
photocatalytic performance, impact of sol–gel parameters on 107
photocatalytic persulfate oxidation 329
photocatalytic processes 193
  antibacterial properties 350–354
  ecotoxicity from wastewater treatment *347*

photocatalytic reactions 91
photocatalytic reactor design 277–285
  categories of 277
  double sheet skin photocatalytic reactor 280, *281*
  drum reactor 279, *279*
  fixed bed reactor 278, *278*
  fluidized bed reactor 278, *278*
  issues of 277
  lake water treatment system 283, *284*
  Purifics Photo-Cat 285, *285*
  solar photocatalytic reactor 280, *282*
  technology readiness level (TRL) method 282–284, *284*
  thin film photo reactor 278–279, *279*
  UV photo microreactors 279–280, *280*
  V-trough photo-reactor 282, *283*
photocatalytic reactors 483–484
  carbon dioxide reduction 489–492
  dyes 484–485
  natural organic matter (NOM) 488–489, *489*
  pharmaceuticals 486–488
  phenol 485–486, *485*
photocatalytic water treatment 271–291
  digitalization and process optimization 289, *289*
  economic viability 289–290, *290*
  environmental and health concerns 288
  future perspectives 289–290
  innovative reactor designs 289
  integration with existing processes 288
  kinetic analysis 285–287, *286*
  modified TiO_2 catalysts, research on 275–277, **276–277**
  photodegradation mechanism 273–275, *274*
  reactor design 277–285
  technical challenges 287–288, *288*
photochemical reactors 471–473
  design of 471–472
  evolution of 471, *472*
  industrial settings 472
  technology advancements 472–473
photocurrent density 165
photodegradation mechanism 273–275, *274*
photodegradation of organic pollutions 217–219, *218*
photoelectrocatalysis (PEC) 330–331
photoelectrochemical (PEC) 423–424
  detection of biomolecules 451–458, *453, 454, 456, 457*
  detection of harmful ions 458–462, *459, 461*
  detection of organic pollutants and oxygen demand 462–464
  fundamentals of 450–451
  sensing 449–464, *450*
photo-Fenton 328
photogenerated electrons 390–391
photo-generated superhydrophilic self-cleaning property 499–500, *499, 500*
photoinduced electronic processes
  cognition of time consumption for 142
  light-intensity-dependent electronic process 142–144, *143*
  time constant estimation for SC interfacial electron transfer 144–146, *145*
photolysis and photocatalysis 350
photoreactor architecture 473–483
photoreforming (PR) 419, 420
photosystem I (PS I) 189
photosystem II (PS II) 189
phthalic acid esters (PAEs) 462
piezoelectric materials 332
pipemidic acid (PA) 308
piperazine ring 310
plants, toxic impact of TiO_2 on 235–236, *235*
plasma-assisted chemical vapor deposition method 373
plasma-assisted deposition 98–99, *98*
plasma treatment 100
plasmonic metal NPs 207
  sol–gel method for hydrolyzing TiO_2 on surface of 211–212, *212, 213*
  to TiO_2 NPs 214–216, *216*
plasmonic metal–TiO_2 nanocomposites, synthesis strategy of 211
  deposition metal NPs on TiO_2 by reduction 212–214, *213, 214, 215*
  loading single metal atom on surface of TiO_2 217
  plasmonic metal NPs to TiO_2 NPs by linkers 214–216, *216*
  sol–gel method for hydrolyzing TiO_2 211–212, *212, 213*
plasmonic metal–TiO_2 photocatalysts
  applications of 217
  enhanced photocatalytic mechanism based on LSPR 207–211
  photocatalytic CO_2 reduction 221–224, *222, 223*
  photocatalytic hydrogen evolution 219–221, *219, 220, 221*
  photodegradation of organic pollutions 217–219, *218*
  strategies for synthesis of plasmonic metal–TiO_2 nanocomposites 211–217
  surface plasmon resonance 207, *208*
plasmon-induced resonant energy transfer (PIRET) 207, 209–211, *210*
plastic waste (PET) 420
platinum (Pt) 116, 375
plug flow reactor (PFR) 472

p–n heterojunctions 185, 186, 417, *418*
poly(vinyl alcohol) (PVA) 88
poly(vinyl pyrrolidone) (PVP) 45
polychlorinated biphenyls (PCBs) 230, 462
polydimethylsiloxane (PDMS) 109
polylactic acid (PLA) 240
polymeric carbon nitride (PCN) 332
polymeric phenylboronic acid (pPBA) 248
polymeric substrates 95
  substrate coating 95–96
polymers 101
  advantages of CS–TiO_2 composites 103
  cellulose-based TiO_2 photocatalysts 104–110
  chitosan 101–102
  chitosan–TiO_2 composite synthesis techniques 102
  future research directions 103–104
  photocatalytic applications of CS–TiO_2 composites 102–103
  photocatalytic mechanism of CS–TiO_2 composites 103
polyoxometallate (POM) 122–123
  modification for proton-coupled electron transfer 122–124, *123*
polysulfone (PSf) 486
polyvinyl alcohol (PVA) 60
porous silica 69
power conversion efficiency (PCE) 159
primary amines 13–14, *13*
proton-coupled electron transfer (PCET) 122
  polyoxometallate modification for 122–124, *123*
pseudo-steady state (PSS) 286
pure nano-TiO_2 500–502, *502*
Purifics Photo-Cat 285, *285*

q

quantum-dot sensitized solar cells (QDSSCs) 159–160
quantum efficiency of regeneration reaction 176
quartz surfaces, immobilization of TiO_2 91–92, *92*
quasi-Fermi level (QFL) 161
quasi-static approximation 169–170
quenching 324–325
quinolone antibiotics 308–311, *308*, **309**
quinolone ring 308

r

radical-dependent experiments 193–194
Raphidocelis subcapitata 344
reactive oxygen species (ROS) 103, 232, 239, 272, 306, 346
reactor engineering 471–473
  photocatalysis research 492
  photocatalytic reactors 483–492
  photoreactor architecture 473–483
recombination 163
  rate equation 171
  in sensitized TiO_2, 163–177
recycling of heavy metal ions from wastewater 389–390
  arsenic 402–405, *403*
  chromium 396–400
  copper 391–396
  mechanism of photocatalytic reduction recovery 390–391
  photocatalytic treatment of *390*, 391–406, *392*
  uranium 400–402, **400**
reduction, reuse, and recycling (3R principle) 351
reduction photocatalysts (RPs) 191
resonant photon scattering 207, 211
response surface methodology 392
reverse double-beam photoacoustic spectroscopy (RDB-PAS) 126
R-factor 4
rifamycins 315
roxithromycin (RXM) 311
ruthenium (Ru) 116
rutile NRs
  arrays 45–46, *46*
  formation of 37, *37*, *38*
rutile phase formation 9–11, *10*

s

Salmonella typhimurium 351
scanning electron microscopy (SEM) 14, 89, 105
seawater photocatalytic hydrogen production 418–419, *419*
secondary amines 14–17, *14*, *15*, *17*
secondary wastewater (SWW) 352
selective metal (oxide) photo-deposition 196–197, *197*
self-cleaning processes 507–509
self-doping concept 120
self-organization theory from HTCS 72–73
self-template route 72
  hydrous TiO_2 colloidal spheres 72–76
  titanium MOFs 76–77, *77*
semiconductor-cocatalyst (SC) 143
  interfacial electron transfer 144–146, *145*
semiconductor-cocatalyst-solution (SCS) 145
semiconductor I (SC I) 186
semiconductor II (SC II) 186

sensitized nanostructured TiO_2
  carrier transport and recombination in 163
  diffusion model 163–166, *166*
  electron–redox mediator
    recombination 170–176, *172, 173, 175*
  electron–sensitizer recombination and
    sensitizer regeneration 176–177
  multiple trapping model 166–169, *167*
  quasi-static approximation 169–170
  SSCs 160–163, *161, 163*
sensitized solar cells (SSCs) 159–163, *161, 163*
sensitizers
  modification with 117–118
  regeneration 176–177
silica 94
silica-based compounds 506–507
silver-loaded TiO_2 376
silver nanoparticles (Ag NPs) 211, 217
single-atom catalysts (SACs) 416–417
single-atom cocatalyst 129–131, *130*
single metal atom on surface of TiO_2 217
sintering 96
skin contact risk of TiO_2 243
slurry photocatalytic reactor (SPR) 473–475,
    474, 475
slurry suspension 152
slurry system 334
sodium hydroxide (NaOH) 54
  elongated TiO_2 NTs 40–41, *41*
  formation of TiO_2 NTs in 39–41
  Kasuka Method 39–40, *39*
sodium oleate 18
sodium titanate 39
  nanobelts (NBs) 42–43
soft template method 70–72, *71, 72*
solar energy 415–420
solar fuel generation 415
  ammonia from N_2 425–427, *425–427*
  carbonaceous from CO_2 420–424
  hydrogen production 415–420
solar parabolic photocatalytic reactor
    (SPPR) 480–481, *482*
solar photocatalytic reactor 280, *282*
sol–gel method 90, 100
  advantages of 92
  for hydrolyzing TiO_2 on surface of
    plasmonic metal NPs 211–212, *212, 213*
  procedure 92
  thin film production via 92
  TiO_2 immobilization on glass or quartz
    surfaces using 91–92, *92*
sol–gel parameters
  impact on photocatalytic performance 107
  optimization of 106–107
sol–gel reaction
  in acidic condition 3–12
  in basic condition 13–20
  coarsening process 6–9, *8*
  nonhydrolytic sol–gel process 12, *12*
  oleate and stearate, effect of 18–19, *18*
  peptization process 5–6, *6, 7*
  primary amines, effect of 13–14, *13*
  rutile phase formation 9–11, *10*
  secondary and tertiary amines, effect
    of 14–17, *14, 15, 17*
  size control depending on solvent
    system 11–12, *11*
  tetrabutylammonium hydroxide, effect
    of 19–20, *19*
  tetramethylammonium hydroxide, effect
    of 17–18, *18*
solution pH effects 322
solution plasma modification 125–127,
    125, 127
solvent system, size control depending on
    11–12, *11*
solvo/hydrothermal process
sono-assisted liquid-phase exfoliation 54
spin-coating procedure 98
spiramycin (SP) 311
spray pyrolysis 97
S-scheme heterojunctions 318, 417–418, *418*
staircase photocatalytic reactor (SPR)
    480, *481*
stearate 18–19, *18*
stearic acid (SA) 105
step-scheme (S-scheme) 190–192, *192*
  characterization technologies of
    S-scheme heterojunction
    192–197
  density functional theory 194–195, *195*
  radical-dependent experiments
    193–194
  selective metal (oxide) photo-deposition
    196–197, *197*
  in situ-irradiated X-ray photoelectron
    spectroscopy 193, *193, 194*
  surface photovoltage 195–196, *196*
stone degradation 497, *498*
strongly binding capping agents (SBCA) 28
structure-directing agents (SDAs) 73
  for HTCS synthesis 73–74, *74*
submerged ceramic membrane photocatalytic
    reactor (SCMPR) 475–476
substrate(s)
  coating 95
  sintering 96
  substrate surface treatment 96
  surface preparation 99
  techniques for immobilizing TiO_2 on to
    various 91–93
sulfadiazine (SDZ) 306, 308
sulfamerazine (SMZ) 306
sulfamethazine (SMT) 306

sulfamethoxazole (SMX) 306
sulfamethoxydiazine (SMD) 306
sulfisoxazole (SIZ) 306
sulfonamide antibiotics (SAs) 305–308, *306*, **307**
superhydrophilic self-cleaning effect 499–500
superhydrophilic surface coatings 508–509, *508, 509*
superhydrophobic surface coatings 507–509, *508, 509*
superior photocatalytic activity 320, 321
surface capping agents 11
surface doping modification 120–122, *121, 122*
surface-enhanced Raman scattering (SERS) effect 59
surface modification of TiO_2
  general strategies of 116
  with MO*x* Cluster 118–120, *119*
  new direction of 124
  with noble metal cocatalysts 116–117, **117**
  polyoxometallate modification for, proton-coupled electron transfer 122–124, *123*
  with sensitizers 117–118
  single-atom cocatalyst 129–131, *130*
  solution plasma modification 125–127, *125, 127*
  surface doping modification 120–122, *121, 122*
  two-dimensional materials modification 124
  ultrasonic modification 127–129, *129*
surface photovoltage (SPV) 192, 195–196, *196*
surface plasmon resonance (SPR) 207, *208*, 452
surface reaction 201
Systematic Evolution of Ligands by Exponential Enrichment (SELEX) 460

t

TC hydrochloride (TCH) 329, 331, 332
technology readiness level (TRL) method 282–284, *284*, 287
template-free self-assembly method 59–60, *59*
template method 58–59, *58, 59*, 68, 69
  hard template method 69–70
  soft template method 70–72, *71, 72*
tertiary amines 14–17, *14, 15, 17*
tetrabutylammonium (TBA^+) cations 54
tetrabutyl ammonium hydroxide (TBAH/TBAOH) 19–20, *19*, 41, 54–55
tetrabutyl titanate (TBOT) 56, 58
2,3′,5,5′-tetrachlorobiphenyl (PCB72) 462
7,7,8,8-tetracyanoquiondimenthane (TCNQ) 117
tetracycline hydrochloride (TC) 455
tetracyclines 327
  antibiotics 302–305, *302*, **303–304**
tetraethoxysilane/tetraethyl orthosilicate (TEOS) 73, 503
tetrahydrofuran (THF) 56
tetramethylammonium hydroxide (TMAH) 31
  effect of 35–36, *35, 36*
  growth mechanism in presence of 31–33
tetramethylammonium hydroxide, effect of 17–18, *18*
tetramethylpiperidine *N*-oxyl (TEMPO) 434, 437
textile substrates 97–99
  layer-by-layer deposition 97
  pad-dry-cure method 97
  plasma-assisted deposition 98–99, *98*
  spin-coating procedure 98
  TiO_2 immobilization, considerations for 98
thin film photo reactor 278–279, *279*
thin film production 92
3D hierarchical porous TiO_2 spheres 77–80, *78*
  0D→3D Pathway 78–79
  1D→3D Pathway 79–80
  2D→3D Pathway 80
3D layered titanates 55
3D printing 100
3D titanium dioxide
  template method 69–72
  self-template route 72–77
  3D hierarchical porous TiO_2 spheres 77–80, *78*
Ti-based zeolitic imidazolate framework-8 (Ti-ZIF-8) 76
time constants (τSCS) 146
  estimation for SC interfacial electron transfer 144–146, *145*
time-dependent electron continuity equation 167
TiO_2—SiO_2 nanocomposite coatings 506–507
TiO_2-based heterojunction photocatalysts
  classification of heterojunction 186–192
  roles of heterojunctions 197–201, *198, 199–201*
  S-scheme heterojunction 192–197
TiO_2-based protective coatings 498, 499
  degradation of 497, *498*
  dispersions of nano-TiO_2 500–506
  historical monuments and contemporary buildings 501, 510
  Middle Age practices of 498
  nano-TiO_2 embedded, organic/inorganic matrices 506–510
  natural resins 498

TiO_2-based protective coatings (*Cont.*)
photocatalytic nanomaterials 498
photo-generated superhydrophilic self-cleaning property 499–500, *499, 500*
protection treatment 498
TiO_2 Degussa P25 362, 363, 367
TiO_2 nanobelts 42
acid-hydrothermal process 44
alkali-hydrothermal process 42–44, *44*
TiO_2 nanofibers 45, *45*
TiO_2 nanorices (TNRs) 217
TiO_2 nanorods 27
anatase [100] direction, selective growth to 29–33
formation of brookite NR 37–39, *38*
formation of rutile NRs 37, *37, 38*
fundamental principle of 28
oriented aggregation mechanism 28–29, *29*
preparation of anatase TiO_2 NRs 33–36
TiO_2 nanotubes (TiNT) 394
formation in NaOH 39–41
structure 39–42
TiO_2 microtubes 41–42, *42*
TiO_2 P25 UVA photocatalysis 348
TiO_2–Poly(Lactic Acid) (PLA) composites 101
tissue engineering, application of TiO_2 in 249–250
Titanium alkoxide [$Ti(OR)_4$], 72–73
titanium bis(ammonium lactate) dihydroxide (TALH) 37
titanium dioxide (TiO_2) 87–88, 197
accumulation in organisms 238
algae and plants, toxicity impact on 235–236, *235*
antibacterial applications of TiO_2 241–242
antibacterial properties 350–354, *351–353*
antidotal effects of 238
antimicrobial mechanisms of 238–240
antimicrobial properties of 238
aquatic organisms, toxicity impact on 233–235, *234*
based hydrophilic coatings 499–500
cellulose composites 101
chitosan composites 101
contaminants of emerging concern (CECs) 343
deposition metal NPs on TiO_2 by reduction 212–214, *213, 214, 215*
drug delivery systems and therapy, application in 247–249, *247*
ecotoxicity from wastewater treatment 347–350, *347*
ecotoxicological effects of 232–238
edible risk of 244–245
gel formation 108
immobilization 98
immobilized biodegradable polymers 99–101
impact on health 242, *242*
inhalation risk of 243–244
medical devices and products, application in 245–247, *246*
medical potential of 245
on mice and humans, toxicity impact on 237
microorganisms, toxicity impact on 236–237, *236*
microplasma printer for TiO_2 deposition 94–95, *95*
microtubes 41–42, *42*
nanoparticles 95
nanotube walls 245
nanowires 221
particles ecotoxic impact 345–346, *346*
photocatalysis mechanism *230*
photocatalysts 89, 239
photocatalytic water treatment, applications of 271–291
purification and decontamination effects of 240–241
single metal atom on surface of 217
skin contact risk of 243
species commonly used, ecotoxicity tests 344, **344**
tissue engineering, application in 249–250
toxicity mechanisms and modes of action of 232–233, *232*
use of environment friendly 250–251
titanium dioxide nanoparticles ($nTiO_2$) 235–237, 240–241, 243, 246
titanium halide (TiX_4) 4
titanium ions (Ti^{4+}) 239
titanium MOFs 76–77, *77*
titanium oxide nanoparticles (TiO_2 NPs) 3
attaching plasmonic metal NPs to 214–216, *216*
shape control with capping agent under neutral pH 20–22
sol–gel reaction in basic condition 13–20
synthesis by sol–gel reaction under acidic condition 3–12
synthesis of extremely small TiO_2 NPs using dendrimer 23–24, *23*
titanium tetrachloride ($TiCl_4$) 61
titanium tetraisopropoxide (TTIP) 5, 11, 41, 92, 212
toluene 2,4-diisocyanate (TDI) 118
top-down methods for 2D material synthesis 53
electrochemical exfoliation 56–57, *57*
liquid-phase exfoliation 53–56, *55, 56*
mechanical exfoliation 53

topochemical method 60–61, *61*
total organic carbon (TOC) 330
toxicity 343–354
  mechanisms of TiO_2 232–233, *232*
toxicological bioassays 344, **344**
transient photovoltage (TPV) 120
transition metal dichalcogenides (TMDs) 52, 124
transition metals 417, 503
transmission electron microscopy (TEM) 7, 28
transparent conducting oxides (TCOs) 161–162
tri-doping 320
triethanolamine (TEOA) 13–14
triethylamine (TEA) 14, 440
trimethylamine (TMA) 14, 440
trioctylphosphine oxide (TOPO) 12
  effect of 33–34, *34*
Tris(2-chloropropyl) phosphate (TCPP) 238
tumor necrosis factor (TNF) α 237
tungstosilicic acid (HSiW) 123
2D anatase NSs 54
two-dimensional (2D) materials 51, 124
2D-layered materials 58
2D nanosheets 53
2D→3D Pathway 80
2D TiO_2 nanosheets 54
two-dimensional TiO_2 nanostructures
  bottom-up methods 57–61
  synthesis of 52, *52*
  top-down methods 53–57
tylosin 311
type-I, II, III heterojunctions 186–187, *187*, 190, 318, 417–418, *418*

u

ultrasonic irradiation 332
ultrasonic modification 127–129, *129*
ultraviolet (UV)
  absorption 229
  active titanium dioxide 87
  irradiation 75
  light 185
  photo microreactors 279–280, *280*
uranium 400, **400**
  modifying TiO_2, elemental doping/functional additives 400–401
  recovery and elution of U(VI) 402
  TiO_2-based heterojunction 401–402, *401*
US Environmental Protection Agency 230

v

valence band (VB) 118, 186, 318–319, 391
vertically aligned TiO_2 nanotube arrays 213
visible-light-induced selective oxidation of organic molecules 431–432, *433*
  of alcohols 432–435, *434, 435*
  of amines 435–439, *436–439*
  C—C coupling reactions 441–442, *442*
  of organic sulfides 439–441, *440, 441*
visible light photocatalysis 432
V-trough photo-reactor 282, *283*

w

waxes 497–498
weakly binding capping agents (WBCA) 28
wet chemical methods 57
  hydro/solvothermal methods 57–58
  template-free self-assembly method 59–60, *59*
  template method 58–59, *58, 59*
wet-on-dry techniques 503
wet-on-wet techniques 503
World Health Organization (WHO) 301, 362

x

X-ray absorption near edge structure (XANES) 376
X-ray photoelectron spectroscopy (XPS) 376

z

0D→3D Pathway 78–79
zero-dimensional titanium dioxide nanoparticles 3
Z-scheme family 187
  all-solid-state 188–189, *188*
  direct 189–190
  liquid-phase 187–188
  S-scheme 190–192, *192*
Z-type heterojunctions 318, 417–418, *418*